Intelligent Imaging and Analysis

Intelligent Imaging and Analysis

Special Issue Editors

DaeEun Kim
Dosik Hwang

MDPI • Basel • Beijing • Wuhan • Barcelona • Belgrade

Special Issue Editors

DaeEun Kim
Yonsei University
Korea

Dosik Hwang
Yonsei University
Korea

Editorial Office
MDPI
St. Alban-Anlage 66
4052 Basel, Switzerland

This is a reprint of articles from the Special Issue published online in the open access journal *Applied Sciences* (ISSN 2076-3417) from 2018 to 2019 (available at: https://www.mdpi.com/journal/applsci/special_issues/Intelligent_Imaging_and_Analysis).

For citation purposes, cite each article independently as indicated on the article page online and as indicated below:

LastName, A.A.; LastName, B.B.; LastName, C.C. Article Title. *Journal Name* **Year**, *Article Number*, Page Range.

ISBN 978-3-03921-920-9 (Pbk)
ISBN 978-3-03921-921-6 (PDF)

Cover image courtesy of DaeEun Kim.

Contents

About the Special Issue Editors

DaeEun Kim, Professor, received his B.E. and M.S. from the Department of Computer Science and Engineering of Seoul National University, South Korea and the University of Michigan at Ann Arbor, USA, respectively. He received his Ph.D. degree from the University of Edinburgh, UK, in 2002. From 2002 to 2006, he was Research Scientist at the Max Planck Institute for Human Cognitive and Brain Sciences in Munich, Germany. He is currently Professor at Yonsei University in Seoul, Korea. His research interests are in the areas of biorobotics, autonomous robots, artificial intelligence, artificial life, neural networks, and neuroethology.

Dosik Hwang, Professor, received his B.S. and M.S. degrees from Yonsei University, Seoul, Korea, in 1997 and 1999, respectively, and Ph.D. degree from the University of Utah, Salt Lake City, Utah, USA in 2006. From 2006 to 2008, he was associated with the Brain Imaging Center, University of Colorado Health Science Center, USA. He is currently Professor in the Department of Electrical and Electronic Engineering, Yonsei University, Seoul, Korea. He is mainly engaged in the research fields of artificial intelligence and medical imaging. He is currently working on deep learning-based medical image reconstruction, multicontrast image conversion, and automatic measurements and diagnosis. He has published 40 papers in various international journals as well as 26 international and domestic patents. He received a Magna Cum Laude Merit Award and Summa Cum Laude Merit Award from the International Society for Magnetic Resonance in Medicine in 2012 and 2017, respectively. He has made 3 technology transfers in the fields of medical imaging, diagnosis, and intelligent image analysis.

Editorial

Special Features on Intelligent Imaging and Analysis

Dosik Hwang * and DaeEun Kim *

School of Electrical and Electronic Engineering, Yonsei University, Seoul 03722, Korea
* Correspondence: dosik.hwang@yonsei.ac.kr (D.H.); daeeun@yonsei.ac.kr (D.K.)

Received: 29 October 2019; Accepted: 4 November 2019; Published: 10 November 2019

---

1. Introduction

Intelligent imaging and analysis have been studied in various research fields, including medical imaging, biomedical applications, computer vision, visual inspection and robot systems. Recently, artificial intelligence (AI) technologies with deep learning, machine learning and image processing have been applied to many difficult tasks, including challenging issues in image generation, reconstruction, de-noising, segmentation, and defect detection.

This Special Issue handles various applications of intelligent imaging technologies, covering medical imaging, visual detection, segmentation, medical diagnosis, image retrieval, image reconstruction and texture mapping. It shows the trend and the latest developments of intelligent imaging and analysis techniques. Theories of image processing, learning schemes and state-of-art artificial intelligence techniques for imaging and analysis are introduced.

2. Intelligent Imaging and Analysis

This section provides overall summaries of the papers included in this Special Issue for quick guidance for readers. The first half introduces non-medical or general applications, and the second half focuses on medical applications of the intelligent imaging and analysis techniques.

As an application of image analysis, face sketch synthesis has been tested by Wan and Lee [1], using the joint training method on face photos and sketches. Thus, more detailed information can be recorded in the synthesized sketches. Autonomous underwater vehicles (AUVs) have been a challenging subject requiring the underwater location accuracy. The work suggested by Wang et al. [2] shows the PL-SLAM (point and line simultaneous localization and mapping) to improve the accuracy of localization in the underwater environment. To measure volumetric tooth wear, a set of sequence images over sprocket teeth in a scraper conveyer can be collected. A focused morphology restoration algorithm has been applied to the image set by Ding et al. [3]. The method uses image filtering with a normal distribution operator to improve the accuracy of an evaluation function. Statistical body shape models can be used to estimate 3D human pose. However, self-intersection often occurs on images with a pattern of poses, especially between body parts. Wu et al. [4] introduced a self-intersection penalty term for statistical body shape models for 3D pose estimation, thus improving the accuracy of the pose estimation.

Many imaging techniques have been recently developed with CNN (convolutional neural network). A new approach with a depth module has been applied to human parsing by Jiang and Chi [5]. The method integrates a depth estimation module and a segmentation module as a variation of CNN (convolutional neural network) for image analysis, thus improving the performance for human parsing. Another CNN approach was used for scene parsing in the road scene context by Li et al. [6]. They provided a fast 3D semantic mapping system with monocular vision by combining localization, mapping and scene parsing. The semantic segmentation runs on selected key frames and their depth information, reducing the computational cost and also improving the accuracy of semantic mapping. The CNN model was also applied to geological structure image classification. As Zhang et al. [7]

showed, transfer learning based on a deep learning model is effective to extract features of geological structure data. Oil spill detection on the ocean surface has been a hot issue, and a refined technique is required to detect dark spots on SAR (synthetic aperture radar) images. A deep convolution neural network, called Segnet has been tested for this application by Guo et al. [8]. It has the basic framework of encoder and decoder for image semantic segmentation. It improves accuracy performance to extract an oil spill location and area.

Three-dimensional image modeling reconstructs the 3D model of an object with multiple 2D images as well as maintaining its texture. Lai et al. [9] proposed a texture mapping method to use mesh partitioning, mesh parameterization and packing, texture transferring, and texture correction and optimization. It forms a high-resolution texture map of a 3D model for application in e-commerce. Image super-resolution technique is one of the most promising CNN based image processing techniques, which can be applied to various kinds of images. Chen et al. [10] propose a dual-channel CNN super-resolution network to extract the detailed texture information (deep channel) and the overall outline of the original image (shallow channel). Image quality assessment is an important process to maintain good quality of images for various types of imaging system. Jang et al. [11] proposed a new method of automatically assessing image quality when several enhancement techniques are applied, by using feature sets derived from high dynamic range images.

Detection-specific AI techniques such as the region-based CNN (R-CNN) can be used for various industrial purposes with specific configuration of a vision system. Wang et al. [12] proposed a novel one-camera–five-mirror system for cavitation bubble cluster study and have used a faster R-CNN method to detect bubbles in their system. The concept of residual network has widely been adopted in various deep learning architectures. Guo et al. [13] proposed an efficient deep residual network approach by using sparse feedbacks, which improves the convergence speed and the training stability in their image restoration applications.

Image segmentation has been tackled with a novel active contour model by Sun et al. [14]. It uses an improved SPF (signed pressure function) and a LIF (local image fitting) model. A weight function of the grayscale mean values around the contour curve was introduced to segment blurred images and weak gradient images, and another metric function was used to check local image information to segment intensity-inhomogeneous images. Image segmentation plays a role of monitoring the weld pool surface affecting the weld quality. The reflected laser lines used for arc welding contain the weld pool surface information, and Wang et al. [15] proposed that various image processing methods are combined for image segmentation approach, applied to the reflected laser lines. An effective image search is demanding work in the field of content-based image retrieval. A novel approach to encode the relative spatial information for histogram-based representation of the visual worlds has been introduced by Zafar et al. [16]. It computes the geometric relationship in the visual worlds and enhances the performance of image retrieval.

Surface defect detection is a challenging problem in industrial product manufacturing. Vision-based defect detection of steel sheets focuses on finding the salient characteristics of the defects. Zhou et al. [17] processed the defect segmentation with a double low-rank and sparse decomposition model to obtain high-quality defect images. Another style of surface defect detection can be found in the railway surface. A novel visual inspection approach based on UAV (unmanned aerial vehicle) images has been tested by Wu et al. [18]. It characterizes the defective sub-regions and defect-free background sub-regions, and highlights the critical defect regions in the image analysis. Their approach has two key techniques for UAV-based rail images: image enhancement and surface defect segmentation.

The inspection of a PCB (printed circuit board) is a kind of surface defect detection problem with visual image in manufacturing. Yuk et al. [19] extracted robust features in the visual image with various types of defect patterns such as scratches and improper etching, applying a random forest method. They also used probabilistic kernel density estimation to improve the detection performance. Visual inspection technology in the manufacturing process in the iron and steel industry has been investigated by Sun et al. [20]. Coverage of cameras or light source information can differentiate the hardware selection of visual inspection. The inspection or detection algorithm depends on filtering, statistics and learning methods. Basic theories and foundations of image processing highly contribute to the visual inspection technologies, according to this review paper.

The following part introduces medical applications of intelligent imaging and analysis techniques on various types of image datasets including magnetic resonance images (MRI), computed tomography (CT), optical scan data, photographic images and digital videos.

Segmentation of medical images is one of the most important tasks for many medical image analyses. However, the boundaries of the object segmentation have not been well addressed so far. Kim et al. [21] proposed a method to overcome the poor performance issues encountered on the boundaries of objects. Their boundary-specific U-net improved the segmentation performance on the boundaries of the intervertebral discs in MR spine images. Human lumbar spine diagnosis can be assisted by accurate segmentation of the vertebral bodies. Kim et al. [22] also proposed a semi-automatic segmentation technique for vertebral bodies in MR images which can reduce the user's role while achieving good segmentation accuracy.

A data balancing problem is an important issue to address for some deep learning applications, especially when there are not enough datasets for training. It is often noticed that non-balanced datasets lead to a biased estimation model. Zhang et al. [23] proposed a pre-training strategy to address this problem encountered in the liver segmentation from computed tomography (CT) scans. Zheng et al. proposed an intelligent evaluation system for strabismus diagnosis with a sequence of digital video analyses [24]. Each image in the recorded video is analyzed to localize important features, which are combined to evaluate the strabismus. Updating and automatic training can be an important step for the practical application and subsequence maintenance of an AI-based system. Sugimori [25] discussed the overall accuracy of additional learning and automatic training system for CT classification task.

Intelligent design and manufacturing can help medical procedures. Kim et al. [26] proposed a computer-aided design and manufacturing technology for patient-specific optimization of the Nuss procedure for minimally invasive surgery of pectus excavatum. This can be an example of how the intelligent imaging and analysis techniques can change the shape of the medical procedures in the future. Automatic alignment of images and volumes in medical imaging data is an important step for further subsequent analysis. Rehman and Lee [27] proposed an efficient automatic midsagittal plane extraction in brain MRI images. This method can be useful for image registration, asymmetric analysis and tilt correction encountered in the analysis of brain images.

Registration between volume data and surface data is also an important application in several medical procedures. Jung et al. [28] proposed an effective registration technique for dental tomographic volume data and scan surface data by using a dynamic segmentation technique. Another registration task can be found as a point cloud registration for 3D datasets. Liu et al. [29] presented a 3D point cloud registration algorithm based on a greedy projection triangulation method to address the 3D problem.

3. Future Intelligent Imaging and Analysis

Various technological advancements of intelligent imaging and analysis have been introduced in this Special Issue. With the advent of the deep learning and related machine learning techniques, many of the conventional imaging and analysis techniques are being improved or substituted by the intelligent learning-based methods. It is expected that robust and reliable intelligent techniques will soon be deployed for many practical applications in industries, sciences, medicine and arts.

Author Contributions: The authors of this editorial served as Guest Editors for this Special Issue. Both the authors contributed equally to this editorial and proofread for publication.

Funding: The first author was supported by Basic Science Research Program through the National Research Foundation of Korea (NRF) funded by the Ministry of Science and ICT (2019R1A2B5B01070488), Brain Research Program through NRF (2018M3C7A1024734), and Bio & Medical Technology Development Program of NRF

(NRF-2018M3A9H6081483). The second author was supported by the NRF funded by the Korea government (MSIT) (No. 2017R1A2B4011455).

Acknowledgments: We would like to thank all the authors who contributed to this Special Issue. Their contributions have made this Special Issue a very valuable one. We also thank all reviewers for their time and commitment to providing comments and suggestions. This issue would not be possible without their efforts and time. Finally, we would like to express our special thanks to the editorial team and Daria Shi, the managing editor of *Applied Sciences* for their kind and endless help to keep us focused on this Special Issue.

Conflicts of Interest: The authors declare no conflict of interest.

References

1. Wan, W.; Lee, H.J. A Joint Training Model for Face Sketch Synthesis. *Appl. Sci.* **2019**, *9*, 1731. [CrossRef]
2. Wang, R.; Wang, X.; Zhu, M.; Lin, Y. Application of a Real-Time Visualization Method of AUVs in Underwater Visual Localization. *Appl. Sci.* **2019**, *9*, 1428. [CrossRef]
3. Ding, H.; Liu, Y.; Liu, J. Volumetric Tooth Wear Measurement of Scraper Conveyor Sprocket Using Shape from Focus-Based Method. *Appl. Sci.* **2019**, *9*, 1084. [CrossRef]
4. Wu, Z.; Jiang, W.; Luo, H.; Cheng, L. A Novel Self-Intersection Penalty Term for Statistical Body Shape Models and Its Applications in 3D Pose Estimation. *Appl. Sci.* **2019**, *9*, 400. [CrossRef]
5. Jiang, Y.; Chi, Z. A CNN Model for Human Parsing Based on Capacity Optimization. *Appl. Sci.* **2019**, *9*, 1330. [CrossRef]
6. Li, X.; Wang, D.; Ao, H.; Belaroussi, R.; Gruyer, D. Fast 3D Semantic Mapping in Road Scenes. *Appl. Sci.* **2019**, *9*, 631. [CrossRef]
7. Zhang, Y.; Wang, G.; Li, M.; Han, S. Automated Classification Analysis of Geological Structures Based on Images Data and Deep Learning Model. *Appl. Sci.* **2018**, *8*, 2493. [CrossRef]
8. Guo, H.; Wei, G.; An, J. Dark Spot Detection in SAR Images of Oil Spill Using Segnet. *Appl. Sci.* **2018**, *8*, 2670. [CrossRef]
9. Lai, J.-Y.; Wu, T.-C.; Phothong, W.; Wang, D.W.; Liao, C.-Y.; Lee, J.-Y. A High-Resolution Texture Mapping Technique for 3D Textured Model. *Appl. Sci.* **2018**, *8*, 2228. [CrossRef]
10. Chen, Y.; Wang, J.; Chen, X.; Sangaiah, A.K.; Yang, K.; Cao, Z. Image Super-Resolution Algorithm Based on Dual-Channel Convolutional Neural Networks. *Appl. Sci.* **2019**, *9*, 2316. [CrossRef]
11. Jang, J.; Jang, H.; Eo, T.; Bang, K.; Hwang, D. No-reference Automatic Quality Assessment for Colorfulness-Adjusted, Contrast-Adjusted, and Sharpness-Adjusted Images Using High-Dynamic-Range-Derived Features. *Appl. Sci.* **2018**, *8*, 1688. [CrossRef]
12. Wang, H.; Xu, H.; Pooneeth, V.; Gao, X.-Z. A Novel One-Camera-Five-Mirror Three-Dimensional Imaging Method for Reconstructing the Cavitation Bubble Cluster in a Water Hydraulic Valve. *Appl. Sci.* **2018**, *8*, 1783. [CrossRef]
13. Guo, Z.; Sun, Y.; Jian, M.; Zhang, X. Deep Residual Network with Sparse Feedback for Image Restoration. *Appl. Sci.* **2018**, *8*, 2417. [CrossRef]
14. Sun, L.; Meng, X.; Xu, J.; Tian, Y. An Image Segmentation Method Using an Active Contour Model Based on Improved SPF and LIF. *Appl. Sci.* **2018**, *8*, 2576. [CrossRef]
15. Wang, Z.; Zhang, C.; Pan, Z.; Wang, Z.; Liu, L.; Qi, X.; Mao, S.; Pan, J. Image Segmentation Approaches for Weld Pool Monitoring during Robotic Arc Welding. *Appl. Sci.* **2018**, *8*, 2445. [CrossRef]
16. Zafar, B.; Ashraf, R.; Ali, N.; Iqbal, M.K.; Sajid, M.; Dar, S.H.; Ratyal, N.I. A Novel Discriminating and Relative Global Spatial Image Representation with Applications in CBIR. *Appl. Sci.* **2018**, *8*, 2242. [CrossRef]
17. Zhou, S.; Wu, S.; Liu, H.; Lu, Y.; Hu, N. Double Low-Rank and Sparse Decomposition for Surface Defect Segmentation of Steel Sheet. *Appl. Sci.* **2018**, *8*, 1628. [CrossRef]
18. Wu, Y.; Qin, Y.; Wang, Z.; Jia, L. A UAV-Based Visual Inspection Method for Rail Surface Defects. *Appl. Sci.* **2018**, *8*, 1028. [CrossRef]
19. Yuk, E.H.; Park, S.H.; Park, C.-S.; Baek, J.-G. Feature-Learning-Based Printed Circuit Board Inspection via Speeded-Up Robust Features and Random Forest. *Appl. Sci.* **2018**, *8*, 932. [CrossRef]
20. Sun, X.; Gu, J.; Tang, S.; Li, J. Research Progress of Visual Inspection Technology of Steel Products—A Review. *Appl. Sci.* **2018**, *8*, 2195. [CrossRef]

21. Kim, S.; Bae, W.C.; Masuda, K.; Chung, C.B.; Hwang, D. Fine-Grain Segmentation of the Intervertebral Discs from MR Spine Images Using Deep Convolutional Neural Networks: BSU-Net. *Appl. Sci.* **2018**, *8*, 1656. [CrossRef] [PubMed]
22. Kim, S.; Bae, W.C.; Masuda, K.; Chung, C.B.; Hwang, D. Semi-Automatic Segmentation of Vertebral Bodies in MR Images of Human Lumbar Spines. *Appl. Sci.* **2018**, *8*, 1586. [CrossRef]
23. Zhang, Y.; Wang, Y.; Wang, Y.; Fang, B.; Yu, W.; Long, H.; Lei, H. Data Balancing Based on Pre-Training Strategy for Liver Segmentation from CT Scans. *Appl. Sci.* **2019**, *9*, 1825. [CrossRef]
24. Zheng, Y.; Fu, H.; Li, R.; Lo, W.-L.; Chi, Z.; Feng, D.D.; Song, Z.; Wen, D. Intelligent Evaluation of Strabismus in Videos Based on an Automated Cover Test. *Appl. Sci.* **2019**, *9*, 731. [CrossRef]
25. Sugimori, H. Evaluating the Overall Accuracy of Additional Learning and Automatic Classification System for CT Images. *Appl. Sci.* **2019**, *9*, 682. [CrossRef]
26. Kim, Y.-J.; Heo, J.-Y.; Hong, K.-H.; Lim, B.-Y.; Lee, C.-S. Computer-Aided Design and Manufacturing Technology for Identification of Optimal Nuss Procedure and Fabrication of Patient-Specific Nuss Bar for Minimally Invasive Surgery of PectusExcavatum. *Appl. Sci.* **2019**, *9*, 42. [CrossRef]
27. Rehman, H.Z.U.; Lee, S. An Efficient Automatic Midsagittal Plane Extraction in Brain MRI. *Appl. Sci.* **2018**, *8*, 2203. [CrossRef]
28. Jung, K.; Jung, S.; Hwang, I.; Kim, T.; Chang, M. Registration of Dental Tomographic Volume Data and Scan Surface Data Using Dynamic Segmentation. *Appl. Sci.* **2018**, *8*, 1762. [CrossRef]
29. Liu, J.; Bai, D.; Chen, L. 3-D Point Cloud Registration Algorithm Based on Greedy Projection Triangulation. *Appl. Sci.* **2018**, *8*, 1776. [CrossRef]

Article

Intelligent Evaluation of Strabismus in Videos Based on an Automated Cover Test

Yang Zheng [1,2,3], Hong Fu [3,*], Ruimin Li [1,2,3], Wai-Lun Lo [3], Zheru Chi [4], David Dagan Feng [5], Zongxi Song [1] and Desheng Wen [1]

[1]   Xi'an Institute of Optics and Precision Mechanics of CAS, Xi'an 710119, China; zhengyang2015@opt.cn (Y.Z.); liruimin2015@opt.cn (R.L.); songxi@opt.ac.cn (Z.S.); ven@opt.ac.cn (D.W.)
[2]   University of Chinese Academy of Sciences, Beijing 100049, China
[3]   Department of Computer Science, Chu Hai College of Higher Education, Hong Kong 999077, China; wllo@chuhai.edu.hk
[4]   Department of Electronic and Information Engineering, The Hong Kong Polytechnic University, Hong Kong 999077, China; enzheru@polyu.edu.hk
[5]   School of Computer Science, The University of Sydney, Sydney 2006, Australia; dagan.feng@sydney.edu.au
*   Correspondence: hfu@chuhai.edu.hk, Tel.: +852-29727250

Received: 30 December 2018; Accepted: 12 February 2019; Published: 20 February 2019

Abstract: Strabismus is a common vision disease that brings about unpleasant influence on vision, as well as life quality. A timely diagnosis is crucial for the proper treatment of strabismus. In contrast to manual evaluation, well-designed automatic evaluation can significantly improve the objectivity, reliability, and efficiency of strabismus diagnosis. In this study, we have proposed an innovative intelligent evaluation system of strabismus in digital videos, based on the cover test. In particular, the video is recorded using an infrared camera, while the subject performs automated cover tests. The video is then fed into the proposed algorithm that consists of six stages: (1) eye region extraction, (2) iris boundary detection, (3) key frame detection, (4) pupil localization, (5) deviation calculation, and (6) evaluation of strabismus. A database containing cover test data of both strabismic subjects and normal subjects was established for experiments. Experimental results demonstrate that the deviation of strabismus can be well-evaluated by our proposed method. The accuracy was over 91%, in the horizontal direction, with an error of 8 diopters; and it was over 86% in the vertical direction, with an error of 4 diopters.

Keywords: intelligent evaluation; automated cover tests; deviation of strabismus; pupil localization

---

1. Introduction

Strabismus is the misalignment of the eyes, that is, one or both eyes may turn inward, outward, upward, or downward. It is a common ophthalmic disease with an estimated prevalence of 4%, in adulthood [1], 65% of which develops in childhood [2]. Strabismus could have serious consequences on vision, especially for children [3,4]. When the eyes are misaligned, the eyes look in different directions, leading to the perception of two images of the same object, a condition called diplopia. If strabismus is left untreated in childhood, the brain eventually suppresses or ignores the image of the weaker eye, resulting in amblyopia or permanent vision loss. Longstanding eye misalignment might also impair the development of depth perception or the ability to see in 3D. In addition, patients with paralytic strabismus might turn their face or head to overcome the discomfort and preserve the binocular vision for the paralyzed extraocular muscle, which might lead to a skeletal deformity in children, such as scoliosis. More importantly, it has been shown that people with strabismus show higher levels of anxiety and depression [5,6] and report a low self-image, self-esteem, and self-confidence [7,8], which brings adverse impact on a person's life, including education, employment, and social communication [9–14]. Thus, timely quantitative evaluation of strabismus is essential, in order to get a suitable treatment for strabismus. More specifically, accurate measurement of the deviation in strabismus is crucial in planning surgery and other treatments.

Currently, several tests need to be performed, usually, to diagnose strabismus in a clinical context [15]. For example, the corneal light reflex is conducted by directly observing the displacement of the reflected image of light from the center of the pupil. Maddox rod is a technique that utilizes filters and distorting lenses for quantifying eye turns. Another way to detect and measure an eye turn is to conduct a cover test, which is the most commonly used technique. All these methods require conduction and interpretation by the clinician or ophthalmologist, which is subjective to some extent. Taking the cover test as an example, the cover procedures and assessments are conducted manually, in the existing clinical systems, and well-trained specialists are needed for the test. Therefore, this limits the effect of strabismus assessment in two aspects [16,17]. With respect to cover procedure, the cover is given manually, so the covering time and speed of occluder movement depend on the experience of the examiners and can change from time to time. These variations of the cover may influence the assessment results. With respect to assessment, the response of subject is evaluated subjectively, which leads to more uncertainties and limitations in the final assessment. First, the direction of eye movement, the decision of whether or not moving and the responding speed for recovery, rely on the observation and determination of the examiners. The variances of assessment results, among examiners, cannot be avoided. Second, the strabismus angle has to be measured with the aid of a prism, in a separate step and by trial-and-error. This strings out the diagnosis process. Being aware of these clinical disadvantages, researchers are trying to find novel ways to improve the process of strabismus assessment.

With the development of computer technology, image acquisition technology, etc., researchers have made some efforts to utilize new technologies and resources to aid ophthalmology diagnostics. Here, we give a brief review on the tools and methodologies that support the detection and diagnosis of strabismus. These methods can be summarized into two categories, namely the image-based or video-based method, and the eye-tracking based method.

The image-based or video-based method uses image processing techniques to achieve success in diagnosing strabismus [18–22]. Helveston [18] proposed a store-and-forward telemedicine consultation technique that uses a digital camera and a computer to obtain patient images and then transmits them by email, so the diagnosis and treatment plan could be determined by the experts, according to the images. This was an early attempt to apply new resources to aid the diagnosis of strabismus. Yang [19] presented a computerized method of measuring binocular alignment, using a selective wavelength filter and an infrared camera. Automated image analysis showed an excellent agreement with the traditional PCT (prism and alternate cover test). However, the subjects who had an extreme proportion that fell out of the normal variation range, could not be examined using this system, because of its software limitations. Then in [20], they implemented an automatic strabismus examination system that used an infrared camera and liquid crystal shutter glasses to simulate a cover test and a digital video camera, to detect the deviation of eyes. Almedia et al. [21] proposed a four-stage methodology for automatic detection of strabismus in digital images, through the Hirschberg test: (1) finding the region of the eyes; (2) determining the precise location of eyes; (3) locating the limbus and the brightness; and (4) identifying strabismus. Finally, it achieved a 94% accuracy in classifying individuals with or without strabismus. However, the Hirschberg test was less precise compared to other methods like the cover test. Then in [22], Almeida presented a computational methodology to automatically diagnose strabismus through digital videos featuring a cover test, using only a workstation computer to process these videos. This method was recognized to diagnose strabismus with an accuracy value of 87%. However, the effectiveness of the method was considered only for the horizontal strabismus and it could not distinguish between the manifest strabismus and the latent strabismus.

The eye-tracking technique was also applied for strabismus examination [23–27]. Quick and Boothe [23] presented a photographic method, based on corneal light reflection for the measurement of binocular misalignment, which allowed for the measurement of eye alignment errors to fixation targets presented at any distance, throughout the subject's field of gaze. Model and Eizenman [24] built up a remote two-camera gaze-estimation system for the AHT (Automated Hirschberg Test) to measure binocular misalignment. However, the accuracy of the AHT procedure has to be verified with a larger sample of subjects, as it was studied on only five healthy infants. In [25], Pulido proposed a new method prototype to study the eye movement where gaze data were collected using the Tobii eye tracker, to conduct ophthalmic examination, including strabismus, by calculating the angles of deviation. However, the thesis focused on the development of the new method to provide repeatability, objectivity, comprehension, relevance, and independence and lacked an evaluation of patients. In [26], Chen et al. developed an eye-tracking-aided digital system for strabismus diagnosis. The subject's eye alignment condition was effectively investigated by intuitively analyzing gaze deviations, but only a strabismic person and a normal person were asked to participate in the experiments. Later, in [27], Chen et al developed a more effective eye-tracking system to acquire gaze data for strabismus recognition. Particularly, they proposed a gaze deviation image to characterize eye-tracking data and then leveraged the Convolutional Neural Networks to generate features from gaze deviation image, which finally led to an effective strabismus recognition. However, the performance of the proposed method could be further evaluated with more gaze data, especially data with different strabismus types.

In this study, we have proposed an intelligent evaluation system for strabismus. Intelligent evaluation of strabismus, which could also be termed an automatic strabismus assessment, assesses strabismus without ophthalmologists. We developed a set of intelligent evaluation systems in digital videos based on a cover test, in which an automatic stimulus module, controlled by chips, was used to generate the cover action of the occluder; the synchronous tracking module was used to monitor and record the movement of eyes; and the algorithm module was used to analyze the data and generate the measurement results of strabismus.

The rest of paper is organized as follows. A brief introduction of the system is given in Section 2. The methodology exploited for strabismus evaluation is described in detail in Section 3. Then, in Section 4, the results achieved by our methodology are presented, and in Section 5, some conclusions are drawn and a prospect of future work is discussed.

2. System Introduction

In our work, we have developed a set of intelligent evaluation systems of strabismus, in which the subject needs to sit on the chair with his chin on the chin rest and fixate on the target. The cover tests are automatically performed and finally a diagnosis report is generated, after a short while. The system, as shown in Figure 1, can be divided into three parts, i.e., the automated stimulus module for realizing the cover test, the video acquisition module for motion capture, and the algorithm module for detection and measurement of strabismus. More details of the system have been presented in our previous work [28].

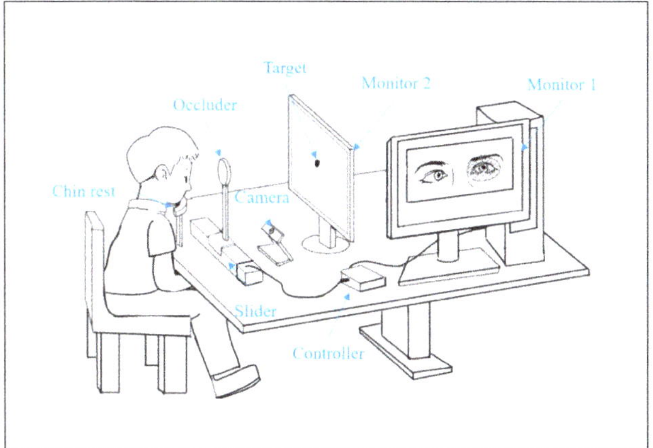

Figure 1. The proposed intelligent evaluation system of strabismus.

2.1. Hardware Design

The automated stimulus module is based on a stepper motor connected to the controller, a control circuit, which makes the clinical cover test automated. The stepper motor used in the proposed system is iHSV57-30-10-36, produced by JUST MOTION CONTROL (Shenzhen, China). The occluder is hand-made cardboard, 65 mm wide, 300 mm high, and 5 mm thick. The subject's sight is completely blocked when the occluder occludes the eye so that our method can properly simulate the clinical cover test. XC7Z020, a Field Programmable Gate Array (FPGA), is the core of the control circuit. The communication between the upper computer and the FPGA is via a Universal Serial Bus (USB). The motor rotates at a particular speed in a particular direction, clockwise or counterclockwise, to drive the left and right movement of the occluder on the slider, once the servo system receives the control signals from the FPGA.

As for the motion-capture module, the whole process of the test is acquired by the high-speed camera RER-USBFHD05MT with a 1280 × 720 resolution at 60 fps. A near-infrared led array with a spectrum of 850 nm and a near-infrared lens were selected to compensate for the infrared light illumination and separately reduce the noise from the visible light. AMCap is used to perform the control of the camera, such as the configuration of the frame rate and resolution, exposure time, the start and end of a recording, and so on.

Being ready to execute the strabismus evaluation, the subject is told to sit in front of the workbench with chin on the chin rest and fixate on the given target. The target is a cartoon character displayed on a MATLAB GUI on Monitor 2, for the purpose of attracting attention, especially for children. The experimenter sends a code "0h07" (the specific code for automatic cover test) to the system, and the stimulus module reacts to begin the process of the cover test. Meanwhile, the video acquisition application AMCap automatically starts recording. When the cover test ends, AMCap stops recording and saves the video in a predefined directory. Then the video is fed into the proposed intelligent algorithm performing the strabismus evaluation. Finally, a report is generated automatically which contains the presence, type, and degree of strabismus.

2.2. Data Collection

In the cover test, the examiner covers one of the subject's eyes and uncovers it, repeatedly, to see whether the non-occluded eye moves or not. If the movement of the uncovered eye can be observed, the subject is thought to have strabismus. The cover test can be divided into three subtypes—the unilateral cover test, the alternating cover test, and the prism alternate cover test. The unilateral cover test is used principally to reveal the presence of a strabismic deviation. If the occlusion time is extended, it is also called the cover-uncover test [29]. The alternating cover test is used to quantify the deviation [30]. The prism alternate cover test is known as the gold standard test to obtain the angle of ocular misalignment [31]. In our proposed system, we sequentially performed the unilateral cover test, the alternate cover test, and the cover-uncover test, for each subject, to check the reliability of the assessment.

The protocol of the cover tests is as follows. Once the operator sends out the code "0h07", the automatic stimulus module waits for 6 s to let the application "AMCap" react, and then the occlusion operation begins. The occluder is initially held in front of the left eye. The first is the unilateral cover test for the left eye—the occluder moves away from the left eye, waiting for 1 s, then moves back to cover the left eye for 1 s. This process is repeated for three times. The unilateral cover test for the right eye is the same as that of the left eye. When this procedure ends, the occluder is at the position of occluding the right eye. Then the alternate cover test begins. The occluder moves to the left to cover the left eye for 1 s and then moves to the right to cover the right eye. This is considered as one round, and it needs to be repeated for three rounds. Finally, the cover-uncover test is performed for both eyes. The only difference from the above unilateral cover test is that the time of the occluder's occluding eyes is increased to 2 s. Eventually, the occluder returns to the initial position.

We cooperated with the Hong Kong Association of Squint and Double Vision Sufferers to collect strabismic data. In total 15 members of the association consented to participate in our experiments. In addition to the 15 adult subjects, 4 children were invited to join our study. The adults and children, including both male and female, were within the age ranges of 25 to 63 years and 7 to 10 years, respectively. The camera was configured to capture a resolution of 1280 × 720 pixels at a frame rate of 60 fps. The distance between the target and eyes was 33 cm. If wearing corrective lenses, the subject was requested to perform the tests twice, once wearing it and once without it. After ethics approval and informed consent, the 19 subjects followed the data acquisition procedure introduced above, to participate in the experiments. Finally, 24 samples were collected, five of which were with glasses.

3. Methodology

To assess the strabismus, it is necessary to determine the extent of unconscious movement of eyes when applying the cover test. To meet the requirement, a method consisting of six stages is proposed, as shown in Figure 2. (1) The video data are first processed to extract the eye regions, to get ready for the following procedures. (2) The iris measure and template is detected to obtain its size for the further calculations and segment region for the template matching. (3) The key frame is detected to locate the position at which the stimuli occur. (4) The pupil localization is performed to identify the coordinates of the pupil location. (5) Having detected the key frame and pupil, the deviation of eye movements can be calculated. (6) This is followed by the strabismus detection stage that can obtain the prism diopter of misalignment and classify the type of strabismus. Details of these stages of the method are described below.

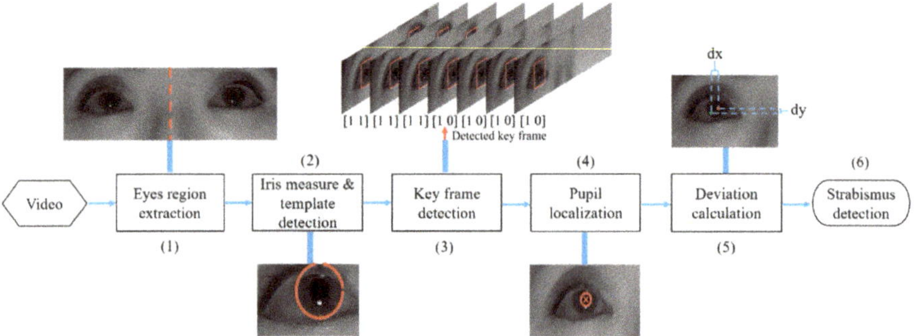

Figure 2. The workflow of the proposed method for the assessment of strabismus.

3.1. Eye Region Extraction

At this stage, a fixed sub-image containing the eye regions, while excluding regions of no interest (like nose and hair), is extracted to reduce the search space for the subsequent steps. In our system, the positions of the subject and the camera remain the same so that the data captured by the system show a high degree of consistency, that is, half of the face from the tip of the nose to the hair occupies the middle area of the image. This information, known as a priori, together with the anthropometric relations, can be used to quickly identify the rough eye regions. The boundary of the eye regions can be defined as

$$p_x = 0.2 \times W, \ p_y = 0.4 \times H, \ w = 0.6 \times W, \ h = 0.3 \times H, \tag{1}$$

where W and H are the width and height of the image, w and h are the width and height of the eye regions, and $\left(p_x, p_y\right)$ defines the top-left position of the eye regions, respectively, as shown in Figure 3.

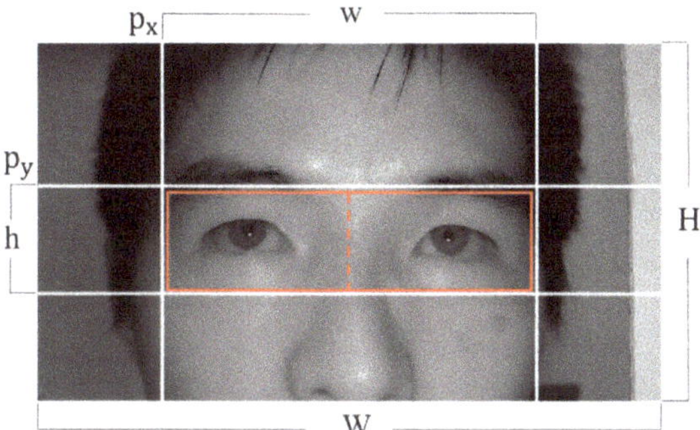

Figure 3. The rough localization of eye regions of the image in our database. Half of the face occupies the middle of the image in our database. The eye region is highlighted by the red window, which is slightly smaller than the actual eye region, in order to obtain a clearer display. The right and left eye region can be obtained by dividing the eye region into two parts.

Thus, the eye regions are extracted, and the right and left eye can be distinguished by dividing the area into two parts, of which the area with smaller x coordinate corresponds to the right eye and vice versa, by comparing the x coordinates of the left upper corner of both eye regions.

3.2. Iris Measure and Template Detection

During this stage, the measure and template of the iris are detected. To achieve this, it is necessary to locate the iris boundary, particularly, the boundary of iris and sclera. The flowchart of this stage is shown in Figure 4. (1) First, the eye image is converted from RGB to grayscale. (2) Then the Haar-like feature is applied to the grayscale image to detect the exact eye region with the objective of further narrowing the area of interest. This feature extraction depends on the local feature of the eye, that is, the eye socket appears much darker in grayscale than the area of skin below it. The width of the rectangle window is set to be approximately the horizontal length of the eye, while the height of the window is variable within a certain range. The maximum response of this step corresponds to the eye and the skin area below it. (3) The Gaussian filter is applied to the result of (2), with the goal of smoothing and reducing noise. (4) Then, the canny method is applied as an edge-highlighting technique. (5) The circular Hough transform is employed to locate the iris boundary, due to its circular character. In performing this step, only the vertical gradients (4) are taken for locating the iris boundary [32]. This is based on the consideration that the eyelid edge map will corrupt the circular iris boundary edge map, because the upper and lower iris regions are usually occluded by the eyelids and the eyelashes. The eyelids are usually horizontally aligned. Therefore, the technique of excluding the horizontal edge map reduces the influence of the eyelids and makes the circle localization more accurate and efficient. (6) Subsequently, we segment a region with dimensions of 1.2× radius, horizontally, and 1.5× radius, vertically, on both sides from the iris center in the original gray image. The radius and iris center used in this step are the radius and center coordinates of the iris region detected in the last step. These values were chosen so that a complete iris region could be segmented without any damage. This region will be used as a template for the next stage.

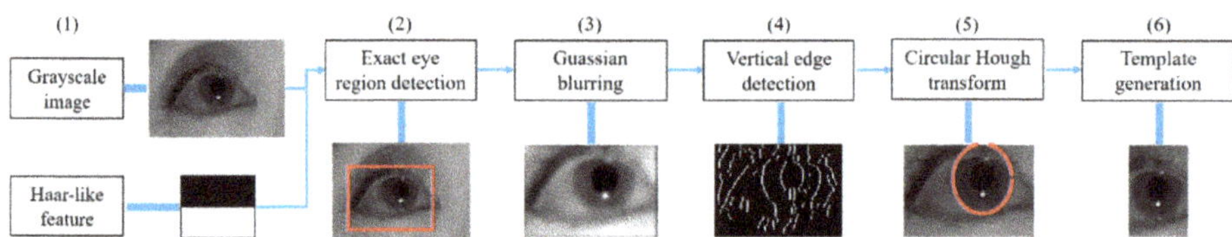

Figure 4. The flowchart of iris boundary detection. The result of the exact eye region detection is marked by the red window on the original grayscale image. The red circle represents the iris boundary.

The above operations are applied on the right and left eye regions, respectively, in a ten-frame interval, and ten pairs of iris radius values are extracted. The interval should be chosen to meet two conditions. First, the radius should be accurately determined in the interval. Second, the interval should not influence the next stage because the segmented region will be used for template matching. By the end of the interval, the iris radius value with the largest frequency is determined as the final iris radius. Thus, we have the right iris and left iris, with the radius of R_r and R_l, respectively.

3.3. Key Frame Detection

At this stage, the key frame detection is performed with the template matching technique on the eye region. The cover test is essentially a stimulus-response test. What we are interested in is whether the eye movements occur when a stimulus occurs. In the system, an entire process of tests is recorded in the video, which contains nearly 3000 frames at a length of about 50 s. We examined all frames between two stimuli. The stimuli we focused on are the unilateral cover test for the left and right eye, the alternating cover test for left and right eye, and the cover-uncover test for the left and right eye. In total, 18 stimuli are obtained with 6 types of stimuli for 3 rounds. The useful information accounts for about two-fifth of the video. Therefore, it is more efficient for the algorithm to discard these redundant information. The key frame detection is for the purpose of finding the frame where the stimulus occurs.

The right iris region segmented in Section 3.2 is used as a template, and the template matching is applied to the eye regions. The thresholds TH_1, TH_2 are set for the right eye region and the left region, respectively, and TH_2 is smaller than TH_1, as the right iris region is used as a template. The iris region is detected if the matching result is bigger than the threshold. In the nearby region of the iris, there may be many matching points which present the same iris. The repeated point can be removed by using the distance constraint. Therefore, the number of the matching template is consistent with the actual number of irises. The frame number, the number of iris detected in the right eye region, the number of the iris detected in the left region are saved in memory. Then we search the memory to find the position of the stimulus. Taking the unilateral cover test for the left eye as an example, the number of iris detected is [1 1], separately, before the left eye is covered and then [1 0], after covering the left eye. Therefore, we can use the state change from [1 1] to [1 0] to determine the corresponding frame of the stimuli. The correspondence between state changes and stimulus is shown in Table 1. Thus, the frame number of the eighteen stimulus can be obtained.

Table 1. State changes and stimulus.

State Change	Stimuli
[1 1] → [1 0]	Covering the left eye in unilateral cover test
[1 1] → [0 1]	Covering the right eye in unilateral cover test
[0 1] → [1 0]	Uncovering the right eye in alternate cover test
[1 0] → [0 1]	Uncovering the left eye in alternate cover test
[1 0] → [1 1]	Uncovering the left eye in cover-uncover test
[0 1] → [1 1]	Uncovering the right eye in cover-uncover test

3.4. Pupil Localization

The pupil localization process is used to locate the pupil, which is the dark region of the eye controlling the light entrance. The flowchart of this stage is shown in Figure 5. (1) First, the eye image is converted into grayscale. (2) The Haar-like rectangle feature, same as that in Section 3.2, is applied to narrow the eye region. (3) Then another Haar-like feature, the center-surround feature with the variable inner radius of r and outer radius of 3r, is applied to the detected exact eye region of step 2. This feature makes use of the pupil being darker than the surrounding area. Therefore, the region corresponding to the maximum response of the Haar feature is a superior estimate of the iris region. The center coordinates and radius of the Haar feature is obtained and a region can be segmented with a dimension of 1.2× radius, horizontally and vertically, on both sides from the center of the detected region, to make sure the whole pupil is in the segment. Then we perform the following techniques. (4) Gaussian filtering is used to reduce noise and smooth the image while preserving the edges. (5) The morphology opening operation is applied to eliminate small objects, separate small objects at slender locations, and restore others. (6) The edges are detected through the Canny filter, and the contour point is obtained.

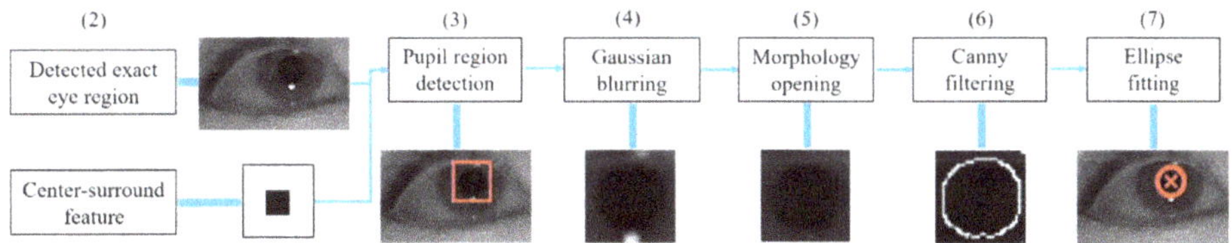

Figure 5. The flowchart of pupil localization. Steps 1 and 2 are omitted here since these two steps are the same as Steps 1 and 2 of Section 3.2, and Step 2 represents the detected exact eye region of Step 2. In (3) The segmented pupil region is marked by the red window on the detected exact region. In (4), (5), and (6), the presented images are enlarged and bilinearly interpolated for a good display. In (7), an ellipse is fit to the contour of the pupil and the red ellipse and the cross separately marks the result of fitting and the center of the pupil.

Given a set of candidate contour points of the pupil, the next step of the algorithm is to find the best fitting ellipse. (7) We applied the Random Sample Consensus (RANSAC) paradigm for ellipse fitting [33]. RANSAC is an effective technique for model fitting in the presence of a large but unknown percentage of outliers, in a measurement sample. In our application, inliers are all of those contour points that correspond to the pupil contour and outliers are contour points that correspond to other contours, like the upper and the lower eyelid. After the necessary number of iterations, an ellipse is fit to the largest consensus set, and its center is considered to be the center of the pupil. The frame number and pupil center coordinates are stored in memory.

3.5. Deviation Calculation

In order to analyze the eye movement, the deviation of the eyes during the stimulus process needs to be calculated. During a stimulus process, the position of the eye remains motionless before the stimulus occurs; after stimuli, the eye responds. The response can be divided into two scenarios. For the simpler case, the eyeball remains still. For the more complex case, the eyeball starts to move after a short while and then stops moving and keeps still until the next stimuli. Based on the statistics in the database, the eyes complete the movement within 17 frames and the duration of the movement is about 3 frames.

The schematic of deviation calculation is shown in Figure 6. The pupil position data within the region from the 6 frames before the detected key frame, to 60 frames after the key frames, are selected as a data matrix. Each row of the matrix corresponds to the frame number, the x, y coordinate of the pupil center. Next, an iterative process is applied to filter out the singular values of pupil detection. The current line of the matrix is subtracted from the previous line of the matrix, and the frame number of the previous line, the difference between the frame numbers Δf, and the difference between the coordinates Δx, Δy are retained. If $\Delta x > \Delta f.v$, where v is the statistical maximum of the offset of pupil position of two adjacent frames, then this frame is considered to be wrong and the corresponding row in the original matrix is deleted. This process iterates until no rows are deleted or the number of loops exceeds 10. Finally, we use the average of the coordinates of the first five rows of the reserved matrix as the starting position and the average of the last five rows as the ending position, thus, obtaining the deviation of each stimulus, as expressed by the equations:

$$
\begin{aligned}
\mathrm{Dev}_{\mathrm{p}}^{(x)} &= \left| x_e - x_s \right|; \\
\mathrm{Dev}_{\mathrm{p}}^{(y)} &= \left| y_e - y_s \right|,
\end{aligned}
\tag{2}
$$

where x_e and y_e are the ending positions of the eye in a stimulus, the x_s and y_s are the starting positions of the eye, and $\mathrm{Dev}_{\mathrm{p}}^{(x)}$, $\mathrm{Dev}_{\mathrm{p}}^{(y)}$ are the horizontal and vertical deviations in pixels, respectively.

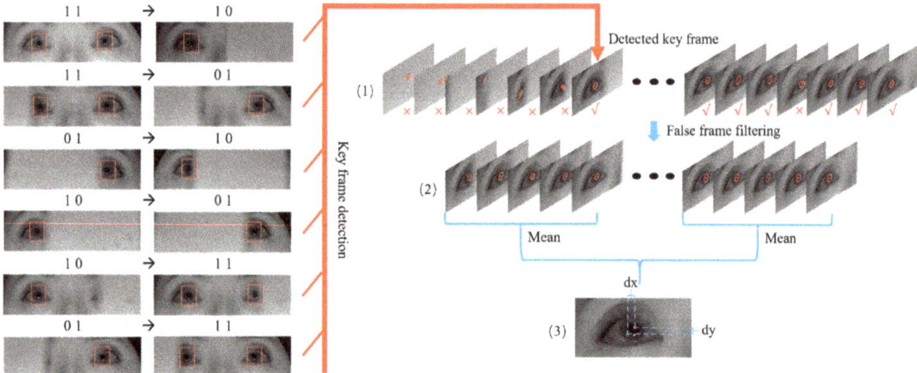

Figure 6. The schematic of deviation calculation. (1) The input data consist of the interval from 6 frames before the key frame to 60 frames after the key frame. The frame pointed to by the red arrow represents the key frame detected, and the symbol "×" "√" below the image indicates the abnormality or normality of the pupil detection. (2) With the false frame filtering completed, the abnormal frames are deleted while the normal frames are reserved. The average of the pupil locations of the first five frames is calculated as the starting position, while the average of the pupil locations of the last five frames is the ending position. (3) The deviations in pixels are calculated. For an intuitive show, the starting position represented by the green dot and the ending position represented by the red dot are matched into one eye image, indicating the size of the image. "dx", "dy" represent the deviation in horizontal and vertical directions, respectively.

3.6. Strabismus Detection

Obtained deviation of each stimulus, the deviation value in pixel Dev_p can be converted into prism diopters Dev_Δ, which is calculated out using the equation:

$$Dev_\Delta = \left(\frac{DE_{mm}}{DE_p} \right) \cdot dpMM \cdot Dev_p, \tag{3}$$

where DE_{mm} is the value of the mean diameter of iris boundary of adult patients and $DE_{mm} = 11$ mm [34], DE_p is the diameter value of the iris boundary detected in pixels, dpMM is a constant in millimeter conversion for prismatic diopters (Δ/mm) and dpMM $= 15\Delta$ [35]. Finally, we have the deviation Dev_Δ expressed in diopter.

The subject's deviation values are calculated separately for different cover tests. The subject's deviation value for each test is the average of the deviations calculated for both eyes. These values are used to detect the presence or absence of strabismus.

The types of the strabismus can first be classified as manifest strabismus or latent strabismus. According to the direction of deviation, it can be further divided into—horizontal eso-tropia (phoria), exo-tropia (phoria), vertical hyper-tropia (phoria), or hypo-tropia (phoria). The flowchart of the judgment of strabismus type is shown in Figure 7. If the eyes move in the unilateral cover test, the subject will be considered to have manifest strabismus and the corresponding type can be determined too, so it is unnecessary to consider the alternate cover test and the cover-uncover test and assessment ends. Nevertheless, if the eye movement does not occur in the unilateral test stage, there is still the possibility of latent strabismus, in spite of the exclusion of the manifest squint. We proceed to explore the eye movement in the alternating cover test and cover-uncover test. If the eye movement is observed, the subject is determined with heterophoria and then the specific type of strabismus is given on the basis of the direction of eye movement. If not, the subject is diagnosed as normal.

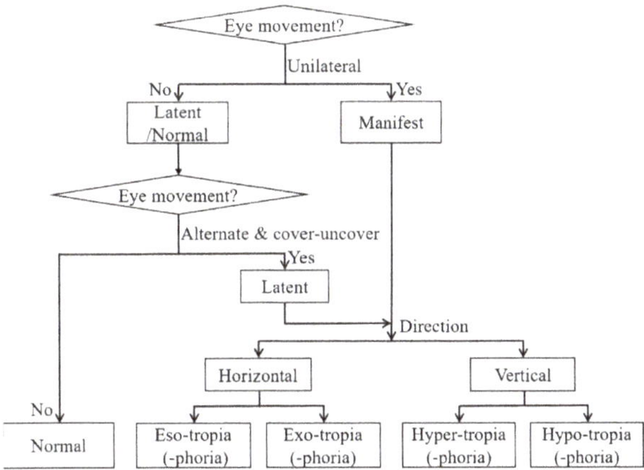

Figure 7. The flowchart of the judgment of strabismus types.

4. Experimental Results

In this section, the validation results of the intelligent strabismus evaluation method are presented, including the results of iris measure, key frame detection, pupil localization, and the measurement of the deviation of strabismus. In order to verify the effectiveness of the proposed automated methods, the ground truths of deviations in prism diopters were provided by manually observing and calculating the deviations of eyes for all samples. The measures of the automated methods have been compared with the ground truths.

4.1. Results of Iris Measure

With the eye regions extracted, an accuracy of 100% was achieved in detecting the iris measure. The range of values that defines the minimum and maximum radius size for Hough transform was empirically identified to be between 28 and 45 pixels, for our database. Due to our strategy for choosing the radius with the largest frequency in the interval, the radius of the iris could be accurately obtained even if there were individual differences or errors. An example of the iris measure in 10 consecutive frames is shown in Figure 8. As can be seen from the figure, in an interval of 10 frames, there were 8 frames detected with a radius of "33" and 2 frames detected with a radius of "34", so the radius of the iris was determined to be 33.

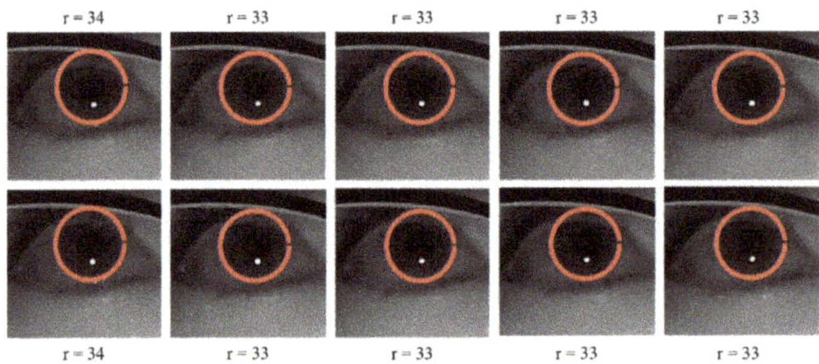

Figure 8. An example of the iris measure in 10 consecutive frames. The iris boundary located by the methodology is marked by the red circle. The radius of the iris was determined to be 33 pixels, according to this strategy.

4.2. Results of Key Frame Detection

In order to measure the accuracy of the key frame detection, the key frames of all samples observed and labeled, manually, were regarded as the ground truths. The distance $D^{(f)}$ of the key detected frame f_p and the manual ground truths f_g, was calculated using the equation:

$$D^{(f)} = \left| f_p - f_g \right| \tag{4}$$

The accuracy of the key frame detection could be measured by calculating the percentage of the key frames for which the distance $D^{(f)}$ was within a threshold in the frames. The accuracy of the key frame detection for each cover test was given, as shown in Table 2.

Table 2. The accuracy of the key frame detection for cover tests.

	Unilateral Cover Test	Alternating Cover Test	Cover-Uncover Test
$D^{(f)} \leq 2$	93.1%	62.5%	85.4%
$D^{(f)} \leq 4$	97.9%	88.2%	89.6%
$D^{(f)} \leq 6$	97.9%	97.2%	91.0%

Taking the unilateral cover test as an example, the detection accuracy was 93.1%, 97.9%, and 97.9%, at a distance of within 2, 4, and 6 frames, separately. As we can see, the detection rate in the alternating cover test was lower than that in others within the 2 and 4 frames intervals. This could be attributed to the phantom effects which might occur with the rapid motion of the occluder. It might interfere with the detection in the related frames, as the residual color of the trace left by the occluder merges with the color of the eyes. The movement of the occluder between two eyes brings more perturbation than that on one side. The detection rate appears good results for each cover test when the interval is set within 6 frames. As the deviation calculation method (Section 3.5) relaxes the reliance on key frame detection, our method could get a promising result.

4.3. Results of Pupil Localization

The accuracy of the proposed pupil detection algorithm was tested on static eye images on the dataset we built. The dataset consists of 5795 eye images with a resolution of 300×150 pixels for samples without wearing corrective lenses and 400×200 pixels for samples wearing lenses. All images were from our video database. The pupil location was manually labeled as the ground truth data for analysis.

In order to appreciate the accuracy of the pupil detection algorithm, the Euclidean distance $D_P^{(E)}$ between the detected and the manually labeled pupil coordinates, as well as the distance $D_P^{(x)}$ and $D_P^{(y)}$ on both axes of the coordinate system was calculated for the entire dataset. The detection rate measured in individual directions had a certain reference value, as the mobility of eyes has two degrees of freedom. The accuracy of the pupil localization could be measured by calculating the percentage of the eye pupil images for which the pixel error was lower than a threshold in pixels. We compared our pupil detection method with the classical Starburst [33] algorithm and circular Hough transform (CHT) [36]. The performance of pupil localization with different algorithms is illustrated in Table 3. The accuracy rates of the following statistical indicators were used: "$D_P^{(E)} < 5$" and "$D_P^{(E)} < 10$" corresponded to the detection rate, at 5 and 10 pixels, in Euclidean distance; "$D_P^{(x)} < 4$" or "$D_P^{(y)} < 2$" represented the percentage of the eye pupil images for which the pixel error was lower than 4 pixels in horizontal direction or 2 pixels in vertical direction.

Table 3. Performance of pupil localization with different algorithms on our dataset.

Method	$D_P^{(E)} < 5$	$D_P^{(E)} < 10$	$D_P^{(x)} < 4$	$D_P^{(y)} < 2$
Starburst [33]	27.0%	44.2%	39.9%	27.8%
CHT [36]	84.6%	85.0%	84.6%	83.4%
Ours	86.6%	94.3%	90.6%	80.7%

As we can see, the performance of the Starburst algorithm was much poorer, which was due to the detection of the pupil contour points, using the iterative feature-based technique. In this step, the candidate feature points that belonged to the contour points were determined along the rays that extended radially away from the starting point, until a threshold $\varnothing = 20$ was exceeded. For our database, there were many disturbing contour points detected, especially the limbus. This could cause the final stage to find the best-fitting ellipse for a subset of candidate feature points by using the RANSAC method to misfit the limbus. The performance of the CHT was acceptable, but it was highly dependent on the estimate of the range of the radius of the pupil. There might have been overlaps between the radius of the pupil and the limbus for different samples, which made the algorithm invalid for some samples. While our method shows a good detection result overall.

Actually, the overall detection rate was an average result. Poor detection in some samples had lowered the overall performance. Listed below, are some cases in locating the pupil, as shown in Figure 9. Some correct detection results are shown in Figure 9a, which shows that our algorithm could get a good detect effect in most cases, even if there was interference from glasses. Some typical examples of errors are described in Figure 9b. The errors could be attributed to the following factors—(1) a large part of the pupil was covered by the eyelids so that the exposed pupil contour, together with a part of eyelid contour, were fitted to an ellipse when the model was fitted, as shown in set 1 of Figure 9b; (2) the pupil was extremely small, so the model fitting was prone to be similar to the result discussed in factor 1, as shown in set 2 of Figure 9b; (3) the difference within the iris region was not so apparent that the canny filter could not get a good edge of the pupil, thus, leading to poor results, as shown in set 3 of Figure 9b; (4) the failure detection caused by the phantom effects when the fast-moving occluder was close to the eyeball, as shown in set 4 of Figure 9b.

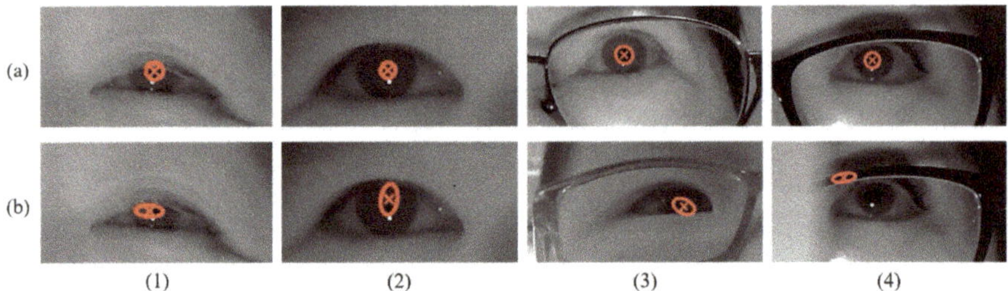

Figure 9. The pupil detection cases: (**a**) Some examples of correctly-detected images; (**b**) some typical errors.

4.4. Results of the Deviation Measurement

For analyzing the accuracy of the deviation calculated by the proposed method, the deviation of each stimulus was calculated as ground truth, by manually determining the starting position and ending position of each excitation for all samples, and labeling the pupil position of the corresponding frames, and then calculating the strabismus degrees in prism diopters. The deviations of the automated method were compared with the ground truths. The accuracy of the deviation measurement was measured by calculating the percentage of deviations for which the error of the deviation detected and manual ground truths was lower than a threshold.

The accuracy rate using different indicators are shown in Tables 4 and 5. For example, "*error* < 2" represents the percentage of the deviation calculation for which the error in prism diopters was lower than the threshold 2 in certain axes, and so on. The indicators for vertical direction was set to be more compact as the structure of the eye itself causes it to have a smaller range of motion in the vertical direction than that in the horizontal direction. The calculation accuracy was acceptable when the error was set to be 8Δ in the horizontal direction or 4Δ in the vertical direction. This conclusion could also be seen from Figure 10, which shows the correlation of deviation between the ground truth and the predicted results. Each point represents the average of three stimuli. It can be seen that most of the points were within the 8Δ or 4Δ error, and it could be considered an error as the points were outside the range. The results demonstrated a high consistency between the proposed method and the manual measurement of deviation, and that the proposed methods were effective for automated evaluations of strabismus.

Table 4. The accuracy rate of deviation calculation in prism diopters (Δ) for different cover test stages in the horizontal direction.

(Δ)	*Error* < 4	*Error* < 8	*Error* < 12
Unilateral	81.3%	95.8%	97.2%
Alternate	85.4%	93.8%	97.9%
Uncover	74.3%	91.7%	96.5%

Table 5. The accuracy rate of deviation calculation in prism diopters (Δ) for different cover test stages in the vertical direction.

(Δ)	*Error* < 2	*Error* < 4	*Error* < 6
Unilateral	70.1%	88.2%	94.4%
Alternate	71.5%	93.8%	96.5%
Uncover	68.8%	86.1%	88.2%

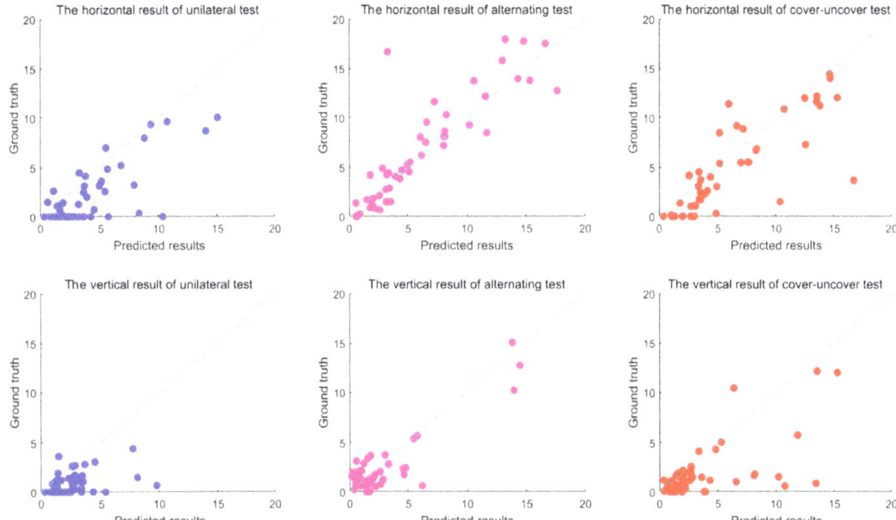

Figure 10. The correlation of the deviation between the ground truth and predicted results. The first row shows the horizontal axis for different cover tests, and the second row shows the vertical direction.

5. Conclusions and Future Work

In this paper, we proposed and validated an intelligent measurement method for strabismus deviation in digital videos, based on the cover test. The algorithms were applied to video recordings by near-infrared cameras, while the subject performed the cover test for a diagnosis of strabismus. In particular, we focused on the automated algorithms for the identification of the extent to which the eyes involuntarily move when a stimulus occurs. We validated the proposed method using the manual ground truth of deviations in prism diopters, from our database. Experimental results suggest that our automated system can perform a high accuracy of evaluation of strabismus deviation.

Although the proposed intelligent evaluation system for strabismus could achieve a satisfying accuracy, there are still some aspects to be further improved in our future work. First, for the acquisition of data, there are obvious changes in the video brightness, due to the cover of the occluder. For example, almost half of the light was blocked when one eye was covered completely. This might bring a perturbation for the algorithm, especially for the pupil detection. Therefore, our system needed to be further upgraded to reduce this interference. Second, the subjects were required to remain motionless while the cover test is performed. In fact, a slight movement of the head that is not detectable to humans will cause a certain deviation in the detection of eye position, thus, reducing the accuracy of the final evaluation. To develop a fine eye localization, eliminating slight movements would improve the result. Additionally, our system can also be used for an automatic diagnosis of strabismus, in the future.

Author Contributions: Y.Z. and H.F. conceived and designed the experiments. Y.Z. and R.L. performed the experiments. Y.Z. and H.F. analyzed the data and wrote the manuscript. Y.Z., H.F., W.-L.L., Z.C., D.D.F., Z.S. and D.W. contributed to revising the manuscript.

Funding: The research was fully funded by a grant from the Research Grants Council of the Hong Kong Special Administrative Region, China (Project Reference No.: UGC/FDS13/E01/17).

Acknowledgments: We would like to thank the Hong Kong Association of Squint and Double Vision Sufferers for their cooperation in collecting the experimental data.

Conflicts of Interest: The authors declare no conflict of interest.

References

1. Beauchamp, G.R.; Black, B.C.; Coats, D.K.; Enzenauer, R.W.; Hutchinson, A.K.; Saunders, R.A. The management of strabismus in adults-I. clinical characteristics and treatment. *J. AAPOS* **2003**, *7*, 233–240. [CrossRef]
2. Graham, P.A. Epidemiology of strabismus. *Br. J. Ophthalmol.* **1974**, *58*, 224–231. [CrossRef] [PubMed]
3. Castanes, M.S. Major review: The underutilization of vision screening (for amblyopia, optical anomalies and strabismus) among preschool age children. *Binocul. Vis. Strabismus Q.* **2003**, *18*, 217–232. [PubMed]
4. Mojon-Azzi, S.M.; Kunz, A.; Mojon, D.S. The perception of strabismus by children and adults. *Graefes Arch. Clin. Exp. Ophthalmol.* **2010**, *249*, 753–757. [CrossRef] [PubMed]
5. Jackson, S.; Harrad, R.A.; Morris, M.; Rumsey, N. The psychosocial benefits of corrective surgery for adults with strabismus. *Br. J. Ophthalmol.* **2006**, *90*, 883–888. [CrossRef] [PubMed]
6. Klauer, T.; Schneider, W.; Bacskulin, A. Psychosocial correlates of strabismus and squint surgery in adults. *J. Psychosom. Res.* **2000**, *48*, 251–253.
7. Menon, V.; Saha, J.; Tandon, R.; Mehta, M.; Sudarshan, S. Study of the Psychosocial Aspects of Strabismus. *J. Pediatr. Ophthalmol. Strabismus* **2002**, *39*, 203–208. [CrossRef] [PubMed]
8. Merrill, K.; Satterfield, D.; O'Hara, M. Strabismus surgery on the elderly and the effects on disability. *J. AAPOS* **2009**, *14*, 196–198. [CrossRef] [PubMed]
9. Bez, Y.; Coskun, E.; Erol, K.; Cingu, A.K.; Eren, Z.; Topcuoglu, V.; Ozerturk, Y.; Erol, M.K. Adult strabismus and social phobia: A case-controlled study. *J. Am. Assoc. Pediatr. Ophthalmol. Strabismus* **2009**, *13*, 249–252. [CrossRef] [PubMed]
10. Nelson, B.A.; Gunton, K.B.; Lasker, J.N.; Nelson, L.B.; Drohan, L.A. The psychosocial aspects of strabismus in teenagers and adults and the impact of surgical correction. *J. Am. Assoc. Pediatr. Ophthalmol. Strabismus* **2008**, *12*, 72–76.e1. [CrossRef] [PubMed]
11. Egrilmez, E.D.; Pamukçu, K.; Akkin, C.; Palamar, M.; Uretmen, O.; Köse, S. Negative social bias against children with strabismus. *Acta Ophthalmol. Scand.* **2003**, *81*, 138–142.
12. Mojon-Azzi, S.M.; Mojon, D.S. Strabismus and employment: The opinion of headhunters. *Acta Ophthalmol.* **2009**, *87*, 784–788. [CrossRef] [PubMed]
13. Mojon-Azzi, S.M.; Potnik, W.; Mojon, D.S. Opinions of dating agents about strabismic subjects' ability to find a partner. *Br. J. Ophthalmol.* **2008**, *92*, 765–769. [CrossRef] [PubMed]
14. McBain, H.B.; Au, C.K.; Hancox, J.; MacKenzie, K.A.; Ezra, D.G.; Adams, G.G.; Newman, S.P. The impact of strabismus on quality of life in adults with and without diplopia: a systematic review. *Surv. Ophthalmol.* **2014**, *59*, 185–191. [CrossRef] [PubMed]
15. Douglas, G.H. The Oculomotor Functions & Neurology CD-ROM. Available online: http://www.opt.indiana.edu/v665/CD/CD_Version/CONTENTS/TOC.HTM (accessed on 1 September 2018).
16. Anderson, H.A.; Manny, R.E.; Cotter, S.A.; Mitchell, G.L.; Irani, J.A. Effect of Examiner Experience and Technique on the Alternate Cover Test. *Optom. Vis. Sci.* **2010**, *87*, 168–175. [CrossRef]
17. Hrynchak, P.K.; Herriot, C.; Irving, E.L. Comparison of alternate cover test reliability at near in non-strabismus between experienced and novice examiners. *Ophthalmic Physiol. Opt.* **2010**, *30*, 304–309. [CrossRef] [PubMed]
18. Helveston, E.M.; Orge, F.H.; Naranjo, R.; Hernandez, L. Telemedicine: Strabismus e-consultation. *J. Am. Assoc. Pediatr. Ophthalmol. Strabismus* **2001**, *5*, 291–296. [CrossRef]
19. Yang, H.K.; Seo, J.-M.; Hwang, J.-M.; Kim, K.G. Automated Analysis of Binocular Alignment Using an Infrared Camera and Selective Wavelength Filter. *Investig. Ophthalmol. Vis. Sci.* **2013**, *54*, 2733–2737. [CrossRef]
20. Min, W.S.; Yang, H.K.; Hwang, J.M.; Seo, J.M. The Automated Diagnosis of Strabismus Using an Infrared Camera. In Proceedings of the 6th European Conference of the International Federation for Medical and Biological Engineering, Dubrovnik, Croatia, 7–11 September 2014; Volume 45, pp. 142–145.
21. De Almeida, J.D.S.; Silva, A.C.; De Paiva, A.C.; Teixeira, J.A.M.; Paiva, A. Computational methodology for automatic detection of strabismus in digital images through Hirschberg test. *Comput. Biol. Med.* **2012**, *42*, 135–146. [CrossRef]
22. Valente, T.L.A.; De Almeida, J.D.S.; Silva, A.C.; Teixeira, J.A.M.; Gattass, M. Automatic diagnosis of strabismus in digital videos through cover test. *Comput. Methods Prog. Biomed.* **2017**, *140*, 295–305. [CrossRef]

23. Quick, M.W.; Boothe, R.G. A photographic technique for measuring horizontal and vertical eye alignment throughout the field of gaze. *Investig. Ophthalmol. Vis. Sci.* **1992**, *33*, 234–246.
24. Model, D.; Eizenman, M. An Automated Hirschberg Test for Infants. *IEEE Trans. Biomed. Eng.* **2011**, *58*, 103–109. [CrossRef] [PubMed]
25. Pulido, R.A. Ophthalmic Diagnostics Using Eye Tracking Technology. Master's Thesis, KTH Royal Institute of Technology, Stockholm, Sweden, 2012.
26. Chen, Z.; Fu, H.; Lo, W.L.; Chi, Z. Eye-Tracking Aided Digital System for Strabismus Diagnosis. In Proceedings of the IEEE International Conference on Systems, Man, and Cybernetics 2016, Budapest, Hungry, 9–12 October 2016.
27. Chen, Z.; Fu, H.; Lo, W.-L.; Chi, Z. Strabismus Recognition Using Eye-Tracking Data and Convolutional Neural Networks. *J. Healthc. Eng.* **2018**, *2018*, 1–9. [CrossRef] [PubMed]
28. Zheng, Y.; Fu, H.; Li, B.; Lo, W.L.; Wen, D. An Automatic Stimulus and Synchronous Tracking System for Strabismus Assessment Based on Cover Test. In Proceedings of the International Conference on Intelligent Informatics and Biomedical Sciences, Bangkok, Thailand, 21–24 October 2018; Volume 3, pp. 123–127.
29. Barnard, N.A.S.; Thomson, W.D. A quantitative analysis of eye movements during the cover test—A preliminary report. *Ophthalmic Physiol. Opt.* **1995**, *15*, 413–419. [CrossRef]
30. Peli, E.; McCormack, G. Dynamics of Cover Test Eye Movements. *Optom. Vis. Sci.* **1983**, *60*, 712–724. [CrossRef]
31. Wright, K.; Spiegel, P. *Pediatric Ophthalmology and Strabismus*, 2nd ed.; Springer Science and Business Media: Berlin, Germany, 2013.
32. Wildes, R.; Asmuth, J.; Green, G.; Hsu, S.; Kolczynski, R.; Matey, J.; McBride, S. A system for automated iris recognition. In Proceedings of the Second IEEE Workshop on Applications of Computer Vision 1994, Sarasota, FL, USA, 5–7 December 1994; pp. 121–128.
33. Winfield, D.; Li, D.; Parkhurst, D.J. Starburst: A hybrid algorithm for video-based eye tracking combining feature-based and model-based approaches. In Proceedings of the 2005 IEEE Computer Society Conference on Computer Vision and Pattern Recognition (CVPR'05)–Workshops, San Diego, CA, USA, 21–23 September 2005.
34. Khng, C.; Osher, R.H. Evaluation of the relationship between corneal diameter and lens diameter. *J. Cataract Refract. Surg.* **2008**, *34*, 475–479. [CrossRef] [PubMed]
35. Schwartz, G.S. *The Eye Exam: A Complete Guide*; Slack Incorporated: Thorofare, NJ, USA, 2006.
36. Cherabit, N.; Djeradi, A.; Chelali, F.Z. Circular Hough Transform for Iris localization. *Sci. Technol.* **2012**, *2*, 114–121. [CrossRef]

 *applied sciences*

Article

Application of a Real-Time Visualization Method of AUVs in Underwater Visual Localization

Ran Wang, Xin Wang *, Mingming Zhu and Yinfu Lin

School of Mechanical Engineering and Automation, Harbin Institute of Technology (Shenzhen), Shenzhen 518055, China; wangran@stu.hit.edu.cn (R.W.); zhumingminghit@outlook.com (M.Z.); linyinfu@stu.hit.edu.cn (Y.L.)
* Correspondence: wangxinsz@hit.edu.cn

Received: 4 March 2019; Accepted: 1 April 2019; Published: 4 April 2019

Abstract: Autonomous underwater vehicles (AUVs) are widely used, but it is a tough challenge to guarantee the underwater location accuracy of AUVs. In this paper, a novel method is proposed to improve the accuracy of vision-based localization systems in feature-poor underwater environments. The traditional stereo visual simultaneous localization and mapping (SLAM) algorithm, which relies on the detection of tracking features, is used to estimate the position of the camera and establish a map of the environment. However, it is hard to find enough reliable point features in underwater environments and thus the performance of the algorithm is reduced. A stereo point and line SLAM (PL-SLAM) algorithm for localization, which utilizes point and line information simultaneously, was investigated in this study to resolve the problem. Experiments with an AR-marker (Augmented Reality-marker) were carried out to validate the accuracy and effect of the investigated algorithm.

Keywords: underwater visual localization method; line segment features; PL-SLAM

1. Introduction

Underwater research has been evolving rapidly during the last few decades and autonomous underwater vehicles (AUVs), as an important part of underwater research, are widely used in harsh underwater environments instead of human exploration. However, it is a tough challenge to guarantee the underwater location accuracy of AUVs. Currently, many methods are used to position AUVs, such as inertial measurement units (IMUs), Doppler velocity logs (DVLs), pressure sensors, sonar and visual sensors. For example, Fallon et al. used a side-scan sonar and a forward-look sonar as perception sensors in an AUV for mine countermeasures and localization [1]. The graph was initialized by pose nodes from a GPS, and a nonlinear least square optimization was performed for the dead-reckoning (DVL and IMU) sensor and sonar images. However, sonars are susceptible to interference from the water surface and other sources of sound reflection in shallow water areas. To localize the AUV, inertial units (from an accelerometer or gyroscope), DVLs and pressure sensors are fused by Kalman filter [2]. However, this approach is prone to generating drift without a periodical correction based on a loop closing detection mechanism.

AUVs needs to have good positioning accuracy in shallow water areas, and this paper introduces a periodic correction based on a closed-loop detection mechanism to further improve positioning accuracy. Therefore, the visual simultaneous localization and mapping (SLAM) localization method was chosen in this study. In shallow water areas, visual sensors are better than sonar because they cannot be affected by reflection. The SLAM [3] technique has proved to be one of the most popular and available methods to perform precise localization in unknown environments. When the AUV reaches a position that it has been to before, SLAM provides a loop detection mechanism that eliminates cumulative errors and drift.

Although the quality of the picture will be seriously affected by the scattering and attenuation of underwater light and poor illumination conditions, cameras have larger spatial and temporal resolutions compared to acoustic sensors, which makes cameras more suitable for certain applications, such as surveying, object identification and intervention [4]. Hong and Kim proposed a computationally efficient approach that could be applied in visual simultaneous localization and mapping for the autonomous inspection of underwater structures using monocular vision [5]. Jung et al. proposed a vision-based simultaneous localization and mapping of AUVs in which underwater artificial landmarks were used to help the visual sensing of forward- and downward-looking cameras [6]. Kim and Eustice performed a visual SLAM using monocular video images and utilized a special saliency method using local and global saliency for feature detection in hull inspection [7]. Carrasco et al. proposed the stereo graph-SLAM algorithm for the localization and navigation of underwater vehicles, which optimized the vehicle trajectory and processed the features from the graph [8]. Negre et al. proposed a novel technique to detect loop closings visually in underwater environments to increase the accuracy of vision-based localization systems [9].

All of the above-mentioned visual SLAM methods are based on point feature localization. However, the above methods would cause instability of the system because of the low-texture in many underwater environments, which contain a small number of point features. However, there are rich planar elements in the linear shapes in many low-texture environments, from which line segment features can be extracted. Based on the oriented FAST (Features From Accelerated Segment Test) and rotated brief SLAM (ORB-SLAM) [10] framework, point and line SLAM PL-SLAM [11] can simultaneously utilize point and line information. As suggested in [12], lines are parameterized by their endpoints, the precise locations of which are estimated by following a two-step optimization process in the image plane. In this representation, lines were integrated within the SLAM machinery as if they were points and were hence able to be processed by reusing the ORB-SLAM [10] architecture.

The line segment detector (LSD) method [13] was applied to extract line segments, as it has high precision and repeatability. For stereo matching and frame-to-frame tracking, line segments are augmented with a binary descriptor provided by the line band descriptor (LBD) method [14], which is useful to find correspondences among lines based on their local appearance. The characteristics of LBD were used in the closed-loop detection mechanism outlined below.

In this paper, a visual-based underwater location approach is presented. In Section 2, the PL-SLAM method is presented for real-time underwater localization, including estimation of the real-time position of the AUV, closed loop detection and a closed-loop optimization algorithm based on point-line features. Section 3 is the experimental section, in which the errors of positioning for linear and arbitrary trajectory motions are evaluated. Section 4 summarizes the experimental results and analyzes the problems.

2. Methodology

In the underwater stereo visual localization algorithm, both ORB [10] feature points and LSD [13] line features were selected in this study. The system is composed of three main modules, including tracking, local mapping and loop closing. In the tracking thread, the camera's position is estimated and the timing of when to add new key-frames is decided. In the local mapping thread, the new key-frame information is added into the map and it is optimized using bundle adjustment (BA). In the loop closing thread, loops are checked and corrected constantly.

2.1. Line Segment Feature Algorithm

Point features are mainly corner points in the image. At present, there are a lot of widely used point feature detection and extraction algorithms such as scale invariant feature transform (SIFT), speeded up robust features (SURF) and oriented FAST and rotated brief (ORB). Compared to other algorithms, ORB has a high extraction speed and good real-time performance while maintaining the invariance of feature sub-rotation and scale. Since it is likely that sufficient point features of the splicing registration algorithm cannot be effectively extracted in underwater environments, the robustness of the algorithm decline. In contrast, line features are mainly the edges of objects in the image. The depth information in the line segment feature changes less and thus the line segment feature is easier to extract underwater. Thus, the robustness of the algorithm is improved.

Line segment detection is an important and frequently used application in computer vision. In the traditional method, the Canny edge detector is used to extract the edge information and then the line segments consisting of edge points exceeding the set threshold are extracted by Hough transform. Finally, the length thresholds are used to select these line segments. There are serious defects in extracting straight lines by Hough transformation, and error detection will occur at a high edge density. In addition, this method has high time complexity and cannot be used as the line segment extraction algorithm for underwater real-time positioning.

As shown in Figure 1, LSD is an extraction algorithm that extracts sub-pixel precision line segments in linear time. It uses heuristic search and inverse verification methods to achieve sub-pixel precision in linear time without setting any parameters. It defines a line segment as the image region, called the line-support region, which is a straight region the points of which have roughly the same image gradient angle. Finally, the judgement as to whether the line-support region is a line segment is conducted by counting the number of aligned points.

Figure 1. Feature map of underwater extraction lines.

2.2. Motion Estimation

After extracting the ORB point features and LSD line segment features, feature matching between continuous frames is carried out by the traditional violent matching method. After the correspondences between two stereo frames are established, the key points and line segments of the first frame is projected to the next frame. In order to estimate the movements, robust Gauss–Newton minimization is used to reduce the error of the line and key-point projections [11].

The difference between the transformed coordinates of the 3D point in the first frame image and in the second frame image is denoted as the re-projection error of the point. It can be solved by Equation (1):

$$\Delta p_i(\xi) = \hat{p}_i(\xi) - p_i'$$

(1)

where p_i' is the coordinates of the three dimensional points on the second frame, and p_i is the coordinates of the three dimensional points on the first frame. The traditional violent matching method indicates the information that features p_i in frame n and that corresponds to feature p_i' in frame $n + 1$. $\hat{p}_i(\xi)$ is the coordinates of the point projected onto the second frame from the first frame image after the transformation of ξ. ξ is the motion transformation matrix between two frames, including rotation and displacement. i is the sequence number of the points feature. The first frame and the second frame are consecutive in time.

The sum of the distances between the two transformed corresponding endpoints of the two frame images is the re-projection error of the line segment (see Figure 2). It can be solved by Equation (2):

$$\Delta l_j(\xi) = \left[l_j'^T \cdot \left[\; \hat{p}_j[\xi] \quad \hat{q}_j[\xi] \; \right] \right]^T \tag{2}$$

where l_j' is the corresponding line segment feature on the second frame image, and $l_j' = p_j' \times q_j'$. p_j' and q_j' are the coordinates of two endpoints of a line segment feature transformed on the second frame image. $\hat{p}_j(\xi)$ and $\hat{q}_j(\xi)$ are the coordinates of two endpoints of a line segment feature transformed from the first frame image to the second frame image. $l_j'^T \cdot \hat{p}_j(\xi)$ is the distance from $\hat{p}_j(\xi)$ to $l_j'^T$. $l_j'^T \cdot \hat{q}_j(\xi)$ is the distance from $\hat{q}_j(\xi)$ to $l_j'^T$. j is the sequence number of the line segment feature.

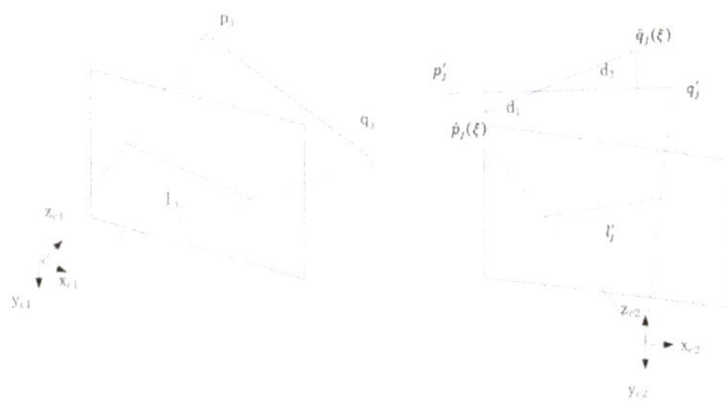

Figure 2. The re-projection error of the line segments.

For the point-line-feature pairs on the two frames, the pose transformation of two frames can be obtained by minimizing the sum of the re-projection errors. It can be solved by Equation (3):

$$\xi^* = argmin \left\{ \sum_i^{N_p} \Delta p_i\{\xi\}^T \sum_{\Delta p_i}^{-1} \Delta p_i\{\xi\} + \sum_j^{N_l} \Delta l_i\{\xi\}^T \sum_{\Delta l_i}^{-1} \Delta l_i\{\xi\} \right\} \tag{3}$$

where $\sum_{\Delta p_i}^{-1} \Delta p_i(\xi)$ is the inverse of the covariance matrix of the point re-projection error. $\sum_{\Delta l_i}^{-1} \Delta l_i(\xi)$ is the inverse of the covariance matrix of the re-projection error of a line.

Although the motion transformation matrix ξ can be obtained according to Equations (1)–(3), it still has some errors because of the mistaken matching of point features and line segment features. Thus we called it an "estimate". These errors are eliminated in the closed-loop optimization in the following section.

2.3. Closed Loop Detection and Closed-Loop Optimization Algorithm Based on Point-Line Features

Closed loop detection is mainly used to judge whether the AUV is in the area that has been visited before on the basis of the current observation data [9]. If it is in such an area, a complete graph structure is constructed and redundant constraints are added.

The main purpose of the closed loop test is to eliminate the cumulative error caused by inter-frame registration and the basic idea is to compare the current frame with all the key frames in the system. If they are similar, a closed loop is generated. The methods for judging similarity are presented below.

In contrast to the previous closed loop algorithm, or feature dictionary, the characteristic dictionary tree used in this paper is generated by the off-line training of point-line features. The basic training method consists of three steps, as follows.

- For the collected environmental image data, the features of point and line segments are extracted, including the corresponding 256 bits ORB and LBD [14] feature description vectors. The two characteristic description vectors can be used to establish the later feature dictionary (see Figure 3).

- A simple K-means clustering method is used to obtain each leaf node at the bottom of the tree structure. Then the nodes of each layer are obtained in turn. Thus the training process of the feature tree can be completed.
- When the new data is collected, the point-line features are extracted first, and the corresponding number of words in the dictionary are obtained using these features. Thus each picture can be described by the vectors, which are composed of the number of words in the characteristic dictionary tree. The vectors are called bags of words vectors (BOWV).

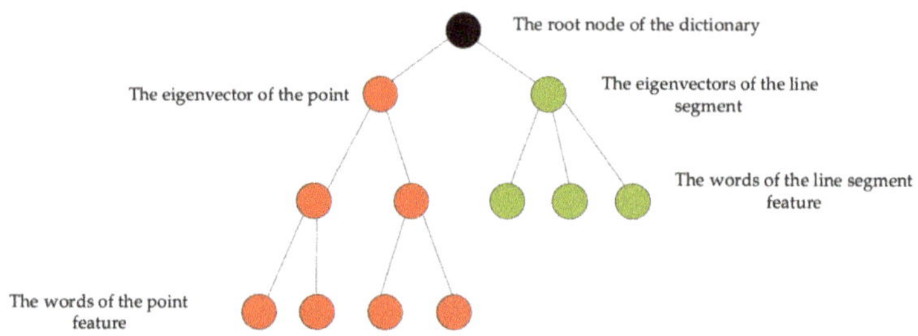

Figure 3. Dictionary tree based on point-line features.

For each word vector in each image, it is important to calculate the distance between two vectors as it is necessary to be able to determine the similarity between the two images. The shorter the distance, the more similar the two images are. A shorter distance also indicates that the AUV may reach the previous positions.

The specific approach is as follows. For the two vectors and definitions, the evaluation scores based on the L1 norm $s(v_1, v_2)$ can be defined as follows:

$$s(v_1, v_2) = 1 - \frac{1}{2}\left| \frac{v_1}{|v_1|} - \frac{v_2}{|v_2|} \right| \tag{4}$$

where v_1 is the eigenvector of the current frame, and v_2 is the eigenvector of the dictionary and the previous frame.

The higher the evaluation score, the greater the similarity between the two images and the greater the possibility of a closed loop. When the above score exceeds the set value, the algorithm enters the closed-loop optimization state.

After estimating all consecutive loop closures in the trajectory, both sides of the loop closure are fused and the error distributed along the loop is corrected. Usually, the pose graph optimization (PGO) method is required in this process. The main purpose of pose graph optimization is to optimize the previous pose of a robot using the redundant constraints. The redundant constraints are obtained by closed-loop detection. When the robot walks to a position where it has walked before, the pose matrix changes due to the accumulation of errors. At this time, the redundant constraint is to reduce the error of the pose matrix. The process can be explained as follows: $u = \begin{pmatrix} u_1 & u_2 & \cdots & u_t \end{pmatrix}$, which represents the state quantity of the pose matrix of the robot at t moment. Where $u_i = \prod_{k=0}^{i} \varsigma_{(k-1,k)}$ and $\varsigma_{(k-1,k)}$ is the pose transition matrix between frames $k-1$ and k, ς_{ij} and Ω_{ij} are the observed transformation matrix and information matrix (representing the weight of noise) from the state of i time to the state of j time. $\hat{\varsigma}_{ij}$ is a real transformation matrix. The error function is shown in Equation (5).

$$e_{ij}(u_i, u_j) = \varsigma_{ij} - \hat{\varsigma}_{ij} \tag{5}$$

The sum of the overall error functions is:

$$F(u) = \sum_{(i,j)\in c} \underbrace{e_{ij}^T \Omega_{ij} e_{ij}}_{F_{ij}} \tag{6}$$

The goal of optimization is to find the u that minimizes $F(u)$:

$$\hat{u} = arg \min_u F(u) \tag{7}$$

Then a first-order Taylor expansion on the error function is performed:

$$\begin{aligned} F_{ij}(u + \Delta u) &= e_{ij}(u + \Delta u)^T \Omega_{ij} e_{ij}(u + \Delta u) \approx \left(e_{ij} + J_{ij}\Delta u\right)^T \Omega_{ij}\left(e_{ij} + J_{ij}\Delta u\right) \\ &= \underbrace{e_{ij}^T \Omega_{ij} e_{ij}}_{c_{ij}} + \underbrace{2e_{ij}^T \Omega_{ij} J_{ij}\Delta u}_{b_{ij}^T} + \underbrace{\Delta u^T J_{ij}^T \Omega_{ij} J_{ij}\Delta u}_{H_{ij}^T} \\ &= c_{ij} + 2b_{ij}^T \Delta u + \Delta u^T H_{ij}^T \Delta u \end{aligned} \tag{8}$$

Then let $c = \sum c_{ij}$, $b = \sum b_{ij}$, $H = \sum H_{ij}$, so we can obtain Formula (9):

$$F(u + \Delta u) = \sum_{(i,j)\in c} F_{ij}(u + \Delta u) \approx \sum_{(i,j)\in c}\left[c_{ij} + 2b_{ij}^T\Delta u + \Delta u^T H_{ij}\Delta u\right] = c + 2b^T\Delta u + \Delta u^T H\Delta u \tag{9}$$

In order to find the minimum of Equation (9), we can use the following formula:

$$\frac{\partial F}{\partial(\Delta u)} = 2b + 2H\Delta u \tag{10}$$

Thus,

$$H\Delta u = -b \tag{11}$$

Through the above formula, the increment Δu of each iteration is calculated using the L-M iterative algorithm. This kind of problem can be optimized through the g2o (General Graph Optimization) library [15], which can greatly simplify the operation.

3. Results

In this section, three experiments were performed, including walking along the wall of a pool, walking along a linear route, and walking along an irregular route. The robustness of the algorithm is verified and the accuracy is evaluated. Finally, a description of the experimental environment and AUV is provided.

3.1. Experiment in an Artificial Pool

An AUV was controlled to walk along the wall of a pool to build an underwater three-dimensional map composed of spatial points and lines (as shown in Figure 4). The AUV performed well with the PL-SLAM method. However, the other visual SLAM algorithms were prone to collapse. The phenomenon of system collapse due to the lack of features was greatly reduced, and the actual frame rate remained at around 20 fps. In order to display the trajectory of the AUV in real time, the pose information was translated into the path topic of the ROS (Robot Operating System). In Figure 5, the color axis represents the AUV's pose, and the yellow line represents the real time trajectory.

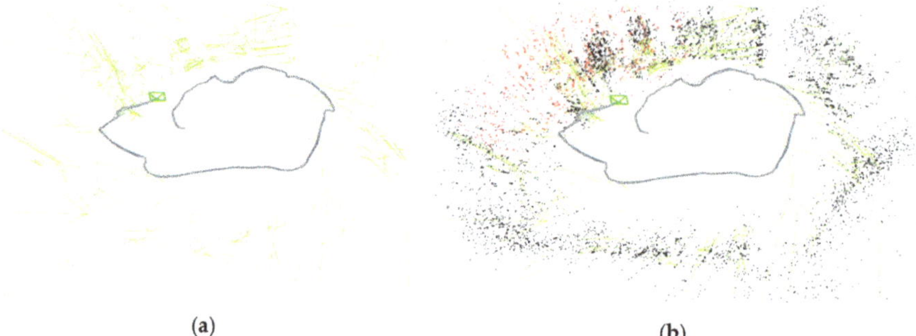

(a)   (b)

Figure 4. Stereo PL-SLAM performed in an artificial pool. (**a**) The green line represents the line features in the global map. The green quadrilateral box represents the current camera. The grey rough line represents the trajectory; (**b**) The red and black points represent the current local map points and global map points.

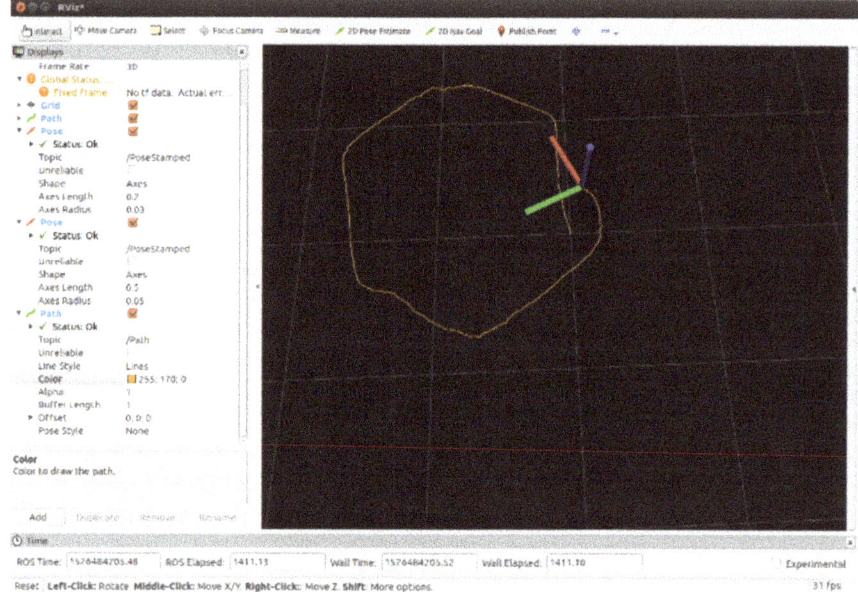

Figure 5. Trajectory of the AUV launched by RVIZ (a 3D visualizer for the Robot Operating System (ROS) framework) released by ROS kinetic.

3.2. Comparison of Pose Measurement Experiments with AR-Markers

In order to further verify the accuracy of the underwater algorithm, the algorithm was compared with an AR-marker, because there is no cumulative error and drift in the attitude information measured by an AR-marker. An AR-marker with a length of 150 mm was arranged in the experimental scene, so that the AR-marker always appeared in the field of vision of the AUV. For convenient measurement, the ARToolKitPlus [16] library was used to measure the attitude of the AR-marker. ARToolKitPlus is a software library for calculating camera position and orientation relative to physical markers in real time (see Figure 6)

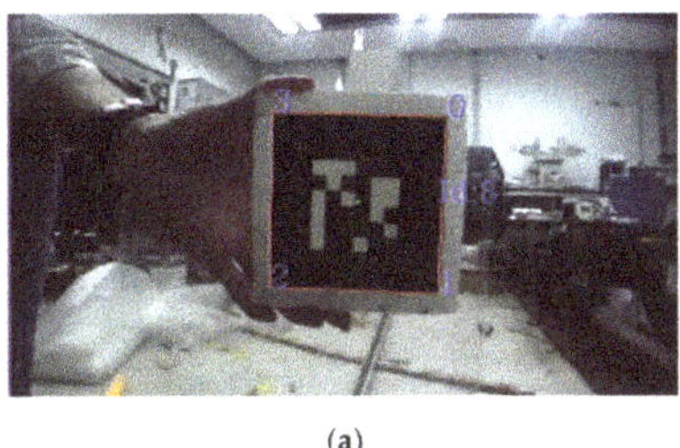
(**a**)

(**b**)

Figure 6. Comparison of pose measurement experiments with the AR-marker. (**a**) The red square locking AR-marker indicated that the recognition of the AR-marker was successful, which allows us to obtain the transformation matrix between the camera coordinate system and the AR-marker coordinate system. (**b**) Using the AR-marker recognition program and PL-SLAM simultaneously.

The AUV was set to walk along a linear route. The process of the experiment is shown in Figure 7. The AR-marker recognition program and PL-SLAM were used simultaneously. The straightness of the AUV's walking trajectory as found using the PL-SLAM was analyzed and the accuracy of the pose measurement using PL-SLAM was compared with the AR-marker. The results of the experiment are shown in Figure 8. The deviation in the y,z-direction was within 5 cm. Because the AUV was not able to be strictly aligned to the center of the y,z-plane, the deviation was within the acceptable range. As shown in Figure 9, the motion trajectories of the two were approximately close, and the final error was less than 3 cm in the x-direction.

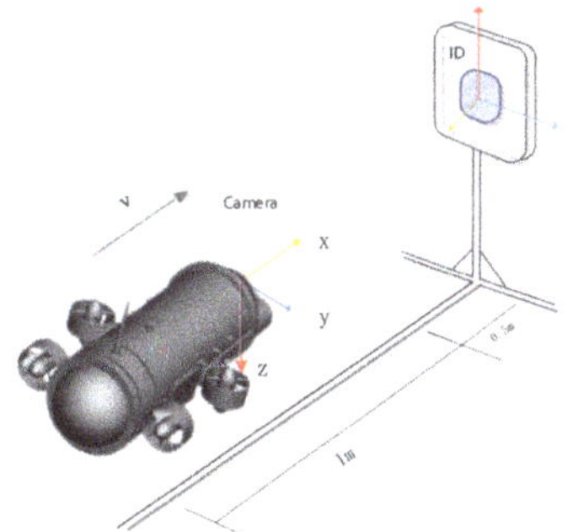

Figure 7. The linear motion experiment.

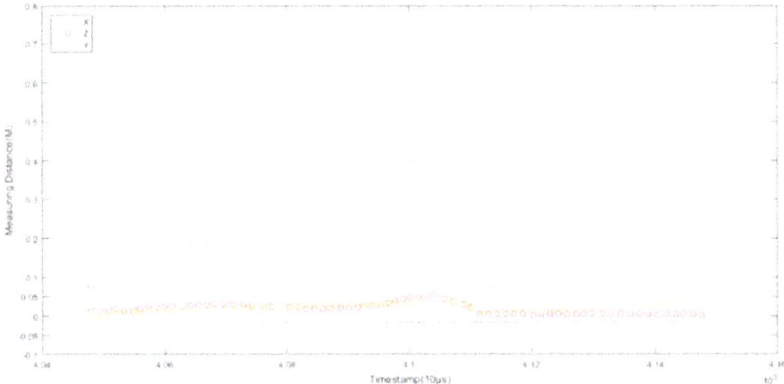

Figure 8. Results of the linear motion experiments. The experimental diagram is shown in Figure 7. The red curve and the yellow curve indicate that the Z-direction and Y-direction pose changes measured by PL-SLAM were small. The blue curve indicates that the change of the X-direction pose measured by PL-SLAM was proportional to time.

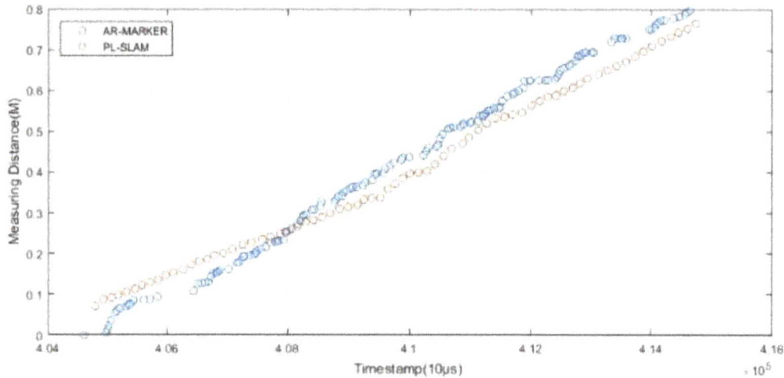

Figure 9. Result of the linear motion experiments in the X-direction, using the AR-marker recognition program and PL-SLAM simultaneously. The experimental diagram is shown in Figure 7.

As shown in Figure 10, in order to analyze the cumulative error generated by the algorithm, the fixed AR-marker was arranged to appear in view of the camera at the beginning and end of the experiment, as described in Section 3.1. Thus, the accurate termination attitude of the AUV was obtained. After the algorithm finished running, the AUV's termination attitude with cumulative error was also able to be obtained. These two termination attitudes were compared and the error rates are shown in Table 1.

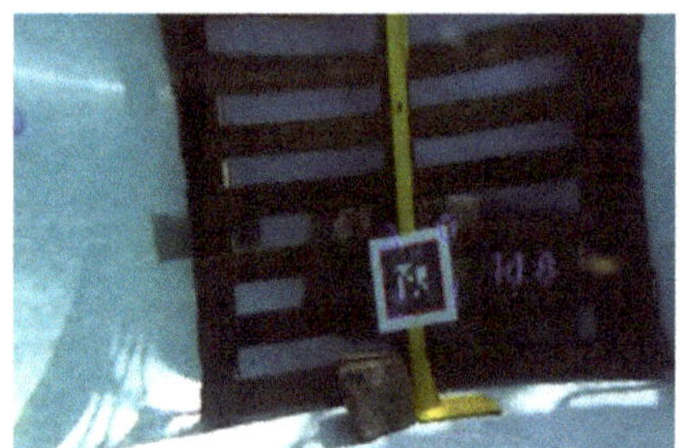

(a) Camera vision at the beginning

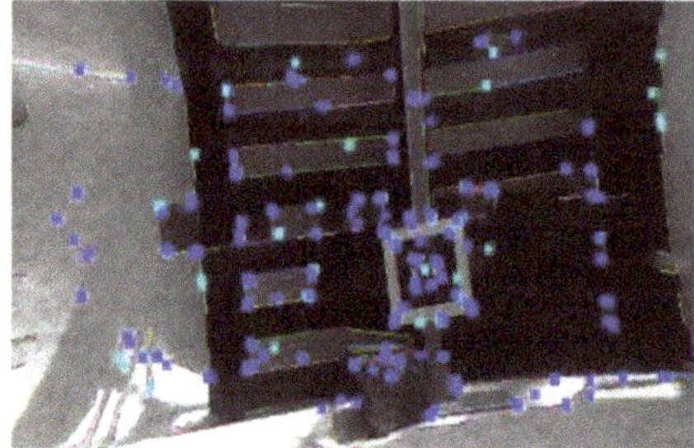

(b) Camera vision at the end

Figure 10. Camera vision in the experiment described in Section 3.1. (a) Camera vision at the beginning; (b) Camera vision at the end.

Table 1. Error comparison of termination attitude.

	Attitude_z (m)	Attitude_y (m)	Attitude_x (m)	Attitude_Error
AR-marker	−0.4523	0.1434	−1.056	
PL-SLAM	−0.3651	0.2180	−1.151	2.98%
Error	0.0872	0.0746	−0.095	

According to Table 1, the attitude error was $\sqrt{Error\_x^2 + Error\_y^2 + Error\_z^2} = 0.14897$ m. The AUV walked 4.98 m during the experiment described in Section 3.1. Thus, the relative error was 2.98%.

The AUV was then controlled to walk along an irregular route. The irregular route is shown in Figure 11. The accuracy of the pose measurement by PL-SLAM was then compared with the AR-marker. The results of the experiment are shown in Figure 12.

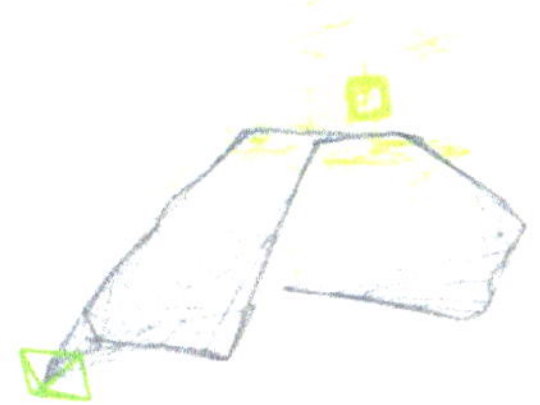

Figure 11. Comparison of the pose measurement experiments with the AR-marker.

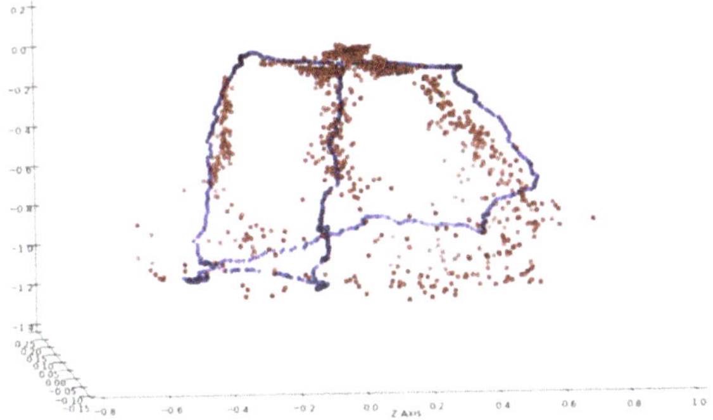

Figure 12. Analysis and contrast diagram of the experimental results. The blue scatter points Pb (Xb, Yb, Zb) represent the PL-SLAM positions. The red scatter points Pr (Xr, Yr, Zr) represent the AR-marker positions.

3.3. Experiment Setup

Nvidia Jetson TX2 was chosen as the platform, with an Ubuntu 16.04 system and ROS kinetic. The embedded development board was located in the sealing bin of the AUV. The ZED stereo camera (forward-looking) was located in the head of the AUV. The two parts were connected through the USB (Universal Serial Bus) 3.0 interface (see Figure 13a). The length of the AUV was approximately 730 mm and the diameter of the cabin was approximately 220 mm. Because of the limited experimental conditions, the experiment was only carried out in an artificial pool with a diameter of 4 m and a depth of 1.5 m. Scenes were placed in the pool to simulate the natural environment of the shallow water area (see Figure 13b). In the experiment, the speed of the AUV was set to 0.6 m/s.

(a)   (b)

Figure 13. The developed shark AUV. (**a**) The appearance of the AUV; (**b**) Experiment in an artificial pool.

Due to the large distortion of the underwater scene, the stereo camera should have been calibrated first. However, camera calibration based on rigorous underwater camera calibration models is time-consuming and cannot achieve real-time performance. Thus, the traditional pinhole model in air was selected to calibrate the AUV. The calibration experiment shows that the refraction effect was considered to be absorbed by the focal length and radial distortion, and the conclusion was that when the camera is in an underwater environment, the focal length is approximately 1.33 times of that in air.

4. Discussion

The experimental results showed that the algorithm was highly robust in underwater low-texture environments due to the inclusion of line segments. At the same time, the algorithm achieved a high accuracy of location effectively, which means that it can be implemented in the navigation and path planning of AUVs in the future. The current positioning accuracy was 2.98%. The experimental environment simulated a shallow water area well, and also verified that the visual positioning method can be applied in shallow water areas. However, there were some problems with the experiment. The added line features undoubtedly increased the complexity of the algorithm, which is more time-consuming than ORB-SLAM. Secondly, it was difficult to calibrate the underwater cameras, especially using the stereo underwater calibration model, however this fact is beyond the scope of our discussion. Due to the fact that high distortion cannot be neglected in underwater environments, the matching of line features occasionally failed. However, the ORB-SLAM algorithm also often fails due to the small number of point features in an underwater environment. The PL-SLAM algorithm combined with the point line feature method had a high success rate. Finally, the accuracy of the pose measured by the AR-marker is not very high and therefore can only be used as a reference.

5. Conclusions

Future work will focus on the following two points: eliminating point features and line features near the edges of the image in a high underwater distortion environment in order to reduce mismatch caused by distortion, and considering matching two stereo cameras, one looking forward and the other looking down, and improving the real-time performance of the algorithm.

Author Contributions: Conceptualization, X.W. and R.W.; methodology, M.Z.; software, R.W.; validation, X.W., R.W. and Y.L.; formal analysis, Y.L.; investigation, M.Z.; resources, Y.L.; data curation, R.W.; writing—original draft preparation, M.Z.; writing—review and editing, R.W.; supervision, X.W.; project administration, X.W.; funding acquisition, X.W.

Funding: This research was funded by the Shenzhen Bureau of Science Technology and Information, grant number No. JCYJ20170413110656460 and grant number No. JCYJ20180306172134024.

Conflicts of Interest: The authors declare no conflict of interest.

References

1. Fallon, M.F.; Folkesson, J.; McClelland, H.; Leonard, J.J. Relocating Underwater Features Autonomously Using Sonar-Based SLAM. *IEEE J. Ocean. Eng.* **2013**, *38*, 500–513. [CrossRef]
2. Karimi, M.; Bozorg, M.; Khayatian, A. Localization of an Autonomous Underwater Vehicle using a decentralized fusion architecture. In Proceedings of the 2013 9th Asian Control Conference (ASCC), Istanbul, Turkey, 23–26 June 2013.
3. Durrant-Whyte, H.; Bailey, T. Simultaneous localization and mapping (SLAM): Part I. *IEEE Robot. Autom. Mag.* **2006**, *13*, 99–110. [CrossRef]
4. Bonin, F.; Burguera, A.; Oliver, G. Imaging systems for advanced underwater vehicles. *J. Mariti. Res.* **2011**, *8*, 65–86.
5. Hong, S.; Kim, J. Selective image registration for efficient visual SLAM on planar surface structures in underwater environment. *Auton. Robots* **2019**, 1–15. [CrossRef]

6.   Jung, J.; Lee, Y.; Kim, D.; Lee, D.; Myung, H.; Choi, H.-T. AUV SLAM using forward/downward looking cameras and artificial landmarks. In Proceedings of the 2017 IEEE Underwater Technology (UT), Busan, Korea, 21–24 February 2017.

7.   Kim, A.; Eustice, R.M.R. Real-Time Visual SLAM for Autonomous Underwater Hull Inspection Using Visual Saliency. *IEEE Trans. Robot.* **2013**, *29*, 719–733. [CrossRef]

8.   Carrasco, P.L.N.; Bonin-Font, F.; Codina, G.O. *Stereo Graph-SLAM for Autonomous Underwater Vehicles*; Intelligent Autonomous Systems 13; Springer International Publishing: Cham, Switzerland, 2016; pp. 351–360.

9.   Negre, P.L.; Bonin-Font, F.; Oliver, G. Cluster-based loop closing detection for underwater slam in feature-poor regions. In Proceedings of the IEEE International Conference on Robotics and Automation, Stockholm, Sweden, 16–21 May 2016; pp. 2589–2595.

10.   Mur-Artal, R.; Montiel, J.M.M.; Tardos, J.D. ORB-SLAM: A versatile and accurate monocular slam system. *TRO* **2015**, *31*, 1147–1163. [CrossRef]

11.   Pumarola, A.; Vakhitov, A.; Agudo, A.; Sanfeliu, A.; Moreno-Noguer, F. PL-SLAM: Real-time monocular visual SLAM with points and lines. In Proceedings of the IEEE International Conference on Robotics and Automation, Singapore, 29 May–3 June 2017.

12.   Vakhitov, A.; Funke, J.; Moreno-Noguer, F. Accurate and linear time pose estimation from points and lines. In *ECCV 2016*; Springer: Cham, Switzerland, 2016.

13.   Von Gioi, R.G.; Jakubowicz, J.; Morel, J.M.; Randall, G. LSD: A line segment detector. *IPOL* **2012**, *2*, 35–55. [CrossRef]

14.   Zhang, L.; Koch, R. An efficient and robust line segment matching approach based on LBD descriptor and pairwise geometric consistency. *JVCIR* **2013**, *24*, 794–805. [CrossRef]

15.   Kümmerle, R.; Grisetti, G.; Strasdat, H.; Konolige, K.; Burgard, W. G^{2}o: A general framework for graph optimization. In Proceedings of the 2011 IEEE International Conference on Robotics and Automation (ICRA), Shanghai, China, 9–13 May 2011; pp. 3607–3613.

16.   Available online: https://github.com/paroj/artoolkitplus (accessed on 1 June 2018).

 *applied sciences*

Article

Volumetric Tooth Wear Measurement of Scraper Conveyor Sprocket Using Shape from Focus-Based Method

Hua Ding [1,2,*], Yinchuan Liu [1,2] and Jiancheng Liu [3]

[1] College of Mechanical and Vehicle Engineering, Taiyuan University of Technology, Taiyuan 030024, China; liuyinchuan@tyut.edu.cn

[2] Shanxi Key Laboratory of Fully Mechanized Coal Mining Equipment, Taiyuan 030024, China

[3] School of Engineering and Computer Science, University of the Pacific, 3601 Pacific Ave., Stockton, CA 95211, USA; jliu@PACIFIC.EDU

*   Correspondence: dinghua@tyut.edu.cn; Tel.: +86-188-3512-3666

Received: 18 February 2019; Accepted: 11 March 2019; Published: 14 March 2019

Abstract: Volumetric tooth wear measurement is important to assess the life of scraper conveyor sprocket. A shape from focus-based method is used to measure scraper conveyor sprocket tooth wear. This method reduces the complexity of the process and improves the accuracy and efficiency of existing methods. A prototype set of sequence images taken by the camera facing the sprocket teeth is collected by controlling the fabricated track movement. In this method, a normal distribution operator image filtering is employed to improve the accuracy of an evaluation function value calculation. In order to detect noisy pixels, a normal operator is used, which involves with using a median filter to retain as much of the original image information as possible. In addition, an adaptive evaluation window selection method is proposed to address the difficulty associated with identifying an appropriate evaluation window to calculate the focused evaluation value. The shape and size of the evaluation window are autonomously determined using the correlation value of the grey scale co-occurrence matrix generated from the measured pixels' neighbourhood pixels. A reverse engineering technique is used to quantitatively verify the shape volume recovery accuracy of different evaluation windows. The test results demonstrate that the proposed method can effectively measure sprocket teeth wear volume with an accuracy up to 97.23%.

Keywords: shape from focus; wear measurement; sprocket teeth; normal distribution operator image filtering; adaptive evaluation window; reverse engineering

---

1. Introduction

A scraper conveyor is the primary production and transportation equipment in a fully mechanized mining face [1]. In modern coal mining, the conveyor transports coal and provides hydraulic support and a walking track for the shearer. Therefore, its reliability directly affects the safety and production efficiency of modern coal mines. Sprockets are the core components of the chain drive system, which is the most important subsystem in the scraper conveyor [2]. Sprocket's performance is directly related to the transport performance and service life of the scraper conveyor [3]. Sprockets contact chains directly; consequently, friction causes wear and excessive wear is the main form of sprocket failure and the main cause of scraper conveyor failure [4]. The sprocket conveyor chain may jump when it engages with the excessively worn sprocket; worn sprocket teeth may break, which affects the safe and efficient production of the coal mine, therefore, sprocket teeth wear analysis is required. Conventional wear measurement methods for scraper conveyor sprocket teeth include a weighing method, water volume measurement method, ANSYS analysis method [5] and wear monitoring [6]. Wang et al. [7]

discussed the wear condition of a driving sprocket and the influence of wear on the sliding distance by taking the sliding speed and sliding distance of the meshing process as the index. Wang et al. [8] also analyzed the relationship between the deformation of a ring chain and driving sprocket wear by combining numerical analysis with experiments. However, these methods are not only tedious and time-consuming, they are also not sufficiently accurate or efficient.

The research of computer vision in industry field has attracted more and more attention of many researchers. Alverdi et al. [9] proposed a new way of using images to model the kerf profile in abrasive water jet milling. Qian et al. [10] presented an algorithm to compute the axis and generatrix focus on complex surfaces or irregular surfaces. A new monitoring technique for burr detection was proposed for the optimization of drill geometry and process parameters [11]. In addition, as a relatively simple and practical 3D reconstruction technology, shape from focus (SFF) has been applied to tool wear measurements [12,13], LCD/TFT (Liquid Crystal Display/Thin-Film Technology) display manufacturing [14] and grinding wheel surface morphology [15], etc.

To realize 3D surface topography restoration, in 1988, Darrell et al. [16] proposed using a Laplace operator-Gauss fitting method to search the clear frame of pixels in the sequence partial focus image according to image focusing information. In the 1990s, Nayar et al. [17,18] proposed an SFF-based method and obtained the height information of the corresponding surface of the window image by searching the image position corresponding to the maximum value of the focus evaluation function in the evaluation window. However, SFF suffers from some technical defects in pro-processing images and choosing evaluation function window size, thus developing methods to improve SFF accuracy has been the focus of ongoing research.

Many studies have proposed image pre-processing methods. For example, for wavelet transform, Karthikeyan et al. [19] introduced an effective denoising method for grey images using joint bilateral filtering. Khan et al. [20] introduced a new impulse noise detection algorithm that is based on Noise Ratio Estimation and a combination of K-means clustering and Non-Local Means based filter. An adaptive type-2 fuzzy filter is used to remove salt-and-pepper noise from images [21]. To improve the processing performance of image texture-free regions, Fan et al. [22] presented a shape focusing method combined with a 3D adjustable filter that considered edge response and image blurring. Liu et al. [23] proposed a graph Laplacian regularizer to preserve the inherent piecewise smoothness of depth, and this method demonstrated effective filtering. An iterative algorithm that combines stationary wavelet transform, bilateral filtering, Bayesian estimation and anisotropic diffusion filtering was used to reduce speckle noise in SAR images [24]. Khan et al. [25] designed a meshfree algorithm (Kansa technique) that uses a DTV method and a radial basis function approximation method to solve DTV-based model numerically to eliminate multiplicative noise in measurements. However, although the above methods can remove image noise to some extent, they change the grey-level information of non-noise areas of the image and affect the accuracy of 3D morphology restoration.

Mahmood et al. [26] analyzed the influence of different evaluation window sizes and noise types on the focusing evaluation function and concluded that, for different resolutions, the best evaluation window size for the same evaluation function was no single. Lee et al. [27,28] studied the focusing evaluation function window size. To determine the focusing evaluation function value, different standard window sizes were used to analyze the evaluation results of the size and shape of the focusing evaluation function window. Muhammad et al. [29] conducted 3D morphology restoration experiments on images collected using imaging equipment with different parameters and formulated the selection of the evaluation window. However, most of the above studies are based on the optimal size selection of a fixed square evaluation window without simultaneously optimizing both the shape and size of the window.

This paper presents an SFF-based method to measure scraper conveyor sprocket teeth wear efficiently. A specially designed device was used to collect a set of sprocket tooth wear sequence images. Normal distribution operator filtering, adaptive window evaluation and a Laplacian focusing evaluation function are applied to the obtained images. We obtain an initial depth map of the entire tooth wear surface. Then, a 3D shape recovery map is constructed to calculate the wear volume. This method improves measurement accuracy, can be operated remotely, and can be used to predict the life of the sprocket. More importantly, it is an efficient, fast and safe measurement method that provides data and technical support for coal mine production safety.

2. Measurement Scheme of Sprocket Teeth Wear of Scraper Conveyor

The scraper conveyor sprocket tooth wear measurement system based on SFF primarily comprises hardware and software. The hardware includes industrial cameras and control tracks, and the software includes 3D topography recovery and calculation of wear volume. The measurement process is summarised as follows. First, the sequence images of the tooth are collected using the hardware device. Then, the images are transmitted to the computer. Finally, the wear volume and geometric position of the sprocket teeth are obtained via 3D topography recovery and wear volume calculation.

2.1. Structure and Wear of Sprocket Teeth of Scraper Conveyor

A scraper conveyor sprocket [30] comprises a hub and sprocket teeth. The shape of the teeth is a geometric polygon, and each sprocket generally has five or seven teeth. The structure of the sprocket is shown in Figure 1.

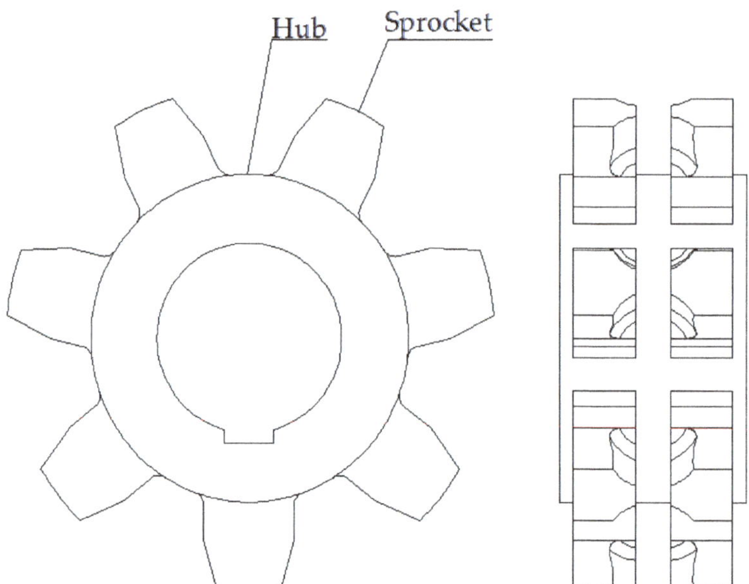

Figure 1. Structure diagram of scraper conveyor sprocket.

The working principle of the sprocket is to rotate the drive shaft to drive the hub to rotate, and the sprocket teeth engage with the circle chain. The different wear degrees of sprocket teeth are shown in Figure 2.

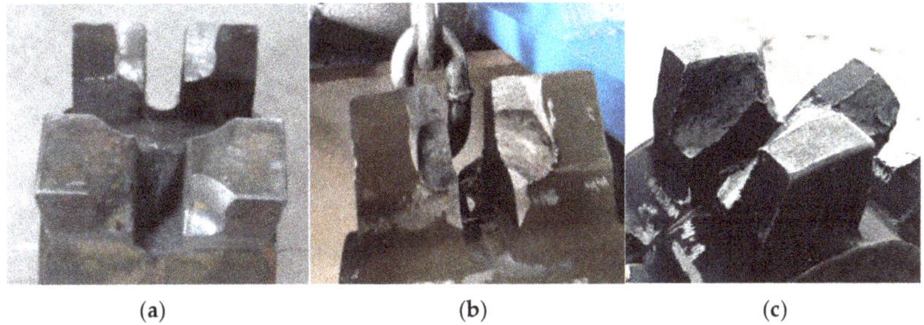

(**a**)   (**b**)   (**c**)

Figure 2. Scraper conveyor sprocket teeth structure shape: (**a**) Before wear; (**b**) After wear; (**c**) After failure.

Figure 3a,b show the hardware device's design, where 1: box; 2: circular track; 3: tooth radial motion module; 4: circular track slider; 5: circular slider driving module; 6: circular slider auxiliary track; 7: light receiver; 8: linear light; 9: ring light; 10: industrial lens; 11: industrial camera; 12: longitudinal motion module; 13: slider connection plate; 14: box connection plate.

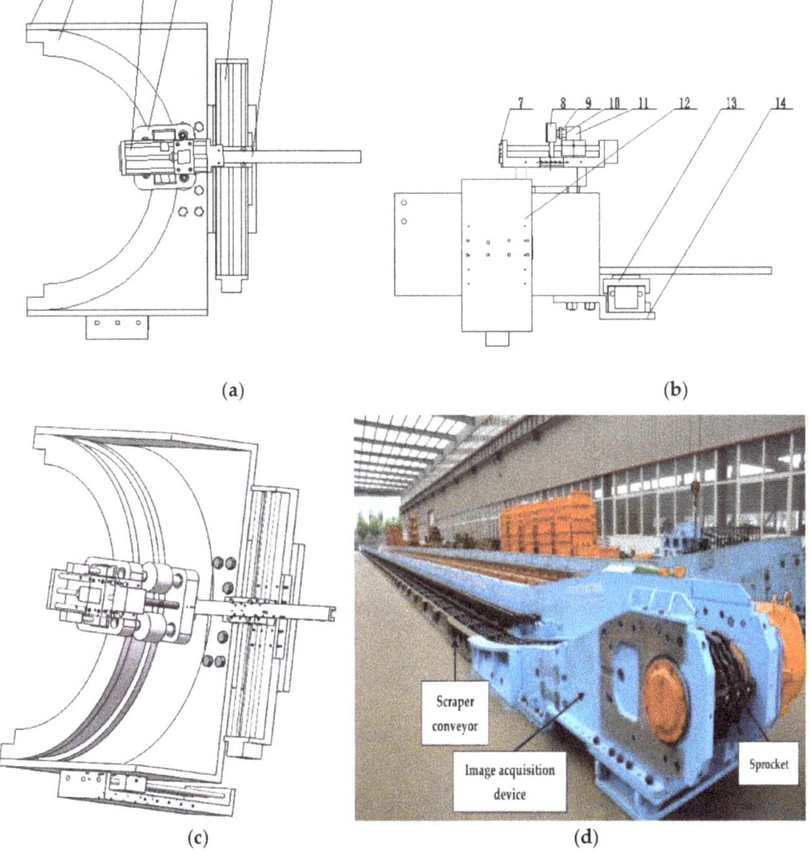

Figure 3. Hardware structure of wear measurement device based on SFF: (**a**) Front view; (**b**) Top view; (**c**) 3D model; (**d**) Installation position of sequence image acquisition device

The hardware device that measures sprocket wear includes position control, motion, centering, sequence image acquisition and other modules. The position control module primarily comprises a PLC unit and a driver unit, where the PLC unit includes different sub-units, such as longitudinal motion control, circumferential motion control, sprocket teeth radial motion control, camera control, linear light switch control, light receiver monitoring and ring light control. The motion module comprises a longitudinal motion unit, a circumferential motion unit and a radial sprocket teeth movement unit. The longitudinal movement unit includes a circular arc guide, a slider, a slider auxiliary guide rail and a slider drive module, and the centering module comprises a linear light unit and a light receiver unit. The sequence image acquisition module comprises a Charge Coupled Device CCD camera unit, a lens unit and an auxiliary light unit, and the other modules include support units and connection units. The structure of the sprocket wear measurement device is shown in Figure 3.

2.2. Wear Measurement Process Flow Chart

Figure 4 illustrates wear measurement process, which is addressed as follows.

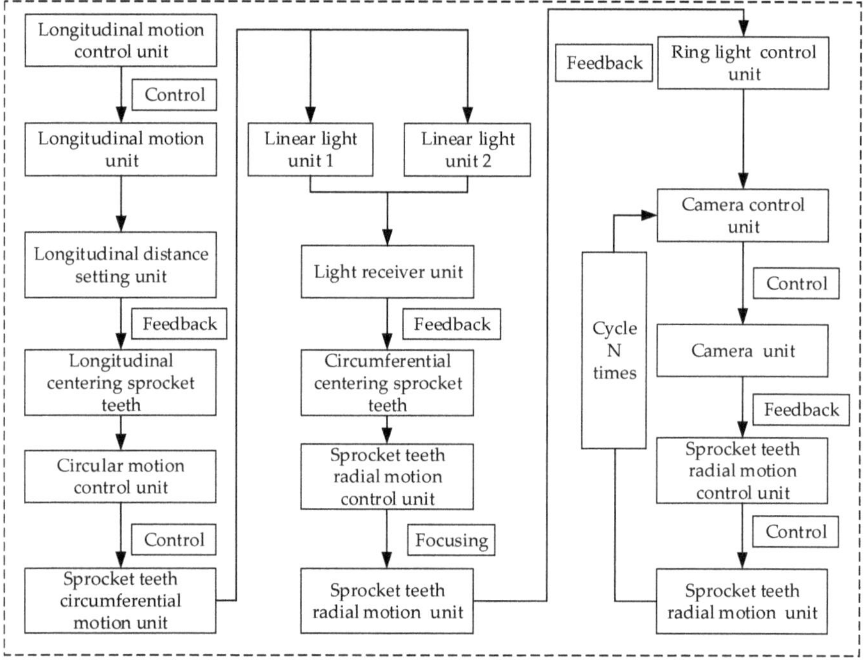

Figure 4. Technical route of sequence images acquisition.

Firstly, the longitudinal motion unit is driven to move in the longitudinal direction by the longitudinal motion control unit and stops when the moving distance reaches the set distance of the longitudinal distance unit. As a result, the camera unit is aligned with the longitudinal row of teeth. Secondly, the two linear light switches are turned on using linear light control unit, and the circular slider driving module is driven to move with the help of the circumferential control unit. When the light receiver unit simultaneously receives the signals from the two lights reflected back through the tooth surface, the circular slider driving module stops moving, which means that the camera unit is aligned with one of the teeth in the circumferential direction. Thirdly, the radial motion unit is driven to the designated focal length position by the sprocket tooth radial control unit, and the step distance is set to one N of the sprocket tooth height. The camera is driven by the camera control unit. Each step forward, the camera takes a picture, which cycles N times. The radial motion unit stops moving and returns to the original position, and then sequence image acquisition is completed.

The technical flow chart of the focused morphology restoration algorithm is shown in Figure 5.

First of all, the collected sequence images are read into the computer, and the field of view and resolution and cropping of all N-frame sequence images are transformed according to the proportional relationship of the target region of the N-frame sequence image. Normal distribution operator image filtering is used to filter the sequence images to obtain the same resolution. The pre-processed sequence images have N frames of the field of view. Therefore, N-frame pre-processed sequence image of the same resolution and field of view are obtained. Then, the clear pixel points of each pre-processed sequence image are extracted in order to construct a full-focus image. Then, the proposed adaptive method is used to select the focus evaluation window of any pixel in the full-focus image. The focus factor of each pixel in the pre-processed sequence image is calculated and the sequence image number corresponding to the maximum focus factor of all pixels in the full-focus image is obtained. The sequence image number is taken as the depth value of the corresponding pixels to form the initial depth map of the full-focus image. Next, a full-focus image is obtained with the help of image binarization, inversion, filling and contour recognition, and the object contour is extracted. The extracted object contour is applied to the initial depth map and the region outside the object contour is hollowed out to obtain a three-dimensional shape recovery map of the object. Lastly, the wear volume is calculated. The pixel equivalent and actual depth value of each pixel in the complete 3D topography is calculated, the tooth volume is determined using the limit method and the volume difference and wear volume between the recovered tooth model and the actual tooth model are calculated using the difference method.

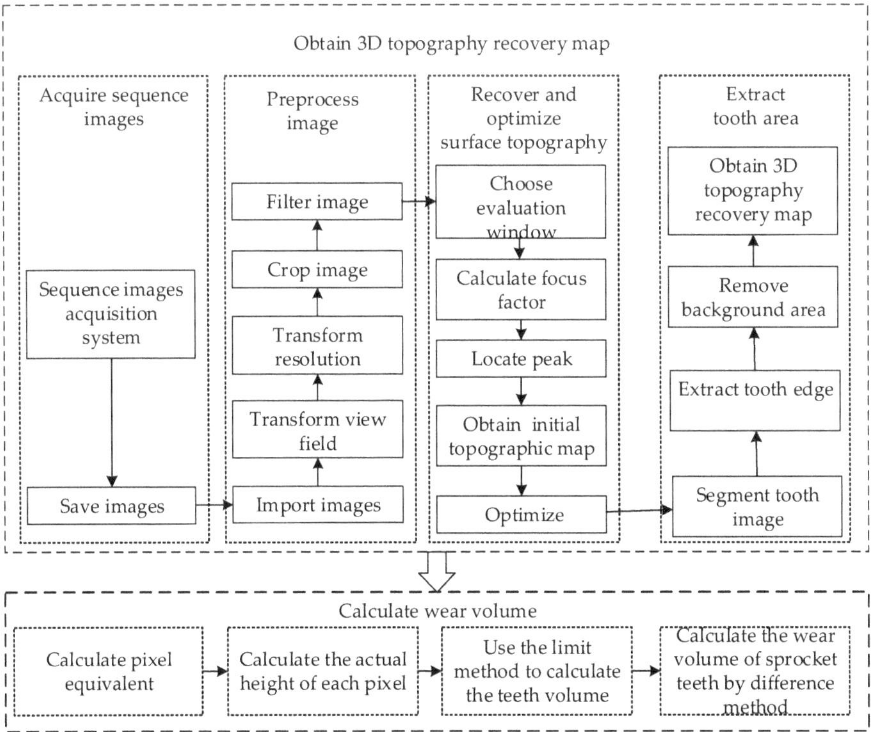

Figure 5. Flow chart of the focused morphology restoration algorithm.

3. Improved SFF

3.1. Principle of SFF

SFF is a method to recover a 3D topography from 2D sequence images [31]. SFF collects a series of partially-focused sequence images and obtains the depth information of each pixel based on focus information. Figure 6 shows a schematic diagram of an ideal optical system imaging principle. The object distance u, focal length f and distance v satisfy the relationship $1/f = 1/u + 1/v$ in an ideal optical imaging system. For a fixed-focus lens, the object point P forms a clear image point P_f on the focus plane through the optical system when the image sensor coincides with the focus plane. The object point P forms a blur circle of radius R on the image sensor when the image sensor does not coincide with the focus plane. Moreover, a greater distance between the image sensor and focus plane is results in greater R and the image points become more blurred. SFF must collect K-frame partial focus images I_k ($k = 1, 2, \ldots, K$) of the measured surface along the optical axis, and these images contain the depth information of the entire measured surface.

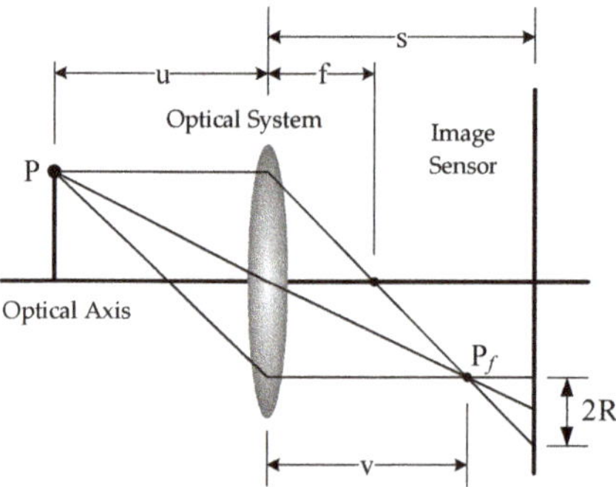

Figure 6. Schematic diagram of ideal optical system imaging principle.

To increase the robustness of the focus measure, the neighbourhood window $U(x, y)$ of the pixel (x, y) is usually selected, rather than the pixel as the calculation object, and its size is $w \times w$, this variable is expressed as follows.

$$U(x,y)_k = \{(\xi, \eta) | |\xi - x| \le w \land |\eta - y| \le w\} \tag{1}$$

where (ξ, η) represents the pixels in the neighbourhood $U(x, y)$, and k is the image sequence number.

Focused images have more high-frequency components than blurred images. Therefore, the focusing degree is usually characterized by the sharpness of the pixel points and quantified using the focus measure in SFF.

When an evaluation function is selected, the evaluation function value sequence $I_k(i, j)$ of pixel (i, j) can be obtained.

$$F_k(x,y) = \sum_{(\xi,\eta)\in(x,y)_k} F_k(\xi, \eta) \tag{2}$$

Since the clearest pixel can provide depth information of the corresponding surface element of the pixel, the depth of each pixel corresponding to the surface element can be obtained by obtaining each pixel in the image corresponding to the maximum focus volume. In this manner, the initial depth map of the measured surface is obtained. The formula is as follows:

$$D(x,y) = \arg_k\max[F_k(x,y)] \tag{3}$$

Then, an approximation technique method is applied to refine the initial depth map.

3.2. Normal Distribution Operator Image Filtering

The influence of many factors like the image capturing hardware, surface texture and light inevitably introduces noise during data acquisition and transmission. Image noise greatly affects the accuracy of the value of the focus measure; it is necessary to use filtering techniques to eliminate them. During image acquisition, two main types of noise are produced: Gaussian noise and salt-and-pepper noise. Salt-and-pepper noise has a greater impact on the accuracy of the value of the focus measure [32]. The most prominent feature of salt-and-pepper noise is that the grey value of the noise pixels is different from those of its neighbourhood pixels. Therefore, the median filter is the best filtering method for this type of application. Median filtering is the most common filtering method; however, median filtering will change the grey value of all pixels in the image. To maintain the grey value of non-noise pixels in the image, the normal distribution operator is used to detect noise points. Then, median filtering is employed for the noise pixels, thereby retaining more of the original information contained in the image.

3.2.1. Principle of Normal Distribution

The normal distribution operator is a filtering algorithm based on normal distribution, which is defined as the probability distribution of random variable X obeying position parameter μ and scale parameter σ. The probability density formula is given as follows:

$$f(x) = \frac{1}{\sqrt{2\pi}\sigma} \exp\left(-\frac{(x-\mu)^2}{2\sigma^2}\right) \tag{4}$$

This random variable is referred to as a normal random variable, and the distribution it obeys is called a normal distribution, expressed as $X \sim N(\mu, \sigma^2)$.

Figure 7 plots the normal distribution. According to the principle that salt-and-pepper noise is distant from the mean value, in the normal distribution with a mean of 5 and variance of 2, there are nine pixels in the 3×3 evaluation window, of which seven normal points are concentrated near the mean while the other noise points are distant from the mean. Use the median filter to replace the two abnormal points under the help of $(\mu - K\sigma, \mu + K\sigma)$.

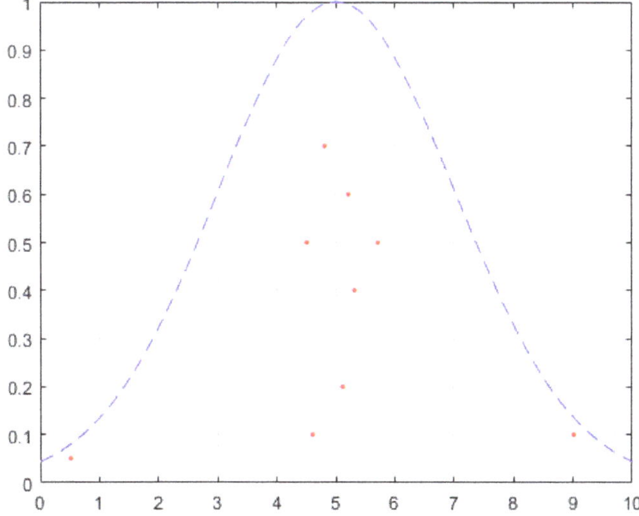

Figure 7. Schematic diagram of normal distribution.

3.2.2. Noise Point Detection

In the 3×3 filter evaluation window, the grey value of pixel centre point (x, y) is $f(x, y)$, and the pixel points in the centre pixel and its neighbourhood are represented as $f_{11}, f_{12}, f_{13}, f_{21}, f_{22}, f_{23}, f_{31}, f_{32}$ and f_{33}, as shown in Figure 8.

f_{11}	f_{12}	f_{13}
f_{21}	f_{22}	f_{23}
f_{31}	f_{32}	f_{33}

Figure 8. 3×3 filter evaluation window.

According to the normal distribution principle and the Figure 8, the maximum and minimum of the nine points in the 3×3 filter evaluation window are removed. The mean and variance of the remaining seven points are taken as μ and σ, which are expressed as follows.

$$F_1 = \max \left(f_{11} + f_{12} + f_{13} + f_{21} + f_{22} + f_{23} + f_{31} + f_{32} + f_{33} \right) \tag{5}$$

$$F_2 = \min \left(f_{11} + f_{12} + f_{13} + f_{21} + f_{22} + f_{23} + f_{31} + f_{32} + f_{33} \right) \tag{6}$$

$$\mu = \left(f_{11} + f_{12} + f_{13} + f_{21} + f_{22} + f_{23} + f_{31} + f_{32} + f_{33} - F_1 - F_2 \right) \tag{7}$$

$$\sigma^2 = \left(\begin{array}{c} (f_{11} - \mu)^2 + (f_{12} - \mu)^2 + (f_{13} - \mu)^2 + (f_{21} - \mu)^2 + (f_{22} - \mu)^2 + (f_{23} - \mu)^2 \\ + (f_{31} - \mu)^2 + (f_{32} - \mu)^2 + (f_{33} - \mu)^2 - (F_1 - \mu)^2 - (F_2 - \mu)^2 \end{array} \right) / 7 \tag{8}$$

From the above formula, the average of the seven points is calculated as the mean and variance as the variance of the normal distribution, where K is the threshold. The centre pixel is a non-noise point when the absolute value of the difference between the centre pixel and the mean is in the $K\sigma$ range and do not change its grey value. By taking the centre pixel as a non-noise point, the absolute value of the difference between the centre pixel and the mean is not in the $K\sigma$ range. After median filtering, the original grey value is replaced with *Med*. The grey value obtained after filtering is $F(x, y)$, and the formula is given as follows.

$$F(x, y) = \begin{cases} f(x, y) & |f(x, y) - \mu| < K\sigma \\ Med & |f(x, y) - \mu| \geq K\sigma \end{cases} \tag{9}$$

An experiment proved that the filtering effect was best when threshold K was set to 2.2.

3.2.3. Algorithm Verification and Evaluation Analyzes

To verify the feasibility of the operator filtering, three images of a vegetable, a ball and a human, respectively, were selected as test objects. (These images, shown in Figure 9, were chosen from a book named Detailed Explanation of Image Processing Example in MATLAB.) Salt-and-pepper noise with a density of 0.01 was added to the images, and we performed median filtering and normal distribution operator filtering on the noise images (threshold value K: 2.2).

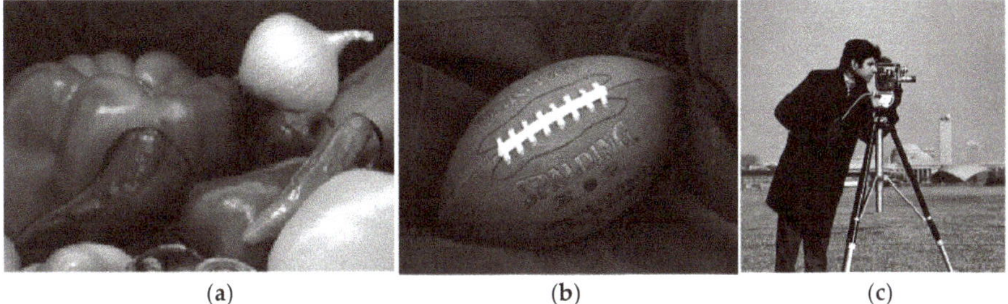

Figure 9. Test objects: (**a**) Vegetable; (**b**) Ball; (**c**) Human.

To evaluate the quality of median filtering and normal distribution operator filtering quantitatively, we selected correlation and the peak signal-to-noise ratio (PSNR) as quantitative assessment criteria. Correlation was used to evaluate the similarity between the reference and real data, where a greater correlation value indicates that the reference data are more consistent with the real data. PSNR was used to measure image quality after filtering, where a greater PSNR value indicates less image distortion. The formulas for correlation and PSNR are given follows.

$$Cor = \frac{\sum\limits_{i=1}^{M}\sum\limits_{j=1}^{N}\left\{\left[I'(x,y) - \overline{I'(x,y)}\right] \times \left[I(x,y) - \overline{I(x,y)}\right]\right\}}{\sqrt{\sum\limits_{i=1}^{M}\sum\limits_{j=1}^{N}\left[I'(x,y) - \overline{I'(x,y)}\right]^2 \times \sum\limits_{i=1}^{M}\sum\limits_{j=1}^{N}\left[I(x,y) - \overline{I(x,y)}\right]^2}} \tag{10}$$

$$PSNR = 10 \times \log\left(\frac{255^2}{MSE}\right) \tag{11}$$

$$MSE = \frac{1}{M \times N}\sum\limits_{i=1}^{M}\sum\limits_{j=1}^{N}\left[I'(x,y) - I(x,y)\right] \tag{12}$$

In Equations (10)–(12), M represents the width of the image, N denotes the height of the image, $I'(x, y)$ stands for the actual grey value of the pixel point (x, y), and $\overline{I'(i,j)}$ represents the estimated average grey value for all pixels in the image. While $I(x, y)$ denotes the grey value of the pixels (x, y) after filtering, and $\overline{I(i,j)}$ stands for the estimated average grey value for all pixels after filtering in the image, MSE represents the mean square error.

Figure 10 shows the noise and filter processing results with a density of 0.01. Here, the first, second and third columns show images with density of 0.01, images processed using median filtering and images processed using normal distribution operator filtering, respectively. As shown in Figure 10a, the images contain large number of errors. Note that these errors are reduced significantly by filtering the noise image. In addition, the median filtered images (Figure 10b) are more blurred than the original image and the filtering effect is poor. However, images obtained via normal operator filter processing are closer in appearance to the corresponding original images.

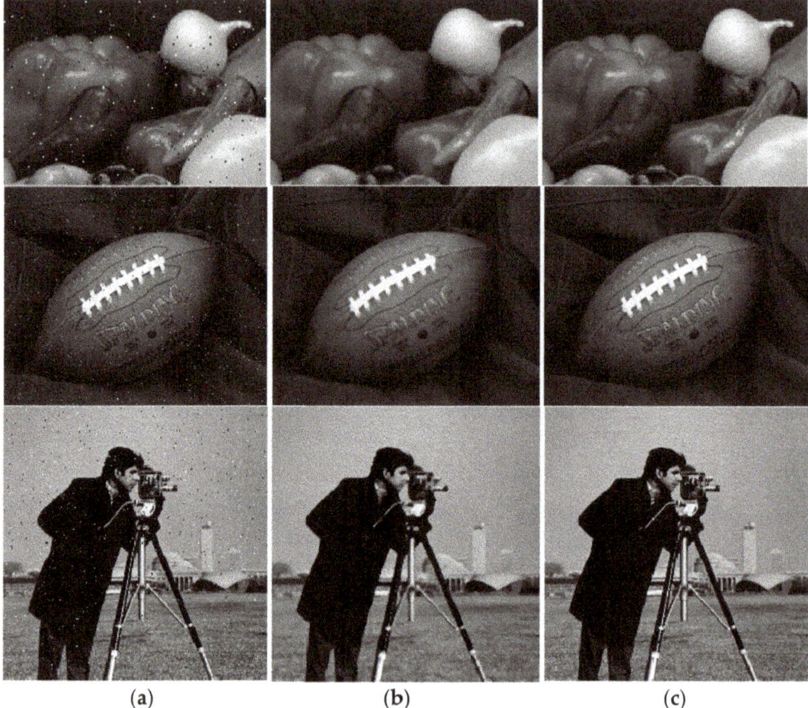

Figure 10. Noise images (density: 0.01) and filtered images: (**a**) Noise Image with density of 0.01; (**b**) Image processed by median filtering; (**c**) Image processed by normal distribution operator filtering

Table 1 shows the correlation and root mean squared error (RMSE) data for three sequence images processed by adding noise and by applying median filtering and normal distribution operator filtering. As can be seen, the correlation and RMSE values obtained by the two filtering methods are greater than those obtained with the noisy image sequence. Furthermore, the increase to these values is more obvious with normal distribution operator filtering. Both filtering methods improve the accuracy of image filtering; however, the results demonstrate that normal distribution operator filtering is better.

Table 1. Correlation and RMSE values of filtering effects of different filtering methods.

Test Object	Vegetable			Ball			Human		
Type	Noise Image	Median Filter	Normal Distribution Operator Filtering	Noise Image	Median Filter	Normal Distribution Operator Filtering	Noise Image	Median Filter	Normal Distribution Operator Filtering
Correlation	0.9669	0.9977	0.9997	0.9440	0.9854	0.9956	0.9223	0.9842	0.9970
RMSE	58.9862	84.9033	104.6258	60.6358	74.1619	86.1075	46.8312	62.4062	79.2773

3.3. Proposed Adaptive Window Selection Method

3.3.1. Grey-Level Co-Occurrence Matrix and Its Correlation Features

The grey-level co-occurrence matrix [33] is a matrix function of the distance and angle between pixels. This measure reflects the comprehensive information on the direction, interval, amplitude and speed of the image through the correlation between a certain distance of the image and the two-pixel grey of a certain direction.

The galactic co-occurrence matrix [34–36] is defined as the probability from grey-level i to a fixed position $d = (Dx, Dy)$ to the grey-level j. The grey-level co-occurrence matrix is denoted by $P_d(i, j)(i, j = 0, 1, 2, \dots , L-1)$, where i and j represent the grey scale values of two pixels, respectively, and L denotes the grey level of the image. The spatial relationship d between two pixels is shown in Figure 11, where θ is the direction of generation of the grey-level co-occurrence matrix.

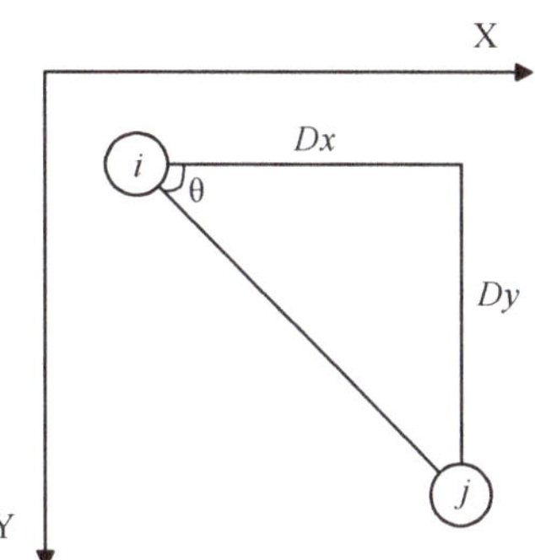

Figure 11. Position relation of the pixel pair of a grey-level co-occurrence matrix.

When d is selected, the grey-level co-occurrence matrix P_d under a certain relation d is generated.

$$
P_d = \begin{bmatrix}
P_d(0,0) & P_d(0,1) & \dots & P_d(0,j) & \dots & P_d(0,L-1) \\
P_d(1,0) & P_d(1,1) & \dots & P_d(1,j) & \dots & P_d(1,L-1) \\
\dots & \dots & \dots & \dots & \dots & \dots \\
P_d(i,0) & P_d(i,1) & \dots & P_d(i,j) & \dots & P_d(i,L-1) \\
\dots & \dots & \dots & \dots & \dots & \dots \\
P_d(L-1,0) & P_d(L-1,1) & \dots & P_d(L-1,j) & \dots & P_d(L-1,L-1)
\end{bmatrix}
\tag{13}
$$

Usually, scalars can be used to describe the characteristics of the grey-level co-occurrence matrix. The correlation features are used to measure the degree of similarity in the horizontal or vertical direction of the grey level of the image, and the magnitude of the value reflects the approximate degree of the local grey level correlation. The larger the correlation value is, the larger the correlation of the local grey level as shown in Equation (14).

$$
Cor = \frac{\sum\limits_{i=0}^{L-1} \sum\limits_{j=0}^{L-1} (i,j)P_d(i,j) - \mu_1\mu_2}{\sigma_1^2 \sigma_2^2}
\tag{14}
$$

Here, μ_1, μ_2, σ_1, and σ_2 are respectively defined as follows:

$$
\mu_1 = \sum\limits_{i=0}^{L-1} \sum\limits_{j=0}^{L-1} iP_d(i,j)
\tag{15}
$$

$$
\mu_2 = \sum\limits_{i=0}^{L-1} \sum\limits_{j=0}^{L-1} jP_d(i,j)
\tag{16}
$$

$$
\sigma_1^2 = \sum\limits_{i=0}^{L-1} \sum\limits_{j=0}^{L-1} (i - \mu_1)^2 P_d(i,j)
\tag{17}
$$

$$
\sigma_2^2 = \sum\limits_{i=0}^{L-1} \sum\limits_{j=0}^{L-1} (j - \mu_2)^2 P_d(i,j)
\tag{18}
$$

where i and j represent the grey values of two pixels, L is the grey level of the image, and d represents the spatial position relationship of two pixels.

3.3.2. Calculation of the Shape and Size of the Evaluation Window

For any pixel (x, y) in the image, the horizontal left neighbourhood $N\,h1\,(x, y)$, the horizontal right neighbourhood $N\,h2\,(x, y)$, the vertical neighbourhood $N\,v1\,(x, y)$ and the vertical lower neighbourhood $N\,v2\,(x, y)$ are respectively in Equations (19)–(22).

$$N^{h1}_{(x,y)} = \{x_{k1}|x_{k1} = R(x + (k1 - (m+1)), y), 1 \le k1 < m+1\} \tag{19}$$

$$N^{h2}_{(x,y)} = \{x_{k2}|x_{k2} = R(x + (k2 - (m+1)), y), 1 \le k2 < m+1\} \tag{20}$$

$$N^{v1}_{(x,y)} = \{x_{k3}|x_{k3} = R(x + (k3 - (m+1)), y), 1 \le k3 < m+1\} \tag{21}$$

$$N^{v2}_{(x,y)} = \{x_{k4}|x_{k4} = R(x + (k4 - (m+1)), y), 1 \le k4 < m+1\} \tag{22}$$

For an overly large m, the neighbourhood of the pixel exceeds the range of the acceptable evaluation window. Taking $m = 3$, the corresponding maximum evaluation window size is 7×7 pixels. In the horizontal direction, a grey-level co-occurrence matrix $P_{d1}(k1)$ with a distance $D1 = (m+1) - k1$ from the centre pixel point (x, y) and an angle of $180°$ is generated with the pixel horizontal left neighbourhood $N\,h1\,(x, y)$, and the correlation eigenvalue $Cor(k1)$ corresponding to the grey-level co-occurrence matrix is obtained. A grey-level co-occurrence matrix $P_{d2}(k2)$ with a distance $D2 = k2 - (m+1)$ from the centre pixel point (x, y) and an angle of $0°$ is generated with the pixel horizontal right neighbourhood $N\,h2\,(x, y)$, and the correlation eigenvalue $Cor(k2)$ corresponding to the grey-level co-occurrence matrix is obtained. Similarly, a grey-level co-occurrence matrix $P_{d3}(k3)$ with a distance from the centre pixel point (x, y) and an angle of $270°$ is generated with the pixel vertical upper neighbourhood $N\,v1\,(x, y)$, and the correlation eigenvalue $Cor(k3)$ corresponding to the grey-level co-occurrence matrix is obtained. A grey-level co-occurrence matrix $P_{d4}(k4)$ with a distance from the centre pixel point (x, y) and an angle of $90°$ is generated with the pixel vertical lower neighbourhood $N\,v2\,(x, y)$, and the correlation eigenvalue $Cor(k4)$ corresponding to the grey-level co-occurrence matrix is obtained.

To find the maximum correlation pixels of the four neighbourhoods of the centre pixels (x, y), the pixels corresponding to the maximum correlation eigenvalues of the grey-level co-occurrence matrix in each direction are taken as the maximum relevant pixel, and the maximum correlation distances $D1$, $D2$, $D3$ and $D4$ of the centre pixel in the four squares' directions are calculated by Equations (23)–(26).

$$D1 = (m+1) - \mathrm{argmax}(Cor(k1)) \tag{23}$$

$$D2 = \mathrm{argmax}(Cor(k2)) - (m+1) \tag{24}$$

$$D3 = (m+1) - \mathrm{argmax}(Cor(k3)) \tag{25}$$

$$D4 = \mathrm{argmax}(Cor(k4)) - (m+1) \tag{26}$$

The maximum correlation distances $D1$, $D2$, $D3$ and $D4$ of the centre pixels in four directions can be used to determine the shape of the rectangular evaluation window of the pixel, and the width $Lx = D1 + D2 + 1$ and height $Ly = D3 + D4 + 1$ of the neighbourhood window are obtained. A diagram of the neighbourhood window is shown in Figure 12.

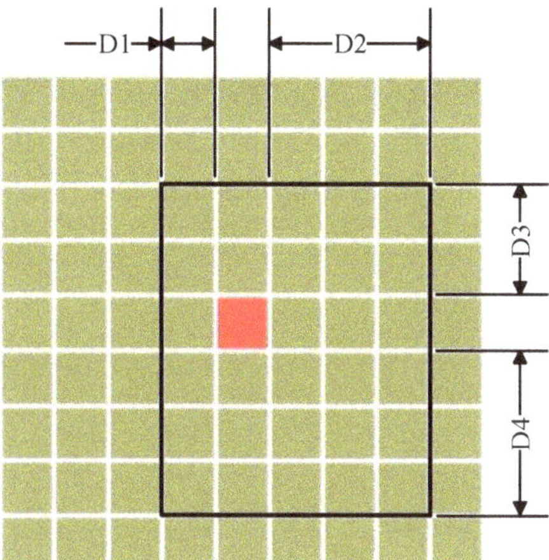

Figure 12. Schematic of the evaluation window.

3.4. Main Procedures of the Improved SFF Algorithm

According to the above stated method, the process of the improved SFF algorithm is shown in Figure 13. The improved SFF algorithm has three main steps, consisting of the original sequence image de-noising, the initial depth map calculation, and the initial depth map refining. First, the image sequence I_k is detected by using the normal distribution operator, and a new grey value is assigned to a pixel determined as noise by using the median filtering method. Otherwise, the original pixel grey level is kept unchanged, and the preprocessed image sequence I'_k is obtained (the threshold value K is 2.2.) Then, the Laplacian operator is used to extract the clear pixels in the preprocessed sequence image I'_k of each frame to construct an all-focus image I_f. The adaptive evaluation window selection method is used to determine the evaluation window $W(i, j)$ of each pixel (i, j) in the fully focused image; then, it calculates the focus measure value of each pixel of the image sequence I'_k, and find the image number corresponding to the maximum focus measure value of each pixel (i, j) to obtain the initial depth map. Lastly, using the depth valued of all pixels, the 3D topography is reconstructed via interpolation. The pixel equivalent is calculated according to the width and height of the pixel and the tooth size of the 3D topography. The limit method is employed to obtain the 3D volume of the worn sprocket teeth and is combined with difference method to obtain the tooth wear volume.

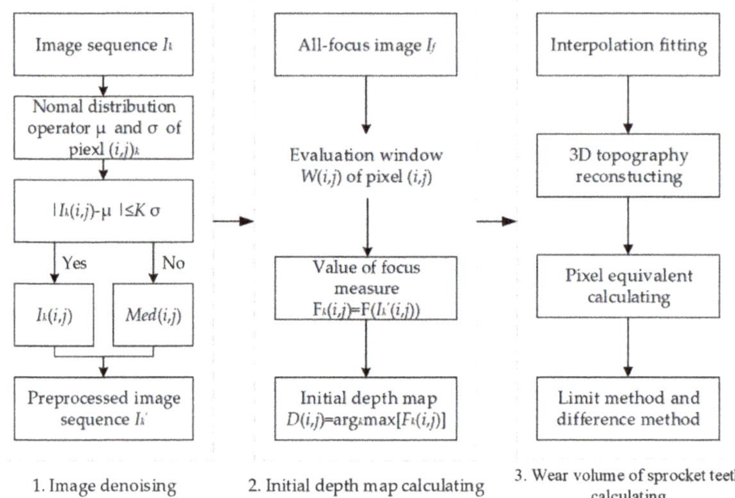

Figure 13. Process of the improved SFF algorithm.

3.5. Test Results and Analysis

To verify the effectiveness of the algorithm, three synthetic objects, i.e., spherical surface, a complex surface and a simple surface, were used as test virtual objects, as shown in Figure 14. In addition, an analogue camera imaging mathematical model was used to create differently focused image sequences of 100 frames, corresponding to the three virtual models, which are 360×360 [37].

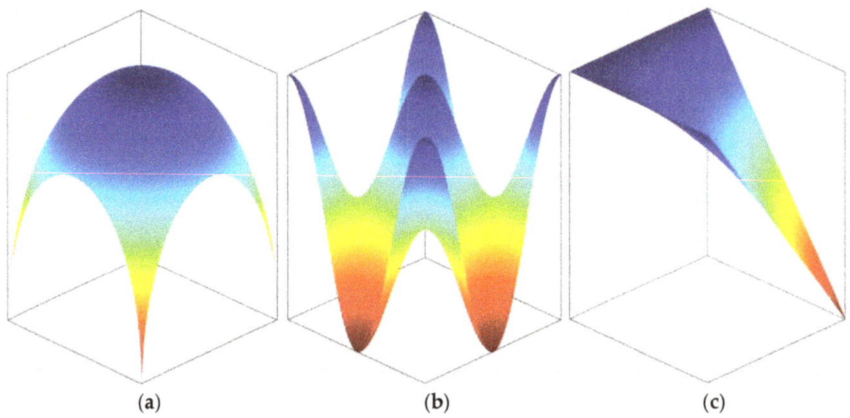

Figure 14. Test virtual objects: (**a**) Spherical surface; (**b**) Complex surface; (**c**) Simple surface.

Salt-and-pepper noise with a density of 0.01 was added to the images of the 10th, 20th, 30th, 40th, 50th, 60th, 70th, 80th, 90th, and 100th frames of the spherical, complex, and simple surface models. The three focus measures F_{SML}, F_{TEN} [38], and F_{GLV} [39] were selected as the test measures. Windows with a size of 3×3, 5×5, 7×7, and the adaptive evaluation window proposed in this paper were used to conduct a 3D morphological recovery test for the image sequences generated using the three models.

Figures 15–17 show the initial depth maps of the three models when F_{SML}, F_{TEN}, and F_{GLV} were chosen as the focus evaluation functions, and different evaluation windows were applied. Columns one to four in the figure are the three models of the 3D morphologies that were recovered and generated by using 3×3, 5×5, and 7×7 windows and the adaptive evaluation window proposed in this paper. Figures show that when the evaluation window is 3×3, the 3D surface topography of all three recovered models appears to have more error values. In comparing the recovery results of the three models, when the surface of the recovery object is smooth and we appropriately increase the evaluation window, the three types of evaluation functions can obtain accurate 3D topographic images. As the surface geometry tends to be complex, increasing the evaluation window does not obviously reduce the error effect. As can be seen from the recovery results of the spherical surfaces in figures, the error of the adaptive evaluation window is obviously less than that of the other evaluation windows, which indicates that the adaptive evaluation window in this algorithm is also effective for noisy image sequences.

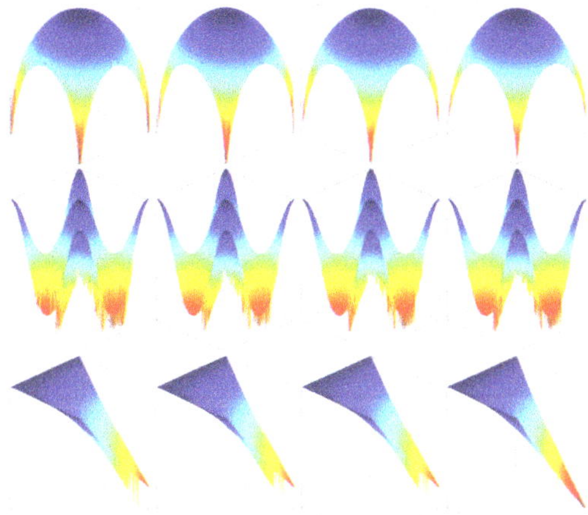

Figure 15. Initial depth map of the three models reconstructed using different evaluation windows of F_{SML}.

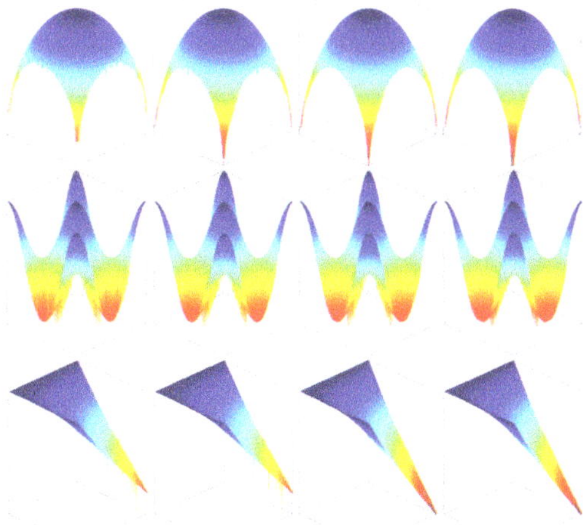

Figure 16. Initial depth map of the three models reconstructed using different evaluation windows of F_{TEN}.

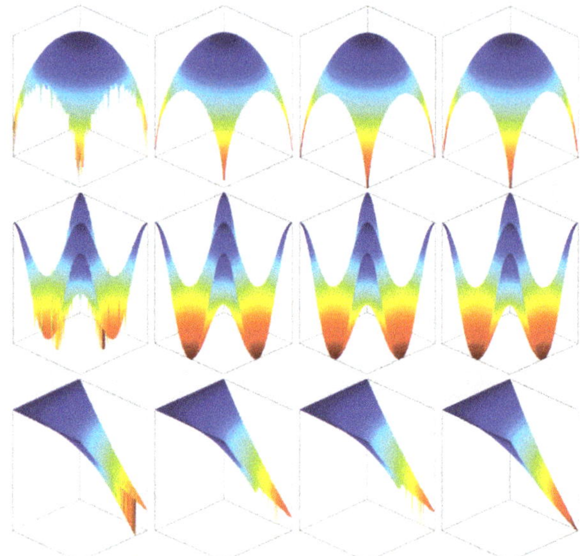

Figure 17. Initial depth map of the three models reconstructed using different evaluation windows of F_{GLV}.

The test compares the actual morphology with the morphology obtained through the test using qualitative observation. Then, the recovery is quantitatively evaluated using the assessment criteria, RMSE and correlation [40]. RMSE and correlation are used to evaluate the error and similarity between the reference data and the real data, respectively. The smaller the RMSE value, the smaller the error between the reference data and the real data. The greater the correlation value, the more consistent the reference data is with the real data. The calculation methods of RMSE and correlation are addressed in Equations (27) and (28).

$$RMSE = \sqrt{\frac{1}{M \times N}\sum_{i=1}^{M}\sum_{j=1}^{N}[D'(i,j) - D(i,j)]^2} \tag{27}$$

$$Cor = \frac{\sum_{i=1}^{M}\sum_{j=1}^{N}\left[D'(i,j) - \overline{D'(i,j)}\right] \times \left[D(i,j) - \overline{D(i,j)}\right]}{\sqrt{\sum_{i=1}^{M}\sum_{j=1}^{N}\left[D'(i,j) - \overline{D'(i,j)}\right]^2 \times \sum_{i=1}^{M}\sum_{j=1}^{N}\left[D(i,j) - \overline{D(i,j)}\right]^2}} \tag{28}$$

In Equations (27) and (28), M represents the number of rows of the image, N denotes the columns of the image, $D'(i,j)$ stands for the actual depth of the pixel point (i,j), and $\overline{D'(i,j)}$ represents the average depth value for all pixels in the image. While $D(i,j)$ stands for the estimated depth of the pixel point (i,j), and $\overline{D(i,j)}$ represents the estimated average depth value for all pixels in the image.

It can be seen from the 3D morphology diagram of the three models reconstructed by the three evaluation functions in the Figures 15–17, Tables 2–4 show the RMSE and correlation data of the models' morphological recovery results of the spherical surface, complex surface and simple surface when F_{SML}, F_{TEN}, and F_{GLV} are chosen as evaluation functions.

Table 2. Evaluation result of the recovery effect of the three models reconstructed using different evaluation windows of F_{SML}.

Size	Cor			RMSE		
	Spherical Surface	Complex Surface	Simple Surface	Spherical Surface	Complex Surface	Simple Surface
3×3	0.9823	0.9897	0.9972	0.0129	0.0323	0.0167
5×5	0.9899	0.9927	0.9978	0.0100	0.0302	0.0157
7×7	0.9942	0.9930	0.9979	0.0074	0.0286	0.0149
Adaptive window	0.9985	0.9936	0.9981	0.0036	0.0256	0.0121

Table 3. Evaluation result of the recovery effect of the three models reconstructed using different evaluation windows of F_{TEN}.

Size	Cor			RMSE		
	Spherical Surface	Complex Surface	Simple Surface	Spherical Surface	Complex Surface	Simple Surface
3×3	0.9983	0.9947	0.9985	0.0075	0.0315	0.0165
5×5	0.9989	0.9954	0.9987	0.0058	0.0296	0.0156
7×7	0.9990	0.9954	0.9988	0.0045	0.0280	0.0148
Adaptive window	0.9994	0.9981	0.9991	0.0023	0.0209	0.0131

Table 4. Evaluation result of the recovery effect of the three models reconstructed using different evaluation windows of F_{GLV}.

Size	Cor			RMSE		
	Spherical Surface	Complex Surface	Simple Surface	Spherical Surface	Complex Surface	Simple Surface
3×3	0.9940	0.9934	0.9803	0.0095	0.0337	0.0198
5×5	0.9990	0.9972	0.9859	0.0073	0.0312	0.0171
7×7	0.9990	0.9976	0.9933	0.0059	0.0295	0.0158
Adaptive window	0.9992	0.9981	0.9966	0.0028	0.0280	0.0128

According to the RMSE data in the table, when the evaluation window is 3×3, the RMSE values restored by the three evaluation functions are the largest, and the larger the evaluation window, the smaller the RMSE value. For example, in the surface morphology recovery using the F_{SML} evaluation function, the RMSE values of the conical surface, simple surface and complex surface obtained by the adaptive window are 0.0036, 0.0256 and 0.0121, in contrast of 3×3, 5×5, 7×7, window, the RMSE value of the adaptive window is minimal. This finding shows that, when the evaluation window is smaller, there are more error values in the recovery results. Furthermore, as the evaluation window increases, the error values gradually decrease, and the effect of the adaptive window is better when the image tends to be smooth. When the evaluation window is 3×3, the correlation values restored by the three evaluation functions are significantly smaller than those of other evaluation windows. In addition, the correlation value of the adaptive evaluation window in this algorithm is larger compared to a fixed-size evaluation window. This phenomenon is more obvious when surface geometry is spherical. For instance, in the surface topography recovery of spherical surfaces using the F_{SML} evaluation function, the Cor values of 3×3, 5×5, 7×7 and adaptive windows are 0.9823, 0.9899, 0.9942 and 0.9985 respectively. The Cor value of adaptive windows is at least 0.4% higher. This finding shows that the 3D topographic map reconstructed with the adaptive evaluation window of this algorithm is closer to the original surface. The more complex the surface topography, the greater the advantage of an adaptive evaluation window.

On the basis of the qualitative observation and comparison and the quantitative data analysis, when we restore the 3D image sequence with noise, compared to a fixed-size evaluation window, regardless of the evaluation function we choose, the error value of the recovery result, the coincidence degree of the 3D image or the original surface morphology, employing the adaptive evaluation window provides good results. Therefore, this algorithm is also feasible for the restoration of the 3D topography of noisy images.

4. Application Example

The above is a test of three virtual models, which shows that the algorithm is effective for virtual models. In addition, to verify the effectiveness of this algorithm in the physical 3D surface morphology restoration of actual solids, a scraper conveyor sprocket tooth is selected as recovery objects, and image acquisition device designed in this paper is used to sequentially collect 100 frames of 1980 × 1114 object images. Figure 18 is an actual entity of sprocket teeth and the 3D model of the sprocket teeth.

(a)   (b)

Figure 18. Test object and 3D model: (**a**) an actual entity of sprocket teeth; (**b**) 3D model of the sprocket teeth.

Firstly, image cropping and filtering are performed on the collected 100 frame sequential image. Then, different evaluation windows, evaluation function and peak positioning technology are used to acquire the initial depth map of sprocket tooth. Finally, the background area is removed by image segmentation technology, and the 3D recovery map of sprocket tooth is obtained. Figure 19a,b are the partial original images and pre-processed images of 100 frame sequential images respectively. Figure 19c,d are the initial depth map and the 3D recovery map of sprocket teeth respectively.

To compare the recovery accuracy of adaptive window with other fixed-size windows, the morphology recovery test is carried out. F_{SML} focus measures is selected in this test, and 3 × 3, 5 × 5, and 7 × 7 windows and the adaptive evaluation window proposed in this paper are used to reconstruct the 3D image of the sprocket tooth. The recovery effect is qualitatively evaluated based on observations, and the actual appearance and the shape obtained are compared in the test. Figure 19 shows the initial depth map of the sprocket tooth. The row from left to right are restored by F_{SML} focus evaluation operator using the evaluation windows of size 3 × 3, 5 × 5 and 7 × 7 and the adaptive window, respectively.

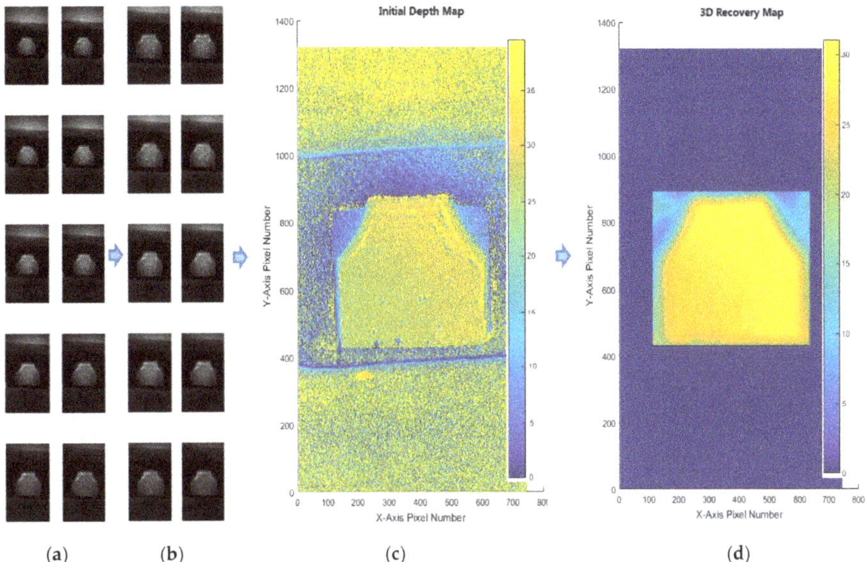

(a)　　　　　(b)　　　　　　　(c)　　　　　　　　　(d)

Figure 19. Image processing of the test: (**a**) the partial original images; (**b**) the partial pre-processed images; (**c**) the initial depth map; (**d**) the 3D recovery map.

We can observe form Figure 20, when the evaluation window is 3 × 3, on the surface of the part, there are many error values in the 3D morphologies restored using F_{SML} evaluation functions. With the increase in the evaluation window, the overall image tends to be smooth, the error value is gradually reduced, and the surface morphology is closer to the surface of the original part. When we select the adaptive evaluation window, the surface profile of the part is the clearest and the surface is smooth. In particular, the pits on the surface of the part are retained, and compared with the other four evaluation windows, the recovery effect is the best. In summary, the evaluation window size has a great influence on the result of the appearance recovery when the 3D surface morphology of the sprocket tooth is restored. An undersized evaluation window is not conducive to recovery results. Compared to the traditional fixed-size evaluation window, the adaptive evaluation window can effectively reduce the error value and preserve the surface texture details.

Figure 20. 3D topographic recovery map reconstructed by FSML focus evaluation operator using different evaluation windows.

To further quantitatively verify the accuracy of the adaptive evaluation window, the degree of similarity between the reconstructed 3D model and the original model was deeply analyzed, and a further experiment was carried out. First, MATLAB (The MathWorks Inc., Natick, MA, ver.2015b) software was used to extract the 3D surface point data in Figure 20, and the extracted point data was saved in the txt folder; then, the IMAGEWARE software was used to read the point data under the txt folder to form a point cloud image, and reverse engineering was applied to restore the point cloud to a curved surface. Finally, the surface was formed to a 3D model in the SolidWorks software, as shown in Figure 21; from left to right, the reconstructed 3D entities of the $3 \times 3, 5 \times 5, 7 \times 7$ windows and the adaptive evaluation window were obtained by the F_{SML} focus evaluation operator.

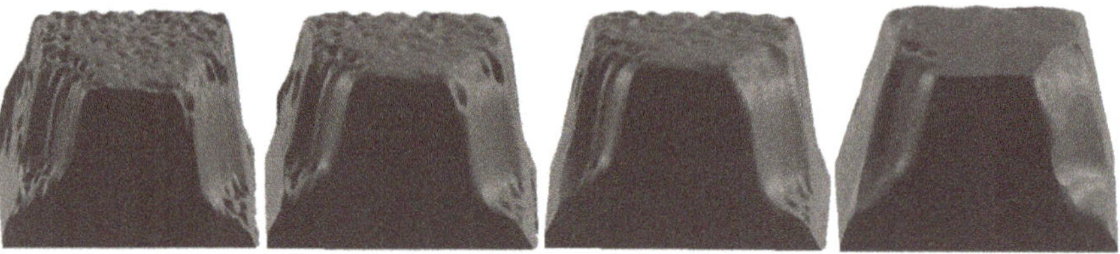

Figure 21. 3D entities using different evaluation windows reconstructed by FSML focus evaluation operator.

The reconstructed 3D entity coincides with the centre of gravity of the original 3D model established in Figure 18 and performs a Boolean operation to obtain a public part reconstructed from the original 3D entity, as shown in Figure 22, from left to right; the public part of the 3×3, $5 \times 5, 7 \times 7$. windows and the adaptive evaluation window are obtained from the F_{SML} focus evaluation operator.

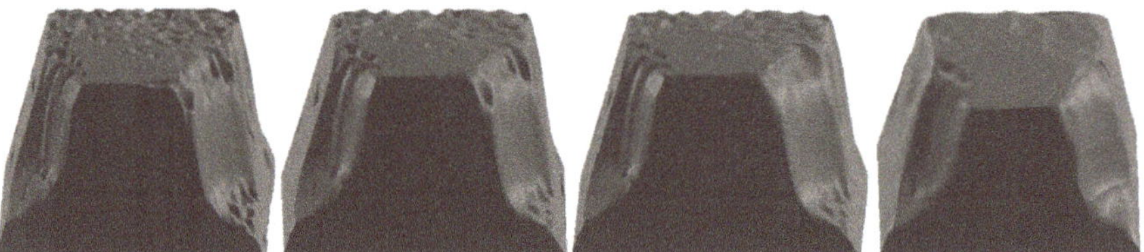

Figure 22. FSML focusing evaluation operator adopts different evaluation windows and original 3D entities of public part.

It can be seen from the above figure that, when the evaluation window is 3×3, the surface of the common part has the largest number of pits and the largest error value, and with the increase of evaluation window, the surface of public entities tends to be smooth, while the adaptive evaluation window has the best smoothness and the lowest error value.

The larger the volume of the public part, the greater the accuracy of recovery; therefore, the accuracy of recovery β can be expressed by the following formula:

$$\beta = \left(1 - \frac{(V_i - V_{0i}) + (V_0 - V_{0i})}{V_0}\right) \times 100\% \tag{29}$$

i takes values from 1–4, SolidWorks software measures the volume of original model ($V_0 = 32511.83$ mm^3), V_i represents the volume of 3D entities reconstructed by different evaluation windows by F_{SML} focusing evaluation operator, and V_{0i} represents the volume of the 3D entities reconstructed by different evaluation windows and common entities of original models by the F_{SML} focusing evaluation operator and calculates the volume evaluation results of metal entity reconstructed by F_{SML} in different evaluation windows, as shown in Table 5.

Table 5. Volume evaluation results of metal cones reconstructed by F_{SML} at different evaluation windows/mm^3.

Evaluation Window	3×3	5×5	7×7	Adaptive Window
V_i	32,021.26	32,037.06	32,089.74	32,056.08
V_{0i}	31,468.01	31,488.91	31,550.28	31,833.45
β	95.09%	95.17%	95.38%	97.23%

It can be seen from Table 5 that the evaluation window is $5 \times 5, 7 \times 7$ and the adaptive evaluation window obtains an image volume, which is basically the same as the volume of the common part. With the increase of the evaluation window, the volume of the overlapping part increases, the volume of the common part of the adaptive evaluation window reaches the maximum, i.e., 31833.45mm^3, and the recovery accuracy of the adaptive evaluation window reaches 97.23%.

According to the 3D shape restoration test results of the scraper conveyor sprocket tooth, the focus value obtained using the adaptive evaluation window is more accurate than the traditional fixed-size square evaluation window when we qualitatively and quantitatively analyze the test results, and it is also feasible to combine the normal distribution operator filtering method in this algorithm.

5. Conclusions

An SFF-based method was proposed in order to effectively measure the wear volume of sprocket teeth in a scraper conveyor; the following conclusions were drawn:

1. A hardware device for volumetric tooth wear measurement was designed and assembled to collect sequential images of sprocket teeth, which provides a way for images acquisition of measuring the wear volume of sprocket teeth in a scraper conveyor.
2. A normal distribution operator image filtering method was presented, which only filters the noise points in the image without changing the grey value of the non-noise point pixels. Therefore, compared with the traditional filtering method, more original information of the image is retained to a large extent.
3. An adaptive evaluation window selection method was proposed. A focused morphology restoration algorithm based on the normal distribution operator-region pixel reconstruction was formed, which not only effectively eliminates the error of restoration accuracy caused by noise interference, but also satisfies the requirement of peak location. Therefore, both the accuracy and effectiveness of morphology restoration has been improved.
4. Compared to other focused 3D restoration methods, the proposed methods can effectively measure the wear volume of sprocket teeth with a recovery accuracy of up to 97.23%.
5. In order to further improve the accuracy of this method and expand the scope of application, we will consider the advantages of structured light [41] for further research.

Author Contributions: H.D. proposed the method; Y.L. performed the experiments and analyzed the data; J.L. contributed method guidance and language modification; H.D. wrote the paper.

Funding: This work was funded by the Shanxi Science and Technology Foundation Condition Platform Project, grant 201805D141002 and the Joint Training Base for Postgraduate Students in Shanxi Province, grant 2018JD15.

Acknowledgments: Thanks are due to Shanxi Coal Mine Machinery Manufacturing Co., Ltd. for providing experimental equipment and application environment.

References

1. Dolipski, M.; Remiorz, E.; Sobota, P. Determination of dynamic loads of sprocket drum teeth and seats by means of a mathematical model of the longwall conveyor. *Arch. Min. Sci.* **2012**, *57*, 1101–1119.

2. Jiang, S.B.; Zeng, Q.L.; Wang, G.; Gao, K.D.; Wang, Q.Y.; Hidenori, K. Contact analysis of chain drive in scraper conveyor based on dynamic meshing properties. *Int. J. Simul. Model.* **2018**, *17*, 81–91. [CrossRef]

3. Sobota, P. Determination of the friction work of a link chain interworking with a sprocket drum. *Arch. Min. Sci.* **2013**, *58*, 805–822.

4. Ren, F.; Shi, A.Q.; Yang, Z.J. Research on load identification of mine hoist based on improved support vector machine. *Trans. Can. Soc. Mech. Eng.* **2018**, *42*, 201–210. [CrossRef]

5. Põdra, P.; Andersson, S. Finite element analysis wear simulation of a conical spinning contact considering surface topography. *Wear* **1999**, *224*, 13–21. [CrossRef]

6. China University of Ming and Technology. A Monitoring Device and Method for Abrasion of Scraper Conveyor Sprocket Tooth Wear. CN Patent CN201610325560, 17 August 2016.

7. Wang, S.P.; Yang, Z.J.; Wang, X.W. Wear of driving sprocket for scraper convoy and mechanical behaviors at meshing progress. *J. China Coal Soc.* **2014**, *39*, 166–171.

8. Wang, S.; Yang, Z.; Wang, X. Relationship between Round Link Chain Deformation and Worn Sprocket. *China Mech. Eng.* **2014**, *25*, 1586–1590.

9. Alberdi, A.; Rivero, A.; López de Lacalle, L.N.; Etxeverria, I.; Suárez, A. Effect of process parameter on the kerf geometry in abrasive water jet milling. *Int. J. Adv. Manuf. Technol.* **2010**, *51*, 467–480. [CrossRef]

10. Qian, X.; Huang, X. Reconstrction of surfaces of revolution with partial sampling. *J. Comput. Appl. Math.* **2004**, *163*, 211–217. [CrossRef]

11. Peña, B.; Aramendi, G.; Rivero, A.; López de Lacalle, L.N. Monitoring of drilling for burr detection using spindle torque. *Int. J. Mach. Tools Manuf.* **2005**, *45*, 1614–1621. [CrossRef]

12. Xiong, G.X.; Liu, J.C.; Avila, A. Cutting tool wear measurement by using active contour model based image processing. In Proceedings of the IEEE International Conference on Mechatronics and Automation, Beijing, China, 7–10 August 2011; pp. 670–675.

13. Liu, J.C.; Xiong, G.X. Study on Volumetric tool wear measurement using image processing. *Appl. Mech. Mater. Manuf.* **2014**, *670–671*, 1194–1199. [CrossRef]

14. Ahmad, M.; Choi, T.S. Application of three dimensional shape from Image focus in LCD/TFT displays Manufacturing. *IEEE Trans. Consum. Electr.* **2007**, *53*, 1–4. [CrossRef]

15. Tang, J.; Qiu, Z.; Li, T. A novel measurement method and application for grinding wheel surface topography based on shape from focus. *Measurement* **2018**, *133*, 495–507. [CrossRef]

16. Darrell, T.; Wohn, K. Pyramid based depth from focus. In Proceedings of the Computer Vision and Pattern Recognition, Ann Arbor, MI, USA, 5–9 June 1988; pp. 504–509.

17. Nayar, S.K.; Nakagawa, Y. Shape from Focus. *IEEE Trans. Pattern Anal. Mach. Intell.* **1994**, *16*, 824–831. [CrossRef]

18. Nayar, S.K.; Nakagawa, Y. Shape from Focus: An Effective Approach for Rough Surfaces. In Proceedings of the IEEE International Conference on Robotics and Automation, Cincinnati, OH, USA, 13–18 May 1990.

19. Karthikeyan, P.; Vasuki, S. Multiresolution joint bilateral filtering with modified adaptive shrinkage for image denoising. *Multimed. Tools Appl.* **2016**, *75*, 1–18.

20. Khan, A.; Waqas, M.; Ali, M.R.; Altalhi, A.; Alshomrani, S.; Shim, S.O. Image de-noising using noise ratio estimation, K-means clustering and non-local means-based estimator. *Comput. Electr. Eng.* **2016**, *54*, 370–381. [CrossRef]

21. Singh, V.; Dev, R.; Dhar, N.K.; Agrawal, P.; Verma, N.K. Adaptive Type-2 Fuzzy Approach for Filtering Salt and Pepper Noise in Greyscale Images. *IEEE Trans. Fuzzy Syst.* **2018**, *26*, 3170–3176. [CrossRef]

22. Fan, T.; Yu, H. A novel shape from focus method based on 3D steerable filters for improved performance on treating textureless region. *Opt. Commun.* **2018**, *410*, 254–261. [CrossRef]

23. Liu, X.; Zhai, D.; Chen, R.; Ji, X.; Zhao, D.; Gao, W. Depth super-resolution via joint color-guided internal and external regularizations. *IEEE Trans. Image Process.* **2018**, *28*, 1636–1645. [CrossRef]

24. Saravani, S.; Shad, R.; Ghaemi, M. Iterative adaptive Despeckling SAR image using anisotropic diffusion filter and Bayesian estimation denoising in wavelet domain. *Multimed. Tools Appl.* **2018**, *77*, 31469–31486. [CrossRef]

25. Khan, M.A.; Chen, W.; Fu, Z.J.; Khalil, A.U. Meshfree digital total variation based algorithm for multiplicative noise removal. *J. Inf. Sci. Eng.* **2018**, *34*, 1441–1468.

26. Mahmood, M.T.; Majid, A.; Choi, T.S. Optimal depth estimation by combining focus measures using genetic programming. *Inf. Sci.* **2011**, *181*, 1249–1263. [CrossRef]

27. Lee, I.; Mahmood, M.T.; Choi, T.S. Adaptive windows election for 3D shape recovery from image focus. *Opt. Laser Technol.* **2013**, *35*, 21–31. [CrossRef]

28. Lee, I.H.; Shim, S.O.; Choi, T.-S. Improving focus measurement via variable window shape on surface radiance distribution for 3D shape reconstruction. *Opt. Laser Eng.* **2013**, *51*, 520–526. [CrossRef]

29. Muhammad, M.S.; Mutahira, H.; Choi, K.W.; Kim, W.Y.; Ayaz, Y. Calculation accurate window size for shape from focus. In Proceedings of the IEEE International Conference on Information Science & Applications, Seoul, South Korea, 6–9 May 2014; Computer Society Press: Washington, DC, USA, 2014; pp. 1–4.

30. Thipprakmas, S. Improving wear resistance of sprocket parts using a fine-blanking process. *Wear* **2011**, *271*, 2396–2401. [CrossRef]

31. Billiot, B.; Cointault, F.; Journaux, L.; Simon, J.-C.; Gouton, P. 3D image acquisition system based on shape from focus technique. *Sensors* **2013**, *13*, 5040–5053. [CrossRef]

32. Shim, S.-O.; Malik, A.S.; Choi, T.-S. Noise reduction using mean shift algorithm for estimating 3D shape. *Imaging Sci. J.* **2011**, *59*, 267–273. [CrossRef]

33. Huang, X.; Liu, X.; Zhang, L. A Multichannel Gray Level Co-Occurrence Matrix for Multi/Hyperspectral Image Texture Representation. *Remote Sens.* **2014**, *6*, 8424–8445. [CrossRef]

34. Zhang, X.; Cui, J.; Wang, W.; Lin, C. A Study for Texture Feature Extraction of High-Resolution Satellite Images Based on a Direction Measure and Grey Level Co-Occurrence Matrix Fusion Algorithm. *Sensors* **2017**, *17*, 1474. [CrossRef]

35. Zheng, G.; Li, X.; Zhou, L.; Yang, J.; Ren, L.; Chen, P. Development of a Grey-Level Co-Occurrence Matrix-Based Texture Orientation Estimation Method and Its Application in Sea Surface Wind Direction Retrieval From SAR Imagery. *IEEE Trans. Geosci. Remote Sens.* **2018**, *56*, 5244–5260. [CrossRef]

36. Varish, N.; Pal, A.K. A novel image retrieval scheme using grey level co-occurrence matrix descriptors of discrete cosine transform based residual image. *Appl. Intell.* **2018**, *12*, 1–24.

37. Subbarao, M.; Lu, M.-C. Image sensing model and computer simulation for CCD camera systems. *Mach. Vis. Appl.* **1994**, *7*, 277–289. [CrossRef]

38. Xia, X.; Yao, Y.; Liang, J.; Fang, S.; Yang, Z.; Cui, D. Evaluation of focus measures for the autofocus of line scan cameras. *Opt. Int. J. Light Electron Opt.* **2016**, *127*, 19–7762. [CrossRef]

39. Krotkow, E. Focusing. *Int. J. Comput. Vis.* **1988**, *1*, 223–237. [CrossRef]

40. Malik, A.S.; Choi, T.S. A novel algorithm for estimation of depth map using image focus for 3D shape recovery in the presence of noise. *Pattern Recogn.* **2008**, *41*, 2200–2225. [CrossRef]

41. Scharstein, D.; Szeliski, R. High-accuracy stereo depth maps using structured light. In Proceedings of the IEEE conference on computer vision and pattern recognition., Madison, WI, USA, 18–20 June 2003; pp. 195–202.

 *applied sciences*

Article

A Novel Self-Intersection Penalty Term for Statistical Body Shape Models and Its Applications in 3D Pose Estimation

Zaiqiang Wu [1], Wei Jiang [1,*], Hao Luo [1] and Lin Cheng [2]

[1] College of Control Science and Engineering, Zhejiang University, Hangzhou 310007, China; wuzaiqiang@zju.edu.cn (Z.W.); haoluocsc@zju.edu.cn (H.L.)
[2] 2012 Lab, Huawei Technologies, Hangzhou 310028, China; chenglin17@huawei.com
* Correspondence: jiangwei_zju@zju.edu.cn; Tel.: +86-150-6818-8247

Received: 29 December 2018; Accepted: 21 January 2019; Published: 24 January 2019

Abstract: Statistical body shape models are widely used in 3D pose estimation due to their low-dimensional parameters representation. However, it is difficult to avoid self-intersection between body parts accurately. Motivated by this fact, we proposed a novel self-intersection penalty term for statistical body shape models applied in 3D pose estimation. To avoid the trouble of computing self-intersection for complex surfaces like the body meshes, the gradient of our proposed self-intersection penalty term is manually derived from the perspective of geometry. First, the self-intersection penalty term is defined as the volume of the self-intersection region. To calculate the partial derivatives with respect to the coordinates of the vertices, we employed detection rays to divide vertices of statistical body shape models into different groups depending on whether the vertex is in the region of self-intersection. Second, the partial derivatives could be easily derived by the normal vectors of neighboring triangles of the vertices. Finally, this penalty term could be applied in gradient-based optimization algorithms to remove the self-intersection of triangular meshes without using any approximation. Qualitative and quantitative evaluations were conducted to demonstrate the effectiveness and generality of our proposed method compared with previous approaches. The experimental results show that our proposed penalty term can avoid self-intersection to exclude unreasonable predictions and improves the accuracy of 3D pose estimation indirectly. Further more, the proposed method could be employed universally in triangular mesh based 3D reconstruction.

Keywords: statistical body shape model; self-intersection penalty term; 3D pose estimation

1. Introduction

Estimating a 3D human pose from a single 2D image, and more generally, reconstructing the 3D model from 2D images is one of the fundamental and challenging problems in 3D computer vision due to the inherent ambiguity in inferring 3D from 2D. Choosing the appropriate 3D representation is vital for 3D reconstruction. There are many types of 3D representations for 3D modeling. Voxels, point clouds and polygon meshes are commonly used 3D formats for 3D representation. Voxels can be fed directly to convolutional neural networks (CNNs), therefore a lot of works applied voxels for classification [1,2] and 3D reconstruction [3–5]. However voxels are poor in memory efficiency. To avoid this drawback of voxel representation, Fan et al. [6] proposed a method to generate point clouds from 2D images. But since there are no connections between points in the point cloud representation, the generated point cloud is often not close to a surface. Polygon mesh is promising due to its high memory efficiency when compared to voxels and point clouds [7]. Polygon mesh is also convenient to visualize since it is compatible with most existing rendering engines.

There are many works using polygon meshes, especially triangular meshes, to represent 3D pose estimation results. Anguelov et al. [8] proposed the first statistical body shape model called SCAPE, represented as a triangular mesh. Loper et al. [9] proposed another statistical body shape model with higher accuracy called SMPL which is also represented as a triangular mesh. Guan et al. [10,11] employed SCAPE to estimate 3D poses based on manually marked 2D joints. Bogo et al. [12] employed the SMPL human model and minimized the error between the projected human model joints and 2D joints detected by DeepCut [13] to estimate the 3D pose of the human body automatically, this method is iterative optimization-based which results in high accuracy but is time consuming. Pavlakos et al. [14] used a variant of Hourglass [15] to predict 2D joints and 2D masks simultaneously, then the 2D joints and 2D masks were used to regress pose parameters and shape parameters of SMPL separately in a direct prediction way. This approach is much faster than method proposed in [12], however self-intersection occurs on images with pattern of poses that never appeared in the training set.

Other work utilized triangle meshes to represent 3D reconstruction results of objects. Kar et al. [16] trained a mesh deformable model to reconstruct 3D shapes limited to the popular categories. Kato et al. [7] deformed a predefined mesh to approximate the 3D object by minimizing the silhouette error. Wang et al. [17] proposed an end-to-end deep learning architecture which represents a triangular mesh in a graph-based convolutional neural network to estimate the 3D shape of objects from a single image.

However, one of the main disadvantages of representing 3D shapes as triangular meshes is that self-intersection is difficult to prevent. Some examples of 3D pose estimation results with self-intersection are shown in Figure 1. It is impossible for objects in the real world to have surfaces with self-intersection. Therefore previous work paid great attention to avoiding the self-intersection of triangular meshes. Since it is difficult to derive a differentiable expression of the intersection volume directly, approximation is often taken to simplify the derivation. Although approximation provides great convenience to deriving a differentiable penalty term, self-intersection can not be removed strictly since approximation can not describe the original surface accurately.

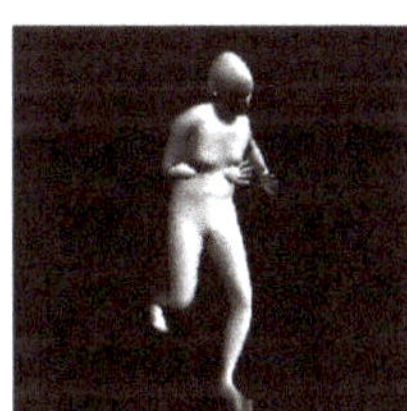

Figure 1. Examples of model-based 3D pose estimation results with self-intersection between body parts.

To overcome the weakness of previous methods of preventing self-intersection, this paper proposed a novel self-intersection penalty term (SPT) which is able to avoid self-intersection strictly. Unlike previous approaches, our proposed self-intersection penalty term is defined as the volume of intersection regions which is expensive to compute however, we managed to work around this problem. Notably, no approximation was taken in this paper to derive a differentiable expression. Besides, it is not worthwhile to derive a differentiable expression of intersection volume since calculating the exact volume is not our intention. Moreover, the partial derivatives can be easily derived even without the expression of intersection volume. Inspired by [18], we developed an algorithm to detect vertices of self-intersection regions quickly by only going through triangles in the mesh once. This process is similar to rasterization in computer graphic with linear time complexity. A linked list is applied to store depth values and orientations of triangles intersected with the same detection ray. Then vertices in self-intersection region can be easily detected by going through the linked list. The partial derivatives of the self-intersection term with respect to the coordinate of each vertex are easy to derive from the perspective of geometry. The partial derivatives with respect to vertices not in the region of self-intersection is obviously zero, while the partial derivatives with respect to vertices in the region of self-intersection could be obtained from the normal vectors of neighbouring triangles. The value of the penalty term is assigned as the ratio of number of vertices in self-intersection to number of vertices not in self-intersection. In this way the value of penalty term is easy to compute and indicates the degree of self-intersection in some degree. The experimental results show that the proposed penalty term avoids self-intersection strictly and works effectively. The main contributions of this paper are summarized as follows:

- We proposed a novel self-intersection penalty term which does not require deriving a differentiable expression and generally applies to triangular mesh-based representations.
- We performed 3D pose estimation from 2D joints and compared our method with other state-of-the-art approaches qualitatively and quantitatively to demonstrate the practical value and the superiority of our method.
- To the best of our knowledge, this is the first time the conception of self-intersection in relation to disconnected meshes in the field of 3D reconstruction has been generalized.

The content of this paper consists of five sections. In Section 2 an overview of related work is provided. In Section 3, the details of the proposed self-intersection penalty term are presented. In Section 4, the results of experiments and analysis of the proposed self-intersection penalty term are given. The conclusions are presented in Section 5.

2. Related Work

The work presented in this section is closely related to our work and involves avoiding self-intersection with triangular mesh.

In computer graphics, it is common to use proxy geometries to prevent self-intersection [19,20]. In computer vision, recent works followed this approach to prevent self-intersection of 3D reconstruction results represented as triangular meshes. Sminchisescu et al. [21] defined an implicit surface as a approximation of body shape to avoid self-intersection. Pons et al. [22] applied a set of spheres to approximate the interior volume of body mesh, and used the radius of each sphere to define a penalty term of self-intersection. These approaches are not accurate since the shape of human body can not be described exactly by spheres. To improve the accuracy of this approach, Bogo et al. [12] trained a regressor to generate capsules with minimum error to the body surface, then the authors further defined the penalty term as a 3D isotropic Gaussian derived from the capsule radius. It is worth mentioning that these approaches mentioned above do not strictly avoid self-intersection as approximations were applied to derive a differentiable penalty term. In [17] the authors employed a Laplacian term to prevent the vertices from moving too freely, this penalty term avoids self-intersection to some degree. However this method still does not strictly avoid self-intersection since the Laplacian term acts just like a surface smoothness term preventing the 3D mesh from deforming too much.

Our work differs from previous works by identifying that a differentiable self-intersection penalty term is not necessary and the gradients can be calculated manually. We demonstrated appealing results in 3D pose estimation based on a statistical body shape model.

3. Self-intersection Penalty Term

In this section, the details of our proposed self-intersection penalty term are discussed. We employed the SMPL human body shape model [9] to evaluate our method. Essential notations are provided here. SMPL model defines a function $\mathcal{M}(\beta, \theta; \Phi)$, where θ are the pose parameters, β are the shape parameters and Φ are fixed parameters of the model. The output of this function is a triangular mesh $P \in \mathbb{R}^{N \times 3}$ with $N = 6890$ vertices $P_i \in \mathbb{R}^3 (i = 1, \ldots, N)$. The shape parameters β represents the coefficients of linear combination of a low number of principal body shapes learned from a dataset containing body scans [23].

3.1. Definition and Description

Our method generally applies to meshes satisfying the conditions described below:

- The mesh is a two-dimensional manifold.
- The mesh describes an orientable surface.
- The mesh is a closed surface.

Since the two-dimensional manifolds do not have to be connected, we can say a mesh with several disconnected parts also satisfies the conditions above. We demonstrated that our proposed method also works with a disconnected mesh in Section 4.

To remove the self-intersection, the triangle mesh should be iteratively deformed by moving each vertex in a specific direction. The moving directions of vertices are obtained by computing the partial derivatives of the self-intersection penalty term which is defined as the volume of the self-intersection region in this paper. This penalty term is denoted as $E_{SPT}(V)$, where V is the coordinates of all the vertices. An ideal self-intersection penalty term should satisfy the following conditions:

- When there is no self-intersection, both the penalty term and the gradient of the penalty term should be zero.
- When there is self-intersection, the value of penalty term indicates the degree of intersection.
- When there is self-intersection, the gradient of penalty term offers meaningful direction for optimization.

Leaving the strategy of computing the value of penalty term aside, the method of computing gradient is discussed first.

The first step of computing the partial derivatives is separating the vertices into two sets: (1) vertices in the self-intersection region and (2) vertices not in the self-intersection region. We implemented this separation by emitting a beam of detection rays, the density of rays is manually set according to the number of triangles in the mesh. An appropriate setting of density of rays guarantees the accuracy of classification and low memory consumption. There are two typical ways of intersection shown in Figure 2. According to the type of self-intersection, the set of vertices in self-intersection could be further separated into two sets. Overall, the vertices are divided into three sets: (a) vertices in the self-intersection region due to interpenetration of the outer surface, denoted as V_{out}; (b) vertices in the self-intersection region due to interpenetration of the inner surface, denoted as V_{in} and (c) vertices not in the self-intersection region, denoted as V_0; Based on the classification result, the partial derivative of the penalty term with respect to coordinate of a vertex can be obtained according to the normal vector and which set the vertex is belonging to.

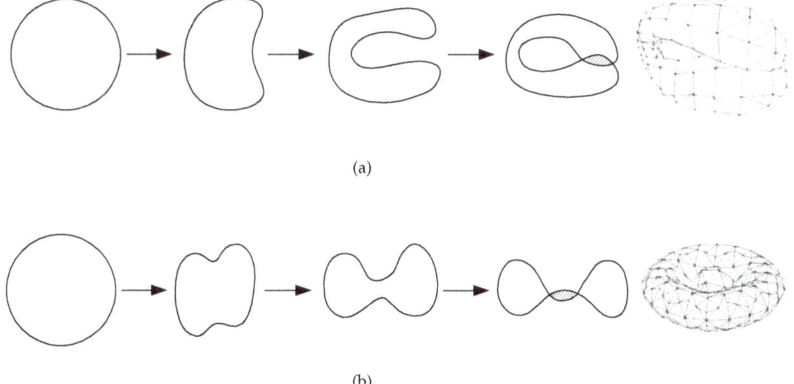

(a)

(b)

Figure 2. Two typical way of self-intersection: (**a**) Self-intersection due to interpenetration of the outer surface. (**b**) Self-intersection due to interpenetration of the inner surface.

It is worth pointing out that self-intersection described in Figure 2b rarely occurs for statistical body shape model. To maintain the generality of our method, both of these two type of self-intersection are considered in the following discussion.

3.2. Detection and Classification of Vertices

To compute the partial derivatives of our proposed self-intersection penalty term, it is necessary to divide the vertices into three sets: V_0, V_{in} and V_{out}. A camera screen with pixels arranged in a square with H rows and W columns is set in front of the 3D mesh such that the orthogonal projection of the 3D mesh falls totally inside the screen. It is worth noting that the camera mentioned here is used only to emit detection rays, not for rendering and visualization. Detection rays are emitted from the center of each pixel to detect self-intersection, as is shown in Figure 3. The detail of the detection and classification is presented in Algorithm 1. To make our algorithm more intuitive, a schematic representation is given in Figure 4.

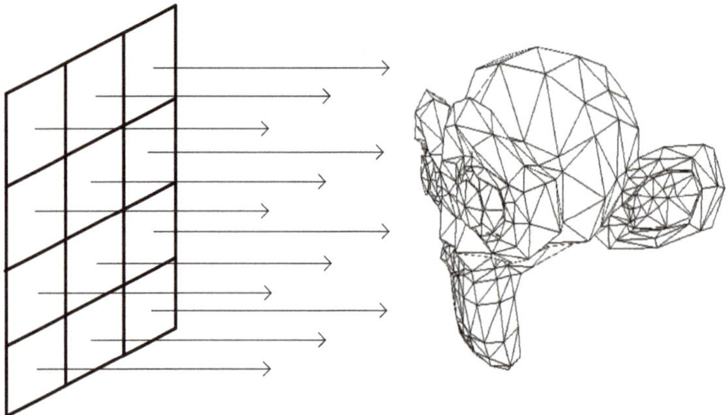

Figure 3. Schematic representation of our approaches to detect self-intersection (the image of triangular mesh is rendered by Blender).

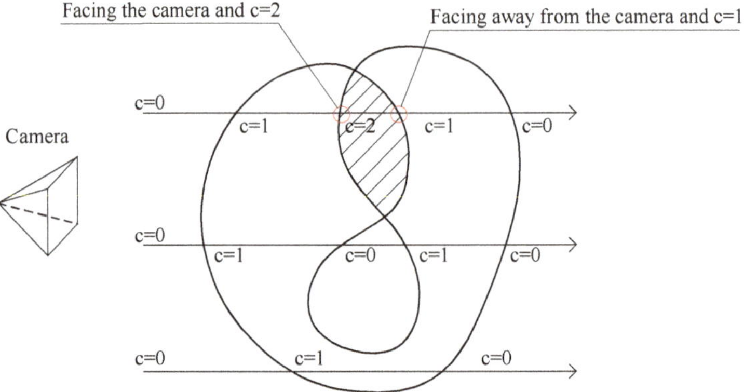

Figure 4. An intuitive representation of our algorithm to detect self-intersection. The detection rays are emitted from the camera, the dashed area represents the self-intersection region. The vertices in red circle were detected to be in self-intersection according to the counter and orientation.

Algorithm 1: Illustration of self-intersection detection and classification of vertices

---

Input: the height H and width W of the screen, V containing coordinates of all vertices, F containing vertex indexes of all triangles.

Output: V_0 containing vertices not in self-intersection, V_{out} containing vertices in self-intersection of the outer surface, V_{in} containing vertices in self-intersection of the inner surface.

1 **initialize** V_0, V_{out} and V_{in} as empty sets
2 **initialize** the linked list of each pixel as empty list
3 **for** *each triangle in F* **do**
4   project the triangle onto the screen by orthogonal projection
5   **for** *each pixel in the screen* **do**
6   **if** *this pixel located inside the triangle* **then**
7   insert this triangle into the linked list of this pixel in the order of smallest to largest in terms of depth

8 **for** *each pixel in the screen* **do**
9   **if** *the linked list of this pixel is empty* **then**
10   **continue**

11   **initialize** *counter* $\leftarrow 0$
12   **for** *each triangle in the linked list* **do**
13   **if** *this triangle is facing the camera* **then**
14   *count* $\leftarrow$ *counter* $+ 1$

15   **else**
16   *count* $\leftarrow$ *counter* $- 1$

17   **if** *this triangle is facing the camera* **then**
18   **if** *counter* $= 1$ **then**
19   append the three vertices of this triangle to V_0

20   **else if** *counter* $= 2$ **then**
21   append the three vertices of this triangle to V_{out}

22   **else if** *counter* $= 0$ **then**
23   append the three vertices of this triangle to V_{in}

24   **else**
25   **if** *counter* $= 0$ **then**
26   append the three vertices of this triangle to V_0

27   **else if** *counter* $= 1$ **then**
28   append the three vertices of this triangle to V_{out}

29   **else if** *counter* $= -1$ **then**
30   append the three vertices of this triangle to V_{in}

31 **return** V_0, V_{out} and V_{in}

3.3. Gradients Calculation

The self-intersection penalty term $E_{SPT}(V)$ is defined as:

$$E_{SPT}(V) = V_{intersect} \tag{1}$$

where V denotes the set of coordinates of all vertices, $V_{intersection}$ is the volume of the self-intersection region. For the sake of further discussion, a vertex $p_t(t = 1, \ldots, N)$ is randomly chosen from the triangular mesh with N vertices. Coordinates of all vertices are frozen, except for p_t. The coordinate of p_t is denoted as (x_t, y_t, z_t).

Under the preceding assumptions, the penalty term $E_{SPT}(V)$ can be regarded as a function of x_t, y_t and z_t.

$$E_{SPT}(V) = f(x_t, y_t, z_t) \tag{2}$$

To compute $\frac{\partial E_{SPT}(V)}{\partial x_t}$, $\frac{\partial E_{SPT}(V)}{\partial y_t}$ and $\frac{\partial E_{SPT}(V)}{\partial z_t}$, p_t is displaced from (x_t, y_t, z_t) to $(x_t + \Delta x, y_t + \Delta y, z_t + \Delta z)$. Then the change in $E_{STP}(V)$ due to the displacement could be represented as:

$$\Delta E_{SPT}(V) = f(x_t + \Delta x, y_t + \Delta y, z_t + \Delta z) - f(x_t, y_t, z_t) \tag{3}$$

Further computation is hard to continue without imposing constraints on p_t. First, for the simplest case, p_t is assumed to belong to V_0, that is to say $p_t \in V_0$. Since tiny displacement of p_t brings no effect to the volume of self-intersection, it is obvious that:

$$\Delta E_{SPT}(V) = \Delta f = 0, p_t \in V_0 \tag{4}$$

The gradient of $E_{SPT}(V)$ could be represented as:

$$\begin{pmatrix} \frac{\partial E_{SPT}(V)}{\partial x_t} \\ \frac{\partial E_{SPT}(V)}{\partial y_t} \\ \frac{\partial E_{SPT}(V)}{\partial z_t} \end{pmatrix} = \begin{pmatrix} \lim_{\Delta x \to 0} \frac{\Delta f}{\Delta x} \\ \lim_{\Delta y \to 0} \frac{\Delta f}{\Delta y} \\ \lim_{\Delta z \to 0} \frac{\Delta f}{\Delta z} \end{pmatrix} = \begin{pmatrix} 0 \\ 0 \\ 0 \end{pmatrix}, p_t \in V_0 \tag{5}$$

For the second case $p_j \in V_{out}$, more assumptions need to be made for a detailed discussion. We assume that there are n neighboring triangles sharing p_t as a common vertex. One of the neighboring triangles is denoted as $T_l(l = 1, \ldots, n)$, the area of T_l is represented as S_l and the unit normal vector of T_l is denoted as $\boldsymbol{n}_l$. An intuitive representation of this situation is shown in Figure 5.

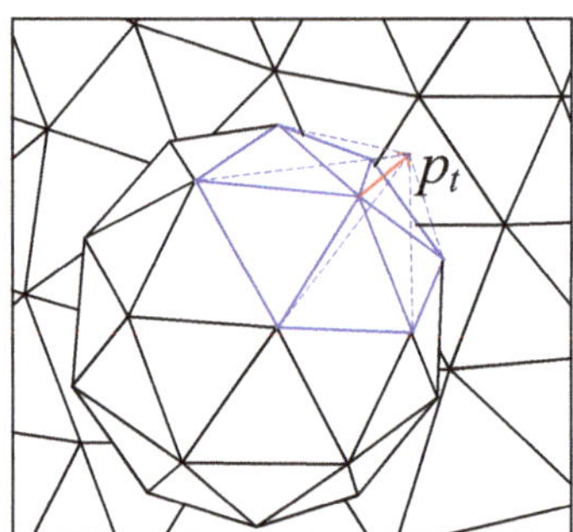

Figure 5. A tiny displacement on the vertex p_t. The change in the volume of self-intersection is equal to the sum of volume of several neighboring tetrahedrons.

The change in the volume of self-intersection due to tiny displacement of vertex p_t can be represented as:

$$\Delta f = \Delta V_{intersection} = \frac{1}{3} \sum_{l=1}^{n} S_l \boldsymbol{n}_l (\Delta x \boldsymbol{i} + \Delta y \boldsymbol{j} + \Delta z \boldsymbol{k}), p_t \in V_{out} \tag{6}$$

where $\boldsymbol{i}, \boldsymbol{j}$ and $\boldsymbol{k}$ are unit vectors in the same directions as the positive directions of x, y and z axes.

The gradient can be obtained as:

$$\begin{pmatrix} \frac{\partial E_{SPT}(V)}{\partial x_t} \\ \frac{\partial E_{SPT}(V)}{\partial y_t} \\ \frac{\partial E_{SPT}(V)}{\partial z_t} \end{pmatrix} = \begin{pmatrix} \lim_{\Delta x \to 0} \frac{\Delta f}{\Delta x} \\ \lim_{\Delta y \to 0} \frac{\Delta f}{\Delta y} \\ \lim_{\Delta z \to 0} \frac{\Delta f}{\Delta z} \end{pmatrix} = \begin{pmatrix} \frac{1}{3} \sum_{l=1}^{n} S_l \boldsymbol{n}_l \boldsymbol{i} \\ \frac{1}{3} \sum_{l=1}^{n} S_l \boldsymbol{n}_l \boldsymbol{j} \\ \frac{1}{3} \sum_{l=1}^{n} S_l \boldsymbol{n}_l \boldsymbol{k} \end{pmatrix}, p_t \in V_{out} \tag{7}$$

The equation above can be simplified as:

$$\begin{pmatrix} \frac{\partial E_{SPT}(V)}{\partial x_t} \\ \frac{\partial E_{SPT}(V)}{\partial y_t} \\ \frac{\partial E_{SPT}(V)}{\partial z_t} \end{pmatrix} = \frac{1}{3} \sum_{l=1}^{n} S_l \boldsymbol{n}_l, \, p_t \in V_{out} \tag{8}$$

For the last case $p_t \in V_{in}$, the derivation process of partial derivative is similar to the process described in the second case, and the results are same in magnitude but opposite in sign. The gradient in this case can be obtained as:

$$\nabla E_{SPT}(V) = -\frac{1}{3} \sum_{l=1}^{n} S_l \boldsymbol{n}_l, \, p_t \in V_{in} \tag{9}$$

In summary, the gradient of self-intersection penalty term with respect to the coordinate of vertex p_t can be represented as:

$$\nabla E_{SPT}(V) = \begin{cases} \vec{0} & p_t \in V_0 \\ \frac{1}{3} \sum_{l=1}^{n} S_l \boldsymbol{n}_l & p_t \in V_{out} \\ -\frac{1}{3} \sum_{l=1}^{n} S_l \boldsymbol{n}_l & p_t \in V_{in} \end{cases} \tag{10}$$

Employing the equation above for a gradient-based optimization algorithm to remove self-intersection works well in most cases. But when there are great differences between the areas of triangles, the process of optimization tends to be unstable. To solve this problem, a modified version of the gradient formula is presented:

$$\nabla E_{SPT}(V) = \begin{cases} \vec{0} & p_t \in V_0 \\ \frac{\sum_{l=1}^{n} S_l \boldsymbol{n}_l}{\left\| \sum_{l=1}^{n} S_l \boldsymbol{n}_l \right\|_2} & p_t \in V_{out} \\ -\frac{\sum_{l=1}^{n} S_l \boldsymbol{n}_l}{\left\| \sum_{l=1}^{n} S_l \boldsymbol{n}_l \right\|_2} & p_t \in V_{in} \end{cases} \tag{11}$$

For the value of $E_{SPT}(V)$, it is difficult and unnecessary to calculate the exact volume of the intersection region by coordinates of vertices. Instead, $E_{SPT}(V)$ is assigned with the ratio of the number of vertices in self-intersection to the number of vertices not in self-intersection. This ratio is easy to calculate and significant for indicating the degree of self-intersection. The expression of $E_{SPT}(V)$ is presented as:

$$\begin{aligned} E_{SPT}(V) &= \frac{|V_{out}| + |V_{in}|}{|V_{out}| + |V_{in}| + |V_0|} \\ &= \frac{|V_{out}| + |V_{in}|}{N} \end{aligned} \tag{12}$$

where $|\cdot|$ denotes the number of elements in a set.

So far, the forward and backward processes of our proposed self-intersection penalty term have been defined.

4. Experimental Results and Discussion

In this section, experiments were conducted to evaluated the effectiveness of our proposed self-intersection penalty term (SPT). We employed a statistical body shape model SMPL [9] to show the effectiveness of our method.

4.1. Self-Intersection Removal on a Single SMPL Model

To evaluate the validity of our method, we set the pose parameters θ to generate triangular meshes with self-intersection deliberately. Then we performed gradient descent algorithm to optimize the pose parameters θ to remove the self-intersection, the shape parameters β were the ground truth values from dataset and were fixed during iterations. The learning rate of gradient descent was set to 1.0×10^{-4}. The number of detection rays was set to 512×512.

The process of iteration is visualized in Figure 6. We can see that the gradient calculated via proposed method works effectively in a gradient descent algorithm to minimize the value of SPT, that is to say, minimize the number of vertices in self-intersection region.

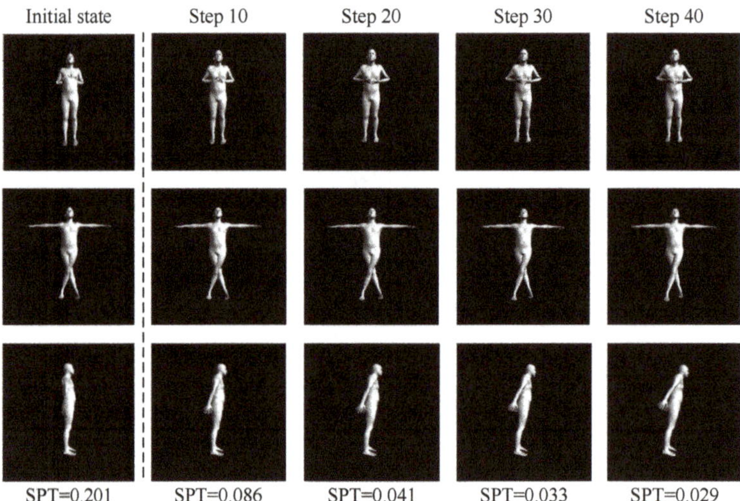

Figure 6. Images rendered from the iterative process. First column: Images rendered from initial meshes with self-intersection. Second through fourth columns: Images rendered from optimized meshes every 10 iterations. The self-intersection penalty term (SPT) values in the bottom of each column denote the average SPT of three SMPL models.

In order to show the necessity of gradient normalization which is presented in Equation (11), an experiment with same conditions as the experiment described above, but without gradient normalization, was conducted. The visualized result is shown in Figure 7. Since there are great differences between the areas of triangles in the body meshes, gradients of vertices differ greatly in magnitudes. This often leads to unstable iterations and unpredictable results and we demonstrated that this problem can be solved by gradient normalization.

As can be seen from Figures 6 and 7, gradient normalization improves the stability of optimization. Therefore it can be concluded that gradient normalization is significant for 3D pose estimation since unstable iterations often lead to failure predictions.

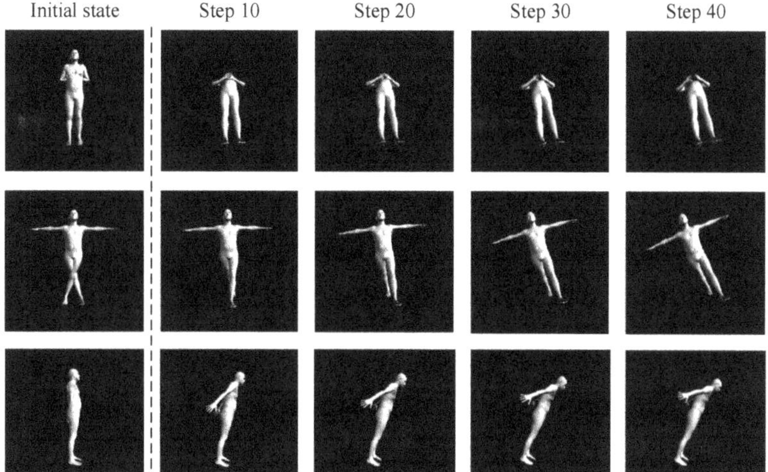

Figure 7. Images rendered from the iterative process without gradient normalization. First column: images rendered from initial meshes with self-intersection. Second through fourth columns: images rendered from optimized meshes every 10 steps.

4.2. Intersection and Self-Intersection Removal on Multiple SMPL Models

To demonstrate that our method applies to general closed surfaces, an experiment on mesh with two disconnected surfaces was carried out. In the experiment, two SMPL mesh models were generated and were regarded as one mesh, gradient descent was employed to remove the intersection between the two SMPL models and the self-intersection of themselves. We visualized the result in Figure 8.

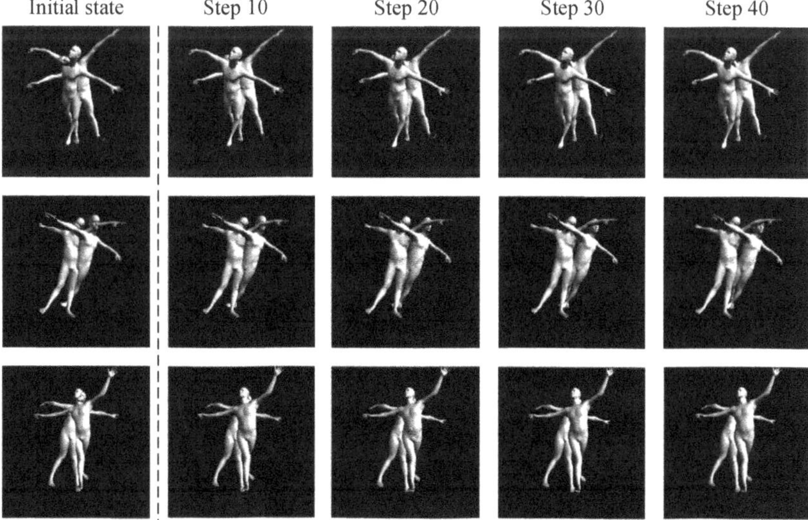

Figure 8. Images rendered from the iterative process with two SMPL models. First column: Images rendered from initial meshes with self-intersection. Second through fourth columns: Images rendered from optimized meshes every 10 steps.

Figure 8 shows that our proposed self-intersection penalty term can remove both the intersection between different body meshes and the self-intersection of each body mesh. This property of our method is of great significance for multi-person 3D pose estimation.

4.3. 3D Pose Estimation from 2D Joints

We tested our proposed self-intersection penalty term (SPT) on UP-3D [24] whose sample images were labeled with ground truth 2D joints. To estimate 3D pose from 2D joints, we defined an objective function as:

$$E(\boldsymbol{\theta}) = E_J(\boldsymbol{\theta}, \boldsymbol{\beta}, K, J_{gt}) + E_{SPT}(\mathcal{M}(\boldsymbol{\beta}, \boldsymbol{\theta}; \boldsymbol{\Phi})) \tag{13}$$

where K are camera parameters and J_{gt} is the ground truth of 2D joints. E_J represents the error between projected joints of the SMPL model and the ground truth 2D joints.

The shape parameters $\boldsymbol{\beta}$ are fixed during optimization. Since the objective function defined in Equation (13) is differentiable, gradient descent can be directly applied to optimize the pose parameters $\boldsymbol{\theta}$ by minimizing the objective function. To demonstrate that SPT can improve the accuracy of 3D pose estimation by excluding unreasonable predictions, we also performed the optimization of objective function without SPT which can be represented as:

$$E'(\boldsymbol{\theta}) = E_J(\boldsymbol{\theta}, \boldsymbol{\beta}, K, J_{gt}) \tag{14}$$

Figure 9 visually compares the results of two different objective functions on a few images from UP-3D dataset. It is obvious that minimizing the error between projected joints and the ground truth 2D joints directly without self-intersection penalty term tends to result in body meshes with self-intersection. The fitting results with SPT are more natural and more reasonable compared with the results without SPT. This experiment demonstrated that it is effective to add our proposed SPT into the objective function to avoid self-intersection of body meshes in optimization-based 3D pose estimation.

Figure 9. Visualized results of 3D pose estimation. Images in row (**a**) are the input images, images in row (**b**) are fitting results without SPT, images in row (**c**) are fitting results with SPT.

4.4. Comparison with State-of-the-Art

To compare our proposed method with other state-of-the-art methods, we conducted an experiment on UP-3D dataset. Since our iterative optimization-based method often fails because of local minima, we used a reduced test set of 139 images selected by Tan et al. [25] to limit the range for the global rotation of body shape model.

We implemented two state-of-the-art methods for qualitative and quantitative comparisons. In the baseline method, the objective function is defined as the reprojection error of 2D joints only. These methods are described in more detail below:

- **Reprojection error of 2D joints (RE) only**
  This method only minimizes the error between ground truth 2D joints and projected 2D joints to estimate the 3D pose.
- **RE + Laplacian regularization (LR)**
  Laplacian regularization is proposed by Wang et al. [17] to prevent the vertices from moving too freely and potentially avoids mesh self-intersection in triangular mesh based 3D reconstruction. We employed this method in 3D pose estimation and the objective function is defined as the sum of reprojection error of 2D joints and the Laplacian regularization term.
- **RE + Sphere approximation (SA)**
  Pons-Moll et al. [22] built a set of spheres as a coarse approximation to the body shape model and derived a differentiable penalty term via calculating the intersection between spheres. To implement this method, We designed a set of spheres to approximate the surface of human body, as is shown in Figure 10. The objective function is defined as the sum of reprojection error of 2D joints and the intersection between spheres.
- **RE + SPT**
  This is our proposed method whose objective function is defined as the sum of the reprojection error of 2D joints and the self-intersection penalty term proposed in this paper.

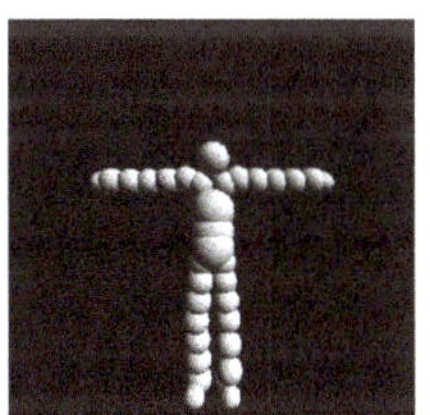

Figure 10. Spheres designed to approximate the human body are kept in the same pose with the body shape model.

4.4.1. Qualitative Comparison

We implemented two state-of-the-art methods and one baseline method for qualitative comparison. Figure 11 presents a part of results from the reduced test set of UP-3D by our proposed method and other methods. These results demonstrate that our proposed method can remove the self-intersection of the statistical body shape model effectively and produces more reasonable results.

As is can seen from Figure 11, the results obtained by the baseline method without any self-intersection penalty tends to intersects with itself. The Laplacian regularization can not strictly avoid self-intersection and often leads to unnatural results. We can see that it is not suitable to employ Laplacian regularization in statistical body shape model because of the fact that this laplacian term brings negative effect to 3D pose estimation. The method of sphere approximation is very competitive in removing the self-intersection of body mesh, however method requires designing a appropriate set of spheres and we found that it is an excessive trivial procedure to set the radius and the coordinate of each sphere appropriately. In addition, since the body mesh can not be approximated accurately by spheres, this method may lead to fail results by simply removing the intersection between spheres.

Compared with other approach, our proposed method can obtain more visually appealing results from two points. One is that when there is no self-intersection, our proposed SPT will have no effect on the body mesh, this means no side effects on 3D pose estimation. The other is that our proposed approach can avoid the self-intersection of body mesh strictly without taking any approximation.

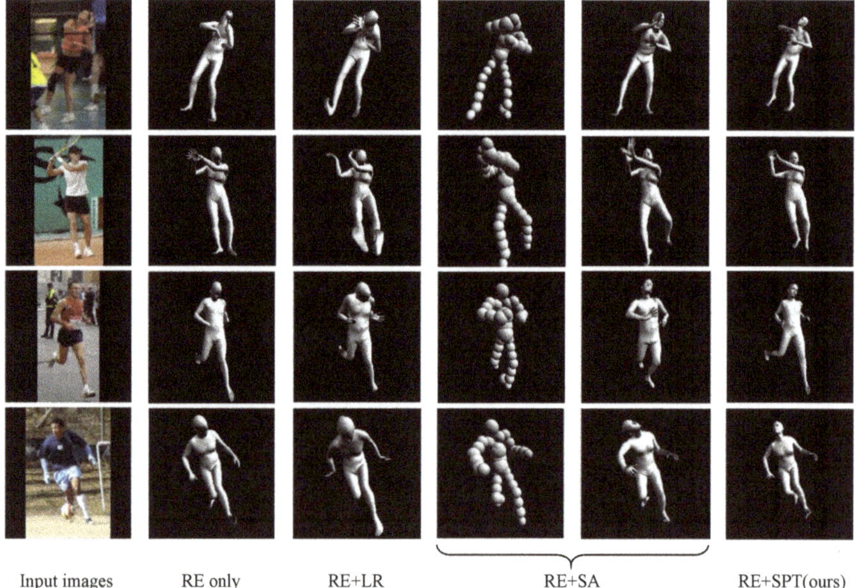

Figure 11. 3D pose estimation results by four different methods. First column: Input images. Second column: Results obtained by minimizing reprojection (RE) error of 2D joints only. Third column: Results obtained by minimizing the sum of reprojection error and Laplacian regularization (LR). Fourth column through fifth column: Results obtained by minimizing the sun of reprojection error and the intersection between spheres. Sphere Approximation (SA). Sixth column: Results obtained by our proposed approach.

4.4.2. Quantitative Comparison

To the best of our knowledge, there is no commonly used evaluation metric for methods preventing self-intersection of triangular mesh. In order to compare our method with other state-of-the-art approaches quantitatively, we adopt the per vertex errors as 3D pose estimation metric and we used the percentage of vertices in region of self-intersection computed by our proposed algorithm to evaluate the performance of our method and other state-of-the-art approaches. It should be noted that it is inappropriate to evaluate these methods only by per vertex error or percentage of vertices in self-intersection because an ideal approach of avoiding self-intersection should achieve both minimum per vertex error and minimum percentage of vertices in self-intersection region. Therefore these two evaluation metrics, per vertex error and percentage of vertices in self-intersection, were adopted to conduct a quantitative comparison.

The quantitative results of the baseline method, the other state-of-the-art approaches and our proposed method is shown in Table 1. The Laplacian regularization method achieved a lower percentage of vertices in self-intersection but a higher per vertex error compared to the baseline method, which demonstrates that the Laplacian regularization does work in avoiding self-intersection but the side effect leads to loss of precision in 3D pose estimation. The approach of approximating body shape by a set of spheres outperforms the baseline method both in per vertex error and percentage of vertices in self-intersection. It is undeniable that this method may perform better with more carefully designed spheres, but it will be extremely tedious to implement this method. Our proposed approach outperforms the baseline method and two state-of-the-art methods and avoids the tedious procedure required for the sphere approximation.

Table 1. Results of baseline method, other state-of-the-art methods and our proposed approach on reduced test set of UP-3D [24].

Methods	Per Vertex Error (mm)	Percentage of Vertices in Self-Intersection
RE only	257.54	15.22%
RE+LR [17]	452.72	7.64%
RE+SA [22]	186.98	0.87%
RE+SPT (ours)	**140.31**	**0.23%**

4.5. Analysis of Time Efficiency

In order to evaluate the time efficiency of our method, we carried out experiments with a different number of detection rays and SMPL models. All experiments in this section were done on a laptop with Intel(R) Core(TM) i5-3230M processer. The most time-consuming part of our technique is self-intersection detection as is described in Algorithm 1. The highlight of our method is that the time complexity is linear to the number of triangles. The number of detection rays has almost no effect on the elapsed time as is shown in Table 2. As can be seen in the Table 2, the number of detection rays was increased from 128×128 to 2048×2048 but the elapsed time remained approximately constant.

Table 2. Time consumed by an iteration with different number of detection rays. The number of SMPL models is 1.

Number of Detection Rays	Elapsed Time (ms)
128×128	53.84
256×256	54.96
512×512	56.76
1024×1024	58.32
2048×2048	59.28

In the next experiment, we fixed the number of detection rays and changed the number of SMPL models to test the performance of our method with growing number of triangles. A single SMPL model has about 13,000 triangles. As is shown in Table 3, the elapsed time grows about 30 ms for each additional SMPL model added to the mesh. Result of this experiment demonstrated that the time complexity of our proposed algorithm is linear to the number of triangles.

Table 3. Time consumed by an iteration with different number of SMPL models. The number of detection rays is 512×512.

Number of SMPL Models	Elapsed Time (ms)
1	56.76
2	89.80
3	119.25
4	149.58
5	186.76

Our proposed method is obviously more computational compared with traditional methods, but it is worthwhile to apply this method because of the accuracy and generality of our approach. Moreover, the time efficiency of our method is totally acceptable according to the experimental results.

5. Conclusions

In this paper, we proposed a novel self-intersection penalty term for statistical body shape models to remove the self-intersection of the mesh by gradient-based optimization. Unlike most traditional approaches, our method does not require a hard-to-obtain differentiable penalty term, but instead gradients are manually calculated. In the course of analysis, we have demonstrated that it is not necessary to derive differentiable expressions of a penalty term and gradients can be manually calculated from the perspective of geometry. Since no approximation is used in our method, self-intersection can be strictly removed. The highlight of our work is that our method applies to general meshes with different shapes and topology without the need to design a set of appropriate proxy geometries. Despite the fact that our proposed self-intersection penalty term is more time consuming than traditional approaches, the elapsed time of one iteration is totally acceptable according to the experimental results. The applications of our method are not limited to the statistical body shape models presented in this paper. Our proposed self-intersection penalty term can be incorporated into other 3D reconstruction problems based on a triangular mesh, such as the mesh-based 3D reconstruction described in [7,16,17].

Our proposed approach has its limitations. When there are some triangles happened to be parallel to the detection rays, these triangles will not intersect with any detection rays no matter how dense the detection rays are. That is to say, vertices of triangles parallel to the detection rays may be mistakenly classified, and further the gradients with respect to these vertices will be incorrect. We assume that each triangle and its vertices are located in the same region, this assumption does not apply to the situation where the mesh is sparse and in this situation undesirable consequences may be caused. Another limitation is that it is difficult to manually set the number of detection rays appropriately, such that all vertices in self-intersection can be detected and classified correctly with minimum memory consumption.

Future research directions of this work may include modifying the way detection rays are emitted to avoid incorrect results when there are triangles parallel to the detection rays. It may also include developing a strategy to set the number of detection rays appropriately and automatically. We are also interested in reducing the time complexity of our proposed method to make this approach more suitable for gradient-based optimization.

Author Contributions: Conceptualization, Z.W. and H.L.; Methodology, Z.W., H.L. and W.J.; Writing—original draft preparation, Z.W.; Writing—review and editing, Z.W., H.L. and W.J.; Visualization, Z.W. and H.L.; Supervision, W.J.; Funding acquisition, W.J., H.L. and L.C.

Funding: This research was funded by [the National Natural Science Foundation of China] grant number [61633019], [the Public Projects of Zhejiang Province, China] grant number [LGF18F030002] and [Huawei innovation research program] grant number [HO2018085209].

Conflicts of Interest: The authors declare no conflict of interest.

References

1. Maturana, D.; Scherer, S. Voxnet: A 3D convolutional neural network for real-time object recognition. In Proceedings of the 2015 IEEE/RSJ International Conference on Intelligent Robots and Systems (IROS), Hamburg, Germany, 28 September–2 October 2015; IEEE: Piscataway, NJ, USA, 2015; pp. 922–928.
2. Qi, C.R.; Su, H.; Nießner, M.; Dai, A.; Yan, M.; Guibas, L.J. Volumetric and multi-view cnns for object classification on 3D data. In Proceedings of the IEEE Conference on Computer Vision and Pattern Recognition, Las Vegas Valley, NV, USA, 26 June–1 July 2016; pp. 5648–5656.

3. Choy, C.B.; Xu, D.; Gwak, J.; Chen, K.; Savarese, S. 3D-R2N2: A unified approach for single and multi-view 3d object reconstruction. In Proceedings of the European Conference on Computer Vision, Amsterdam, The Netherlands, 8–16 October 2016; Springer: Berlin, Germany, 2016; pp. 628–644.

4. Tatarchenko, M.; Dosovitskiy, A.; Brox, T. Octree generating networks: Efficient convolutional architectures for high-resolution 3D outputs. In Proceedings of the IEEE International Conference on Computer Vision (ICCV), Venice, Italy, 22–29 October 2017; Volume 2, p. 8.

5. Tulsiani, S.; Zhou, T.; Efros, A.A.; Malik, J. Multi-view supervision for single-view reconstruction via differentiable ray consistency. In Proceedings of the 2017 Conference on Computer Vision and Pattern Recognition (CVPR), Honolulu, HI, USA, 21–26 July 2017; Volume 1, p. 3.

6. Fan, H.; Su, H.; Guibas, L.J. A Point Set Generation Network for 3D Object Reconstruction from a Single Image. In Proceedings of the 2017 Conference on Computer Vision and Pattern Recognition (CVPR), Honolulu, HI, USA, 21–26 July 2017; Voume 2, p. 6.

7. Kato, H.; Ushiku, Y.; Harada, T. Neural 3D mesh renderer. In Proceedings of the IEEE Conference on Computer Vision and Pattern Recognition, Salt Lake City, UT, USA, 18–23 June 2018; pp. 3907–3916.

8. Anguelov, D.; Srinivasan, P.; Koller, D.; Thrun, S.; Rodgers, J.; Davis, J. SCAPE: Shape completion and animation of people. *ACM Trans. Graph.* **2005**, *24*, 408–416. [CrossRef]

9. Loper, M.; Mahmood, N.; Romero, J.; Pons-Moll, G.; Black, M.J. SMPL: A skinned multi-person linear model. *ACM Trans. Graph.* **2015**, *34*, 248. [CrossRef]

10. Guan, P.; Weiss, A.; Balan, A.O.; Black, M.J. Estimating human shape and pose from a single image. In Proceedings of the 2009 IEEE 12th International Conference on Computer Vision, Kyoto, Japan, 27 September–4 October 2009; IEEE: Piscataway, NJ, USA, 2009; pp. 1381–1388.

11. Guan, P. Virtual human bodies with clothing and hair: From images to animation. Ph.D. Thesis, Brown University, Providence, RI, USA, 2012.

12. Bogo, F.; Kanazawa, A.; Lassner, C.; Gehler, P.; Romero, J.; Black, M.J. Keep it SMPL: Automatic estimation of 3D human pose and shape from a single image. In Proceedings of the European Conference on Computer Vision, Amsterdam, The Netherlands, 8–16 October 2016; Springer: Berlin, Germany, 2016; pp. 561–578.

13. Pishchulin, L.; Insafutdinov, E.; Tang, S.; Andres, B.; Andriluka, M.; Gehler, P.V.; Schiele, B. DeepCut: Joint subset partition and labeling for multi person pose estimation. In Proceedings of the IEEE Conference on Computer Vision and Pattern Recognition, Las Vegas Valley, NV, USA, 26 June–1 July 2016; pp. 4929–4937.

14. Pavlakos, G.; Zhu, L.; Zhou, X.; Daniilidis, K. Learning to Estimate 3D Human Pose and Shape from a Single Color Image. *arXiv* **2018**, arXiv:1805.04092.

15. Newell, A.; Yang, K.; Deng, J. Stacked hourglass networks for human pose estimation. In Proceedings of the European Conference on Computer Vision, Amsterdam, The Netherlands, 8–16 October 2016; Springer: Berlin, Germany, 2016; pp. 483–499.

16. Kar, A.; Tulsiani, S.; Carreira, J.; Malik, J. Category-Specific Object Reconstruction From a Single Image. In Proceedings of the IEEE Conference on Computer Vision and Pattern Recognition (CVPR), Boston, MA, USA, 7–12 June 2015.

17. Wang, N.; Zhang, Y.; Li, Z.; Fu, Y.; Liu, W.; Jiang, Y.G. Pixel2Mesh: Generating 3D Mesh Models from Single RGB Images. *arXiv* **2018**, arXiv:1804.01654.

18. Crow, F.C. Shadow algorithms for computer graphics. In Proceedings of the 4th Annual Conference on Computer Graphics and Interactive Techniques, SIGGRAPH'77, San Jose, CA, USA, 20–22 July 1977; ACM: New York, NY, USA, 1977; Volume 11, pp. 242–248.

19. Ericson, C. *Real-Time Collision Detection (The Morgan Kaufmann Series in Interactive 3-D Technology)*; CRC Press: Boca Raton, FL, USA, 2004.

20. Thiery, J.M.; Guy, É.; Boubekeur, T. Sphere-meshes: Shape approximation using spherical quadric error metrics. *ACM Trans. Graph.* **2013**, *32*, 178. [CrossRef]

21. Sminchisescu, C.; Triggs, B. Covariance scaled sampling for monocular 3D body tracking. In Proceedings of the 2001 IEEE Computer Society Conference on Computer Vision and Pattern Recognition, Kauai, HI, USA, 8–14 December 2001; IEEE: Piscataway, NJ, USA, 2001; p. 447.

22. Pons-Moll, G.; Taylor, J.; Shotton, J.; Hertzmann, A.; Fitzgibbon, A. Metric regression forests for correspondence estimation. *Int. J. Comput. Vis.* **2015**, *113*, 163–175. [CrossRef]

23. Robinette, K.M.; Blackwell, S.; Daanen, H.; Boehmer, M.; Fleming, S. *Civilian American and European Surface Anthropometry Resource (CAESAR), Final Report. Volume 1. Summary*; Technical Report; Sytronics Inc.: Dayton, OH, USA, 2002.

24. Lassner, C.; Romero, J.; Kiefel, M.; Bogo, F.; Black, M.J.; Gehler, P.V. Unite the people: Closing the loop between 3D and 2D human representations. In Proceedings of the IEEE Conference on Computer Vision and Pattern Recognition (CVPR), Honolulu, HI, USA, 21–26 July 2017; Volume 2, p. 3.

25. Tan, J.; Budvytis, I.; Cipolla, R. Indirect deep structured learning for 3D human body shape and pose prediction. In Proceedings of the BMVC, London, UK, 4–7 September 2017; Volume 3, p. 6.

Article

A CNN Model for Human Parsing Based on Capacity Optimization

Yalong Jiang * and Zheru Chi

Department of Electronic and Information Engineering, The Hong Kong Polytechnic University,
Hong Kong 999077, China; enzheru@polyu.edu.hk
* Correspondence: yalong.jiang@connect.polyu.hk

Received: 28 February 2019; Accepted: 21 March 2019; Published: 29 March 2019

Abstract: Although a state-of-the-art performance has been achieved in pixel-specific tasks, such as saliency prediction and depth estimation, convolutional neural networks (CNNs) still perform unsatisfactorily in human parsing where semantic information of detailed regions needs to be perceived under the influences of variations in viewpoints, poses, and occlusions. In this paper, we propose to improve the robustness of human parsing modules by introducing a depth-estimation module. A novel scheme is proposed for the integration of a depth-estimation module and a human-parsing module. The robustness of the overall model is improved with the automatically obtained depth labels. As another major concern, the computational efficiency is also discussed. Our proposed human parsing module with 24 layers can achieve a similar performance as the baseline CNN model with over 100 layers. The number of parameters in the overall model is less than that in the baseline model. Furthermore, we propose to reduce the computational burden by replacing a conventional CNN layer with a stack of simplified sub-layers to further reduce the overall number of trainable parameters. Experimental results show that the integration of two modules contributes to the improvement of human parsing without additional human labeling. The proposed model outperforms the benchmark solutions and the capacity of our model is better matched to the complexity of the task.

Keywords: human parsing; depth-estimation; computational efficiency; capacity optimization

---

1. Introduction

Semantic segmentation and human parsing are critical tasks in visually describing humans under various scenes. Deep Convolutional Neural Networks (CNNs) have brought significant improvements to human parsing tasks [1–3] thanks to the availability of an increased amount of training data. Existing works in this field include Path Aggregation (PA) [4], Large Kernel Matters (LKM) [5], Mask RCNN (MRCNN) [6], holistic models for human parsing [7], and joint pose estimation and part segmentation [8] with spatial pyramid pooling [9]. Moreover, human parsing aligns well with other tasks such as group behavior analysis [10], person re-identification [11], e-commerce [12], image editing [13], video surveillance [14], autonomous driving [3], and virtual reality [15].

However, the performance of existing human parsing methods is still far from robust due to the heavy reliance on the limited training data. In real-world scenarios, one image is very likely to contain multiple people with various human interactions, poses, and occlusion. However, very few of the scenarios can be included in common datasets. For instance, the Pascal Person Part Dataset [7] contains annotations of less than 10 classes and no more than 4,000 images for training and validation, which is far from enough to train complex CNNs [16,17] with over 100 layers. What is worse, is that data augmentation is challenging because labeling an image pixel-by-pixel takes 239.7 s on average [18].

To reduce the cost in labelling, image-level annotations and bounding boxes have been adopted by weakly supervised methods to improve segmentation [19–21]. Additionally, scribbles and points have been introduced in [18,22] as auxiliary supervision. Unlike most weakly supervised methods, depth information is utilized in this paper as guidance to help distinguish foregrounds from backgrounds and use the limited capacity of a CNN model more on foreground areas. A module for depth estimation was trained firstly, then the concatenation of the depth predictions and RGB images composed the input to the segmentation module during both training and testing processes. For simplicity, we used the Depth-Module (DM) and Segmentation-Module (SM) to represent the two modules. The depth annotations from RGB image pairs with overlapping viewpoints could be obtained automatically [23] with multi-view stereo (MVS). The two modules composed the Overall Model (OM).

The advantage of integrating a Depth-Model with Segmentation-Module comes in two ways. Firstly, the training data for depth estimation can cover the variations which seldomly appear in the training data for human parsing. Robustness is improved in this way. Secondly, depth estimation and segmentation are closely correlated. The former assigns continuous depth values to pixels while the latter assigns discrete categorical labels to pixels. The predicted depth maps facilitate hierarchical descriptions of images which are helpful for segmentation. The learning process is divided into two stages: (1) To train the DM on the large-scale MegaDepth Dataset [24] collected from Internet photos. (2) Both the training and testing of SM were based on the predictions from the DM and original RGB images [2,7]. The strategy introduced in Section 3.3 was applied. As is shown in Figure 1, DM helped to focus the SM's limited capacity on the qualified regions and boost the performance of segmentation. Each input image of the SM had four channels, RGB and the depth prediction from DM.

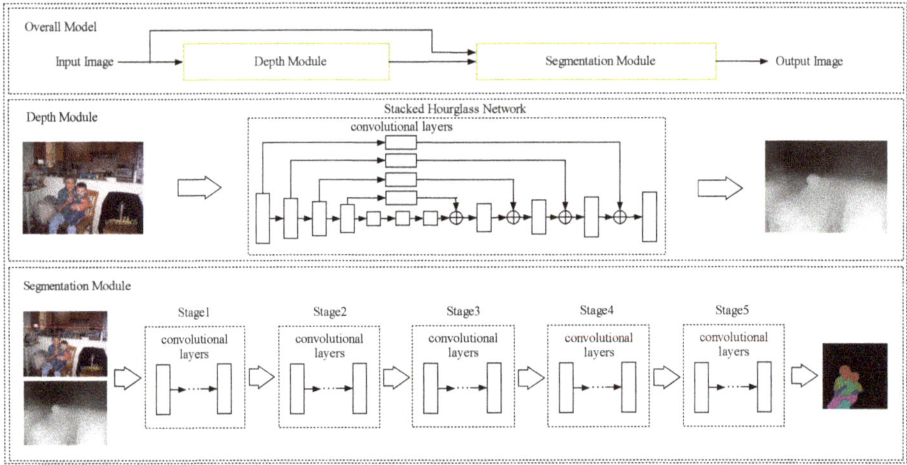

Figure 1. The proposed model for human parsing. The Overall Model (OM) is composed of a Depth Module (DM) and a Segmentation Module (SM). The DM is based on the hourglass network proposed in [25] and is trained on the MegaDepth Dataset proposed in [24]. The trained DM is used to pre-process the training and test images. The training data includes the Pascal Person Part Dataset [7] and the LIP Dataset [26]. The SM is composed of five stages of convolutional layers.

Existing works on optimizing the capacity of CNNs are divided into three categories. The first type of work [27,28] explores the sparseness of feature representations within a CNN and keeps only branches which are useful to tasks. The second type of work [29] makes use of the favorable properties of a shallow network and improves the performance of the CNN based on novel training strategies. The third type of work [30] maximizes the representational power of a CNN by maximizing the number of inter-connections between the features in different levels. Although the existing methods have improved computational efficiency, none of them has explored the relationship between depth and accuracy. This paper proposes a way to match the representational power and capacity of a CNN to a task by adjusting the depth and width of the CNN and training the CNN with a novel scheme. Both quantitative and qualitative experiments are conducted on the LIP Dataset [26] and the Pascal Person Part Dataset [7] to show that the proposed model conducts human parsing in a more effective and time-efficient manner.

In summary, the contributions of this paper are in three aspects: (1) A model for human parsing is proposed with integrated DM and SM modules. Moreover, a novel scheme is proposed for training the modules. The DM is trained on a large amount of automatically labeled images and provides information which is complementary to the features in the SM. As a result, the OM outperforms the SM only, especially at the boundaries. (2) A new algorithm is proposed to train the SM with 24 layers, achieving a similar performance as the baseline model with over 100 layers trained on the currently largest dataset for human parsing. (3) The influences of depth and width on capacity are studied. Two methods are proposed to build a SM which is deeper with performance improvement but uses less parameters. As a result, the performance-complexity ratio is improved and the capacity of the CNN model is better utilized.

The rest of the paper is organized as follows. Section 2 introduces related work. Section 3 discusses the details of our proposed model and the method for adjusting the CNN's capacity. Section 4 shows the details of implementation as well as experimental results. Concluding remarks are drawn in Section 5.

2. Related Work

Human Parsing Approaches. Human parsing has become an active research topic in the last few years [7,9,21,26,31–33]. The JPPNet [8] and the Nested Adversarial Network [34] represent the current state-of-the-art methods. The improvements in the methods, such as those in Reference [8,23], over traditional methods [9] are achieved by combining pose estimation with semantic part segmentation. The estimated poses provide the shape prior, which is necessary for segmentation. Similarly, the authors of Reference [35] proposed to integrate parsing with optical flow estimation. The authors of Reference [36] incorporated a self-supervised joint loss to ensure the consistency between parsing and pose. However, the guidance from poses cannot improve borders. As a result, it is still quite difficult to delineate the boundaries. In our proposed method, it is shown that depth information can improve the classification of pixels near boundaries. Other work, such as Reference [34], proposed to integrate three sub-nets which perform semantic saliency prediction, instance-agnostic parsing, and instance-aware clustering, respectively. The authors of Reference [37] proposed a framework integrating a human detector and a category-level segmentation module. However, both methods involve multiple stages. The outputs from earlier stages compose the only inputs to later stages and misleading outputs from the earlier stages disable the later stages. In our proposed method, the input to the SM composes not only the output from the DM, but also the original RGB images. The SM is less dependent on the DM and can function even when the DM fails.

Weakly Supervised Methods. To tackle the lack in training data, three types of research works have been conducted. The first type involves learning based on bounding boxes, scribbles, image tags or mixing multiple types of annotations. The labels in the form of bounding boxes and scribbles indicate the locations and sizes of objects. The BoxSup proposed in [20] and DeepCut proposed in [38] trained the segmentation model based on iterating between bounding box generation and training the CNN. The 3D U-Net proposed in Reference [39] performed volumetric segmentation with a semi-automated setup or a fully-automated setup. The multi-task learning proposed in Reference [40] adopted image-level and point-level supervision. Image-level supervision shows whether certain objects are present in an image. Point-level supervision indicates the locations and rough boundaries of objects. Bounding boxes and scribbles were mixed in Reference [21,41] to facilitate better training. However, the methods cannot deal with images containing multiple people or those with complex backgrounds.

The second type of work focuses on predicting the weights of a model in a target task, such as image classification using the weights from a source task, such as natural language descriptions or few-shot examples [42–44]. Among vision tasks, the authors of Reference [45] proposed to transform the weights in a detection model to those in a segmentation model. Similarly, the LSDA (Large Scale Detection Through Adaptation) proposed in Reference [46] introduced a way to transform a classification model to a detection model. The rationality lies in the fact that more training data are available in source domains. However, the transformation of weights is based on a parameterized function which is learned on the limited annotations from a target domain. The taskonomy proposed in Reference [47] re-used the supervision among related tasks. It trains higher-order transfer functions to map the feature representations from a source task to a target task. However, different types of annotations need to be present on the same set of images. The requirements on annotations limits the scale of training data. Different from the above-mentioned methods, our proposed human parsing model utilizes depth information without learning a parameterized function. The DM can be trained on large datasets with only depth annotations and provide robust predictions on images with multiple people or those with complex backgrounds. The SM benefits from the complementary features provided by the DM and the segmentation performance is improved.

Capacity Optimization. The definition of capacity is introduced in Reference [48]. The term capacity tends to relate to volumes, quantities or memorization. It measures how complex a function a neural network can model. Existing work on capacity optimization includes pruning feature representations [49–51], exploring favorable properties of a shallow network [29], and maximizing the expressive power of a fixed-size network [16,30]. The first type of work focuses on pruning convolutional kernels to obtain the most compressed sets of feature representations required for a task. However, most of the related methods improved the processing speed at the cost of accuracy. The second type of method replaces the end-to-end training scheme by a sequentially training scheme. Accuracy is improved without increasing depth. However, the depth of a CNN is not yet well matched to a task. Our proposed training scheme outperforms the scheme in Reference [29], as will be shown in Section 4.3. The third type of work [30] tried to improve the performance-complexity ratio, but the added connections significantly reduced the efficiency in memory accessing. In our proposed scheme, the depth of a CNN is better matched to the complexity of a human parsing task. Moreover, a deeper but more efficient module is built to optimize the capacity of a CNN without dropping in accuracy. Experimental results will demonstrate the superiority of our proposed methods.

3. Methodology

Our proposed model is shown in Figure 1. The complementary nature of the SM and DM is explored. Besides color, the depth prediction provided by the DM facilitates an extra way to understand images to improve segmentation. To be more specific, nearby regions belonging to different instances are predicted to share the same label by the SM because of similar colors and textures. However, the regions can be differentiated from each other by the DM because of the difference in their depth values.

As is shown in Figure 2, the results of using the SM only for segmentation and those of integrating the SM with the DM are compared. It is shown in the first row that the heads of different identities cannot be distinguished because of the similarity in colors. However, the depth predictions help to differentiate the instances. In the second row, foreground instances share the same color as backgrounds. The depth predictions help to segment out foreground instances. Similarly, the lower arms in the third row can hardly be distinguished from the background with color information only. Successful segmentation results from depth predictions.

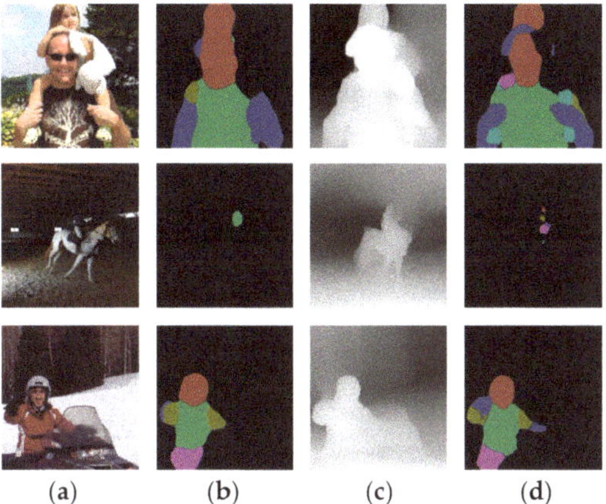

(a)   (b)   (c)   (d)

Figure 2. The improvement on segmentation is brought by the DM: (**a**) RGB images which are the input to both DM and SM; (**b**) The results of using the SM only for segmentation; (**c**) The predictions from the DM; (**d**) The results of integrating the SM with the DM for segmentation.

3.1. Depth Module

Similar to other network models which output the same resolution as the inputs [37,52], the DM is also trained end-to-end. It processes and passes information across multiple scales. The design of the DM is based on the stacked hour-glass network proposed in Reference [25,53]. The symmetric structure consists of convolutions, pooling layers which are followed by up-sampling layers and convolutions. The detailed structure was discussed in Reference [54].

The loss function for training the network is the weighted sum of three terms. The first term denotes the mean square error of predicted depth values.

$$D_{\mathrm{MSE}} = \frac{1}{n}\sum_{i=1}^{n} d_i^2 - \frac{1}{n^2}\left(\sum_{i=1}^{n} d_i\right)^2 \tag{1}$$

where d_i denotes the difference between the prediction at the i-th pixel and the corresponding ground truth depth value. n denotes the number of pixels. The training images are from the large-scale MegaDepth dataset [24]. D_{MSE} is invariant to the shifts on the mean values of images. The second term takes into consideration the gradients on the difference map:

$$D_{grad} = \frac{1}{n}\sum_{i=1}^{n} \left(|\nabla_x d_i| + |\nabla_y d_i|\right) \tag{2}$$

This term improves the performance on sharp discontinuities and makes depth predictions smoother. The third term enforces the ordinal depth relations between foreground super-pixels and background super-pixels:

$$D_{ord} = \sum_{k=1}^{K} \log\left(1 + \exp\left(-abs\left(z_{i_k} - z_{j_k}\right)\right)\right) \tag{3}$$

K pairs of points are sampled from the depth predictions and ground truth depth maps. i_k denotes the k-th point sampled from the largest foreground super-pixel while j_k denotes the k-th point sampled from the surrounding background super-pixels. z_{i_k} denotes the depth at point i_k and z_{j_k} the depth at point j_k. The weight of D_{grad} in the loss function is set to 0.5 while the weight of D_{ord} is set to 0.1.

The existence of term Equation (3) enforces the predicted depth values of neighboring instances to be different and ordered. The overall lost function is defined as

$$D = D_{MSE} + 0.5D_{grad} + 0.1D_{ord} \tag{4}$$

3.2. Segmentation Module

In this section, an SM with 24 layers is introduced, which is compared with the baseline model Deeplab-V2 [9] on the segmentation task.

Figure 3 shows our proposed SM. The architecture is based on the backbone of VGG-16 [55]. The blocks in red denote residual blocks [16].

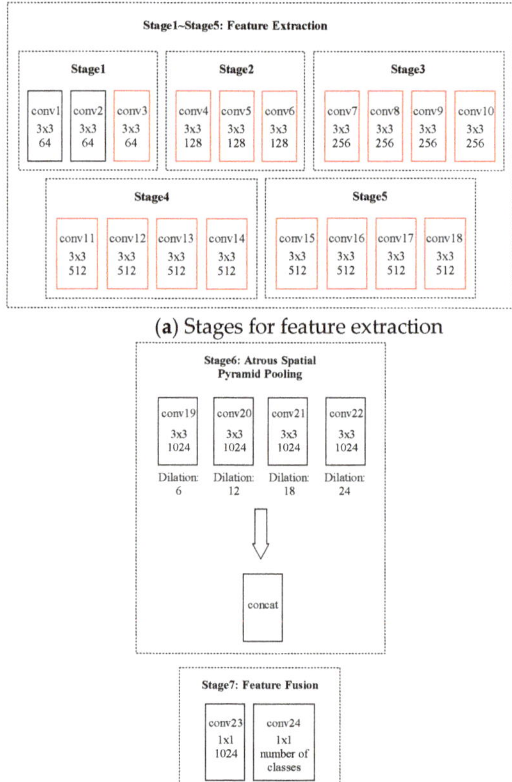

(**a**) Stages for feature extraction

(**b**) ASPP (Atrous Spatial Pyramid Pooling) and feature fusion

Figure 3. Architecture of an SM. Each block denotes one convolutional layer. ReLU nonlinearities are used throughout, and max pooling occurs between adjacent groups of convolutional layers. The first line within each block denotes the name of the layer, the second line shows the kernel size, and the third line shows the number of output channels. (**a**) The five stages for extracting basic features. (**b**) Task-specific stages. A mechanism known as ASPP [9] is adopted to enable 3×3 filters to have different field-of-views. Four parallel filters with different field-of-views are adopted to extract the features for pixel classification. The kernels in the layers from conv19 to conv22 are with size 3×3 but differ in the distance between weights in the kernels. The heat-maps generated by the four parallel convolutional layers are fused in Stage 7.

3.3. Strategy of Combining DM with SM

Solely integrating the predictions from the DM with RGB images during training and testing the SM only brings slight improvements. However, better strategies can be adopted to fully explore the complementary nature of color and depth information to contribute more to performance improvement. In this section, we propose a strategy to better utilize depth information. The strategy is divided into two steps. An example is shown in Figure 4.

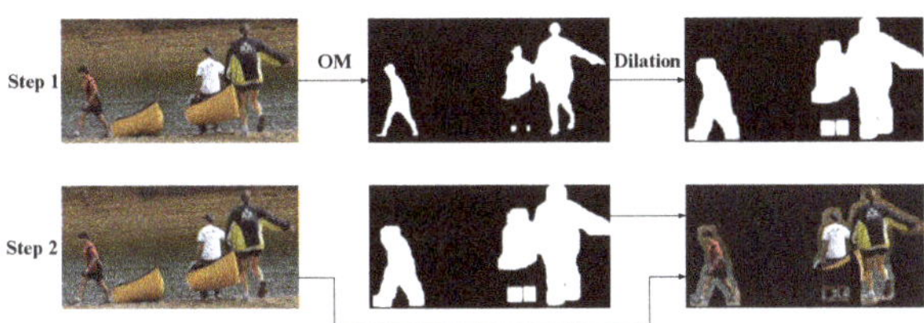

Figure 4. The procedure for combining DM with SM.

In the first step, the DM is trained and used to augment the training data and test data of the SM. OM segments out foreground objects as one class. To reduce false negatives, dilation is conducted on the masks produced by the OM. In the second step, the regions which are predicted by Step 1 as foregrounds are kept unchanged, with remaining parts set to zero. The images produced by Step 2 are then used for re-training and testing SM. Table 1 compares the performance of three cases: Only using the SM, direct training and testing of the SM based on the predictions from the DM and applying the strategy in this section to combine DM with the SM.

Table 1. Mean pixel IOU (mIOU) of human parsing on the PASCAL Person Part Dataset.

Method	mIOU (%)
Attention [56]	56.39%
HAZN [57]	57.54%
LG-LSTM [58]	57.97%
Joint pose estimation and part segmentation (with Resnet-101 as backbone) [8]	64.39%
SM pre-trained on ImageNet [59]	61.57%
Our overall model (OM) pre-trained on ImageNet [59]. Directly concatenate DM's predictions with RGB images for training and test SM.	62.49%
Our overall model (OM) pre-trained on ImageNet [59]. Combine DM with SM based on the strategy introduced in Section 3.3.	65.03%

3.4. Capacity Optimization of SM

Each input image is mapped by a CNN from the image space χ to the feature space $\mathbb{F}$, a CNN learns the low-dimensional structures of data and represents them using a parametric polyhedral manifold which is then partitioned into pieces [60]. The more pieces there are, the higher the representation capability of the CNN becomes. For a ReLU deep neural network, each neuron functions as a hyperplane and partitions the input manifold into multiple polyhedra. As a result, the number of pieces is decided by the number of ReLU operations. The bound of the encoding or representation capability of a ReLU DNN is measured by Rectified Linear Complexity (RL Complexity) $\mathbb{N}(N)$. For a neural network with k hidden layers of widths $\{w_i\}_{i=1}^{k}$, the upper bound of RL Complexity is given by

$$\mathbb{N}(N) \leq \Pi_{i=1}^{k+1} C(w_{i-1}, w_i). \tag{5}$$

where

$$C(d, n) = \binom{n}{0} + \binom{n}{1} + \binom{n}{2} + \cdots + \binom{n}{d} \tag{6}$$

Only when the RL complexity of a neural network is no less than that of the manifold can the data be encoded by the neural network. It can be inferred from Equations (5) and (6) that the depth of neural network contributes much more significantly to the capacity of a neural network than width. As a result, we have made the CNN in Figure 3a deeper than the backbone [55]. The contribution to performance improvement from additional depth will be shown in Sections 4.2 and 4.4.

Moreover, we have also developed a scheme for training the CNN in Figure 3a, as is shown in Algorithm 1. The training is conducted layer by layer because it was discussed in Reference [29] that a CNN with a fixed number of layers can perform better if it is built sequentially layer by layer instead of trained end-to-end.

Algorithm 1: Scheme of training SM

Step 1	Train the backbone network [55] without layers conv3, conv6, conv10, conv14, conv18 until convergence.
Step 2	Add layer conv18 which is initialized with Gaussian weight matrices.
Step 3	Use the pre-trained layers conv1-conv17 to extract the feature vectors from images and use the feature vectors as the input to train conv18.
Step 4	Freeze the weights in all layers in Stage 6 and Stage 7 and train conv18 until convergence.
Step 5	Re-train SM until convergence with all layers un-frozen.
Step 6	Add layer conv14, conv10, conv6, conv3 and go through the same operations from Step 2 to Step 5.

Different from Reference [29] which only trained the added layer each time, for each added layer in Algorithm 1, the network is trained for two times. In the first time the added layer only is trained while in the second time, the overall network is trained. As will be shown in Section 4.3 the strategy of our adding layers has an advantage over both adding all layers together at once and applying the method in Reference [29]. Moreover, the involvement of the five additional layers has brought significant improvements in accuracy over the backbone network, as will be shown in Section 4.2.

Different from traditional segmentation models, the feature representations at Stage 6 which correspond to 4 point-of-views are concatenated and fused by the 1×1 convolutions at Stage 7, as compared to the direct summation in Reference [8]. Feature fusion offers a much more flexible scheme of combining the features from different point-of-views. The network can learn to add up the features or combine the features in more complex ways.

Besides improving depth which leads to the increase in computational complexity, we also propose to exchange width for depth to obtain further improvements in accuracy while reducing the number of parameters. Two methods have been proposed to reduce the complexity of convolutional layers by replacing one conventional convolutional layer with a stack of simplified layers. The overall complexity of stacked simplified layers is less than that of one original convolutional layer. The mechanism is shown in Figure 5. Figure 5b shows our first proposed way of exchanging width for depth. In one conventional layer, C_{in} independent 3×3 convolutional kernels function on C_{in} channels to obtain one output channel. In one simplified sub-layer, N ($N \ll C_{out}$) kernels function on each of the C_{in} input channels and the intermediate representation has NC_{in} output channels which are processed by 1×1 convolutions. M is the number of sub-layers which are stacked to replace one conventional layer. Even if one sub-layer is less expressive than one conventional layer with the same C_{in} and C_{out}, the stack of M sub-layers achieves a higher expressive power than one conventional layer without increasing the overall number of parameters for proper choices of M and N. In Figure 5c, the second way of exchanging width for depth is introduced. Each conventional layer is replaced by two sub-layers. The difference lies in that Figure 5b uses 4 sub-layers to replace one conventional layer, while Figure 5c uses two sub-layers.

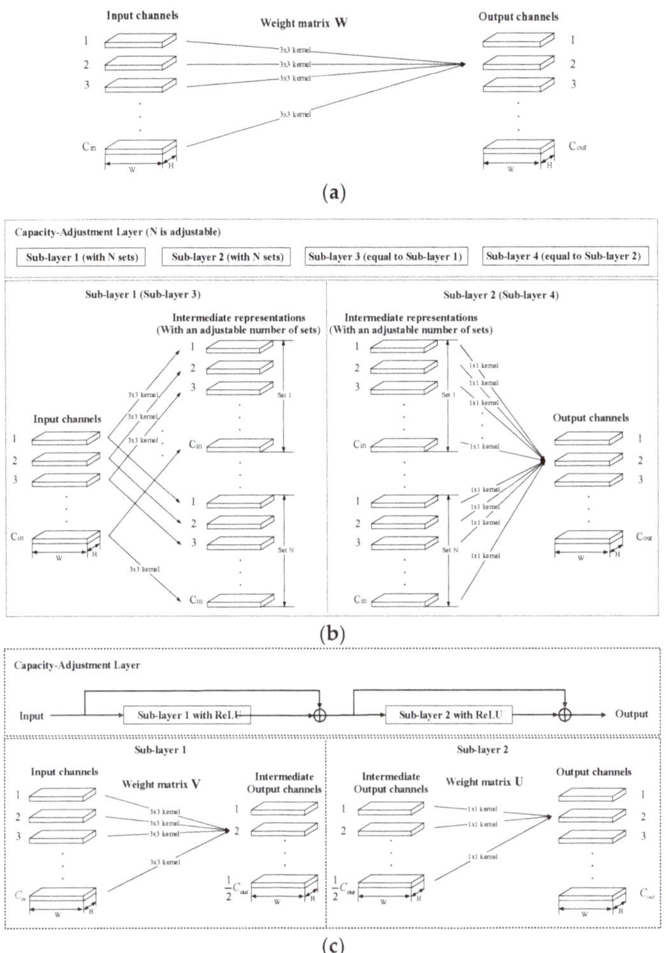

Figure 5. The proposed way of exchanging width for depth: (**a**) A conventional convolutional layer; (**b**) simplify one conventional layer to sub-layers and concatenate $M(M = 4)$ sub-layers to replace one conventional layer. Each sub-layer is obtained by decomposing one conventional layer and increasing the dependency between convolutional kernels; (**c**) simplify one conventional layer to sub-layers and stack 2 sub-layers to replace one conventional layer.

The structures introduced in Figure 5b and c will be used to simplify conv17 and conv18 shown in Figure 3. As will be shown in the experiments, accuracy is kept almost the same with the overall computational complexity reduced. Deeplab-V2 [9] and joint pose estimation and part segmentation [8] which is based on Resnet-101 [9] are also trained on the segmentation datasets for comparison.

The initialization of weights in Figure 5b was discussed in Reference [61] and the initializations of weights in Figure 5c is based on minimizing the reconstruction error:

$$\underset{\mathbf{U},\mathbf{V}}{\text{argmin}} \sum_{i=1}^{C_{out}} \left\| W_i - \sum_{j=1}^{C_{out}/2} U_{ij} V_j \right\|_2 \tag{7}$$

where **W** denotes the weights matrix in one layer shown in Figure 5a with size $C_{out} \times C_{in}k_hk_w$, with $\mathbf{W}_i$ denoting the i-th row of **W**. **V** denotes the weights in Sub-layer 1 with $\mathbf{V}_i$ being the i-th row of the matrix. **U** denotes the weights in Sub-layer 2 with $\mathbf{U}_{ij}$ being the (i, j)-th entry of the matrix.

3.5. Domain Randomization

To reduce the generalization error, it is necessary to bridge the gap between the source domain (training data) and the target domain (test data). Some methods have been proposed to discuss the problem [62]. However, these methods mainly focus on other tasks, such as object detection [62], in which all types of simulated variability at training time are utilized, including positions, textures, orientations, field-of-views, and lightening conditions. Too many variations may result in a low convergence rate.

The LIP dataset includes the variations in poses and lightening conditions in the training set. As a result, we are only concerned with the variations in backgrounds which contribute to the divergence between domains. We propose to crop the predicted backgrounds from test images, which are then used to fill the background regions in training images. In this way, an augmented training dataset is produced to help the SM develop more generalizable representations. The detailed scheme is introduced in Algorithm 2.

Algorithm 2: Scheme of domain randomization

Step 1	Train SM until convergence and use it to segment out the background regions of test images.
Step 2	For each training image, find the image from test set with the most similar aspect ratio. Resize the test image to be with the same size as the training image.
Step 3	Crop the background regions from the test image and replace the background regions on the training image with those from the test image.
Step 4	Re-train SM and return to Step 1. (Two iterations are adopted.)

Figure 6 shows two examples of domain randomization. The backgrounds in training images are replaced by those in test images.

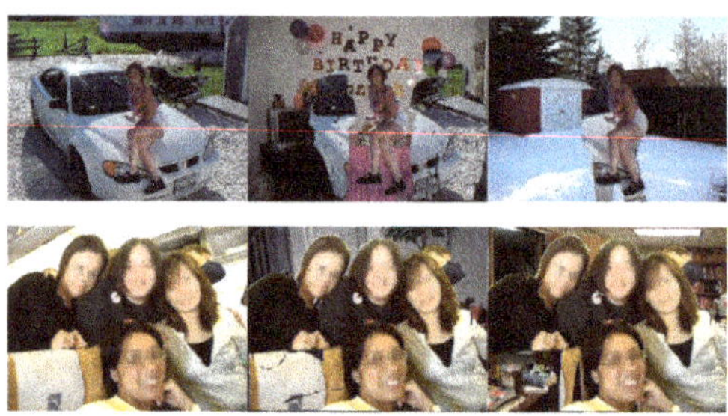

Figure 6. Two examples of domain randomization. In each row, the left image denotes the original image, the middle and the right ones show two modified images with different backgrounds.

4. Results

4.1. Datasets and Implementation Details

The two modules in our proposed model were trained on two datasets. The DM was trained on the MegaDepth Dataset [24] which involves 130K images from 200 different landmarks. The depth information in images with over-lapping viewpoints is automatically obtained with SFM (Structure from Motion) and MVS (Multi-View Stereo). The SM was trained on the images from the PASCAL VOC 2010 Person Part Dataset for body part segmentation [2,63]. The Person Part Dataset includes annotations on 3,533 images where 1,716 images are used for training while the other 1,817 images are for testing. The ground truth labels are in the form of segmentation masks. There are six annotated semantic types, that is, head, torso, upper arm, lower arm, upper leg, lower leg, and background. To evaluate on larger datasets and to demonstrate the improvement on capacity, the SM was also trained on the LIP (Look into Person) Dataset [26] with 30,462 images for training, 10,000 images for validation and 10,000 test images. There are 19 annotated semantic types, that is, face, upper clothes, hair, right arm, pants, left arm, right shoe, left shoe, hat, coat, right leg, left leg, glove, socks, sunglasses, dress, skirt, jumpsuits, scarf. The capacity of the SM was optimized using the two methods shown in Figure 5.

4.2. Integration of the Two Modules

As is shown in Figure 1, the DM is used to preprocess an image and predict depth masks. The predicted masks are concatenated with corresponding input RGB images to produce the input of the SM during both training and testing. The experiments were conducted to show that the combination of color and depth information during training and inference improves the performance of human parsing. The metric for evaluating the performance of human parsing is mean Intersection Over Union (mIOU) which is proposed in [7]. mIOU is computed by dividing the number of true positive samples by the summation of true positive, false negative, and false positive samples:

$$mIOU = \frac{1}{N} \sum_{i=1}^{N} \frac{n_{ii}}{t_i + \sum_{j \neq i} n_{ji}} \tag{8}$$

where n_{ji} is the number of pixels of class j which are predicted to class i, and $t_j = \sum_i n_{ji}$ is the total number of pixels belonging to class j. The metric takes into account both false positives and false negatives. For the Pascal VOC 2010 Person Part Dataset, mIOU is computed for each of the seven classes and averaged. For instance, the mIOU of head is obtained by regarding head as the foreground and other six types as the backgrounds. Table 1 shows the results on the test set.

For the LIP Dataset, mIOU is computed in the same way and the experimental results are shown in Table 2.

Table 2. Mean pixel IOU (mIOU) of human parsing on Look into Person (LIP) Dataset.

Method	mIOU (%)
Deep Lab-V2 (VGG-16) [9]	41.56%
Deep Lab-V2 (Resnet-101) [9]	44.96%
SM (24 layers) pre-trained on ImageNet [59]	44.89%
Our overall model (OM) pre-trained on ImageNet [59]. The scheme introduced in Section 3.3 is applied.	46.73%

By comparing the last three rows in Table 1 and the last two rows in Table 2, it can be inferred that the integration of the DM and the SM outperforms the SM on both large and small datasets. Table 1 shows that the contribution is mainly attributed to the scheme proposed in III-C, which fully explores the complementary nature of the DM and the SM.

Moreover, it is discussed in Section 3.2 that the backbone of the SM is Deep Lab-V2 (VGG-16) [9] with 18 layers. Figure 7 shows the changes in accuracy upon increasing the depth of SM from 18 to 24. With the six added layers, the SM not only outperforms the backbone, but also performs as well as the baseline model with over 100 layers which is shown in the second row in Table 2. Note that the number of parameters in the OM is less than that in the baseline model [9]. More importantly, LIP is the currently largest dataset for human parsing and the number of parameters in SM is much less than that in the baseline model. The improvement demonstrates that by adjusting the depth of a CNN model, its capacity is better matched to a task than Reference [8,9].

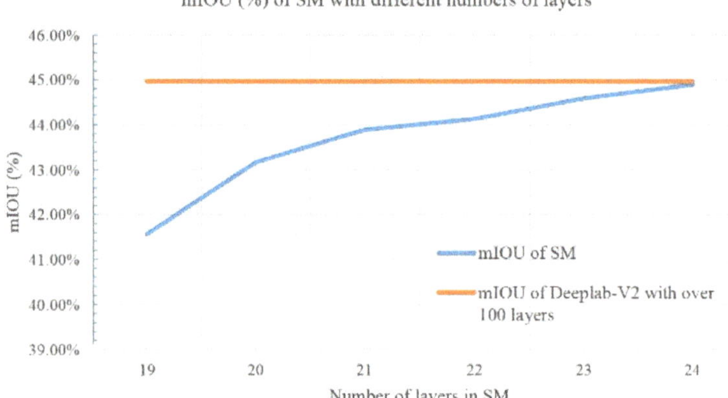

Figure 7. The changes in accuracy upon increasing the depth of SM.

4.3. The Advantage of Layer-Wise Training

As is demonstrated by the second and third rows in Table 2, the SM can perform as well as a model that is much more complex. The advantage results from the layer-wise training scheme which is proposed in Section 3.4 and shown in Algorithm 1. To demonstrate the advantage of Algorithm 1 over adding all the layers to the backbone at once, we compare the performance of the SMs trained with the two schemes and show the results in Table 3. In the scheme where all the layers are added at once and the SM was trained for one time, the number of iterations during training is 600,000. In the scheme proposed in Algorithm 1, the SM was trained for 100,000 iterations upon the addition of each layer. The overall time cost during training is the same. The performance is evaluated on the LIP Dataset.

Table 3. Mean pixel IOU (mIOU) of the Segmentation Module (SM) trained with different schemes.

Scheme	mIOU (%)
Directly adding layers to SM and train at once	42.76%
Train using the layer-wise scheme in [29]	44.37%
Train SM with Algorithm 1	44.89%

It can be inferred from Table 3, that layer-wise training significantly outperforms direct training all layers at once. Different from the layer-wise training in Reference [29], we firstly train each added layer while keeping other layers fixed for 50,000 iterations and then re-training all layers for another 50,000 iterations. Compared with training each added layer for 100,000 iterations, Algorithm 1 performs better, as is shown by the last two rows in Table 3.

4.4. The Exchange Between Width and Depth for Capacity Optimization

We have also tried to replace traditional convolutional layers with the stacking of simplified convolutional layers shown in Figure 5b,c. The N in Figure 5b is chosen to be 3 and the number of sub-layers M is selected to be 4. We have replaced conv18 in Figure 3 with the stacked simplified layers. Upon replacing the layer, the SM is re-trained for 100,000 iterations. The changes in performance and the drop in computational burden is shown in Table 4. The performance is evaluated on the LIP dataset.

The initializations of the weights in the layers in Figure 5b,c are discussed in Section 3.4.

It can be inferred from Table 4 that the methods shown in Figure 5b,c reduce computational burdens while maintaining or slightly improving performance.

Table 4. The influence of exchanging width for depth on the SM. Performance is evaluated with mIOU.

Method	mIOU (%)	Deduction on the Number of Floating-Point Multiplications (%)
SM shown in Figure 3	46.73%	-
SM with conv18 converted to the stack of layers shown in Figure 5b	46.85%	1.75%
SM with conv18 converted to the stack of layers shown in Figure 5c	46.79%	2.42%
SM with conv18 and conv17 converted to the stack of layers shown in Figure 5b	46.89%	3.51%
SM with conv18 and conv17 converted to the stack of layers shown in Figure 5c	46.54%	4.85%

4.5. Domain Randomization

We augmented the training data with domain randomization where the backgrounds in training images were replaced by the counterparts in test images. In the implementation of Algorithm 2, we cropped the backgrounds from two test images to replace the background of each training image. As a result, the augmented dataset includes 60,924 images.

It can be inferred from Table 5 that iterative domain randomization improves the generalization of the SM.

Table 5. The influence of domain randomization on the SM.

Method	mIOU (%)
SM trained on the original images from the LIP Dataset	46.73%
SM trained on the augmented LIP Dataset	47.21%

4.6. Examples of Segmentation Results

Besides objective results, some results are shown in Figures 8 and 9 to show the advantages of integrating the DM with the SM which can be judged subjectively. A comparison between the SM and the OM is made on multiple cases, including samples with complex gestures in identities, images with occlusions, and those suffering from darkness.

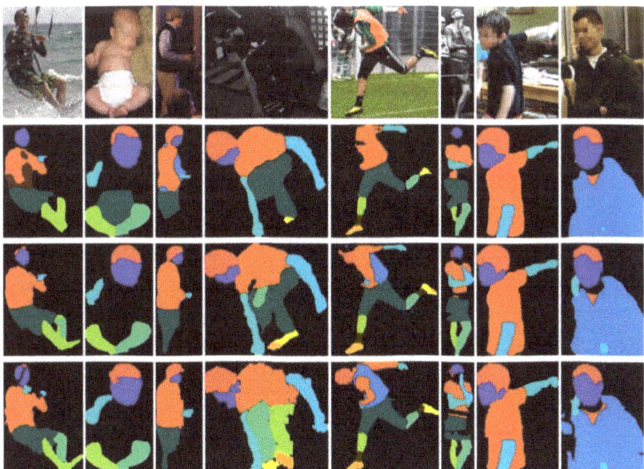

Figure 8. *Cont.*

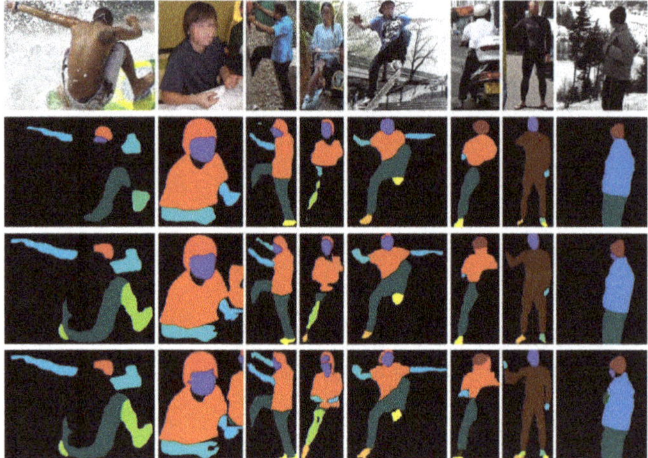

Figure 8. Performance comparison between the OM and the SM on the validation set. From the top row to the bottom row are input images, predictions from the SM, predictions from the OM and the ground truth labels.

Figure 9. Performance comparison between the OM and the SM on the test set. From the top row to the bottom row are input images, predictions from the SM and predictions from the OM.

5. Discussion

In this paper, depth information is combined with color using a novel strategy. The performance of human parsing is significantly improved. Moreover, depth information is obtained by a module which is trained on automatically acquired labels, thus saving human labor cost. Secondly, the SM with 24 layers, which is trained using the scheme in Algorithm 1 achieves a similar performance as the baseline model with over 100 layers on the currently largest dataset for human parsing. The number of parameters in the OM is less than that in the baseline model. Thirdly, two methods have been proposed to optimize the capacity of the SM by increasing depth while reducing parameters, achieving a more efficient solution with a better performance. Both quantitative and subjective results have shown the effectiveness of our proposed methods.

Author Contributions: Conceptualization, Y.J. and Z.C.; methodology, Y.J.; software, Y.J.; writing—original draft preparation, Y.J.; writing—review and editing, Y.J. and Z.C.; supervision, Z.C.

Acknowledgments: The work described in this paper was partially supported by a Natural Science Foundation of China (NSFC) grant (Project Code: 61473243). Yalong Jiang would like to acknowledge the financial support from The Hong Kong Polytechnic University for his PhD study.

Conflicts of Interest: The authors declare no conflict of interest.

References

1. Lin, T.; Maire, M.; Belongie, S.; Hays, J. Microsoft coco: Common objects in context. In Proceedings of the European Conference on Computer Vision, Zurich, Switzerland, 6–12 September 2014.
2. Everingham, M.; Eslami, S.A.; Van Gool, L. The pascal visual object classes challenge a retrospective. *Int. J. Comput. Vis.* **2015**, *111*, 98–136. [CrossRef]
3. Cordts, M.; Omran, M.; Ramos, S.; Rehfeld, T. The cityscapes dataset for semantic urban scene understanding. In Proceedings of the IEEE Conference on Computer Vision and Pattern Recognition, Las Vegas, NV, USA, 26 June–1 July 2016.
4. Liu, S.; Qi, L.; Qin, H.; Shi, J.; Jia, J. Path aggregation network for instance segmentation. *arXiv* **2018**, arXiv:1803.01534.
5. Peng, C.; Zhang, X.; Yu, G.; Luo, G.; Sun, J. Large Kernel Matters—Improve Semantic Segmentation by Global Convolutional Network. *arXiv* **2017**, arXiv:1703.02719.
6. He, K.; Gkioxari, G.; Dollár, P.; Girshick, R. Mask r-cnn. *arXiv* **2017**, arXiv:1703.06870.
7. Chen, X.; Mottaghi, R.; Liu, X.; Fidler, S. Detect what you can: Detecting and representing objects using holistic models and body parts. In Proceedings of the IEEE Conference on Computer Vision and Pattern Recognition (CVPR), Columbus, OH, USA, 24–27 June 2014.
8. Xia, F.; Wang, P.; Chen, X.; Yuille, A. Joint Multi-person Pose Estimation and Semantic Part Segmentation. In Proceedings of the IEEE Conference on Computer Vision and Pattern Recognition, Honolulu, HI, USA, 22–25 July 2017.
9. Chen, L.C.; Papandreou, G.; Kokkinos, I. Deeplab: Semantic image segmentation with deep convolutional nets, atrous convolution, and fully connected crfs. *IEEE Trans. Pattern Anal. Mach. Intell.* **2018**, *40*, 834–848. [CrossRef] [PubMed]
10. Gan, C.; Lin, M.; Yang, Y.; de Melo, G. Concepts not alone: Exploring pairwise relationships for zero-shot video activity recognition. In Proceedings of the AAAI 2016, Phoenix, AZ, USA, 12–17 February 2016.
11. Zhao, R.; Ouyang, W.; Wang, X. Unsupervised salience learning for person re-identification. In Proceedings of the IEEE Conference on Computer Vision and Pattern Recognition, Portland, OR, USA, 25–27 June 2013.
12. Turban, E.; King, D.; Lee, J.; Viehland, D. *Electronic Commerce: A Managerial Perspective 2002*; Prentice Hall: Upper Saddle River, NJ, USA, 2002; Volume 2.
13. Xu, N.; Price, B.; Cohen, S.; Yang, J.; Huang, T. Deep interactive object selection. In Proceedings of the IEEE Conference on Computer Vision and Pattern Recognition, Las Vega, NV, USA, 26 June–1 July 2016.
14. Collins, R.; Lipton, A.; Kanade, T. *A System for Video Surveillance and Monitoring*; VSAM Final Report; Robotics Institute, Carnegie Mellon University: Pittsburgh, PA, USA, 2000; Volume 2, pp. 1–68.
15. Lin, J.; Guo, X.; Shao, J.; Jiang, C.; Zhu, Y. A virtual reality platform for dynamic human-scene interaction. In Proceedings of the SIGGRAPH, San Jose, CA, USA, 24–28 July 2016.

16. He, K.; Zhang, X.; Ren, S.; Sun, J. Deep residual learning for image recognition. *Comput. Vis. Pattern Recognit. arXiv* **2016**, arXiv:1512.03385.

17. Szegedy, C.; Ioffe, S.; Vanhoucke, V.; Alemi, A. Inception-v4, inception-resnet and the impact of residual connections on learning. In Proceedings of the AAAI, San Francisco, CA, USA, 4–9 February 2017.

18. Bearman, A.; Russakovsky, O.; Ferrari, V.; Li, F. What's the point: Semantic segmentation with point supervision. In Proceedings of the European Conference on Computer Vision, Amsterdam, The Netherlands, 8–16 October 2016.

19. Pathak, D.; Krahenbuhl, P.; Darrell, T. Constrained convolutional neural networks for weakly supervised segmentation. In Proceedings of the IEEE International Conference on Computer Vision, Santiago, Chile, 11–18 December 2015.

20. Dai, J.; He, K.; Sun, J. Boxsup: Exploiting bounding boxes to supervise convolutional networks for semantic segmentation. In Proceedings of the IEEE International Conference on Computer Vision, Santiago, Chile, 11–18 December 2015.

21. Papandreou, G.; Chen, L.; Murphy, K.; Yuille, A. Weakly-and semi-supervised learning of a DCNN for semantic image segmentation. *arXiv*, 2015; arXiv:1502.02734.

22. Sun, J.; Lin, D.; Dai, J.; Jia, J.; He, K. Scribblesup: Scribble-supervised convolutional networks for semantic segmentation. In Proceedings of the IEEE Conference on Computer Vision and Pattern Recognition, Las Vegas, NV, USA, 26 June–1 July 2016.

23. Snavely, N.; Seitz, S.; Szeliski, R. Photo tourism: Exploring photo collections in 3D. *ACM Trans. Graph.* **2006**, *25*, 835–846. [CrossRef]

24. Snavely, N.; Li, Z. MegaDepth: Learning Single-View Depth Prediction from Internet Photos. In Proceedings of the IEEE Conference on Computer Vision and Pattern Recognition, Salt Lake City, UT, USA, 19–21 June 2018.

25. Newell, A.; Yang, K.; Deng, J. Stacked hourglass networks for human pose estimation. In Proceedings of the European Conference on Computer Vision, Amsterdam, The Netherlands, 8–16 October 2016.

26. Liang, X.; Gong, K.; Shen, X.; Lin, L. Look into Person: Joint Body Parsing & Pose Estimation Network and A New Benchmark. *arXiv*, 2018; arXiv:1804.01984v1.

27. Ierusalem, A. Catastrophic Importance of Catastrophic Forgetting. *arXiv* **2018**, arXiv:1808.07049.

28. He, Y.; Lin, J.; Liu, Z.; Wang, H. Amc: Automl for model compression and acceleration on mobile devices. In Proceedings of the European Conference on Computer Vision (ECCV), Munich, Germany, 8–14 September 2018.

29. Shallow learning for deep networks. *Int. Conf. Learn. Represent.* **2019**, Under review.

30. Huang, G.; Liu, Z.; Van Der Maaten, L. Densely Connected Convolutional Networks. In Proceedings of the IEEE Conference on Computer Vision and Pattern Recognition, Honolulu, HI, USA, 22–25 July 2017.

31. Ronneberger, O.; Fischer, P.; Brox, T. U-net: Convolutional networks for biomedical image segmentation. In Proceedings of the International Conference on Medical Image Computing and Computer-Assisted Intervention, Munich, Germany, 5–9 October 2015.

32. Pohlen, T.; Hermans, A.; Mathias, M.; Leibe, B. Full-resolution residual networks for semantic segmentation in street scenes. *arXiv* **2017**, arXiv:1611.08323.

33. Chen, L.; George, P.; Florian, S.; Hartwig, A. Rethinking atrous convolution for semantic image segmentation. *arXiv* **2017**, arXiv:1706.05587.

34. Zhao, J.; Li, J.; Cheng, Y.; Zhou, L.; Sim, T. Understanding Humans in Crowded Scenes: Deep Nested Adversarial Learning and A New Benchmark for Multi-Human Parsing. *arXiv* **2018**, arXiv:1804.03287.

35. Liu, S.; Wang, C.; Qian, R.; Yu, H.; Bao, R. Surveillance video parsing with single frame supervision. In Proceedings of the IEEE Conference on Computer Vision and Pattern Recognition, Honolulu, HI, USA, 22–25 July 2017.

36. Zhao, J.; Li, J.; Nie, X.; Zhao, F.; Chen, Y.; Wang, Z.; Feng, J.; Yan, S. Self-supervised neural aggregation networks for human parsing. In Proceedings of the IEEE Conference on Computer Vision and Pattern Recognition, Honolulu, HI, USA, 22–25 July 2017.

37. Li, Q.; Arnab, A.; Torr, P. Holistic, instance-level human parsing. *arXiv* **2017**, arXiv:1709.03612.

38. Oktay, O.; Kamnitsas, K.; Passerat-Palmbach, J. Deepcut: Object segmentation from bounding box annotations using convolutional neural networks. *IEEE Trans. Med. Imag.* **2017**, *36*, 674–683.

39. Çiçek, Ö.; Abdulkadir, A.; Lienkamp, S.; Brox, T. 3D U-Net: Learning dense volumetric segmentation from sparse annotation. In Proceedings of the International Conference on Medical Image Computing and Computer-Assisted Intervention, Athens, Greece, 17–21 October 2016.

40. Vezhnevets, A.; Buhmann, J. Towards weakly supervised semantic segmentation by means of multiple instance and multi-task learning. In Proceedings of the IEEE Conference on Computer Vision and Pattern Recognition, San Francisco, CA, USA, 13–18 June 2010.

41. Xu, J.; Schwing, A.G.; Urtasun, R. Learning to segment under various forms of weak supervision. In Proceedings of the IEEE Conference on Computer Vision and Patter Recognition, Boston, MA, USA, 7–12 June 2015.

42. Ha, D.; Dai, A.; Le, Q. Hypernetworks. *arXiv* **2016**, arXiv:1609.09106.

43. Elhoseiny, M.; Saleh, B.; Elgammal, A. Write a classifier: Zero-shot learning using purely textual descriptions. In Proceedings of the IEEE International Conference on Computer Vision, Sydney, Australia, 1–8 December 2013.

44. Wang, Y.; Hebert, M. Learning to learn: Model regression networks for easy small sample learning. In Proceedings of the European Conference on Computer Vision, Amsterdam, The Netherlands, 8–16 October 2016.

45. Hu, R.; Dollár, P.; He, K.; Darrell, T. Learning to Segment Every Thing. *arXiv* **2017**, arXiv:1711.10370.

46. Hoffman, J.; Guadarrama, S.; Tzeng, E.; Hu, R. LSDA: Large scale detection through adaptation. In Proceedings of the Advances in Neural Information Processing Systems, Montreal, Canada, 8–12 December 2014.

47. Zamir, A.; Sax, A.; Shen, W.; Guibas, L.; Malik, J. Taskonomy: Disentangling Task Transfer Learning. In Proceedings of the IEEE Conference on Computer Vision and Pattern Recognition, Salt Lake City, UT, USA, 19–21 June 2018.

48. Goodfellow, I.; Bengio, Y.; Courville, A. *Deep Learning*; MIT Press: Cambridge, MA, USA, 2016.

49. He, Y.; Zhang, X.; Sun, J. Channel pruning for accelerating very deep neural networks. In Proceedings of the International Conference on Computer Vision, Honolulu, HI, USA, 22–25 July 2017.

50. Ma, N.; Zhang, X.; Zheng, H.T.; Sun, J. Shufflenet v2: Practical guidelines for efficient cnn architecture design. *arXiv* **2018**, arXiv:1807.11164.

51. Sandler, M.; Howard, A.; Zhu, M.; Zhmoginov, A. MobileNetV2: Inverted Residuals and Linear Bottlenecks. In Proceedings of the IEEE Conference on Computer Vision and Pattern Recognition, Salt Lake City, UT, USA, 19–21 June 2018.

52. Long, J.; Shelhamer, E.; Darrell, T. Fully convolutional networks for semantic segmentation. In Proceedings of the IEEE Conference on Computer Vision and Pattern Recognition, Boston, MA, USA, 7–12 June 2015.

53. Chen, W.; Fu, Z.; Yang, D.; Deng, J. Single-image depth perception in the wild. In Proceedings of the Advances in Neural Information Processing Systems, Barcelona, Spain, 5–10 December 2016.

54. Szegedy, C.; Liu, W.; Jia, Y.; Rabinovich, A. Going deeper with convolutions. In Proceedings of the IEEE Conference on Computer Vision and Pattern Recognition, Boston, MA, USA, 7–12 June 2015.

55. Simonyan, K.; Zisserman, A. Very deep convolutional networks for large-scale image recognition. In Proceedings of the ICLR 2015, San Diego, CA, USA, 7–9 May 2015.

56. Oliveira, G.; Valada, A.; Bollen, C.; Burgard, W. Deep Learning for human part discovery in images. In Proceedings of the IEEE International Conference on Robotics and Automation (ICRA), Stockholm, Sweden, 17–20 May 2016.

57. Chen, L.; Yang, Y.; Wang, J.; Yuille, A. Attention to scale: Scale-aware semantic image segmentation. In Proceedings of the IEEE Conference on Computer Vision and Pattern Recognition (CVPR), Las Vegas, NV, USA, 26 June–1 July 2016.

58. Xia, F.; Wang, P.; Chen, L.C.; Yuille, A. Zoom better to see clearer: Human part segmentation with auto zoom net. In Proceedings of the European Conference on Computer Vision, Amsterdam, The Netherlands, 8–16 October 2016.

59. Liang, X.; Shen, X.; Xiang, D.; Feng, J.; Lin, L. Semantic object parsing with local-global long short-term memory. In Proceedings of the IEEE Conference on Computer Vision and Pattern Recognition, Las Vegas, NV, USA, 26 June–1 July 2016.

60. Deng, J.; Berg, A.; Satheesh, S.; Su, H. Large Scale Visual Recognition Challenge 2012 (ILSVRC2012). 2012. Available online: http://www.image-net.org/challenges/LSVRC/2012/ (accessed on 1 September 2012).

61. Lei, N.; Luo, Z.; Yau, S.; Gu, D. Geometric Understanding of Deep Learning. *arXiv* **2018**, arXiv:1805.10451.
62. Jiang, Y.; Chi, Z. A CNN Model for Semantic Person Part Segmentation with Capacity Optimization. *IEEE Trans. Image Process.* **2018**. [CrossRef] [PubMed]
63. Tobin, J.; Fong, R.; Ray, A.; Schneider, J. Domain Randomization for Transferring Deep Neural Networks from simulation to the real world. In Proceedings of the IEEE/RSJ International Conference on Intelligent Robots and Systems (IROS), Vancouver, BC, Canada, 24–28 September 2017.

Article

Fast 3D Semantic Mapping in Road Scenes [†]

Xuanpeng Li [1,*], Dong Wang [1], Huanxuan Ao [1], Rachid Belaroussi [2] and Dominique Gruyer [2]

[1] School of Instrument Science and Engineering, Southeast University, Nanjing 210096, Jiangsu, China; kingeast16@seu.edu.cn (D.W.); aohuanxuan@163.com (H.A.)

[2] COSYS/LIVIC, IFSTTAR, 25 allée des Marronniers, 78000 Versailles, France; rachid.belaroussi@ifsttar.fr (R.B.); dominique.gruyer@ifsttar.fr (D.G.)

* Correspondence: li_xuanpeng@seu.edu.cn

† This paper is an extended version of our paper published in LI, Xuanpeng, et al. Fast semi-dense 3D semantic mapping with monocular visual SLAM. In Proceedings of the 2017 IEEE 20th International Conference on Intelligent Transportation Systems, Yokohama, Japan, 16–19 October 2017; pp. 385–390.

Received: 30 December 2018; Accepted: 8 February 2019; Published: 13 February 2019

Abstract: Fast 3D reconstruction with semantic information in road scenes is of great requirements for autonomous navigation. It involves issues of geometry and appearance in the field of computer vision. In this work, we propose a fast 3D semantic mapping system based on the monocular vision by fusion of localization, mapping, and scene parsing. From visual sequences, it can estimate the camera pose, calculate the depth, predict the semantic segmentation, and finally realize the 3D semantic mapping. Our system consists of three modules: a parallel visual Simultaneous Localization And Mapping (SLAM) and semantic segmentation module, an incrementally semantic transfer from 2D image to 3D point cloud, and a global optimization based on Conditional Random Field (CRF). It is a heuristic approach that improves the accuracy of the 3D semantic labeling in light of the spatial consistency on each step of 3D reconstruction. In our framework, there is no need to make semantic inference on each frame of sequence, since the 3D point cloud data with semantic information is corresponding to sparse reference frames. It saves on the computational cost and allows our mapping system to perform online. We evaluate the system on road scenes, e.g., KITTI, and observe a significant speed-up in the inference stage by labeling on the 3D point cloud.

Keywords: 3D semantic mapping; incrementally probabilistic fusion; CRF regularization; road scenes

1. Introduction

Scene understanding plays a key background role in most vision-based mobile robots. For example, autonomous navigation in indoor/outdoor scenes asks for a rapid and comprehensive understanding of surroundings for obstacle avoidance and path planning [1–3]. Vehicle movement in limited temporal and spatial contexts always requires knowledge of what there are around ego-vehicle and where it is located. Robotic maps, such as Occupancy grid map and OctoMap, traditionally provide geometric presentation of the environment. However, they lack the correlation in data between map points and semantic knowledge; thus, they could not be directly utilized for scenes understanding.

Scene parsing/semantic segmentation is an important and promising step to address this issue. It has been an active topic for a long time, and it benefits from the state-of-the-art Deep Convolutional Neural Networks (DCNNs), which contributes to better performance of 2D pixel labeling than traditional methods. Then, combined with the SLAM technology, an automobile could locate itself and meanwhile recognize surrounding objects on a pixel-wise level. For instance, it could make autonomous vehicle accomplish certain high-level tasks, such as "parking on the right free space" and "stopping at the crosswalk". This form of 3D semantic representation provides mobile robots with functions of understanding, interaction, and navigation in various scenes.

Most semantic segmentation methods focus on increasing the accuracy of the semantic segmentation, and have seen major improvements [4,5]. However, they usually asks for high-power computing resources, which is not suitable for embedded platforms. Several recent research try to make a balance between the computing cost and the accuracy of object detection, classification, and 2D pixel labeling [6,7]. They achieve better performance on the embedded and mobile platforms.

Visual SLAM is a promising technology, especially based on monocular vision, which is flexible, inexpensive, and widely equipped on most recent vehicles. Although scaled sensors like stereo cameras and RGB-Depth (RGB-D) cameras could provide reliable measurement in their specific ranges, they lack the capability of seamless switch between various-scaled scenes. In addition, they normally need large storage resources. Most man-made environments, e.g., road scenes, usually exhibit distinctive spatial relations among varied classes of objects. Employing theses relations could enhance semantic segmentation performance in the 3D semantic mapping [8]. In this paper, we exploit a monocular SLAM method that provides cues of 3D spatial information and utilize state-of-the-art DCNN to build a 3D scene understanding system towards road scenes. Moreover, a Bayesian 2D-3D transfer and a map regularization process are utilized to generate a consistent reconstruction in the spatial and semantic context.

In our monocular mapping system, the map is incrementally reconstructed with a sequence of automatically selected keyframes and corresponding semantic information. There is no need to label each frame in a sequence, which could save a considerable amount of computation cost. We refer readers to Figure 1. Different from the frame skipping strategy proposed by Hermans et al. [9] and McCormac et al. [10], our method could work well under fast motions. Since the 3D map should have global consistent depth information, it should be regularized in term of spatial structures. The regularization is aimed at removing distinctive outliers and it makes components more consistent in the point cloud map, i.e., local points with same semantic label should be approached in 3D space. Two datasets, Cityscapes [11] and KITTI [12], are used to evaluate our approach. This work is an extension of our previous work [13]. We not only modify the 2D semantic segmentation module, but also revise the offline regularization module with new potential constraint. More experiments and theoretical details are involved in this work. The main contributions involve the improvement of 2D semantic segmentation model, the associative hierarchical Conditional Random Field (CRF) with *High Order Potential* towards the point cloud, the extended experiments and the quantitative evaluation of the performance including accuracy and runtime.

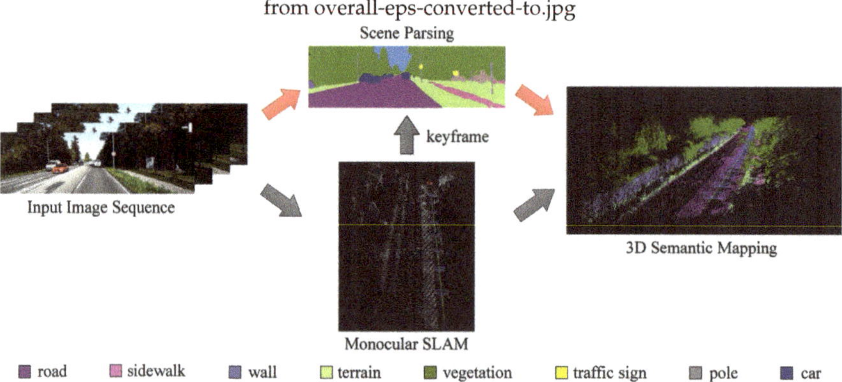

Figure 1. Overview of our system: From monocular image sequence, keyframes are selected to obtain the 2D semantic information, which then transfer to the 3D reconstruction, and then incrementally build the 3D semantic map.

This paper is presented as follows. In the Section 2, a review of the related work is given. The problem formulation is presented in Section 3. The 3D semantic mapping is described in Section 4, including the semantic segmentation, the monocular visual SLAM, the Bayesian incremental fusion, and the global regularization. Section 5 includes the results of 2D semantic inference and 3D semantic mapping. Finally, Section 6 concludes the paper and discusses possible extensions of our work.

2. Related Work

Our work is motivated by [10] which contributes an indoor 3D semantic SLAM from the RGB-D input. It aims towards a dense 3D map based on ElasticFusion SLAM [14] with semantic labeling. Pixel-wise semantic information is acquired from a Deconvolutional semantic segmentation network [15] using the scaled RGB information and the depth as the input. Depth information is also used to update surfel's depth and normal information to construct 3D dense map during loop closure. In addition, a previous work, SLAM++ [16], creates a map with semantically defined objects, but it is limited to predefined database and hand-crafted template models. In this paper, we make use of an incremental Bayesian fusion strategy with state-of-the-art visual SLAM and semantic segmentation.

Visual SLAM usually contains sparse, semi-dense, and dense types depending on the methods of image alignment. Feature-based methods only exploit limited feature points—typically image corners and blobs or line segments, such as classic MonoSLAM [17] and ORB-SLAM [18,19]. They are not suitable for 3D semantic mapping due to rather sparse feature points. In order to better exploit image information and avoid the cost on calculation of features, direct dense SLAM system, such as the surfel-based dense slam, ElasticFusion [14], and Dense Visual SLAM [20], have been proposed recently. Whereas, direct image alignment from these dense methods is well-established for monocular, RGB-D and stereo sensors. Semi-dense methods like Large-Scale Direct-SLAM (LSD-SLAM) [21] and Semi-direct Visual Odometry (SVO) [22] provide possibility to build a synchronized 3D semantic mapping system.

Deep CNNs have proven to be effective in the field of image semantic segmentation. Long et al. [23] firstly introduces an inverse convolution layer to realize an end-to-end training and inference. Then, the encoder-decoder architectures with specified upsampling layers, such as max unpooling and deconvolutional layer, are proposed to avoid the problem of separate step training in the Fully Convolutional Network (FCN) and improve the accuracy [15,24]. Zhao et al. [4] exploit the capability of global context information through embedding various scenery context feature in a pyramid structure. The fusion of varied scaled feature has been a popular strategy in the recent deep CNNs. The cutting-edge method, namely, DeepLab series [5,7], combines atrous convolutions and atrous spatial pyramid pooling (ASPP) to achieve a state-of-the-art performance on semantic segmentation. The early DeepLab models have a reasonable accuracy but require much computation overhead. Recently proposed efficient convolution neural network, such as MobileNets [25,26] boosts real-time performance of semantic segmentation without losing the accuracy too much. The state-of-the-art DeepLab-v3+ [7] contains a simple effective decoder module to refine the segmentation results especially along object boundaries. Furthermore, combining the encoder part of MobileNet-v2 in its structure, DeepLab-v3+ could achieve a better trade-off between precision and runtime.

In the topic of scene understanding and mapping, recent researchs employ 3D priors of objects increasingly. Salas-Moreno et al. [16] project 3D mesh of objects to the RGB-D frame in a graphical SLAM framework. Valentin et al. [27] propose a triangulated meshed representation of the scene from multiple depth measurements and exploit the CRF to capture the consistency of 3D object mesh. Kundu et al. [28] exploit the CRF for joint voxels to infer the semantic information and occupancy. Sengupta and Sturgess [29] use stereo camera, estimated pose, and CRF to infer the semantic octree presentation of the 3D scene. Vineet et al. [30] propose an incremental dense stereo reconstruction and semantic fusion technique to handle dynamic objects in the large-scale outdoor scenes. Kochanov et al. [31] employ scene flow measurements to incorporate temporal updates into the mapping of dynamic environment. Landrieu et al. [32] introduce a regularization framework to obtain spatially smooth semantic labeling of 3D point clouds from a point-wise classification, considering the uncertainty associated with each label. Gaussian Process (GP) is another popular method for map inference. Jadidi et al. [33] exploit GP to learn the structural and semantic correlation between map points. This technique also incorporates OcotoMap to handle sparse measurements and missing labels. In order to improve the training and query time complexities of the GP-based semantic mapping, Gan et al. [34] further introduce a Relevance Vector Machine (RVM) inference technique for efficient map query at any resolution.

Our semi-dense approach is also inspired by dense 3D semantic mapping methods [8,9,35,36] in both indoor and outdoor scenes. The major contributions from these work involve the 2D-3D transfer and the map regularization. Especially, Hermans et al. [9] propose an efficient 3D CRF to regularize 3D semantic mapping consistently considering influence between neighbors of 3D points (voxels). In this work, we adopt a similar strategy to improve the performance of the 3D semantic reconstruction in the road scenes. The key concepts are

- a 3D semantic mapping system based on monocular vision,
- integration of monocular SLAM [21] and scene parsing [7] into 3D semantic representation,
- exploiting the correlation between semantic information and geometrical information to enforce spatial consistency,
- active sequence downsampling and sparse semantic segmentation so that to achieve a real-time performance and reduce the storage.

Following the comparison in [30], we list the characteristics of our approach and some related works in Table 1.

Table 1. Comparison with some related work: M = monocular camera, S/D = stereo/depth camera, L = lidar, O = outdoor, I = incremental, SDT = sparse data structures, RT = real time.

Method	M	S/D	L	O	I	SDT	RT
Hu et al. [37]			√	√	√	√	√
Sengupta et al. [35]		√		√			
Hermans et al. [9]		√			√		√
Kundu et al. [28]	√			√		√	
Vineet et al. [30]		√		√	√	√	√
Wolf et al. [8]		√					√
McCormac et al. [10]		√			√	√	√
Ours	√			√	√	√	√

3. Problem Formulation

3.1. Notation

The target is to estimate the 3D semantic map $\mathcal{M}$ comprising of a pose-graph of keyframes with semantic map from a monocular camera. The 3D map $\mathcal{M}$ is reconstructed by the estimation of depth and poses, where each 3D point $\mathbf{P}$ can be labeled as one of the solid semantic objects in the label space $\mathcal{L} = \{l_1, l_2, \ldots, l_k\}$ like *Road, Building, Tree*, etc. We use $\mathbf{X} = \{X_1, X_2, \ldots, X_M\}$ to denote the set of random variables corresponding to the 3D points $\mathbf{P}_i : i \in \{1, \ldots, M\}$, where each variable $X_i \in \mathbf{X}$ take a value l_i from the predefined label space $\mathcal{L}$.

Let $I_i : \Omega_i \to \mathbb{R}^3$ symbolize an H × W RGB image of an input sequence at the frame indexed by i. Keyframes are extracted from the sequence in light of camera's pose $\mathbf{T}_i^j$ at the jth frame with respect to the previous ith keyframe. We define the ith keyframe to be a tuple $\mathcal{K}_i = (I_i, D_i, V_i, S_i)$, where $D_i : \Omega_{D_i} \to \mathbb{R}$ is the full-resolution inverse depth map associated with image I_i, and $V_i : \Omega_{V_i} \to \mathbb{R}$ is the variance associated with the inverse depth map. The inverse depth model is a better description for visual depth estimation, which assumes normally distributed [38]. The inverse depth map and the variance map are defined in the subset of pixels as $\Omega_{D_i}, \Omega_{V_i} \subset \Omega_i$, which means semi-dense, available for certain image regions of large intensity gradient. The symbol $S_i : \Omega_{S_i} \to \mathbb{R}^{|\mathcal{L}|}$ represents the full-resolution semantic map with the probability of each object class.

The keyframes are consecutively stacked in a pose-graph $\mathcal{G} = (\mathcal{V}, \mathcal{E})$, where $\mathcal{V} = \{\mathcal{K}_0, \ldots, \mathcal{K}_n\}$ is the set of keyframes and $\mathcal{E} = \{\mathbf{S}_i^j \in \text{Sim}(3) : \mathcal{K}_i, \mathcal{K}_j \in \mathcal{V}\}$ is the set of constraint factors. Each $\mathbf{S}_i^j = (\mathbf{T}_i^j, s_i^j)$ consists of a camera's pose $\mathbf{T}_i^j = \binom{R\,t}{0\,1}$ from the keyframe ith to the keyframe jth, and scale factor $s_i^j > 0$. In reference to world frame W, regarded as the pose of the first keyframe $\mathcal{K}_0$, the pose of the keyframe ith is denoted as $\mathbf{T}_W^i$. For a sequence of keyframes (n keyframes), we get the nth keyframe's pose $\mathbf{T}_W^n = \prod_1^n \mathbf{T}_{k-1}^k$.

3.2. Framework

Our target is to build a 3D semantic map with semi-dense and consistent label information online while the image sequences are captured by a forward-moving monocular camera. Given an image sequence, the inference of the 3D semantic map is regarded as:

$$\mathcal{M}^* = \text{argmax}_{\mathcal{M}} P(\mathcal{M}|\mathcal{G}), \tag{1}$$

which can be estimated by the maximum a-posterior (MAP). Compared to the model used in [28], our measures are continually updated with new keyframes. Thus, we adopt an incremental fusion strategy to estimate the 3D semantic map by incorporating new estimation of pose, depth, and semantic information. Correspondingly, the approach is decoupled into three separately running processes as shown in Figure 2.

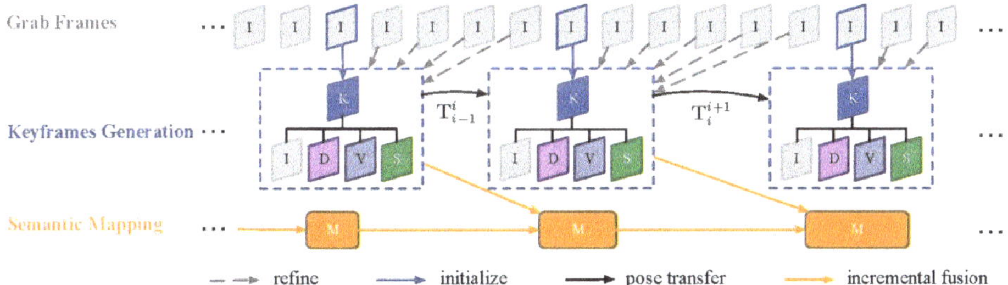

Figure 2. Framework of our method: The input is the sequence of RGB frames I. There are three separate processes, a keyframe selection process, a 2D semantic segmentation process, and a 3D reconstruction with semantic optimization process. Keyframe, denoted as K, is conditionally extracted from the sequence based on the distance between the poses **T**. The following frames refine the inverse depth map D and the variance map V of each keyframe until new keyframe is extracted. The 2D semantic segmentation module predicts the pixel-level class with scores S of the new arriving keyframe. Finally, the keyframes are incrementally explored to reconstruct the 3D map with semantic labeling and then it is regularized by a dense Conditional Random Field (CRF).

In the system, the monocular SLAM process maintains and tracks on a global map of the environment, which contains a number of keyframes connected by pose-pose constraints with associated probabilistic depth maps. It runs in real-time on a CPU. Represented as point clouds, the map gives a semi-dense and highly accurate 3D reconstruction of the environment. Meanwhile, the second process of the 2D semantic segmentation generates the pixel-level classification on the extracted keyframes. A fast deep CNN model is explored to predict the semantic information on a GPU. In addition, an incremental fusion process for the semantic label optimization is operated in a parallel way. It builds a local optimal correspondence between semantic labeling and 3D points in the point cloud. To obtain a globally optimal 3D semantic segmentation, we attempt to make use of information of neighboring 3D points, involving the distance, color similarity, and semantic label. It optimizes the point cloud and semantic labels to generate a globally consistent 3D semantic map.

4. 3D Semantic Mapping

4.1. 2D Scene Parsing

We explore the DeepLab-v3+ deep neural network proposed by Chen et al. [7]. Two important components in the DeepLab series are the atrous convolution and atrous spatial pyramid pooling (ASPP), which enlarge the field of view of filters and explicitly combine the feature maps at multiple scales. The improvement in the DeepLab-v3+ involves the encoder-decoder structure and the augmentation of ASPP module with image-level feature. The former is able to capture sharper object boundaries by regaining the spatial information, while the latter encodes multi-scale contextual information to capture long range information. These contributions make DeepLab successfully handle both large and small objects and achieve a better trade-off between precision and run-time.

For the semantic segmentation of road scenes, we exploit the Cityscapes dataset and the KITTI dataset and adopt the predefined 19-class label space $\mathcal{L} = \{l_1, l_2, \ldots, l_{19}\}$, which contains *Road*, *Sidewalk*, *Building*, *Wall*, and so on. We use all semantic annotated images in the Cityscapes dataset for training and fine-tune the model with the KITTI dataset. Note that there is not any depth information involved in the training process. In the inference, we keep the original resolution of input image according to different datasets.

4.2. Semi-Dense SLAM

We explore LSD-SLAM to track camera's trajectory and build consistent, large-scale maps of the environment. LSD-SLAM is a real-time, semi-dense 3D mapping method. It has several advantages: firstly, it is a scale-aware image alignment algorithm to directly estimate the similarity transform between two keyframes against different scale environments, such as office rooms (indoor) and urban roads (outdoor). The second one is that it is a probabilistic approach to incorporate noise on the estimated inverse depth maps into the tracking based on the propagation of uncertainty. Moreover, it could easily integrate with various kinds of sensors like monocular, stereo and panoramic cameras for various applications. Thus, it is able to make a reliable trajectory estimation and map reconstruction even in challenging surroundings.

LSD-SLAM has three major components: tracking, depth estimation and map optimization. Spatial regularization and outlier removal are incorporated in the depth estimation with small-baseline stereo comparisons. In addition, a direct, scale-drift aware image alignment is carried on these existing keyframes to detect scale-drift and loop closures. Due to the inherent correlation between the depth and the tracking accuracy, depth residual is used to estimate the similarity transform $\mathrm{sim}(3)$ constraints between keyframes. Consequently, a 3D point cloud map is built based on a set of keyframes with the estimated inverse depth maps via minimizing the error of image alignment. The map is continuously optimized in the background using a *g2o* pose-graph optimization. The approach runs at 25 Hz on an Intel i7 CPU. More details like keyframe selection and depth estimation can be referred to the work [21].

4.3. Incremental Fusion

There might be a large amount of inconsistent 2D semantic labels between consecutive frames, due to the noise of sensors, the complexity of environments in the real world and the failure of scene parsing model. Incremental fusion of semantic label from the stacked keyframes allows associating probabilistic label in a Bayesian way, when combining with the inverse depth map propagation between keyframes in the LSD-SLAM. We give the details about the incremental semantic fusion as follows.

The camera projection transformation function $\pi(\cdot) : \mathbb{R}^3 \to \mathbb{R}^2$ is defined as

$$\mathbf{p} = \pi(\mathbf{P}) = [\alpha\frac{x}{z} + c_x, \beta\frac{y}{z} + c_y]^T, \tag{2}$$

which maps a point $\mathbf{P} = [x, y, z]^T$ in 3D space into a 2D point $\mathbf{p} = [x', y']^T$ on the digital image plane I_i in the camera coordinate system. Since this projection function is nonlinear, for the computation efficiency, the transformation should be augmented into the homogeneous coordinate system, which is defined as

$$\mathbf{p}_h = \begin{bmatrix} x'_h \\ y'_h \\ z'_h \end{bmatrix} = \begin{bmatrix} \alpha & 0 & c_x & 0 \\ 0 & \beta & c_y & 0 \\ 0 & 0 & 1 & 0 \end{bmatrix} \begin{bmatrix} x \\ y \\ z \\ 1 \end{bmatrix} = \mathbf{K}[\mathbf{I}\ 0]\mathbf{P}_h, \tag{3}$$

where $\mathbf{K}$ is referred to as the camera matrix. Given a 3D point $\mathbf{P}_W$ in the world reference system, the mapping to image plane I_i in the homogeneous reference system is calculated as

$$\mathbf{p}_h = \mathbf{K}\mathbf{T}^i_W\mathbf{P}_{Wh}, \tag{4}$$

where $\mathbf{T}^i_W$ is the pose of the camera in the world reference system. Then, we get Euclidean coordinates $\mathbf{p} = [x'_h/z'_h, y'_h/z'_h]^T$ from the homogeneous coordinates. From this point on, any point $\mathbf{p}$ and $\mathbf{P}$ is assumed to be in homogeneous coordinates and thus we drop the h index, unless stated otherwise.

Correspondingly, given the inverse depth estimation $\hat{d}$ for a pixel $\mathbf{p} = [x', y']^T$ in the image I_i of the keyframe $\mathcal{K}_i$, we also have an inverse projection function from 2D pixel point into the 3D point in the current camera coordinate system as:

$$\mathbf{P} = \pi^{-1}(\mathbf{p}, \hat{d}) = [\frac{x'/\hat{d} - c_x/\hat{d}}{\alpha}, \frac{y'/\hat{d} - c_y/\hat{d}}{\beta}, \frac{1}{\hat{d}}]^T, \tag{5}$$

where $\hat{d} = D_i(\mathbf{p}) \sim \mathcal{N}(\mu, V_i(\mathbf{p}))$ corresponds to the inverse depth of the point $\mathbf{p}$, which is normally distributed. The inverse depth estimation of each existing keyframe is continuously refined using its following frames until new keyframe is selected. In reference to Equations (4) and (5), we can derive the normally distributed 3D points in the world reference system as follows:

$$\hat{\mathbf{P}}_W = {T^i_W}^{-1}\pi^{-1}(\mathbf{p}, D_i(\mathbf{p}), V_i(\mathbf{p})), \tag{6}$$

where the homogeneous transformation matrix has the property: ${T^i_W}^{-1} = T^W_i$.

Once a new frame is chosen to become a keyframe $\mathcal{K}_j$, its inverse depth map D_j is initialized by projecting points from previous keyframe into it. The information of existing, close-by keyframes is propagated to new keyframe for its initialization and semantic probabilistic refinement. The corresponding point $\hat{\mathbf{p}}$ in the image I_j of new keyframe is located by

$$\hat{\mathbf{p}} = \mathbf{K}\mathbf{T}^i_W\mathbf{T}^j_i\hat{\mathbf{P}}_W \in I_j. \tag{7}$$

Here, since the estimation of the inverse depth map is normally distributed, we have a one-to-many transform between keyframes, which involves a couple of estimated 2D/3D points, regarded as $\mathbf{p} \in I_i \to \hat{\mathbf{P}}_W \to \hat{\mathbf{p}} \subset I_j$, as shown in Figure 3.

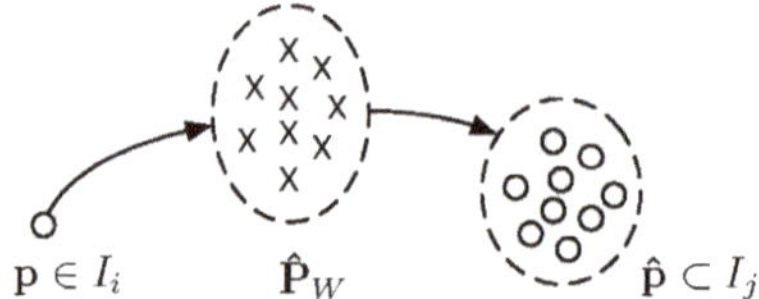

Figure 3. The Gaussian translation from a pixel $\mathbf{p}$ in the image I_i of the keyframe $\mathcal{K}_i$ to estimated pixels $\hat{\mathbf{p}}$ in the image I_j of the keyframe $\mathcal{K}_j$.

The class label corresponding to a couple of 3D points $\hat{\mathbf{P}}_W$ in the world reference is denoted as $X : \hat{\mathbf{P}}_W \rightarrow l \in \mathcal{L}$. Note that the label *Sky* is removed from $\mathcal{L}$ for the 3D semantic mapping. Our target is to obtain the independent probability distribution of each 3D point over the class labels $P(X|\mathcal{K}_0^i)$ given a sequence of existing keyframes $\mathcal{K}_0^i = \{\mathcal{K}_0, \mathcal{K}_1, \dots, \mathcal{K}_i\}$ in the pose-graph $\mathcal{G}$.

We explore a recursive Bayesian fusion to refine the corresponding probability distribution of 3D points with new keyframe's update:

$$P(X|\mathcal{K}_0^i) = \frac{1}{Z_i} P(\mathcal{K}_i|\mathcal{K}_0^{i-1}, X) P(X|\mathcal{K}_0^{i-1}), \tag{8}$$

with $Z_i = P(\mathcal{K}_i|\mathcal{K}_0^{i-1})$. Applying the first-order Markov assumption to $p(\mathcal{K}_i|\mathcal{K}_0^{i-1}, X)$, then we have:

$$P(X|\mathcal{K}_0^i) = \frac{1}{Z_i} P(\mathcal{K}_i|X) P(X|\mathcal{K}_0^{i-1}) = \frac{1}{Z_i} \frac{p(\mathcal{K}_i)P(X|\mathcal{K}_i)}{P(X)} P(X|\mathcal{K}_0^{i-1}). \tag{9}$$

We assume that $P(X)$ does not change over time and there is no need to calculate the normalization factor $P(\mathcal{K}_i)/Z_i$ explicitly.

According to the formulations above, the semantic probability distribution of all given keyframes can be recursively updated as follows:

$$P(X|\mathcal{K}_0^i) \propto P(X|\mathcal{K}_i) P(X|\mathcal{K}_0^{i-1}), \tag{10}$$

where a couple of 2D pixels matching between $\mathcal{K}_0^i$ can be calculated with the Equations (4) and (5). The semantic map in $\mathcal{K}_0^i$ contributes to the accumulated probabilistic estimation of object class. For example, given a pixel $\mathbf{p}$ in the image I_i of the keyframe $\mathcal{K}_i$, its corresponding scores (probabilities) of object classes are $S_i(\mathbf{p}) = \{P(Road|\mathbf{p}) = p_1, P(Sidewalk|\mathbf{p}) = p_2, P(Building|\mathbf{p}) = p_3, \dots, P(Bicycle|\mathbf{p}) = p_{19}\}$ with $\sum p_i = 1$. Then, at each fusion step, the predicted labels of 3D point $\hat{\mathbf{P}}_W$ is the label with maximum probabilities as

$$\max_{k=1}^{N} S_j(\hat{\mathbf{p}}_k) S_i(\mathbf{p}), \tag{11}$$

where there are N possible projected 3D points and pixels $\hat{\mathbf{p}}$ in the image I_j of the keyframe $\mathcal{K}_j$.

The incremental fusion can refine the semantic label of the points in the 3D space based on the pose-graph of keyframes. It could handle the inconsistent 2D semantic labels, even though its performance relies on the depth estimation. In addition, map geometry is another useful feature which could improve the performance of the 3D semantic mapping further. The following section describes how we use the dense CRF to regularize the 3D semantic map by exploring the map geometry, which could propagate semantic information between spatial neighbors.

4.4. Map Regularization

The dense CRF is widely used in the 2D semantic segmentation to enhance the performance of semantic segmentation. Some previous works [8,9,35] seek its application on the 3D map to model contextual relations between various class labels in a fully connected graph. It is a heuristic approach that assumes the influence between neighbors should be proportional to their distance, visual, and geometrical similarity [9].

The CRF model is defined as a graph composed of unary potentials as nodes and pairwise potentials as edges, but the size of the model makes traditional inference algorithms impractical. Thanks to Krahenbuhl and Koltun's work [39], a highly efficient approximate inference algorithm is proposed to handle this issue by defining the pairwise edge potentials as a linear combination of Gaussian kernels. We apply the efficient inference of the dense CRF to maximize label agreement between similar 3D points as follows.

Assume the 3D semantic map $\mathcal{M}$ containing M 3D points is defined as a random field. A CRF $(\mathcal{M}, \mathbf{X})$ is characterized by a Gibbs distribution as follows:

$$P(\mathbf{X}|\mathcal{M}) = \frac{1}{Z(\mathcal{M})} \exp(-E(\mathbf{X}|\mathcal{M})), \tag{12}$$

where $E(\mathbf{X}|\mathcal{M})$ is the Gibbs energy and $Z(\mathcal{M})$ is the partition function. The maximum a posteriori (MAP) labeling of the random field is

$$\mathbf{X}^* = \operatorname{argmax}_{l \in \mathcal{L}} P(\mathbf{X}|\mathcal{M}) = \operatorname{argmin}_{l \in \mathcal{L}} E(\mathbf{X}|\mathcal{M}), \tag{13}$$

which is converted into minimizing the Gibbs energy by the mean-field approximation and message passing scheme.

We employ the associative hierarchical CRF [35,40] which integrates the unary potential ψ_i, the pairwise potential $\psi_{i,j}$, and the higher order potential ψ_c into the Gibbs energy at different levels of the hierarchy (voxels and supervoxels) given by:

$$E(\mathbf{X}|\mathbf{C};\theta) = \sum_i \psi_i(X_i|\mathbf{C}) + \sum_{i<j} \psi_{i,j}(X_i, X_j|\mathbf{C};\theta) + \sum_c \psi_c(X_c|\mathbf{c}) \tag{14}$$

by the indexes $i, j \in \{1, \ldots, M\}$ correspond to different 3D points $\mathbf{P}_i, \mathbf{P}_j$ in the 3D map $\mathcal{M}$.

Unary Potential: The unary potential $\psi_i(\cdot)$ is defined as the negative logarithm of the probabilistic label for a given 3D point:

$$\psi_i(X_i|\mathbf{C}) = -\log(P(X_i \to l|\mathcal{K}_0^t)). \tag{15}$$

This term means the cost of 3D point P_i taking an object label $l \in \mathcal{L}$ based on the incremental semantic probabilistic fusion above. The output of the unary potential for each point is produced independently, and thus, the MAP labeling produced by the unary potential alone is generally inconsistent.

Pairwise Potentials: The pairwise potential $\psi_{i,j}(\cdot)$ is modeled to be a log-linear combination of m Gaussian edge potential kernels:

$$\psi_{i,j}(X_i, X_j|\mathbf{C};\theta) = \mu(X_i, X_j) \sum_m \omega^{(m)} k^{(m)}(\mathbf{f}_i, \mathbf{f}_j;\theta), \tag{16}$$

where $\mu(\cdot)$ is a label compatibility function corresponding to the Gaussian kernel functions $k^{(m)}(\mathbf{f}_i, \mathbf{f}_j)$. $\mathbf{f}$ denotes the feature vector for the 3D point $\mathbf{P}$ including the position, the RGB appearance and the surface normal vector of the reconstructed surface. Furthermore, $\mu(\cdot)$ is defined as a Potts model given by:

$$\mu(l, l') = [l \neq l'] = \begin{cases} 1 & l \neq l' \\ 0 & l = l' \end{cases}. \tag{17}$$

This term is defined to encourage the consistency over pairs of neighboring points for the local smoothness of the 3D semantic map. We employ two Gaussian kernels for the pairwise potentials following the previous work [9]. The first one is an appearance kernel as follows:

$$k^{(1)}(\mathbf{f}_i, \mathbf{f}_j; \theta) = \exp\left(-\frac{|\mathbf{P}_i - \mathbf{P}_j|^2}{2\theta_{\mathbf{P},c}^2} - \frac{|\mathbf{c}_i - \mathbf{c}_j|^2}{2\theta_c^2}\right), \tag{18}$$

where $\mathbf{c}$ is the RGB color vector of the corresponding 3D points. This kernel is used to build long range connections between 3D points with a similar appearance.

The second one, a spatial smoothness kernel, is defined to enforce a local, appearance-agnositc smoothness among 3D points with similar normal vectors.

$$k^{(2)}(\mathbf{f}_i, \mathbf{f}_j; \theta) = \exp\left(-\frac{|\mathbf{P}_i - \mathbf{P}_j|^2}{2\theta_{\mathbf{P},n}^2} - \frac{|\mathbf{n}_i - \mathbf{n}_j|^2}{2\theta_n^2}\right), \tag{19}$$

where $\mathbf{n}$ are the respective surface normals. The surface normal are computed using the Triangulated Meshing using Marching Tetrahedra (TMMT) proposed in [35]. Note that the original method is towards producing a dense labeling with the stereo vision. Since the LSD-SLAM only generates semi-dense 3D point clouds, we modify the TMMT to extract a triangulated mesh within limited ranges of short distance between 3D points.

High Order Potential: The higher order term $\psi_c(X_c|\mathbf{c})$ encourages the 3D points (voxels) in the given segment take the same label and penalizes partial inconsistency of supervoxels as described in [40]. It is defined as

$$\psi_c(X_c|\mathbf{c}) = \min_{l\in\mathcal{L}}(\gamma_c^{\max}, \gamma_c^l + k_c^l N_c^l), \tag{20}$$

where γ_c^l represents the cost if all voxels in the segment take the label l. $N_c^l = \sum_{i\in\mathbf{c}}\delta$ is the number of inconsistent 3D points with the label l which is penalized with a factor k_c, regarded as the inconsistency cost.

All parameters $\theta_{\mathbf{P},c}, \theta_c, \theta_{\mathbf{P},n}, \theta_n, \theta_{\mathbf{P},s}, \theta_s$ specify the range in which points with similar features affect each other, respectively. They can be obtained using piece-wise learning.

5. Experiments and Results

We demonstrate the performance of our approach on the KITTI dataset [12], which contains a variety of urban scene sequences involving lots of moving objects in various lighting conditions. It consists of various datasets, such as the semantic dataset, the odometry dataset, and the detection dataset. Thus, it is very challenging for the 3D reconstruction. The KITTI dataset contains a 2D semantic segmentation data of 200 labeled training images and 200 test images (http://www.cvlibs.net/datasets/kitti/eval_semseg.php?benchmark=semantics2015) . Its data format and metrics conform with the Cityscapes dataset [11]. The Cityscapes dataset involves 19 classes within high quality pixel-level annotations of 5000 images with a resolution of 2048×1024, including 2975 training images, 500 validation images, and 1525 testing images. In our experiment, we train the model on the Cityscapes and then tune it on the KITTI taking the volume size of dataset into account.

For the training of 2D semantic segmentation model, various encoder models in the DeepLab-v3+ are evaluate including *ResNet* [41], *Xception* [42], and *MobileNet-v2* [26]. We find that the "poly" stochastic gradient descent is better than the "step" one on these datasets. The *TensorFlow* library is employed to do the training and inference on the workstation with 4 Nvidia Titan X GPU cards. The hyper-parameters used in training are set corresponding to the datasets and models as shown in Table 2.

We benchmark the performance of our semantic mapping system on the KITTI odometry dataset (http://www.cvlibs.net/datasets/kitti/eval_odometry.php). There are 22 sequences with the consecutive RGB frames, in which there are 11 sequences with the ground-truth poses for evaluation. These scenes involves serious illumination change, moving objects like persons and vehicles, and some turns as shown in Figure 4. These road-scene frames involves two resolutions 1242×375 and 1226×370. Our system runs on an Intel Core i7-5960K CPU and a NVIDIA Titan X GPU for online process.

Table 2. Hyper-parameters used in the training step.

Dataset	Encoder	Learning Rate	Learning Power	Momentum	Weight Decay	Batch	Steps
	ResNet_50	0.003	0.9	0.9	0.0001	8	20,000
	ResNet_101	0.003	0.9	0.9	0.0001	8	20,000
	Xception_41	0.01	0.9	0.9	0.00004	8	10,000
Cityscapes	Xception_65	0.01	0.9	0.9	0.00004	8	10,000
	Xception_71	0.01	0.9	0.9	0.00004	8	10,000
	MobileNet_v2	0.001	0.9	0.9	0.00004	64	10,000
	ResNet_50	0.003	0.9	0.9	0.0001	8	20,000
	ResNet_101	0.003	0.9	0.9	0.0001	8	20,000
	Xception_41	0.01	0.9	0.9	0.00004	8	10,000
KITTI	Xception_65	0.01	0.9	0.9	0.00004	8	10,000
	Xception_71	0.01	0.9	0.9	0.00004	8	10,000
	MobileNet_v2	0.001	0.9	0.9	0.00004	64	10,000

Since the KITTI sequences are mostly captured in 10 Hz, it is highly below the normal speed requirements of LSD-SLAM about 60 Hz. In addition, the LSD-SLAM is hard to handle severe turning when the platform moves. Due to the limit of the monocular LSD-SLAM, we choose certain sequences to evaluate.

In the following sections, we show some qualitative results for our approach in Section 5.1 and the quantitative results of our evaluation are presented in Section 5.2, in which we also make the runtime analysis on our semantic mapping approach.

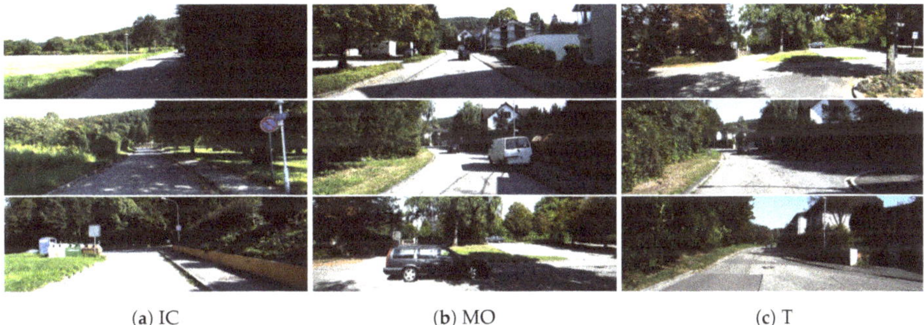

(a) IC   (b) MO   (c) T

Figure 4. Instances in the *odometry_03* sequence. IC: Illumination Change, MO: Moving Objects, T: Turns.

5.1. Qualitative Results

First, we present some qualitative results of the KITTI semantic dataset in Figure 5. Then, we use the trained model to make prediction on the KITTI odometry dataset, and the results are exemplified as shown in Figure 6.

Take the sequence *odometry_03* as an example of our semantic mapping approach. The sequence consists of 801 RGB frames on a urban road of about 560m. Figure 7 shows the semantic reconstruction with a close-up view including large-scale annotations such as *road*, *building*, and even small-scale objects like *traffic signs*. Note we discard some keyframes at the beginning, due to random initialization of LSD-SLAM. A close-up view is exemplified to illustrate the offline CRF processing as shown in Figure 8.

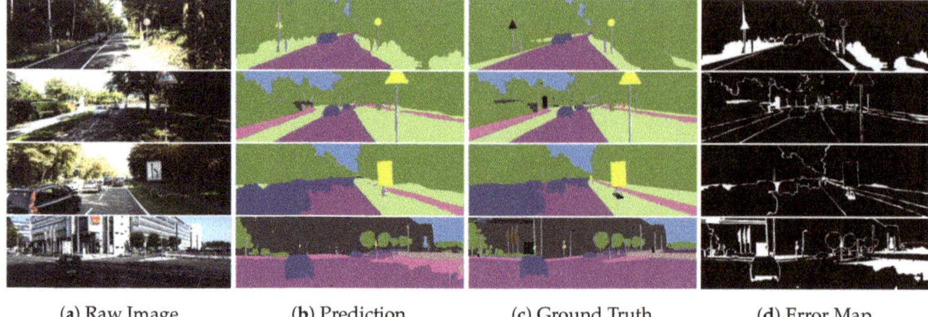

(**a**) Raw Image   (**b**) Prediction   (**c**) Ground Truth   (**d**) Error Map

Figure 5. Qualitative results of 2D semantic segmentation.

Figure 6. Instances of 2D semantic segmentation in the KITTI odometry set.

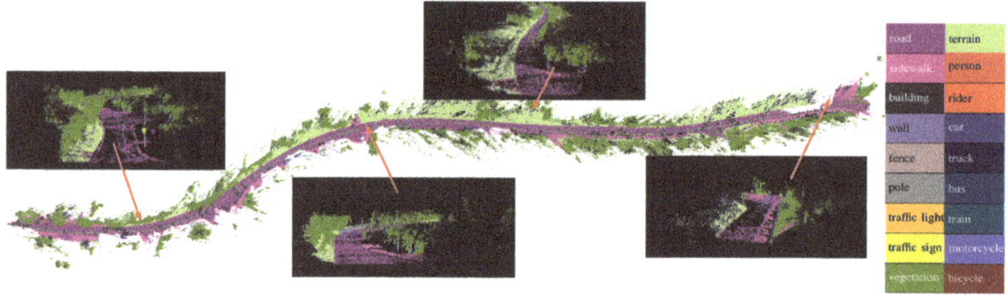

Figure 7. Qualitative results of 3D semantic mapping from the sequence KITTI *odometry_03*. Our approach not only reconstructs and labels entire outdoor scenes that include roads, sidewalks, and buildings, but also accurately recovers thin objects such as traffic signs and trees. The close-up views show the details of the map.

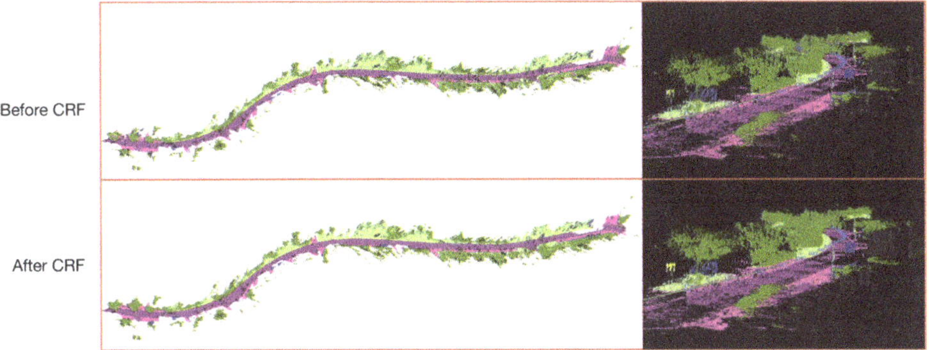

Figure 8. Comparison between the before CRF processing and the after CRF processing from different views of the 3D semantic map on the KITTI *odometry_03*.

Another possible qualitative comparison on the KITTI *odometry_05* as used in Kundu et al.'s work [28] is illustrated in Figure 9. Whereas monocular LSD-SLAM is not resistant to strong rotation in the sequence, we present the qualitative result based on the subset (500 frames) of this data.

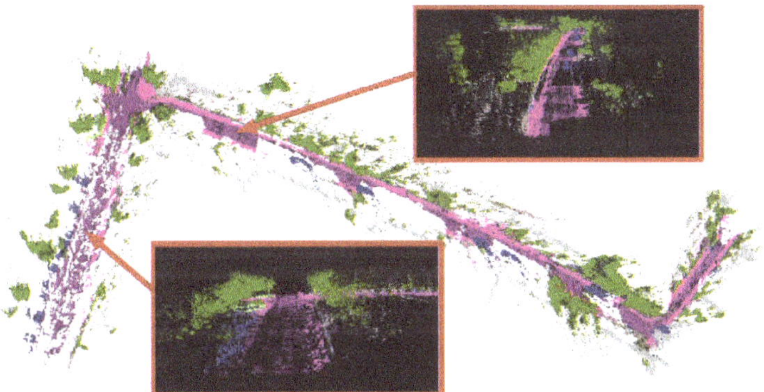

Figure 9. A qualitative result of 3D semantic mapping on the KITTI *odometry_05*.

5.2. Quantitative Results

For the quantitative performance of our approach, we focus on the 2D semantic segmentation and the runtime of the entire system, since the 3D reconstruction mainly depends on the SLAM module.

Semantic Segmentation: Table 3 shows the quantitative results of 2D semantic segmentation based on different DeepLab-v3+ models on the KITTI datasets. We evaluate these models by the mean intersection/union (mIOU) score, the model size, and the computational runtime. The mIOU score is defined as

$$\mathrm{mIOU} = \frac{1}{|\mathcal{L}|} | \sum_{i=1}^{|\mathcal{L}|} \mathrm{TP}_i / (\mathrm{TP}_i + \mathrm{FP}_i + \mathrm{FN}_i) \tag{21}$$

in terms of the True/False Positives/Negatives for a given class i. We do not resize the image to evaluate the models here. Whereas, for the 3D semantic mapping process, we need to half resize the input images in order to make a trade-off between accuracy and computational speed.

Table 3. Quantitative results of various encoder parts of DeepLab-v3+ on the Cityscapes and the KITTI. I: ImageNet, M: MS-COCO, C: Cityscapes.

Dataset	Encoder	Crop Size	mIOU[0.5:0.25:1.75]	Pb Size (MB)	Runtime (ms)	I	M	C
Cityscapes	ResNet_50	769	63.9	107.8	-	√		
	ResNet_101	769	69.9	184.1	-	√		
	Xception_41	769	68.5	113.4	-	√		
	Xception_65	769	78.7	165.7	1800	√		
	Xception_71	769	80.2	167.9	2000	√	√	
	MobileNet_v2	513	70.7	8.8	400		√	
	MobileNet_v2	769	70.9	8.8	400		√	
KITTI	ResNet_50	769	51.4	107.8	120	√		√
	ResNet_101	769	57.1	184.1	140	√		√
	Xception_41	769	54.2	113.4	140	√		√
	Xception_65	769	64.8	165.6	160	√		√
	Xception_71	769	66.2	167.9	170	√	√	√
	MobileNet_v2	513	57.7	8.8	80		√	√
	MobileNet_v2	769	60.7	8.8	80		√	√

During the training process, these models are initialized with the checkpoints pre-trained from various datasets including ImageNet [43] and MS-COCO [44]. In the training step on the Cityscapes dataset, we directly use the ImageNet-pretrained checkpoints as the initialization. Note we employ the *MobileNet_v2* based model which has been pre-trained on MS-COCO dataset, and the *Xception_71* based model has been pre-trained on both ImageNet and MS-COCO datasets. These pre-trained models can be accessed from the github (https://github.com/tensorflow/models/blob/master/research/deeplab/g3doc/model_zoo.md).

Then we fine-tune the models on the KITTI dataset by using the pre-trained Cityscapes model. The *Xception_71* based model performs the best mIOU performance but a rather slow computational speed. The *MobileNet_v2* based model has a moderate mIOU, the smallest file size and the fastest speed. Note the *MobileNet_v2* based model does not employ ASPP and decoder modules for fast computation. Considering the balance between computational speed and accuracy, we choose the *MobileNet_v2* based model to carry out the 2D semantic segmentation in our approach. Table 4 shows the performance of the *MobileNet_v2* based model on the VAL/TEST split of the KITTI dataset.

Table 4. Results of our selected model on the val/test of the KITTI datasets.

Method	Road	Sidewalk	Building	Wall	Fence	Pole	Traffic Light	Traffic Sign	Vegetation	Terrain	Sky	Person	Rider	Car	Truck	Bus	Train	Motorcycle	Bicycle	IoU
VAL	95.7	73.9	87.1	38.1	44.2	42.7	48.6	60.3	89.1	52.3	90.1	70.1	36.5	89.1	44.6	62.2	37.4	36.1	67.7	60.3
TEST	96.1	73.7	86.2	37.9	41.4	40.1	50.3	58.3	90.2	66.8	91.3	72.4	40.3	91.8	33.7	46.4	37.1	46.0	62.4	60.9

We also make the test regarding to the effect of pre-training on the Cityscapes dataset. In Table 5, the salience has been illustrated on training the *Xception_65* and *MobileNet_v2* models. The Cityscapes pre-trained models could greatly improve the performance of 2D semantic segmentation on the KITTI dataset.

Note that towards the 3D semantic mapping, since we use a novel monocular 3D mapping different from the other related work, it is not easy to make quantitative comparison here. Kundu et al.'s work [28] proposes a joint semantic segmentation and 3D reconstruction from monocular video, but it is an offline approach with different 3D representation in the form of a 3D volumetric semantic and occupancy map.

Table 5. Performance of 2D semantic segmentation with/without the Cityscapes. Using the pre-trained Cityscapes model, the accuracy of 2D semantic segmentation could be greatly improved on the KITTI semantic data.

Encoder	mIOU[0.5:0.25:1.75]	WITH Cityscapes
ResNet_101	52.5	
ResNet_101	57.1	√
Xception_65	56.0	
Xception_65	64.8	√
MobileNet_v2	51.8	
MobileNet_v2	60.7	√

Runtime and Storage: As shown in Table 6, the SLAM module in our system runs about 40 ms on average to process each frame, extract the keyframes and update the map. The semantic segmentation process requires about 100 ms to infer 2D semantic information parallel upon the keyframes, and the incremental fusion process needs 50 ms on average. In the experiments, the SLAM process at least selects a keyframe every 4 or 5 frames. It keeps enough timing for the 2D semantic segmentation and the incremental fusion during the 3D semantic mapping. Thus, our approach could run in real-time. Moreover, considering the speed of moving platform, in case of the speed of 60 km/h, the semantic segmentation process on selected keyframes corresponds to a distance about 2 meters, which is not too sparse for an urban scene.

Table 6. Timing results. The table lists the operation time for different components of our system. Times of three core components are averaged over all sequences and the Conditional Random Field (CRF) timings depends on the iterations and the point cloud sizes.

Component	Average Consumed Time
Semantic segmentation	$\simeq$100 ms
SLAM	$\simeq$40 ms
Incremental fusion	$\simeq$50 ms
3D CRF 1 Iter.	800–2000 ms
3D CRF 2 Iter.	1200–2400 ms
3D CRF 3 Iter.	1500–3000 ms
3D CRF 4+ Iter.	>2000 ms
Kundu et al. [28]	$\simeq$20 min/800 frames
Our 'baseline'	$\simeq$200 s/800 frames
Our	$\simeq$80 s/800 frames

The lower part of this table shows the ranges of the CRF timing with different configurations due to the different size of point clouds when testing various sequences in the experiments. The CRF update runs offline due to slow inference speed on the CPU. Thus, it is only applied once at the end of the sequence. Optimized GPU implementation could be studied in future to realize the online CRF update.

Taking the *odometry_03* sequence as example, our approach acquires 114 keyframes with 28 million 3D points from the sequence of 801 frames, which utilizes only about 1/7 frames for mapping. Note that smaller values of the parameters *KFDistWeight* and *KFUsageWeight* could give more constraints between keyframes so that to achieve more accurate mapping. But it has a rather limited influence on the number of keyframes, the number of 3D points, and the size of storage. Compared to the system [28], it costs around 20 minutes on a standard desktop machine for 800 images long sequence involving about 20 million 3D points. Our system is a fast monocular vision mapping, even though it uses an offline CRF optimization.

Lastly, we test the system with semantic segmentation on all frames as the 'baseline' pipeline. We find that for one thing it is hard to say the accuracy of 3D semantic mapping is improved. Because for the LSD-SLAM, the current keyframe is refined with its following frames until new keyframe is selected. The depth map of the current keyframe is more accurate than the depth measure on each frame. If we use the 'baseline' pipeline, we need the depth information on each frame; even though more semantic information is used in the incremental fusion, the noisy depth would lead to inaccurate semantic map. Besides, since the visual SLAM process runs faster than the semantic segmentation at present, the untreated frames would quickly exceed the buffer limit, leading to new frames blocked. The entire system cannot run in real-time and it would not simultaneously generate the semantic map.

6. Conclusions

We have presented a fast monocular 3D semantic mapping system which runs on a CPU coupled with a GPU. An incremental fusion method is introduced to combine 2D semantic segmentation and 3D reconstruction online. We exploit a state-of-the-art deep CNN to accomplish scene parsing in the road context. Direct monocular visual SLAM provides a quick 3D mapping based on selected keyframes and corresponding depth estimation. Since the semantic segmentation only runs and propagates on the keyframes, this reduces the computational cost and improves the accuracy of semantic mapping. The offline regularization with a CRF model can enhance the mapping further.

Since the original LSD-SLAM is hard to handle in the case of sharp turns which are frequent in ordinal driving, our system is not stable in such conditions. In addition, semi-dense 3D reconstruction should be replaced by a dense model. In future work, we plan to introduce several state-of-the-art SLAM methods to improve the initialization and resistance to serious movements. Research on how labeling boosts 3D reconstruction of SLAM would be an interesting direction. The optimization of the regularization module would be another effective effort on the wide-range mapping.

Author Contributions: Conceptualization, X.L.; methodology, X.L.; software, X.L. and H.A.; validation, X.L., D.W. and H.A.; formal analysis, X.L.; investigation, X.L. and H.A.; resources, X.L.; data curation, X.L.; writing—original draft preparation, X.L.; writing—review and editing, R.B. and D.G.; visualization, X.L.; supervision, X.L.; project administration, X.L. and D.W.; funding acquisition, X.L. and D.W.

Funding: This research was funded by the Natural Science Foundation of Jiangsu Province grant number No. BK20160700 and No. BK20170681.

Conflicts of Interest: The authors declare no conflict of interest. The funders had no role in the design of the study; in the collection, analyses, or interpretation of data; in the writing of the manuscript, or in the decision to publish the results.

References

1. Chow, J.C.; Lichti, D.D.; Hol, J.D.; Bellusci, G.; Luinge, H. Imu and multiple RGB-D camera fusion for assisting indoor stop-and-go 3D terrestrial laser scanning. *Robotics* **2014**, *3*, 247–280. [CrossRef]
2. Alzugaray, I.; Sanfeliu, A. Learning the hidden human knowledge of UAV pilots when navigating in a cluttered environment for improving path planning. In Proceedings of the 2016 IEEE/RSJ International Conference on Intelligent Robots and Systems, Deajeon, Korea, 9–14 October 2016; pp. 1589–1594.
3. Alzugaray, I.; Teixeira, L.; Chli, M. Short-term UAV path-planning with monocular-inertial SLAM in the loop. In Proceedings of the 2017 IEEE International Conference on Robotics and Automation, Singapore, 29 May–3 June 2017; pp. 2739–2746.
4. Zhao, H.; Shi, J.; Qi, X.; Wang, X.; Jia, J. Pyramid scene parsing network. In Proceedings of the 2017 IEEE Conference on Computer Vision and Pattern Recognition, Honolulu, HI, USA, 21–26 July 2017; pp. 2881–2890.
5. Chen, L.C.; Papandreou, G.; Kokkinos, I.; Murphy, K.; Yuille, A.L. Deeplab: Semantic image segmentation with deep convolutional nets, atrous convolution, and fully connected crfs. *IEEE Trans. Pattern Anal. Mach. Intell.* **2018**, *40*, 834–848. [CrossRef] [PubMed]

6. Zhao, H.; Qi, X.; Shen, X.; Shi, J.; Jia, J. ICNet for Real-Time Semantic Segmentation on High-Resolution Images. In Proceedings of the 2018 European Conference on Computer Vision, Munich, Germany, 8–14 September 2018; pp. 418–434.

7. Chen, L.C.; Zhu, Y.; Papandreou, G.; Schroff, F.; Adam, H. Encoder-decoder with atrous separable convolution for semantic image segmentation. In Proceedings of the 15th European Conference on Computer Vision, Munich, Germany, 8–14 September 2018; pp. 833–851.

8. Wolf, D.; Prankl, J.; Vincze, M. Fast semantic segmentation of 3D point clouds using a dense CRF with learned parameters. In Proceedings of the 2015 IEEE International Conference on Robotics and Automation, Seattle, WA, USA, 26–30 May 2015; pp. 4867–4873.

9. Hermans, A.; Floros, G.; Leibe, B. Dense 3d semantic mapping of indoor scenes from rgb-d images. In Proceedings of the 2014 IEEE International Conference on Robotics and Automation, Hong Kong, China, 31 May–7 June 2014; pp. 2631–2638.

10. McCormac, J.; Handa, A.; Davison, A.; Leutenegger, S. SemanticFusion: Dense 3D semantic mapping with convolutional neural networks. In Proceedings of the 2017 IEEE International Conference on Robotics and Automation, Singapore, 29 May–3 June 2017; pp. 4628–4635.

11. Cordts, M.; Omran, M.; Ramos, S.; Rehfeld, T.; Enzweiler, M.; Benenson, R.; Franke, U.; Roth, S.; Schiele, B. The cityscapes dataset for semantic urban scene understanding. In Proceedings of the 2016 IEEE Conference on Computer Vision and Pattern Recognition, Las Vegas, NV, USA, 26 June–1 July 2016; pp. 3213–3223.

12. Geiger, A.; Lenz, P.; Urtasun, R. Are we ready for autonomous driving? the kitti vision benchmark suite. In Proceedings of the 2012 IEEE Conference on Computer Vision and Pattern Recognition, Providence, RI, USA, 18–20 June 2012; pp. 3354–3361.

13. LI, X.; Ao, H.; Belaroussi, R.; Gruyer, D. Fast semi-dense 3D semantic mapping with monocular visual SLAM. In Proceedings of the 2017 IEEE 20th International Conference on Intelligent Transportation Systems, Yokohama, Japan, 16–19 October 2017; pp. 385–390.

14. Whelan, T.; Salas-Moreno, R.F.; Glocker, B.; Davison, A.J.; Leutenegger, S. ElasticFusion: Real-time dense SLAM and light source estimation. *Int. J. Rob. Res.* **2016**, *35*, 1697–1716. [CrossRef]

15. Noh, H.; Hong, S.; Han, B. Learning Deconvolution Network for Semantic Segmentation. In Proceedings of the 2015 IEEE International Conference on Computer Vision, Santiago, Chile, 11–18 December 2015; pp. 1520–1528.

16. Salas-Moreno, R.F.; Newcombe, R.A.; Strasdat, H.; Kelly, P.H.; Davison, A.J. Slam++: Simultaneous localisation and mapping at the level of objects. In Proceedings of the 2013 IEEE Conference on Computer Vision and Pattern Recognition, Portland, OR, USA, 23–28 June 2013; pp. 1352–1359.

17. Davison, A.J.; Reid, I.D.; Molton, N.D.; Stasse, O. MonoSLAM: Real-time single camera SLAM. *IEEE Trans. Pattern Anal. Mach. Intell.* **2007**, *29*, 1052–1067. [CrossRef] [PubMed]

18. Mur-Artal, R.; Montiel, J.M.M.; Tardos, J.D. ORB-SLAM: A versatile and accurate monocular SLAM system. *IEEE Trans. Robot.* **2015**, *31*, 1147–1163. [CrossRef]

19. Mur-Artal, R.; Tardós, J.D. Orb-slam2: An open-source slam system for monocular, stereo, and rgb-d cameras. *IEEE Trans. Robot.* **2017**, *33*, 1255–1262. [CrossRef]

20. Kerl, C.; Sturm, J.; Cremers, D. Dense visual SLAM for RGB-D cameras. In Proceedings of the 2013 IEEE/RSJ International Conference on Intelligent Robots and Systems, Tokyo, Japan, 3–7 November 2013; pp. 2100–2106.

21. Engel, J.; Schöps, T.; Cremers, D. LSD-SLAM: Large-scale direct monocular SLAM. In Proceedings of the 2014 European Conference on Computer Vision, Zurich, Switzerland, 6–12 September 2014; pp. 834–849.

22. Forster, C.; Zhang, Z.; Gassner, M.; Werlberger, M.; Scaramuzza, D. Svo: Semidirect visual odometry for monocular and multicamera systems. *IEEE Trans. Robot.* **2017**, *33*, 249–265. [CrossRef]

23. Long, J.; Shelhamer, E.; Darrell, T. Fully convolutional networks for semantic segmentation. In Proceedings of the 2015 IEEE Conference on Computer Vision and Pattern Recognition, Boston, MA, USA, 7–12 June 2015; pp. 3431–3440.

24. Badrinarayanan, V.; Kendall, A.; Cipolla, R. SegNet: A Deep Convolutional Encoder-Decoder Architecture for Image Segmentation. *IEEE Trans. Pattern Anal. Mach. Intell.* **2017**, *39*, 2481–2495. [CrossRef] [PubMed]

25. Howard, A.G.; Zhu, M.; Chen, B.; Kalenichenko, D.; Wang, W.; Weyand, T.; Andreetto, M.; Adam, H. Mobilenets: Efficient convolutional neural networks for mobile vision applications. *arXiv* **2017**, arXiv:1704.04861.

26. Sandler, M.; Howard, A.; Zhu, M.; Zhmoginov, A.; Chen, L.C. MobileNetV2: Inverted Residuals and Linear Bottlenecks. In Proceedings of the 2018 IEEE Conference on Computer Vision and Pattern Recognition, Salt Lake City, UT, USA, 18–22 June 2018; pp. 4510–4520.

27. Valentin, J.P.; Sengupta, S.; Warrell, J.; Shahrokni, A.; Torr, P.H. Mesh based semantic modelling for indoor and outdoor scenes. In Proceedings of the 2013 IEEE Conference on Computer Vision and Pattern Recognition, Portland, OR, USA, 23–28 June 2013; pp. 2067–2074.

28. Kundu, A.; Li, Y.; Dellaert, F.; Li, F.; Rehg, J.M. Joint semantic segmentation and 3d reconstruction from monocular video. In Proceedings of the 2014 European Conference on Computer Vision, Zurich, Switzerland, 6–12 September 2014; pp. 703–718.

29. Sengupta, S.; Sturgess, P. Semantic octree: Unifying recognition, reconstruction and representation via an octree constrained higher order MRF. In Proceedings of the 2015 IEEE International Conference on Robotics and Automation, Seattle, WA, USA, 26–30 May 2015; pp. 1874–1879.

30. Vineet, V.; Miksik, O.; Lidegaard, M.; Nießner, M.; Golodetz, S.; Prisacariu, V.A.; Kähler, O.; Murray, D.W.; Izadi, S.; Pérez, P. Incremental dense semantic stereo fusion for large-scale semantic scene reconstruction. In Proceedings of the 2015 IEEE International Conference on Robotics and Automation, Seattle, WA, USA, 26–30 May 2015; pp. 75–82.

31. Kochanov, D.; Ošep, A.; Stückler, J.; Leibe, B. Scene flow propagation for semantic mapping and object discovery in dynamic street scenes. In Proceedings of the 2016 IEEE/RSJ International Conference on Intelligent Robots and Systems, Deajeon, Korea, 9–14 October 2016; pp. 1785–1792.

32. Landrieu, L.; Raguet, H.; Vallet, B.; Mallet, C.; Weinmann, M. A structured regularization framework for spatially smoothing semantic labelings of 3D point clouds. *ISPRS J. Photogramm. Remote Sens.* **2017**, *132*, 102–118. [CrossRef]

33. Jadidi, M.G.; Gan, L.; Parkison, S.A.; Li, J.; Eustice, R.M. Gaussian processes semantic map representation. *arXiv* **2017**, arXiv:1707.01532.

34. Gan, L.; Jadidi, M.G.; Parkison, S.A.; Eustice, R.M. Sparse Bayesian Inference for Dense Semantic Mapping. *arXiv* **2017**, arXiv:1709.07973.

35. Sengupta, S.; Greveson, E.; Shahrokni, A.; Torr, P.H. Urban 3d semantic modelling using stereo vision. In Proceedings of the 2013 IEEE International Conference on Robotics and Automation, Karlsruhe, Germany, 6–10 May 2013; pp. 580–585.

36. Martinovic, A.; Knopp, J.; Riemenschneider, H.; Van Gool, L. 3d all the way: Semantic segmentation of urban scenes from start to end in 3d. In Proceedings of the 2015 IEEE Conference on Computer Vision and Pattern Recognition, Boston, MA, USA, 7–12 June 2015; pp. 4456–4465.

37. Hu, H.; Munoz, D.; Bagnell, J.A.; Hebert, M. Efficient 3-d scene analysis from streaming data. In Proceedings of the 2013 IEEE International Conference on Robotics and Automation, Karlsruhe, Germany, 6–10 May 2013; pp. 2297–2304.

38. Civera, J.; Davison, A.J; Montiel, J.M.M. Inverse depth parametrization for monocular SLAM. *IEEE Trans. Robot.* **2008**, *24*, 932–945. [CrossRef]

39. Krähenbühl, P.; Koltun, V. Efficient inference in fully connected crfs with gaussian edge potentials. In Proceedings of the 2011 Conference in Neural Information Processing Systems, Granada, Spain, 12–17 December 2011; pp. 109–117.

40. Russell, C.; Kohli, P.; Torr, P.H. Associative hierarchical crfs for object class image segmentation. In Proceedings of the 2009 IEEE International Conference on Computer Vision, Kyoto, Japan, 29 September–2 October 2009; pp. 739–746.

41. He, K.; Zhang, X.; Ren, S.; Sun, J. Deep residual learning for image recognition. In Proceedings of the 2016 IEEE Conference on Computer Vision and Pattern Recognition, Las Vegas, NV, USA, 26 June–1 July 2016; pp. 770–778.

42. Chollet, F. Xception: Deep Learning With Depthwise Separable Convolutions. In Proceedings of the 2017 IEEE Conference on Computer Vision and Pattern Recognition, Honolulu, HI, USA, 21–26 July 2017; pp. 1251–1258.

43. Russakovsky, O.; Deng, J.; Su, H.; Krause, J.; Satheesh, S.; Ma, S.; Huang, Z.; Karpathy, A.; Khosla, A.; Bernstein, M. Imagenet large scale visual recognition challenge. *Int. J. Comput. Vis.* **2015**, *115*, 211–252. [CrossRef]
44. Lin, T.Y.; Maire, M.; Belongie, S.; Hays, J.; Perona, P.; Ramanan, D.; Dollár, P.; Zitnick, C.L. Microsoft coco: Common objects in context. In Proceedings of the 2014 European Conference on Computer Vision, Zurich, Switzerland, 6–12 September 2014; pp. 740–755.

 *applied sciences*

Article

Automated Classification Analysis of Geological Structures Based on Images Data and Deep Learning Model

Ye Zhang [1], Gang Wang [2], Mingchao Li [1,*] and Shuai Han [1]

[1] State Key Laboratory of Hydraulic Engineering Simulation and Safety, Tianjin University, Tianjin 300354, China; jgzhangye@tju.edu.cn (Y.Z.); hs2015205039@tju.edu.cn (S.H.)
[2] Chengdu Engineering Corporation Limited, PowerChina, Chengdu 610072, China; wgcd126a@126.com
* Correspondence: lmc@tju.edu.cn

Received: 1 November 2018; Accepted: 29 November 2018; Published: 4 December 2018

Featured Application: This work aims to build a robust model with a comparison of machine learning, convolutional neural network and transfer learning. The model can be combined with an unmanned aerial vehicle (UAV) to act as a tool in geological surveys in the future.

Abstract: It is meaningful to study the geological structures exposed on the Earth's surface, which is paramount to engineering design and construction. In this research, we used 2206 images with 12 labels to identify geological structures based on the Inception-v3 model. Grayscale and color images were adopted in the model. A convolutional neural network (CNN) model was also built in this research. Meanwhile, K nearest neighbors (KNN), artificial neural network (ANN) and extreme gradient boosting (XGBoost) were applied in geological structures classification based on features extracted by the Open Source Computer Vision Library (OpenCV). Finally, the performances of the five methods were compared and the results indicated that KNN, ANN, and XGBoost had a poor performance, with the accuracy of less than 40.0%. CNN was overfitting. The model trained using transfer learning had a significant effect on a small dataset of geological structure images; and the top-1 and top-3 accuracy of the model reached 83.3% and 90.0%, respectively. This shows that texture is the key feature in this research. Transfer learning based on a deep learning model can extract features of small geological structure data effectively, and it is robust in geological structure image classification.

Keywords: OpenCV; machine learning; transfer learning; Inception-v3; geological structure images; convolutional neural networks

---

1. Introduction

The primary objective of a geological survey is to identify geological structures in the field and, this is also important for project schedule management and safety guarantees. In construction, engineers search for the exposure of geological structures to the Earth's surface in field surveys, then explore geological structures that partly extend below the Earth's surface with boreholes, adits, etc. Some geological structures should be given special attention because of their poor properties. Anticline and ptygmatic folds weather easily; xenoliths, boudins, and dikes usually have low strength at the contact surface because they contain rocks with different properties; ripple marks, mudcracks, and concretion always indicate there is an ancient river course; faults and scratches mean broken structures in engineering; basalt columns have a low strength because of the columnar joints; a gneissose structure also has a low shear strength at the direction of schistosity. The geological structures have a significant influence on project site selection, general layout and schedule management, which is also crucial to construction quality. The identification of geological structures can help engineers make a better choice in construction. On the other hand, geological structures, such as faults [1] and folds [2], are connected to hazards. Vasu and Lee [3] applied an extreme learning machine to build the landslide susceptibility model with 13 selected features (including geological structure features). The performance of prediction was better, with the accuracy of 89.45%. Dickson and Perry [4] explored three machine learning methods, namely maximum entropy models, classification and regression trees and boosted regression trees, to make identification of coastal cliff landslide control based on geological structure parameters. The final result showed a high performance with 85% accuracy. The researches proved geological structures were connected to geologic hazard prediction and prevention. However, the machine learning methods are applied in structured data. For unstructured data, such as image, audio, text, we need to extract the features of the unstructured data and input them to train machine learning models. In the process of data type transformation, other algorithms are going to be selected.

3D visualization techniques and data interpolation are also used in geological structure detection. Zhong et al. [5] made a 3D model of the complex geological structure based on borehole data. Discrete fractures in rocks were also computed and estimated by 3D model construction [6,7]. Spatial relationships of geological data were easy to understand in a 3D model. As a result, a 3D model was built to explore geological conditions under the Earth's surface. With limited geological data because of cost controls, such as several boreholes, we were able to build the whole plane with spatial interpolation methods. It is a significant and easy way to show the distribution of the discrete points. However, it is mostly used for underground data analysis. The geological situation on the Earth's surface is often explored by geological engineers in the geological survey. It requires many computation resources because of the rendering in the 3D visualization model. It is a time-consuming method with low accuracy in some cases.

Image powered methods have become increasingly popular recently. These provide a novel method in geological structure identification. Vasuki et al. [8] captured rock surface images with unmanned aerial vehicles (UAVs) and detected rock features from the photos. According to the features detected on UAVs images, 3D models were built to show folds, fractures, cover rocks and other geological structures [9,10]. Furthermore, the automatic classification of geological images was also studied. The geological image, as a kind of unstructured data, contains much information including the object features. Młynarczuk et al. [11] applied four methods to make a classification of microscopic rock images automatically. The result of the nearest neighbor (NN) analysis showed high recognition level with 99.8%. Li et al. [12] also proposed a transfer learning method for sandstone identification automatically based on microscopic images. Shu et al. [13] used an unsupervised machine learning method to classify rock texture images. The experimental results indicated the outstanding performance of self-taught learning. Geological structures identification has many similarities with rock classification, which indicates what features we should extract from geological structures images. The color and texture are both critical in rock images to both micro and regular images. In some cases, the rock mineral was able to be classified just by color. While the geological structures data has unique characters. The texture is addressed more by the geomorphometric analysis [14].

Deep learning is prominent in image processing. It was proposed by Hinton [15] and was further developed recently [16]. Because of the positive performance, it was used to analyze unstructured data in many areas, such as image classification [17] and semantic analysis [18]. In medicine, deep learning is also popular [19,20]. However, Kim et al. [21] thought the input was massive in deep learning. Deep learning was also applied in remote-sensing image scene classification [22] and terrain features recognition [23]. It was able to extract features from unstructured data and make a classification with high accuracy. The convolutional neural network (CNN) is an essential method in deep learning. Scholars were able to build different CNN models by adding different kernels, pooling layers, and fully connected layers. Palafox et al. [24] used different CNN architectures in detecting landforms on Mars which proved convenient to extract features from images. Nogueira et al. [25] also used convolutional neural networks to extract features of images then built a support vector machine (SVM) linear model based on the features. Xu et al. [26] also used transfer learning to predict geo-tagged field photos based on convolutional neural networks. All the results showed good performance of the convolutional neural network in extracting images features. While we should consider that the CNN model depends on a large dataset to avoid overfitting.

In this research, we established a transfer learning model based on Inception-v3 for geological structures with 12 labels. The test result showed a high accuracy of the model. Then we made a comparison between the identification model and the other four models, namely K nearest neighbors (KNN), artificial neural network (ANN), extreme gradient boosting (XGBoost) and CNN. The result showed the transfer learning model had a high accuracy on a small dataset. The machine learning method's accuracy was poor because it is hard to extract accurate features of images from a pixel vector or histogram. CNN was overfitting strongly. Transfer learning based on deep learning model was an effective method for geological structure images classification. Moreover, Wu et al. [27] applied the UAV and recognition model to detect rail surface defects. The retrained model in this research can also be combined with a UAV, which can be an assistant tool in the geological survey in further study.

2. Data Collection

2.1. Data Information

In this research, we collected 2206 geological structures images with 12 labels, including anticline, ripple marks, xenolith, scratch, ptygmatic folds, fault, concretion, mudcracks, gneissose, boudin, basalt columns, and dike. The dataset was collected from the geological survey and the internet [28]. In data collection, we tried to make each category cover images with different scales and sizes as many as possible. The resolution of the image is not limited. All the images are going to be processed at the same size before training. The numbers of images in each label are listed in Table 1. Figure 1 shows samples of the data.

Table 1. Information of geological structures images dataset.

Geological Structure	No.	Geological Structure	No.	Geological Structure	No.
Anticline	179	Ptygmatic folds	162	Gneissose structure	206
Ripple marks	221	Fault	127	Boudin	190
Xenolith	208	Concretion	181	Basalt columns	196
Scratch	164	Mudcracks	181	Dike	191

(a)   (b)   (c)
(d)   (e)   (f)

Figure 1. *Cont.*

Figure 1. Samples of geological structure images: (**a**) Anticline; (**b**) Ripple marks; (**c**) Xenolith; (**d**) Scratch; (**e**) Ptygmatic folds; (**f**) Fault; (**g**) Concretion; (**h**) Mudcracks; (**i**) Gneissose structure; (**j**) Boudin; (**k**) Basalt columns; (**l**) Dike.

2.2. Data Preprocessing

It is necessary to make data preprocessing in images classification. Some feature pre-processing methods [29,30] were adopted to improve the performance of the model. We applied two direct and straightforward preprocessing methods to extract features in images as the input of KNN, ANN, and XGBoost. The first method is to convert pixels in each image into a row vector directly; the second method is to build the histogram of pixels based on their statistical characteristics, as shown in Figure 2. In Figure 2b,d, x-axis means the range of pixels, which is [0, 225]; y-axis means the numbers of pixels at each level. The color images have three channels, namely red, green, and blue; the grayscale images just have one channel-grayscale. In Figure 2b, the red, green, and blue lines refer to the numbers of the pixels at R, G, and B levels. In Figure 2d, The red line refers to the numbers of the pixels at the grayscale level. In Figure 2a,c, the x- and y-axis measure the size of the photo.

The features extracted from color, and grayscale images were both set as the input in KNN, ANN, and XGBoost, which was able to show the influence of image color. In CNN and transfer learning based on Inception-v3, the convolutional neural network was applied to extract the features of the images.

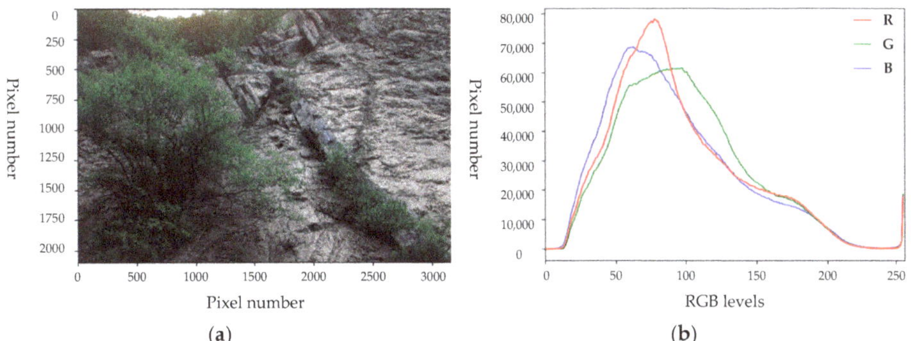

Figure 2. *Cont.*

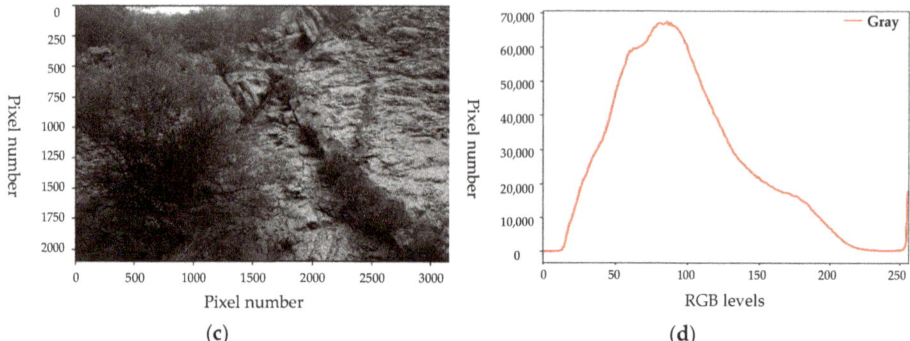

Figure 2. The histogram features extracted from color and grayscale geological structure image: (a) Color image; (b) Color image histogram; (c) Grayscale image; (d) Grayscale image histogram.

The raw data is not enough for a training model using CNN and transfer learning. As a consequence, some data augmentation methods, such as channel shift, shear, flip from left to right and flip from top to bottom, were adopted to raw data, as shown in Figure 3. A channel shift means to change the general color of the whole image; shear means to keep the horizontal axis (or vertical axis) stable and translate the other axis at a ratio. The translation distance is proportional to the distance to the horizontal axis.

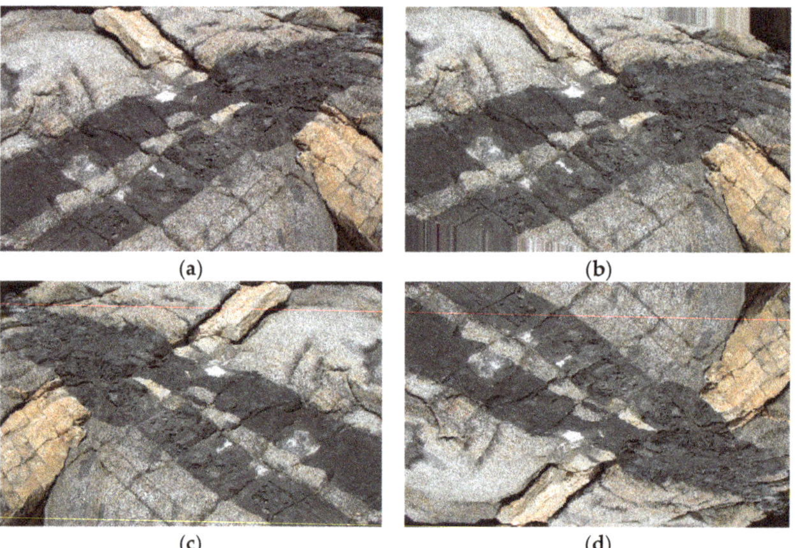

Figure 3. Data augmentation: (a) Channel shift; (b) Shear; (c) Flip from left to right; (d) Flip from top to bottom.

3. Key Techniques and Methods

3.1. Machine Learning

KNN is a lazy algorithm with no parameters, and ANN is a kind of supervised learning method. The two methods are used widely in prediction. XGBoost [31] is an improved gradient boosting decision tree (GBDT) method. In GBDT, the weak learners are combined to be a strong learner, as in the following equation:

$$f_i^{(t)} = \sum_{k=1}^{t} f_k(x_i) = f_i^{(t-1)} + f_t(x_i), \tag{1}$$

where $f_t(x_i)$ is the weak learner, namely a single decision tree; $f_t(x_i)$ is the sum of the weak learners. In each iteration, the new decision tree was added to the model. XGBoost improves the loss function, and also regularizes objective function, as shown in the following equation:

$$\begin{cases} L = \sum_{k=1}^{n} l(\bar{y}_i, y_i) + \sum_{k=1}^{t} \Omega(f_i) \\ \Omega(f) = \gamma T + \frac{1}{2}\lambda\|\omega\|^2 \end{cases}, \tag{2}$$

where l is the loss function, which is used to measure the difference between prediction $\hat{y}_i$ and target y_i. Ω is used to control the complexity of the model. ω is the score of the leaves, λ is the parameter for regularization, which is used to evaluate the node split.

3.2. Convolutional Neural Network (CNN)

CNN is a kind of feedforward neural network. The neurons in CNN can respond to the specific region in an image to extract features, which makes it outstanding in processing large unstructured data. A convolutional neural network includes convolutional layer, pooling layer, and fully connected layer.

There are three critical concepts in CNN, namely receptive field, parameter sharing, and pooling layers. Altenberger and Lenz [32] explained them in detail. The receptive field is a square region. It is a local subset of the neurons that the kernel connected to. The size of this square is the receptive field. Neurons of the same kernel should get the same pattern of the image regardless of their positions. As a result, the parameters should be shared by all the neurons of the same kernel. This concept is called parameter sharing. Pooling layers are also connected to a square region of the previous layer. However, polling layers are different from the convolutional layers. They are not determined by the weights or bias in the learning process, and the result just depends on the input. The max pooling is the common type in CNN. The maximum value that the neurons return is taken as the feature of the image. The average pooling can be interpreted in a similar way. Pooling reduces the complexity and dimensions of the feature map and improves the result to lead to less overfitting. At the same time, the features can keep translation invariance after pooling, which means if there are some translations, such as rotation, scale, distortion, in images, the pooling features are also effective.

As shown in Figure 4, the sizes of the image and convolutional layers are 5×5 and 3×3, respectively. There are nine parameters in the convolutional layer, namely the weight matrix. The nine parameters mean nine neurons. According to the sizes of image and kernel, the output is going to be a 3×3 matrix, which is called the feature map. In the first step, the neurons were connected to the receptive field on the image; then it slides to the next region by one stride in the second step, as shown in Figure 4b. The computation was processed in each neuro as follows.

$$f(x) = act\left(\sum_{i,j}^{n} \theta_{(n-i)(n-j)} x_{ij} + b\right), \tag{3}$$

where $f(x)$ was the output, *act* is the activation function, θ is the weight matrix, x_{ij} is the input, b is the bias. The softmax activation function is selected in this research.

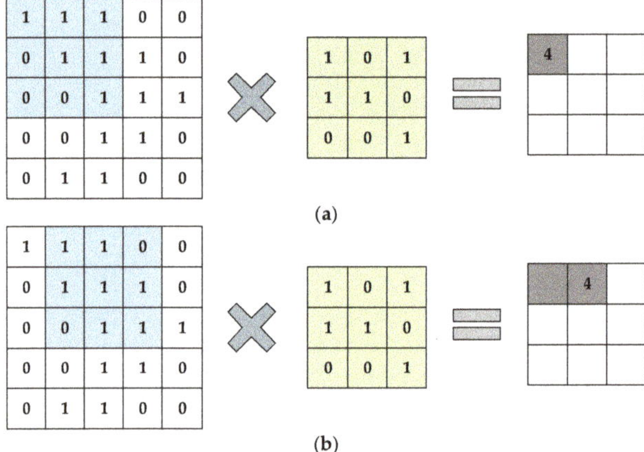

Figure 4. The process of convolutional neural network computation: (**a**) Computation in the first step; (**b**) Computation in the second step.

3.3. Transfer Learning

Even if there is an established model in a similar area, the model is going to be established from scratch using a machine learning method. It costs much manpower and time to solve problems individually in a similar domain. Considering the similarity between different tasks, we are able to build the model based on the knowledge obtained using transfer learning method. The knowledge obtained can be used again in a related domain with small change. If the gained knowledge can work in most cases or the data is hard to collect in the new task, we can make the most of the gained knowledge with transfer learning to build the new model. It benefits much in reducing dependency on big data and establishing a new model using a transfer learning method.

Furthermore, it is necessary to have a high-performance computer and time to train big data. However, it can reduce time cost and dependency on big data using transfer learning based on the pre-trained model [26,33,34]. We can apply a pre-trained model, which contains parameters trained by another big dataset, in training a new model in a similar domain. The kernels in the convolutional neural network can extract features of images automatically and effectively.

In this research, we adopted Inception-v3 [35] as the pre-trained model. The dataset which is used to train Inception-v3 contains 1.2 million images with more than 1000 labels. In the result of recognition, the top-5 accuracy in Inception-v3 is 96.5%, which is better than humans, with an accuracy of 94.9%. The convolutional layers and pooling layers in Inception-v3 can extract features from images as 2048 dimensional vectors. We removed the softmax layer in Inception-v3 and retained the new layer in our own domain. All the convolutional layers and pooling layers in Inception-v3 were used in extracting features from images, a process shown in Figure 5.

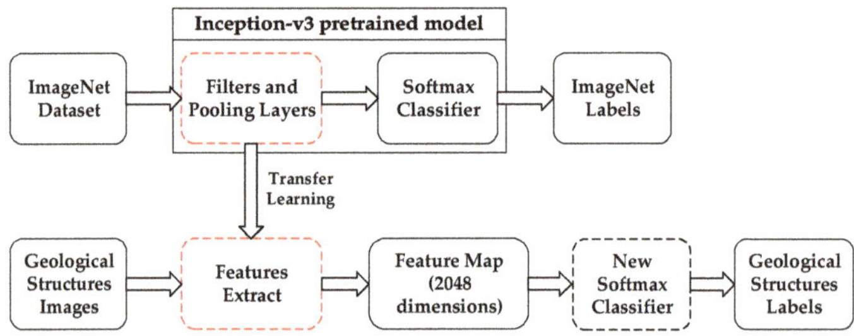

Figure 5. The process of retraining Inception-v3.

4. Model Establishment

4.1. Parameters Set

In the process of Inception-v3 retraining, iteration was set as 20,000; learning rate was set 0.01. In each iteration, 100 images were selected randomly to train the model, namely the training batch size equals 100. Batch size is limited by computer performance; 10% of the data was set as the test dataset. After 10 iterations, the model was evaluated. All the images were going to be used in the training process. Training accuracy, validation accuracy, and cross-entropy were used to evaluate the model in the training process. Training accuracy is gained by testing the trained dataset with the model; validation accuracy is gained by testing the validation dataset with the model; cross-entropy shows the model performance in identification. The smaller cross-entropy indicates a better performing model. In each training step, the prediction value and target value were measured to update the weight matrix. The geological structures images were cut as the same size before training. As a consequence, there were no strict limitations on the size and resolution of the images. The color images and grayscale images were both used in training, which was able to show the influence of color on the model.

In the training of CNN, we set two, three, and four convolutional layers and two fully connected layers to establish the model. The convolutional layers were set as 5×5 and 3×3, respectively. There were 64 neurons in each convolutional layer. While there were 128 neurons in fully connected layers. The learning rate was set as 10^{-4}. Batch size was set as 32. The data was split into training data and validation data; 80% of the data was set as training data; 10% data was set as validation data and test data.

In the model establishment of KNN, ANN, and XGBoost, we used OpenCV [36] to process the raw data into two datasets, namely color images and grayscale images, with the size of 128×128. The two datasets were used to build images features with origin pixels and pixels histogram. Then we took the pixel vectors and histogram features as the input of KNN, ANN, and XGBoost. The python package Scikit-learn [37] was used in the research to build the three models, and all the parameters of the models were set as in Table 2.

Table 2. Parameters in K nearest neighbors (KNN), artificial neural network (ANN), and extreme gradient boosting (XGBoost).

Method	Parameters	Value
KNN	n_neighbors	1
	p	2
XGBoost	colsample_bytree	0.8
	learning_rate	0.1
	eval_metric	mlogloss
	max_depth	5
	min_child_weight	1
	nthread	4
	seed	407
	subsample	0.6
	objective	multi:softprob
ANN	hidden_layer_sizes	50
	max_iter	1000
	alpha	10^{-4}
	solver	sgd
	tol	10^{-4}
	random_state	1
	learning_rate_init	0.1

4.2. Model Train and Test

Figure 6 showed the training process of the transfer learning. The model was evaluated by train accuracy, validation accuracy, and cross-entropy. In Figure 6, train accuracy, and validation accuracy both increased gradually. Then train accuracy converged to about 97.0% and validation accuracy converged to about 90.0%. Cross-validation decreased gradually and converged to about 0.2. Finally, the test accuracy based on grayscale and color images were 91.0% and 92.6%, respectively. The small difference between the two models indicates that color had little influence on the model identification for geological structures, which means textures are more important in identification.

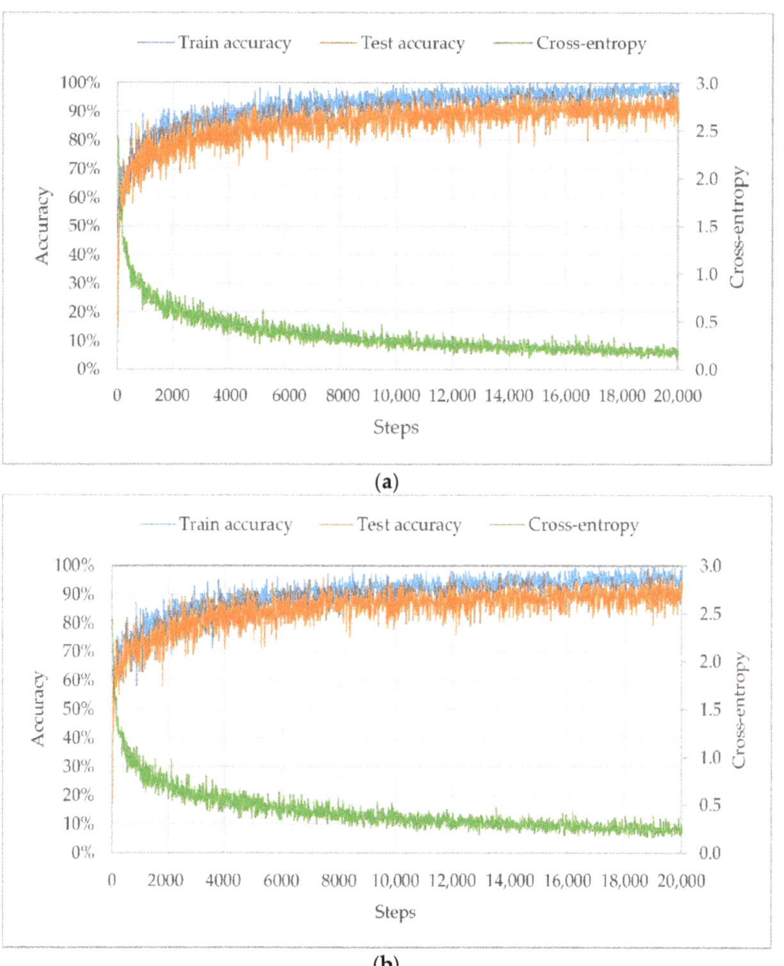

Figure 6. Train accuracy, validation accuracy and cross-entropy variation in transfer learning process: (**a**) Grayscale dataset; (**b**) Color dataset.

The patches from the same image are similar using data augmentation. At the same time, we want to apply the model to identify geological structure images from the geological survey. So we chose another 60 images which were from an engineering project to test the model accuracy. Top-1 and top-3 accuracy were used in model evaluation. Top-1 accuracy means the prediction with the highest probability matches the right label. Top-3 accuracy means any one of predictions with the three highest probability matches the right label. In the model testing, the top-1 accuracy was 83.3% and the top-3 accuracy was 90.0%. The test images in Figure 7 are the same as those in Figure 1. In Figure 7, the result showed top-3 prediction probability. In the identification of faults, the top-1 and top-3 result were wrong. The number of fault images should be increased. In the identification of boudin images, the top-1 prediction was wrong. However, we found that the probability of boudin was 15.7%, which ranked third. As a consequence, it is better to apply top-1 and top-3 accuracy to evaluate the result comprehensively in predictions.

Figure 7. Identification of geological structures images results.

The training process of CNN was shown in Figure 8. Train accuracy, validation accuracy, and cross-entropy were also used to evaluate the model. Figure 8a–c are the results of the CNN with two, three and four convolutional layers on the color dataset; Figure 8d is the result of the CNN with three convolutional layers on the grayscale dataset. The effects of the three-layer CNN was the best, and the grayscale data was also trained by the CNN architecture, as shown in Figure 8d. The train accuracy was almost 100.0%, while the validation accuracy was about 85.0%, which indicated the model was overfitting.

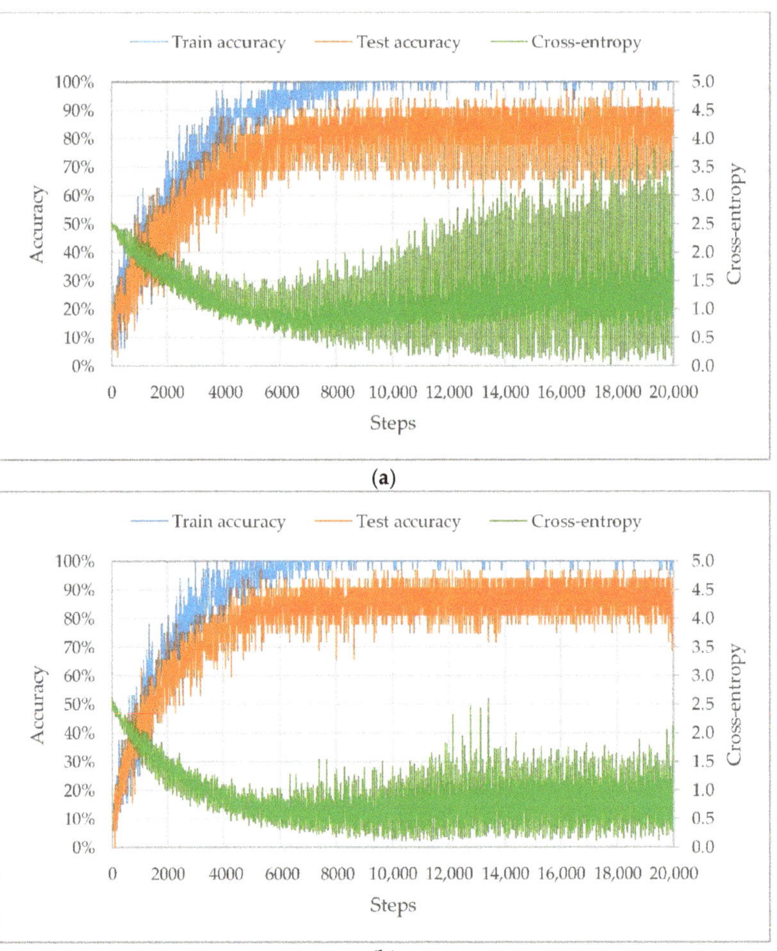

Figure 8. *Cont.*

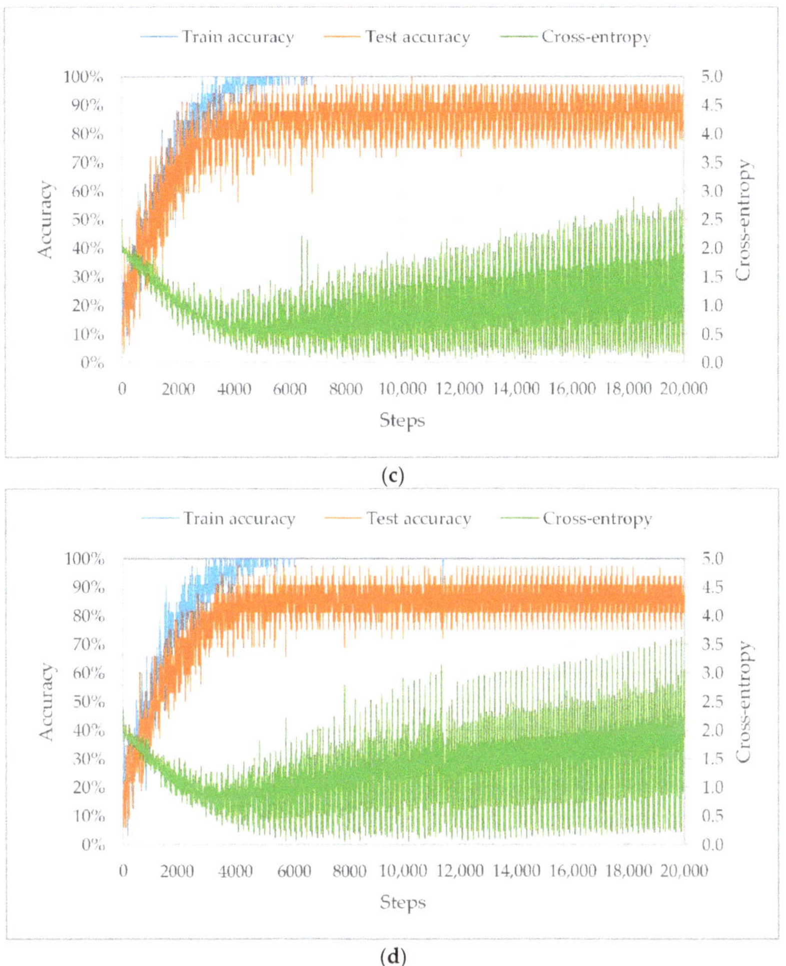

(c)

(d)

Figure 8. Train accuracy, validation accuracy and cross-entropy variation in the convolutional neural network (CNN) training process: (**a**) Two layers using color dataset; (**b**) Three layers using color dataset; (**c**) Four layers using color dataset; (**d**) Three layers using grayscale dataset.

The results of the five methods were summarized in Table 3. The accuracies of KNN, ANN, and XGBoost models were less than 40%, while the KNN and XGBoost also showed better performance in color images with histogram features. The test accuracy of CNN model based on grayscale and color images was 80.1% and 83.3%, but the model was overfitting; the test accuracy of the transfer learning model based on Inception-v3 model reached 91.0% and 92.6%, which was the best in all the methods. The result indicated the convolutional layers and pooling layers in Inception-v3 were able to extract features from geological structures images effectively. As a consequence, the transfer learning method was chosen to identify the geological structure image from engineering. The top-1 and top-3 accuracy were 83.3% and 90.0%, respectively.

Table 3. Comparison result between the five methods.

	Grayscale Image Feature		Color Image Feature	
	Pixel	Histogram	Pixel	Histogram
KNN	20.4%	19.6%	20.4%	33.4%
ANN	9.1%	19.3%	9.4%	31.4%
XGBoost	25.2%	20.7%	33.4%	34.8%
Three-layer CNN	80.1%		83.3%	
Transfer Learning	91.0%		92.6%	

4.3. Discussion

In this research, we built a geological structure identification model based on Inception-v3. In a comparison between KNN, ANN, XGBoost, and CNN, the convolutional layers and pooling layers in Inception-v3 were effective in extracting features from images of the small dataset. Actually, the small geological structures dataset we used in the research has its own characters. For example, the boudins in Figure 9a–c are very different, even though they have the same label. On the other hand, boudin and xenolith are with different labels; however, they are similar and easy to be mixed in some cases, as shown in Figure 9c,d. In Figure 7, the identification result of boudin also proved that. Meanwhile, the prediction shows the probability of xenolith is 29.4%, which is higher than that of boudin. The result shows that the features of boudin and xenolith are similar in some cases.

Figure 9. (a–c) Boudin; (d) Xenolith.

We also built a CNN model to establish identification. Because the dataset was small, we designed a simple net with two convolutional layers and two fully connected layers, while the model was still overfitting. The single architecture of CNN did not work in any cases. In ConvNets, Palafox et al. [24] also designed several CNN models with different architectures in different cases. The retrained model based on Inception-v3 can extract image features effectively with the convolutional layers and pooling layers. It is not necessary to redesign model architecture in the transfer learning model, and it worked well on the small dataset in this research. It was hard to extract real image features based on pixel vector and histogram in which translation invariance could not be kept. The background noise also interfered the features extraction significantly. As a result, the inaccurate input led to a low accuracy in KNN, ANN, XGBoost. Actually, some feature pre-processing methods can be applied in model training. The restricted Boltzmann machines (RBM) pre-processing method [29] and the mixture of handcrafted and learned features [30] can improve the model performance. Model ensemble [38] is also a robust method to enhance the feature extraction from images. Model ensemble and feature engineering are going to be applied in the model establishment in the further study.

5. Conclusions

In this research, we built the geological structure identification model based on a small dataset and the Inception-v3 model. The grayscale and color images datasets were both trained to construct different models. The two models had an accuracy of 91.0% and 92.6%, respectively. At the same time, we used 60 engineering images to test the model. The top-1 and top-3 accuracies were 83.3% and 90.0%, which showed the kernels and pooling layers in the Inception-v3 model could extract image features effectively. CNN models with different layers were built as well, while the model was overfitting in training even just with two convolutional layers and two fully connected layers. Three convolutional layers were adopted to establish the model in our study. The best parameters in CNN are hard to reach because it depends on experience. We also used OpenCV to build pixel feature based on origin pixel information and a pixel's histogram. However, the images features could not be extracted accurately in this way, which led to low accuracy in KNN, ANN, and XGBoost models. More feature engineering methods should be considered in the future. The retrained models based on Inception-v3 were trained using transfer learning method with color and grayscale datasets and had a small difference in accuracy, which indicated that color had little influence on geological structure identification.

There are also some weaknesses in the model trained by a small dataset. Test data is small and overfitting still exists in the training process. Even though data augmentation was adopted, some patterns and features were not learned by the model. In this research, we proved the feasibility of transfer learning for geological structures classification. If the model is applied in practice in the future, more data should be added.

Transfer learning based on the Inception-v3 model has strong adaptability for a small dataset. In the future, we are going to extend our dataset and combine the model with a UAV, which can be applied as a tool in geological surveys.

Author Contributions: Y.Z. wrote the code and the paper; G.W. gave professional geological guidance and the image data; M.L. provided the idea and edited the manuscript; S.H. collected and analysed the data.

Funding: This research was funded by the Tianjin Science Foundation for Distinguished Young Scientists of China, Grant no. 17JCJQJC44000 and the National Natural Science Foundation for Excellent Young Scientists of China, Grant no. 51622904.

Conflicts of Interest: The authors declare no conflict of interest.

References

1. Fleming, R.W.; Johnson, A.M. Structures associated with strike-slip faults that bound landslide elements. *Eng. Geol.* **1989**, *27*, 39–114. [CrossRef]
2. Fisher, M.A.; Normark, W.R.; Greene, H.G.; Lee, H.J.; Sliter, R.W. Geology and tsunamigenic potential of submarine landslides in Santa Barbara Channel, Southern California. *Mar. Geol.* **2005**, *224*, 1–22. [CrossRef]
3. Vasu, N.N.; Lee, S.R. A hybrid feature selection algorithm integrating an extreme learning machine for landslide susceptibility modeling of Mt. Woomyeon, South Korea. *Geomorphology* **2016**, *263*, 50–70. [CrossRef]
4. Dickson, M.E.; Perry, G.L.W. Identifying the controls on coastal cliff landslides using machine-learning approaches. *Environ. Model. Softw.* **2016**, *76*, 117–127. [CrossRef]
5. Zhong, D.H.; Li, M.C.; Song, L.G.; Wang, G. Enhanced NURBS modeling and visualization for large 3D geoengineering applications: An example from the Jinping first-level hydropower engineering project, China. *Comput. Geosci.* **2006**, *32*, 1270–1282. [CrossRef]

6. Li, M.; Zhang, Y.; Zhou, S.; Yan, F. Refined modeling and identification of complex rock blocks and block-groups based on an enhanced DFN model. *Tunn. Undergr. Space Technol.* **2017**, *62*, 23–34. [CrossRef]

7. Li, M.; Han, S.; Zhou, S.; Zhang, Y. An Improved Computing Method for 3D Mechanical Connectivity Rates Based on a Polyhedral Simulation Model of Discrete Fracture Network in Rock Masses. *Rock Mech. Rock Eng.* **2018**, *51*, 1789–1800. [CrossRef]

8. Vasuki, Y.; Holden, E.J.; Kovesi, P.; Micklethwaite, S. Semi-automatic mapping of geological Structures using UAV-based photogrammetric data: An Image analysis approach. *Comput. Geosci.* **2014**, *69*, 22–32. [CrossRef]

9. Bemis, S.P.; Micklethwaite, S.; Turner, D.; James, M.R.; Akciz, S.; Thiele, S.T.; Bangash, H.A. Ground-based and UAV-based photogrammetry: A multi-scale, high-resolution mapping tool for structural geology and paleoseismology. *J. Struct. Geol.* **2014**, *69*, 163–178. [CrossRef]

10. Vollgger, S.A.; Cruden, A.R. Mapping folds and fractures in basement and cover rocks using UAV photogrammetry, Cape Liptrap and Cape Paterson, Victoria, Australia. *J. Struct. Geol.* **2016**, *85*, 168–187. [CrossRef]

11. Młynarczuk, M.; Górszczyk, A.; Ślipek, B. The application of pattern recognition in the automatic classification of microscopic rock images. *Comput. Geosci.* **2013**, *60*, 126–133. [CrossRef]

12. Li, N.; Hao, H.; Gu, Q.; Wang, D.; Hu, X. A transfer learning method for automatic identification of sandstone microscopic images. *Comput. Geosci.* **2017**, *103*, 111–121. [CrossRef]

13. Shu, L.; McIsaac, K.; Osinski, G.R.; Francis, R. Unsupervised feature learning for autonomous rock image classification. *Comput. Geosci.* **2017**, *106*, 10–17. [CrossRef]

14. Trevisani, S.; Rocca, M. MAD: Robust image texture analysis for applications in high resolution geomorphometry. *Comput. Geosci.* **2015**, *81*, 78–92. [CrossRef]

15. Hinton, G.E.; Salakhutdinov, R.R. Reducing the dimensionality of data with neural networks. *Science* **2006**, *313*, 504–507. [CrossRef] [PubMed]

16. LeCun, Y.; Bengio, Y.; Hinton, G. Deep learning. *Nature* **2015**, *52*, 436. [CrossRef] [PubMed]

17. He, K.; Zhang, X.; Ren, S.; Sun, J. Deep residual learning for image recognition. In Proceedings of the IEEE Conference on Computer Vision and Pattern Recognition, Las Vegas, NV, USA, 26 June–1 July 2016; pp. 770–778.

18. Hinton, G.; Deng, L.; Yu, D.; Dahl, G.E.; Mohamed, A.; Jaitly, N.; Senior, A.; Vanhoucke, V.; Nguyen, P.; Sainath, T.N.; et al. Deep neural networks for acoustic modeling in speech recognition: The shared views of four research groups. *IEEE Signal Process. Mag.* **2012**, *29*, 82–97. [CrossRef]

19. Kim, S.; Bae, W.C.; Masuda, K.; Chung, C.B.; Hwang, D. Fine-Grain Segmentation of the Intervertebral Discs from MR Spine Images Using Deep Convolutional Neural Networks: BSU-Net. *Appl. Sci.* **2018**, *8*, 1656. [CrossRef]

20. Greenspan, H.; van Ginneken, B.; Summers, R.M. Guest editorial deep learning in medical imaging: Overview and future promise of an exciting new technique. *IEEE Trans. Med. Imaging* **2016**, *35*, 1153–1159. [CrossRef]

21. Kim, S.; Bae, W.C.; Masuda, K.; Chung, C.B.; Hwang, D. Semi-Automatic Segmentation of Vertebral Bodies in MR Images of Human Lumbar Spines. *Appl. Sci.* **2018**, *8*, 1586. [CrossRef]

22. Han, W.; Feng, R.; Wang, L.; Cheng, Y. A semi-supervised generative framework with deep learning features for high-resolution remote sensing image scene classification. ISPRS-J. *Photogramm. Remote Sens.* **2017**, *145*, 23–43. [CrossRef]

23. Li, W.; Zhou, B.; Hsu, C.Y.; Li, Y.; Ren, F. Recognizing terrain features on terrestrial surface using a deep learning model: An example with crater detection. In Proceedings of the 1st Workshop on Artificial Intelligence and Deep Learning for Geographic Knowledge Discovery, Los Angeles, CA, USA, 7–10 November 2017; pp. 33–36.

24. Palafox, L.F.; Hamilton, C.W.; Scheidt, S.P.; Alvarez, A.M. Automated detection of geological landforms on Mars using Convolutional Neural Networks. *Comput Geosci.* **2017**, *101*, 48–56. [CrossRef] [PubMed]

25. Nogueira, K.; Penatti, O.A.B.; dos Santos, J.A. Towards better exploiting convolutional neural networks for remote sensing scene classification. *Pattern Recognit.* **2017**, *61*, 539–556. [CrossRef]

26. Xu, G.; Zhu, X.; Fu, D.; Dong, J.; Xiao, X. Automatic land cover classification of geo-tagged field photos by deep learning. *Environ. Model. Softw.* **2017**, *91*, 127–134. [CrossRef]

27. Wu, Y.; Qin, Y.; Wang, Z.; Jia, L. A UAV-Based Visual Inspection Method for Rail Surface Defects. *Appl. Sci.* **2018**, *8*, 1028. [CrossRef]

28. He, J. Learn Boudin. Available online: http://blog.sina.com.cn/s/blog_69fb738a0102vmkh.html (accessed on 3 July 2015).
29. Mousas, C.; Anagnostopoulos, C.N. Learning Motion Features for Example-Based Finger Motion Estimation for Virtual Characters. *3D Res.* **2017**, *8*, 25. [CrossRef]
30. Nanni, L.; Ghidoni, S.; Brahnam, S. Handcrafted vs. non-handcrafted features for computer vision classification. *Pattern Recognit.* **2017**, *71*, 158–172. [CrossRef]
31. Chen, T.; Guestrin, C. Xgboost: A scalable tree boosting system. In Proceedings of the 22nd ACM SIGKDD International Conference on Knowledge Discovery and Data Mining, San Francisco, CA, USA, 13–17 August 2016; pp. 785–794.
32. Altenberger, F.; Lenz, C. A Non-Technical Survey on Deep Convolutional Neural Network Architectures. *arXiv*, **2018**; arXiv:1803.02129.
33. Qureshi, A.S.; Khan, A.; Zameer, A.; Usman, A. Wind power prediction using deep neural network based meta regression and transfer learning. *Appl. Soft Comput.* **2017**, *58*, 742–755. [CrossRef]
34. Zhang, Y.; Li, M.; Han, S. Automatic identification and classification in lithology based on deep learning in rock images. *Acta Petrol. Sin.* **2018**, *34*, 333–342.
35. Szegedy, C.; Vanhoucke, V.; Ioffe, S.; Shlens, J.; Wojna, Z. Rethinking the inception architecture for computer vision. In Proceedings of the IEEE Conference on Computer Vision and Pattern Recognition, Seattle, WA, USA, 27–30 November 2017; pp. 2818–2826.
36. Bradski, G.; Kaehler, A. The OpenCV library. *Dr. Dobbs J.* **2000**, *25*, 120.
37. Pedregosa, F.; Varoquaux, G.; Gramfort, A.; Michel, V.; Thirion, B.; Grisel, O.; Blondel, M.; Prettenhofer, P.; Weiss, R.; Dubourg, V.; et al. Scikit-learn: Machine learning in Python. *J. Mach. Learn. Res.* **2011**, *12*, 2825–2830.
38. Nanni, L.; Ghidoni, S.; Brahnam, S. Ensemble of Convolutional Neural Networks for Bioimage Classification. *Appl. Comput. Inform.* **2018**. [CrossRef]

 *applied sciences*

Article

Dark Spot Detection in SAR Images of Oil Spill Using Segnet

Hao Guo *, Guo Wei * and Jubai An

Information Science and Technology College, Dalian Maritime University, Dalian 116026, China; jubaian@dlmu.edu.cn
* Correspondence: guohao0512@dlmu.edu.cn (H.G.); weiguo@dlmu.edu.cn (G.W.); Tel.: +86-130-8413-6921 (H.G.); +86-178-2485-5992 (G.W.)

Received: 15 November 2018; Accepted: 16 December 2018; Published: 18 December 2018

Abstract: Damping Bragg scattering from the ocean surface is the basic underlying principle of synthetic aperture radar (SAR) oil slick detection, and they produce dark spots on SAR images. Dark spot detection is the first step in oil spill detection, which affects the accuracy of oil spill detection. However, some natural phenomena (such as waves, ocean currents, and low wind belts, as well as human factors) may change the backscatter intensity on the surface of the sea, resulting in uneven intensity, high noise, and blurred boundaries of oil slicks or lookalikes. In this paper, Segnet is used as a semantic segmentation model to detect dark spots in oil spill areas. The proposed method is applied to a data set of 4200 from five original SAR images of an oil spill. The effectiveness of the method is demonstrated through the comparison with fully convolutional networks (FCN), an initiator of semantic segmentation models, and some other segmentation methods. It is here observed that the proposed method can not only accurately identify the dark spots in SAR images, but also show a higher robustness under high noise and fuzzy boundary conditions.

Keywords: image segmentation; deep learning; synthetic aperture radar (SAR); oil slicks; segnet

1. Introduction

Due to the influence of short gravity waves and capillary waves on the sea surface, Bragg scattering of the sea surface is greatly weakened, causing the oil film to produce dark spots on synthetic aperture radar (SAR) images [1]. Solberg et al. pointed out that SAR oil spill detection includes three steps: (1) dark spot detection; (2) feature extraction; and (3) discrimination of oil slicks and lookalikes [2]. Among them, the accuracy of dark spot detection is bound to affect the extraction of oil spill location and area. However, some natural phenomena (such as waves, ocean currents, and low wind belts, as well as human factors) may change the backscatter intensity on the surface of the sea, thus leading to an uneven intensity, high noise, or blurred boundaries of oil slicks or lookalikes, making the automatic segmentation of the oil spill area sometimes very difficult. Therefore, a robust and accurate segmentation method plays a crucial role in monitoring oil spills.

There are many studies on dark spot detection on SAR images of oil spill, among which the most widely used method is based on pixel grayscale threshold segmentation, such as a manual single threshold segmentation [3], an adaptive threshold segmentation method [4], and some double threshold segmentation methods [5]. Those methods have simple principles and fast implementation speeds, but they are easily affected by speckle noise and global gray unevenness, thus reducing the accuracy of dark spot recognition. The active contour models (ACM) are another common image segmentation method [6,7]. Compared with traditional segmentation methods, the smooth and closed contours can be obtained by ACM. The most famous and widely used region-based ACM is the borderless ACM proposed by Chan and Vese [7]. The Chan-Vese model performs well in processing

Appl. Sci. **2018**, *8*, 2670; doi:10.3390/app8122670

128

www.mdpi.com/journal/applsci

images with weak edge and noise, but it cannot process images with uneven intensity and high speckle noise.

With the popularization of neural networks and machine learning algorithms, some studies have used these methods for dark spot detection. Topouzelis et al. proposed a fully connected feed forward neural network to monitor the dark spots in an oil spill area, and obtained a very high detection accuracy at that time [8]. Taravat et al. used a Weibull multiplication filter to suppress speckle noise, enhance the contrast between target and background, and used a multi-layer perceptron (MLP) neural network to segment the filtered SAR images [9]. Taravat et al. also proposed a new method to distinguish dark spots from the combination of the Weibull multiplication model (WMM) and pulse coupled neural networks (PCNN) [10]. Singha used artificial neural networks (ANN) to identify the characteristics of oil slicks and lookalikes [11]. Although this method improved the segmentation accuracy to some extent and suppressed the influence of speckle noise on dark spot extraction, it still cannot obtain high segmentation accuracy and robustness. Jing et al. discussed the application of fuzzy c-means (FCM) clustering in SAR oil spill segmentation, in which it is easy to generate fragments in the segmentation process due to speckle noise images [12]. To suppress the influence of speckle noise on SAR image segmentation of an oil spill, Teng et al. proposed a hierarchical clustering-based SAR image segmentation algorithm, which effectively maintained the shape characteristics of oil slicks in SAR images using the idea of multi-scale segmentation [13]. However, its ability to suppress speckle noise was not good, and the segmentation of weak boundary region was also not ideal.

In recent years, deep learning methods have been successfully applied in extracting high level feature representations of images, especially in semantic segmentation. Long et al. changed the full connection layer of the traditional convolution neural networks (CNN) for pixel-based classification [14]. Persello et al. used fully convolutional networks (FCN) to improve the detection accuracy of informal residential areas in high-resolution satellite images [15]. Huang et al. successfully applied the FCN model to weed identification in paddy fields [16]. However, FCN is not sensitive to the details in the images, and its up-sampling results are often blurred. Badanlayan et al. proposed a classic deep learning method (i.e., Segnet) for image semantic segmentation, which was used for automatic driving or intelligent robots [17]. The model has obvious advantages over FCN in storage, calculation time, and segmentation accuracy.

Inspired by the great success of Segnet in image semantic segmentation [17,18], we used Segnet as a segmentation model to detect dark spots in oil spill areas. The proposed method is applied to a data set of 4200 from five original SAR images of oil spill. Each scene image is cropped according to four different window sizes, and samples containing oil slicks and seawater are selected from the cropped pieces as data sets. Four hundred and twenty samples were selected from each oil slick scene, with a total of 2100 sample data. To enhance the robustness of the training model, 21 samples in each oil slick scene were added with 10 noise levels of multiplicative and additive noise, respectively, totaling 2100 noisy images. The training set consisted of 1800 original samples and 1800 noisy samples, totaling 3600. The testing set consisted of 600 samples, including 300 original samples and 300 sets of noisy samples (20 noise level data corresponding to three samples randomly selected in each oil slick scene). The effectiveness of the method is demonstrated through the comparison with FCN and some classical segmentation methods (such as support vector machine (SVM), classification and regression tree (CART), random forests (RF), and Otsu, etc.). The segmentation accuracy based on Segnet can reach 93.92% under high noise and weak boundary conditions. It is here observed that the proposed method can not only accurately identify the dark spots in SAR images, but also show higher robustness.

The rest of this paper is organized as follows. Section 2 focuses on the preparation process of the data set, which includes description, preprocessing, and sampling of five SAR oil slick scenes acquired by C-band Radarsat-2. In Section 3, we describe the segmentation based on the Segnet model and the parameter selection in the training process. The validity of the algorithm is verified through analysis and compared with the experimental results of the semantic segmentation model FCN8s. In Section 4, we analyze the validity and stability of the model. The conclusions and outlooks are discussed in the final section.

2. Study Area and Data Sets

2.1. Study Area and Pretreatment

Five SAR oil slick scenes acquired by Radarsat-2 (fine quad-polarized mode) are described in Table 1, and some information on those data (e.g., wind direction, water temperature, etc.) is described in detail in Guo's studies [19]. Radarsat-2 images of the Mexico Bay area (No.1 and No.4) were acquired on 8 May 2010 and 24 August 2011, respectively. The dark spots in Figure 1 were interpreted as crude oil. North Sea of Europe area data sets (No.2 and No.3) were acquired from 6–9 June 2011. There are three substances (i.e. crude oil, oil emulsion, and plant oil) in the two scenes, and the acquisition interval of No.2 and No.3 was about 12 hours. The data No.5 was obtained in the South China Sea on 18 September 2009. The experimental data contain a small amount of crude oil and plant oil, which were poured with 15-minute intervals.

Table 1. Information of the five quad-polarization SAR images of oil spill.

Scene ID	No.1	No.2.	No.3	No.4	No.5
Location of Center	26°49′N/92°01′W	59°59′N/2°25′E	60°09′N/2°19′E	27°54′N/90°55′W	18°06′N/109°25′E
Wind Speed	~6.5 m/s	1.6–3.3 m/s	1.6–3.3 m/s	~15 m/s	~10 m/s
Types of Dark Spot	Crude oil	Plant oil/oil emulsion	Plant oil/oil emulsion/crude oil	Crude oil	Oil emulsion/crude oil
Sample size	420	420	420	420	420

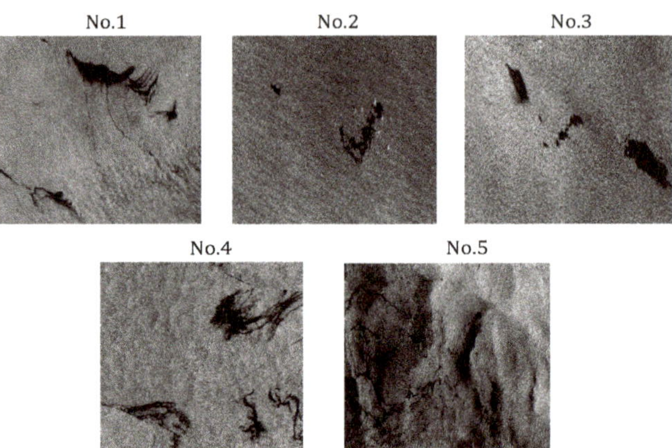

Figure 1. Five oil spill RadarSat-2 scenes.

Quad-polarization SAR images are susceptible to noise. Pauli decomposition has the advantages of anti-interference and a general high adaptability [20]. In general, the Pauli decomposition images are clearer than original quad-polarization SAR images. The image preprocessing stages are as follows:

(1)   The original quad-polarization SAR data are decomposed by Pauli.
(2)   The obtained Pauli decomposition images are filtered by 3 × 3 Boxcar filtering.

2.2. Sampling Process

The Vapnik-Chervonenkis (VC) dimension is usually used to predict the probability of testing errors of models. Vapnik [21] proves that the probability of the upper bound of the testing error is given by (1):

$$\Pr\left(testing\,error \le training\,error + \sqrt{\frac{1}{N}\left[D\left(\log\left(\frac{2N}{D}\right) + 1 \right) - \log\left(\frac{\eta}{4}\right) \right]} \right) = 1 - \eta \qquad (1)$$

where D is the VC dimension of the classification model. N is the number of training samples. $\sqrt{\frac{1}{N}\left(\left[D\left(\log\left(\frac{2N}{D}\right) + 1 \right) - \log\left(\frac{\eta}{4}\right) \right] \right)}$ is also called the model complexity penalty. When the testing error is less than the training error plus the model complexity penalty, the probability is $1 - \eta$. The bigger the D, the bigger the model complexity penalty, and the bigger the N, the smaller the model complexity penalty. Generally speaking, a deep learning model needs enough samples. Otherwise, the generalization of the model would be limited, i.e., over-fitting.

The proposed method was applied to a data set of 4200 samples from five original SAR images of oil spill. Here, the data set was processed by the following steps and was called the OIL_SPILL_DATASET:

(1) In order to ensure that each sampling window includes oil slicks and seawater, the window size cannot be too small or too large, and the window sizes were selected to be 500×500, 1000×1000, 1500×1500, and 2000×2000 for each scene of the quad-polar SAR image, respectively. Samples, including oil slicks and seawater, were selected from those sub-images, 420 samples were selected from each scene data, totaling 2100 samples. The boundary complexity and weak boundary were the main factors affecting the segmentation accuracy. The boundary complexity and boundary strength of 420 samples selected from each scene data are shown in detail in Table 2.

(2) To ensure the balance of the sample distribution, 105 samples (21 samples in each scene) in 2100 samples were added with multiplicative noise and additive noise, respectively, among which multiplicative noise had 10 levels (peak signal-to-noise ratio (PSNR) was between 50 and 30) and additive noise had 10 levels (PSNR was between 50 and 30). A total of 20 different levels of noise were applied to each sample. In this way, the number of samples per scene was extended from 420 to 840, and the total number of samples was up to 4200.

(3) Due to the limitation of the graphics processing unit (GPU) capabilities, the samples with different sizes obtained in Steps (1) and (2) were resized into 256×256.

(4) Segnet is a supervised pre-training process, and a label should be made for each sample. In Figure 2b, the black area represents the background (seawater) and the red region represents the target (oil slicks or lookalikes).

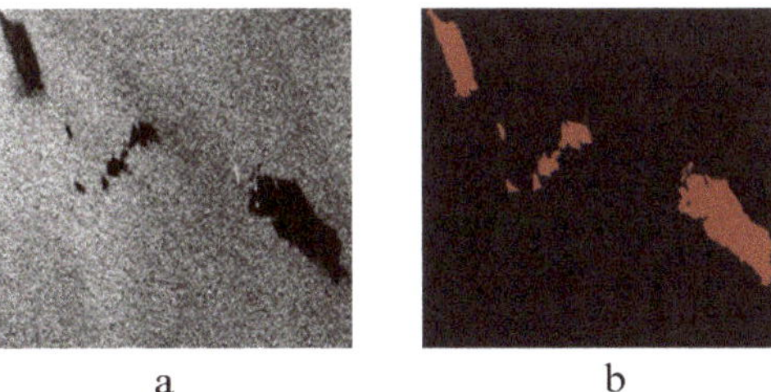

a   b

Figure 2. Label for each sample. (**a**) An input image; (**b**) the label of the image.

(5) 4200 samples were randomly divided into a testing set or a training set according to the ratio of 1:6. To ensure the same distribution of data in the training set and the testing set, 15 samples were selected from 105 samples added with noise, of which three were contained in each oil slick

scene. 300 samples of each noise level corresponding to the 15 test samples were selected from the remaining 2100 noisy samples. The testing set contained 600 samples and the training set contained 3600 samples.

Table 2. Sample collection of No.1–No.5.

Boundary Complexity - Boundary Strength	No.1	No.2	No.3	No.4	No.5
Low - Low	-	-	-	-	190
Low - Moderate	-	-	-	-	50
Low - High	320	50	420	-	30
Moderate - Low	-	-	-	-	130
Moderate - Moderate	-	140	-	-	20
Moderate - High	100	-	-	60	-
High - Low	-	-	-	-	-
High - Moderate	-	230	-		-
High - High	-	-	-	360	-

3. Dark Spot Detection Using Segnet

3.1. Introduction to Segnet

Segnet is a deep convolution neural network with a sound performance of image semantic segmentation [17]. The basic framework of Segnet is an encoder and a decoder. The most important components of Segnet include a convolution layer, pooling layer, up-sampling layer, and softmax layer, see Figure 3. The encoder consists of the convolution layer, batch normalization layer, and rectified linear unit (ReLU), and its structure is similar to the visual geometry group (VGG)-16 network [22]. The convolution layer is the main component of the encoder, and each output pixel is only linked to the local area of the next input layer, thus forming a local receptive field [23]. The decoder consists of a transposed convolutional layer and an up-sampling layer, and its structure is symmetrical to that of the encoder. The convolution layer corresponds to the transposed convolution layer and the max pooling layer corresponds to the up-sampling layer [24]. At the end of the decoder, the category of each pixel is output through a softmax layer.

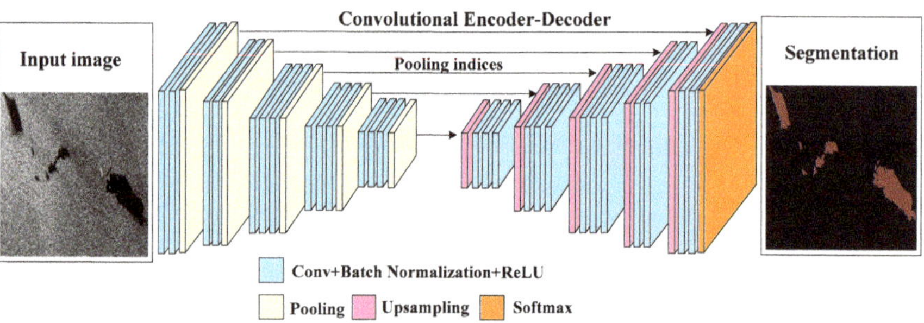

Figure 3. The framework of Segnet.

The training process of Segnet can be summarized as follows:

(1) Each sample in the training set and its corresponding label are input into the Segnet in sequence.
(2) The cross loss entropy is used as the objective function of the training model, and its value is calculated by weighted average for all pixels in each training sample [17].
(3) Through the back propagation algorithm, the weights are updated according to the minimum error.

The information in Steps (1) and Steps (2) is forward propagating, and the outputs are obtained by convolution of inputs and weights. Step (3) is the backward propagation process. According to the results of Step (2), the weights are passed to the previous layers through the backward propagation algorithm, and the weights are updated.

3.2. Image Segmentation of Oil Spill Using Segnet

Segnet is an end-to-end training process. In this experiment, the training process of the Segnet was based on 3600 training samples, of which 1800 were original images, 900 were additive noise images, and 900 were multiplicative noise images. Due to limitations of the GPU capabilities, one sample at a time was input during the training process. Here, the epoch was set at 30, and the fixed learning rate was 0.01.

The structure and parameters of Segnet are shown in Figure 4. The application of Segnet excluding the batch normalization layer is shown in Figure 5. The training performance diagram of Segnet is shown in Figure 6. In our study, the weight initialization process of the encoder and decoder was based on the research of He et al. [25]. When the learning rate is 0.0001, the training loss would be relatively large. When the learning rate is 0.001, the training would be relatively stable soon, but it is not as good as the training performance when the learning rate is 0.01.

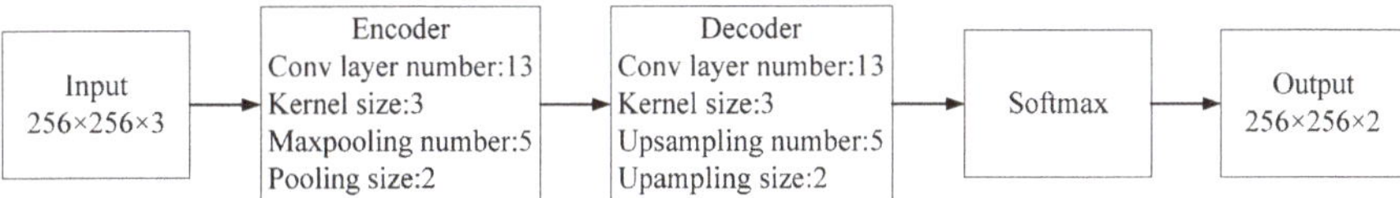

Figure 4. The structure and parameters of Segnet.

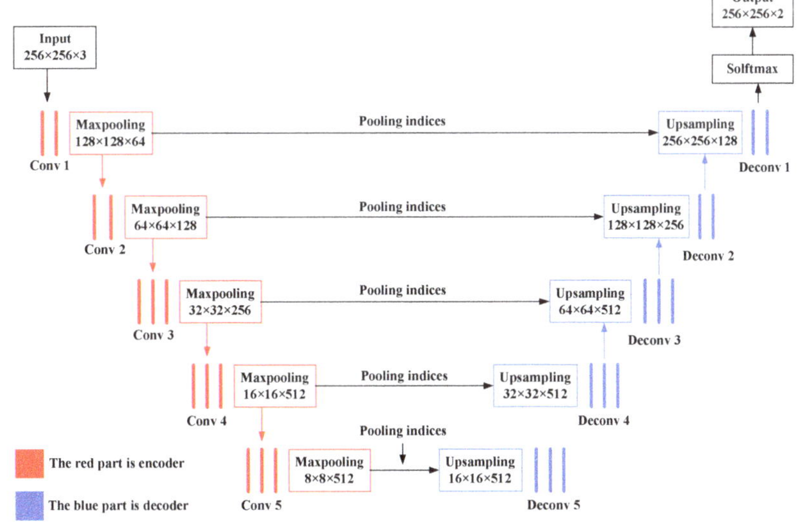

Figure 5. The application of Segnet without the batch normalization layer.

Parts of of the test results of Segnet model based on the OIL_SPILL_DATASET are shown in Figure 7. Where a, b, c, d, and e are representative samples of five boundary statuses, respectively, and a brief description of the five boundary statuses is shown in Table 3. It can be seen from Figure 7 that Status-a (medium boundary complexity) and Status-c (ideal boundary) achieved the best segmentation results in the five boundary statuses, and Status-b (strong noise) and Status-d (complex boundary) were slightly inferior. For Status-e (weak boundary), the Segnet can still effectively segment dark spots in general, although some backgrounds were incorrectly segmented into dark spots.

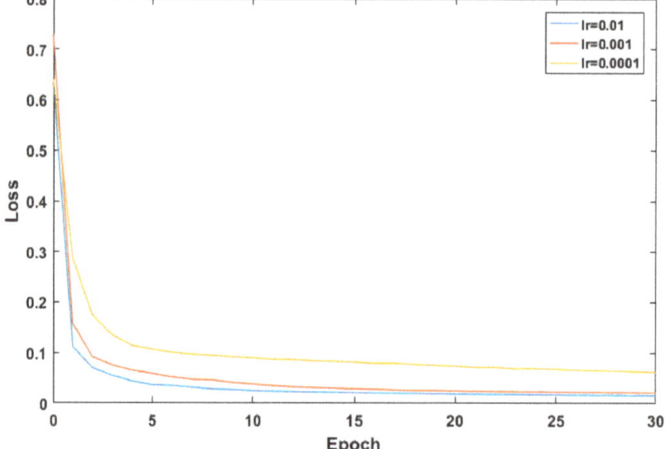

Figure 6. Training performance of Segnet based on different learning rates.

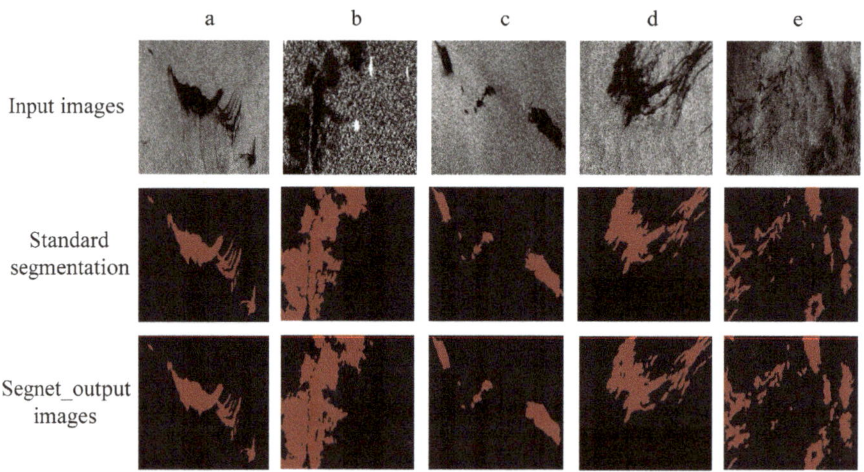

Figure 7. The segment effect of some test samples with different boundary status.

Table 3. Five boundary statuses of oil slicks.

Status ID	a	b	c	d	e
Status of boundary	Medium boundary complexity, low noise	Strong noise	Ideal boundary	High boundary complexity	Weak boundary strength
Number of samples	30	30	30	30	30

We used a trained model to test samples with a learning rate of 0.01. The segmentation results did not achieve the expected results (seeing Figure 8). For the samples without noise or with low multiplicative and additive noise in the testing set, the segmentation effect was good. However, almost all pixels with high additive noise were predicted as the background. To reduce the computational and storage pressure of GPU, we chose a Segnet's batch size of 1 (i.e., inputting one sample at a time), and found that in this case, the Segnet can achieve a better segmentation effect without using the batch normalization layer [26]. Finally, the learning rate was 0.01, and the batch normalization layer was removed based on the basic structure of Segnet.

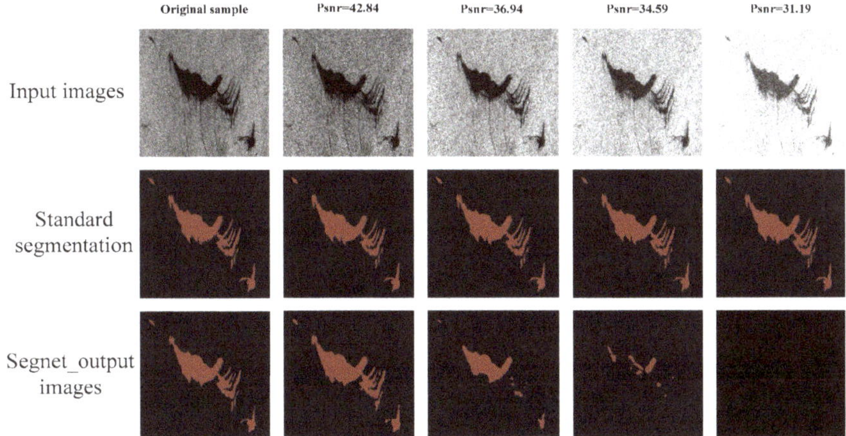

Figure 8. Comparison of segmentation results with different levels of additive noise.

3.3. Comparison of Segnet to FCN

Three end-to-end FCN models (i.e., FCN32s, FCN16s, and FCN8s) were proposed by Long et al. [14], among which FCN8s (8-step sampling) was considered the best one. The FCN8s encoder includes convolution layers with a 3×3 convolution kernel. The convolution layers changed from the last three full connection layers are convolution kernels of 7×7, 1×1, 1×1, and the convolution layers of layers 6–7 are all characteristic images $1 \times 1 \times 4096$. The last de-convolution layer can be considered as an up-sampling process, which can be used to obtain the segmentation image with the same size as the original image. The up-sampling process of FCN8s is a jump architecture, which performs up-sampling on the results of different pool layers of pool 3, pool 4, and pool 5, and then optimizes the output according to these results. The size of the output picture was the same as that of the input, and its number of channels was 2, which indicates that the output prediction picture contained two categories (seawater and oil slicks). A schematic diagram of FCN8s' model operation is given in Figure 9.

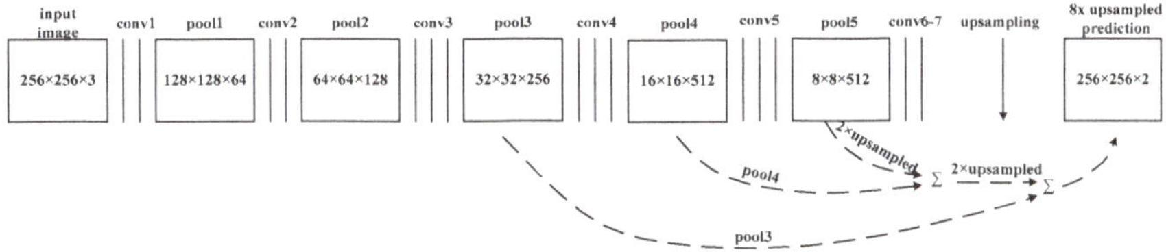

Figure 9. The structure and parameters of FCN8s (8-step sampling).

In our study, the FCN8s' structure did not include the batch normalization layer. We tested some learning rates (0.01, 0.001, 0.0001) during FCN8s' training and found that the required accuracy could be obtained when the learning rate was 0.001, but the cost was a longer training period than Segnet. The training parameters of FCN8s and Segnet are shown in Table 4. The training performance of FCN8s is shown in Figure 10. When epoch reached 10, the training loss was close to 0.06 and tended to be stable.

Table 4. Training parameters of Segnet and FCN8s.

Methods	Learning Rate	Batch Size	Batch Normalization	Volatile GPU-Until	GPU Training Memory	Model Size
Segnet	0.01	1	off	99%	1119 MB	114 MB
FCN8s	0.001	1	off	100%	5319 MB	537 MB

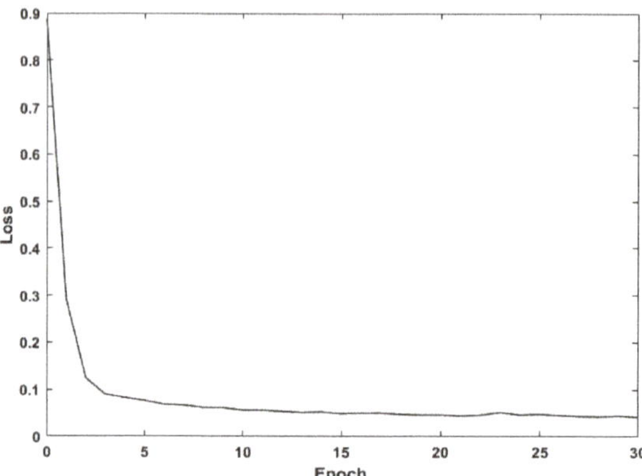

Figure 10. Training performance of FCN8s (Learning rate is 0.001).

The comparison between Segnet and FCN8s is shown in Figure 11, and the five samples (a–e) represent the boundary statuses, respectively (see Table 3). The results show that FCN8s has a good overall segmentation effect. However, the performance needs to be improved in oil spill images with weak boundaries and high boundary complexity. The Sample-d and Sample-e are both high wind speed regions in Figure 11. Due to the high wind speed, the oil slick boundary complexity in Sample-d was high, and the segmentation results were not ideal.

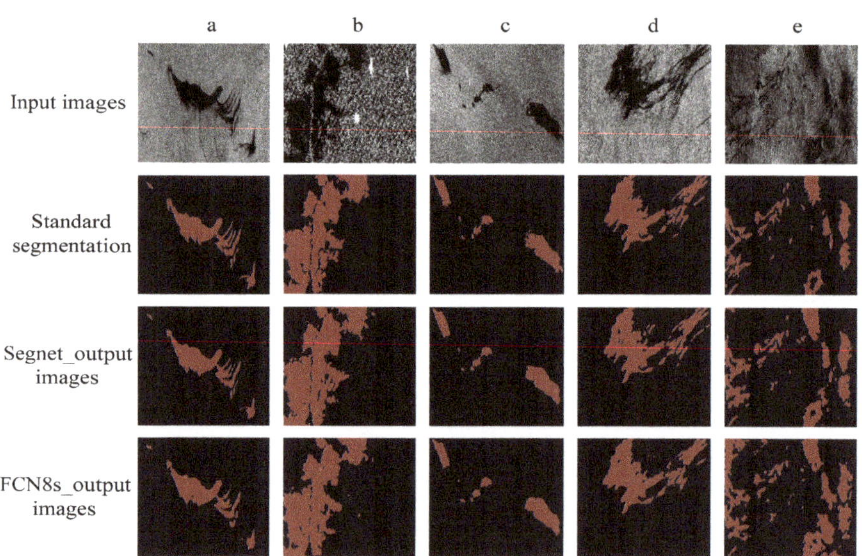

Figure 11. Comparison of segmentation results of FCN8 and Segnet for five boundary statuses.

The receiver operating characteristic (ROC) analysis was used to evaluate the proposed algorithm with the pixel classification accuracy of FCN8s. Since our model input was a sample containing both oil slicks and seawater, it was difficult to ensure that the pixel ratio of oil slicks and seawater was 1:1. To ensure that the pixel ratio of oil slicks and seawater was as close as possible to 1:1, we re-selected the training set and the testing set. According to the label of each sample in the training set, the number of pixels of oil slicks and seawater in each sample was calculated, respectively, and finally, 2800 pieces of data with the pixel-to-pixel ratio of oil slicks and seawater of 0.998:1 were selected. The testing set selected 200 test data with a ratio of 0.989: 1 from 600 testing sets. The ratio of the training set to testing set was still 6:1. The ROC curves for Segnet and FCN8s are shown in Figure 12, and we can see that the ROC curves of Segnet and FCN8s are very close to the upper left corner, but there are still some differences. Under the condition of a high false positive rate (FPR), both showed a higher true positive rate (TPR). However, under the condition of low FPR, the TPR of FCN was lower than that of Segnet. The results show that Segnet has achieved a moderate TPR in the whole range of FPR.

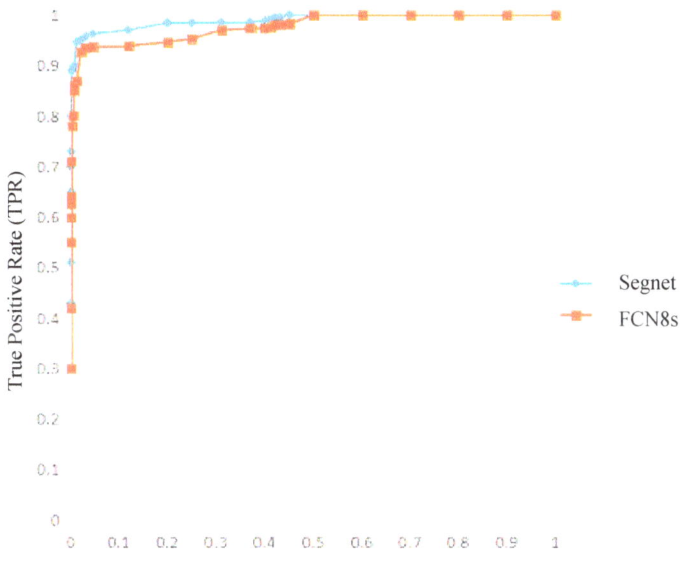

Figure 12. ROC curves of Segnet and FCN8s.

4. Analysis of Segmentation Experiments

4.1. Efficiency Analysis

We compared the performance of FCN8s and Segnet from the following four aspects: Pixel-classification accuracy (PA), mean accuracy (MA), mean intersection over union (MIoU), and frequency weighted intersection over union (FWIoU). The comparison of the four standard values for FCN8s and Segnet with five boundary statuses (see Table 3) is shown in Figure 13. It can be observed that the performance of Segnet and FCN8s for the first four boundary statuses was almost the same, and the PA was above 95%. However, for Status-e (Weak boundary strength), Segnet was superior to FCN8s in the segmentation effect, and the PA of Segnet reached 93.92%. FCN8s performed slightly worse for weak boundary segmentation, achieving 87.53% of PA. Thus, that Segnet can effectively detect dark spots (oil slicks or lookalikes) in SAR images.

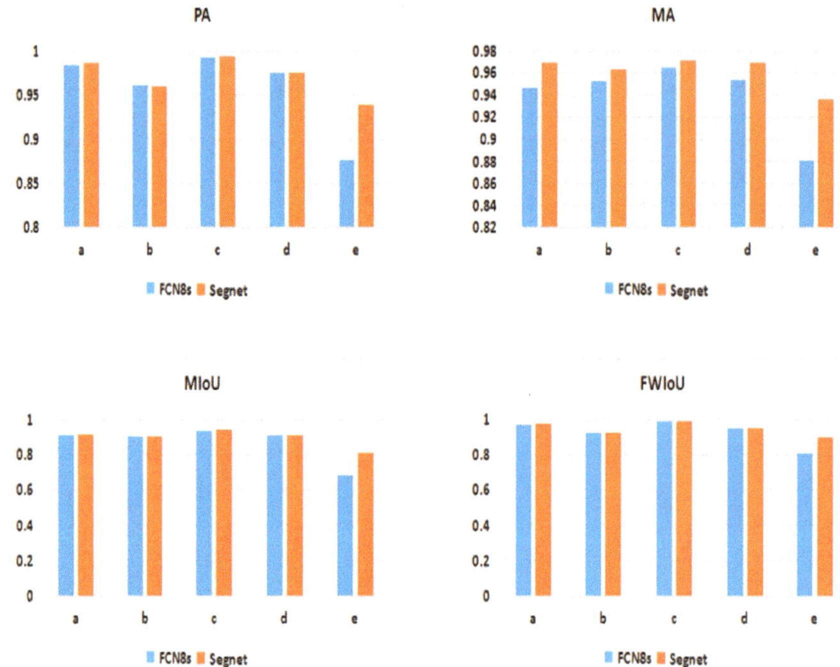

Figure 13. Comparison of the four evaluation parameters with five boundary statuses.

4.2. Stability Analysis

Due to the influence of the sea surface environment (such as waves, ocean currents, and low wind belts) and the characteristics of SAR sensors, high noise and weak boundaries are commonly found in SAR images of oil spill. Figure 14 shows an example of the segmentation effect of a test sample at five additional noise levels. The first row is a SAR test sample with five different peak signal-to-noise ratio (PSNR), and the second row is the label of each sample. The outputs of Segnet and FCN8s are listed in the third and fourth rows, respectively. Figure 15 shows the segmentation effect of the same test sample at five multiplicative noise levels.

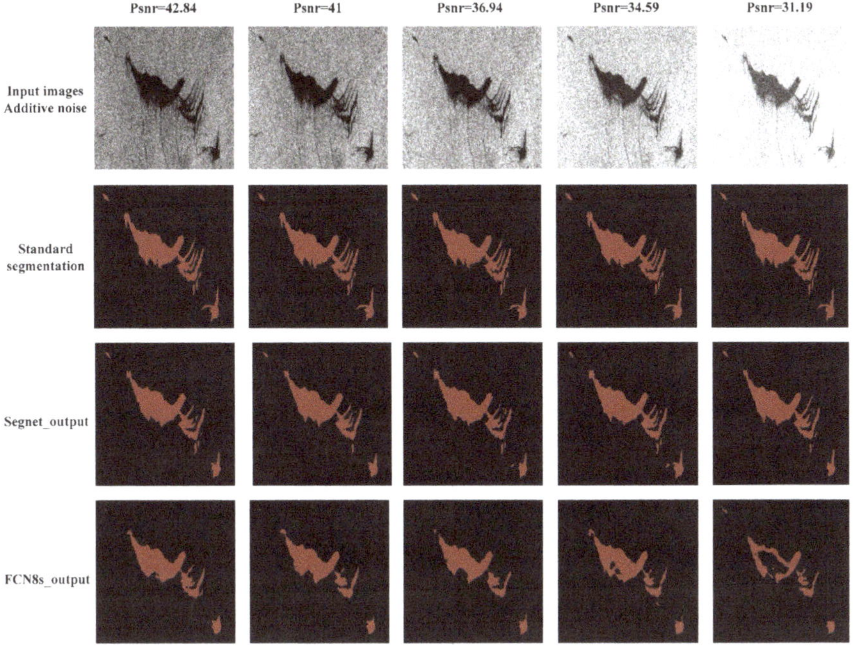

Figure 14. Comparison of Segnet and FCN8 with five levels of additive noise.

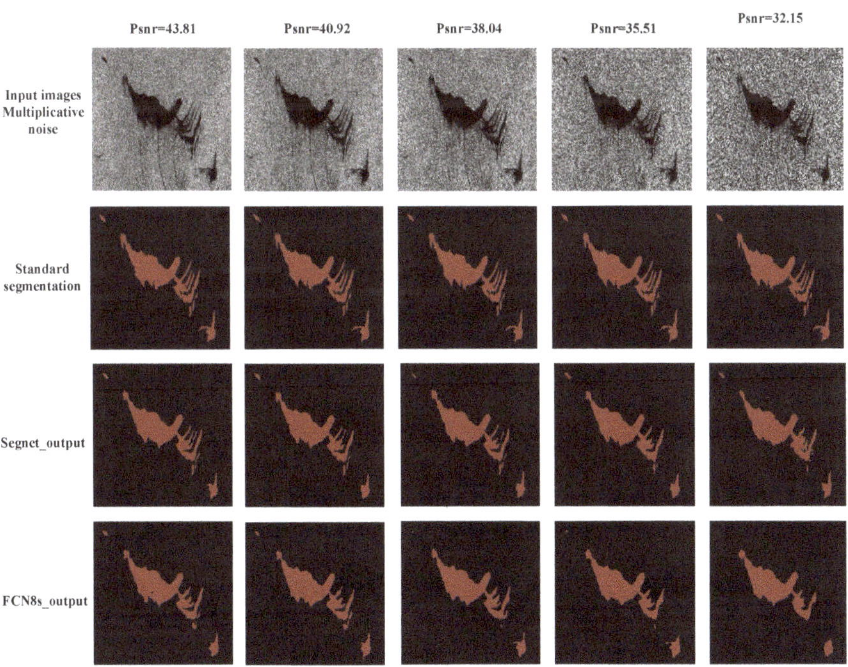

Figure 15. Comparison of Segnet and FCN8 with five levels of multiplicative noise.

Correspondingly, the effectiveness of the proposed method is demonstrated through the analysis of some experimental results. The same training set and test set were applied to FCN8s and some classical image segmentation methods, such as SVM, CART [27], RF [28], and Otsu. The four aforementioned evaluation parameters (PA, MA, MIoU, and FWIoU) at 10 additive noise levels are shown in Figure 16, and a comparison at 10 multiplicative noise levels is shown in Figure 17, where the X coordinate is PSNR. We can see the following trend:

(1) In addition to some fluctuations of MA of Segnet where PSNR is relatively large in Figure 16, the other three parameters (PA, MIoU, and FWIoU) are basically on a horizontal line, which proves that Segnet shows high robustness in terms of additive noise.

(2) When PSNR is generally less than 35 in Figure 16, all four indicators of FCN8s have a clear downward trend, which indicates that FCN8S is not as stable as Segnet when the additive noise is relatively high.

(3) In Figure 16, the four classical segmentation methods (SVM, CART, RF, and Otsu) are sensitive to additive noise (especially when PSNR is generally less than 35), and the comparison of the three indicators (MA, MIoU, and FWIoU) shows that they are not as good as Segnet and FCN8s. Although these three methods (SVM, CART, and RF) seem to have a similar performance with Segnet and FCN8s based on PA, this phenomenon should be related to PA's defects. It is very difficult to ensure that oil slicks and seawater have the same initial probability in the testing set.

(4) In Figure 17, Segnet and FCN8s show high stability and tolerance to multiplicative noise, although the overall performance of FCN8s is not as good as that of Segnet. When PSNR is less than 35, the PA of FCN8s is obviously decreased.

(5) The four classical segmentation methods (SVM, CART, RF, and Otsu) are much more sensitive to multiplicative noise than Segnet and FCN8s, especially when the noise is high, and the performance of those classification methods drops sharply. In addition, Otsu's performance is significantly worse than the other three methods.

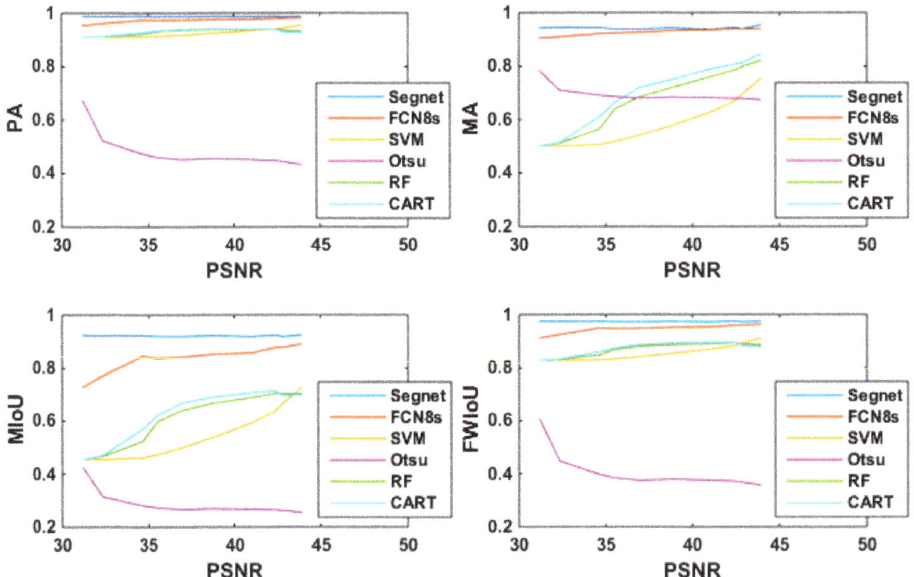

Figure 16. Four image segmentation standard values under 10 additive noise levels.

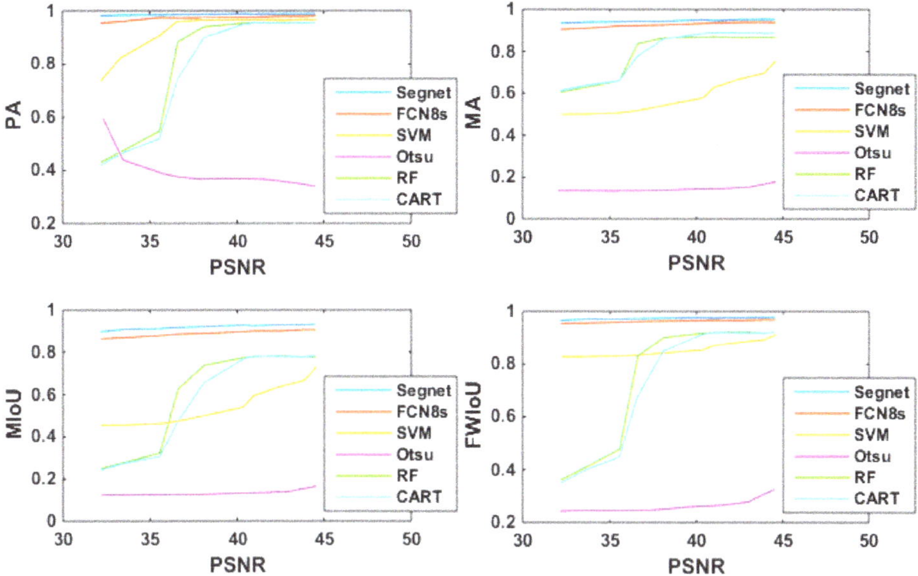

Figure 17. Four image segmentation standard values under 10 multiplicative noise levels.

Overall, by comparing the four parameters (PA, MA, MIoU, and FWIoU) of the additive and multiplicative noise in Figures 16 and 17, the traditional machine algorithm performed poorly in detecting dark spots compared with semantic segmentation algorithms. Table 5 shows the comparison of segmentation accuracy (averages values of PA, MA, MIoU, and FWIoU) and running time (GPU time) using the same test set (600 samples). Due to the complex structure of the deep learning model, its running time was much longer than that of the classical machine learning model.

Table 5. Comparison of the six segmentation methods.

Methods	Time (s)	PA	MA	MIoU	FWIoU
Segnet	1.639×10^4	0.939	0.895	0.801	0.914
FCN8s	1.703×10^4	0.884	0.805	0.783	0.823
SVM	1.353×10^2	0.854	0.693	0.568	0.795
Otsu	2.192×10^{-3}	0.573	0.543	0.375	0.508
RF	1.183	0.867	0.735	0.612	0.814
CART	2.023×10^{-2}	0.855	0.749	0.605	0.810

4.3. Overfitting Analysis

The initial probability of the data in the overfitting experiment was equal. For the first experimental model, the training samples were from the first three SAR oil slick scenes (No.1–No.3). The number of samples from each scene was 720, and the total number of training samples was 2160. The training data of the second experimental model included 720 samples selected from the first SAR oil slick scene (No.1) only. The two models were tested using the same test data, and those 120 test samples here were selected from the SAR oil slick scenes (No.1). The training set of the first model contained 1080 original samples and 1080 noise samples. Accordingly, the training set of the second training model included 360 original samples and 360 noise samples, which have the same distribution as that of the first model. The average values of each parameter based on the first and second models are shown in Table 6, and the average values of the four parameters of the second model were higher than that of the first model. It can be seen that there is indeed an over-fitting phenomenon when the sample space is insufficient.

Table 6. Evaluation of Segnet and FCN8s based on the four image segmentation standard values.

Training Set	Average PA		Average MA		Average MIoU		Average FWIoU	
	Segnet	FCN8s	Segnet	FCN8s	Segnet	FCN8s	Segnet	FCN8s
No.1–No.3	0.9845	0.9589	0.9437	0.8644	0.9174	0.8419	0.9724	0.9481
No.1	0.9858	0.9692	0.953	0.8952	0.9208	0.8637	0.9734	0.9547

4.4. K-Fold

K-fold cross validation (K-CV) can effectively avoid over-fitting and under-fitting [29]. The data set is randomly divided into K groups to verify the validity of the training model. Each subset of the data set is used as a testing set and the remaining K-1 groups are used as training sets. On the basis of K-CV, we verified the performance of the model by using the mean and variance of PA. K = 3, 5, 7, and 9 are shown in Table 7. With the improvement of PA, the stability of the model would be improved. When K increased to 7, the increase of PA and variance tended to be stable. Here, K was set to 7 in consideration of statistical stability and calculation costs. Therefore, the ratio of the testing set to training set was 1:6, the total data set had 4200 samples, the testing set had 600 samples, and the training set contained 3600 samples.

Table 7. Mean and variance of pixel-classification accuracy (PA) based on K-CV.

K	Average	Variance
3	0.966617469	0.010223167
5	0.975305136	0.001727448
7	0.985242757	0.000513106
9	0.986190631	0.000499049

5. Conclusions and Outlooks

The current research used Segnet to extract dark spots in SAR images of an oil spill. To reduce the computational and storage pressure of GPU, we chose a Segnet's batch size of 1 (i.e., inputting one sample at a time), and found that in this case, the Segnet achieved a better segmentation effect without using the batch normalization layer. The proposed method effectively distinguished between oil slicks and seawater based on the data set (OIL_SPILL_DATASET), and high accuracy segmentation results were obtained for SAR images with high noise and weak boundaries.

The OIL_SPILL_DATASET was also applied to FCN8s and some other classical segmentation methods. By comparing the four parameters (PA, MA, MIoU, and FWIoU) of different addition and multiplication noise levels, the following trends were found:

- Segnet and FCN8s showed high stability and tolerance to addition and multiplicative noise, although the overall performance of FCN8s was not as good as that of Segnet. In addition, Segnet was obviously superior to FCN8s in weak boundary regions.
- Some classical segmentation methods (such as SVM, CART, RF, and Otsu) were much more sensitive to addition and multiplicative noise than the deep learning models.

However, Segnet's training process was supervised, and its training relies on a large number of label images. The production of labels was not only time-consuming and laborious in the data preparation stage, but also the training effect could be easily affected by human factors. In the future, we hope to shift to a weak or unsupervised training process to improve the convenience of application.

Appl. Sci. **2018**, *8*, 2670

Author Contributions: H.G. conceived and designed the algorithm, and constructed the outline for the manuscript; G.W. performed the experiments, analyzed the data, and made the first draft of the manuscript; J.A. was responsible for determining the overall experimental plan, and polished the manuscript.

Funding: The work was carried out with the supports of the National Natural Science Foundation of China (Grant 61471079) and State Oceanic Administration of China for ocean nonprofit industry research special funds (No.2013418025).

Conflicts of Interest: The authors declare no conflict of interest. The founding sponsors had no role in the design of the study; in the collection, analyses, or interpretation of data; in the writing of the manuscript, and in the decision to publish the results.

References

1. Yin, J.; Yang, J.; Zhou, Z.S. The Extended Bragg Scattering Model-Based Method for Ship and Oil-Spill Observation Using Compact Polarimetric SAR. *IEEE J. Sel. Top. Appl. Earth Obs. Remote Sens.* **2015**, *8*, 3760–3772. [CrossRef]

2. Solberg, A.H.S.; Brekke, C.; Husoy, P.O. Oil Spill Detection in Radarsat and Envisat SAR Images. *IEEE Trans. Geosci. Remote Sens.* **2007**, *45*, 746–755. [CrossRef]

3. Bern, T.I.; Wahl, T.; Andersen, T. Oil Spill Detection Using Satellite Based SAR: Experience from a Field Experiment. *Photogramm. Eng. Remote Sens.* **1993**, *59*, 423–428.

4. Solberg, A.H.S.; Storvik, G.; Solberg, R. Automatic detection of oil spills in ERS SAR images. *IEEE Trans. Geosci. Remote Sens.* **1999**, *37*, 1916–1924. [CrossRef]

5. Kanaa, T.F.N.; Tonye, E.; Mercier, G. Detection of oil slick signatures in SAR images by fusion of hysteresis thresholding responses. In Proceedings of the 2003 IEEE International Geoscience and Remote Sensing Symposium, Toulouse, France, 21–25 July 2003; pp. 2750–2752.

6. Li, C.; Kao, C.Y.; Gore, J.C. Minimization of Region-Scalable Fitting Energy for Image Segmentation. *IEEE Trans. Image Process.* **2008**, *17*, 1940–1949. [PubMed]

7. Chan, T.F.; Vese, L.A. *Active Contours without Edges*; IEEE Press: New York, NY, USA, 2001.

8. Topouzelis, K.; Karathanassi, V.; Pavlakis, P. Oil spill detection: SAR multiscale segmentation and object features evaluation. *Int. Symp. Remote Sens.* **2003**, *4880*, 77–87.

9. Taravat, A.; Oppelt, N. Adaptive Weibull Multiplicative Model and Multilayer Perceptron neural networks for dark-spot detection from SAR imagery. *Sensors* **2014**, *14*, 22798–22810. [CrossRef] [PubMed]

10. Taravat, A.; Latini, D.; Frate, F.D. Fully Automatic Dark-Spot Detection from SAR Imagery with the Combination of Nonadaptive Weibull Multiplicative Model and Pulse-Coupled Neural Networks. *IEEE Trans. Geosci. Remote Sens.* **2014**, *52*, 2427–2435. [CrossRef]

11. Singha, S.; Bellerby, T.J.; Trieschmann, O. Satellite Oil Spill Detection Using Artificial Neural Networks. *IEEE J. Sel. Top. Appl. Earth Obs. Remote Sens.* **2013**, *6*, 2355–2363. [CrossRef]

12. Jing, Y.; An, J.; Liu, Z. A Novel Edge Detection Algorithm Based on Global Minimization Active Contour Model for Oil Slick Infrared Aerial Image. *IEEE Trans. Geosci. Remote Sens.* **2011**, *49*, 2005–2013. [CrossRef]

13. Teng-Fei, S.U.; Meng, J.M.; Zhang, X. Segmentation Algorithm for Oil Spill SAR Images Based on Hierarchical Agglomerative Clustering. *Adv. Mar. Sci.* **2013**, *31*, 256–265.

14. Long, J.; Shelhamer, E.; Darrell, T. Fully convolutional networks for semantic segmentation. In Proceedings of the IEEE Conference on Computer Vision and Pattern Recognition, Boston, MA, USA, 7–12 June 2015; pp. 3431–3440.

15. Persello, C.; Stein, A. Deep Fully Convolutional Networks for the Detection of Informal Settlements in VHR Images. *IEEE Geosci. Remote Sens. Lett.* **2017**, *14*, 2325–2329. [CrossRef]

16. Huang, H.; Deng, J.; Lan, Y. A fully convolutional network for weed mapping of unmanned aerial vehicle (UAV) imagery. *PLoS ONE* **2018**, *13*, e0196302. [CrossRef] [PubMed]

17. Badrinarayanan, V.; Kendall, A.; Cipolla, R. SegNet: A Deep Convolutional Encoder-Decoder Architecture for Image Segmentation. *IEEE Trans. Pattern Anal. Mach. Intell.* **2017**, *39*, 2481–2495. [CrossRef] [PubMed]

18. Mesbah, R.; McCane, B.; Mills, S. Deep convolutional encoder-decoder for myelin and axon segmentation. In Proceedings of the International Conference on Image and Vision Computing New Zealand, Palmerston North, New Zealand, 21–22 November 2016; pp. 1–6.

19. Guo, H.; Wu, D.; An, J. Discrimination of Oil Slicks and Lookalikes in Polarimetric SAR Images Using CNN. *Sensors* **2017**, *17*, 1837. [CrossRef] [PubMed]

Appl. Sci. **2018**, *8*, 2670

20. Yin, H. Classification of polSAR Images Based on Polarimetric Decomposition. Master's Thesis, University of Electronic Science and Technology of China, Chengdu, China, 2013.
21. Vapnik, V.N. *The Nature of Statistical Learning Theory*; Springer: Berlin, Germany, 2000.
22. Simonyan, K.; Zisserman, A. Very Deep Convolutional Networks for Large-Scale Image Recognition. *arXiv* **2014**; arXiv:1409.1556.
23. Wang, J.; Lin, J.; Wang, Z. Efficient Hardware Architectures for Deep Convolutional Neural Network. *IEEE Trans. Circuits Syst. I Regul. Pap.* **2018**, *65*, 1941–1953. [CrossRef]
24. Zhang, M.; Hu, X.; Zhao, L. Learning Dual Multi-Scale Manifold Ranking for Semantic Segmentation of High-Resolution Images. *Remote Sens.* **2017**, *9*, 500. [CrossRef]
25. He, K.; Zhang, X.; Ren, S. Delving Deep into Rectifiers: Surpassing Human-Level Performance on ImageNet Classification. In Proceedings of the IEEE International Conference on Computer Vision, Santiago, Chile, 11–18 December 2015; pp. 1026–1034.
26. Deep Learning. Available online: http://www.deeplearningbook.org (accessed on 8 April 2016).
27. Loh, W.-Y. *Classification and Regression Trees*; John Wiley & Sons: New York, NY, USA, 2011; Volume 1, pp. 14–23.
28. Svetnik, V.; Liaw, A.; Tong, C. Random Forest: A Classification and Regression Tool for Compound Classification and QSAR Modeling. *J. Chem. Inf. Comput. Sci.* **2003**, *43*, 1947–1958. [CrossRef] [PubMed]
29. Cross-Validation. Available online: http://blog.sina.com.cn/s/blog_688077cf0100zqpj.html (accessed on 14 October 2011).

Article

A High-Resolution Texture Mapping Technique for 3D Textured Model

Jiing-Yih Lai [1,*], Tsung-Chien Wu [1], Watchama Phothong [1], Douglas W. Wang [2], Chao-Yaug Liao [1] and Ju-Yi Lee [1]

[1] Department of Mechanical Engineering, National Central University, Taoyuan 32001, Taiwan; rabbit94577@gmail.com (T.-C.W.); p_watchama@hotmail.com (W.P.); cyliao@ncu.edu.tw (C.-Y.L.); juyilee@ncu.edu.tw (J.-Y.L.)

[2] Ortery Technologies, Inc., New Taipei City 22052, Taiwan; dwmwang@gmail.com

* Correspondence: jylai@ncu.edu.tw

Received: 1 October 2018; Accepted: 8 November 2018; Published: 12 November 2018

Abstract: We proposed a texture mapping technique that comprises mesh partitioning, mesh parameterization and packing, texture transferring, and texture correction and optimization for generating a high-quality texture map of a three-dimensional (3D) model for applications in e-commerce presentations. The main problems in texture mapping are that the texture resolution is generally worse than in the original images and considerable photo inconsistency exists at the transition of different image sources. To improve the texture resolution, we employed an oriented boundary box method for placing mesh islands on the parametric (UV) map. We also provided a texture size that can keep the texture resolution of the 3D textured model similar to that of the object images. To improve the photo inconsistency problem, we employed a method to detect and overcome the missing color that might exist on a texture map. We also proposed a blending process to minimize the transition error caused by different image sources. Thus, a high-quality 3D textured model can be obtained by applying this series of processes for presentations in e-commerce.

Keywords: conformal mapping; mesh parameterization; mesh partitioning; pixel extraction; texture mapping

---

1. Introduction

Two-dimensional (2D) images are commonly used for product presentations in e-commerce because they can reveal the object's texture and are easy to process. However, as 2D images can display only limited views of an object, it may be possible to capture hundreds of 2D images and orient an image at any viewing angle via a web viewer [1]. However, storing and displaying so many images while maintaining high image quality would have huge memory requirements. In addition, the actual three-dimensional (3D) shape and dimensions of an object cannot be obtained in this representation. 3D image-modeling technology is a technique for reconstructing the 3D model of an object by using multiple 2D images while maintaining its texture on the model (called 3D textured model hereafter). If its texture quality can be comparable to that of 2D images, this technology could be used to replace 2D images for product presentations, because a 3D textured model requires less memory and can freely be oriented in 3D space.

Product presentation usually requires a dedicated photography device to catch high-quality object images with known position and orientation in 3D space. The object images can be obtained using a single-camera device that applies a digital single-lens reflex (DSLR) camera to capture an object placed on a turntable, or a multi-camera device that applies several DSLR cameras mounted on an arm to capture an object placed on a turntable from different angles. These devices can position the camera precisely such that the camera information can be calibrated. The object on the turntable can also be oriented to capture object images in different views. These devices also provide a controlled environment, for example, single background color and adjustable lighting, such that the object images and the background color can easily be separated. As these devices are already used in the field of product presentation, we use them as the image source of the 3D-image modeling technology.

3D image-modeling technology primarily involves the generation of two kinds of information, the 3D model of an object and its texture map. The former employs triangular meshes to describe the object's surface geometry, and the latter describes its color information. There is a mapping between the 3D model and the texture map such that when the model is displayed in 3D space, accurate object texture can be displayed accordingly. Approaches to generating 3D models from multiple images can be classified into two groups: shape-from-silhouette (SFS) and shape-from-photoconsistency (SFP). The SFP approach has received extensive attention because it can simultaneously yield a 3D geometric model of an object and its texture map. The main idea of this method is to generate photo-consistent models that can reduce some measure of the discrepancy between different image projections of their surface vertices [2–4]. The main advantage of the SFP approach is that it can generate fine surface details by using photometric and geometric information. However, the reliability of the SFP approach remains a problem because the texture quality can easily be affected by environmental factors such as noise in the colors, inaccuracies in camera calibration, non-Lambertian surfaces, and homogeneous object color.

However, the SFS approach is a common method used to estimate an object's shape from images of its silhouettes [5–7]. This method is essentially based on a visual hull concept in which the object's shape is constructed by the intersection of multiple sets of polygons from the silhouettes of multiple 2D images. With a sufficient number of images from different views, this method can yield an approximate model to describe the outline shape of an object. However, this model is not yet suitable for visualization due to the following two reasons. First, the SFS method can produce visual features on the 3D model, such as sharp edges and artifacts, which do not exist on the real object surface; some virtual features may be sufficient large to affect the outline shape. Second, concavities on the object surface are often formed as convex shapes because these are invisible on image silhouettes. Therefore, a quality improvement method must be implemented to remove virtual features while recovering the smoothness of the model [8]. The removal of artifacts is particularly important because they are difficult to detect and eliminate.

Texture mapping generally includes multiple techniques, such as mesh partitioning, mesh parameterization, texture transferring, and correction and optimization, which are related to each other and affect the texture quality. Research in mesh partitioning can be summarized using several different approaches. Shamir [9] categorized several methods of mesh partitioning according to segmentation type, partitioning technique, and segmentation criterion. Segmentation type refers to surface-type and part-type. Surface-type mesh partitioning is commonly used in texture mapping [10–12] because it can prevent large distortion in mesh parameterization. Mangan et al. [13,14] and Lavoué et al. [15] proposed a constant curvature watershed method to separate a mesh model into several regions. Other applications of surface-type partitioning include remeshing and simplification [16], mesh morphing, and mesh collision detection [17]. Part-type mesh partitioning is commonly used for part recognition on a mesh model composed of multiple parts. Mortara et al. [18,19] proposed a partitioning method by applying the curvature information at the transition of different parts to decompose a mesh model. Funkhouser et al. [20] proposed another method by establishing the database of some known parts for the separation of a mesh model. Partitioning techniques include region growing, hierarchical clustering, iterative clustering, and inferring from a skeleton, which can be implemented either alone or together. Segmentation criterion approaches include dihedral angle or normal angle, geodesic distance, and topological relationship, which can also be implemented either alone or together.

Mesh parameterization was classified in accordance with distortion minimization, boundary condition, and numerical complexity [21,22]. Distortion minimization can be summarized based on three types: angle, area, and distance. For angle minimization, an objective function is formulated to minimize the distortion of 2D meshes on the UV domain. Several methods can be employed for angle minimization. Lévy et al. [11] proposed a least-squares approximation of the Cauchy-Riemann equations to minimize both angle and area distortion on 2D meshes. Desbrun et al. [23] presented an instinct parameterization to minimize angle distortion. These two methods allow free boundaries and linear numerical complexity. Sheffer et al. [24] optimized the angles on the UV domain based on angle-base flattening. This method sets constraints on the topology of triangular meshes to preserve the correctness of 2D meshes. Sheffer et al. [25] proposed a hierarchical algorithm to improve the optimization efficiency for the case of huge triangular meshes, and Zayer et al. [26] proposed a method to solve the optimization problem for a set of linear equations that were derived based on the angle-base flattening approach with a set of constraints specified. In addition, the barycentric mapping is commonly used for mapping 3D meshes onto the UV domain in mesh processing. Tutte [27,28] proposed an algorithm to embed a 3D mesh onto the UV domain by evaluating the barycentric position in terms of its neighboring meshes. Eck et al. [29] proposed an algorithm to calculate the multiresolution form of a mesh via a barycentric map. Floater [30] applied a "shape-preserve" condition for the barycentric map to preserve the shape of 2D meshes on the UV domain. Floater [31] and Floater et al. [32] further applied mean-value weights for the barycentric map to preserve the shape of 2D meshes. For all above-mentioned barycentric mapping, the boundary is fixed and the numerical complexity is linear, which is not suitable for texture editing. For texture mapping, a method of free boundary is more appropriate as it can ensure that the boundary of each island of 2D meshes is close to the real profile, making the texture editing easy. Some other approaches have focused on minimizing the area distortion [33] and distance [34].

For texture map generation, the main idea is to deal with the texture transferring problem. Niem et al. [35] proposed a texture transferring method by identifying the most appropriate image source for a group of meshes. They also minimized the color inconsistency at the transition of two different groups and synthesized the invisible meshes using the color of neighboring pixels. Genç et al. [36] proposed a method to extract and render the texture dynamically. The extraction was implemented by horizontally scanning the pixels and rendering every color onto the meshes. Baumberg [37] proposed a blending method to handle the color difference between two different images. The images were separated into high and low bands; the low band images were averaged to minimize the color difference, whereas the high band images were kept to preserve the outline profile. In addition, texture synthesizing is commonly used to improve the transition between different textures. Efros et al. [38] proposed an image quilting method to quilt together different texture patterns. They extended the boundary of each original pattern and calculated the minimum color difference on the overlapping area to find the new boundary between two patterns. Wei et al. [39] proposed an algorithm to synthesize the texture pattern based on deterministic searching and use tree-structured vector quantization to improve the efficiency. These two approaches focus mainly on the transition synthesis between two texture patterns.

2. Problem Statement

For product presentations in e-commerce, texture quality is the most crucial issue to investigate because it directly affects the visualization effect. Ideally, the texture quality at any view in 3D space should perfectly match that of the corresponding 2D image. Actual texture on the 3D model, however, is usually worse than that of 2D images, mainly because individual texture on the 3D model comes from different image sources. A 3D model reconstructed using multiple images of an object is only an approximation of the object geometry. The camera model and calibration method used to estimate the camera parameters might yield additional errors in the position and orientation of the object images. These errors, combined with errors caused by the texture mapping process, might lead to discrepancy between the texture of the 3D model and the real object. Any defect in the 3D texture could negatively impact perceptions of the product being presented.

The following are typical problems involving the 3D texture:

1. Reduced texture resolution: The texture resolution at any view in 3D space is worse than that of the corresponding object image, primarily because of inappropriate scaling of the pixels between the real image domain and the texture mapping image domain.
2. Missing color on some mesh regions: All 2D meshes on the texture domain should ideally be color-filled, but some may be missed if they are beyond the boundary of the object image, primarily because of insufficient accuracy of the 3D model, especially for those meshes near the image silhouette.
3. Photo inconsistency at the transition of different image sources: Photo inconsistency usually occurs along the boundary of different groups of meshes, with each group textured by different image sources. This problem is the combined effect of insufficient accuracy of the 3D model and the camera parameters.

Thus, we develop a texture mapping algorithm that focuses on detecting and removing these problems.

The objective of this study is to develop a high-quality texture mapping algorithm that can be combined with a 3D modeling algorithm to generate the 3D textured model of an object for use in e-commerce product presentation. High-quality texture here indicates that the texture at any view in 3D space should be as close as possible to that of the corresponding 2D image, which mainly requires maintenance of the resolution on the texture and elimination of photo-inconsistent errors at the transition of different image sources. A general texture mapping process comprising the following three techniques is proposed: mesh partitioning, mesh parameterization and packing, and texture transferring. Specific efforts are made at each step to initially eliminate problems that might affect the texture of the 3D model. To further reduce the discrepancy of the texture owing to insufficient inaccuracy of the 3D model and camera parameters, a correction and optimization algorithm is presented. The entire texture mapping process is fully automatic and is intended to be used for all kinds of objects.

The main contribution of the proposed texture mapping method is as follows. First, we enhance the techniques of converting 3D meshes onto the UV domain so that the shape of most 2D meshes can be preserved and the finest resolution can be obtained in texture transferring. Three main techniques in converting 3D meshes onto the UV domain are mesh partitioning, mesh parameterization and packing. In the proposed mesh partitioning algorithm, a novel chart growth method is proposed to partition 3D meshes iteratively so that each chart of 3D meshes can be as flat (disk type) as possible, which can reduce the error of 2D meshes in mesh parameterization. In the proposed mesh parameterization algorithm, a novel conformal mapping method is proposed to preserve the shape of 2D meshes as close to that of 3D meshes as possible. In the proposed packing method, all regions of 2D meshes are tightly packed in a rectangular area to acquire the finest resolution. Second, we propose an optimized texture transferring algorithm for generating the texture map, emphasizing the elimination of erroneous texture mapping owing to insufficient accuracy of the 3D model as well camera parameters, and the improvement of the texture resolution as close to that of 2D object images as possible. The strategies used in the proposed algorithm include: (1) increase overall texture size in pixels; (2) increase the number of pixels occupied by each 2D mesh; (3) detect and fill in void meshes; and (4) perform texture blending at the boundaries of mesh islands. The first two operations can improve the resolution of the final texture map, whereas the last two operations can eliminate erroneous texture mapping. Several realistic examples are presented to verify the feasibility of the proposed texture mapping method. The results are also compared with those form commercial software.

3. Overview of the Proposed Method

The 3D textured model is created by covering a 3D model with a texture map that stores the color information of the object. The main idea of direct texture mapping is to generate the texture of the 3D model by directly using the object images. Figure 1 shows the overall flowchart of the proposed texture mapping method. The input data are the 3D model of an object and multiple object images from different views (Figure 1a). The original 3D model was generated from silhouettes of the object images using an SFS method. However, the surface quality of the original meshes was not satisfactory because of artifacts and virtual features affecting the outline shape, as well as the surface smoothness. A mesh optimization algorithm combining re-meshing, mesh smoothing, and mesh reduction was employed to eliminate the effect of the above-mentioned phenomena and yield an optimized mesh model [8]. The model after mesh optimization served as the input of the proposed texture mapping algorithm.

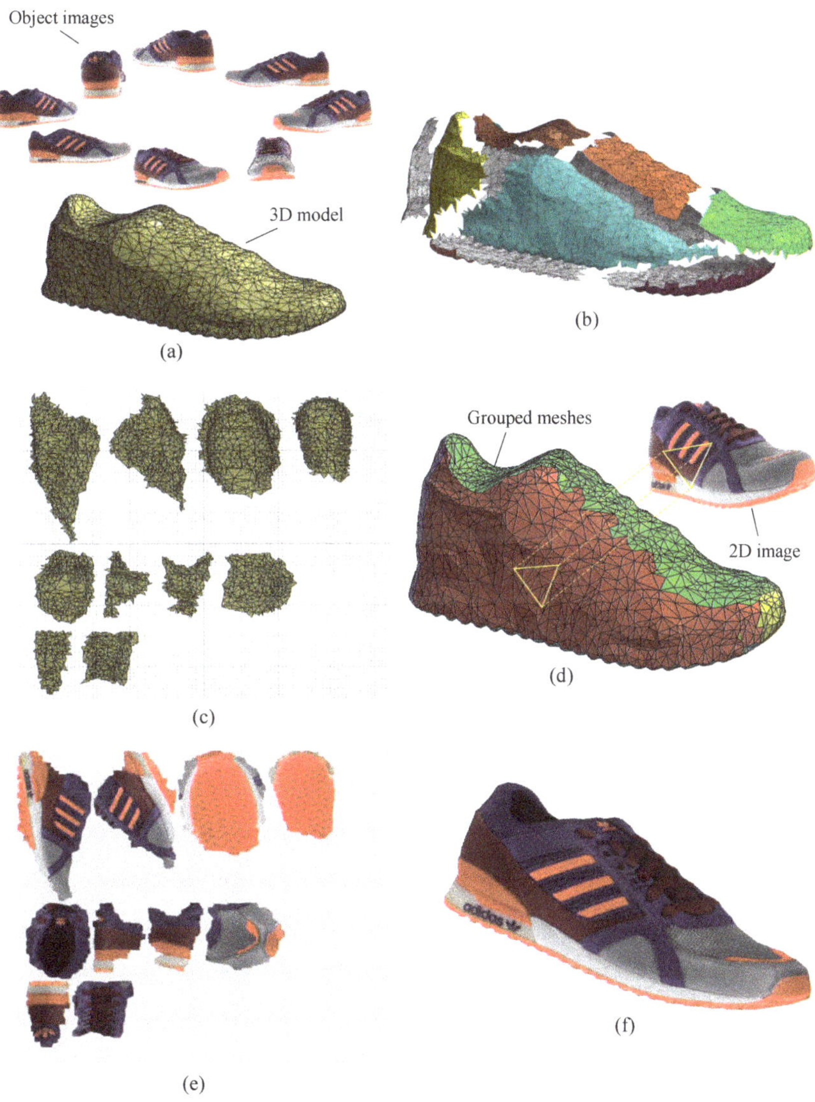

Figure 1. Overall flowchart of the proposed texture mapping method: (**a**) input data, (**b**) mesh partitioning, (**c**) mesh parameterization and packing, (**d**) texture transferring, (**e**) correction and optimization, and (**f**) output object file.

In the proposed texture mapping algorithm, mesh partitioning is first implemented to subdivide the 3D model into several charts (Figure 1b), each of which is later individually mapped onto the

UV domain. Mesh partitioning is based on a chart growth method to assign a weight to each mesh on the model, and grow each chart of meshes one by one from a set of initial seed meshes. The seed meshes are optimized in an iteration process until all meshes have been clustered. This ensures that all charts are flat and compact in the boundary for easy mapping in the mesh parameterization. Mesh parameterization and packing is then implemented to map the meshes on each chart and to pack all 2D meshes on the UV domain (Figure 1c). An angle-preserving algorithm is proposed to optimize the mapping between the 3D and 2D domains, which can preserve the shape of most 2D meshes. Furthermore, all 2D meshes are tightly packed in a rectangular area to acquire the finest resolution when mapping the pixels from the image domain to the texture domain.

Next, texture transferring is implemented to extract pixels from the image domain, and place them on the texture domain appropriately (Figure 1d). This procedure comprises three main steps: grouping the 3D meshes, extracting pixels from the image domain, and placing pixels onto the texture domain. We also propose a method to analyze the texture resolution. The proposed texture transferring algorithm ensures that the texture resolution can be set to the equivalent of the 2D images. Finally, we implement correction and optimization of the texture to eliminate erroneous color mapping that might occur due to the insufficient accuracy of the 3D model and camera parameters and to improve the photo consistency at the boundary of different image sources (Figure 1e). Several photo inconsistent problems are detected and solved one by one. The output texture map is saved as a universal data format (*.obj), which can be displayed with a website viewer (Figure 1f).

4. The Proposed Texture Mapping Method

4.1. Mesh Partitioning

The purpose of mesh partitioning is to partition 3D meshes into several charts, where a chart denotes a group of meshes that are tightly connected to each other and form a boundary loop only. When a chart is mostly flat and compact in boundaries, it is easy to preserve the shape in mesh parameterization. By contrast, when a chart is bent too much or closed on both sides, that is, two boundary loops, the shape distortion in the mesh parameterization increases, thereby reducing the texture resolution in some regions. A conventional approach to dealing with this issue is to map each mesh on the 3D model onto the UV domain independently, which can accurately preserve the shape of all 2D meshes and pack them all tightly row by row [40]. However, this approach might result in an un-editable texture map, because all 2D meshes are independently projected and distributed irregularly.

The proposed mesh partitioning technique essentially assigns a cost to each mesh, which denotes a mesh's weight calculated by considering the flatness and distance of the mesh with respect to a chart. An iterative procedure combining chart growth and seed mesh upgrades is implemented to expand and modify charts as well as seed meshes in sequence. The chart growth is a process to cluster all meshes into charts in accordance with each mesh's cost. When a closed chart is detected as possibly occurring, a new seed mesh is added to separate the chart into two. The seed mesh upgrading is a process to upgrade the seed mesh of each chart that has been expanded. Whenever a chart is grown, its seed mesh is recomputed by putting it near the center of the new chart.

Two costs are defined and used in chart growth and seed mesh upgrading. The cost used in chart growth is defined as

$$Cost1(F, F') = (1 - (N_C \cdot N_{F'}))(|P_{F'} - P_F|), \tag{1}$$

where $Cost1(F, F')$ denotes the weight of a candidate mesh F' neighboring a chart C, F is the neighboring mesh of F' that has been in C, N_C is the normal vector of C evaluated by the average of all normal vectors of the meshes in C, $N_{F'}$ is the normal vector of the candidate mesh, and $P_{F'}$ and P_F are the centroids of F' and F, respectively. Equation (1) indicates that the cost $Cost1(F, F')$ considers both the flatness and distance of F' with respect to the chart C. The cost used in seed mesh upgrading is defined as

$$Cost2(F, F') = |P_{F'} - P_F|,\tag{2}$$

which is used to determine the mesh F that is closest to the candidate mesh F'.

Figure 2 depicts the flowchart of the proposed mesh partitioning algorithm, which has three main steps: initial seed meshes, chart growth, and seed mesh upgrading. In step 1, a set of seed meshes are initially assigned on the input 3D meshes. A default value of 10 is typically used and the seed meshes are randomly selected from the 3D meshes. Each seed mesh is initially assigned as a chart.

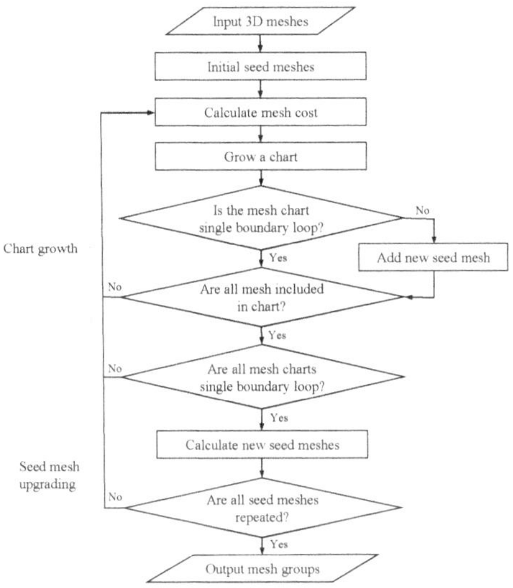

Figure 2. Flowchart of mesh partitioning.

In step 2, chart growth, all meshes neighboring the charts are found by using the topological data of the mesh model. Equation (1) is then employed to evaluate a cost for each of these meshes. The costs are sorted from minimum to maximum. The mesh with the minimum cost is selected to cluster with its neighboring chart. Three criteria are then checked in sequence. First, is this chart (which has just grown) a single boundary loop? If yes, go to the next criterion. If not, a new seed mesh is added. The last mesh added to this chart is regarded as the new seed mesh. Second, are all meshes clustered into charts? If yes, go to the next criterion. If not, go back to the beginning of this step. Third, are all charts a single boundary loop? If yes, this step is finished. If not, go back to the beginning of this step. Notably, after step 2, all meshes are clustered into charts.

In step 3, seed mesh upgrading, the seed mesh on each chart is recomputed. The upgraded seed mesh is located near the center of the chart, which is achieved by a reverse searching process from the boundary of the chart. Starting from a mesh on the boundary, Equation (2) is repeatedly employed to find a loop of meshes around the boundary of the chart. The same search is repeated from outside to inside to yield several layers of loops. The final mesh on the last loop is regarded as the upgraded seed mesh. If all upgraded seed meshes are identical to the ones in the previous iteration, this indicates that all charts obtained are converged, and the entire process is finished. Otherwise, we return to the beginning of step 2 to regenerate all charts with the upgraded seed meshes. Table 1 lists the process (CPU) time required vs. number of meshes for the case "Shoe 1". It is noted that the number of meshes used in this study is only 4500 as the model is to be used on a web viewer. Therefore, the computational time in this case is sufficiently fast.

Table 1. Time consuming for mesh partitioning.

Object	Number of Meshes	No. of Initial Seeds	No. of Final Charts	Total Time (s)
	4500 *	10	10	0.246
	10,000	10	30	1.812
	20,000	10	66	7.075
Shoe 1	30,000	10	111	18.431
	40,000	10	148	31.603
	50,000	10	202	44.574

* used in the case study herein.

4.2. Mesh Parameterization and Packing

After mesh partitioning, the 3D model can be separated into several disk-type mesh groups. This series of mesh groups is flattened onto the 2D domain based on an angle-preserving and conformal mesh parameterization. The main idea of this parameterization method is to make the difference between angles in the 3D and 2D domains as small as possible. Several topological constraints are also applied during the optimization of the angles to ensure the topological correctness on the 2D domain. The proposed angle-based flattening method sets three kinds of mesh-topology constraints, namely, triangle, vertex and wheel consistencies, as shown in Figure 3a–c. This series of topological constraints can be formulated as the following objective function in a linear system:

$$\begin{bmatrix} 10101 & \cdots & 0 \\ \vdots & & \\ 10110 & \ddots & \vdots \\ \vdots & & \\ cot(\varphi)0cot(\omega)\,0 & \cdots & 0 \end{bmatrix} \begin{bmatrix} \varepsilon_1 \\ \varepsilon_2 \\ \varepsilon_3 \\ \vdots \\ \varepsilon_{n\times3} \end{bmatrix} = \begin{bmatrix} 180 - (\alpha + \beta + \gamma) \\ \vdots \\ 360 - (\theta_1 + \ldots \theta_d) \\ \vdots \\ (log(sin(\varphi)) - log(sin(\omega))) + \ldots \end{bmatrix}, \quad (3)$$

where n is the number of meshes; ε_i is the error of the angle on the ith mesh; α, β, and γ are the angles on each mesh; θ_d is the angle around the inner vertex; d is the number of angles around the inner vertex; and φ and ω are the angles on two adjacent meshes, respectively, corresponding to the common edge. Equation (3) is essentially $Ax = b$, where the errors ε_i, $i = 1 \ldots n \times 3$ can be minimized. The optimized angles on the 2D domain can then be obtained by adding the errors and the original angles together.

The new vertices on the 2D domain must be calculated in accordance with the optimized angles. Let three vertices of a triangle be e_1, e_2 and e_3, and the corresponding angles be α_1, α_2, and α_3, respectively. The calculation of the new vertices on the 2D domain uses the following least-squares approximation:

$$Q_{obj} = min \sum_j \left[\left(e_3^j - e_1^j \right) - \frac{sin\alpha_2^j}{sin\alpha_3^j} R_{\alpha_1^j} \left(e_2^j - e_1^j \right) \right]^2, \quad (4)$$

where R is a rotation matrix with angle α_1, and j is the jth iteration. Assume that the two vertices e_1 and e_2 of a triangle are known. Equation (4) employs the known vertices e_1 and e_2 to optimize the unknown vertex e_3, where Q_{obj} is the objective function for the optimization. For all 2D meshes, if the first two vertices on a mesh can be determined, the remaining vertices can be evaluated by using the least-squares approximation [11], which is formulated as a set of linear equations. The topology of all vertices on the UV domain can be maintained correctly.

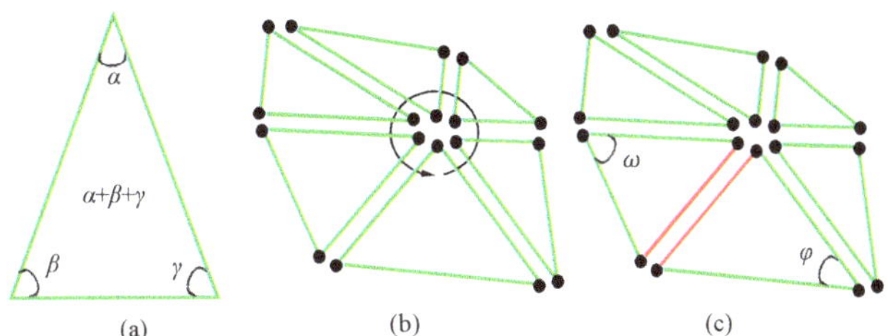

Figure 3. Three kinds of mesh-topology constraints in mesh parameterization: (**a**) triangle consistency, (**b**) vertex consistency, and (**c**) wheel consistency.

The parameterized mesh islands are all independent. This series of mesh islands needs to be packed together onto the UV map. The UV map is essentially a kind of image that records all 2D meshes and is of the same image size as the texture map. The process of collecting all mesh-islands and converting them into the UV map is called packing. The objective in packing is to let each mesh island occupy as much space as possible, thereby maintaining the resolution of the texture as close to that of the 2D images as possible. Therefore, we consider how to efficiently arrange the mesh islands on the UV map. First, an oriented boundary box (OBB) method [41] is employed to construct a best-fit boundary box for each mesh island, as shown in Figure 4a. Using the OBB method to arrange the islands ensures that less space on the UV map is wasted compared to when using the axis aligned bounding box (AABB) method shown in Figure 4b. Next, the mesh islands are arranged together according to their OBB lengths on the UV map, as shown in Figure 4c.

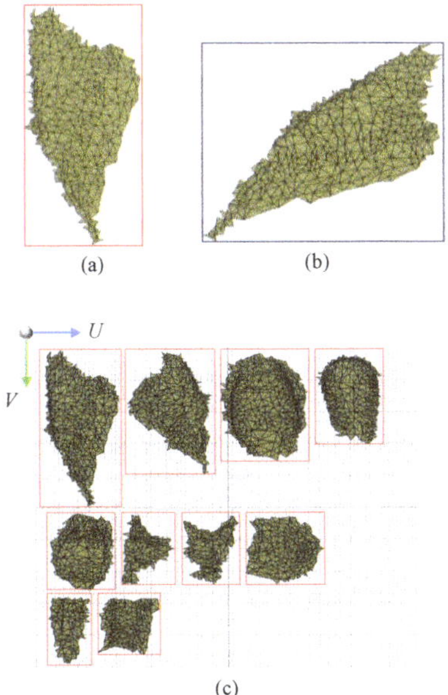

Figure 4. Mesh-islands packing: (**a**) the oriented boundary box (OBB) method to determine the boundary box, (**b**) the axis aligned bounding box (AABB) method to determine the boundary box, and (**c**) packing of all mesh-islands.

4.3. Texture Transferring

Texture transferring is essentially a process which yields a texture map by filling in each pixel on and inside the mesh islands on the UV map with a color extracted from the object images. The following sentences describe the basic idea of this algorithm (see Figure 5). For each 3D mesh, we allocate the most appropriate object image (called front image hereafter) and extract a triangular range of pixels and color information for this mesh. We can also find a triangular range of pixels on the UV map for the same mesh. However, two pixel ranges might not be the same. Therefore, we perform a transformation for pixel mapping between these two domains. The texture transferring algorithm has three main steps: grouping the 3D meshes, extracting the pixels from object images, and placing the pixels onto the UV map. A detailed description for each step is given below.

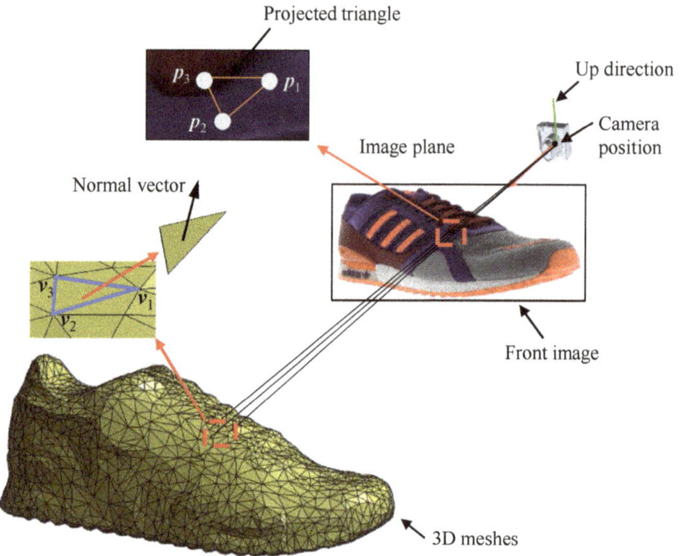

Figure 5. Texture transferring.

4.3.1. Grouping the 3D Meshes

The purpose of this step is to allocate each mesh to a front image and put all meshes that use the same front image in a group. Each mesh can be projected onto several candidate images. The candidate image that yields the largest projected area and hence the highest texture resolution is chosen as the front image. Ideally, all object images could be regarded as the candidate images and selected by all meshes. However, erroneous texture mapping might occur owing to insufficient inaccuracy of the 3D model, as well as camera parameters. A seam line is a photo-inconsistent phenomenon that often occurs at the transition of two different image sources. As the number of candidate images increases, so does the possibility of seam lines. Therefore, to reduce the occurrence of seam lines, we only select some object images as the candidate images and perform mesh grouping.

The algorithm of grouping is as follows. A series of pieces of camera information corresponding to the object images and the 3D meshes are the input. One of the important parameters is the looking vector, which represents the camera viewing direction and is perpendicular to the image plane. In addition, each of the meshes has its own surface normal. The grouping criterion is based on the angle between the looking vector of an image and the surface normal of a mesh. The front image of a mesh is defined as the image with the minimum angle among a set of candidate images. It can yield the largest projected area when projecting the mesh onto the front image. All meshes that use the same front image can thus be grouped.

Visibility should be considered when grouping meshes. The following two criteria are checked to detect the visibility of a mesh. First, the angle between an image and a mesh must be less than $90°$. This criterion is employed to ensure that the image faces the front side of the mesh. Second, this mesh cannot be obstructed by other meshes that use the same front image. An obstruction check in terms of the above two criteria could be developed by comparing each mesh with all other meshes. However, it would require substantial computational time. A cell subdivision algorithm [42] is employed to check the possibility of mesh obstruction, which can save the computational time efficiently. The visibility check can prevent the occurrence of mesh obstruction for all meshes on the same group. When the visibility problem occurs on an image, the front image can be selected from one of its two neighboring images.

After these processes, the meshes are grouped. The existence of isolated meshes may result in additional seam lines. An isolated mesh is a small mesh island, which has a front image different to its surrounding meshes. As the boundary of the mesh island represents two different image sources, seam lines easily occur around the boundary of the mesh island. Therefore, when a mesh island is detected, its image source is changed to that of its surrounding meshes. Figure 6 depicts the grouping result of an example using six candidate images.

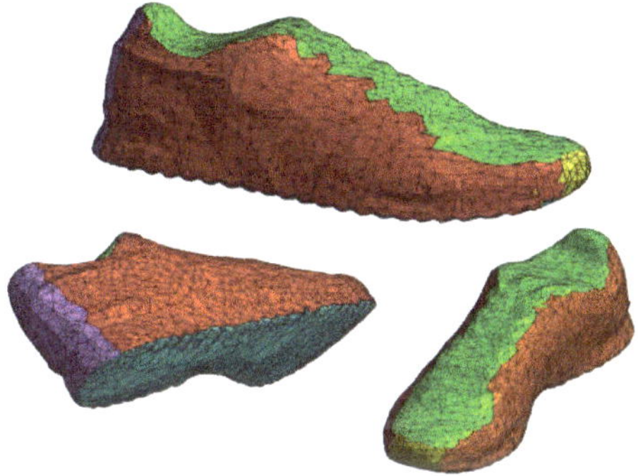

Figure 6. The grouping result of a shoe example using six candidate images.

4.3.2. Extraction of Pixels from the Object Images

The purpose of this step is to extract pixels from the front image with respect to a 3D mesh. A prospective projection is performed to project 3D meshes back to the image domain. As Figure 5 depicts, the triangle $\Delta p_1 p_2 p_3$ denotes the projection of a 3D mesh $\Delta v_1 v_2 v_3$ onto the image domain. All pixels and color information on and inside this triangle represent the corresponding texture for the 3D mesh. The extraction of a pixel inside a triangle is explained below. The image is made up of pixels in a grid plane containing horizontal and vertical lines, which gives each pixel a unique coordinate. A scanline method is implemented to compute all pixels inside a triangle. The scanline shown in Figure 7 intersects two triangle edges, which yields the two endpoints of the line segment inside the triangle. All pixels on this line segment can then be evaluated in sequence. An endpoint of the line segment can be evaluated by using the following equation:

$$\delta_x = X_2 - \left[\frac{(Y_2 - Y)\cdot(X_2 - X_1)}{(Y_2 - Y_1)}\right],$$

(5)

where (X_1, Y_1) and (X_2, Y_2) denote two vertices of an edge on the triangle, Y is the vertical coordinate value of the current scanline, and δ_x is the horizontal coordinate value of the endpoint on this edge. Equation (5) is applied twice on the left and right edges, respectively, for each scanline.

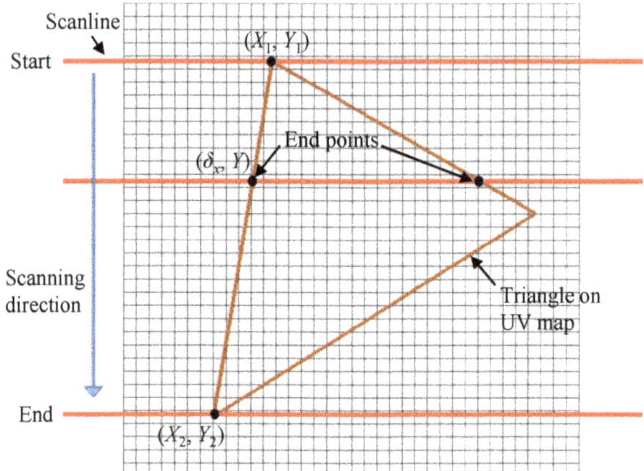

Figure 7. A scanline method to evaluate all pixels inside a triangle.

4.3.3. Placement of Pixels onto the UV Domain

The final step is the placement of pixels onto the UV map. The pixels with respect to each 2D mesh are evaluated in the previous step. However, as each 2D mesh on the UV domain is different from the projected mesh on the image domain, the pixels on these two-pixel domains do not have a one to one correspondence. Therefore, a transformation algorithm must be employed to map the pixels between these two domains. The proposed algorithm is explained below. The three vertices of a mesh on the image domain are respectively mapped onto the corresponding three vertices on the UV domain by using the following equation:

$$aX + bY + c = X',$$ (6)

$$dX + eY + f = Y',$$ (7)

where X and Y denote the coordinates of a vertex on the image domain, and X' and Y' denote the coordinates of the corresponding vertex on the UV domain. The parameters a to f can be evaluated as all three pairs of vertices on the image and UV domains are given. Once a to f corresponding to a triangle are obtained, the colors of all pixels within this triangle can thus be interpolated by using Equations (6) and (7). Therefore, all pixels of different triangles on the UV domain can be filled in with correct colors, which yield the texture map for all 2D meshes.

4.4. Texture Correction and Optimization

The purpose of this study is to generate a high quality texture for a 3D mesh model. Thus, the texture correction and optimization need to be investigated to ensure that the texture quality is similar to that of original 2D images. There are four key issues to study: packing the meshes on the UV domain efficiently, arranging the pixel resolution on the texture map, eliminating the influence of geometric error on the 3D model, and blending the texture at the transition of different images. For the first issue, the main idea has already been described in Section 4.2. The meshes can be packed efficiently on the UV map by applying the OBB method to each mesh island, which can yield a smaller boundary box for each mesh island packed on the UV map as compared with the AABB method. In this way, the overall space required for the OBB method is more compact than that without applying the OBB method. Hence, each 2D mesh can allocate more pixels on the UV map, which is especially useful for small meshes with respect to preserving the texture resolution.

The next issue is arranging the overall resolution of the texture map. An object image only partially covers the texture of an object. However, a texture map must cover the entire object texture.

If the texture size of a texture map is the same as that of an object image, the image resolution of the texture map is worse than that of the object image. The texture size of an object image used is 5184×3456, whereas the original texture size for a texture map is 4096×4096. After a careful comparison of several kinds of image resolution, the texture size of the texture map is expanded to 8192×8192, with a texture space four times larger than before. This kind of texture size ensures that the pixel number within a mesh on the texture map is close to that of the same mesh projected onto an object image. The original high-quality image information can therefore be kept on the final texture map.

The texture information is extracted from an object image by projecting a 3D mesh onto the corresponding image plane. Normally, a projected mesh is completely inside an image silhouette, and the corresponding range of pixels can be extracted from the projection. However, due to the insufficient accuracy of the 3D model, some of the meshes could be wrongly projected and are partially or completely outside the image silhouette, such as the example in Figure 8a. When a projected mesh is not completely inside the image silhouette, no matching texture can be obtained, and hence the corresponding color is void. To deal with this kind of problem, it is necessary to detect each occurrence of this kind of mesh, and change the front image for each of them. The detection is based on the background removal of object images. First, the object image is converted into a binary image by verifying the foreground and background information. An alpha channel, which records the transparency of each pixel on an object image, is saved and associated with the object image after background removal. This process can be used to verify the foreground and background information of the object image. We convert the object image into black and white in accordance with the data on the alpha channel, such as the example in Figure 8b, where the pixels in white and black denote inside and outside the object, respectively. This additional image is used to check if a projected mesh is outside the image silhouette during the texture transferring process. Since the transferring is scanned pixel by pixel, the black color can be detected and the mesh that covers the black color can be marked for further correction later. Figure 8c depicts the meshes covering pixels of black color, and are marked to individually change their front images. For each of this kind of mesh, the new front image is determined by choosing one of the two neighboring images of the original front image.

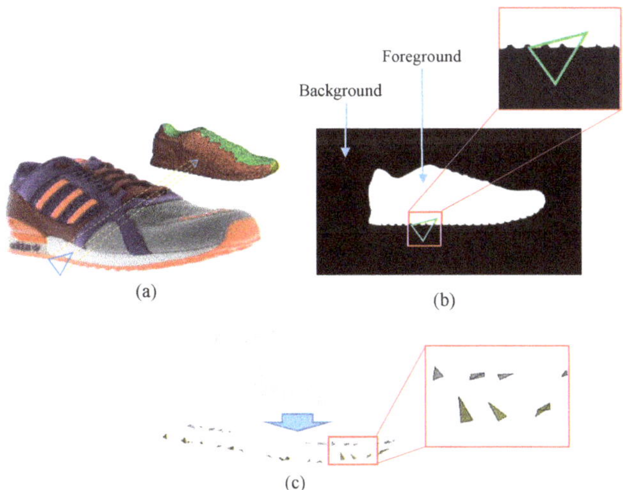

Figure 8. Detection and removal of meshes with missing color: (**a**) the projected mesh outside the image silhouette, (**b**) the image converted into foreground (white) and background (black) in accordance with the alpha channel, and (**c**) meshes detected outside the image silhouette.

The final optimization is to blend the texture at the transition of different images. The texture is extracted from different front images. However, the texture between different image sources may be inconsistent in color. This difference will cause seam lines on the 3D textured model. The blending between two texture sources can be performed to optimize the color consistency on the model. The boundary meshes should be detected first. The blending is based on the pixel distance to the boundary edge. The equation of color blending is

$$P'(i) = (P_m(i) \times D_f + P_n(i) \times (D_f - D_c))/(2 \times D_f - D_c),\qquad(8)$$

where $P'(i)$ denotes the blending pixel color of the mesh, $P_m(i)$ denotes the main pixel color of the mesh, $P_n(i)$ denotes the neighboring pixel color of the mesh, D_f denotes the farthest pixel of the mesh, and Dc denotes the current pixel of the mesh. Figure 9a depicts the blending of two pixel colors on two neighboring meshes. A linear variation on the weight for blending is applied so that when the distance of the pixel is close to the boundary edge, the weight is larger; whereas, when the distance of the pixel is further from the boundary edge, the weight decreases linearly. That is, the original color information on each mesh is kept if the pixel is far from the boundary edge. In this way, the seam lines on the model can be eliminated to support the consistency of the 3D textured model. Figure 9b shows one example to illustrate the effect of blending, where the left and right plots indicate the results before and after blending, respectively.

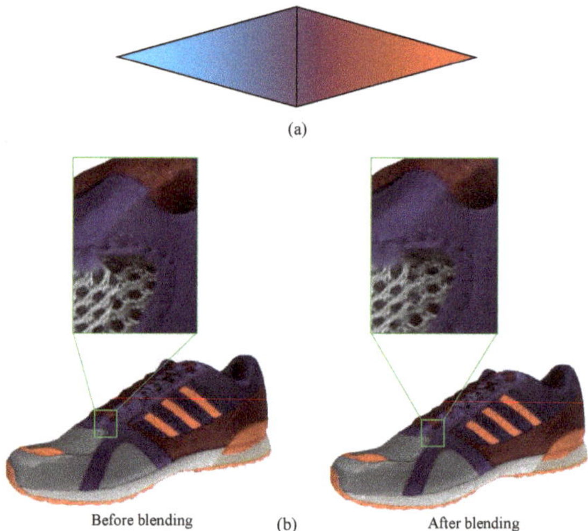

Figure 9. Texture blending at the transition of different images: (**a**) the blending of two neighboring meshes, and (**b**) a shoe example before and after blending.

5. Result and Discussion

The results of the texture map and 3D textured model for six examples are depicted in Figure 10a–f, where the left and right images in each figure panel denote the 3D textured model and the texture map, respectively. The entire texture mapping process is done automatically, with a 3D model and 16 object images in different views as inputs, and the corresponding texture map as the output. The texture size for all six examples is 8912 × 8912. The proposed process includes the following key procedures: mesh partitioning, mesh parameterization and packing, texture transferring, and correction and optimization of the texture. The initial number of seeds on mesh partitioning is set to 10, and the final number of mesh islands generated for all six examples is 10–13. Each of the results in Figure 10 can be demonstrated as a high-quality 3D textured model by applying the texture correction and

optimization during the texture generation process. The results with and without texture correction
and optimization are further discussed below.

(a)

(b)

(c)

Figure 10. *Cont.*

(d)

(e)

(f)

Figure 10. The results of the texture map and 3D textured model for six examples: (**a**) shoe 1, (**b**) microphone, (**c**) shoe 2, (**d**) cup, (**e**) shoe 3, and (**f**) statue.

The first optimization process is mesh island packing. When the AABB method is employed (Figure 11a), the bounding box of each mesh island is larger, and the empty space inside each boundary box is also larger. When all these boundary boxes are packed onto a UV map of fixed size, each mesh island is over-compressed and loses the texture resolution that it should have. By contrast, when the OBB method is employed (Figure 11b), each boundary box can best fit its mesh island so that the space that a mesh island occupies is more compact. In addition, the previous resolution of the texture map was 4096 × 4096 pixels. To maintain the resolution 5184 × 3456 of the original image, the larger resolution 8192 × 8192 has been applied to enhance the quality of the final texture. The texture space is four times larger than before. Therefore, each mesh island can be allocated more pixel space when all boundary boxes are packed on the same UV map. Figure 12 depicts the distribution of the mesh number on each range of pixel numbers for the following four cases: 8192 × 8192/OBB, 8192 × 8192/AABB, 4096 × 4096/OBB, 4096 × 4096/AABB and commercial (3DSOM) software [43], where 3DSOM is commercial software. When the number of meshes with fewer pixels is reduced, the texture resolution is closer to that of the original images. It is evident that the texture resolution of the case 8192 × 8192/OBB is the best among the five cases because it has the minimum number of meshes with fewer pixels. In addition, the texture resolution of 3DSOM software is the worst as most of meshes have pixels less than 2000. Therefore, the texture resolution of the proposed method is better than that of 3DSOM software. Figure 13 depicts a local region of the texture for three cases, 3DSOM software, 4096 × 4096/AABB and 8192 × 8192/OBB. The result clearly indicates that the sharpness of the texture in Figure 13c is better than that in Figure 13a,b. The 3DSOM software blends the color with a low-pass filtered image, which will result in a loss on the texture resolution. This result indicates that the proposed method can yield a better texture resolution than 3DSOM software.

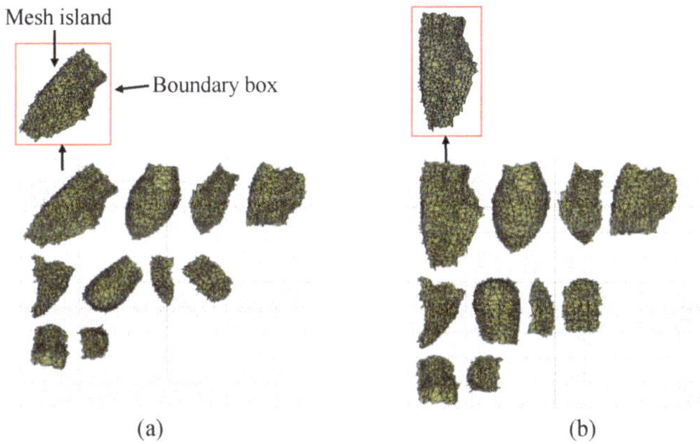

Figure 11. The results of mesh-island packing for two methods: (**a**) AABB method and (**b**) OBB method.

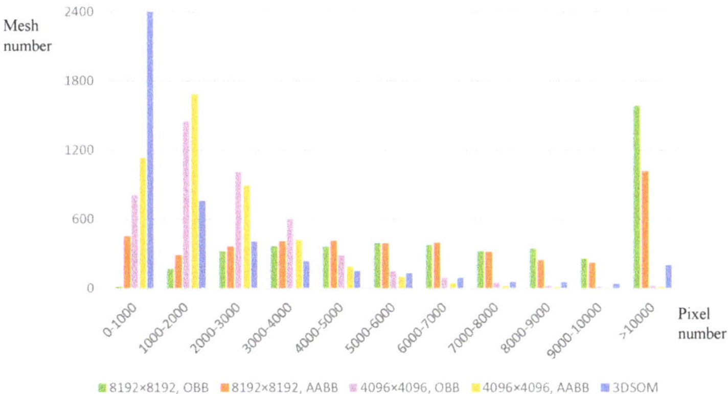

Figure 12. The bar chart of mesh number vs. pixel number for five cases: 8192 × 8192/OBB, 8192 × 8192/AABB, 4096 × 4096/OBB, 4096 × 4096/AABB and 3DSOM software.

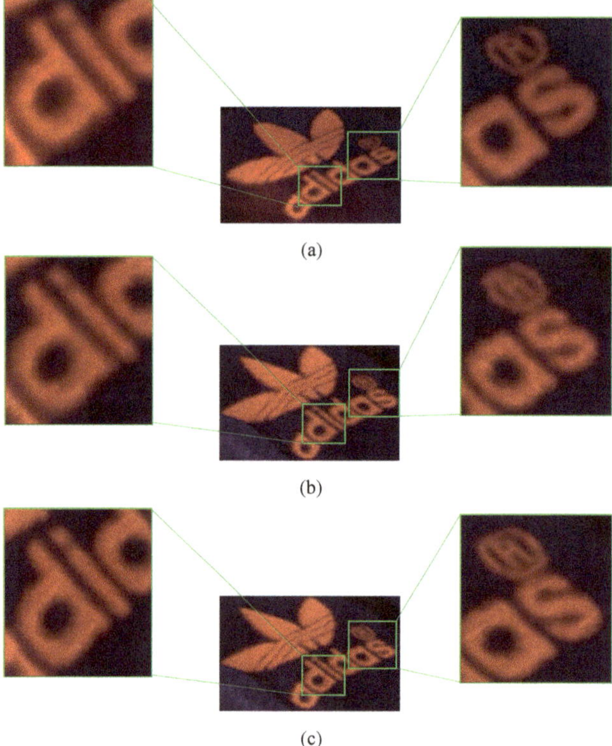

Figure 13. The comparison of texture quality for three cases: (**a**) 3DSOM software, (**b**) 4096 × 4096/AABB and (**c**) 8192 × 8192/OBB.

The next optimization process is the elimination of the texture defects caused by the geometric error. The background color of the image might be wrongly extracted for some meshes near the image silhouette, resulting in white spots on the 3D textured model. The incorrect extraction is caused by the meshes that are located outside the image silhouette when they are projected onto the front image. Thus, we wish to eliminate the influence of the error. Figure 14 depicts the comparison of 3DSOM software, the previous result, and the proposed result where, for the previous result, no action was taken to deal with this problem, and for the proposed result, the data on the alpha channel of each object image was employed to detect this problem, and then its front image was replaced where necessary. It is evident that white background spots appear both on the result of 3DSOM software and previous result, they have been eliminated on the proposed result and the color is more consistent on the boundary area. For the e-commerce presentation, the color correctness is increased and the entire model viewing experience is improved.

The final optimization process is blending the texture information on the image transition area. The texture information is extracted from different front images. The boundary between two image sources might be inconsistent in color. The results before and after the implementation of the proposed blending algorithm for a shoe and a cup are shown in Figure 15a,b, respectively. The texture quality on the transition area has been improved. The quality of the entire 3D textured model can therefore be improved for the purpose of e-commerce presentation.

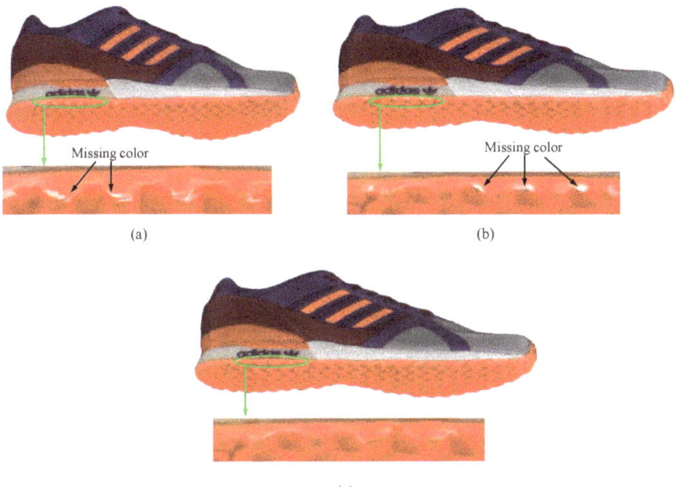

Figure 14. Implementation of the proposed algorithm to remove missing colors: (**a**) 3DSOM software (**b**) before and (**c**) after.

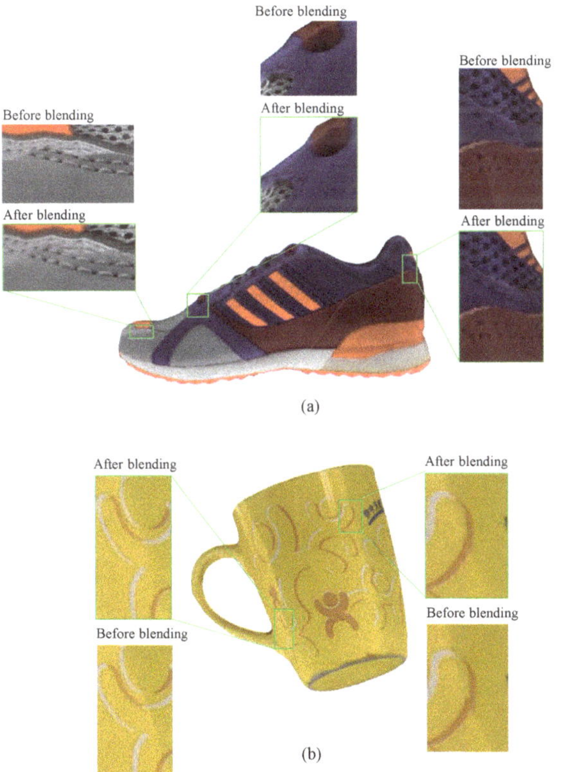

Figure 15. Results before and after the implementation of the proposed blending algorithm: (**a**) shoe and (**b**) cup.

6. Conclusions

In this study, we proposed a texture mapping technique that incorporates mesh partitioning, mesh parameterization and packing, texture transferring, and texture correction and optimization. The proposed mesh partition minimizes the growing cost to find the optimized mesh group. The mesh parameterization was based on an angle-based flattening to yield the optimized angles for 2D meshes, and a least-squares approximation to obtain all vertices. The texture transferring was implemented by projecting 3D meshes onto the image domain, and then extracting the pixels to map onto the UV map. However, to maintain the original quality of the texture information, a correction and optimization process was proposed. The OBB method was applied to allocate the UV map space more efficiently in the packing stage. The resolution of the texture map was increased to sufficiently include the original extracted pixels. Additional images were also employed to correct the error extraction of the background color by applying the alpha channel onto the object image. Finally, a blending process was proposed to minimize the transition error caused by different image sources. A high-quality 3D textured model can be obtained by applying this series of processes for presentations in e-commerce. However, the photo consistency of the 3D textured model is still not as good as that of 2D images. The color information from different image sources for the same point may differ slightly. This error is caused by the inaccuracy of 3D vertices and the calibration error; it can affect the projection accuracy of the vertices onto different texture sources. To further improve the quality of the 3D textured model, the photo inconsistency problem should be studied further.

Author Contributions: Conceptualization, J.-Y.L., T.-C.W., W.P., D.W.W., C.-Y.L. and J.-Y.L.; methodology, J.-Y.L. and T.-C.W.; writing—original draft preparation, J.-Y.L. and T.-C.W.; writing—review and editing, J.-Y.L. and T.-C.W.

Funding: This research received no external funding.

Conflicts of Interest: The authors declare no conflicts of interest.

References

1. Ortery. Available online: https://www.ortery.com/ (accessed on 1 October 2018).
2. Kutulakos, V.; Seitz, S. A theory of shape by space carving. *Int. J. Comput. Vis.* **2000**, *38*, 199–218. [CrossRef]
3. Sinha, S.; Pollefeys, M. Multi-view reconstruction using photo-consistency and exact silhouette constraints: A maximum flow formulation. In Proceedings of the Tenth IEEE International Conference on Computer Vision, Washington, DC, USA, 17–21 October 2005; Volume 1, pp. 349–356. [CrossRef]
4. Lazebnik, S.; Furukawa, S.; Ponce, J. Projective visual hulls. *Int. J. Comput. Vis.* **2007**, *74*, 137–165. [CrossRef]
5. Mulayim, A.Y.; Yilmaz, U.; Atalay, V. Silhouette-based 3D model reconstruction from multiple images. *IEEE Trans. Syst. Man Cybern. Part B Cybern.* **2003**, *34*, 582–591. [CrossRef] [PubMed]
6. Franco, J.S.; Boyer, E. Exact polyhedral visual hulls. In Proceedings of the British Machine Vision Conference, Norwich, UK, 9–11 September 2003; Volume 1, pp. 329–338. [CrossRef]
7. Yous, S.; Laga, H.; Kidode, M.; Chihara, K. Gpu-based shape from silhouettes. In Proceedings of the 5th International Conference on Computer Graphics and Interactive Techniques in Australia and Southeast Asia ACM, Perth, Australia, 1–4 December 2007; pp. 71–77. [CrossRef]
8. Phothong, W.; Wu, T.C.; Lai, J.Y.; Yu, J.Y.; Wang, D.W.; Liao, C.Y. Quality improvement of 3D models reconstructed from silhouettes of multiple images. In Proceedings of the CAD'17, Okayama, Japan, 10–12 August 2017. [CrossRef]
9. Shamir, A. A survey on mesh segmentation techniques. In *Computer Graphics Forum*; Blackwell Publishing: Oxford, UK, 2008; Volume 27, pp. 1539–1556. [CrossRef]
10. Sander, P.; Snyder, J.; Gortler, S.; Hoppe, H. Texture mapping progressive meshes. In Proceedings of the 28th Annual Conference on Computer Graphics and Interactive Techniques, Los Angeles, CA, USA, 12–17 August 2001; pp. 409–416. [CrossRef]
11. Lévy, B.; Petitjean, S.; Ray, N.; Maillot, J. Least squares conformal maps for automatic texture atlas generation. In Proceedings of the 29th Annual Conference on Computer Graphics and Interactive Techniques, San Antonio, TX, USA, 23–26 July 2002; Volume 21, pp. 362–371. [CrossRef]

12. Sander, P.; Wood, Z.; Gortler, S.; Snyder, J.; Hoppe, H. Multi-chart geometry images. In Proceedings of the 2003 Eurographics/ACM SIGGRAPH Symposium on Geometry Processing, Aachen, Germany, 23–25 June 2003; pp. 146–155. [CrossRef]

13. Mangan, A.P.; Whitaker, R.T. Surface segmentation using morphological watersheds. In Proceedings of the IEEE Visualization, 18–23 October 1998.

14. Mangan, A.P.; Whitaker, R.T. Partitioning 3D surface meshes using watershed segmentation. *IEEE Trans. Vis. Comput. Graph.* **1999**, *5*, 308–321. [CrossRef]

15. Lavoué, G.; Dupont, F.; Baskurt, A. A New cad mesh segmentation method, based on curvature tensor analysis. *Comput. Aided Des.* **2005**, *37*, 975–987. [CrossRef]

16. Mortara, M.; Patan'e, G.; Spagnuolo, M.; Falcidieno, B.; Rossignac, J. Blowing bubbles for multi-scale analysis and decomposition of triangle meshes. *Algorithmica* **2004**, *38*, 227–248. [CrossRef]

17. Mortara, M.; Patan'e, G.; Spagnuolo, M.; Falcidieno, B.; Rossignac, J. Plumber: A method for a multi-scale decomposition of 3d shapes into tubular primitives and bodies. In Proceedings of the Ninth ACM Symposium on Solid Modeling and Applications, Genoa, Italy, 9–11 June 2004; pp. 139–158. [CrossRef]

18. Funkhouser, T.; Kazhdan, M.; Shilane, P.; Min, P.; Kiefer, W.; Tal, A.; Rusinkiewicz, S.; Dobkin, D. Modeling by example. *ACM Trans. Graph.* **2004**, *23*, 652–663. [CrossRef]

19. Sheffer, A. Model simplification for meshing using face clustering. *Comput. Aided Des.* **2001**, *33*, 925–934. [CrossRef]

20. Garland, M.; Willmott, A.; Heckbert, P. Hierarchical face clustering on polygonal surfaces. In Proceedings of the 2001 Symposium on Interactive 3D Graphics, New York, NY, USA, 19–21 March 2001; pp. 49–58. [CrossRef]

21. Sheffer, A.; Praun, E.; Rose, K. Mesh parameterization methods and their applications. *Found. Trends Comput. Graph. Vis.* **2006**, *2*, 105–171. [CrossRef]

22. Hormann, K.; Lévy, B.; Sheffer, A. Mesh parameterization: Theory and practice. In *ACM SIGGRAPH 2007 Courses on–SIGGRAPH 07*; ACM: New York, NY, USA, 2007; Volume 1. [CrossRef]

23. Desbrun, M.; Meyer, M.; Alliez, P. Intrinsic parameterizations of surface meshes. *Comput. Graph. Forum* **2002**, *21*, 209–218. [CrossRef]

24. Sheffer, A.; De Sturler, E. Parameterization of Faceted Surfaces for Meshing using Angle-Based Flattening. *Eng. Comput.* **2001**, *17*, 326–337. [CrossRef]

25. Sheffer, A.; Lévy, B.; Mogilnitsky, M.; Bogomyakov, A. ABF++: Fast and robust angle based flattening. *ACM Trans. Graph.* **2005**, *24*, 311–330. [CrossRef]

26. Zayer, R.; Lévy, B.; Seidel, H.P. Linear angle based parameterization. In Proceedings of the Fifth Eurographics Symposium on Geometry Processing-SGP, Eurographics Association, Barcelona, Spain, 4–6 July 2007; pp. 135–141. [CrossRef]

27. Tutte, W.T. *Convex Representations of Graphs*; London Mathematical Society: London, UK, 1960; Volume 3, pp. 304–320.

28. Tutte, W.T. *How to Draw a Graph*; London Mathematical Society: London, UK, 1963; Volume 3, pp. 743–767.

29. Eck, M.; DeRose, T.D.; Duchamp, T.; Hoppe, H.; Lounsbery, M.; Stuetzle, W. Multiresolution analysis of arbitrary meshes. In Proceedings of the 22nd Annual Conference on Computer Graphics and Interactive Techniques, Los Angeles, CA, USA, 6–11 August 1995; pp. 173–182. [CrossRef]

30. Floater, M.S. Parametrization and smooth approximation of surface triangulations. *Comput. Aided Geom. Des.* **1997**, *14*, 231–250. [CrossRef]

31. Floater, M.S. Mean value coordinates. *Comput. Aided Geom. Des.* **2003**, *20*, 19–27. [CrossRef]

32. Floater, M.S.; Hormann, K.; Kós, G. A general construction of barycentric coordinates over convex polygons. *Adv. Comput. Math.* **2006**, *24*, 311–331. [CrossRef]

33. Zigelman, G.; Kimmel, R.; Kiryati, N. Texture mapping using surface flattening via multidimensional scaling. *Vis. Comput. Graph.* **2002**, *8*, 198–207. [CrossRef]

34. Degener, P.; Jan, M.; Reinhard, K. An Adaptable Surface Parameterization Method. *IMR* **2003**, *3*, 201–213.

35. Niem, W.; Buschmann, R. Automatic Modelling of 3D Natural Objects from Multiple Views. In *Image Processing for Broadcast and Video Production*; Springer: London, UK, 1995; pp. 181–193.

36. Genç, S.; Atalay, V. Texture extraction from photographs and rendering with dynamic texture mapping. In Proceedings of the 10th International Conference on Image Analysis and Processing, Venice, Italy, 27–29 September 1999; pp. 1055–1058. [CrossRef]

37. Baumberg, A. Blending Images for Texturing 3D Models. In Proceedings of the BMVC, Cardiff, UK, 2–5 September 2002; Volume 3, p. 5. [CrossRef]

38. Efros, A.A.; Freeman, W.T. Image quilting for texture synthesis and transfer. In Proceedings of the 28th Annual Conference on Computer Graphics and Interactive Techniques ACM, Los Angeles, CA, USA, 12–17 August 2001; pp. 341–346. [CrossRef]

39. Wei, L.Y.; Levoy, M. Fast texture synthesis using tree-structured vector quantization. In Proceedings of the 27th Annual Conference on Computer Graphics and Interactive Techniques ACM, New Orleans, LA, USA, 23–28 July 2000; pp. 479–488. [CrossRef]

40. Maruya, M. Generating a Texture Map from Object-Surface Texture Data. In *Computer Graphics Forum*; Blackwell Science Ltd.: Edinburgh, UK, 1995; Volume 14, pp. 397–405.

41. Gottschalk, S.; Lin, M.C.; Manocha, D. OBBTree: A hierarchical structure for rapid interference detection. In Proceedings of the 23rd annual conference on Computer graphics and interactive techniques ACM, New Orleans, LA, USA, 4–9 August 1996; pp. 171–180. [CrossRef]

42. Lai, J.Y.; Shu, S.H.; Huang, Y.C. A cell subdivision strategy for r-nearest neighbors computation. *J. Chin. Inst. Eng.* **2006**, *29*, 953–965. [CrossRef]

43. 3DSOM Software. Available online: https://www.3dsom.com/ (accessed on 1 October 2018).

Article

Image Super-Resolution Algorithm Based on Dual-Channel Convolutional Neural Networks

Yuantao Chen [1], Jin Wang [1,2,*], Xi Chen [1], Arun Kumar Sangaiah [3], Kai Yang [4] and Zhouhong Cao [5]

[1] School of Computer and Communication Engineering & Hunan Provincial Key Laboratory of Intelligent Processing of Big Data on Transportation, Changsha University of Science and Technology, Changsha 410114, China; chenyt@csust.edu.cn (Y.C.); chentianjun@163.com (X.C.)

[2] School of Information Science and Engineering, Fujian University of Technology, Fujian 350118, China

[3] School of Computing Science and Engineering, Vellore Institute of Technology University, Vellore 632014, India; arunkumarsangaiah@gmail.com

[4] Technical Quality Department, Hunan ZOOMLION Heavy Industry Intelligent Technology Corporation Limited, Changsha 410005, China; yangkai@zoomlion.com

[5] School of Hydraulic Engineering & Hunan Provincial Science and Technology Innovation Platform of Key Laboratory of Dongting Lake Aquatic Eco-Environmental Control and Restoration, Changsha University of Science and Technology, Changsha 410114, China; caozhouhong@csust.edu.cn

* Correspondence: jinwang@csust.edu.cn; Tel.: +86-180-1484-9250

Received: 17 March 2019; Accepted: 30 May 2019; Published: 5 June 2019

Abstract: For the image super-resolution method from a single channel, it is difficult to achieve both fast convergence and high-quality texture restoration. By mitigating the weaknesses of existing methods, the present paper proposes an image super-resolution algorithm based on dual-channel convolutional neural networks (DCCNN). The novel structure of the network model was divided into a deep channel and a shallow channel. The deep channel was used to extract the detailed texture information from the original image, while the shallow channel was mainly used to recover the overall outline of the original image. Firstly, the residual block was adjusted in the feature extraction stage, and the nonlinear mapping ability of the network was enhanced. The feature mapping dimension was reduced, and the effective features of the image were obtained. In the up-sampling stage, the parameters of the deconvolutional kernel were adjusted, and high-frequency signal loss was decreased. The high-resolution feature space could be rebuilt recursively using long-term and short-term memory blocks during the reconstruction stage, further enhancing the recovery of texture information. Secondly, the convolutional kernel was adjusted in the shallow channel to reduce the parameters, ensuring that the overall outline of the image was restored and that the network converged rapidly. Finally, the dual-channel loss function was jointly optimized to enhance the feature-fitting ability in order to obtain the final high-resolution image output. Using the improved algorithm, the network converged more rapidly, the image edge and texture reconstruction effect were obviously improved, and the Peak Signal-to-Noise Ratio (PSNR) and structural similarity were also superior to those of other solutions.

Keywords: super-resolution; dual-channel; residual block; convolutional kernel parameter; long-term and short-term memory blocks

1. Introduction

Because images are affected by both the image processing system and the transmission environment during the process of acquisition, the resolution of the original image is typically low; moreover, since key information is missing from these original low-resolution images, they are generally not capable of meeting many actual user needs. Accordingly, the use of high-resolution images is required in some areas and fields of research. In order to solve the problems caused by low image quality, Single Image Super Resolution (SISR) technology is used to transform a single Low-Resolution (LR) image into a High-Resolution (HR) image containing rich high-frequency information. There are wide applications for this technology in the research fields of object detection, satellite image, medical image and face recognition [1–4].

Traditional *SISR* methods have included interpolation methods based on the sample extraction theory, such as Bicubic Interpolation [5] and Bilinear Interpolation [6]. The image reconstruction is based on methods including the Iterative Back Projection (IBP) method [7], the Projection Onto method (PO) [8], the Maximum A Posteriori method (MAP) [9], and so on. Based on learning methods such as embedded neighborhood [10], the regression or mapping relationship between HR and LR blocks has been understood by using the concept of geometric similarity. In sparse representation based on the interrelated approach, Yang et al. [11] and Yang et al. [12] reconstructed HR image blocks and HR images by strengthening the similarity between LR and HR image blocks and their corresponding real dictionaries, so that the sparse representation of the LR block and the super-completed HR dictionary can be used to reconstruct HR image blocks and then connect HR images. A complete high-resolution image is obtained like a block [13–16].

In recent years, deep learning has achieved remarkable results in the research field of image super resolution, benefiting from the powerful feature characterization [17] of deep learning, which is more effective than traditional methods. Dong et al. [18] first proposed the application of the Super Resolution using Convolutional Neural Networks (SRCNN) algorithm to super-resolution images. Compared with traditional methods, the simple network structure obtains the ideal super-resolution; however, there are limitations of the simple network structure. Firstly, it is dependent on the context information of small image blocks. Secondly, the training convergence is too slow, and the time complexity is high. Thirdly, the simple network only can be used for a single-scale super resolution (SR) procedure. Dong et al. [19] proposed the Fast Super-Resolution Convolutional Neural Network (FSRCNN) by reducing the speed training of the network parameters. FSRCNN used eight layers of network structure, making it deeper than SRCNN; moreover, instead of Bicubic Interpolation, the anti-coiling layer was used on the last layer of the network. Finally, FSRCNN has achieved success in the convergence and super-resolution reconstruction field. Considering the slow convergence and shallow network of SRCNN and FSRCNN networks, Wang et al. [20] proposed an image super-resolution algorithm (EEDS) based on end-to-end and shallow convolutional neural networks that has achieved better performance than others. However, because the deep network cannot fully extract the features of an *LR* image in the feature extraction stage, the loss of useful information and long-term memory content during the reconstruction process becomes serious when the feature of the up-sampling process is nonlinear mapping, as this causes the effect of super resolution to be reduced by the deep network. However, generally speaking, the shallow network master is the main problem. Moreover, when restoring the main components of LR images, the fast convergence of the network can be limited if too many parameters are used. Kim et al. [21] proposed a highly accurate single-image super-resolution method named Very Deep Networks for Super Resolution (VDSR). By using a very deep convolutional network of *VGG-net* [22] in image classification, the model employs cascaded small filters in a deep-network structure, using 20 weight-layers to efficiently utilize the context information of the large image region.

Moreover, Kim et al. [23] proposed the Deep Recursive Convolutional Network (DRCN) for image super resolution. The network uses a very deep recursive layer (as many as 16 recursions), as increasing the recursion depth can improve the performance without the need to introduce additional parameters to additional convolutions. In order to prevent the explosion and disappearance of the gradient, as well as to reduce the difficulty of training, the recursive monitoring and skipping connection methods are far more effective than previous methods. Recently, Ke et al. [24] proposed the Gradual Up-Sampling Network (GUN) method, which is based on a deep convolutional neural network. This method uses a gradual process to simplify the direct *SR* problem into a multi-step sampling task that employs very small magnification at each step. The Enhanced Deep Residual Networks for Single-Image Super Resolution (EDSR) and the Multi-Scale Deep Super-Resolution (MDSR) network were proposed by Lim et al. [25] among others. The model is optimized by removing unnecessary modules from the residual network to significantly enhance the performance of the model. Moreover, by extending the size of the model to further improve the performance, *MDSR* can reconstruct HR images with different magnification factors using a network model. Tai et al. [26] proposed a very deep Memory Network (MemNet) for image restoration, which introduces memory blocks consisting of a recursive unit and a gate control unit that mine persistent memory through an adaptive learning process. The representation and output from previous memory blocks are connected and sent to the gate control unit. The gate control unit is adaptive to control memory [27,28] and controls how many previous states should be retained and how many current states should be stored to achieve superior performance in super-resolution tasks [29,30].

By exploring the above methods and combining them with MemNet [26] and Deep Residual Network (ResNet) [31], Nair et al. [32] proposed an enhanced algorithm of image super-resolution based on Dual-Channel Convolution Neural Network (DCCNN), related to SRCNN and EEDS to solve the above problems. The shallow channel is mainly used to restore the overall outline of the original image and to achieve fast convergence performance. By adjusting the parameters of the three-layer network from the shallow channel, it can quickly converge while ensuring the restoration of the main components from the image. By contrast, deep channels are used to extract detailed texture information from LR images. Deep channels are divided into three steps: feature extraction and mapping, up-sampling, and long-term and short-term memory block reconstruction. Because there are fewer network layers in the extraction stage of the original model, the local sensing field of the image is too small, and the full LR image feature extraction will lead to the final SR effect. In order to avoid loss of important high-frequency content, the proposed model increases the residual layer on the original basis by increasing the number of network layers in the process; it also reduces the LR feature mapping dimension, such that the residual layer can learn edge and texture information of the image better than the common stacked convolution, and the increased network depth avoids the network. It is difficult to train the problem, meaning that the feature can be directly transmitted to the lower level so as to optimize the gradient vanishing problem and make it easier for the network to enhance the training performance. During the up-sampling phase, because the sampling operation is an important part of the model, the goal is to increase the space span to the target of the HR size. In order to get good results, a 1×1 filter is used to increase the number of dimensions to 64 after the mapping is complete. In addition, deconvolution is used to achieve the sampling rather than manual designing. During memory block reconstruction in the long-term and short-term period, because the reconstruction stage directly determines the HR reconstruction effect of the deep channel, the long-term and short-term memory blocks made up of the residual block are used after up-sampling to further reduce the loss of high-frequency information, such that the reconstructed *HR* image texture information is more abundant. Finally, the deep and shallow passages are jointly optimized to obtain the final HR image. Experimental results show that the effect of network super resolution is better than that of bicubic interpolation, A+ [11], SRCNN [18], and EEDS [20] super-resolution reconstruction algorithms.

2. Related Works

2.1. The SCRNN Model

In the learning-based super-resolution image algorithm, *SRCNN* applied a convolutional neural network to the task of image super resolution for the first time. Compared with traditional methods, the method can directly learn the mapping relationship between *LR* images and *HR* images.

As shown in Figure 1, the process of the proposed algorithm was divided into three stages. The data are pre-processed, the training dataset of 91 images is taken to make up the image block of 14, and the LR image block after the bicubic interpolation pre-processing procedure is used as the input for the network. The first layer uses 64 filters and the convolutional core of the size of 3 channels of image block performs feature extraction and representation; at this time, the number of channels is expanded from 3 to 64. The second layer uses 64 filters, and the convolution nucleus (of size 1×1) conducts the nonlinear mapping to extract features; at this time, the number of channels is reduced from 64 to 3. The third layer uses a convolution nuclear size of 5×5 to reconstruct the *HR* image block at this time; the number of channels decreases from 64 to 3. Finally, the mean squared error (MSE) corresponding to the original image and HR output image is constructed to optimize the model's parameters. In SRCNN, the experimental results show that the super-resolution effect is improved by using a large scale of dataset for ImageNet.

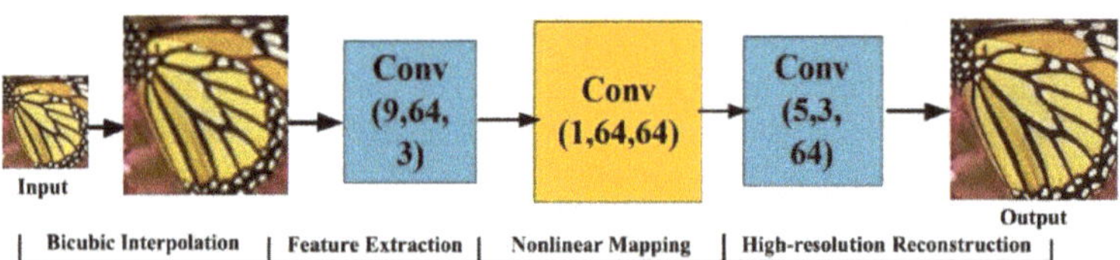

Figure 1. The processing procedure of Super Resolution using Convolutional Neural Networks (SRCNN) construction.

2.2. Image Super-Resolution Algorithm Based on Dual-Channel Convolutional Neural Networks

The image super-resolution algorithm is based on dual-channel convolutional neural networks, such as EEDS [21], and is also a learning-based SISR algorithm. The EEDS algorithm works to improve SRCNN and FSRCNN: its structure is deeper than those of SRCNN and FSRCNN, and the residual block with jump layer, the residual network because of the existence of the fast connection. Data transmission between the network is smoother, and the gradient is improved, resulting in the loss of fitting and making it easier for the network to converge. The network structure of EEDS is divided into two parts: the deep layer and the shallow layer. The deep network contains 13 layers, including a feature extraction layer, an up-sampling layer, and q multi-scale reconstruction layer. The shallow network contains three layers. The design idea comes from the three-layer model of SRCNN, in which the anti-coiling layer replaces the original SRCNN nonlinear mapping layer. Finally, using a deep network combined with the output of the shallow network, the final output of the HR image is obtained.

In the training process, 91 training images are first scaled, rotated, and fragmented, and then sampled according to the required ratio. The obtained *LR* image blocks are input into the double-layer network. The MSE corresponds to the original image, the output HR image is constructed, and the model parameters are optimized.

The shortcomings of the network are as follows: because the deep network cannot fully extract the features of the *LR* image in the feature extraction stage, the nonlinear feature mapping of the up-sampling process leads to the loss of useful information and of long-term memory content in the reconstruction process, which causes the deep network to discount the effect of super-resolution. The shallow network is mainly used to restore the main components of LR images. Too many parameters will limit the fast convergence of the network.

3. Dual-Channel Convolutional Neural Networks

3.1. The Improved Ideas

Because the shallow network cannot adequately extract the features of the *LR* image, the effect produced by super resolution is not ideal. Although the deep network is superior to the shallow network in depth, the deepening architecture of the network will also cause the network to be difficult to train, and the gradient disappearance/explosion will affect the stability of the network. Therefore, by combining with the two factors of width and depth of the network, the SRCNN and EEDS have been improved.

The network structure of SRCNN and EEDS is that of a three-layer network, that of the DCCNN is "deep and shallow", using 13 tiers and 3 tiers, respectively, in a dual-channel network; the shallow channel is used to restore the overall outline of the image, while the deep channel is used to restore rich texture information. Therefore, the combination of these two channels can effectively improve the efficiency of training, enhance the feature-fitting ability, and reduce the computational complexity of the whole model. On the basis of the above factors, the present paper selects the parameters of the shallow channel with a convolutional kernel while adjusting the depth of the deep channel network, so that the shallow channel is mainly responsible for the convergence performance of the network to reduce the time complexity of the model; moreover, the deep channel is mainly responsible for more detailed texture recovery and for improving the restoration precision of the network. It is more efficient to learn the texture information at a high level, and the feature of fitting the image is more accurate, considering that the use of the residual block and the jump layer results in faster convergence than a simple increase in the number of network layers and also reduces the gradient dispersion and the loss of features.

Therefore, residual blocks and skip layers in deep channels are used in this paper. At the same time, as the depth increases, it is more difficult for the model to achieve long-term dependence at each stage; this leads to the reduction of dependence during the up-sampling component of the reconstruction phase and the increased loss of the important, higher-frequency information in the up-sampling stage. Accordingly, in this paper, three residual blocks are selected in the reconstruction stage to grow the short-term memory blocks to the up-sampling feature. The space is rebuilt. Finally, this paper proposes an image super-resolution algorithm based on DCCNN with a deep channel of 19 layers and a shallow channel of 3 layers.

3.2. The Network Structure of DCCNN

The image super-resolution algorithm based on DCCNN fully considers the nonlinear mapping relationship between the low-resolution image and the super-resolution image, and the characteristics of the dual-channel are equal to those of the proposed model. The corresponding weights for each channel are not shared in DCCNN. The shallow channel is mainly used to restore the overall outline of the image. The deep channel is used to extract detailed texture information of the LR image. In the phase of feature extraction and mapping with the deep channel, the input layer of the proposed network is the three-channel LR image, which is 48×48 size of units.

Figure 2 presents the dual-channel network constructed in this paper, which is divided into two sub-channels: the deep channel and the shallow channel. Firstly, the number of channels is increased to 64 through the convolutional kernel of 3×3 size, then entered into the residual block (see Figure 3). It is composed of *Conv*, *ReLU*, and *Conv*, and the *Conv* residual block size is 3×3, while the step length is one and the padding is two. After three residual blocks, the output has 16 48×48 characteristic graphs. At this time, the semantic information in the feature map is richer than it was previously. In the up-sampling phase, as the most important part of the network, the goal is to increase the spatial span of the *LR* images to *HR* size. After the mapping, the dimension of 1×1 is compressed from 16 to 4. Instead of using the manual interpolation method, we used deconvolution (*DeConv*) to achieve the up-sampling. The size of *DeConv* is 9×9. For two times, three times and four times, the different scales of up-sampling by setting different steps. After deconvolution, the feature map is increased to 64. Finally, the dimension of the 1×1 filter is mapped from 64 to 4, the parameters of the 1×1 filter are effectively reduced, and the calculation complexity is also reduced. In the stage of reconstructing long-term and short-term memory, as the last stage of the network, the up-sampling phase is also the most important part, as it determines the quality of the texture information recovery from the network.

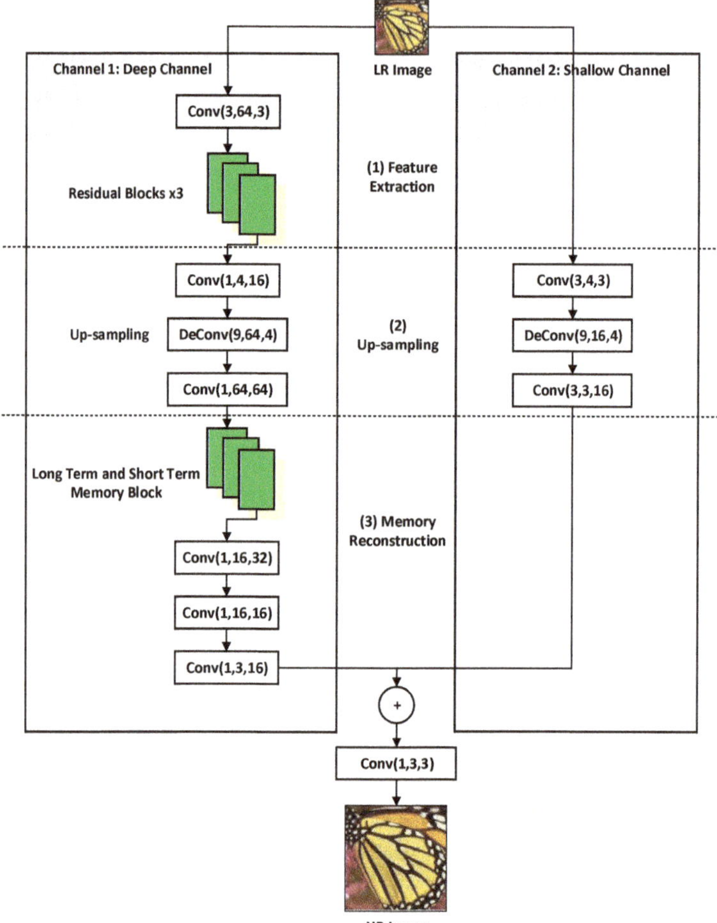

Figure 2. The processing procedure of Dual-Channel Convolutional Neural Networks (DCCNN).

Considering that it is difficult for the model to achieve long-term dependence at each stage, we used the multi-scale convolution to reconstruct the up-sampling feature space using long-term and short-term memory blocks at the beginning of the reconstruction. The long-term and short-term memory blocks, which are shown in Figure 4, consist of three residual blocks. The dimensions of the feature map range from 64 to 32, which further reduces the dimensions and enhances the nonlinear mapping ability. The size of *Conv* in the long-term and short-term memory block is 3×3, the step length is one, the padding is two, and the output is 32 feature graphs. The 1×1 filter is then used to compress the dimensions from 32 to 16, so that the high-dimensional feature is extracted and the computational complexity is reduced. In order to effectively aggregate the local information of the 16 feature maps, multi-scale convolution is used for reconstruction. The multi-scale coiling layer contains four filters of different sizes, namely, $1 \times 1, 3 \times 3, 5 \times 5$ and 7×7. The four filters' convolutions in the layer are parallel. Each filter has 4 outputs in the feature graph, and then the 16 feature graphs are combined. Finally, the 1×1 filter is used as the weighted combination of multi-scale texture features. At this time, the dimension of the feature map is from 16 to 1. Correspondingly, the use of deconvolutional networks to complete the up-sampling operation involves the use of a three-layer structure similar to that of SRCNN in shallow channels. The specific process is as follows: the three-channel LR image input to 48×48 is input from the input layer to the network, and the number of channels is increased to four through the 3×3 filter. The space span of the LR image is increased to the HR size by means of deconvolution. The size of *DeConv* convolutional kernel is the same as that of the 9×9 deep learning network, and the feature map after the deconvolution is increased to 16. Finally, using the convolutional kernel of 3×3 size, the step size is one, the padding is two, and the output is a three-channel feature graph.

In order to avoid the problem of gradient disappearance, our proposed structure is deeper than that of the improved network; moreover, the characteristics of both the feature graph and the feature map of the lower layer are also different. Furthermore, in this paper, the *ReLU* [33] activation function is used in all convolution operations to improve the *PReLU* [34] activation function. All convolutional operations utilized in this paper can improve the high network's nonlinear modeling ability. At this point, the output feature graph of the shallow and deep network is optimized, the output of the two networks is added, the effective component is retained, and the texture information of the feature map is enriched. The feature graph is then input to a convolution layer of 1×1 size. Finally, the image output of HR is obtained, with the result that the image quality has greatly improved.

3.3. Residual Blocks and Long-Term and Short-Term Memory Block

(1) Residual Blocks

The residual blocks' network design is inspired by the 152-level ResNet network proposed by He et al. [31]. The recognition performance on the ImageNet dataset was improved with the increase of the number of network layers, and its performance on computer vision problems [22,23,26] from low to high tasks is excellent.

The original residual block, which is shown in Figure 3a, is composed of a feed-forward convolutional network and a jump around a number of layers. The stacked residuals form the final residual networks. Compared with a smooth network, the residual network exhibits lower convergence loss and a lack of overfitting due to the disappearance of the gradient, which makes the network easier to optimize. The dimensions of the feature map progressively increase to ensure the ability to express the output features.

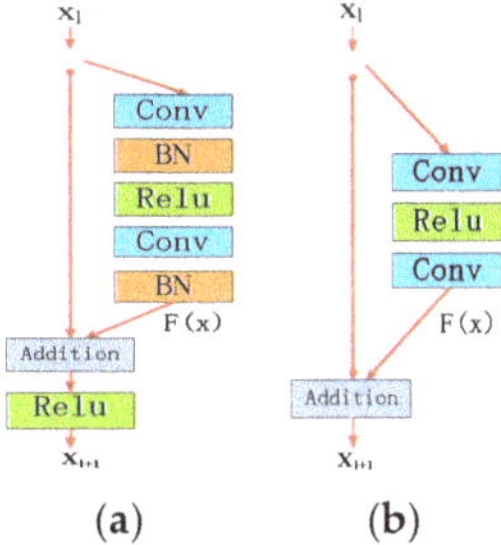

Figure 3. The processing construction with improved residual blocks described in the paper. (**a**) Original Residual Blocks; (**b**) Improved Residual Blocks in the Paper.

Since the original batch normalization layer (BN) [35] is used to normalize the characteristics of the coiling output layer, this will affect the distribution of features learned by the convolution layer and cause the loss of important information from the feature graph. Moreover, the batch positive layer has the same number of parameters as the previous convolutional layer and thus consumes a lot of memory. In their image deblurring task, Zeiler et al. [36] deleted the *BN* layer in the residual block, with the result that the network performance was greatly improved. Therefore, in this paper, we used the residual block to delete the batch regularization layer in order to reduce the color's offset in the output, while maintaining the training stability. Each residual block in the present paper contained two 3×3 convolutional layers and the ReLU layer. The structure of the residual block in the present paper is shown in Figure 3b.

The residual block can be expressed by Equation (1):

$$X_{l+1} = X_l + F(X) \tag{1}$$

Here, X_l and X_{l+1} represent the input and output vectors of residual blocks, respectively. The function $F(X)$ denotes residual mapping. The residual block in this paper contained only the convolutional layer and the ReLU layer. The modified linear unit (ReLU) has unilateral suppression and sparsity. In most cases, the ReLU gradient is a constant term, avoiding the problem of gradient disappearance to a certain extent. The relevant mathematical expression can be expressed by Equation (2):

$$f(x) = \max(0, x_i) \tag{2}$$

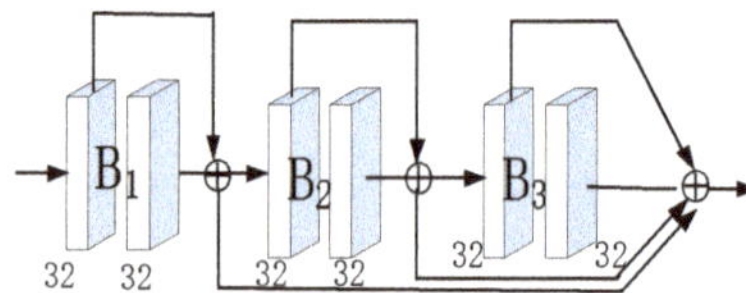

Figure 4. The processing construction of long-term and short-term memory blocks.

In this paper, the activation function of the convolutional layer outside the residual block is the Parametric Rectified Linear Unit (PReLU) [34]. The use of PReLU is mainly designed to avoid the "dead angle" [37] caused by the zero gradient in the ReLU. It is increased by the correction of parameters to a certain extent. It can have a regularizing effect and can also improve the generalization ability of the model. The difference between the proposed model and *ReLU* is mainly reflected in the negative part, and the mathematical expression is shown in Equation (3):

$$f(x) = \max(0, x_i) + a_i \min(x_i, 0) \tag{3}$$

Here, x_i is the input signal of the ith layer, and a_i is the coefficient of the negative part. In Equation (3), the parameter a_i is set to zero, but the negative part of PReLU can be learned. Finally, the output of the activation function can be expressed by Equation (4):

$$f_l(x) = f(W_l * f_{l-1}(x) + B_l) \tag{4}$$

Here, f_l is the final output feature graph and B_l is the offset of the lth layer.

(2) Long-Term and Short-Term Memory Block

It is difficult to achieve long-term dependence at each stage, resulting in lower dependence on the up-sampling phase in the reconstruction phase and more loss of important high-frequency information in the up-sampling phase. In this paper, three residual blocks (B_1, B_2, B_3) were used to synthesize the long-term and short-term memory block for the up-sampling feature space at the beginning of reconstruction. The design of the long-term and short-term memory block was inspired by He et al. [31], who proposed a very deep persistent MemNet. The construction of our long-term and short-term memory blocks is presented in Figure 4.

In this paper, we used three residual blocks to learn recursively in the long-term and short-term memory blocks. We used the eigenvector x of the up-sampling phase as input; the residual block B_i can be expressed by Equation (5):

$$B_i = F(B_{i-1}, w_i) + B_{i-1} \tag{5}$$

In Equation (5), i is set to one, two, and three. B_1, B_2, and B_3, respectively represent the output of the corresponding residual block. When $i = 1$, $B_{i-1} = x$. F represents the residual mapping, and w_i

represents the weight vector of the residual block to learn. Since each residual block consists of two volume layers and ReLU activation functions, Equation (5) can be further expressed as Equation (6):

$$F(B_{i-1}, w_i) = w_i^2 \text{ReLU}\left(w_i^1 \text{ReLU}(B_{i-1})\right) \tag{6}$$

Here, ReLU represents the activation function, while w^1 and w^2 are the two weight vectors of the volume layer, respectively. In the interest of simplicity, the bias is omitted in the above equations.

Finally, unlike the traditional leveling network, the present paper uses cascading methods to combine the output features of the three residual blocks, which effectively avoids content loss from the previous stage. The process of calculation is shown in Equation (7):

$$B_{out} = [B_1, B_2, B_3] \tag{7}$$

Here, B_{out} represents the final output and passes to the next layer.

3.4. Loss Function and Evaluation Standard

(1) Loss Function

By minimizing the loss cost between the super-resolution image and the real high-resolution image, the network constantly adjusts the network parameters $\Theta = \{w_i, b_i\}$. For a group of real high-resolution images X_j and a group of super-resolution images, $F^j(Y; \Theta)$, is reconstructed by the network. This paper uses MSE as the cost function:

$$L(\Theta) = \frac{1}{n} \sum_{i=1}^{n} \|F^j(Y; \Theta) - X_j\|^2 \tag{8}$$

where n represents the number of training samples. Because the weights of the dual-channel network are not shared, they are converted to a dual-channel cost function problem:

$$L_{EDC} = \min[L_d(\Theta) + L_s(\Theta)] \tag{9}$$

Here, $L_d(\Theta)$ and $L_s(\Theta)$ are the loss costs of the deep channel and shallow channel respectively. The network uses the Adam optimization method and back-propagation algorithm [38] to minimize MSE in order to adjust the network parameters, and the update process of the network weights is as in Equation (10):

$$\Delta_{k+1} = 0.9 \times \Delta_k - \eta \times \frac{\partial L}{\partial W_k^l}, W_{k+1}^l = W_k^l + \Delta_{k+1} \tag{10}$$

Δ_k represents the updating value of the last weight, l represents the number of layers of the network, and k represents the number of iterations from the network; η is the learning rate, W_k^l represents the weight of the kth iteration in level l, $\frac{\partial L}{\partial W_k^l}$ represents the corresponding weight of the cost function and derivation of the derivative. The weights are randomly initialized according to a Gaussian distribution with mean value of zero and variance of 0.001. The model can automatically adjust the learning rate in the range of training, making the learning of the parameters more stable.

(2) Evaluation Standards

In this paper, the difference between the generated image quality and the quality of the original high-resolution image is measured by means of two common evaluation indexes, namely the Peak Signal-to-Noise Ratio (PSNR) and the Structural Similarity Index (SSIM) [29,30].

PSNR is used as an objective evaluation index of image quality, which is measured by calculating the error between corresponding pixels. The PSNR's unit is decibel (dB) [16]. The larger the value, the smaller the image distortion. The calculating equation is Equation (11):

$$PSNR = 10\log_{10}\left(\frac{(2^n-1)^2}{MSE}\right) \tag{11}$$

Here, MSE is the direct Mean Squared Error of the original image and the super-resolution image, $(2^n-1)^2$ is the signal maximum square, and n is the number of bits per sampling value.

The SSIM measures image similarity in terms of three aspects: brightness, contrast ratio, and structure. The range of SSIM is [0,1], and its value is closer to one. The distortion effect is smaller. The calculation equations are as follows:

$$SSIM(X,Y) = l(X,Y) \cdot c(X,Y) \cdot s(X,Y) \tag{12}$$

$$l(X,Y) = \frac{2\mu_X\mu_Y + C_1}{\mu_X^2 + \mu_Y^2 + C_1} \tag{13}$$

$$c(X,Y) = \frac{2\sigma_X\sigma_Y + C_2}{\sigma_X^2 + \sigma_Y^2 + C_2} \tag{14}$$

$$s(X,Y) = \frac{2\sigma_{XY} + C_3}{\sigma_X\sigma_Y + C_3} \tag{15}$$

X is the super-resolution image of the LR image obtained through network training, Y is the original HR image. The variances of μ_X and μ_Y are represented by X and Y, respectively, while σ_X and σ_Y represent the variances of the super-resolution image and of the original high-resolution image, respectively, and σ_{XY} represents the covariance of the super-resolution image and the original high-resolution image. C_1, C_2, C_3 are constant terms. In order to avoid a zero in the denominator, the usual practice is to take $C_1 = (K_1 \times L)^2$, $C_2 = (K_2 \times L)^2$, $C_3 = C_2/2$ and, generally, $K_1 = 0.01$, $K_2 = 0.03$, $L = 255$.

4. Experimental Results and Analysis

4.1. Parameter Settings

The experiment used 91 pictures by Bevilacqua et al. [39] and one hundred $2K$ high-definition images selected from the DIV2K dataset. In short, a total of 191 images were used as training datasets to train the network model. Considering that dataset size directly affects network performance, two methods of data expansion were adopted for the image, based on the original training dataset. The image was amplified in two ways: (1) Scaling: each image was zoomed in proportion to 0.9, 0.8, 0.7, and 0.6; (2) Rotating: each image was rotated by 90 degrees, 180 degrees, and 270 degrees. Each image was used 20 times, such that 3820 images were eventually available for the training process. In this process, the sub-sampling size was 48 × 48, the initial learning rate of the network was set to 0.001, and the Adam optimization method was adopted to automatically adjust the learning rate so that the network parameters could be learned smoothly. The number of images per batch was set to 64, and the network was trained 1000 times. The testing dataset comprised the internationally common datasets "Set5" [40,41] and "Set14" [42,43]. The GPU was NVIDIA GeForce 1080 T_i, the experimental environment was Keras, and Python 3.5 and OpenCV 3.0 were applied to carry out the simulation experiments. The results of the network training were compared with those of existing methods in terms of three aspects: subjective visual effect, objective evaluation index, and efficiency comparison.

4.2. Experimental Results and Comparative Analysis

In order to verify the effectiveness of the proposed image super-resolution algorithm based on DCCNN, the present paper used a trained model to reconstruct the *LR* image at "2×", "3×", and "4×" [44] the super resolution. The performance of the proposed DCCNN method was evaluated on the Set5 dataset and Set14 dataset, and the results were compared with the results of the existing bicubic interpolation, A+ [11], SRCNN [19], and EEDS [20] algorithms.

Because of the different experimental environments of each algorithm, the contrast images could differ from the original ones. However, the overall trend of the comparison results would not be affected. In order to ensure the rationality and objectivity of the experimental results, two representative datasets were selected to test and contrast the images with rich texture details. The testing results are presented in Figures 5–7, which compare the results of the bicubic interpolation, A+, SRCNN, and EEDS methods for different reconstruction times of the butterfly image, zebra image, and comic image, and select the whole panorama and more obvious parts of the wing texture of the butterfly, the head markings of the zebra, and the cheek and shoulder of the comic. A subjective visual evaluation was carried out.

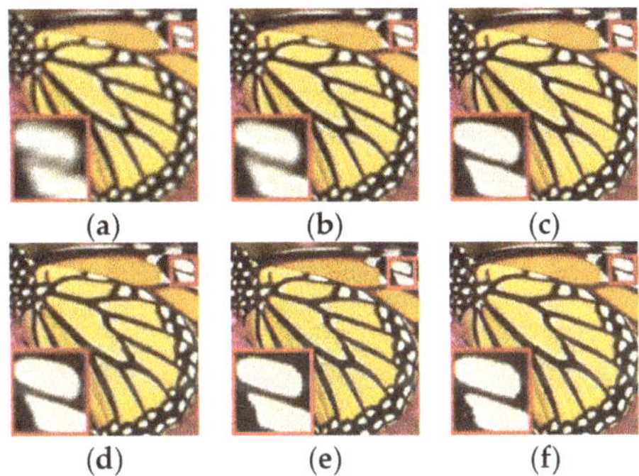

Figure 5. Super-resolution reconstruction results of the image "Butterfly" with "×4" scale factor. (**a**) Bicubic [5]; (**b**) A+ [11]; (**c**) SRCNN [18]; (**d**) EEDS [20]; (**e**) The Proposed Method's; (**f**) Original Image.

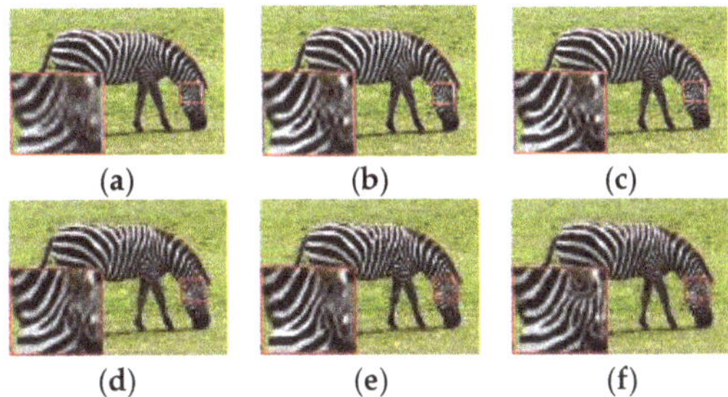

Figure 6. Super-resolution reconstruction results of the image "Zebra" with "×3" scale factor. (**a**) Bicubic [5]; (**b**) A+ [11]; (**c**) SRCNN [18]; (**d**) EEDS [20]; (**e**) The Proposed Method; (**f**) Original Image.

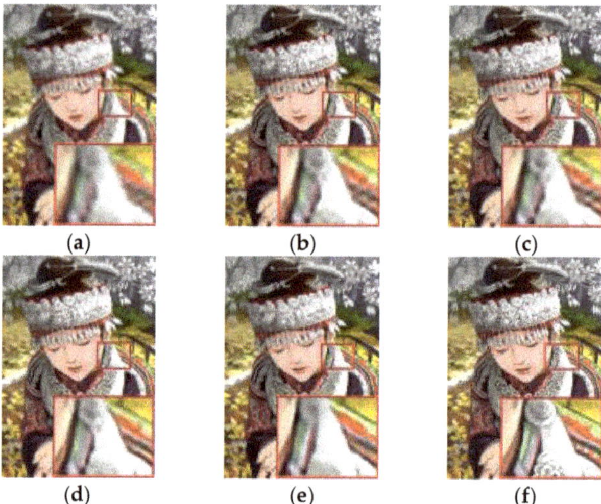

Figure 7. Super-resolution reconstruction results of the image "Comic" with "×3" scale factor. (**a**) Bicubic [5]; (**b**) A+ [11]; (**c**) SRCNN [18]; (**d**) EEDS [20]; (**e**) The Proposed Method; (**f**) Original Image.

Figure 5a–d present four super-resolution images of four contrast models from left to right. Figure 5e is the result of reconfiguration. Figure 5f concentrates on the Set5 testing of the original HR image. The butterfly wing edge of the image produced by the proposed method is sharper relative to the other methods: both the edge and the image are more complete, and the texture is also clearer.

In Figures 6 and 7, from left to right, the reconfiguration of the four contrast models is also three times that of the super-resolution effect diagram. Figure 6e presents the reconstruction result of the proposed method, while Figure 6f is the original HR image of the Set14 testing dataset. It was found that the reconstruction effect of the zebra image was more prominent, the reconstruction of the cheek edge from the comic image was sharper, the edge preservation was better, and the details of the shoulder texture were more abundant. The average PSNR and SSIM objective testing indexes under various experimental conditions are presented in Table 1. The best experimental results in the table are marked in bold.

Table 1. Average Peak Signal-to-Noise Ratio (PSNR) and Structural Similarity Index (SSIM) at different reconstruction scales on Set5 and Set14 datasets.

Dataset	Reconstruction Multiple	Bicubic [5]	A+ [11]	SRCNN [18]	EEDS [20]	Proposed DCCNN
		PSNR/SSIM	PSNR/SSIM	PSNR/SSIM	PSNR/SSIM	PSNR/SSIM
Set5	×2	33.64/0.9296	36.55/0.9543	36.67/0.9541	37.30/0.9578	**37.43/0.9603**
	×3	30.38/0.8681	32.57/0.9089	32.76/0.9091	33.46/0.9190	**33.59/0.9204**
	×4	28.41/0.8106	30.29/0.8602	30.49/0.8627	31.15/0.8782	**31.32/0.8842**
Set14	×2	30.23/0.8687	32.29/0.9058	32.43/0.9062	32.82/0.9104	**32.95/0.9115**
	×3	27.54/0.7743	29.14/0.8187	29.29/0.8208	29.61/0.8283	**29.70/0.8307**
	×4	26.01/0.7028	27.31/0.7492	27.48/0.7502	27.81/0.7625	**28.13/0.7696**

As can be seen from the testing results presented in Table 2 below, the results of the proposed algorithm were better than those of the improved algorithm in terms of average PSNR and SSIM, thereby proving the effectiveness of the proposed algorithm.

Table 2. Comparison of computational complexity with phases.

Method	Feature Extraction/ms	Up-Sampling/ms	Reconstruction/ms	Shallow Channel/ms
EEDS	38,015	**4112**	154,834	7265
DCCNN	**19,151**	24,895	**70,500**	**5231**

4.3. Efficiency Comparison

To further illustrate the effectiveness of the proposed algorithm and evaluate the network performance, the paper analyzed the time complexity [45,46] of the dual channels and compared them in turn with those of the improved network. The specific parameters are shown in Table 2. In the paper, the time complexity of the shallow network is $O\left(f_1^2 n_1\right) + O\left(n_1 f_2^2 n_2\right) + O\left(n_2 f_3^2\right)$, while the time complexity of the deep channel is the same as that of the shallow layer. It can be seen from Table 2 that the amount of parameter computation per iteration was smaller than that of *EEDS*, meaning that a single iteration training consumed less time. With the same number of iterations, the network training of our proposed model was better than those of SRCNN and EEDS, while the computational complexity of our model was also greatly reduced relative to others. In summary, the efficiency of our proposed method is better than that of the EEDS algorithm.

5. Conclusions

This paper proposed the image super-resolution algorithm based on DCCNN. The deep channel was used to extract the detailed texture information of an image and increase the local receptive field of the image. The shallow channel was mainly used to restore the overall outline of the image. Experimental results showed that the simplified model parameters could not only enhance the ability of the network model to fit the model characteristics but also enable the network model to be trained at a higher learning rate, improving the model's convergence speed. At the same time, the long-term and short-term memory blocks constructed by the residual blocks in the network performed better than the single mapping output network using only the residual blocks. The quantity of image recovery was better, and the performance improved, which proves the necessity of using long-term and short-term memory blocks. Improvement could be observed in both subjective visual effect and objective evaluation parameters, as well as in efficiency, which proves the practicability of the proposed method.

Author Contributions: Conceptualization Y.C.; Methodology, J.W.; Software X.C.; Validation, A.K.S.; Formal Analysis, Y.C.; Investigation J.W.; Resources Z.C.; Data Curation, K.Y. and Z.C.; Supervision J.W. and A.K.S.; Funding Acquisition, J.W. and K.Y. Y.C. provided extensive support in the overall research; J.W. conceived and designed the presented idea, developed the theory, performed the simulations, and wrote the paper; K.Y. and X.C. verified the analytical methods and encouraged to investigate various relevant aspects of the proposed research; A.K.S. and Z.C. provided critical feedback and helped shape the research, analysis, and manuscript. All authors discussed the results and contributed to the final manuscript.

Funding: This research was funded by the National Natural Science Foundation of China [61811530332, 61811540410], the Open Research Fund of Hunan Provincial Key Laboratory of Intelligent Processing of Big Data on Transportation [2015TP1005], the Open Research Fund of Hunan Provincial Science and Technology Innovation Platform of Key Laboratory of Dongting Lake Aquatic Eco-Environmental Control and Restoration [2018DT04], the Changsha Science and Technology Planning [KQ1703018, KQ1706064, KQ1703018-01, KQ1703018-04], the Research Foundation of Education Bureau of Hunan Province [17A007, 16A008], Changsha Industrial Science and Technology Commissioner [2017-7], the Junior Faculty Development Program Project of Changsha University of Science and Technology [2019QJCZ011].

Acknowledgments: We are grateful to our anonymous referees for their useful comments and suggestions. The authors also thank Jingbo Xie, Ke Gu, Yan Gui, Jian-Ming Zhang, Runlong Xia and Li-Dan Kuang for their useful advice during this work.

References

1. Gunturk, B.K.; Batur, A.U.; Altunbasak, Y.; Hayes, M.H.; Mersereau, R.M. Eigenface-Domain Super-Resolution for Face Recognition. *IEEE Trans. Image Process.* **2003**, *12*, 597–606. [CrossRef] [PubMed]
2. Li, S.; Fan, R.; Lei, G.Q.; Yue, G.H.; Hou, C.P. A Two-Channel Convolutional Neural Network for Image Super-Resolution. *Neurocomputing* **2018**, *275*, 267–277. [CrossRef]
3. Zhang, L.P.; Zhang, H.Y.; Shen, H.F.; Li, P.X. A Super-Resolution Reconstruction Algorithm for Surveillance Images. *Signal Process.* **2010**, *90*, 848–859. [CrossRef]
4. Shi, W.Z.; Caballero, J.; Ledig, C.; Zhuang, X.H.; Bai, W.J.; Bhatia, K.K.; Marvao, A.M.M.D.; Dawes, T.; O'Regan, D.P.; Rueckert, D. Cardiac Image Super-Resolution with Global Correspondence using Multi-Atlas Patchmatch. In Proceedings of the 2013 Medical image computing and computer-assisted intervention: MICCAI, Nagoya, Japan, 22–26 September 2013; pp. 9–16.
5. Chen, Y.T.; Wang, J.; Xia, R.L.; Zhang, Q.; Cao, Z.H.; Yang, K. The Visual Object Tracking Algorithm Research Based on Adaptive Combination Kernel. *J. Ambient Intell. Humaniz. Comput.* **2019**, 1–19. [CrossRef]
6. Chen, Y.T.; Xiong, J.; Xu, W.H.; Zuo, J.W. A Novel Online Incremental and Decremental Learning Algorithm Based on Variable Support Vector Machine. *Clust. Comput.* **2018**, 1–11. [CrossRef]
7. Zhang, J.M.; Jin, X.K.; Sun, J.; Wang, J.; Sangaiah, A.K. Spatial and Semantic Convolutional Features for Robust Visual Object Tracking. *Multimed. Tools Appl.* **2018**, 1–21. [CrossRef]
8. Wang, J.; Gao, Y.; Liu, W.; Wu, W.B.; Lim, S.J. An Asynchronous Clustering and Mobile Data Gathering Schema based on Timer Mechanism in Wireless Sensor Networks. *Comput. Mater. Contin.* **2019**, *58*, 711–725. [CrossRef]
9. Timofte, R.; De Smet, V.; Van Gool, L. Anchored Neighborhood Regression for Fast Example-Based Super-Resolution. In Proceedings of the 2013 IEEE International Conference Computer Vision, Sydney, Australia, 1–8 December 2013; pp. 1920–1927.
10. Timofte, R.; De Smet, V.; Van Gool, L. A+: Adjusted Anchored Neighborhood Regression for Fast Super-Resolution. In Proceedings of the 2014 Asian Conference Computer Vision, Singapore, Singapore, 1–5 November 2014; pp. 111–126.
11. Yang, J.C.; Wright, J.; Huang, T.S.; Ma, Y. Image Super-Resolution as Sparse Representation of Raw Image Patches. In Proceedings of the 2008 IEEE Conference Computer Vision and Pattern Recognition, Anchorage, AK, USA, 24–26 June 2008; pp. 1–8.
12. Yang, J.; Wright, J.; Huang, T.S.; Ma, Y. Image Super-Resolution Via Sparse Representation. *IEEE Trans. Image Process.* **2010**, *19*, 2861–2870. [CrossRef]
13. Wang, J.; Gao, Y.; Liu, W.; Sangaiah, A.K.; Kim, H.J. An Intelligent Data Gathering Schema with Data Fusion Supported for Mobile Sink in WSNs. *Int. J. Distrib. Sens. Netw.* **2019**, *15*. [CrossRef]
14. Chen, Y.T.; Xia, R.L.; Wang, Z.; Zhang, J.M.; Yang, K.; Cao, Z.H. The Visual Saliency Detection Algorithm Research Based on Hierarchical Principle Component Analysis Method. *Multimedia Tools Appl.* **2019**, *78*. [CrossRef]
15. Yang, Y.; Lin, Z.; Cohen, S. Fast image super-resolution based on in-place example regression. In Proceedings of the 2013 IEEE Conference on Computer Vision Pattern Recognition, Portland, OR, USA, 23–28 June 2013; pp. 1059–1066.
16. Zhou, S.R.; Ke, M.L.; Luo, P. Multi-Camera Transfer GAN for Person Re-Identification. *J. Visual Commun. Image Represent.* **2019**, *59*, 393–400. [CrossRef]
17. Krizhevsky, A.; Sutskever, I.; Hinton, G.E. ImageNet Classification with Deep Convolutional Neural Networks. In Proceedings of the 2012 International Conference Neural Information Processing Systems, Lake Tahoe, NV, USA, 3–6 December 2012; pp. 1097–1105.
18. Dong, C.; Loy, C.C.; He, K.M.; Tang, X.O. Image Super-Resolution using Deep Convolutional Networks. *IEEE Trans. Pattern Anal. Mach. Intell.* **2016**, *38*, 295–303. [CrossRef] [PubMed]
19. Dong, C.; Loy, C.C.; He, K.M.; Tang, X.O. Learning a Deep Convolutional Network for Image Super-Resolution. In Proceedings of the 2014 International European Conference on Computer Vision, Zurich, Switzerland, 6–12 September 2014; pp. 184–199.
20. Wang, Y.F.; Wang, L.J.; Wang, H.Y.; Li, P.H. End-to-End Image Super-Resolution via Deep and Shallow Convolutional Networks. *IEEE Access* **2019**, *7*, 31959–31970. [CrossRef]

21. Kim, J.; Lee, J.K.; Lee, K.M. Accurate Image Super-Resolution using Very Deep Convolutional Networks. In Proceedings of the 2016 IEEE Conference on Computer Vision Pattern Recognition, Las Vegas, NV, USA, 27–30 June 2016; pp. 1646–1654.

22. Yang, J.X.; Zhao, Y.Q.; Chan, J.C.W.; Yi, C. Hyperspectral Image Classification using Two-Channel Deep Convolutional Neural Network. In Proceedings of the 2016 IEEE International Geoscience and Remote Sensing Symposium, Beijing, China, 10–15 July 2016; pp. 5079–5082.

23. Kim, J.; Lee, J.K.; Lee, K.M. Deeply-Recursive Convolutional Network for Image Super-Resolution. In Proceedings of the 2016 IEEE Conference on Computer Vision Pattern Recognition, Las Vegas, NV, USA, 27–30 June 2016; pp. 1637–1645.

24. Ke, R.M.; Li, W.; Cui, Z.Y.; Wang, Y.H. Two-Stream Multi-Channel Convolutional Neural Network (TM-CNN) for Multi-Lane Traffic Speed Prediction Considering Traffic Volume Impact. Available online: https://arxiv.org/abs/1903.01678 (accessed on 4 June 2019).

25. Lim, B.; Son, S.; Kim, H.; Nah, S.; Lee, K.M. Enhanced Deep Residual Networks for Single Image Super-Resolution. In Proceedings of the 2017 IEEE Conference on Computer Vision Pattern Recognition, Workshops, Honolulu, HI, USA, 21–26 July 2017; pp. 1132–1140.

26. Tai, Y.; Yang, J.; Liu, X.M.; Xu, C.Y. MemNet: A Persistent Memory Network for Image Restoration. In Proceedings of the 2017 IEEE International Conference Computer Vision, Venice, Italy, 22–29 October 2017; pp. 4549–4557.

27. Asvija, B.; Eswari, R.; Bijoy, M.B. Security in Hardware Assisted Virtualization for Cloud Computing—State of the Art Issues and Challenges. *Comput. Netw.* **2019**, *151*, 68–92. [CrossRef]

28. Zhou, S.W.; He, Y.; Xiang, S.Z.; Li, K.Q.; Liu, Y.H. Region-Based Compressive Networked Storage with Lazy Encoding. *IEEE Trans. Parallel Distrib. Syst.* **2019**, *30*, 1390–1402. [CrossRef]

29. Min, X.; Ma, K.; Gu, K.; Zhai, G.; Wang, Z.; Lin, W. Unified Blind Quality Assessment of Compressed Natural, Graphic, and Screen Content Images. *IEEE Trans. Image Process.* **2017**, *26*, 5462–5474. [CrossRef] [PubMed]

30. Gu, K.; Zhai, G.T.; Yang, X.K.; Zhang, W.J. Using Free Energy Principle for Blind Image Quality Assessment. *IEEE Trans. Multimed.* **2015**, *17*, 50–63. [CrossRef]

31. He, K.M.; Zhang, X.Y.; Ren, S.Q.; Sun, J. Deep Residual Learning for Image Recognition. In Proceedings of the 2016 IEEE International Conference on Computer Vision Pattern Recognition, Las Vegas, NV, USA, 27–30 June 2016; pp. 770–778.

32. Nair, V.; Hinton, G.E. Rectified Linear Units Improve Restricted Boltzmann Machines. In Proceedings of the 2010 International Conference on Machine Learning, Haifa, Israel, 21–24 June 2010; pp. 807–814.

33. He, K.M.; Zhang, X.Y.; Ren, S.Q.; Sun, J. Delving Deep into Rectifiers: Surpassing Human-Level Performance on Imagenet Classification. In Proceedings of the 2015 IEEE International Conference on Computer Vision, Santiago, Chile, 7–13 December 2015; pp. 1026–1034.

34. Sun, J.; Xu, Z.B.; Shum, H.Y. Image Super-Resolution using Gradient Profile Prior. In Proceedings of the 2008 IEEE International Conference on Computer Vision and Pattern Recognition, Anchorage, AK, USA, 24–26 June 2008. [CrossRef]

35. Nah, S.; Kim, T.H.; Lee, K.M. Deep Multi-Scale Convolutional Neural Network for Dynamic Scene Deblurring. In Proceedings of the 2016 IEEE Conference on Computer Vision Pattern Recognition, Honolulu, HI, USA, 21–26 July 2017; pp. 257–265.

36. Zeiler, M.D.; Fergus, R. Visualizing and Understanding Convolutional Networks. In Proceedings of the 2014 European Conference on Computer Vision, Zurich, Switzerland, 6–12 September 2014; pp. 818–833.

37. Goodfellow, I.; Pouget-Adadie, J.; Mirza, M.; Xu, B.; Farley, D.W.; Ozair, S.; Courville, A.; Bengio, Y. Generative Adversarial Nets. In Proceedings of the 2014 Advances in Neural Information Processing Systems, Montreal, QC, Canada, 8–13 December 2014; pp. 2672–2680.

38. Timofte, R.; Agustsson, E.; Gool, L.V.; Yang, M.H.; Zhang, L.; Lim, B.; Son, S.; Kim, H.; Nah, S.; Lee, K.M.; et al. Ntire 2017 challenge on single image super-resolution: Methods and results. In Proceedings of the 2017 IEEE Conference on Computer Vision Pattern Recognition, Workshops, Honolulu, HI, USA, 21–26 July 2017; pp. 1110–1121. [CrossRef]

39. Bevilacqua, M.; Roumy, A.; Guillemot, C.; Alberi-Morel, M.L. Low-Complexity Single-Image Super-Resolution Based on Nonnegative Neighbor Embedding. In Proceedings of the 2012 British Machine Vision Conference, Northumbria University, Newcastle, UK, 3–6 September 2018; pp. 135.1–135.10.

40. Zeyde, R.; Elad, M.; Protter, M. On Single Image Scale-up using Sparse-Representations. In Proceedings of the 2010 International Conference Curves and Surfaces, Avignon, France, 24–30 June 2010; pp. 711–730.

41. Choi, S.Y.; Dowan, C. Unmanned Aerial Vehicles using Machine Learning for Autonomous Flight; State-of-The-Art. *Adv. Rob.* **2019**, *33*, 265–277. [CrossRef]

42. Gao, G.W.; Zhu, D.; Yang, M.; Lu, H.M.; Yang, W.K.; Gao, H. Face Image Super-Resolution with Pose via Nuclear Norm Regularized Structural Orthogonal Procrustes Regression. *Neural Comput. Appl.* **2018**, 1–11. [CrossRef]

43. Chen, Y.T.; Wang, J.; Chen, X.; Zhu, M.W.; Yang, K.; Wang, Z.; Xia, R.L. Single-Image Super-Resolution Algorithm Based on Structural Self-Similarity and Deformation Block Features. *IEEE Access* **2019**, *7*, 58791–58801. [CrossRef]

44. Hong, P.L.; Zhang, G.Q. A Review of Super-Resolution Imaging through Optical High-Order Interference. *Appl. Sci.* **2019**, *9*, 1166. [CrossRef]

45. Pan, C.; Lu, M.Y.; Xu, B.; Gao, H.L. An Improved CNN Model for Within-Project Software Defect Prediction. *Appl. Sci.* **2019**, *8*, 2138. [CrossRef]

46. Yin, C.Y.; Ding, S.L.; Wang, J. Mobile Marketing Recommendation method Based on User Location Feedback. *Human-Centric Comput. Inf. Sci.* **2019**, *9*, 1–17. [CrossRef]

Article

No-reference Automatic Quality Assessment for Colorfulness-Adjusted, Contrast-Adjusted, and Sharpness-Adjusted Images Using High-Dynamic-Range-Derived Features

Jinseong Jang, Hanbyol Jang, Taejoon Eo, Kihun Bang and Dosik Hwang *

School of Electrical and Electronic Engineering, Yonsei University, Seoul 03722, Korea;
jinseongjang@yonsei.ac.kr (J.J.); hanstar4@yonsei.ac.kr (H.J.); ship9136@naver.com (T.E.);
bangki03@yonsei.ac.kr (K.B.)
* Correspondence: dosik.hwang@yonsei.ac.kr; Tel.: +82-2-2123-5771

Received: 15 August 2018; Accepted: 14 September 2018; Published: 18 September 2018

Abstract: Image adjustment methods are one of the most widely used post-processing techniques for enhancing image quality and improving the visual preference of the human visual system (HVS). However, the assessment of the adjusted images has been mainly dependent on subjective evaluations. Also, most recently developed automatic assessment methods have mainly focused on evaluating distorted images degraded by compression or noise. The effects of the colorfulness, contrast, and sharpness adjustments on images have been overlooked. In this study, we propose a fully automatic assessment method that evaluates colorfulness-adjusted, contrast-adjusted, and sharpness-adjusted images while considering HVS preferences. The proposed method does not require a reference image and automatically calculates quantitative scores, visual preference, and quality assessment with respect to the level of colorfulness, contrast, and sharpness adjustment. The proposed method evaluates adjusted images based on the features extracted from high dynamic range images, which have higher colorfulness, contrast, and sharpness than that of low dynamic range images. Through experimentation, we demonstrate that our proposed method achieves a higher correlation with subjective evaluations than that of conventional assessment methods.

Keywords: image adjustment; colorfulness; contrast; sharpness; high dynamic range

---

1. Introduction

Recently, camera manufacturers and researchers have developed various post-processing methods that enhance image quality. With the development of computer performance, image enhancement techniques have been actively developed for the last 20 years. These image enhancement techniques include denoising that reduces image noise [1,2], sharpening that creates a less blurry image [3], filtering that changes image property [4,5], and histogram equalization that enhances the contrast of the image [6]. Furthermore, various cutting-edge techniques such as super-resolution used to increase image resolutions [7–9], artifact and distortion removal for images degraded by compression [10,11], and methods to adjust the colorfulness, contrast, and sharpness of images have been developed to improve the visual preference of human visual systems (HVS) [12–18].

These methods need parameter adjustments in order to obtain high-quality images. Since the perception of image quality is influenced by HVS properties, these parameter values can be determined through subjective viewer preferences. However, subjective evaluations are time-consuming and expensive because many people are required for the evaluation of test images. Accordingly, objective assessment methods that automatically evaluate no-reference images have been extensively researched, which resulted in developments such as just-noticeable difference (JND)-based techniques [19,20]. Furthermore, natural scene statistic (NSS)-based methods have been developed to improve the correlation with subjective evaluations considering HVS properties [21–25]. The natural image-quality index (NIQE) [24] method uses the statistical difference between the fitted Gaussian functions of the high-quality and low-quality images in spatial domains. NIQE does not require a reference database of human-rated images. Additionally, this method considers a human visual system using a statistics-based difference between a pre-trained high-quality image database. However, the aim of these previous methods has been completed to evaluate images distorted by compression or noise and they are not suitable for evaluating colorfulness-adjusted, contrast-adjusted, and sharpness-adjusted images. Furthermore, they depend on the luminance of images without considering color components.

Several efforts have been made to evaluate contrast-adjusted [26,27] and sharpness-adjusted [27–29] images. Recently, a color quality enhancement evaluation (CQE) method was developed [30], which considered color components. This method considers a color component and is specialized in evaluating the adjusted image. In this method, the overall quality scores of an image are obtained from a linear combination of several feature values that represent the levels of quality for each adjustment. However, this method does not take HVS properties into account and, therefore, its results correlate poorly with the results of subjective evaluations [31].

In this study, to improve the correlation of such no-reference methods with subjective evaluations, we propose in this study a fully automatic method that evaluates colorfulness-adjusted, contrast-adjusted, and sharpness-adjusted images by taking into account HVS preferences using a high dynamic range (HDR) [32–34]—derived features. These HDR-derived features are extracted from HDR images that have higher visual preferences for HVS than that for LDR images. Furthermore, since HDR images have a wider dynamic range, they have more contrast, colorfulness, and sharpness components than LDR images do [18]. To evaluate the performance of our proposed method, we compared its correlation results (with subjective evaluation scores for the colorfulness-adjusted, contrast-adjusted, and sharpness-adjusted images) with those of two conventional methods, a natural image quality evaluator (NIQE) [24], and CQE [30]. Since our method uses HDR-derived features that consider HVS properties, it achieved a higher correlation with the subjective evaluations than the conventional methods did.

1.1. Correlation between the Evaluation Scores and the Adjustment Levels

The feature values for colorfulness, contrast, and sharpness used in the CQE method can effectively represent the levels of each adjustment. As a result, the calculated quality scores obtained by linearly combining these values may represent the overall levels of the image adjustments. However, the subjective evaluation scores for these adjusted images may be different from the quality scores calculated by CQE because HVS perceives images differently than the calculated features and quality scores used in CQE do. As such, we conducted experiments to investigate the correlation between the subjective and calculated scores and the level of image adjustments.

Eight levels of the colorfulness, contrast, and sharpness adjustments were applied to 24 images in the TID2013 database [35], which resulted in 576 adjusted images. The color saturation method [36], linear contrast adjustment [37], and unsharp masking method [3] were used for colorfulness, contrast, and sharpness adjustments, respectively. Eight observers participated in subjective evaluations for these adjusted images and rated the image qualities from 1 to 5 with 1 representing the worst quality and 5 the best. Lastly, we calculated mean opinion scores [38] from these subjective scores.

Figure 1 shows the mean scores of the subjective scores and the calculated feature values used in a previous assessment method, CQE [30], along with the levels of image adjustment. While the calculated feature values keep increasing as the level of adjustment increases, the subjective scores increase only until the level of adjustment of about four or five levels and then they decrease thereafter until the eighth adjustment level. Figure 1 shows that the calculated feature values and the subjective scores have different tendencies depending on the level of adjustment. Generally, the subjective scoring is considered the gold standard for image assessment methods because human perception is the ultimate image receiver. This suggests that excessive adjustment of an image can have adverse impact on the perception of HVS for the image. Therefore, in this study, we propose a new fully automatic assessment method that can consider the relationships between the adjustment levels and the visual preference of HVS. Our assessment algorithm utilizes the features of HDR images, which are better matched with HVS preferences than LDR images. The following Section 2.2 demonstrates that HDR images receive higher subjective evaluation scores than LDR images.

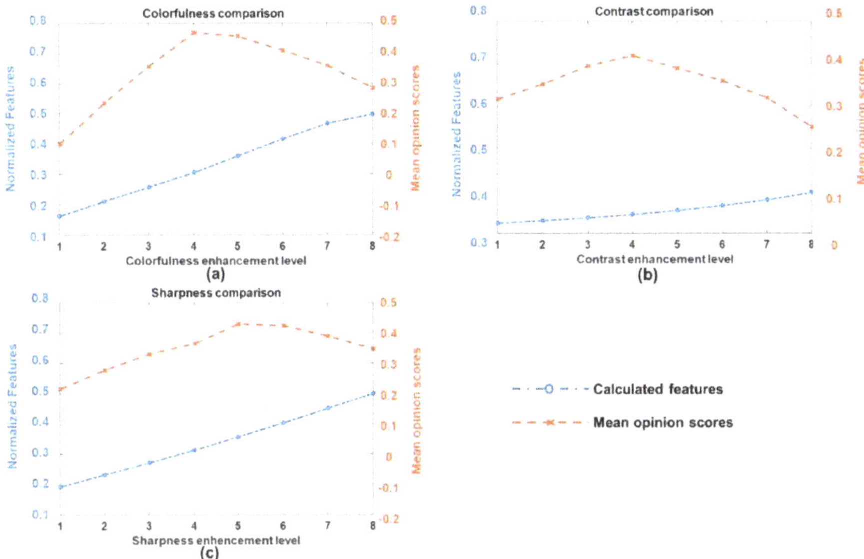

Figure 1. Subjective evaluation scores (x-) and calculated feature values (o-) for different adjustment levels on (**a**) colorfulness adjustment, (**b**) contrast adjustment, and (c) sharpness adjustment. While the calculated feature values keep increasing as the level of adjustment increases, the subjective scores increase only until the level of adjustment of about four or five levels and then they decrease thereafter until the eighth adjustment level.

1.2. Visual Preference Comparison between LDR and HDR Images

To determine whether the visual preference for HDR images is greater than for LDR images, we subjectively evaluated the HDR and LDR images. A total of 190 LDR images and 27 tone-mapped HDR images from the EMPA-HDR database [39] were subjectively evaluated by 10 researchers who specialize in image processing for analysis, image adjustment, and artifact reduction in a dark room display and the same display. The random order images were individually evaluated on a scale of 1 (minimum score) to 5 (maximum score). The subjective scores evaluated by 10 observers of each images were averaged to yield mean opinion scores (MOSs) and the MOSs of all LDR and tone mapped HDR images were also averaged to yield total MOSs of LDR and tone mapped HDR, respectively. Figure 2 shows that the subjective scores for the HDR images are greater than those for the LDR images. While the mean score of the LDR images is 2.44, the mean score of the HDR images is 4.39. This indicates that the HDR images are visually preferred by HVS over the LDR images.

Our proposed assessment method, therefore, uses HDR-derived features extracted from HDR images that are closer to the HVS properties in order to improve our method's correlation with subjective evaluations.

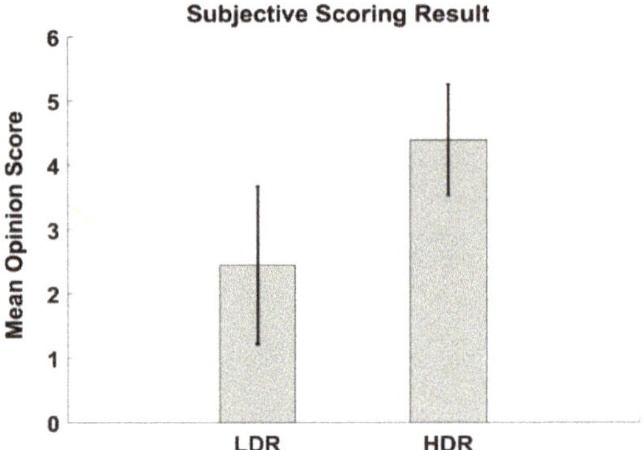

Figure 2. Comparison of the subjective scores for LDR and HDR images. HDR images received significantly higher scores than LDR images.

2. Materials and Methods

2.1. Proposed Assessment Method

Since the visual preference for HDR images is higher than for LDR images, a better quality score can be obtained by using the calculations based on the difference between the HDR-derived features and the test image features. The features used in our method are HDR-derived colorfulness, contrast, and sharpness features. In addition, each colorfulness, contrast, and sharpness feature in our method comprises global and local feature values. Since HVS evaluates images based on the hierarchical visual perception mechanism [40], our proposed method uses both global and local features to consider these HVS properties.

HDR images have different bit depth than LDR images and cannot be directly compared with them. Therefore, the HDR images are first tone-mapped [41] such that the scale of the original HDR images can be matched to those of the LDR test images. 500 HDR images were used to construct the standard HDR-derived features.

2.1.1. Colorfulness

Colorfulness is an aspect of the visual perception, according to which the color of an object is perceived to be more or less chromatic [30]. The colorfulness features can be obtained mainly from color channels a and b of the CIELab space and they are orthogonal to the lightness channel. Two different colorfulness features are used in our proposed method. The first colorfulness feature is calculated by using Equation (1) [30].

$$Col_1 = 0.02 \times \log\left(\frac{\sigma_a^2}{|\mu_a|^{0.2}}\right) \times \log\left(\frac{\sigma_b^2}{|\mu_b|^{0.2}}\right) \tag{1}$$

where σ_a^2 and σ_b^2 are the variance of a and b channels in CIELab and μ_a and μ_b are the mean values of the a and b domains, respectively. This feature value increases as the color adjustment level increases, which is shown in Figure 3a.

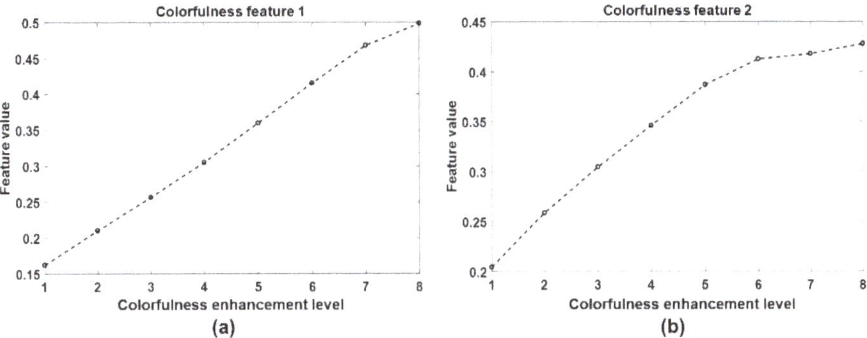

Figure 3. Comparison graph between the color adjustment level and calculated colorfulness features. Colorfulness feature 1 and 2 represent the global and local aspects of the colorfulness of the image, respectively.

Unlike CQE, our proposed method applies a low-pass filter to the RGB space before converting it into the CIELab space. This is done so that the colorfulness features are not influenced by the components of the structural features. The low-pass filter is effective in minimizing the influence of the sharpness features of the images.

The second colorfulness feature in our assessment method is obtained by using Equation (2).

$$Col_2 = \frac{1}{NM} \sum_{i=1}^{N} \sum_{j=1}^{M} \sqrt{\sum_{k=-K}^{K} \sum_{l=-L}^{L} w_{k,l}(R_{k,l}(i,j) - \mu(i,j))^2} \tag{2}$$

where $R = \sqrt{a^2 + b^2}$ is the magnitude of the color components in the chromatic domain, $w_{k,l}$ is the 5×5 sized 2D circularly symmetric Gaussian weighting function, and N and M are the width and height of an image, respectively. This feature also increases as the color adjustment level increases, which is shown in Figure 3b.

The first feature in Equation (1) represents the global colorfulness obtained from the entire image while the second feature in Equation (2) represents the local colorfulness obtained from small image patches. Our proposed method uses these two colorfulness features to consider the global and local aspects of the HVS color perception [40]. These feature values are compared with the HDR colorfulness features that are established from HDR images by using the same Equations (1) and (2) in order to determine the final quality scores that correlate well with the subjective evaluations.

2.1.2. Contrast

Contrast represents the difference in luminance that makes an object distinguishable from other objects within the same field of view [30]. Similar to the colorfulness features, a low-pass filter is also used to minimize the influence of the sharpness component of an image in our study. The first feature is obtained by using Equation (3).

$$Con_1 = \frac{1}{k_1 k_2} \sum_{l=1}^{k_1} \sum_{k=1}^{k_2} \left(\log \left(\frac{L_{max,k,l} + L_{min,k,l}}{L_{max,k,l} - L_{min,k,l}} \right) \right)^{-0.5} \tag{3}$$

where L_{max} and L_{min} are the maximum and minimum luminance values of a 5×5-sized image patches, respectively. This feature increases as the contrast adjustment level increases, which is shown in Figure 4a. The second contrast feature is obtained by using Equation (4).

$$Con_2 = \frac{1}{NM} \sum_{i=1}^{N} \sum_{j=1}^{M} \sqrt{\sum_{k=-K}^{K} \sum_{l=-L}^{L} w_{k,l} (L_{k,l}(i,j) - \mu(i,j))^2} \tag{4}$$

where $w_{k,l}$ is a 15×15 2D circularly symmetric Gaussian weighting function. The second contrast feature also increases as the contrast adjustment level increases, which is shown in Figure 4b.

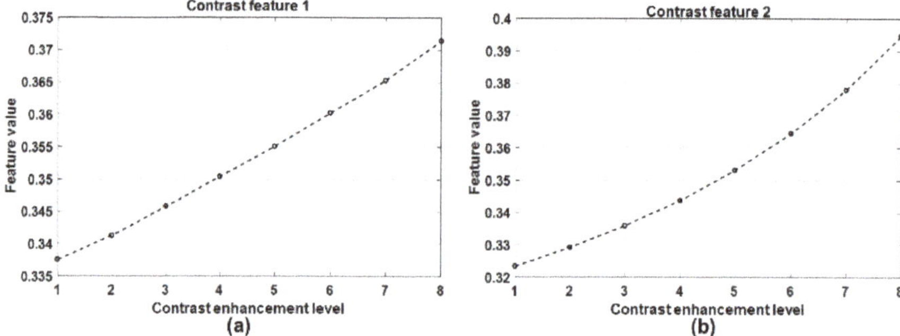

Figure 4. Comparison graph between the contrast adjustment level and the calculated contrast features. Contrast feature 2 and 1 represent the local and global aspects of the contrast of the image, respectively.

The first feature in Equation (3) represents the local contrast feature that is extracted from small-sized image patches. In contrast, the second feature in Equation (4) represents a semi-global contrast feature that is extracted from relatively large-sized image patches. As done with the colorfulness assessment, our proposed method uses two contrast features to consider the HVS properties. These feature values are also compared with the HDR contrast features for the final quality score calculations.

2.1.3. Sharpness

Sharpness represents the aspects of fine details and edge components of an image and it is distributed in the high-frequency band of a Fourier domain [30]. Sharpness features can be extracted from high-pass filtered images, which is outlined in Equations (5) and (6).

$$Sha_1 = \frac{1}{k_1 k_2} \sum_{l=1}^{k_1} \sum_{k=1}^{k_2} \log\left(\frac{E_{max,k,l}}{E_{min,k,l}}\right) \tag{5}$$

The first sharpness feature in Equation (5) is calculated by using 5×5 sized image patches and E_{max} and E_{min} are the maximum and minimum luminance values of the high-pass filtered image patches, respectively. This feature increases as the sharpness adjustment level increases, which is shown in Figure 5a.

$$Sha_2 = \frac{1}{NM} \sum_{i=1}^{N} \sum_{j=1}^{M} \sqrt{\sum_{k=-K}^{K} \sum_{l=-L}^{L} w_{k,l} (E_{k,l}(i,j) - \mu(i,j))^2} \tag{6}$$

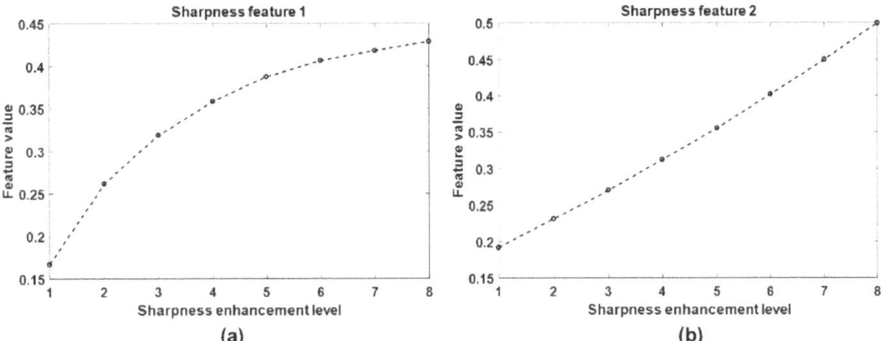

Figure 5. Comparison between the sharpness adjustment level of an image and the calculated sharpness features. Sharpness feature 1 and feature 2 represent the local and global aspects of the colorfulness of the image, respectively.

The second sharpness feature is obtained by using Equation (6). The standard deviation of the high-pass filtered luminance values is used to calculate how widely distributed the high-frequency components were in the images. $w_{k,l}$ is a relatively large-sized 15×15 2D circularly symmetric Gaussian weighting function. This feature increases as the sharpness adjustment level increases, which is shown in Figure 5b. Similar to the assessments of the colorfulness and contrast, our proposed method uses both local (Equation (5)) and global (Equation (6)) sharpness features to consider the HVS properties. These feature values are also compared with the HDR sharpness features for the final quality score calculations.

2.1.4. HDR-Derived Features

A direct assessment using the previously mentioned features obtained from Equations (1)–(6) may lead to erroneous evaluation results that do not correlate well with the subjective evaluations, as previously demonstrated in Figure 1. To overcome this problem, our proposed assessment method uses HDR-derived features as standards when evaluating the features of a test image. The HDR-derived features are extracted by using the same equations (Equations (1)–(6)) applied to many HDR images. Since the HDR images are visually preferred by HVS as shown in Figure 2, HDR-derived features can be used as references, according to which the features of test images are evaluated to determine the final quality scores. Our study demonstrates that this method correlates well with the subjective evaluations.

2.1.5. Assessment Metric Scheme

Figure 6 shows a diagram of our proposed method for determining the quality score of a test image. Six features are extracted from a test image and then compared with the standard HDR-derived features to calculate the quality score of a test image.

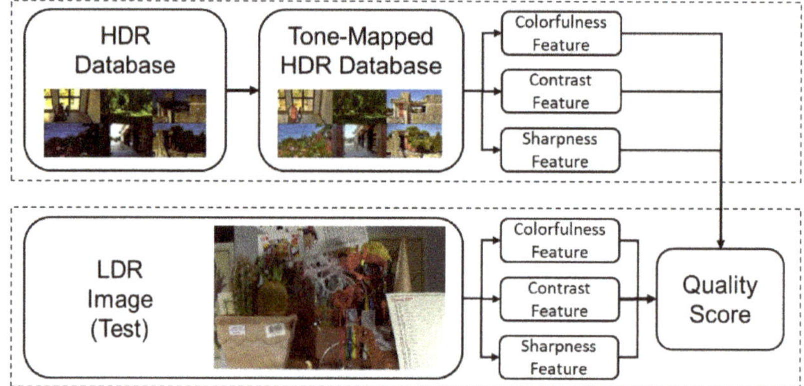

Figure 6. Diagram of the proposed method, which comprises two processes. Above side process: extracting the standard HDR-derived features. Below side process: extracting the features of the test image. These two sets of features are compared with each other to produce the final quality score of the test image.

First, the differences between each feature are calculated by using the following equation.

$$\Delta X = |X_{HDR} - X_{TEST}| \tag{7}$$

where X_{HDR} is a standard HDR-derived single feature vector and the X_{TEST} is a single feature vector of the test image. Many colorfulness, contrast, and sharpness features from HDR images is obtained and averaged to make a single feature vector. Lastly, the single feature vector is used to calculate the quality score based on a weighted combination of the feature differences. In addition, multiple feature vectors from a test image are also averaged to make a single feature vector since multiple feature vectors are obtained from multiple local patches of a single test image. The final quality score, Q, is obtained from a weighted combination of the following feature differences.

$$Q = \vec{C}\vec{D} = [C_1 \ C_2 \ C_3 \ C_4 \ C_5 \ C_6] \times [\Delta Col_1 \ \Delta Col_2 \ \Delta Con_1 \ \Delta Con_2 \ \Delta Sha_1 \ \Delta Sha_2 \ ]^T \tag{8}$$

where $\vec{C}$ is a set of weighting coefficients for each feature. $\vec{C}$ was determined as $[7, 9, 6.1, 8.5, 6.7, 0.54, 1]$ through the training with 576 training image sets.

2.2. Experimental Setup

The total 576 training images were subjectively evaluated by 10 researchers who specialize in image processing for analysis, image adjustment, and artifact reduction in the dark room display and in the same display. The random order images were individually evaluated on a scale of 1 (minimum score) to 5 (maximum score).

A total of 114 adjusted images were assessed by our proposed method as well as by two conventional methods, which were NIQE and CQE. The subjective evaluation was also performed as the ground truth quality scores. Three intact images (traffic, cactus, and basketball) were selected from the LIVE video database [42] and processed with different combinations of three levels of colorfulness adjustment, three levels of contrast adjustment, and four levels of sharpness adjustment, which resulted in 108 adjusted images using MATLAB (R2017a, The Mathworks, Natick, MA, USA) software automatically. The examples of these test image sets are shown in Figure 7. The other six images included the original three images plus enhanced ones by an HDR-toning toolbox in Photoshop CS6. These 114 test images were not used in the training process.

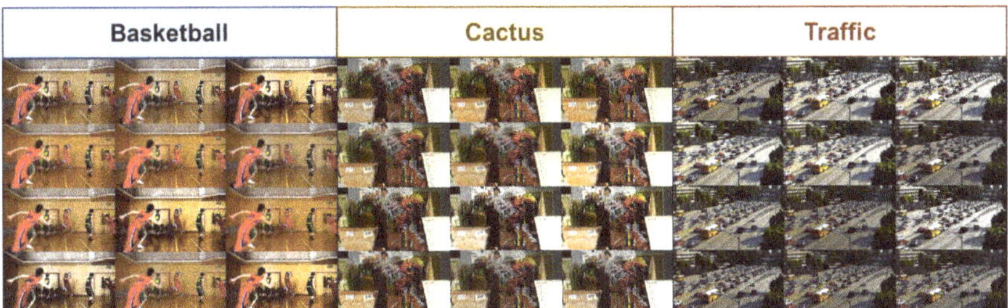

Figure 7. Examples of test images with various levels of colorfulness, contrast, and sharpness adjustments.

3. Evaluation Results

To evaluate the performance of our proposed method, we obtained correlations between the quality scores of our proposed method and the subjective scores (visual preferences of HVS). Ten observers participated in the subjective evaluation. For the purpose of comparison with other methods, correlations between the scores of the conventional methods (NIQE and CQE) and the subjective scores were also obtained. Spearman's rank ordered a correlation coefficient (SROCC) and Pearson's linear correlation coefficient (PLCC), which were used to compare the performances. We used the Spearman and Pearson coefficients methods for performance evaluation because these methods have been widely used for correlation metrics between image quality assessment and subjective scores [21–25].

The results of the correlations with the subjective scores are shown in Table 1. In all the cases, our proposed method achieved higher correlations with the subjective scores than those of the NIQE and CQE. These results demonstrate that our proposed method is more suitable for evaluating the colorfulness-adjusted, contrast-adjusted, and sharpness-adjusted images than the other two conventional methods. This is because the conventional methods were mainly developed for evaluating images distorted by compression or noise corruption (NIQE) and did not consider HVS properties (CQE).

Table 1. Performance comparison between the conventional methods (NIQE, CQE) and the proposed method. Two difference correlation metrics called SROCC and PLCC were calculated between the scores measured by the automatic methods and by the subjective evaluation.

		Basketball	Cactus	Traffic	Average
NIQE	SROCC	0.6465	0.3233	0.1650	0.3283
	PLCC	0.7144	0.3221	0.2373	0.4246
CQE	SROCC	0.6207	0.4241	0.5292	0.5247
	PLCC	0.6225	0.4236	0.5205	0.5222
Proposed Method	SROCC	0.8042	0.9121	0.9669	0.8944
	PLCC	0.7626	0.9284	0.9538	0.8816

4. Discussion and Conclusions

We proposed a fully automatic no-reference quality assessment method for the enhanced images whose colorfulness, contrast, and sharpness were adjusted. HDR-derived features were obtained from HDR images with different scenes. Therefore, the proposed method without using the reference image with the same scene can be a 'no-reference' method.

The proposed method does not require a reference image. It automatically calculates quantitative scores, visual preference, and quality assessment with respect to the level of colorfulness, contrast, and sharpness adjustment. This method considers colorfulness components. Additionally, this method uses HDR-derived features, which have more human visual preference than LDR images. By evaluating the LDR based on the HDR, this method can extract how different the quality of the test images differs from the HDR with a high visual preference and it offers visual preference scores, quantitatively. It shows that the proposed method yielded better performance than other methods.

We investigated the performance of the proposed method depending on the size of image patches in global and local settings and measured the corresponding performance, according to the correlation values with subjective scores. In most cases involving our test image dataset, the highest correlation results were achieved with the sizes of image patches used in our experiments.

Our proposed method currently used linear weighted coefficients trained by 576 HDR images. However, other machine learning metrics such as the support vector machine can be used in the training process. Because SVM-based quality evaluation methods have already been used in the detection on artifacts caused by compression [21–25], it is possible to use the SVM method for our proposed method. In addition, if there are additional features including the six features, neural network methods can be used for the evaluation of adjusted images.

We made an effort to overcome the limitations of conventional methods such as NIQE and CQE methods. NIQE performs well in evaluating the distorted images due to compression, blurring, or noise corruption, but it is not focused on evaluating adjusted images. CQE can evaluate several adjustment effects such as colorfulness and contrast adjustments but does not incorporate HVS properties in its assessment process, which results in a mismatch with the subjective evaluations. These limitations were effectively overcome in our proposed method, which incorporated HVS-favorable HDR-derived features as standards in its evaluation process. HDR images have higher visual preferences for HVS than LDR images and, therefore, the features derived from HDR images are more closely related to the perception properties of HVS than those of LDR images. In addition to the incorporation of the HDR-derived features, both global and local features are extracted and combined to produce the final quality scores for an image assessment, which also considers the hierarchical visual perception mechanism of HVS. Consequently, we found through our experimentation that our new assessment method correlated well with subjective evaluations and outperformed two conventional assessment methods.

Author Contributions: Conceptualization, J.J. and D.H. Methodology, J.J. and H.J. Software, J.J. and T.E. Validation, J.J., H.J., and K.B. Formal Analysis, J.J. and T.E. Writing-Original Draft Preparation, J.J., H.J., and D.H. Writing-Review & Editing, J.J. and D.H.

Funding: This research was supported by Samsung Electronics Company, Suwon, South Korea.

Conflicts of Interest: The authors declare no conflict of interest.

References

1. Danielyan, A.; Katkovnik, V.; Egiazarian, K. BM3D Frames and Variational Image Deblurring. *IEEE Trans. Image Process.* **2012**, *21*, 1715–1728. [CrossRef] [PubMed]
2. Portilla, J.; Strela, V.; Wainwright, M.; Simoncelli, E. Image denoising using scale mixtures of gaussians in the wavelet domain. *IEEE Trans. Image Process.* **2003**, *12*, 1338–1351. [CrossRef] [PubMed]
3. Polesel, A.; Ramponi, G.; Mathews, V. Image enhancement via adaptive unsharp masking. *IEEE Trans. Image Process.* **2000**, *9*, 505–510. [CrossRef] [PubMed]
4. Yu, H.; Zhao, L.; Wang, H. Image Denoising Using Trivariate Shrinkage Filter in the Wavelet Domain and Joint Bilateral Filter in the Spatial Domain. *IEEE Trans. Image Process.* **2009**, *18*, 2364–2369. [PubMed]
5. Kazubek, M. Wavelet domain image denoising by thresholding and Wiener filtering. *IEEE Signal Process. Lett.* **2003**, *10*, 324–326. [CrossRef]
6. Stark, J. Adaptive image contrast enhancement using generalizations of histogram equalization. *IEEE Trans. Image Process.* **2000**, *9*, 889–896. [CrossRef] [PubMed]
7. Park, S.; Park, M.; Kang, M. Super-resolution image reconstruction: A technical overview. *IEEE Signal Process. Mag.* **2003**, *20*, 21–36. [CrossRef]
8. Yang, J.; Wright, J.; Huang, T.S.; Ma, Y. Image super-resolution via sparse representation. *IEEE Trans. Image Process.* **2010**, *19*, 2861–2873. [CrossRef] [PubMed]

9. Zhou, D.; Wang, R.; Lu, J.; Zhang, Q. Depth Image Super Resolution Based on Edge-Guided Method. *Appl. Sci.* **2018**, *8*, 298. [CrossRef]

10. Lai, J.; Liaw, Y.; Lo, W. Artifact reduction of JPEG coded images using mean-removed classified vector quantization. *Signal Process.* **2002**, *82*, 1375–1388. [CrossRef]

11. Lee, R.; Kim, D.; Kim, T. Regression-based prediction for blocking artifact reduction in JPEG-compressed images. *IEEE Trans. Image Process.* **2005**, *14*, 36–48. [PubMed]

12. Lucchese, L.; Mitra, S.; Mukherjee, J. A new algorithm based on saturation and desaturation in the xy chromaticity diagram for enhancement and re-rendition of color images. In Proceedings of the International Conference on Image Processing, Thessaloniki, Greece, 7–10 October 2001; Volume 2, pp. 1077–1080.

13. Naik, S.; Murthy, C. Hue-preserving color image enhancement without gamut problem. *IEEE Trans. Image Process.* **2003**, *12*, 1591–1598. [CrossRef] [PubMed]

14. Agaian, S.; Silver, B.; Panetta, K. Transform Coefficient Histogram-Based Image Enhancement Algorithms Using Contrast Entropy. *IEEE Trans. Image Process.* **2007**, *16*, 741–758. [CrossRef] [PubMed]

15. Panetta, K.; Wharton, E.; Agaian, S. Human Visual System-Based Image Enhancement and Logarithmic Contrast Measure. *IEEE Trans. Syst. Man Cybern. Syst.* **2008**, *38*, 174–188. [CrossRef] [PubMed]

16. Zhang, B.; Allebach, J. Adaptive Bilateral Filter for Sharpness Enhancement and Noise Removal. *IEEE Trans. Image Process.* **2008**, *17*, 664–678. [CrossRef] [PubMed]

17. Panetta, K.; Agaian, S.; Zhou, Y.; Wharton, E. Parameterized Logarithmic Framework for Image Enhancement. *IEEE Trans. Syst. Man Cybern. Syst.* **2011**, *41*, 460–473. [CrossRef] [PubMed]

18. Gu, K.; Zhai, G.; Yang, X.; Zhang, W.; Chen, C. Automatic Contrast Enhancement Technology with Saliency Preservation. *IEEE Trans. Circuits Syst. Video Technol.* **2015**, *25*, 1480–1494.

19. Ferzli, R.; Karam, L. A No-Reference Objective Image Sharpness Metric Based on the Notion of Just Noticeable Blur (JNB). *IEEE Trans. Image Process.* **2009**, *18*, 717–728. [CrossRef] [PubMed]

20. Liu, A.; Lin, W.; Paul, M.; Deng, C.; Zhang, F. Just Noticeable Difference for Images with Decomposition Model for Separating Edge and Textured Regions. *IEEE Trans. Circuits Syst. Video Technol.* **2010**, *20*, 1648–1652. [CrossRef]

21. Moorthy, A.; Bovik, A. Blind Image Quality Assessment: From Natural Scene Statistics to Perceptual Quality. *IEEE Trans. Image Process.* **2011**, *20*, 3350–3364. [CrossRef] [PubMed]

22. Saad, M.; Bovik, A.; Charrier, C. Blind Image Quality Assessment: A Natural Scene Statistics Approach in the DCT Domain. *IEEE Trans. Image Process.* **2012**, *21*, 3339–3352. [CrossRef] [PubMed]

23. Mittal, A.; Moorthy, A.; Bovik, A. No-Reference Image Quality Assessment in the Spatial Domain. *IEEE Trans. Image Process.* **2012**, *21*, 4695–4708. [CrossRef] [PubMed]

24. Mittal, A.; Soundararajan, R.; Bovik, A. Making a "Completely Blind" Image Quality Analyzer. *IEEE Signal Process. Lett.* **2013**, *20*, 209–212. [CrossRef]

25. Zhang, Z.; Wang, H.; Liu, S.; Durrani, T.S. Deep Activation Pooling for Blind Image Quality Assessment. *Appl. Sci.* **2018**, *8*, 478. [CrossRef]

26. Gu, K.; Zhai, G.; Lin, W.; Liu, M. The Analysis of Image Contrast: From Quality Assessment to Automatic Enhancement. *IEEE Trans. Cybern.* **2016**, *46*, 284–297. [CrossRef] [PubMed]

27. Kim, H.; Ahn, S.; Kim, W.; Lee, S. Visual Preference Assessment on Ultra-High-Definition Images. *IEEE Trans. Broadcast.* **2016**, *62*, 757–769. [CrossRef]

28. Feichtenhofer, C.; Fassold, H.; Schallauer, P. A Perceptual Image Sharpness Metric Based on Local Edge Gradient Analysis. *IEEE Signal Process. Lett.* **2013**, *20*, 379–382. [CrossRef]

29. Gu, K.; Zhai, G.; Lin, W.; Yang, X.; Zhang, W. No-Reference Image Sharpness Assessment in Autoregressive Parameter Space. *IEEE Trans. Image Process.* **2015**, *24*, 3218–3231. [PubMed]

30. Panetta, K.; Gao, C.; Agaian, S. No reference color image contrast and quality measures. *IEEE Trans. Consum. Electron.* **2013**, *59*, 643–651. [CrossRef]

31. Panetta, K.; Bao, L.; Agaian, S. A human visual "no-reference" image quality measure. *IEEE Instrum. Meas. Mag.* **2016**, *19*, 34–38. [CrossRef]

32. Reinhard, E. *High Dynamic Range Imaging*; Elsevier Morgan Kaufmann: Amsterdam, The Netherlands, 2010.

33. Ofili, C.; Glozman, S.; Yadid-Pecht, O. Hardware Implementation of an Automatic Rendering Tone Mapping Algorithm for a Wide Dynamic Range Display. *J. Low Power Electron. Appl.* **2013**, *3*, 337–367. [CrossRef]

34. Cauwerts, C.; Piderit, M.B. Application of High-Dynamic Range Imaging Techniques in Architecture: A Step toward High-Quality Daylit Interiors? *J. Imaging* **2018**, *4*, 19. [CrossRef]

Appl. Sci. **2018**, *8*, 1688

35. Ponomarenko, N.; Jin, L.; Ieremeiev, O.; Lukin, V.; Egiazarian, K.; Astola, J.; Vozel, B.; Chehdi, K.; Carli, M.; Battisti, F.; et al. Image database TID2013: Peculiarities, results and perspectives. *Signal Process. Image Commun.* **2015**, *30*, 57–77. [CrossRef]

36. Lübbe, E. *Colours in the Mind-Colour Systems in Reality*; Books on Demand: Norderstedt, Germany, 2010.

37. Al-amri, S.; Kalyankar, N.; Khamitkar, S. Linear and non-linear contrast enhancement image. *Int. J. Comput. Sci. Netw. Secur.* **2010**, *10*, 139–143.

38. Streijl, R.; Winkler, S.; Hands, D. Mean opinion score (MOS) revisited: Methods and applications, limitations and alternatives. *Multimed. Syst.* **2014**, *22*, 213–227. [CrossRef]

39. EMPA Media Technology. Available online: http://www.empamedia.ethz.ch/hdrdatabase/index.php (accessed on 25 January 2017).

40. Wang, Z.; Simoncelli, E.; Bovik, A. Multi-scale structural similarity for image quality assessment. In Proceedings of the Thirty-Seventh Asilomar Conference on Signals, Systems & Computers, Pacific Grove, CA, USA, 9–12 November 2003; Volume 2, pp. 1398–1402.

41. Ashikhmin, M.; Goyal, J. A reality check for tone-mapping operators. *ACM Trans. Appl. Percept.* **2006**, *3*, 399–411. [CrossRef]

42. Seshadrinathan, K.; Soundararajan, R.; Bovik, A.; Cormack, L. Study of Subjective and Objective Quality Assessment of Video. *IEEE Trans. Image Process.* **2010**, *19*, 1427–1441. [CrossRef] [PubMed]

Article

A Novel One-Camera-Five-Mirror Three-Dimensional Imaging Method for Reconstructing the Cavitation Bubble Cluster in a Water Hydraulic Valve

Haihang Wang [1], He Xu [1,*], Vishwanath Pooneeth [1] and Xiao-Zhi Gao [2]

[1] College of Mechanical and Electrical Engineering, Harbin Engineering University, Harbin 150001, China; wanghaihang@hrbeu.edu.cn (H.W.); vpooneeth@ymail.com (V.P.)
[2] School of Computing, University of Eastern Finland, Kuopio FI-70211, Finland; xiao.z.gao@gmail.com
* Correspondence: railway_dragon@sohu.com; Tel.: +1-335-111-7608

Received: 11 August 2018; Accepted: 20 September 2018; Published: 1 October 2018

Abstract: In order to study the bubble morphology, a novel experimental and numerical approach was implemented in this research focusing on the analysis of a transparent throttle valve made by Polymethylmethacrylate (PMMA) material. A feature-based algorithm was written using the MATLAB software, allowing the 2D detection and three-dimensional (3D) reconstruction of bubbles: collapsing and clustered ones. The valve core, being an important part of the throttle valve, was exposed to cavitation; hence, to distinguish it from the captured frames, the faster region-based convolutional neural network (R-CNN) algorithm was used to detect its morphology. Additionally, the main approach grouping the above listed techniques was implemented using an optimized virtual stereo vision arrangement of one camera and five plane mirrors. The results obtained during this study validated the robust algorithms and optimization applied.

Keywords: three-dimensional imaging; optimization arrangement; cavitation bubble; water hydraulic valve

---

1. Introduction

Cavitation is an omnipresent phenomenon observed during flows in valves, pipes, pressure vessels, and so on. Its occurrence favors considerable material losses in complex situations, which requires costly replacements to be made thereby causing significant performance drawbacks to industries. Studies on fluid dynamics about solving issues related to cavitation due to bubbles collapse are being done by researchers from both universities and companies. To be more precise, understanding bubble dynamics in valves requires in-depth investigations as the presence of void fractions in optically dense multiphase flows have been hindering the observation of bubbles [1].

To the best of our knowledge, fewer studies about bubble morphology in throttle valves have been done, albeit with both the invasive techniques (impedance probe and optical fiber probe) and the non-invasive techniques (PIV, PTV, PT) being prevalent. The non-invasive approach has widely been promoted through high speed photography in bubble measurement studies, inclusive of the 2D bubble columns [2–5], the channels [6–9], the flat plates [10], the hydrofoils [11,12], the mixing tanks [13], the liquid-solid interface [14], the dynamically loaded journal bearings [15], the ultrasonic devices [14,16–18], the axisymmetric geometry [10], the throttle orifice, and so on.

To start with, the first ever experiment on capturing the motion of Helium-filled bubbles in an engine using a single camera and multiple mirrors, was done by Kent and Eaton [19] in 1982. Next, Racca and Dawey [20] implemented a measuring method by using a single high speed cine camera through a split field mirror to track small resin beads (tracers) and Belden et al. proposed a "3D synthetic aperture imaging (SA imaging)" by using nine (9) high speed photon cameras to capture the bubbly flow induced by a turbulent circular plunging jet.

Similarly, to study the fluid flow, virtual stereo vision was implemented in this research to observe and reconstruct the 3D bubbles formed in the area between the valve seat, the valve core and the outlet port. Xue et al. [21] applied virtual binocular stereo vision in a glass-made water tank to match and reconstruct the bubble trajectory motion through a "3D polar coordinate homonymy correlation algorithm", thereby determining the analogous relation of alike bubbles from two-half images. Additional studies by Xue et al. [22–24] dealt with the bubble behaviour characteristics in the gas-liquid two-phase flow, modality factors of bubbles, intrinsic and extrinsic parameters of the virtual cameras, and the segmentation of multi-bubbles.

Likewise, by implementing robust algorithms to estimate the velocities and reconstruct the trajectories of bubbles, broad investigations were made by Mitra et al. [25], Acuna et al. [26], Racca et al. [20], Dencks et al. [16], Cheng et al. [27], Krimerman [28], and Bakshi et al. [29], respectively.

For a clearer approach of the stereo vision concept, significant contributions were brought by the following researchers: Feng et al. [30] established a 3D mathematical model to measure a 3D point through a combination of the single camera stereo vision sensor with planar mirror imaging. Upon comparison with the binocular stereo vision, the output in terms of calibration, measurement speed and errors resulted in being more accurate. Figueroa et al. [31] studied a nearly non-dispersed 2D bubbly flow in a thin channel by means of a high speed camera- image processing routine to validate bubble clusters trajectory and sealing arguments to estimate their lifespan. Moreover, bubbly flows were continuously studied by Yucheng et al. [8,32], Chakraborty [33], Lau et al. [3], and Tayler et al. [5] from the last decade. Prolonged observations in dense bubbly flows containing overlapped ellipse-like bubbles through image analysis using algorithms were achieved by de Langlard et al. [34], Honkanen et al. [35,36], and Zhang et al. [37], respectively.

On the other hand, Fujisawa et al. [38] examined erosion caused by bubble implosion and shock waves formed by a cavitating jet. This was carried out using shadowgraph imaging, time-difference analysis, and laser schlieren imaging techniques, thereby giving adequate results.

Similarly, studies about object detection in fluid flow were carried out by Kompella et al. [39] who detected semi-transparent objects in single images, while Hata et al. [40] and Kai et al. [41], emphasized the shape extraction and the dense reconstruction of transparent objects.

On the basis of virtual stereo motion, this paper illustrates a novel approach of using five reflectors (single camera with two-symmetrical reflectors and one stand-alone reflector) in contrast to the literature where single cameras with only two symmetrical reflector sets were used. In addition, compared with using three high-speed cameras, the experimental images reflected from the plane mirrors did not only cut the equipment costs, but also ensured the synchronization of the images from the three directions (x, y, and z). A deep learning method developed by Ren et al. [42] was applied to detect the valve core (opaque) during the bubbly fluid flow. Lastly, the original algorithms written using MATLAB (R2016b, The MathWorks, Inc., Natick, MA, USA) software in this research helped to determine the optimal design of the reflector sets set-up and perform the 3D reconstruction of the bubbles.

The overall structure of the paper is as follows: the introduction is followed by Section 2 in which the overall structure of the 3D imaging experiment system is explained. Next, Section 3 elaborates on the optimized arrangement of the one-camera-five mirror module, while Section 4 illustrates the development of the proposed algorithm used for bubble detection. Finally, Sections 5 and 6 elucidate on the analysis of the results and the conclusion, respectively.

2. Overall Structure of the 3D Imaging Experiment System

Figure 1 shows the experimental setups of the hydraulic system, which consists of a water hydraulic power transmission subsystem, an electric control subsystem and the water hydraulic valve. The dimensions of the transparent valve used in this research were marked in right figure. In addition, the inlet and outlet ports were noted. The "electric control subsystem" consists of a frequency inverter and a component switching panel. The frequency inverter (Schneider Altivar 610, Paris, France) ensures suitable power for the hydraulic test bench through adequate control of the water pumps working frequency (input pressure). Water pump used in the transmission system allows a constant pressure water supply, which eliminates the shortcomings in terms of pressure fluctuations owing to subsequent improvements in the AC frequency conversion technology.

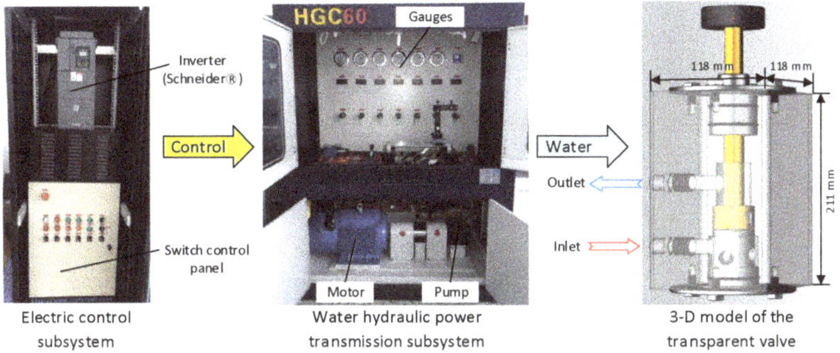

Figure 1. Overall structure of the experimental setups.

Figure 2 clearly demonstrates the optimal arrangement of the one-camera-five-mirror 3D imaging module on the water hydraulics experiment platform. The arrangement allowed the capture and storing of the experimental videos by the high-speed camera and the computer, respectively. High-speed camera can capture thousands of photographs per second. However, because of the very short time interval of two adjacent pictures, the Light Emitting Diode (LED) lamp was an essential device to compensate for the lack of exposure.

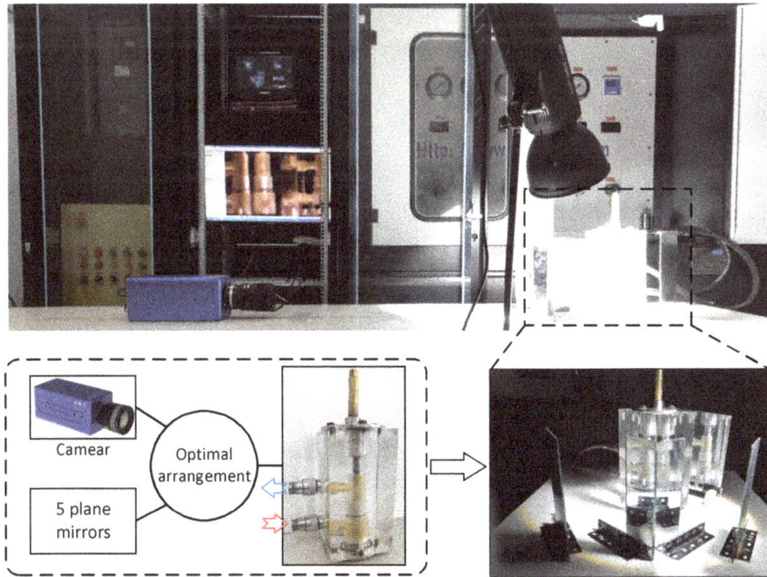

Figure 2. Experimental configuration for the 3D imaging of cavitation bubble.

2.1. Measurement Principle of the 3D Imaging

Viewing cavitating bubbles in the fluid flow from one side might not give satisfying results in terms of the size, the position, and the quantity of bubbles. Hence, to avoid partial invisibility while recording, a 3D approach based on one-camera-five-mirror device was applied in this study.

The 3D virtual stereo vision measurement principle of the bubbles illustrated in Figure 3 explains the arrangement of the reflector sets. To observe the cavitating area in a transparent throttle valve through different view angles, a high-speed camera (Revealer 5KF10, Hefei, China) with a 60 mm Nikkor lens (Tokyo, Japan): resolution and frame rate of 1280 × 860 pixels and 4000 fps, a lighting equipment, and five reflectors (mirror glass) were used. The concept of virtual stereoscopic parallax eased the capturing of the cavitation bubbles around the valve core. With the real camera imaged into virtual cameras: two mirrors positioned symmetrically (L, R) and one mirror placed behind the valve (B). The specular reflection along the inherent optical paths eased mirroring of the real camera. Even with intersecting optical paths, the virtual image planes from the three sides (L, R and B) were distinctively separated on the real camera. Moreover, with the inlet–outlet coupler and the hydraulic hoses hindering the vision, the fourth virtual camera was omitted. As observed during the experiment, this novel method successfully captured clearly both the growth and the collapse of the bubbles.

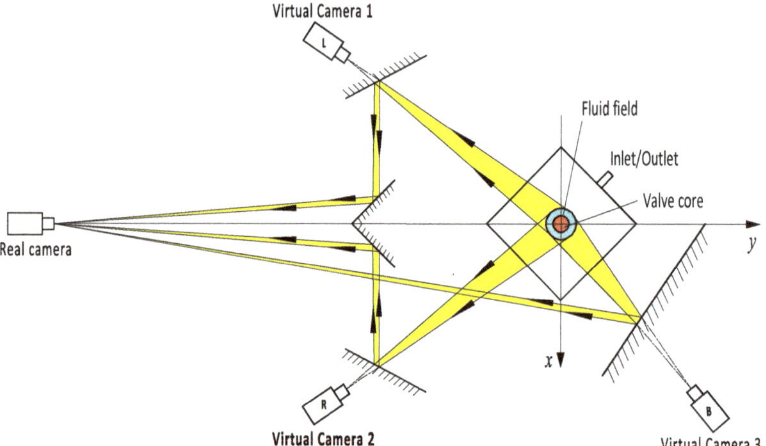

Figure 3. Schematic diagram of the 3D imaging principle.

2.2. Transparent Throttle Valve

In previous studies, the refractive index of glass-made tanks (1.52) [13,21–24] was suitable for analyzing bubbly flows. However, the advent of an easily machined transparent thermoplastic called Polymethylmethacrylate (PMMA, Perspex, acrylic glass), eased the capturing of images by high speed photography. Along with its refractive index (1.490) being relatively closer to the fluid used (water (1.333)) and its ability to withstand higher pressures of 20 bar, PMMA was used to manufacture the throttle valve's body. Operating at 20 °C, no correction factor was required and high light transmission was observed with no substantial image distortion. Having a modular structure, analysis of the bubble features was eased throughout the area under study. To counteract with the effects of high pressure, the chosen material for the valve core was brass while its connecting rod and the valve core were made using stainless steel. In addition, to induce cavitating bubbles in the fluid flow, a pressure difference of 0.2 MPa (the inlet pressure: 0.3 MPa; the outlet pressure: 0.1 MPa) was applied in the valve port area. The bursting effect of the bubbles resulted in the flaking off the materials, thus, eroding the inner area of the valve.

3. Optimization Arrangement of the One-Camera-Five-Mirror Module

The 3D Bubble reconstruction algorithm developed in this paper was based on space rectangular coordinate system. The spatial coordinates of the cavitation bubbles in the valve was provided by the position information of the experiment images from the three directions of left (L), right (R), and back (B). The bubble position coordinates on the horizontal axis from the L and B sides' images were directly used as the x-coordinate values of the bubbles in the spatial location and the R side's image provided the y-coordinate values. In case a virtual camera has an oblique angle with the corresponding observed side of the valve, the transparent surface made by PMMA material will cause image refringence, then the 3D bubble position in the x–y plane will be inaccurate with its actual space position. The larger the oblique angle, the greater the error of the bubble in space position. To eliminate this problem, the light center axes of the three virtual cameras were all defined to be perpendicular to the observed faces.

The resolution of the experiment videos was limited by the capability of the high-speed camera and the distance from the observed field to the camera lens. The resolution capability of the camera and the lens used in the experiment are fixed. The cavitation bubbles in the valve were quite small. In order to ensure the resolution of the bubble images, the distance of the optical path were maintained as short as possible. In addition, the distance of the optical path of the three virtual cameras was set to be equal to ensure the consistent image resolution in the L, R and B three sides.

To optimize the arrangement of the high-speed camera and five plane mirrors to meet the expectation above, the optimization model was built. As shown in Figure 4, the point P and Q are the position of the virtual camera 3 (B) and 2 (R) in Figure 3. Due to the symmetrical relationship of the virtual camera 1 (L) and 2 (R), the optimal design of the two mirrors of virtual camera 1 was omitted. Thus, the optimization design variables in practice were the position parameters of the camera (point H) and the three mirrors (marked as A, B and C).

3.1. Optimization Model

As shown in Figure 4, with regard to the nonlinear constrained optimization in this paper, the optimization objective is defined as $J(\mathbf{x}) = b + c + d$. The objective function is mathematically defined by:

$$\min J(\mathbf{x}), \tag{1}$$

Subject to:

$$\begin{cases} \arctan k_{HC_2} - \arctan k_{HV_0} > \dfrac{1°}{180°}\pi, \\[2mm] x_{A_0} - x_{V_0} > 3, \\[2mm] x_{C_1} > 1, \end{cases}$$

where $\mathbf{x} = [a, b, c, d, \theta_1, \theta_2, \theta_3]^T$.

3.2. Optimization Variables

The four length optimization variables are defined as follows: $l_{OA} = a > 0$, $l_{OB} = b > 0$, $l_{BC} = c > 0$, $l_{CH} = d > 0$. The three angle optimization variables are defined as follows θ_1 (mirror A), θ_2 (mirror B) and θ_3 (mirror C). Hence, the optimal objective can be represented as $l_{OP} = l_{OQ} = b + c + d$ and the coordinates of the virtual cameras 2 and 3 are expressed as follows: $Q(\frac{b+c+d}{\sqrt{2}}, -\frac{b+c+d}{\sqrt{2}})$, $P(\frac{b+c+d}{\sqrt{2}}, \frac{b+c+d}{\sqrt{2}})$. In addition, the coordinates of the plane mirrors A and B can be expressed as $A(\frac{a}{\sqrt{2}}, \frac{a}{\sqrt{2}})$, $B(\frac{b}{\sqrt{2}}, -\frac{b}{\sqrt{2}})$.

All the coordinates in Figure 4 can be deviated and expressed by the optimization variables, $a, b, c, d, \theta_1, \theta_2, \theta_3$, based on the optical and geometrical relationship among them. In addition, the coordinate values are listed in Table 1.

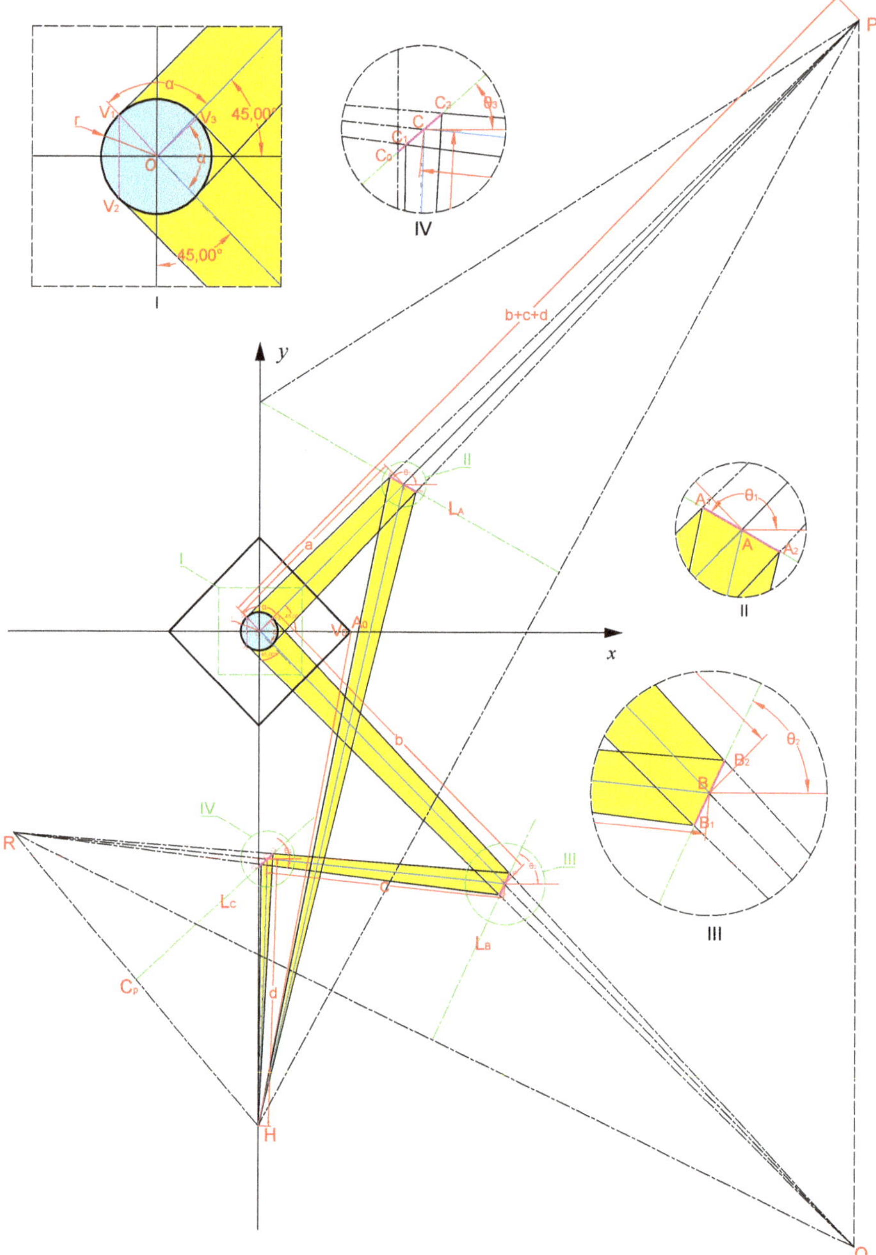

Figure 4. Establishment and parameter setting of the optimization model.

Table 1. Coordinate values of the design points

Point	x coordinate	y coordinate
H	$x_P + y_P \tan\theta_1 - y_H \tan\theta_1$	$\dfrac{2x_P \tan\theta_1 + y_P(\tan^2\theta_1 - 1) + \sqrt{2}a(1 - \tan\theta_1)}{1 + \tan^2\theta_1}$
R	$x_Q + y_Q \tan\theta_2 - y_R \tan\theta_2$	$\dfrac{2x_Q \tan\theta_2 + y_Q(\tan^2\theta_2 - 1) - \sqrt{2}b(1 + \tan\theta_2)}{1 + \tan^2\theta_2}$
C	$\dfrac{\frac{y_B-y_R}{x_B-x_R}x_R - y_R - \frac{x_R+x_H}{2}\tan\theta_3 + \frac{y_R+y_H}{2}}{\frac{y_B-y_R}{x_B-x_R} - \tan\theta_3}$	$\dfrac{y_B - y_R}{x_B - x_R}(x_C - x_R) + y_R$
V_1	$r\cos\left(\arccos\dfrac{r}{\sqrt{x_P^2 + y_P^2}} + 45°\right)$	$r\sin\left(\arccos\dfrac{r}{\sqrt{x_P^2 + y_P^2}} + 45°\right)$
A_1	$\dfrac{y_P - \frac{y_{V_1}-y_P}{x_{V_1}-x_P}x_P - \frac{a}{\sqrt{2}}(1 - \tan\theta_1)}{\tan\theta_1 - \frac{y_{V_1}-y_P}{x_{V_1}-x_P}}$	$x_{A_1}\tan\theta_1 + \dfrac{a}{\sqrt{2}}(1 - \tan\theta_1)$
A_0	$x_H - \dfrac{x_{A_1} - x_H}{y_{A_1} - y_H}y_H$	0
V_0	$\dfrac{l_V}{\sqrt{2}}$	0
V_2	x_{V_1}	$-y_{V_1}$
B_1	$\dfrac{\frac{y_Q-y_{V_2}}{x_Q-x_{V_2}}x_Q - y_Q - \frac{b}{\sqrt{2}}(1 + \tan\theta_2)}{\frac{y_Q-y_{V_2}}{x_Q-x_{V_2}} - \tan\theta_2}$	$x_{B_1}\tan\theta_2 - \dfrac{b}{\sqrt{2}}(1 + \tan\theta_2)$
C_1	$\dfrac{\frac{y_R+y_H}{2} - \frac{x_R+x_H}{2}\tan\theta_3 + \frac{y_R-y_{B_1}}{x_R-x_{B_1}}x_R - y_R}{\frac{y_R-y_{B_1}}{x_R-x_{B_1}} - \tan\theta_3}$	$\tan\theta_3\left(x_{C_1} - \dfrac{x_R + x_H}{2}\right) + \dfrac{y_R + y_H}{2}$
V_3	$r\cos\left(\arccos\dfrac{r}{\sqrt{x_P^2 + y_P^2}} - 45°\right)$	$r\sin\left(\arccos\dfrac{r}{\sqrt{x_P^2 + y_P^2}} - 45°\right)$
B_2	$\dfrac{\frac{b}{\sqrt{2}}(1 + \tan\theta_2) - \frac{y_Q-y_{V_3}}{x_Q-x_{V_3}}x_Q + y_Q}{\tan\theta_2 - \frac{y_Q-y_{V_3}}{x_Q-x_{V_3}}}$	$x_{B_2}\tan\theta_2 - \dfrac{b}{\sqrt{2}}(1 + \tan\theta_2)$
C_2	$\dfrac{\frac{y_R+y_H}{2} - \frac{x_R+x_H}{2}\tan\theta_3 + \frac{y_R-y_{B_2}}{x_R-x_{B_2}}x_R - y_R}{\frac{y_R-y_{B_2}}{x_R-x_{B_2}} - \tan\theta_3}$	$\tan\theta_3\left(x_{C_2} - \dfrac{x_R + x_H}{2}\right) + \dfrac{y_R + y_H}{2}$

The calculation of the following parameters was to define the constraint conditions to meet the requirements of the one-camera-five-mirror 3D imaging module and ensure no interference between the optical paths.

To prevent the optical image reflected by the mirror A from being interrupted by the valve, there should be a certain interval between A_0 and V_0, as expressed in Equation (2):

$$x_{A_0} - x_{V_0} > 3. \tag{2}$$

The left boundary of the optical path reflected by the mirror C is on the positive side of the y-axis, in case of influencing the mirror belonging to the virtual camera 1, which is symmetrical with the mirror C. Thus, the x-coordinate should meet the following constraint:

$$x_{C_1} > 1. \tag{3}$$

The slope of the line HC_2 and HV_0 can be expressed as:

$$k_{HC_2} = \frac{y_H - y_{C_2}}{x_H - x_{C_2}},$$

$$k_{HV_0} = \frac{y_H - y_{V_0}}{x_H - x_{V_0}}.$$

To prevent the optical path from the mirror A and C from interfering with each other, the slopes angle of the line HC_2 and HA_1 should meet:

$$\arctan k_{HC_2} - \arctan k_{HV_0} > \frac{1°}{180°}\pi. \tag{4}$$

3.3. Optimization Solution

The *fmincon* function provided by Matlab optimization toolbox was applied to solve the minimum value of the multi-variable constrained nonlinear function in this paper. The variable initial values was defined as

$$\mathbf{x}_0 = [184.79, 315.54, 216.71, 236.45, 151.03°, 64.68°, 41.25°]^T,$$

which was a quite good arrangement scheme by manual adjustments in the CAD drawing. After the calculation, the optimal solution is obtained as follows:

$$\mathbf{x}_{opt} = [a_{opt}, b_{opt}, c_{opt}, d_{opt}, \theta_{1opt}, \theta_{2opt}, \theta_{3opt}]^T$$
$$= [179.85, 307.36, 197.60, 219.90, 150.75°, 64.71°, 41.25°]^T.$$

4. Algorithm Development

4.1. 2D Bubble Feature Detection

To extract the morphological data of the bubbles from the recorded experimental videos, we used the MATLAB software. The valve port area where the cavitating bubbles appeared was thoroughly analyzed by an image processing algorithm based on the frame differencing method. Obtaining the bubble feature data of bubbles from the images requires the pre-processing and the process is shown in Figure 5. The original images were converted to grayscale image using *rgb2gray* function in MATLAB software. The difference between two frames within a defined internal would be calculated. In addition, the result was converted to binary image (also called as BW image) using *im2bw* function in MATLAB software. As a result, the 2D bubble feature data can be obtained for further calculation.

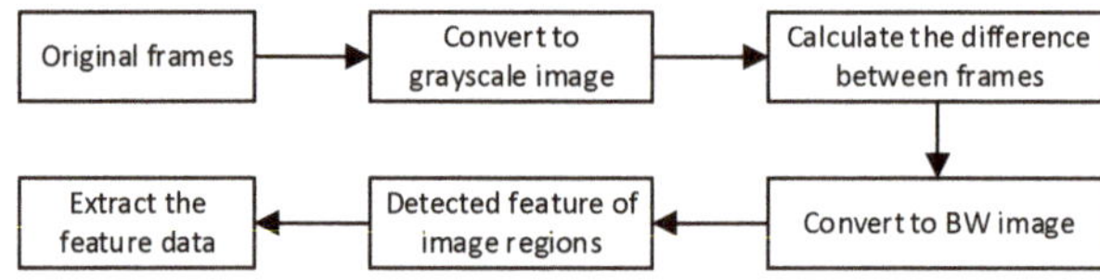

Figure 5. Bubble feature extraction framework.

Figure 6 presents the process for detecting the 2D feature of cavitation bubbles. The semi-major sizes of the cavitation bubble on the x, y, and z axes. The shape of the 2D bubble was defined as ellipse. In addition, the long or short axes of the ellipse are determined by the width (w) and height (h) of the detected area. Furthermore, the center of a bubble is estimated by the detected coordinate (x_{lt}, y_{lt}), w and h:

$$\begin{cases} x = x_{lt} + 0.5 \times w, \\ y = y_{lt} + 0.5 \times h. \end{cases} \tag{5}$$

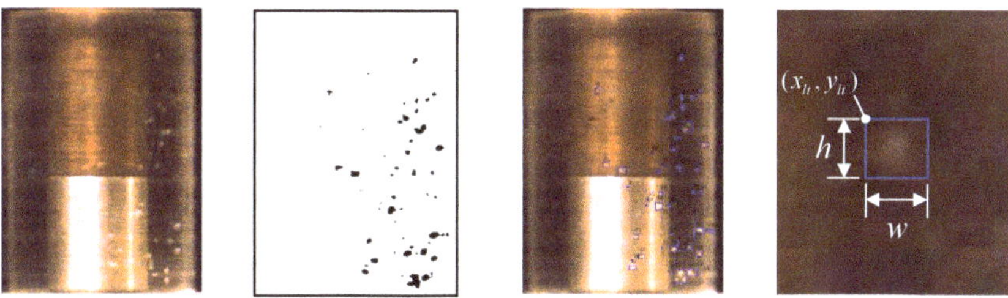

Figure 6. Image processing of the 2D bubble feature detection algorithm.

The center coordinate can be used to reconstruct the 3D motion parameters of bubbles afterwards. The 2D bubble features of the relative motion equalling the previous frame were extracted. Bubble clusters were assumed to be larger bubbles while those in developing or under collapsing mode were presumed smaller bubbles. The motion features in terms of the smallest pixel point were detected by the 2D cavitation bubble detection algorithm; the smallest discernable bubble size was restricted by the resolution of the video.

4.2. Feature Identification of the Opaque Object

The valve seat and core were manufactured by stainless steel materials. The brass material was used to compose the valve rod. Ultimately, it was separately distinguished by its color difference feature at the interface between the valve core and rod. A 3D coordinate system built by the mid-point of the boundary line between the valve core and the its rod to define the coordinates origin of the 3D model. The algorithm of the valve core identification was based on the Faster R-CNN method developed by Ren et al. [42], which is a mainstream deep learning method in object detection. Figure 7 illustrates the object detection model that is based on deep learning, requiring a large number of training samples. The detection accuracy meets the adequate requirements after processing the training model and the coordinates of the valve core contour were generated through the Faster R-CNN model. In addition, getting the valve core's diameter in pixel scale and its real size (r_{core} = 17 mm), it is easy to transform the pixel value of the position of the bubbles into the actual size, whose unit is millimeters (mm).

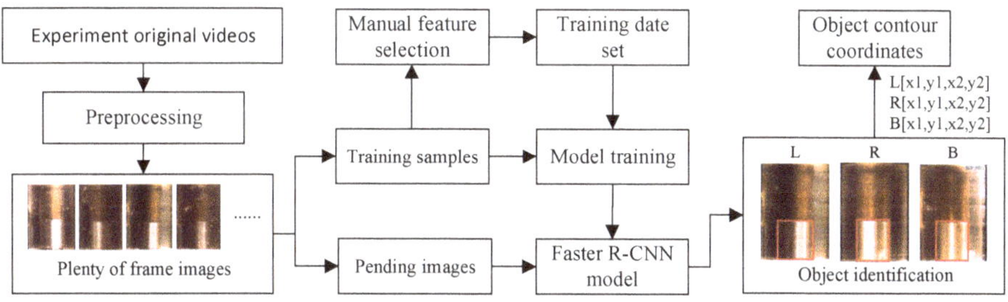

Figure 7. Block diagram of Faster R-CNN model. (R-CNN: region-based convolutional neural network)

4.3. 3D Bubble Cluster Reconstruction

Reconstructing the 3D bubbles was simplified using the 3D Cartesian coordinates system. The origin of the model was determined from the contact surface of the valve core and the valve rod. From Figure 8b, the x and z axes in the left (L) side and back (B) side of the valve body represent the horizontal and vertical directions, respectively. It is worth emphasizing that the type of the space rectangular coordinate system is left-handed cartesian coordinate. And it is different with the x-y coordinate for optimization calculation. Due to different conditions of the mirror reflection, the positive *x*-axis of the L-side and the B-side was analogous. Concerning the right (R) side of the valve body, the *y* and *z*-axes were set in the corresponding horizontal and vertical directions. Additionally, the valve port area in Figure 9 was divided into 15 parts so as to decrease the matching scope and the possible associated errors. The left-handed Cartesian coordinates was used to express the spatial location of the reconstructed bubble.

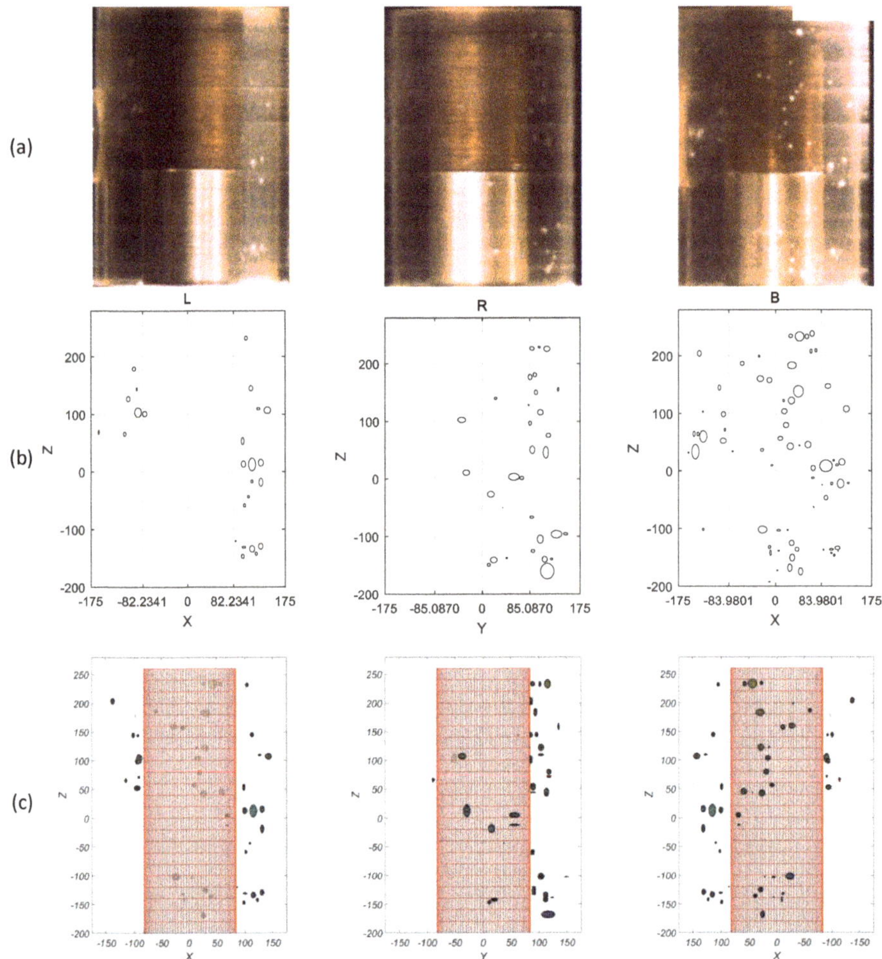

Figure 8. 2D Detection and 3D reconstruction results of the bubble cluster (unit: mm). (**a**) original images; (**b**) 2D detection results; (**c**) 3D reconstuction results.

The requirement of the 3D bubble cluster reconstruction was to match the bubbles from the different sides of the valve with the same bubbles in the 3D space. In addition, the bubbles *y*-axis values on the L and B sides were obtained from the B-side. The lack of vision from the R-side resulted the *y*-axis values of the bubbles belonging to the L2, L1B4, and B3 space to be fixed with a random value. The following equations illustrate the mathematical model of the bubble matching process:

$$D_{ij} = 1 + |Z_i - Z_j|, \tag{6}$$

where i and j are two bubbles on different sides. In addition, Z_i and Z_j are their coordinate values on the z-axis. Z_i and Z_j can be get through the 2D bubble detection results of Equation (5). The Z_i and Z_j are equal to the corresponding value of y in Equation (5). The index D_{ij} is the difference of the z-axes regarding the two matching 2D bubbles:

$$H_{ij} = \frac{h_i}{h_j} + \frac{h_j}{h_i}, \tag{7}$$

where H_{ij} is termed as the height difference. In addition, h_i, h_j are the height of the i and j frames, which can be also obtained according to the calculation results of the 2D bubble detection algorithm, as shown in Figure 8b:

$$M_{ij} = D_{ij}H_{ij}, \tag{8}$$

where M_{ij} is the marching index of the complete judgement of D_{ij} and H_{ij}. If M_{ij} is less than the M_{max}, the bubbles i and j are considered a probable 3D bubble. Prior to the pairs with the minimum M_{ij} being designated as coordinating 3D bubbles, the bubbles from the three sides (L, R, and B) are matched. To simulate the bubbles (small ones, clustered ones) using the reconstruction algorithm, ellipses were used to extract their outer contours, as shown in Figure 8c.

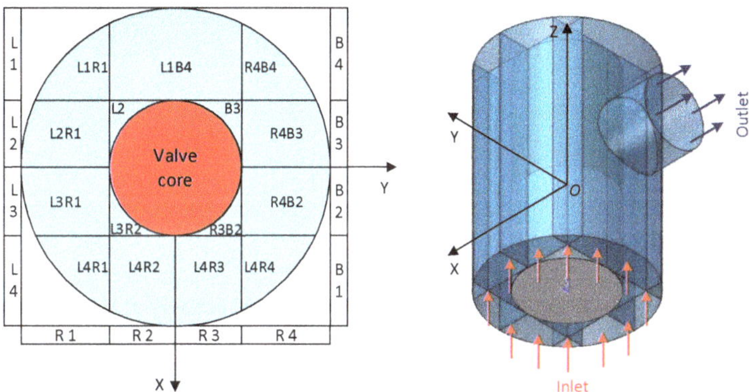

Figure 9. Space partition of the flow field.

5. Results and Analysis

The flow profiles of the bubbles were analyzed based on the calculated results of the 3D bubble reconstruction algorithm. Figure 8c displays the size and distribution of the 3D bubbles recreated by computing and processing a certain frame from the recorded video. Stereoscopic parallax explains the following results: the bubble viewed in the mirror deviated from its actual appearance. Hence, the 3D bubble image from the back (B) side appeared to opposite left and right when compared with the mirror image of the B side in Figure 8a,b. As per the 3D reconstruction results, the bubble algorithm gave satisfying results. Even if not all bubbles were reconstructed, the results of the 3D reconstruction tallied with the original experimental images, thereby concluding with a vivid and solid reference for recreating the bubble morphology.

Figure 10 illustrates the 3D reconstructed bubble flow profile of four adjacent frames from 0 ms to 0.87 ms. The time interval between two adjacent frames was the least time difference captured by the high-speed camera, which was equal to 0.29 ms. Furthermore, the 3D bubble reconstruction algorithm was stable enough to simulate and evaluate the bubble flow profile in each frame.

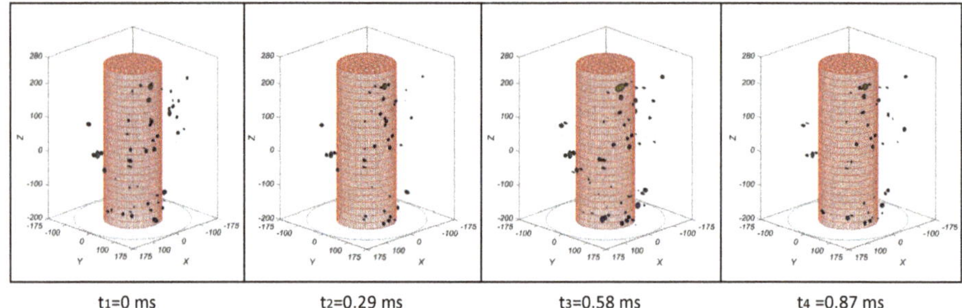

Figure 10. 3D reconstruction results of bubble flow in a short time (unit: mm).

As an example, a short experimental video which consisted of 50 frames of images was used to further analyze the characteristics of the reconstructed 3D cavitation bubble cluster. Its first four frames were the same frames in Figure 10. It is worth noting that the time of 50 frames of experimental images is only 14.5 ms, which exhibits the advantage of the high-camera camera. The number of 3D bubbles fluctuating within 50 frames of experiment images was showed in Figure 11, which can present the size of the bubbles and its number in each time interval The blue-color curve mainly floats between 40 to 70, which indicates that the generation and collapse of the cavitation bubbles can achieve a relative balance in a short time. The average number of bubble is 57.78.

The cavitation volume percentage in each frame is stated as Equation (9):

$$P_v = \frac{\sum\limits_{i=1}^{n} \frac{4}{3}\pi a_i b_i c_i}{\pi(R_{\text{fluid}} - r_{\text{core}})h} \times 100\%, \tag{9}$$

where R_{fluid} is the radius of the fluid field in the valve and r_{core} is the radius of the valve core. In addition, R_{fluid} and r_{core} are constants. a_i, b_i, c_i are the three semi-major sizes of the cavitation bubble on the x, y, and z-axes.

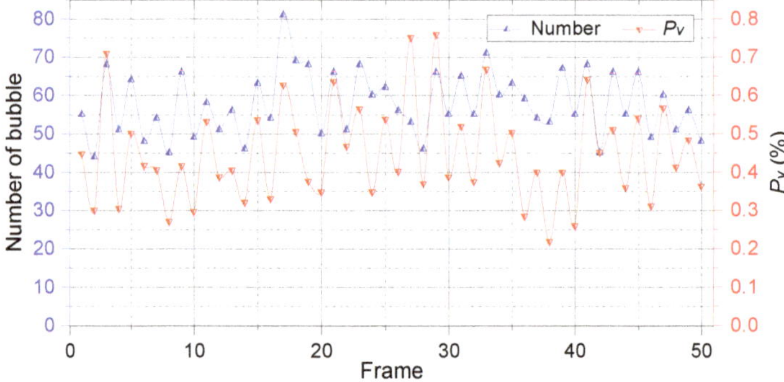

Figure 11. Change of the number of bubble and P_v over time.

The variation of the cavitation volume percentage P_v over time is shown in Figure 11. The change of the P_v and the number of bubbles between the adjacent frames revealed that the growth and collapse of the cavitation bubbles were recurrent and fast. While comparing the P_v curve with the number of bubble curve in Figure 11, an obvious positive correlation between the number of bubbles and the percentage of cavitation can be found. Moreover, the increase in the number of bubble and the P_v did not exactly match; at times, while the cavitation volume percentage was decreasing, the number of bubbles did increase. Through the analysis of both results in Figure 11 and the original experiment images, it was concluded that scattering of the bubble clusters into small bubbles increased the number of bubbles in the cavitating space.

As shown in Figure 12, a short experimental video with 100 frames were analyzed. Various perspectives of the bubble clouds were presented and from the 100th frame (at about 29 ms), 5635 cavitation bubbles were detected and reconstructed, respectively. Based on the scatter diagram, the flow pattern of the 3D bubble cluster could be inferred. For example, the path and direction of the bubble flow were analyzed in Figure 13 through the presence of an obvious bubble headstream on the back side. The bubbles generated by this headstream were then separated into three paths (arrows)as shown in Figure 13. In addition, a large number of bubbles were flown out of the observed area through the outlet port, although there were certainly many that imploded and collapsed in the observed field.

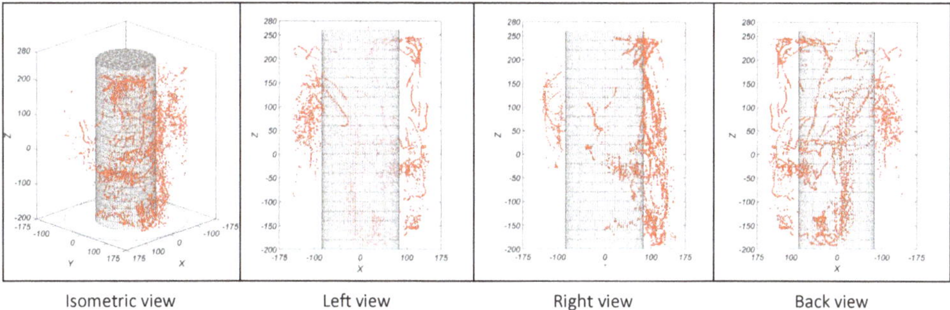

Figure 12. Space distribution of the cavitation bubble cluster of 100 frames (unit: mm).

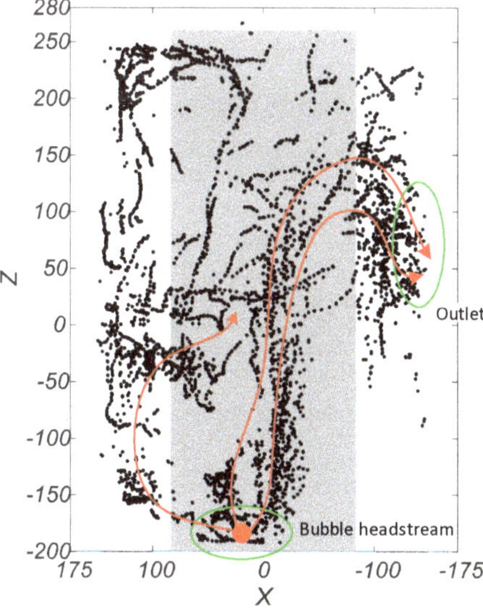

Figure 13. Flow path of the bubble cluster from the back view (unit: mm).

Based on the calculated data of the bubbles' space coordinates, the kernel densities of the bubbles on the x–y plane are shown in Figure 14. Almost all of the cavitation bubbles in the 100 frames are on the positive side of the y-axis. The high density around the point (0,100) validates the analysis of the bubble headstream. A maximum number of bubbles were detected by the algorithm that gave satisfying results as they were easily defined and selected across all the concerned fluid domain, as shown in Figure 8. The regions with the highest kernel densities in Figure 14 (output of the 100th frame) are based on the space partitions R4B4, R4B3, R4B2, and R4B1, respectively. Likewise, partition L1B4 represents the exit of the valve, which directly influences the flow in the partition L1R1 and R4B4, respectively. The bubble cluster rapidly shifted on the right side of the valve core within 29 ms due to the high velocity-low pressure zone (vortices) developed after the vena contracta formed soon after the fluid leaves the valve seat (the valve core opening was set to 1 mm).

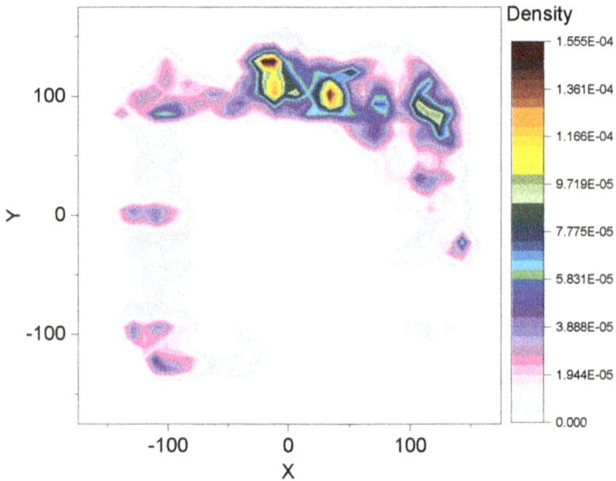

Figure 14. Kernel density of the cavitation bubbles on the x–y plane.

6. Conclusions

Throughout this study, the optimal arrangement of the one-camera-five-mirrors module, the different algorithms applied in detecting the valve core and the bubbles morphology, and the reconstruction of the 3D bubble clusters along with the calculation of the cavitation volume percentage together provided favorable results. In addition, analysis of the recorded bubbly flow using the stable and effective algorithms in the transparent throttle valve proved to be ideal. Lastly, calculating the cavitation volume percentage broadened the concept of bubbles detection in the fluid flow, and it could be applied successfully in other cases as well.

Author Contributions: H.W. implemented the main research, wrote the manuscript and presented the optimization and reconstruction methodology; H.X. provided supervision guidance to this research and conducted a validation of the analysis results; V.P. reviewed and edited the manuscript and searched the literature. X.-Z.G. supported and developed the algorithms. All authors read and approved the final manuscript.

Funding: This work was supported by the Natural Science Foundation of China under Grant 51875113, Natural Science Foundation of the Heilongjiang Province of China under Grant F2016003, "Jinshan Talent" Zhenjiang Manufacture 2025 Leading Talent Project, "Jiangyan Planning" Project in Yangzhong City, and the International Exchange Program of Harbin Engineering University for Innovation-oriented Talents Cultivation.

Conflicts of Interest: The authors declare no conflict of interest.

References

1. Belden, J.; Ravela, S.; Truscott, T.T.; Techet, A.H. Three-dimensional bubble field resolution using synthetic aperture imaging: Application to a plunging jet. *Exp. Fluids* **2012**, *53*, 839–861. [CrossRef]
2. Ahmed, F.S.; Sensenich, B.A.; Gheni, S.A.; Znerdstrovic, D.; Al Dahhan, M.H. Bubble dynamics in 2D bubble column: Comparison between high-speed camera imaging analysis and 4-point optical probe. *Chem. Eng. Commun.* **2015**, *202*, 85–95. [CrossRef]

3. Lau, Y.M.; Deen, N.G.; Kuipers, J.A.M. Development of an image measurement technique for size distribution in dense bubbly flows. *Chem. Eng. Sci.* **2013**, *94*, 20–29. [CrossRef]

4. Zhang, Y.; Liu, M.; Xu, Y.; Tang, C. Three-dimensional volume of fluid simulations on bubble formation and dynamics in bubble columns. *Chem. Eng. Sci.* **2012**, *73*, 55–78. [CrossRef]

5. Tayler, A.B.; Holland, D.J.; Sederman, A.J.; Gladden, L.F. Applications of ultra-fast MRI to high voidage bubbly flow: Measurement of bubble size distributions, interfacial area and hydrodynamics. *Chem. Eng. Sci.* **2012**, *71*, 468–483. [CrossRef]

6. Zhao, L.; Sun, L.; Mo, Z.; Tang, J.; Hu, L.; Bao, J. An investigation on bubble motion in liquid flowing through a rectangular Venturi channel. *Exp. Therm. Fluid Sci.* **2018**, *97*, 48–58. [CrossRef]

7. Xia, G.; Cai, B.; Cheng, L.; Wang, Z.; Jia, Y. Experimental study and modelling of average void fraction of gas-liquid two-phase flow in a helically coiled rectangular channel. *Exp. Therm. Fluid Sci.* **2018**, *94*, 9–22. [CrossRef]

8. Fu, Y.; Liu, Y. Development of a robust image processing technique for bubbly flow measurement in a narrow rectangular channel. *Int. J. Multiphase Flow* **2016**, *84*, 217–228. [CrossRef]

9. Lomakin, V.O.; Kuleshova, M.S.; Kraeva, E.A. Fluid Flow in the Throttle Channel in the Presence of Cavitation. *Procedia Eng.* **2015**, *106*, 27–35. [CrossRef]

10. Gavaises, M.; Villa, F.; Koukouvinis, P.; Marengo, M.; Franc, J.P. Visualisation and les simulation of cavitation cloud formation and collapse in an axisymmetric geometry. *Int. J. Multiphase Flow* **2015**, *68*, 14–26. [CrossRef]

11. Kravtsova, A.Y.; Markovich, D.M.; Pervunin, K.S.; Timoshevskiy, M.V.; Hanjalić, K. High-speed visualization and PIV measurements of cavitating flows around a semi-circular leading-edge flat plate and NACA0015 hydrofoil. *Int. J. Multiphase Flow* **2014**, *60*, 119–134. [CrossRef]

12. Singhal, A.K.; Athavale, M.M.; Li, H.; Jiang, Y. Mathematical Basis and Validation of the Full Cavitation Model. *J. Fluids Eng.* **2002**, *124*, 617–624. [CrossRef]

13. Guevara-Lopez, E.; Sanjuan-Galindo, R.; Cordova-Aguilar, M.S.; Corkidi, G.; Ascanio, G.; Galindo, E. High-speed visualization of multiphase dispersions in a mixing tank. *Chem. Eng. Res. Des.* **2008**, *86*, 1382–1387. [CrossRef]

14. Tan, D.; Mi, J. High speed imaging study of the dynamics of ultrasonic bubbles at a liquid-solid interface. In *Materials Science Forum, Proceedings of the 6th International Light Metals Technology Conference (LMT 2013), Old Windsor, UK, 24–26 July 2013*; Trans Tech Publications Ltd.: Zürich, Switzerland, 2013; Volume 765, pp. 230–234. [CrossRef]

15. Jacobson, B.O.; Hamrock, B.J. High-speed motion picture camera experiments of cavitation in dynamically loaded journal bearings. *J. Lubr. Technol.* **1983**, *105*, 446–452. [CrossRef]

16. Dencks, S.; Ackermann, D.; Schmitz, G. Evaluation of bubble tracking algorithms for super-resolution imaging of microvessels. In Proceedings of the 2016 IEEE International Ultrasonics Symposium (IUS), Tours, France, 18–21 September 2016; pp. 1–4. [CrossRef]

17. Lauterborn, W.; Hentschel, W. Cavitation bubble dynamics studied by high speed photography and holography: Part one. *Ultrasonics* **1985**, *23*, 260–268. [CrossRef]

18. Lauterborn, W.; Hentschel, W. Cavitation bubble dynamics studied by high speed photography and holography: Part two. *Ultrasonics* **1986**, *24*, 59–65. [CrossRef]

19. Kent, J.C.; Eaton, A.R. Stereo photography of neutral density He-filled bubbles for 3D fluid motion studies in an engine cylinder. *Appl. Opt.* **1982**, *21*, 904–912. [CrossRef] [PubMed]

20. Racca, R.G.; Dewey, J.M. A method for automatic particle tracking in a three-dimensional flow field. *Exp. Fluids* **1988**, *6*, 25–32. [CrossRef]

21. Xue, T.; Qu, L.; Wu, B. Matching and 3D Reconstruction of Multibubbles Based on Virtual Stereo Vision. *IEEE Trans. Instrum. Meas.* **2014**, *63*, 1639–1647. [CrossRef]

22. Xue, T.; Xu, L.S.; Zhang, S.Z. Bubble behavior characteristics based on virtual binocular stereo vision. *Optoelectron. Lett.* **2018**, *14*, 44–47. [CrossRef]

23. Xue, T.; Qu, L.; Cao, Z.; Zhang, T. Three-dimensional feature parameters measurement of bubbles in gas-liquid two-phase flow based on virtual stereo vision. *Flow Meas. Instrum.* **2012**, *27*, 29–36. [CrossRef]

24. Xue, T.; Chen, Y.; Ge, P. Multibubbles Segmentation and Characteristic Measurement in Gas-Liquid Two-Phase Flow. *Adv. Mech. Eng.* **2013**, *5*, 143939. [CrossRef]

25. Mitra, A.; Bhattacharya, P.; Mukhopadhyay, S.; Dhar, K.K. Experimental study on shape and path of small bubbles using video-image analysis. In *Proceedings of the 2015 3rd International Conference on Computer, Communication, Control and Information Technology, C3IT 2015, West-Bengal, India, 7–8 February 2015*; Institute of Electrical and Electronics Engineers Inc.: Piscataway, NJ, USA, 2015. [CrossRef]

26. Acuña, C.A.; Finch, J.A. Tracking velocity of multiple bubbles in a swarm. *Int. J. Miner. Process.* **2010**, *94*, 147–158. [CrossRef]

27. Cheng, D.C.; Burkhardt, H. Template-based bubble identification and tracking in image sequences. *Int. J. Therm. Sci.* **2006**, *45*, 321–330. [CrossRef]

28. Krimerman, M. Reconstruction of Bubble Trajectories and Velocity Estimation. Master's Thesis, The University of British Columbia, Vancouver, BC, Canada, February 2013. [CrossRef]

29. Bakshi, A.; Altantzis, C.; Bates, R.B.; Ghoniem, A.F. Multiphase-flow statistics using 3D detection and tracking algorithm (MS3DATA): Methodology and application to large-scale fluidized beds. *Chem. Eng. J.* **2016**, *293*, 355–364. [CrossRef]

30. Feng, X.F.; Pan, D.F. Research on the application of single camera stereo vision sensor in three-dimensional point measurement. *J. Mod. Opt.* **2015**, *62*, 1204–1210. [CrossRef]

31. Figueroa-Espinoza, B.; Mena, B.; Aguilar-Corona, A.; Zenit, R. The lifespan of clusters in confined bubbly liquids. *Int. J. Multiphase Flow* **2018**, *106*, 138–146. [CrossRef]

32. Fu, Y.; Liu, Y. Experimental study of bubbly flow using image processing techniques. *Nucl. Eng. Des.* **2016**, *310*, 570–579. [CrossRef]

33. Chakraborty, S.; Das, P.K. Characterization of bubbly flow through the fusion of multiple features extracted from high speed images. In Proceedings of the 2016 IEEE First International Conference on Control, Measurement and Instrumentation (CMI), Kolkata, India, 8–10 January 2016; pp. 1–5. [CrossRef]

34. De Langlard, M.; Al-Saddik, H.; Charton, S.; Debayle, J.; Lamadie, F. An efficiency improved recognition algorithm for highly overlapping ellipses: Application to dense bubbly flows. *Pattern Recognit. Lett.* **2018**, *101*, 88–95. [CrossRef]

35. Honkanen, M. Reconstruction of a three-dimensional bubble surface from high-speed orthogonal imaging of dilute bubbly flow. *Int. J. Multiphase Flow* **2009**, *63*, 469–480. [CrossRef]

36. Markus, H.; Pentti, S.; Tuomas, S.; Jouko, N. Recognition of highly overlapping ellipse-like bubble images. *Meas. Sci. Technol.* **2005**, *16*, 1760–1770. [CrossRef]

37. Zhang, W.H.; Jiang, X.; Liu, Y.M. A method for recognizing overlapping elliptical bubbles in bubble image. *Patt. Recognit. Lett.* **2012**, *33*, 1543–1548. [CrossRef]

38. Fujisawa, N.; Fujita, Y.; Yanagisawa, K.; Fujisawa, K.; Yamagata, T. Simultaneous observation of cavitation collapse and shock wave formation in cavitating jet. *Exp. Therm. Fluid Sci.* **2018**, *94*, 159–167. [CrossRef]

39. Kompella, V.R.; Sturm, P. Collective-reward based approach for detection of semi-transparent objects in single images. *Comput. Vis. Image Underst.* **2012**, *116*, 484–499. [CrossRef]

40. Hata, S.; Saitoh, Y.; Kumamura, S.; Kaida, K. Shape extraction of transparent object using genetic algorithm. In Proceedings of the 13th International Conference on Pattern Recognition, Vienna, Austria, 25–29 August 1996; Volume 4, pp. 684–688. [CrossRef]

41. Han, K.; Wong, K.Y.K.; Liu, M. Dense reconstruction of transparent objects by altering incident light paths through refraction. *Int. J. Comput. Vis.* **2018**, *126*, 460–475. [CrossRef]

42. Ren, S.; He, K.; Girshick, R.; Sun, J. Faster R-CNN: Towards real-time object detection with region proposal networks. *arXiv* **2016**, arXiv:1506.01497. Available online: https://arxiv.org/abs/1506.01497 (accessed on 23 September 2018).

Article

Deep Residual Network with Sparse Feedback for Image Restoration

Zhenyu Guo [1], Yujuan Sun [1,*], Muwei Jian [2,3] and Xiaofeng Zhang [1]

[1] School of Information and Electrical Engineering, Ludong University, Yantai 264025, China; handlecoding@foxmail.com (Z.G.); iamzxf@126.com (X.Z.)
[2] School of Computer Science and Technology, Shandong University of Finance and Economics, Jinan 250014, China; jianmuweihk@163.com
[3] School of Information Science and Engineering, Linyi University, Linyi 276000, China
* Correspondence: syj_anne@163.com; Tel.: +86-158-5357-8596

Received: 8 October 2018; Accepted: 19 November 2018; Published: 28 November 2018

Abstract: A deep neural network is difficult to train due to a large number of unknown parameters. To increase trainable performance, we present a moderate depth residual network for the restoration of motion blurring and noisy images. The proposed network has only 10 layers, and the sparse feedbacks are added in the middle and the last layers, which are called FbResNet. FbResNet has fast convergence speed and effective denoising performance. In addition, it can also reduce the artificial Mosaic trace at the seam of patches, and visually pleasant output results can be produced from the blurred images or noisy images. Experimental results show the effectiveness of our designed model and method.

Keywords: image restoration; motion deburring; image denoising; sparse feedback

---

1. Introduction

In recent years, many hard problems in computer vision have been well solved, especially in the fields of image classification, object detection [1], and identification [2]. By using the deep learning method, the accuracy and robustness of these issues have been greatly improved.

Until now, many kinds of neural network structures have been proposed, such as five layers LeNet [3], eight layers AlexNet [4], 19 layers VGG [5], 22 layers GoogleNet [6], 152 layers ResNet [7], GAN [8], and so on. These networks are gradually deepened, and the training data set is also getting much larger. Although the technology of deep learning is especially effective for issues of classification, it still includes many problems that cannot be solved in image restoration. The main reason includes two aspects: First, to collect a large number of training data about image degradation is not easy; second, the degradation reason is various, and it is exceptionally difficult to enumerate all cases of image degradation. Therefore, when using a deep neural network to improve these problems, the training set is not usually enough to train a deep neural network model.

In this paper, we improve the network structure of ResNet [7] and propose a sparse feedback residual network, which is called FbResNet. It includes 10 layers with sparse "shortcut connections" (in this article we call the "shortcut connections" forward feedback). Figure 1 shows the structure of FbResNet. It can be seen that FbResNet only includes two forward feedbacks, which are derived from the input layer. One feedback is connected to the middle layer, while the other is connected to the last layer. The two feedbacks can provide an effective constraint of the loss function and help to train reasonable network parameters. Experimental results show FbResNet has a fast convergence speed and effective denoising performance.

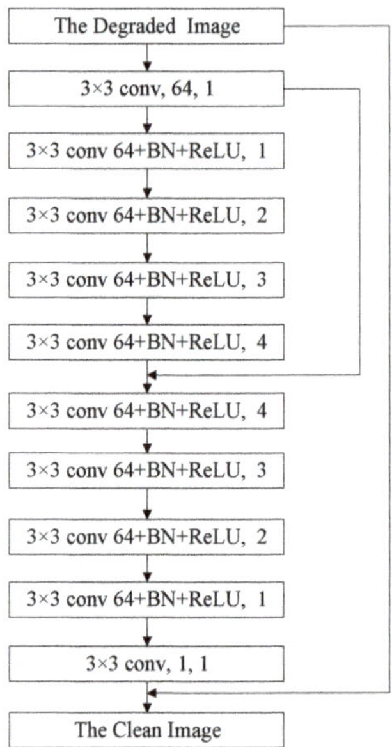

Figure 1. The architecture of proposed network FbResNet.

2. Related Works

Image restoration is a long-standing problem in low-level computer vision. In practice, the obtained images are usually degraded (such as images with noise, blurred images [9], sampled images [10], etc.). Image restoration is used to estimate the original image based on the degenerated image. Since this is an under-constraint problem, the solution is not unique. In order to reduce the solution space of the problem, the prior-based methods had been widely used, which had added the prior information or constraints and could recover the original image from the degraded image [10,11].

However, most of the prior-based methods build the objective function by simplifying the mechanism of the image degradation, and not considering the affection of noises or other factors. Hence, they cannot perfectly restore clean images for severely degraded images. In addition, the prior-based methods involve a complex optimization problem, and most of the prior-based methods can hardly achieve high performance without sacrificing computational efficiency. Furthermore, the prior-based models, in general, are nonconvex and involve several manually chosen parameters [12]. Therefore, there are many limitations in restoring images just using prior-based methods.

Deep convolution neural network (CNN) [13,14] has made a series of breakthroughs in many applications of computer vision [15,16], such as image classification, recognition and target detection, etc. The features of CNN are mainly exacted by increasing the depth of network model. Then, the lower, middle, and the advanced extracted features will be gradually obtained. In general, the advanced features will be used to connect with one or several fully connected layers. The reason for this remarkable achievement in computer vision task is mainly because many rich characteristics of different levels can be extracted by training the deep neural network.

Recent evidence reveals that the depth of neural network is very important. Many visual tasks [17–19], especially the low-level vision problems, have greatly benefited from very deep network. There are several references to perform the denoising problems using deep neural networks.

Reference [18] used a convolutional network as image denoising architecture and claimed that CNNs could provide comparable, and in some cases, superior performance to wavelet and Markov random field methods. Moreover, Reference [20] found that a convolutional network offered similar performance in the blind denoising setting, as compared to other techniques in the non-blind setting. However, training the convolutional network architecture requires substantial computation and many thousands of updates to converge.

Reference [21] combined sparse coding and deep networks pretrained with denoising auto-encoder for the tasks of image denoising and blind inpainting and achieved comparable results to K-SVD [22]. This method could automatically remove complex patterns, such as superimposed text from an image, and improve the performance of unsupervised feature learning. However, the method in Reference [21] also strongly relied on supervised training and could remove only the noise patterns in the training data. In Reference [23], trainable nonlinear reaction diffusion had been proposed and could be used for a variety of image restoration tasks by incorporating appropriate reaction force. In [24], the multi-layer perceptron (MLP) had been used for image denoising.

The model in References [23,24] can achieve promising performance. Reference [24] claimed that training MLP with many hidden layers could lead to problems, such as vanishing gradients and over-fitting. Reference [24] also found back propagation will work well and concluded that deep learning techniques are not necessary.

In a deep network structure, can all these extracted features be fully used? There may be many useless layers or useless parameters, some high-level features that may be actually useless for the low-level applications of image processing. Therefore, a moderate depth neural network has been proposed in this paper, which is a 10-layer deep residual network with sparse feedback loops. The detail of the proposed network will be introduced as following.

3. Deep Residual Network with Sparse Feedback Loops

In the design and application of neural network, the researchers are only required to focus on the input and output, the number of hidden layers, and the initial parameters. As the network depth gradually increases, the parameters of neural network are also difficult to tune. There are also no relevant theories to be presented on how to tune these parameters. Moreover, the updating of neuron parameters depends on the gradient; the more far away from the output layer, the more difficult for updating of neuron parameters. It will be invariant, or will change dramatically, which is called gradient disappearance or gradient explosion problem. Although the dropout strategy, or batch normalization, was adopted to reduce explosion problems, it still often happened. The more layers of neural network, the more obvious the gradient disappearance or explosion problem is.

This question reminds me of the amplifier cascade problem in electronics. When connecting the circuit, the output signal is usually unstable. Single negative feedback or inter-stage negative feedback will generally be added to stabilize the output signal. In the amplifier circuit, the negative feedback is added to enhance the performance of the anti-noise and stability of the circuit, but the feedback will also reduce the amplification factor of the circuit. Usually, sparse feedback with a longer span is adopted to keep a tradeoff between the robustness and amplification factor. We believe that this situation is very similar to the shortcut in residual neural network. We try to adjust the shortcut in ResNet to a sparse longer connection. However, the idea of sparse longer feedback comes from the concept of negative feedback in the circuit; hence it is represented by "feedback."

However, the denoising results of ResNet are not better than that of the convolution neural network without feedback. For example, denoising convolutional neural network (DnCNN) [25] is good at removing Gaussian noise. Therefore, we combine the network structure of ResNet and DnCNN, and propose the deep residual network with sparse feedback for image restoration, which is called FbResNet. The proposed network structure is shown in Figure 1.

There are only two feedbacks in Figure 1; one is a short feedback; the other is a long feedback. The short feedback is connected from layer 1 to layer 5. Because the output dimension of layer 1 and layer 5 is equal, the outputs of layer 1 and layer 5 can be directly added. The dimension of the input layer is the same as that of the last convolutional layer. Hence, the long feedback adds the input image to the output of the last convolutional layer, which can add a constraint to FbResNet to keep the most similarity between the input noisy image and output clean image.

The mean squared error between the clean images and the degraded images can be defined as the loss function to learn the trainable parameters Ξ of FbResNet as follows:

$$f(\Xi) = \frac{1}{N} \sum_{i=1}^{N} \| R(x_i, \Xi) - y_i \|^2 \tag{1}$$

where the input is a noisy image x_i, and FbResNet aims to learn a mapping function $R(x_i) = y_i$, to predict the clean image. $\{(x_i, y_i)\}$ represents N degraded-clean training image (patch) pairs; R represents the network structure, of which parameters Ξ require to be trained.

3.1. Network Structure

Inspired by the residual learning structure, we propose the deep residual network with sparse feedback loops for image restoration, and the structure of FbResNet is shown in Figure 1. It consists of ten layers. "Convolution" block is in the first layer. This layer has no "Batch Normal" and "ReLU," in other words, the information produced by this layer is the original information after filtering the input image, then it is used to estimate the residual information by feeding back to the middle and the last layer. Eight "Convolution + Batch Normalization + ReLU" blocks are in the middle layers. The number behind each middle layer is the dilation factors, which is set to 1, 2, 3, 4, 4, 3, 2 and 1, respectively. By using the increasing dilated factors, the first-half layers can learn the residual information using an enlarged receptive field, and the latter half layers can refine the residual information using the decreasing dilation factors. In order to ensure that the estimated residual information does not deviate greatly, two forward feedbacks from the first layer have been added. The first is connected to the middle of the dilation convolution. The second is connected to the last layer. The main task of FbResNet is to estimate the residual information between the input degraded image and the output clean image.

3.2. Implementation

In order to reduce the size and parameters of the neural network, we cut the input training image into small patches. But the restored image may exist annoying artifact boundary. There are two methods to deal with this problem: Symmetrical padding and zero padding (same padding). To verify the effectiveness of FbResNet in handling boundaries, we use the same padding strategy. Note the dilated convolution with dilation factor 4 pads 4 zeros pixels in the boundaries of each feature map. Batch normalization (BN) is adopted right after each convolution and before activation. We initialize the weights as in Reference [26] and Adam is used as minimizing function with a mini-batch size of 38. The learning rate starts from 0.1 and is divided by 10 when the error plateaus. We use a weight decay of 0.001 and a momentum of 0.9. The dropout is not used in the training phase.

3.3. Comparison

In order to verify the effectiveness of the proposed FbResNet, the comparison with the other network structures has been performed. In our opinion, very deep network architecture requires a huge training set, but in many computer vision tasks, a large number of training samples is not easy to be obtained. Nevertheless, small training samples can be easily constructed. Because our training set is small, for comparison on the same network scale, we reduce the depth of the ResNet and set it to 10. The reformed ResNet is shown in Figure 2. Figure 2 shows two kinds of network structures reformed from ResNet and the network structure of DnCNN. The network structure of (a) is same as that of ResNet except for the depth; besides of the first layer and the last layer, only 4 building blocks are used in the reformed ResNet. In order to compare the performance at the same configuration with Figure 1, Figure 2a is also improved to Figure 2b, which is called ResNet with dilated convolution. Figure 2c shows the structure of DnCNN. The meaning of the parameters on each layer in Figure 2 is similar to that in Figure 1. The experiment setting of different models had been shown in Table 1.

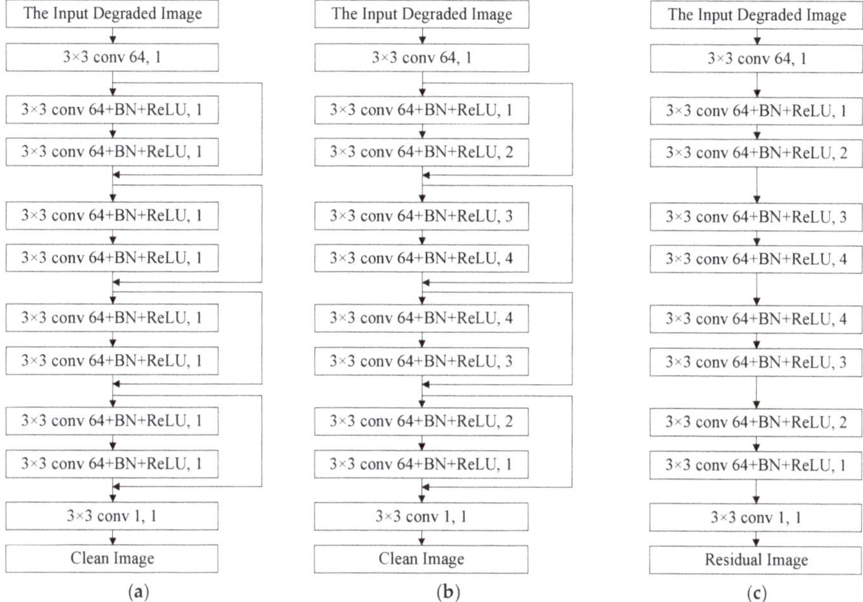

Figure 2. The architectures of the reformed ResNet and denoising convolutional neural network (DnCNN). (**a**) ResNet; (**b**) ResNet with dilated convolution; and (**c**) DnCNN.

Table 1. Experiment setting of different models.

Structure	ResNet	ResNet with Dilated Convolution	DnCNN	FbResNet
Input image	blur image	blur image	blur image	blur image
Number of layers	10	10	10	10
Padding strategy	Same	same	symmetric	same
Dilation convolution	No	yes	yes	yes
Output image	clear image	clear image	residual image	clear image

4. Experiments and Analysis

We evaluate our proposed model and method on two datasets. One is the human face database Facial Recognition Technology (FERET) [27], which includes 1403 human faces with size 64 × 64. The other is 660 images of size 180 × 180, which is a part of images presented in Reference [25], which includes the images of animals, humans, various landscapes, etc.

4.1. Human Face Database

To train FbResNet for motion deblurring with unknown motion direction, we consider four different motion directions: up, down, left, and right. The database used in this paper includes 1403 human faces. These faces are divided into two parts: 1044 images are used to train the parameters of FbResNet, and the remaining 359 images are used to test the network. The size of the human face image is 64 × 64. In this database, the technology of patch cutting is not adopted because the size of the input image is already small.

Figure 3 shows the restored images of human face. The first row shows the images with motion blur; the second row shows the deblurred images of DnCNN; the third row shows the deblurred images of ResNet; the forth row shows the deblurred images of ResNet with dilated convolution; the fifth row shows the deblurred images of FbResNet; and the last row shows the clean images. The restored results of FbResNet are clearer and more similar with the ground truth (clean images) than those of the other models.

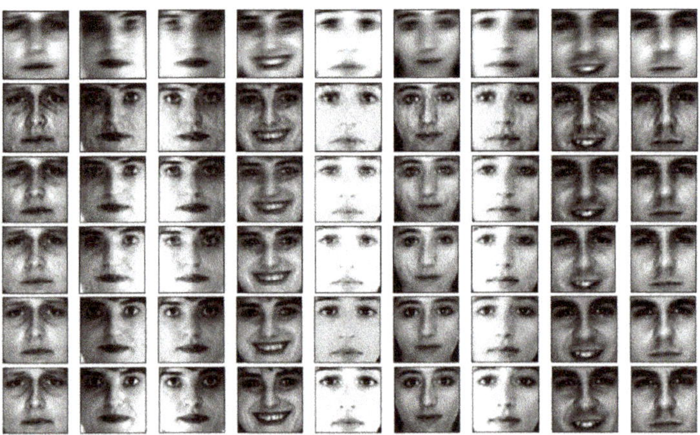

Figure 3. Restored images of human face. The first row shows the images with motion blur; the images from the second row to the fifth row are restored images of DnCNN, ResNet, ResNet with dilated convolution and FbResNet; the last row shows the clean images.

In addition, we find that the restored images using DnCNN and FbResNet are darker than the other ones (the fifth column of Figure 3). The main reason is that there are a few feedbacks or no feedback in these two network structures. Hence, the network structure with a few feedbacks can maintain the average skin color of the restored face image and avoid overexposure, even if the input image has a small exposure effect.

Figure 4 shows the enlarged image of a human face. From left to right, it is the restored image of DnCNN, ResNet, ResNet with dilated convolution, FbResNet, and the ground truth, respectively. It is easier to see the advantages of FbResNet, which has less deformation and higher resolution than those of other models. Figure 5 shows the restored images with motion blur in different directions. The first row shows the blur images and the second row shows the deblured images; the below words describe the movement direction of the produced blur images. We find that the robust performance of FbResNet for motion blur is generally good, but the more mixture of the movement direction, the performance will gradually decline, and the deblurred robustness for various directions needs to be further improved.

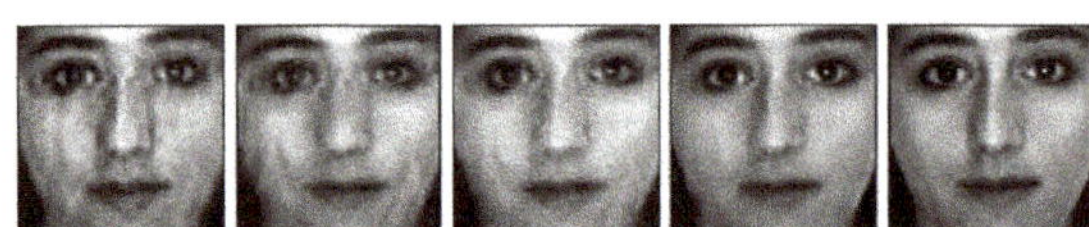

Figure 4. Enlarged restored images. From left to right, it is the restored images of DnCNN, ResNet, ResNet with dilated convolution and FbResNet, and the last is the clean image.

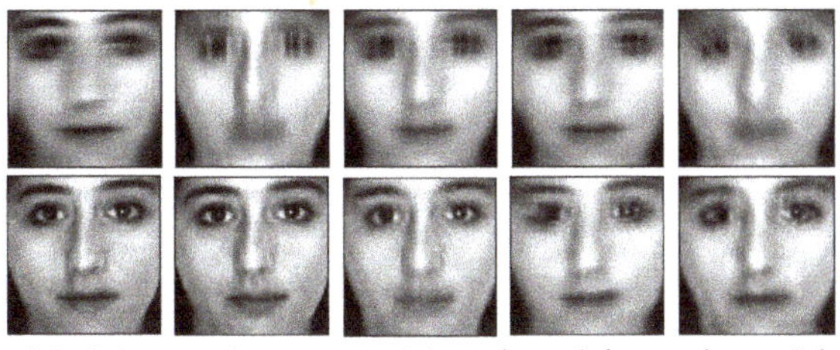

Figure 5. Restored images for input images with motion blur in different directions.

The left figure in Figure 6 shows the average peak signal-to-noise ratio (PSNR) improvement over the other models with respect to different motion direction by FbResNet model. It can be seen that the proposed FbResNet consistently outperforms the other models by a large margin. The right figure in Figure 6 shows the convergence of the loss function. The convergence speed of FbResNet model is faster than that of the other network models.

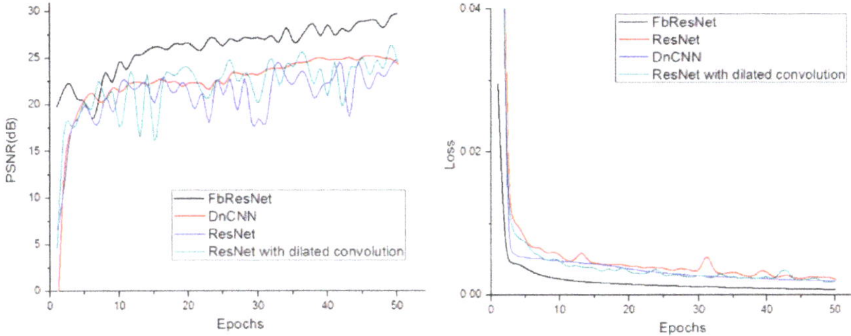

Figure 6. Peak signal-to-noise ratio (PSNR) and loss function for motion deblurring.

4.2. Image Denoising

To train FbResNet for image denoising with different noise level, 660 images of size 180×180 are used. The noise level 15, 35, 45 has been added in the image set. According to the method of Reference [25], we crop each image into small patches, and each patch size is set to 60×60. Then, we obtain 5940 training samples. These samples are divided into two parts, 80% of which is used to train the parameters of FbResNet; the rest is used as the validation set.

In order to validate the effectiveness of FbResNet, 192 images have been used to test the performance of the trained FbResNet and ResNet for image denoising. Figure 7 shows several denoised images with noise level 35. The first column shows the noisy input images; the second is the denoised images of ResNet model and the last column is the results of FbResNet model. Because each test image has been cropped into many small patches, the output patches of the network must be spliced in order to get a complete denoised image.

It can be seen that the images in second column of Figure 7 have obvious artificial stitching trace; nevertheless, it is almost impossible to find the presence of artificial traces from the images in last column. The main reason is that the sparse feedbacks have been added to FbResNet model and can be used to smooth the artificial traces at the seam of patches. Figure 8 shows the enlarged denoised images, and the images from left to right in the first row are input noisy image, restored image of ResNet, restored image of ResNet with "dilation convolution " + " symmetric padding," and restored image of FbResNet. The images in second row, from left to right, are the enlarged images of the red box corresponding to the position in the first row. The restored image of ResNet without symmetric padding has the distinct artificial stitching traces, and the stitching traces have been improved in that of the ResNet with the symmetric padding. However, even if our algorithm is not added symmetric padding, it can also achieve the same effect as that of ResNet with symmetric padding.

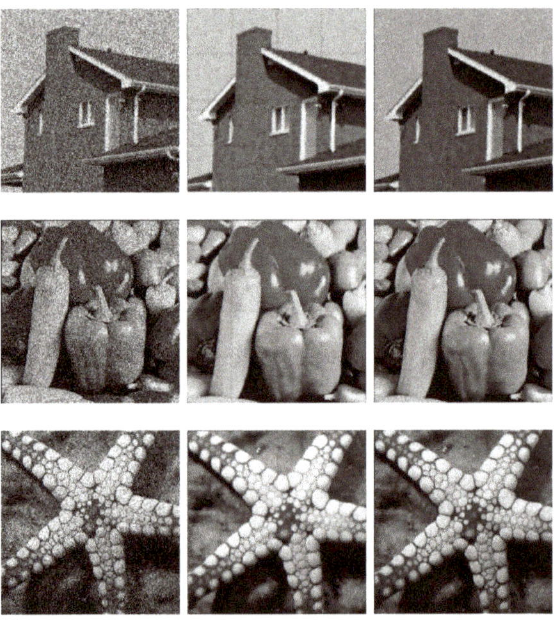

Figure 7. Several denoised image results with noise level 35. From left to right, the first column shows the noisy images, the second column shows the restored image of ResNet, and the last column shows the restored images of FbResNet.

The experiment results on these two datasets demonstrate the feasibility of training FbResNet, which can produce visually pleasant output result for the motion deblurring or image denoising.

Figure 8. Enlarged denoised images. The images from left to right in the first row are input noisy image, restored image of ResNet, restored image of ResNet with "dilation convolution" + "symmetric padding," and restored image of FbResNet; the images in second row, from left to right, are the enlarged images of the red box corresponding to the position in the first row.

5. Conclusions

In this paper, we have designed and trained a deep residual network with sparse feedback loops for image restoration, especially for the restoration of motion deblurring and image denoising. The addition of sparse feedback improves the convergence speed and the training stability of network model. In addition, the proposed FbResNet is good at smoothing artificial stitching trace at the seam of patches, and visually pleasant output results can be produced from the deblurred images or denoised images.

Author Contributions: For this research article with several authors, the author Y.S. conceived and designed the study. Y.S. and M.J. edited and reviewed the manuscript. Z.G. and X.Z. performed the experiments and rendered the figures. All authors have given approval to the final version of the manuscript.

Funding: This research was funded by Natural Science Foundation of Shandong (ZR2016FM13, ZR2016FM40, ZR2016FM21); National Natural Science Foundation of China (61602229, 61601427,61873117); Shandong Provincial Key Research and Development Program of China (NO. 2017CXGC0701, 2017CXGC1504); Fostering Project of Dominant Discipline and Talent Team of Shandong Province Higher Education institutions.

Conflicts of Interest: The authors declare no conflicts of interest.

References

1.   Jian, M.; Lam, K. Simultaneous Hallucination and Recognition of Low-Resolution Faces Based on Singular Value Decomposition. *IEEE Trans. Circuits Syst. Video Technol.* **2015**, *25*, 1761–1772. [CrossRef]

2.   Jian, M.; Lam, K.; Dong, J.; Shen, L. Visual-patch-attention-aware Saliency Detection. *IEEE Trans. Cybern.* **2015**, *45*, 1575–1586. [CrossRef] [PubMed]

3.   Bottou, L.; Cortes, C.; Denker, J.S.; Drucker, H.; Guyon, I.; Jackel, L.D.; LeCun, Y.; Muller, U.A.; Sackinger, E.; Simard, P.; et al. Comparison of classifier methods: a case study in handwritten digit recognition. In Proceedings of the International Conference on Pattern Recognition. IEEE Computer Society, Jerusalem, Israel, Israel, 9–13 October 1994; pp. 77–82.

4.   Krizhevsky, A.; Sutskever, I.; Hinton, G.E. ImageNet classification with deep convolutional neural networks. In Proceedings of the International Conference on Neural Information Processing Systems, Curran Associates Inc, Red Hook, NY, USA, 3–6 December 2012; pp. 1097–1105.

5.   Ma, X.; Dai, Z.; He, Z.; Ma, J.; Wang, Y.; Wang, Y. Learning Traffic as Images: A Deep Convolutional Neural Network for Large-Scale Transportation Network Speed Prediction. *Sensors* **2017**, *17*, 818. [CrossRef] [PubMed]

6.   Szegedy, C.; Liu, W.; Jia, Y.; Sermanet, P.; Reed, S.; Anguelov, D.; Erhan, D.; Vanhoucke, V.; Rabinovich, A. Going deeper with convolutions. In Proceedings of the IEEE Conference on Computer Vision and Pattern Recognition, Boston, MA, USA, 7–12 June 2015; pp. 1–9.

7.   He, K.; Zhang, X.; Ren, S.; Sun, J. Deep Residual Learning for Image Recognition. In Proceedings of the IEEE Conference on Computer Vision and Pattern Recognition, Las Vegas, NV, USA, 27–30 June 2016; pp. 770–778.

8.   Goodfellow, I.J.; Pouget-Abadie, J.; Mirza, M.; Xu, B.; Warde-Farley, D.; Ozair, S.; Courville, A.; Bengio, Y. Generative Adversarial Nets. In *Proceedings of the International Conference on Neural Information Processing Systems*; MIT Press: Cambridge, MA, USA, 2014; pp. 2672–2680.

9.   Park, J.; Min, K.; Chang, S.K.; Lee, K.H. Estimation of motion blur parameters using cepstrum analysis. In Proceedings of the IEEE International Symposium on Consumer Electronics, Singapore, 14–17 June 2011; pp. 406–409.

10.   Jian, M.; Lam, K.; Dong, J. A Novel Face-Hallucination Scheme Based on Singular Value Decomposition. *Pattern Recognit.* **2013**, *46*, 3091–3102. [CrossRef]

11.   Jian, M.; Lam, K.; Dong, J. Facial-Feature Detection and Localization Based on a Hierarchical Scheme. *Inf. Sci.* **2014**, *262*, 1–14. [CrossRef]

12.   Wang, X.; Gao, L.; Song, J.; Shen, H. Beyond Frame-level CNN: Saliency-Aware 3-D CNN with LSTM for Video Action Recognition. *IEEE Signal Process. Lett.* **2017**, *24*, 510–514. [CrossRef]

13.   Chan, T.H.; Jia, K.; Gao, S.; Lu, J.; Zeng, Z.; Ma, Y. PCANet: A Simple Deep Learning Baseline for Image Classification. *IEEE Trans. Image Process.* **2015**, *24*, 5017. [CrossRef] [PubMed]

14. Lecun, Y.; Boser, B.; Denker, J.S.; Henderson, D.; Howard, R.E.; Hubbard, W.; Jackel, L.D. Backpropagation applied to handwritten zip code recognition. *Neural Comput.* **2014**, *1*, 541–551. [CrossRef]

15. Sun, Y.; Jian, M.; Zhang, X.; Dong, J.; Shen, L.; Chen, B. Reconstruction of normal and albedo of convex Lambertian objects by solving ambiguity matrices using SVD and optimization method. *Neurocomputing* **2016**, *207*, 95–104. [CrossRef]

16. Sun, Y.; Zhang, X.; Jian, M.; Wang, S.; Wu, Z.; Su, Q.; Chen, B. An improved genetic algorithm for three-dimensional reconstruction from a single uniform texture image. *Soft Comput.* **2016**, 1–10. [CrossRef]

17. Girshick, R. Fast R-CNN. In Proceedings of the IEEE International Conference on Computer Vision, Santiago, Chile, 7–13 December 2015; pp. 1440–1448.

18. He, K.; Zhang, X.; Ren, S.; Sun, J. Spatial Pyramid Pooling in Deep Convolutional Networks for Visual Recognition. *IEEE Trans. Pattern Anal. Mach. Intell.* **2015**, *37*, 1904–1916. [CrossRef] [PubMed]

19. Shelhamer, E.; Long, J.; Darrell, T. Fully Convolutional Networks for Semantic Segmentation. *IEEE Trans. Pattern Anal. Mach. Intell.* **2017**, *39*, 640–651. [CrossRef] [PubMed]

20. Jain, V.; Seung, H.S. Natural image denoising with convolutional networks. In Proceedings of the International Conference on Neural Information Processing Systems; Curran Associates Inc.: Red Hook, NY, USA, 2008; pp. 769–776.

21. Xie, J.; Xu, L.; Chen, E. Image denoising and inpainting with deep neural networks. In Proceedings of the International Conference on Neural Information Processing Systems; Curran Associates Inc.: Red Hook, NY, USA, 2012; pp. 341–349.

22. Elad, M.; Aharon, M. Image Denoising via Sparse and Redundant Representations Over Learned Dictionaries. *IEEE Trans. Image Process.* **2006**, *15*, 3736–3745. [CrossRef] [PubMed]

23. Chen, Y.; Pock, T. Trainable Nonlinear Reaction Diffusion: A Flexible Framework for Fast and Effective Image Restoration. *IEEE Trans. Pattern Anal. Mach. Intell.* **2015**, *39*, 1256–1272. [CrossRef] [PubMed]

24. Burger, H.C.; Schuler, C.J.; Harmeling, S. Image denoising: Can plain neural networks compete with BM3D. In Proceedings of the IEEE Conference on Computer Vision and Pattern Recognition, Providence, RI, USA, 16–21 June 2012; IEEE Computer Society: Los Alamitos, CA, USA, 2012; pp. 2392–2399.

25. Zhang, K.; Zuo, W.; Chen, Y.; Meng, D.; Zhang, L. Beyond a Gaussian Denoiser: Residual Learning of Deep CNN for Image Denoising. *IEEE Trans. Image Process.* **2017**, *26*, 3142–3155. [CrossRef] [PubMed]

26. He, K.; Zhang, X.; Ren, S.; Sun, J. Delving deep into rectifiers: Surpassing human-level performance on imagenet classification. In Proceedings of the ICCV, Las Condes, Chile, 11–18 December 2015; pp. 1026–1034.

27. Phillips, P.J.; Wechsler, H.; Huang, J.; Rauss, P.J. The FERET database and evaluation procedure for face-recognition algorithms. *Image Vis. Comput.* **1998**, *16*, 295–306. [CrossRef]

Article

An Image Segmentation Method Using an Active Contour Model Based on Improved SPF and LIF

Lin Sun [1], Xinchao Meng [1], Jiucheng Xu [1,*] and Yun Tian [2]

[1]   College of Computer and Information Engineering, Henan Normal University, Xinxiang 453007, China; linsunok@gmail.com (L.S.); 18749607976@163.com (X.M.)

[2]   College of Information Science and Technology, Beijing Normal University, Beijing 100875, China; tianyun@bnu.edu.cn

*   Correspondence: jiuchxu@gmail.com; Tel.: +86-373-332-6190

Received: 14 October 2018; Accepted: 8 December 2018; Published: 11 December 2018

Abstract: Inhomogeneous images cannot be segmented quickly or accurately using local or global image information. To solve this problem, an image segmentation method using a novel active contour model that is based on an improved signed pressure force (SPF) function and a local image fitting (LIF) model is proposed in this paper, which is based on local and global image information. First, a weight function of the global grayscale means of the inside and outside of a contour curve is presented by combining the internal gray mean value with the external gray mean value, based on which a new SPF function is defined. The SPF function can segment blurred images and weak gradient images. Then, the LIF model is introduced by using local image information to segment intensity-inhomogeneous images. Subsequently, a weight function is established based on the local and global image information, and then the weight function is used to adjust the weights between the local information term and the global information term. Thus, a novel active contour model is presented, and an improved SPF- and LIF-based image segmentation (SPFLIF-IS) algorithm is developed based on that model. Experimental results show that the proposed method not only exhibits high robustness to the initial contour and noise but also effectively segments multiobjective images and images with intensity inhomogeneity and can analyze real images well.

Keywords: image segmentation; active contour model; level set; signed pressure force function

---

1. Introduction

Image segmentation is an important task in the field of image analysis and object detection and aims to segment an image into distinctive subregions that are meaningful to analyze [1]. Segmentation is the intermediate step between image processing and image analysis as well as the bridge from low- to high-level research in computer vision. Inhomogeneity, noise, and low contrast in real images have increased the difficulty of image segmentation [2].

Over the past few decades, many segmentation methods have been proposed. The active contour model (ACM), which was proposed by Kass et al. [3], has been proven to be an efficient framework for image segmentation. The fundamental idea of the ACM framework is to control a curve to move toward its interior normal and then stop on the true boundary of an object based on an energy minimization model [4]. The two main shortcomings of ACM algorithms are (1) sensitivity to the initial position and (2) difficulties related to topological changes [5]. Generally, existing ACM methods can be roughly divided into the following types, edge-based models [6–9] and region-based models [10–14].

The geodesic active contour (GAC) model [15] is the most typical of edge-based methods. Owing to the edge-indicator function, the model can stop at high-contrast image gradients [16]. Edge-based models have distinct disadvantages. For example, these methods can effectively segment an object with strong edges; however, they cannot detect the weak edges of an object. Moreover, the methods are sensitive to noise and do not easily obtain satisfactory segmentation results for blurred images [2]. In addition, the contour should initially be set near the object; otherwise, it is difficult to obtain correct segmentation results [17]. Region-based models make full use of image statistical information, whereas edge-based models do not. Thus, region-based models have multiple advantages over edge-based models. For example, because regional information is used, region-based models are less sensitive to contour initialization and noise. Furthermore, these region-based models can easily segment images with weak boundaries or even those without boundaries [18]. One of the most typical region-based methods was proposed by Chan and Vese (C–V) [11], which is based on the Mumford–Shah functional [19]. The C–V model is based on the assumption that image intensities are homogeneous in each region. However, this assumption does not suit the intensity of inhomogeneous images, which limits the method's further applications [20,21].

Recently, hybrid methods have gained popularity among region-based methods. These methods combine region (local or global) and edge information in their energy formulations [22]. Zhang et al. [23] proposed the selective binary and Gaussian filtering regularized level set (SBGFRLS) model. This model combines the advantages of region-based and edge-based active contours and introduces a region-based SPF function, which utilizes the image global intensity means from the C–V method. This method adopts an approach similar to that of the GAC model. However, the edge-indicator function is replaced with a region-based SPF function in the model. Moreover, the traditional regularization function is usually replaced with a Gaussian smoothing function. This traditional method uses only global image intensity information. Therefore, the method is unable to analyze intensity-inhomogeneous images [21,22]. Li et al. [24] investigated a local binary fitting (LBF) model, which is an efficient region-based level set method. The LBF model introduces a local binary fitting energy with a kernel function and uses the intensity of the current pixel to approximate the intensities of the neighboring pixels to obtain accurate segmentation performance; the model can be used to address intensity-inhomogeneous images and has attracted extensive attention due to its satisfactory segmentation performance [25]. However, this model involves high computational complexity. In addition, the model is sensitive to the initialization location and parameters [5,26]. Wang et al. [27] defined an energy functional that combines the merits of the C–V model and the LBF model [21]. Because the new model employs local and global intensity information, it can avoid becoming trapped in a local minimum; however, the result remains partially dependent on the initialization location [21]. Zhang et al. [28] exploited a local image region statistics-based improved ACM method (LSACM) in the presence of intensity inhomogeneity. The LSACM is robust to noise while suppressing intensity overlap to some extent. Yuan et al. [25] offered a model based on global and local regions. The global term takes gradient amplitude into consideration, and the local term adopts local image information by convolving the Gaussian kernel function [29]. This algorithm is sensitive to the initialization location because of the use of gradient information. Similarly, Zhao et al. [30] adopted local region statistical information and gradient information to construct an energy functional and faced the same problem. Zhang et al. [31] introduced a local image fitting (LIF) energy functional to extract local image information and proposed a Gaussian filtering method for a variational level set to regularize the level set function, which can be interpreted as a constraint on the differences between the original image and the fitting image [12,24]. Furthermore, the method used Gaussian kernel filtering to regularize the level set function, and a reinitialization operation was avoided [32]. Unfortunately, the abovementioned methods are sensitive to initialization, and they are also unable to analyze images with intensity inhomogeneity. Hence, these limitations obviously limit their practical applications. Here, we focus on overcoming these drawbacks in this paper.

In this study, to segment the images quickly and accurately, a new image segmentation model is proposed based on an improved SPF and LIF. This method defines a new SPF function, which uses global image information, and the SPF function can segment blurred images and weak gradient images. Then, the LIF model is introduced, which is based on local image information, and this model is used to segment intensity-inhomogeneous images. Moreover, a weight function is established to adjust the weights between the SPF model and the LIF model. Thus, a novel ACM model is presented, and an image segmentation algorithm is investigated. Experimental results demonstrate that our model involves simpler computation, exhibits faster convergence, and can effectively segment multiobjective images and intensity-inhomogeneous images. Furthermore, the proposed method is highly robust to the initial contour and noise.

The remainder of this paper is structured as follows. Section 2 briefly reviews the GAC, C–V, SBGFRLS, and LIF models. In Section 3, by combining the improved SPF function with the LIF model, a novel ACM is presented, and using this model, an image segmentation algorithm is designed. Then, the experimental results and analysis are discussed in Section 4. Section 5 presents the conclusions.

2. Related Work

2.1. The GAC Model

The GAC model uses image gradient information from the boundary of an object [33]. Suppose that $I: \Omega \subset R^2$ is an image domain, $I: \Omega \to R^2$ is an input image, and $C(q)$ is a closed curve. Then, the GAC model is formalized by minimizing the following energy functional as

$$E^{GAC} = \int_0^1 g(|\nabla I(C(q))|)|C'(q)|dq, \tag{1}$$

where g is a strictly decreasing function.

Usually, a satisfactory edge stopping function (ESF) should be defined, which is regular and positive at object boundaries [21], e.g.,

$$g(|\nabla I|) = \frac{1}{1 + |\nabla G_\sigma * I|^2}, \tag{2}$$

where G_σ denotes the Gaussian kernel function and $G_\sigma * I$ describes the convolution operation of I with G_σ.

Using the steepest descent method and the calculus of variations, we obtain the Euler–Lagrange form of Equation (1), which is written as

$$C_t = g(|\nabla I|)k\vec{N} - (\nabla g \cdot \vec{N})\vec{N}, \tag{3}$$

where k is the curvature of the contour and $\vec{N}$ is the inward normal to the curve. A constant velocity term α is typically added to increase the propagation speed [21]. Thus, Equation (3) can be rewritten as

$$C_t = g(|\nabla I|)(k + \varphi)\vec{N} - (\nabla g \cdot \vec{N})\vec{N}. \tag{4}$$

The corresponding level set formulation is described as

$$\frac{\partial \varphi}{\partial t} = g|\nabla \varphi|(div(\frac{\nabla \varphi}{|\nabla \varphi|}) + \alpha) + \nabla g \cdot \nabla \varphi, \tag{5}$$

where ϕ represents the level set function and α is the balloon force that controls the shrinkage or expansion of the contour.

The GAC model utilizes the image gradient to construct an ESF, which can stop the contour evolution on object boundaries. When images have weak boundaries or the initial contour is far from the desired object boundary, the GAC model will fail to find the target [18,22].

2.2. The C–V Model

The C–V model is proposed based on the assumption that the original image intensity is homogeneous. The energy functional of the C–V model [34] is expressed as

$$E^{CV} = \lambda_1 \int_{inside(c)} |I(x) - c_1|^2 \, dx + \lambda_2 \int_{outside(c)} |I(x) - c_2|^2 \, dx, \tag{6}$$

where λ_1 and λ_2 are positive constants that regulate image driving force inside and outside the contour, c_1 represents the mean gray value of the target area and the background area in the evolution curve C, and c_2 represents the mean gray value of the target area and the background area outside the evolution curve C.

By minimizing Equation (6), one has c_1 and c_2, which are described, respectively, as

$$c_1 = \frac{\int I(x)H(\varphi(x))dx}{\int H(\varphi(x))dx}, \tag{7}$$

$$c_2 = \frac{\int I(x)(1 - H(\varphi(x)))dx}{\int (1 - H(\varphi(x)))dx}, \tag{8}$$

where $H(\phi)$ is the Heaviside function.

In practice, the Heaviside function $H(\phi)$ and the Dirac delta function $\delta(\phi)$ must be approximated by smooth functions $H_\varepsilon(\phi)$ and $\delta_\varepsilon(\phi)$ when $\varepsilon \to 0$, which are typically expressed as follows, respectively

$$H_\varepsilon(x) = \frac{1}{2}[1 + \frac{2}{\pi}\arctan(\frac{x}{\varepsilon})], \tag{9}$$

$$\delta_\varepsilon(x) = \frac{1}{\pi} \cdot \frac{\varepsilon}{\varepsilon^2 + x^2}, \tag{10}$$

By incorporating the length and area energy terms into Equation (6) and further minimizing the length and area of the level set curve, the corresponding partial differential equation is described as

$$\frac{\partial \varphi}{\partial t} = \delta(\varphi)\left[\mu\nabla(\frac{\nabla\varphi}{|\nabla\varphi|}) - v - \lambda_1|I - c_1|^2 + \lambda_2|I - c_2|^2\right], \tag{11}$$

where μ, v, λ_1, and λ_2 denote the corresponding coefficients, all of which are positive constants; ∇ is the gradient operator; μ controls the smoothness of the zero level set; v increases the propagation speed; and λ_1 and λ_2 control the image data driving force inside and outside the contour, respectively.

Because c_1 and c_2 are related to the global information inside and outside the curve, this model can segment blurred images and images with weak gradients more effectively than the edge-based model can, and it is insensitive to the initialization location [22,35]. However, when the internal and external intensities of the curve are inhomogeneous, c_1 and c_2 cannot express the local information precisely, which leads to the failure of image segmentation [2].

2.3. The SBGFRLS Model

The SBGFRLS model is proposed based on the traditional C–V model and the GAC model, thereby seizing the advantages of both models [21]. In the SBGFRLS model, an SPF function is used to substitute ESF in the GAC model, and thus the level set formulation of the SBGFRLS can be expressed as

$$\frac{\partial \varphi}{\partial t} = spf(I(x)) \cdot (div(\frac{\nabla\varphi}{|\nabla\varphi|}) + \alpha)|\nabla\varphi| + \nabla spf(I(x)) \cdot \nabla\varphi, \tag{12}$$

where $spf(I(x))$ in Equation (12) is an SPF function, which can be given as

$$spf(I(x)) = \frac{I(x) - \frac{c_1 + c_2}{2}}{\max(|I(x) - \frac{c_1 + c_2}{2}|)}, \tag{13}$$

where c_1 and c_2 represent the gray mean values of regions outside and inside the contour, computed using Equations (7) and (8), respectively.

The SBGFRLS model can reduce the cost of the expensive reinitialization of the traditional level set method and is more efficient than traditional models. The model stops the contour evolution, even with blurred edges, without any a priori training. However, the model assumes that the region to be segmented is homogeneous. This assumption occasionally holds in general clinical cases. When facing heterogeneous intensity distributions, the detection accuracy can fall significantly because the fundamental assumption is violated [36,37]. Moreover, the SBGFRLS model can become trapped in a local minimum without proper initialization, which leads to poor segmentation performance [38–40].

2.4. The LIF Model

The local fitted image (LFI) formulation [31] is defined based on local image information, based on which the LIF model is investigated. This model can segment intensity-inhomogeneous images [41]. The LIF model is expressed as follows

$$E^{LIF}(\varphi) = \frac{1}{2} \int_{\Omega} |I(x) - I^{LFI}(x)|^2 dx, \tag{14}$$

where I^{LFI} is a local fitted image, and any $x \in \Omega$.

It follows that I^{LFI} can be calculated as

$$I^{LFI}(x) = m_1 H_{\varepsilon}(\varphi) + m_2(1 - H_{\varepsilon}(\varphi)), \tag{15}$$

where m_1 and m_2 are expressed, respectively, as

$$\begin{cases} m_1 = mean(I \in (\{x \in \Omega | \varphi(x) > 0\} \cap W_k(x))) \\ m_2 = mean(I \in (\{x \in \Omega | \varphi(x) < 0\} \cap W_k(x))) \end{cases}, \tag{16}$$

ϕ is the zero level set of a Lipschitz function that represents the contour C; $H_{\varepsilon}(\phi)$ is the regularized Heaviside function, as defined in Equation (9); and $W_k(x)$ is a rectangular window function.

In our experiment, $W_k(x)$ is a truncated Gaussian window with a standard deviation of σ and size $(4k + 1) \times (4k + 1)$, where k is the greatest integer that is smaller than σ. Similarly, the segmentation results can be achieved if a constant window is chosen [31].

According to the calculus of variations and the gradient descent method, the following partial differential equation can be obtained by minimizing E^{LIF}:

$$\frac{\partial \varphi}{\partial t} = (I - I^{LFI})(m_1 - m_2)\delta_{\varepsilon}(\varphi), \tag{17}$$

where $\delta_{\varepsilon}(\phi)$ is the regularized Dirac delta function [32], which is calculated as indicated in Equation (10).

According to the complexity analysis and experimental results in [31,32,41,42], the LIF model is more efficient than the LBF model. However, neither model can handle noisy and intensity-inhomogeneous images well [41,42].

3. Proposed Method

3.1. Improved SPF Function

The main strategy of the ACM based on region information is to construct a driving force, which is based on the information of the image region [43]. The region function modulates the sign of the pressure forces using region information such that the contour shrinks when it is outside the object of interest and expands when it is inside the object. For this reason, these external forces are sometimes called SPF [43]. Zhang et al. [22] proposed the SBGFRLS model, which utilizes the statistical information inside and outside the contour to construct a region-based SPF function [37]. However, an SPF function is simply based on image information. Thus, the corresponding model cannot segment intensity-inhomogeneous images or images with weak boundaries [36,41].

In this study, the global information of image I is used to divide the image into two parts, inC and $outC$, and the level set function is then introduced into the new SPF function.

Using global region information and combining c_1 and c_2, a global fitted image formulation is defined as

$$f = H_\varepsilon(\varphi).*(I - c_1) + (1 - H_\varepsilon(\varphi)).*(I - c_2), \tag{18}$$

where $H_\varepsilon(\phi)$ defined in Equation (9) is the regularized Heaviside function and c_1 and c_2 are calculated by Equations (7) and (8), respectively, and .* describes matrix multiplication.

By employing the above-defined global fitted image, a new SPF function is defined as

$$spf(I(x)) = \frac{I(x) - f(x)}{\max(|f(x)|)}. \tag{19}$$

According to the construction approach of the SPF function, a new partial differential equation is defined as

$$\frac{\partial \varphi}{\partial t} = spf(I(x)) \cdot \alpha, \tag{20}$$

where α is the balloon force that controls the shrinkage or the expansion of the contour. In this paper, according to the concept of a balloon force established previously [44], a balloon force is reconstructed to change the evolution rate of the level set function adaptively, which is defined as

$$\alpha_{new} = c_1 + c_2. \tag{21}$$

The new SPF is more efficient than the traditional ACM models because this function avoids the expensive cost of the reinitialization step. Moreover, the SPF is less sensitive to the initialization location. However, the SPF function is constructed with only global image information. Therefore, it appears difficult to handle images with intensity inhomogeneity using this approach.

3.2. Active Contour Model Based on Improved SPF and LIF

Zhang et al. [31] constructed the LIF model, which can effectively process nonhomogeneous images through local image information. Unfortunately, the model is sensitive to the initial curve and noise [2]. To construct a model that can process nonhomogeneous images and reduce the dependence on the location of the initial contour, this subsection combines the new SPF function with the existing LIF model to form a new ACM based on local and global image information.

By combining the new SPF function with the LIF model, the new level set evolution equation is defined as

$$\frac{\partial \varphi}{\partial t} = \lambda(I - I^{LIF})(m_1 - m_2)\delta_\varepsilon(\varphi) + (1 - \lambda)spf(I(x)) \cdot \alpha, \tag{22}$$

where $\delta_\varepsilon(\phi)$, defined in Equation (10), is the regularized Dirac delta function and λ is a new weight coefficient.

Here, λ is a weight function that can be employed to dynamically adjust the ratio between the local and the global term in image segmentation. Namely, the image information term playing a crucial role in segmenting an image can be selected.

Based on the local and global image information, the weight coefficient λ is defined as

$$\lambda = \frac{A}{\max(|B|)}, \tag{23}$$

where A is defined in Equation (21) and B is defined as

$$B = m_1 + m_2, \tag{24}$$

where m_1 and m_2 are defined in Equation (16).

It is noted that the selection of the weight parameter λ is important in controlling the influence of the local and the global terms. Li et al. [45] declared that the local term is critical to the initialization to some extent; a global term is incorporated into the local framework, thereby forming a hybrid ACM. Therefore, with the mutual assistance of the local force and the global force, the robustness to the initialization can be improved, and the global force is dominant if the evolution curve is away from the object. When the contour is placed near the object boundaries, the LIF model plays a dominant role, and fine details can be detected accurately. In contrast, the new SPF model plays a key role when the contour is located far from the object boundaries, and owing to the assistance of the SPF, a flexible initialization is allowed. It follows that the automatic adjustment between the LIF and SPF models in our ACM is very distinct. Furthermore, the objective of the dynamic adjustment is to determine an optimal result for image segmentation.

In general, the new proposed SPFLIF-IS model not only solves the problem that the intensity-inhomogeneous images cannot be accurately segmented by using the global image information but also overcomes the primary shortcoming that the model based on the local image information is sensitive to noise and the initial contour.

3.3. Algorithm Steps

The procedures of image segmentation are illustrated in Figure 1.

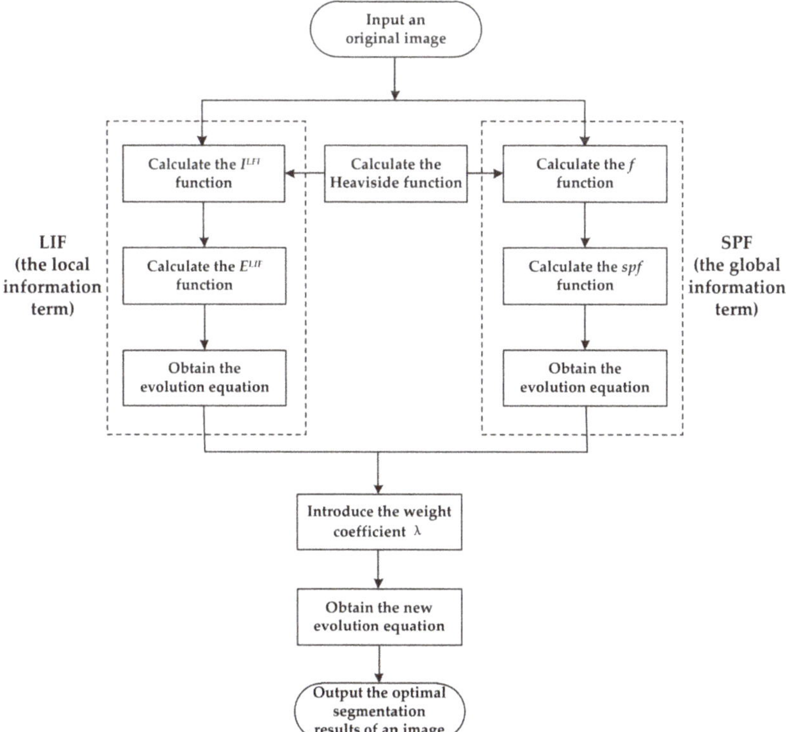

Figure 1. The graphical process of image segmentation.

After the abovementioned image segmentation algorithm has been applied, an improved SPF and LIF-based image segmentation (SPFLIF-IS) algorithm using ACM can be implemented and described as Algorithm 1, which is summarized as follows.

Algorithm 1. SPFLIF-IS

Input: An original image
Output: The result of image segmentation

Step 1:   Initialize the level set function ϕ, and set the coefficients Δt, n, and ε.

Step 2:   Calculate the Heaviside function and the Dirac delta function using Equations (9) and (10), respectively.

Step 3:   **For** $n = 1$: iterNum // iterNum is the total number of iterations.

Step 4:   Compute c_1 and c_2 by Equations (7) and (8), respectively, and obtain f according to Equation (18).

Step 5:   Calculate $spf(I(x))$, according to Equation (19), and obtain the level set evolution equation by Equation (20).

Step 6:   Introduce the LIF model.

Step 7:   Calculate the weight coefficient λ using Equation (23).

Step 8:   Calculate the level set evolution equation using Equation (22).

Step 9:   If the evolution of the curve is stable, then output the segmentation result. Else, return to Step 4.

Step 10: **End for**

It is well known that convolution operations are the most time-consuming with respect to the time complexity of an algorithm. Therefore, it is necessary to explain the complexity of the convolution operation. When an algorithm requires a convolution operation, the time cost is approximately $O(n^2 \times N)$ [46], where N is the image size and n is the Gaussian kernel. The values of N are greater than n^2.

Because the C–V model [34] must be reinitialized in every iteration, its time cost is very high, and the computational complexity is $O(N^2)$ [31]. The LBF model [24] usually needs to perform four convolution operations in each iteration, which greatly increases the computational time complexity. This situation indicates that the time complexity is $O(itr \times 4 \times n^2 \times N)$, where the parameter itr is the number of iterations. In contrast, the SBGFRLS model [23] must perform three convolution operations, two of which are derived by gradient calculation (horizontal and vertical), and the other involves mask image and filter mask. Thus, the total computational complexity of the SBGFRLS model is $O(itr \times 3 \times n^2 \times N)$. The LIF model [31] performs two convolution operations in each iteration. Thus, the total computational time required for the LIF model is $O(itr \times 2 \times n^2 \times N)$. For the SPFLIF-IS algorithm, the computational complexity is mainly concentrated in Step 6. In Algorithm 1, Step 6 is the most time-consuming to calculate in the LIF model. The computational complexity of our proposed method is $O(itr \times 2 \times n^2 \times N)$, where n is the size of the Gaussian kernel function and N is the image size. Since in most cases, $N >> n^2$, the complexity of SPFLIF-IS is $O(N)$ approximately, which is close to that of the LIF model in [31]. It follows that our proposed method is much more computationally efficient than the C–V model [31], the LBF model [24], and the SBGFRLS model [23]. Because the SPFLIF-IS algorithm decreases the number of Gaussian convolution operations required, its time costs and number of iteration operations are drastically reduced. Therefore, the computational complexity of our SPFLIF-IS method is lower than that of the other related ACMs [6,8,11,12,15,17,20,23,24,31,34,43].

4. Experimental Results

4.1. Experiment Preparation

In this section, comprehensive segmentation results for all algorithms compared are presented to validate the performance of our proposed method on various representative synthetic and real images with respect to different characteristics. Following the experimental techniques for image segmentation designed by Ji et al. [42], these selected images are mostly corrupted with one or more degenerative characteristics, including additive noise, low contrast, weak edges, and intensity inhomogeneity. Unless otherwise specified, the same parameters are employed as follows, $\Delta t = 1$, $n = 5$, $\varepsilon = 1.5$, and $\varphi_0(x, y) = 1 : (x, y) \in \mathrm{in}(c)$ or $\phi_0(x, y) = -1 : (x, y) \in \mathrm{out}(c)$. The Gaussian kernel plays an important role in practical applications; the kernel is a scale parameter controlling region scalability from small neighborhoods to the entire image domain [31]. In general, the value of the scale parameter should be appropriately selected from practical images. It is well known that an excessively small value may cause undesirable results, whereas an excessively large value can lead to high computational complexity [31,36]. Thus, the Gaussian kernel size controlling the regularization of the level set function should be chosen according to practical cases [36]. Following the experimental techniques designed in [31,36], the σ selected in our experiments is typically less than 10. All of models compared in this paper are tested in MATLAB R2014a in a Windows 7 environment using a 3.20 GHz Intel (R) Core i5-3470M processor with 4 GB RAM.

4.2. Segmentation Results of Images with Intensity Inhomogeneity

To demonstrate the satisfactory performance and effectiveness of the SPFLIF-IS model, a series of experimental results are presented. We compare our model with the following five existing models: (1) the C–V (The code is available at [47]) model [34], (2) the LBF (The code is available at [48]) model [24], (3) the LIF (The code is available at [47]) model [31], (4) the SBGFRLS (The code is available at [47]) model [23], and (5) the LSACM model [28]. The five representative ACM algorithms are the state-of-the-art level set methods published recently for image segmentation. The algorithms show improvements over the classical ACM and are specially selected based on the level set method for comparison experiments. The chosen parameters for these models can be found in [23,24,28,31,34]. The segmentation results obtained for images with intensity inhomogeneity using the six models are illustrated in Figure 2, where the original images shown in Figure 2a can be found in [2].

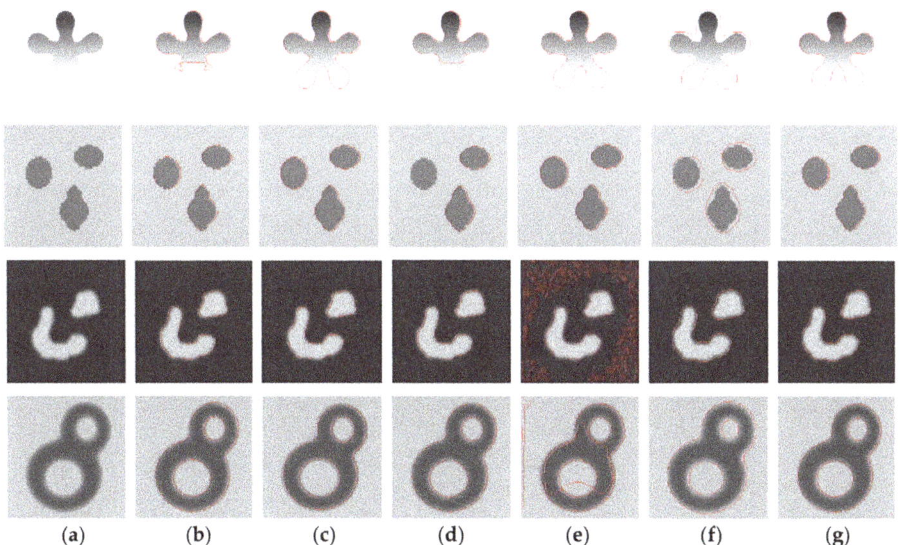

(a) (b) (c) (d) (e) (f) (g)

Figure 2. The segmentation results of images with intensity inhomogeneity for the six models. (a) Original image, (b) C–V model, (c) LBF model, (d) SBGFRLS model, (e) LIF model, (f) LSACM model, and (g) SPFLIF-IS model.

Figure 2b,d,f shows that the C–V model, the SBGFRLS model, and the LSACM model fail to analyze the first image with intensity inhomogeneity. As shown in Figure 2d,f, the SBGFRLS model and the LSACM cannot yield the ideal segmentation results for the second image. The object boundaries of the third image are not identified by the LIF model, and the results are shown in Figure 2e. Figure 2e,f shows that the true boundaries of the fourth image are not accurately extracted by the LIF model or the LSACM model. The SPFLIF-IS model detects the true boundary, and the results are illustrated in Figure 2g. Meanwhile, Figure 2c,g shows that the LBF model perform as well as the SPLIF-IS model.

Note that because the visual evaluations in Figure 2 are partial to subjective measures, to strengthen the objective results of our experiments, the corresponding tables should be added to defend the arguments for all the tested images in the following visual evaluations, in which each failure is clearly labeled to avoid ambiguity. To more clearly illustrate this state, the following symbols are adopted in the tables: F1: fail to detect boundaries, F2: nonideal boundaries detected, F3: fail to detect internal boundaries, and T: true boundaries detected. Table 1 objectively describes the segmentation results of Figure 2 in detail. It can be clearly concluded from Table 1 that the LBF performs as well as the SPLIF-IS, the C–V exhibits slightly bad results, and the LSACM produces the worst results. Therefore, the experimental results shown in Figure 2 and Table 1 indicate that the SPFLIF-IS model can analyze the images with intensity inhomogeneity well.

Table 1. Description of the segmentation results in Figure 2.

Methods	C–V	LBF	SBGFRLS	LIF	LSACM	SPFLIF-IS
	F1	T	F1	F2	F1	T
Segmentation	T	T	F1	T	F1	T
performance	T	T	T	F1	T	T
	T	T	T	F1	F1	T

4.3. Segmentation Results of Multiobjective Images

This portion of our experiment concerns the segmentation results obtained for multiobjective images. The SPFLIF-IS method is consistently compared with the five abovementioned methods (C–V, LBF, SBGFRLS, LIF, and LSACM). The original multiobjective images and the segmentation results of the six models are shown in Figure 3, where the original images shown in Figure 3a are derived from [42,49]. Although our model identifies most of the boundaries of the first image, the boundaries are subtle different when compared with those detected by the LBF model. As shown in Figure 3d,f, the SBGFRLS model and the LSACM model obviously fail to segment the first, second, and fourth multiobjective images. The true boundaries of the third image cannot be extracted by the C–V model, the LBF model, the SBGFRLS model, or the LIF model; the results are shown in Row 3 of Figure 3. Table 2 describes the segmentation results shown in Figure 3. As shown in Table 2, the SPFLIF-IS yields the best results, the C–V performs as well as the LBF, and the LIF exhibits the worst results. Figure 3 and Table 2 clearly show that our proposed SPFLIF-IS method can segment the fourth image, but the other comparison methods cannot. The experimental results indicate that the SPFLIF-IS model can efficiently segment the multiobjective images.

Table 2. Description of the segmentation results of Figure 3.

Methods	C–V	LBF	SBGFRLS	LIF	LSACM	SPFLIF-IS
	F2	T	F3	F2	F1	F2
Segmentation	T	F2	F3	F1	F1	T
performance	F1	F1	F3	F1	T	T
	F1	F1	F1	F1	F1	T

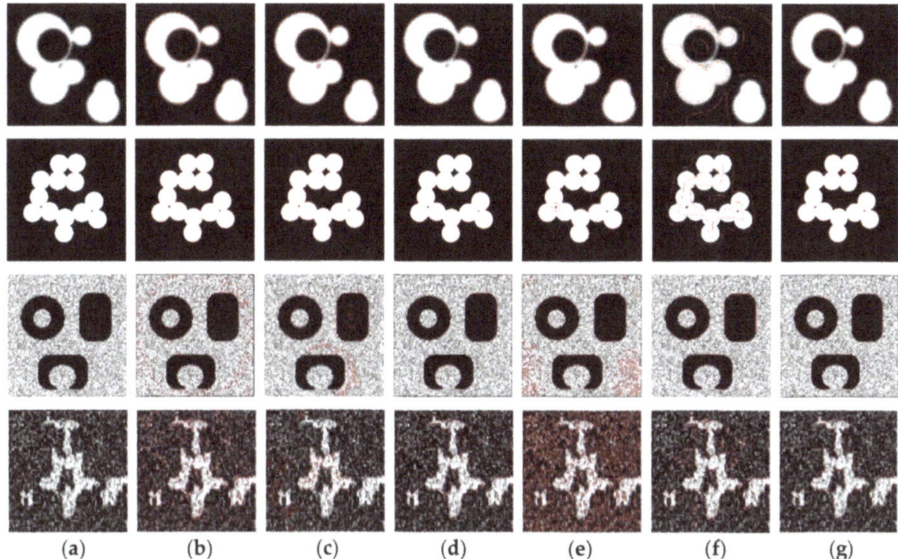

(a)   (b)   (c)   (d)   (e)   (f)   (g)

Figure 3. The segmentation results of the multiobjective images for the six models. (**a**) Original image, (**b**) C–V model, (**c**) LBF model, (**d**) SBGFRLS model, (**e**) LIF model, (**f**) LSACM model, and (**g**) SPFLIF-IS model.

4.4. Segmentation Results of Noisy Images

The following subsection describes the experimental segmentation results obtained for noisy images. The SPFLIF-IS model is still compared with the C–V, LBF, SBGFRLS, LIF, and LSACM models. Figure 4 illustrates the original images with different noise intensities and compares the results of the six state-of-the-art segmentation methods, where the original images without noise in Figure 4a are derived from [46]. In Figure 4, Row 1 shows the original images and the segmentation results. Row 2 to Row 5 show the added Gaussian noise with zero means and different variances (σ = 0.01, 0.02, 0.03, 0.05). Figure 4c,f shows that the LBF model and the LSACM model cannot analyze the five images. Although the C–V model and the SBGFRLS model can segment the first and the second image, neither model performs well when the noise intensity increases; the results are shown in Figure 4b,d. Figure 4e shows that the LIF model could analyze the images without Gaussian noise well. With respect to the segmentation of the images containing Gaussian noise, the LIF model exhibits poor performance. As shown in Figure 4g, the object boundaries are accurately extracted by our proposed SPFLIF-IS model. Table 3 describes the segmentation results of Figure 4. Table 3 shows that the SPFLIF-IS model yields the best results, the C–V model performs as well as the SBGFRLS model, and the LBF model performs as poorly as the LSACM model. The experimental results demonstrate that the SPFLIF-IS model can effectively eliminate the interference of the noise and complete the segmentation of the noisy images.

Table 3. Description of the segmentation results of Figure 4.

Methods	C–V	LBF	SBGFRLS	LIF	LSACM	SPFLIF-IS
	T	F1	T	T	F1	T
	F2	F1	F2	F1	F1	T
Segmentation performance	F1	F1	F1	F1	F1	T
	F1	F1	F1	F1	F1	F2
	F1	F1	F1	F1	F1	F2

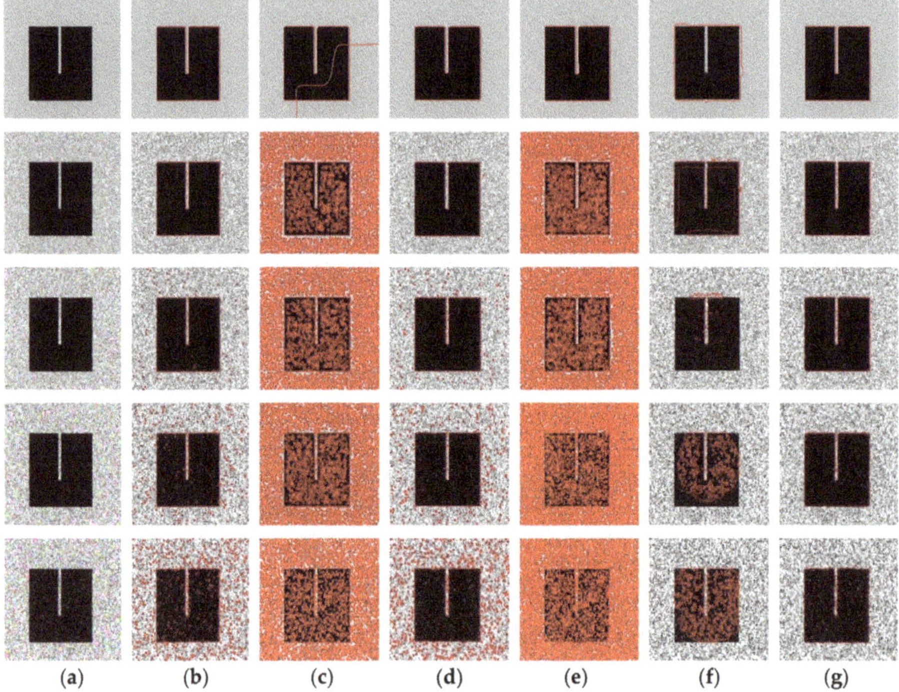

Figure 4. The segmentation results obtained for images with strong noise using the six models. (**a**) Original image, (**b**) C–V model, (**c**) LBF model, (**d**) SBGFRLS model, (**e**) LIF model, (**f**) LSACM model, and (**g**) SPFLIF-IS model.

4.5. Segmentation Results of Texture Image

This part of our experiment tests the segmentation performance of texture images. Figure 5a shows the original texture image, which is derived from [4]. Moreover, the compared models are still the C–V, LBF, SBGFRLS, LIF, and LSACM models. According to Figure 5c,e,f, the object boundaries of the first image are not identified by the LBF, LIF, or LSACM model, respectively. Most of the boundaries are obtained by the SBGFRLS model. However, the internal details are not recognized; the detailed results are illustrated in Figure 5d. Figure 5e shows that the LIF model fails to segment the second image. Although the C–V, LBF, SBGFRLS, and LSACM models recognize the true boundaries of the second image, some boundaries lie in the middle of the image; the results are illustrated in Row 2 of Figure 5. Table 4 describes the segmentation results of Figure 5. Table 4 shows that the SPFLIF-IS model performs the best, the C–V model exhibits the second best performance, and the LIF model shows as poor a performance as the LSACM model. The SPFLIF-IS model can eliminate the interference of the image texture and analyze the texture image well.

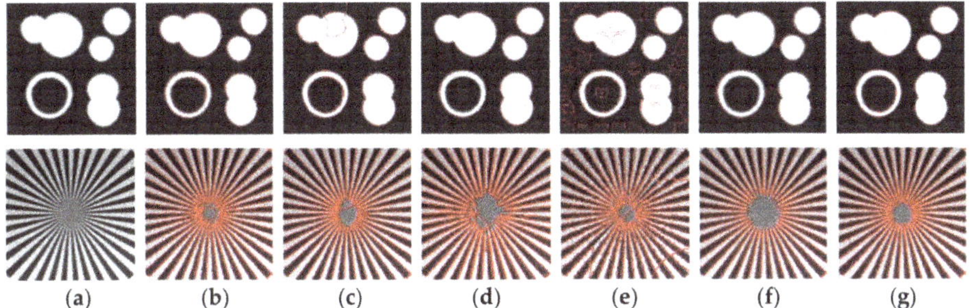

Figure 5. The segmentation results obtained for a texture image using the six models. (**a**) Original image, (**b**) C–V model, (**c**) LBF model, (**d**) SBGFRLS model, (**e**) LIF model, (**f**) LSACM model, and (**g**) SPFLIF-IS model.

Table 4. Description of the segmentation results of Figure 5.

Methods	C–V	LBF	SBGFRLS	LIF	LSACM	SPFLIF-IS
Segmentation	T	F1	F3	F1	F1	T
performance	F3	F3	F3	F1	F1	T

4.6. Segmentation Results of Real Images

In this subsection, we continue testing our algorithms, this time using real images. The SPFLIF-IS method is compared with the same five methods (C–V, LBF, SBGFRLS, LIF, and LSACM). The original real images and the segmentation results of the six models are shown in Figure 6, where the first and second images in Figure 6a can be found in the literature [27,31], and the third and fourth images shown in Figure 6a are selected from Berkeley segmentation data set 500 (BSDS500) (The code is available at [50]). The first image in the third and fourth columns shows that the LBF and SBGFRLS models fail to segment the image; the results are shown in Figure 6c,d. The first image in the fifth and sixth columns shows that most of the boundaries are obtained by the LIF and LSACM models. However, the internal details are not recognized; the results are illustrated in Figure 6e,f. The LBF, LIF, and LSACM models fail to segment the second and third images, as shown in Figure 6c,e,f. The object boundaries of the fourth images are not accurately extracted by the C–V, LBF, SBGFRLS, LIF, and LSACM models, as shown in the fourth rows of Figure 6. As shown in Figure 6g, the object boundaries are accurately extracted by our proposed model. Table 5 objectively offers the segmentation results of Figure 6. Table 5 indicates that the SPFLIF-IS achieves the best results; the C–V exhibits slightly better results than those obtained by the LBF, SBGFRLS, LIF, and LSACM; and the LBF performs as poorly as the LSACM. The segmentation results demonstrate that our SPFLIF-IS model can efficiently analyze real images and yield great segmentation results.

Table 5. Description of the segmentation results of Figure 6.

Methods	C–V	LBF	SBGFRLS	LIF	LSACM	SPFLIF-IS
	T	F1	F1	F3	F1	T
Segmentation	F2	F1	F2	F1	F1	T
performance	T	F1	F2	F1	F1	T
	F1	F1	F2	F1	F1	T

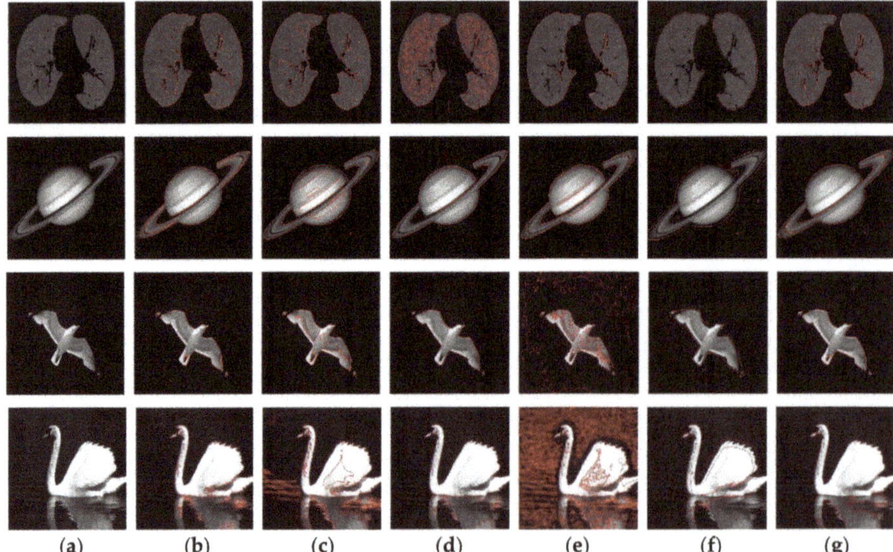

(a)   (b)   (c)   (d)   (e)   (f)   (g)

Figure 6. The segmentation results obtained for real images using the six models. (**a**) Original image, (**b**) C–V model, (**c**) LBF model, (**d**) SBGFRLS model, (**e**) LIF model, (**f**) LSACM model, and (**g**) SPFLIF-IS model.

4.7. Comparative Evaluation Results

In addition to using visual evaluation, the accuracy of the target region segmentation can be assessed quantitatively and objectively using the *DICE* coefficient (*DICE*) [51,52] and the Jaccard similarity index (*JSI*) [53]. Following the experimental techniques designed in [42,54], test images are selected randomly from the BSDS500 database. Note that BSDS500 contains hundreds of natural images whose ground-truth segmentation maps have been generated by multiple individuals [40,55]. To enhance the coherency of our work with the abovementioned algorithms, three comparative experiments are performed on many real-world color images, which are selected from the Berkeley segmentation data set 500 (BSDS500) and consist of a set of natural images.

The first part of this experiment involves evaluating the value of the *DICE* for twenty representative real-world color images, which are chosen from the Berkeley segmentation data set 500 (BSDS500). The algorithms compared are the C–V model [34], the LBF model [24], and the LIF model [31].

The *DICE*, also called the overlap index, is the most frequently used metric for validating image segmentations. The *DICE* measures how well the segmentation results S match the ground truth G. When the value of the *DICE* is close to 1, the segmentation results have high accuracy. The formula for the *DICE* is given as

$$DICE(G, S) = \frac{2|\Omega_G \cap \Omega_S|}{|\Omega_G| + |\Omega_S|},\tag{25}$$

where Ω_S describes the segmented volume and Ω_G denotes the ground truth [56,57]. The *DICE* values of the segmentation results obtained by applying the four models to segment Berkeley color images are listed in Table 6, where the Mean describes the average values of the *DICE* for all test image data. Table 6 shows that the SPFLIF-IS method yields the best values for the *DICE* on the twenty image data, and the corresponding Mean is also the largest. The results indicate that our SPFLIF-IS model outperforms the C–V, LBF, and LIF models. In summary, these results demonstrate that our SPFLIF-IS method is indeed efficient and outperforms these currently available approaches.

Table 6. DICE values of the image segmentation results using the four models for Berkeley color images.

Image ID	C–V	LBF	LIF	SPFLIF-IS
3063	0.9779	0.9576	0.8962	0.9783
8068	0.978	0.9555	0.8673	0.9827
14092	0.9235	0.871	0.8058	0.9257
29030	0.9525	0.9432	0.8106	0.9743
41004	0.9763	0.9565	0.8769	0.9791
41006	0.9625	0.9305	0.8358	0.9643
46076	0.9763	0.9512	0.8392	0.9783
48017	0.9526	0.9066	0.855	0.9562
49024	0.9566	0.9627	0.8531	0.9792
51084	0.9253	0.9401	0.8464	0.9607
62096	0.9641	0.9387	0.8617	0.9734
101084	0.8942	0.8441	0.8103	0.978
124084	0.9578	0.9378	0.8865	0.9616
143090	0.9575	0.9517	0.8633	0.9692
147091	0.9693	0.9387	0.8254	0.9717
207056	0.9677	0.9305	0.8183	0.9826
296059	0.947	0.9276	0.8283	0.9742
299091	0.9708	0.9595	0.8448	0.9759
317080	0.9591	0.9288	0.8665	0.9634
388006	0.9676	0.9452	0.8738	0.9701
Mean	0.9568	0.9339	0.8483	0.9699

The next section of the experiment involves testing the value of the *JSI* coefficient for the twenty representative real-world color images in Table 6. The algorithms compared are still the C–V model [34], the LBF model [24], and the LIF model [31].

The *JSI* is the second statistical measure used for quantitative evaluation in this paper. The *JSI* is calculated by

$$JSI(G, S) = \frac{|\Omega_G \cap \Omega_S|}{|\Omega_G \cup \Omega_S|}. \tag{26}$$

The accuracy of the segmentation results for the Berkeley color images is measured by the *JSI* value, as shown in Figure 7. A *JSI* value close to 1 indicates favorable segmentation results. Figure 7 shows that the SPFLIF-IS method exhibits the best *JSI* values for the twenty image data. For image IDs 3063, 14092, 41006, and 147091, the *JSI* values of the C–V model are very close to those of the SPFLIF-IS model. For image ID 227092, the *JSI* values yielded by the C–V and LBF models are close to those of the SPFLIF-IS model. However, Figure 7 clearly illustrates that the SPFLIF-IS method yields greater *JSI* values than those generated by the C–V, LBF, and LIF models.

In the final part of this experiment, to fully validate the advantages of our SPFLIF-IS method in terms of the *DICE* and *JSI*, the five state-of-the-art methods ((1) the C–V model [34], (2) the LBF model [24], (3) the LIF model [31], (4) the SBGFRLS model [23], and (5) the LSACM model [28]) are applied to eight real color image data selected from the Berkeley segmentation data set 500 (BSDS500). The experimental results are shown in Table 7, where the Mean describes the average values of the *DICE* and *JSI* for all test image data.

The foregoing experimental analysis demonstrates that our proposed method is designed based on an improved SPF function and the LIF method. The model combines the merits of global image information and local image information and can segment noisy images and multiobjective images well. By contrast, the C–V model and the SBGFRLS model are constructed with global image information alone, based on the assumption that the region to be segmented is homogeneous. Unfortunately, this assumption is not suitable for intensity-inhomogeneous images [2,31,35]. The LBF model and LIF model use local information to segment intensity-inhomogeneous images and obtain desirable segmentation results; thus, the models are sensitive to the initial position and image noise [2,36].

The LSACM model is proposed based on the local statistical information of an image; therefore, this model is robust to noise while suppressing intensity overlap to some extent. Nevertheless, this model is assumed that the image gray is separable in a relatively small area, and the offset is smooth in the entire image area. The model is easily trapped in a local minimum and involves high computational complexity [58,59]. It follows from Table 7 that the values of the *DICE* and *JSI* of the SPFLIF-IS method are the highest for the eight real image data, and the Mean is also the largest. Thus, the experimental results obtained for synthetic and real images further demonstrate the superior performance of our method. Therefore, our model is able to obtain better *DICE* and *JSI* values than those yielded by the methods compared.

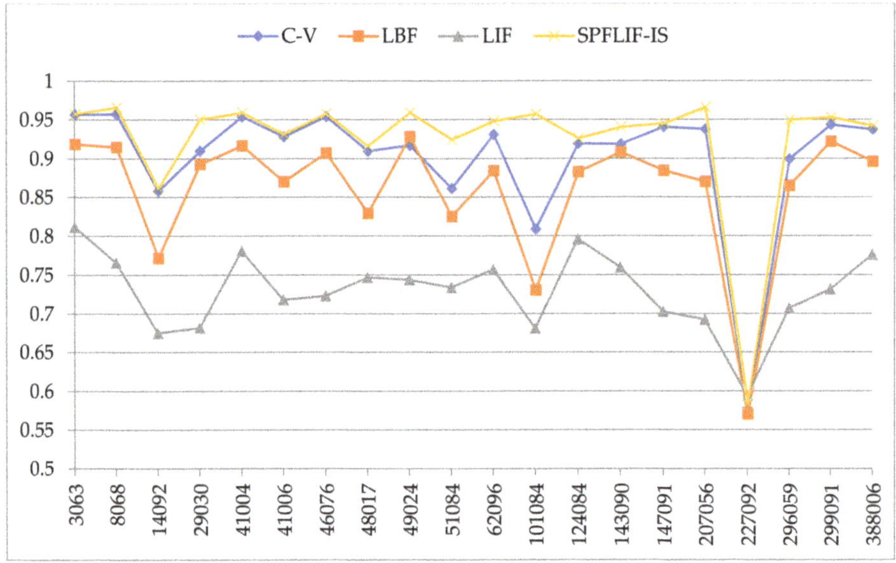

Figure 7. *JSI* values of the image segmentation results using the four models for Berkeley color images.

Table 7. *DICE* and *JSI* values of the results of image segmentation on fifteen Berkeley color images.

Image ID	C–V		LBF		SBGFRLS		LIF		LSACM		SPFLIF-IS	
	DICE	*JSI*	*DICE*	*JSI*	*DICE*	*JSI*	*DICE*	*JSI*	*DICE*	*JSI*	*DICE*	*JSI*
3063	0.9779	0.9568	0.9576	0.9186	0.9728	0.9470	0.8962	0.8119	0.9774	0.9557	0.9783	0.9575
8068	0.9780	0.9570	0.9555	0.9149	0.9785	0.9579	0.8673	0.7657	0.9710	0.9436	0.9827	0.9660
29030	0.9525	0.9093	0.9432	0.8925	0.9709	0.9435	0.8106	0.6815	0.9626	0.9280	0.9743	0.9500
41004	0.9763	0.9537	0.9565	0.9166	0.9775	0.9560	0.8769	0.7808	0.9741	0.9494	0.9791	0.9590
46076	0.9763	0.9537	0.9512	0.9070	0.9753	0.9518	0.8392	0.7230	0.9692	0.9402	0.9783	0.9575
207056	0.9677	0.9375	0.9305	0.8700	0.9774	0.9558	0.8183	0.6924	0.9785	0.9579	0.9826	0.9659
296059	0.9470	0.8994	0.9276	0.8650	0.9684	0.9387	0.8283	0.7069	0.9741	0.9495	0.9742	0.9498
299091	0.9708	0.9433	0.9595	0.9221	0.9750	0.9512	0.8448	0.7313	0.9694	0.9406	0.9759	0.9530
Mean	0.9683	0.9388	0.9477	0.9008	0.9745	0.9502	0.8477	0.7367	0.9720	0.9456	0.9782	0.9573

4.8. Discussion

According to the experimental results and evaluations presented above, the validity and stability of our proposed model are fully verified, and the contributions of the proposed model can be summarized as follows.

(1) The new model is regularized by a Gaussian kernel, which avoids the expensive computation associated with reinitialization. It follows that the model has low computational complexity.

(2) Our proposed model makes the best use of global and local image information. The model solves the problem of not accurately segmenting intensity-inhomogeneous images faced by the traditional image segmentation model and overcomes the shortcoming that local image information is sensitive to the initial contour and noise.

(3) Compared with the existing C–V, LBF, SBGFRLS, LIF, and LSACM models, the SPFLIF-IS model exhibits high robustness to the initial contour and noise and quickly and accurately segments inhomogeneous and multiobjective images.

5. Conclusions

In this paper, to segment intensity-inhomogeneous images quickly and accurately, an image segmentation method using a novel ACM based on an improved SPF function and an LIF model is proposed. The model combines the advantages of global and local information terms in segmenting intensity-inhomogeneous images. Moreover, a weight function is established to adjust the weights between the local information term and the global information term. Thus, a novel ACM model is presented, and an image segmentation algorithm is thereby established. To demonstrate the effectiveness of our proposed model, several experiments are designed in our study. The results indicate that our model not only segments inhomogeneous and multiobjective images effectively but also exhibits high robustness to the initial contour and noise. However, at present, it is difficult to determine a suitable Gaussian kernel size for all the images, and considering the uncertainty of real-world complex images, the proposed method will not be suitable in all the cases. As future work, we plan to accommodate the Gaussian kernel size automatically, which can be used to control region scalability from a small neighborhood to the entire image domain. This approach is considered to be more accurate and efficient in segmenting complex images and reducing computational complexity.

Author Contributions: L.S. and J.X. conceived the algorithm and designed the experiments; X.M. implemented the experiments; Y.T. analyzed the results; and X.M. drafted the manuscript. All authors read and revised the manuscript.

Funding: This research was funded by the National Natural Science Foundation of China (Grants 61772176, 61402153, 61370169, and 61472042), the China Postdoctoral Science Foundation (Grant 2016M602247), the Plan for Scientific Innovation Talent of Henan Province (Grant 184100510003), the Key Scientific and Technological Project of Henan Province (Grant 182102210362), the Young Scholar Program of Henan Province (Grant 2017GGJS041), the Key Scientific and Technological Project of Xinxiang City (Grant CXGG17002), and the Ph.D. Research Foundation of Henan Normal University (Grant qd15132).

Conflicts of Interest: The authors declare no conflicts of interest.

References

1. Mabood, L.; Ali, H.; Badshah, N.; Chen, K.; Khan, G.A. Active contours textural and inhomogeneous object extraction. *Pattern Recognit.* **2016**, *55*, 87–99. [CrossRef]
2. Zhao, L.K.; Zheng, S.Y.; Wei, H.T.; Gui, L. Adaptive active contour model driven by global and local intensity fitting energy for image segmentation. *Opt. Int. J. Light Electron Opt.* **2017**, *140*, 908–920. [CrossRef]
3. Kass, M.; Witki, A.; Terzopoulos, D. Snakes: Active contour models. *Int. J. Comput. Vis.* **1988**, *1*, 321–331. [CrossRef]
4. Hai, M.; Jia, W.; Wang, X.F.; Zhao, Y.; Hu, R.X.; Luo, Y.T.; Xue, F.; Lu, J.T. An intensity-texture model based level set method for image segmentation. *Pattern Recognit.* **2015**, *48*, 1547–1562.
5. Wang, X.F.; Huang, D.S.; Xu, H. An efficient local Chan-Vese model for image segmentation. *Pattern Recognit.* **2010**, *43*, 603–618. [CrossRef]
6. Gao, S.; Bui, T.D. Image segmentation and selective smoothing by using Mumford-Shah model. *IEEE Trans. Image Process.* **2005**, *14*, 1537–1549.
7. Hao, R.; Qiang, Y.; Yan, X.F. Juxta-Vascular pulmonary nodule segmentation in PET-CT imaging based on an LBF active contour model with information entropy and joint vector. *Comput. Math. Methods Med.* **2018**, *2018*, 2183847. [CrossRef] [PubMed]

8. Xie, X.H.; Mirmehdi, M. MAC: Magnetostatic active contour model. *IEEE Trans. Pattern Anal. Mach. Intell.* **2008**, *30*, 632–646. [CrossRef] [PubMed]

9. Ko, M.; Kim, S.; Kim, M.; Kim, K. A novel approach for outdoor fall detection using multidimensional features from a single camera. *Appl. Sci.* **2018**, *8*, 984. [CrossRef]

10. Jing, Y.; An, J.B.; Liu, Z.X. A novel edge detection algorithm based on global minimization active contour model for oil slick infrared aerial image. *IEEE Trans. Geosci. Remote Sens.* **2011**, *49*, 2005–2013. [CrossRef]

11. Chan, T.F.; Vese, L.A. Active contours without edges. *IEEE Trans. Image Process.* **2001**, *10*, 266–277. [CrossRef] [PubMed]

12. Li, C.M.; Kao, C.Y.; Gore, J.C.; Ding, Z.H. Minimization of region-scalable fitting energy for image segmentation. *IEEE Trans. Image Process.* **2008**, *17*, 1940–1949.

13. Han, B.; Wu, Y.Q. A novel active contour model based on modified symmetric cross entropy for remote sensing river image segmentation. *Pattern Recognit.* **2017**, *67*, 396–409. [CrossRef]

14. Li, C.M.; Huang, R.; Ding, Z.H.; Gatenby, J.C.; Metaxas, D.N.; Gore, J.C. A level set method for image segmentation in the presence of intensity inhomogeneities with application to MRI. *IEEE Trans. Image Process.* **2011**, *20*, 2007–2016. [PubMed]

15. Caselles, V.; Kimmel, R.; Sapiro, G. Geodesic active contours. *Int. J. Comput. Vis.* **1997**, *22*, 61–79. [CrossRef]

16. Song, Y.; Wu, Y.Q.; Dai, Y.M. A new active contour remote sensing river image segmentation algorithm inspired from the cross entropy. *Digit. Signal Process.* **2016**, *48*, 322–332. [CrossRef]

17. Cao, G.; Mao, Z.H.; Yang, X.; Xia, D.S. Optical aerial image partitioning using level sets based on modified Chan-Vese model. *Pattern Recognit. Lett.* **2008**, *29*, 457–464. [CrossRef]

18. Li, X.M.; Jiang, D.S.; Shi, Y.H.; Li, W.S. Segmentation of MR image using local and global region based geodesic model. *Biomed. Eng. Online* **2015**, *14*, 8. [CrossRef]

19. Liu, S.G.; Peng, Y.L. A local region-based Chan–Vese model for image segmentation. *Pattern Recognit.* **2012**, *45*, 2769–2779. [CrossRef]

20. Wang, L.; He, L.; Mishra, A.; Li, C.M. Active contours driven by local Gaussian distribution fitting energy. *Signal Process.* **2009**, *89*, 2435–2447. [CrossRef]

21. Zhang, L.; Peng, X.G.; Li, G.; Li, H.F. A novel active contour model for image segmentation using local and global region-based information. *Mach. Vis. Appl.* **2017**, *28*, 75–89. [CrossRef]

22. Soomro, S.; Akram, F.; Munir, A.; Lee, C.H.; Choi, K.N. Segmentation of left and right ventricles in cardiac MRI using active contours. *Comput. Math. Methods Med.* **2017**, *2017*, 1455006. [CrossRef] [PubMed]

23. Zhang, K.H.; Zhang, L.; Song, H.H.; Zhou, W.G. Active contours with selective local or global segmentation: A new formulation and level set method. *Image Vis. Comput.* **2010**, *28*, 668–676. [CrossRef]

24. Li, C.M.; Kao, C.Y.; Gor, J.C.; Ding, Z.H. Implicit active contours driven by local binary fitting energy. In Proceedings of the IEEE Conference on Computer Vision & Pattern Recognition, Minneapolis, MN, USA, 17–22 June 2007; pp. 1–7.

25. Yuan, J.J.; Wang, J.J. Active contours driven by local intensity and local gradient fitting energies. *Int. J. Pattern Recognit. Artif. Intell.* **2014**, *28*, 1455006. [CrossRef]

26. Tu, S.; Su, Y. Fast and accurate target detection based on multiscale saliency and active contour model for high-resolution SAR images. *IEEE Trans. Geosci. Remote Sens.* **2016**, *54*, 5729–5744. [CrossRef]

27. Wang, L.; Li, C.M.; Sun, Q.S.; Xia, D.S.; Kao, C.Y. Active contours driven by local and global intensity fitting energy with application to brain MR image segmentation. *Comput. Med Imaging Graph.* **2009**, *33*, 520–531. [CrossRef] [PubMed]

28. Zhang, K.H.; Zhang, L.; Lam, K.M.; Zhang, D. A level set approach to image segmentation with intensity inhomogeneity. *IEEE Trans. Cybern.* **2016**, *46*, 546–557. [CrossRef]

29. Jiang, X.L.; Li, B.L.; Wang, Q.; Chen, P. A novel active contour model driven by local and global intensity fitting energies. *Opt. Int. J. Light Electron Opt.* **2014**, *125*, 6445–6449. [CrossRef]

30. Zhao, Y.Q.; Wang, X.F.; Shih, F.Y.; Yu, G. A level-set method based on global and local regions for image segmentation. *Int. J. Pattern Recognit. Artif. Intell.* **2012**, *26*, 1255004. [CrossRef]

31. Zhang, K.H.; Song, H.H.; Zhang, L. Active contours driven by local image fitting energy. *Pattern Recognit.* **2010**, *43*, 1199–1206. [CrossRef]

32. Akram, F.; Garcia, M.A.; Puig, D. Active contours driven by local and global fitted image models for image segmentation robust to intensity inhomogeneity. *PLoS ONE* **2017**, *12*, e0174813. [CrossRef]

33. Zhu, H.Q.; Xie, Q.Y. A multiphase level set formulation for image segmentation using a MRF-based nonsymmetric Student's-t mixture model. *Signal Image Video Process.* **2018**, *18*, 1577–1585. [CrossRef]
34. Wang, X.F.; Min, H.; Zou, L.; Zhang, Y.G. A novel level set method for image segmentation by incorporating local statistical analysis and global similarity measurement. *Pattern Recognit.* **2015**, *48*, 189–204. [CrossRef]
35. Cao, J.F.; Wu, X.J. A novel level set method for image segmentation by combining local and global information. *J. Mod. Opt.* **2017**, *64*, 2399–2412. [CrossRef]
36. Tian, Y.; Duan, F.Q.; Zhou, M.Q.; Wu, Z.K. Active contour model combining region and edge information. *Mach. Vis. Appl.* **2013**, *24*, 47–61. [CrossRef]
37. Lok, K.H.; Shi, L.; Zhu, X.L.; Wang, D.F. Fast and robust brain tumor segmentation using level set method with multiple image information. *J. X-ray Sci. Technol.* **2017**, *25*, 301–312. [CrossRef]
38. Sun, Z.; Qi, M.; Lian, J.; Jia, W.K.; Zou, W.; He, Y.L.; Liu, H.; Zheng, Y.J. Image segmentation by searching for image feature density peaks. *Appl. Sci.* **2018**, *8*, 969. [CrossRef]
39. Xu, H.Y.; Jiang, G.Y.; Yu, M.; Luo, T. A global and local active contour model based on dual algorithm for image segmentation. *Comput. Math. Appl.* **2017**, *74*, 1471–1488. [CrossRef]
40. Abdelsamea, M.M. A semi-automated system based on level sets and invariant spatial interrelation shape features for Caenorhabditis elegans phenotypes. *J. Vis. Commun. Image Represent.* **2016**, *41*, 314–323. [CrossRef]
41. Ji, Z.X.; Xia, Y.; Sun, Q.S.; Gao, G.; Chen, Q. Active contours driven by local likelihood image fitting energy for image segmentation. *Inf. Sci.* **2015**, *301*, 285–304. [CrossRef]
42. Xu, C.Y.; Yezzi, A.; Prince, J.L. On the relationship between parametric and geometric active contours. In Proceedings of the IEEE Conference Record of the 34th Asilomar Conference on Signals, Systems and Computers, Pacific Grove, CA, USA, 29 October–1 November 2000; pp. 483–489.
43. Abdelsamea, M.M.; Tsaftaris, S.A. Active contour model driven by globally signed region pressure force. In Proceedings of the IEEE 18th International Conference on Digital Signal Processing, Santorini, Greece, 1–3 July 2013; pp. 1–6.
44. Li, D.Y.; Li, W.F.; Liao, Q.M. Active contours driven by local and global probability distributions. *J. Vis. Commun. Image Represent.* **2013**, *24*, 522–533. [CrossRef]
45. Hanbay, K.; Talu, M.F. A novel active contour model for medical images via the Hessian matrix and eigenvalues. *Comput. Math. Appl.* **2018**, *75*, 3081–3104. [CrossRef]
46. Vese, L.A.; Chan, T.F. A multiphase level set framework for image segmentation using the Mumford and Shah model. *Int. J. Comput. Vis.* **2002**, *50*, 271–293. [CrossRef]
47. Lei Zhang's Homepage. Available online: http://www4.comp.polyu.edu.hk/~{}cslzhang/ (accessed on 10 December 2018).
48. Chunming Li's Homepage. Available online: http://www.engr.uconn.edu/~{}cmli/ (accessed on 10 December 2018).
49. Li, M.; He, C.J.; Zhan, Y. Adaptive regularized level set method for weak boundary object segmentation. *Math. Probl. Eng.* **2012**, *2012*, 369472. [CrossRef]
50. The Berkeley Segmentation Dataset and Benchmark. Available online: https://www2.eecs.berkeley.edu/Research/Projects/CS/vision/bsds/ (accessed on 10 December 2018).
51. Jiang, X.L.; Wang, Q.; He, B.; Chen, S.J.; Li, B.L. Robust level set image segmentation algorithm using local correntropy-based fuzzy c-means clustering with spatial constraints. *Neurocomputing* **2016**, *27*, 22–35. [CrossRef]
52. Dice, L.R. Measures of the amount of ecologic association between species. *Ecology* **1945**, *26*, 297–302. [CrossRef]
53. Jaccard, P. The distribution of the flora in the alpine zone. *New Phytol.* **1912**, *11*, 37–50. [CrossRef]
54. Wang, L.; Chang, Y.; Wang, H.; Wang, Z.Z.; Pu, J.T.; Yang, X.D. An active contour model based on local fitted images for image segmentation. *Inf. Sci.* **2017**, *418*, 61–73. [CrossRef]
55. Arbelaez, P.; Maire, M.; Fowlkes, C.; Malik, J. Contour detection and hierarchical image segmentation. *IEEE Trans. Pattern Anal. Mach. Intell.* **2011**, *33*, 898–916. [CrossRef]
56. Zhao, W.; Fu, Y.; Wei, X.; Wang, H. An improved image semantic segmentation method based on superpixels and conditional random fields. *Appl. Sci.* **2018**, *8*, 837. [CrossRef]
57. Zhou, S.P.; Wang, J.J.; Zhang, M.M.; Cai, Q.; Gong, Y.H. Correntropy-based level set method for medical image segmentation and bias correction. *Neurocomputing* **2017**, *234*, 216–229. [CrossRef]

58. Zhang, Y.C. Research of level set image segmentation based on Rough Set theory and the extended watershed transformation. Ph.D. Thesis, Dalian University of Technology, Dalian, China, 2018; pp. 4–6.
59. Sun, L.; Meng, X.C.; Xu, J.C.; Zhang, S.G. An image segmentation method based on improved regularized level set model. *Appl. Sci.* **2018**, *8*, 2393. [CrossRef]

Article

Image Segmentation Approaches for Weld Pool Monitoring during Robotic Arc Welding

Zhenzhou Wang *, Cunshan Zhang, Zhen Pan, Zihao Wang, Lina Liu, Xiaomei Qi, Shuai Mao and Jinfeng Pan

College of Electrical and Electronic Engineering, Shandong University of Technology, China; zcs@sdut.edu.cn (C.Z.); pz15615732790@163.com (Z.P.); 13110473098@stumail.sdut.edu.cn (Z.W.); linaliu-126@163.com (L.L.); qixiaomei@sdut.edu.cn (X.Q.); maoshuai04965@sdut.edu.cn (S.M.); pjfbysj@163.com (J.P.)
* Correspondence: wangzz@sdut.edu.cn

Received: 3 November 2018; Accepted: 29 November 2018; Published: 1 December 2018

Abstract: There is a strong correlation between the geometry of the weld pool surface and the degree of penetration in arc welding. To measure the geometry of the weld pool surface robustly, many structured light laser line based monitoring systems have been proposed in recent years. The geometry of the specular weld pool could be computed from the reflected laser lines based on different principles. The prerequisite of accurate computation of the weld pool surface is to segment the reflected laser lines robustly and efficiently. To find the most effective segmentation solutions for the images captured with different welding parameters, different image processing algorithms are combined to form eight approaches and these approaches are compared both qualitatively and quantitatively in this paper. In particular, the gradient detection filter, the difference method and the GLCM (grey level co-occurrence matrix) are used to remove the uneven background. The spline fitting enhancement method is used to remove the fuzziness. The slope difference distribution-based threshold selection method is used to segment the laser lines from the background. Both qualitative and quantitative experiments are conducted to evaluate the accuracy and the efficiency of the proposed approaches extensively.

Keywords: Image processing; segmentation; spline; grey level co-occurrence matrix; gradient detection; threshold selection

---

1. Introduction

ARC welding is a widely used process for joining various metals. The electric current is transferred from the electrode to the work piece through the arc plasma, which has been modeled to control the weld quality in the past research [1,2]. However, the direct factor that affects the weld quality is the geometry of the weld pool instead of arc plasma because the skilled welders achieve good weld quality mainly based on the visual information of the weld pool. During arc welding, the incomplete weld pool penetration will reduce the effective working cross-sectional area of the weld bead and subsequently reduce the weld joint strength. It also causes stress concentrations in some cases, e.g., the fillet and T-joints. On the contrary, excessive weld pool penetration might cause melt-through. The skilled welder needs to adjust the position and travelling speed of the weld torch based on the information of the observed weld pool surface to achieve complete penetration. The shortage of skilled welders and a need for welds of a consistently high quality fuels an increasing demand for automated arc welding systems. It is believed that machine vision techniques will lead the development of the next generation intelligent automated arc welding systems.

In recent years, great efforts have been put to develop the automated and high-precision welding equipment to achieve high quality welded joints consistently. The important parts of this kind of automated welding equipment include the seam tracking system [3–6], the weld penetration monitoring system [7–9] and the control system. Desirably, the seam tracking system or the monitoring system and control system are in a closed loop. Thus, the tracking result or the monitoring result could serve as feedback for the control system to ensure high quality of welding. In the past studies, it has been shown that weld pool surface depression has a major effect on weld penetration [10–14]. The geometry of the weld pool surface will affect the convection in the pool. The primary welding parameter, the welding current density is also affected by the geometry of the weld pool greatly. In return, the plasma and the welding current affect the geometry of the weld pool. Both the arc plasma and the molten weld pool are affected by the current density distribution. Therefore, it is fundamental for the automated welding systems to take quantitative measurements of the weld pool surface.

In the past decades, a lot of research has been conducted to measure the shape of the arc welding weld pool by structured light methods [9,15–23]. In Reference [15], the authors pioneered to measure the deformation of the weld pool by inventing a novel sensing system. Their sensing system projects a short duration pulsed laser light through a frosted glass with a grid onto the specular weld pool. The reflected laser stripes are imaged in a CCD camera and the geometry of the reflected stripes contain the weld pool surface information. This method might be the earliest method that made good use of the reflective property of the weld pool surface and achieved state of the art accuracy at that time. In Reference [16], the authors improved the structured light method for gas tungsten arc weld (GTAW) pool shape measurement by reflecting the structured light onto an imaging plane instead of onto the image plane of the camera directly. The imaging plane is placed at a properly selected distance. Thus, the imaged laser patterns are clear enough while the effect of the arc plasma has attenuated significantly, because the propagation of the laser light is much longer than that of the arc plasma. Ever since then, this weld pool sensing technique has become the mainstream of weld pool imaging technology [17–23]. In References [17,18], the authors tried to increase the accuracy of measuring the 3D shape of the GTAW weld pool sensed by the same imaging system as [16]. In Reference [19], the authors came up with an approach for segmentation of the reflected laser lines for pulsed gas metal arc weld (GMAW-P). In References [9,20], the reflected laser lines from the GMAW-P weld pool were segmented manually to measure the weld pool oscillation frequency. In Reference [21], two cameras were used to measure the shape of the weld pool for GMAW-P and an unsupervised approach was proposed to segment and cluster the reflected laser lines. In Reference [22], three cameras are used to measure the specular shape from the projected laser rays and an unsupervised approach was proposed to reconstruct the GTAW weld pool shape with closed form solutions. It achieved on line robust measurement of GTAW weld pool shape with three calibrated cameras.

There is one major difference for the used laser pattern among these structured light methods [9,15–23]. The laser dot pattern is used for the measurement of the GTAW weld pool surface while the laser line pattern is used for the measurement of the GMAW weld pool surface. Compared to the GTAW weld pool surface, the GMAW weld pool surface is much more dynamic and fluctuating, because GMAW process transfers additional metallic and liquid droplets into the weld pool, which increases the fluctuation and dynamics of the pool surface greatly. The position and geometry of the local specular surface changes rapidly and greatly, which causes the reflected rays to change their trajectories rapidly and greatly. If the laser dot pattern is used, the reflected laser dots might interlace irregularly, which makes it impossible for the unsupervised clustering method to identify these dots robustly. Therefore, the laser line patterns are used in measuring the weld pool surface of the GMAW process [19–21]. Robust segmentation and clustering of the reflected laser lines become the most important and challenging part in the whole monitoring system, because of the uncertainty of the weld pool geometry. The quality of the captured image is significantly affected by the welding parameters. Due to the lack of generality and robustness, the proposed image processing methods might work for the images captured in some specifically designed welding experiments, while might not work for the images captured in other experiments. For instance, the reflected laser lines were filtered by the top-hat transform and then segmented by thresholding in Reference [19].

It yields acceptable segmentation results in many cases. However, it fails completely when it is used to segment the images captured in Reference [21] with different welding parameters. The reflected laser lines were filtered by the fast Fourier transform (FFT) and then segmented by a manually specified threshold in Reference [20]. It also fails completely in segmenting the images shown in Reference [21]. To segment the reflected laser lines more robustly, a difference method and an effective threshold selection method are proposed in Reference [21]. The segmented laser lines were then clustered based on their slopes. One big problem that could be seen from the experimental figures in References [19–21] is that quite a few of the laser line parts are missing, because of the limitations of their proposed image processing methods. One direct consequence of missing a significant part of the reflected laser line is the inaccurate characterization of the weld pool shape, since the length of the reflected laser is proportional to the size of the weld pool. Another drawback is that the segmented small part of the laser line might be deleted as noise blobs or some large noise blobs might be recognized as part of the reflected laser line during the unsupervised clustering. As a result, the laser line might be clustered incorrectly.

In Reference [23], a new approach was proposed, and it achieved significantly better segmentation accuracy compared to the past research [19–21]. It comprises several novel image processing algorithms: A difference operation, a two-dimensional spline fitting enhancement operation, a gradient feature detection filter and the slope difference distribution-based threshold selection. The major goal of [23] is to cluster and characterize the laser lines under extremely harsh welding conditions. As a result, it omits the image processing methods for the images captured under less harsh welding conditions with mild welding parameters. In addition, quantitative results and comparisons to validate the effectiveness of the proposed segmentation approach were also not given. Due to the page limit, the reasons why the segmentation approach should contain these image processing algorithms were not explained adequately. One goal of this paper is to complement the research work conducted in Reference [23] and conduct a more thorough experiment to find the most effective solution both qualitatively and quantitatively. Another goal is to come up with more efficient segmentation approaches for images with relatively high quality that are captured under mild welding parameters.

In the past research, the visual inspection of the weld pool surface was mainly used to understand the complex arc welding processes. The obtained data were used to validate and improve the accuracy of the numerical models and to gain insight into the complex arc welding processes. Few are used for in-process welding parameter adjustment and on-line feedback control, due to the lack of a robust and efficient approach that is capable of extracting meaningful feedback information on-line from most of the captured images. The proposed approaches in this paper are promising to accomplish this challenging task in the future.

This paper is organized as follows. Section 2 describes the established monitoring system. In Section 3, state of the art methods for laser line segmentation are evaluated and compared. In Section 4, the combination approach proposed previously is explained theoretically. In Section 5, we propose different segmentation approaches by combining different image processing algorithms. In Section 6, the experimental results and discussions are given. Section 7 concludes the paper.

2. The Structured Light Monitoring System

Figure 1 shows the configuration of the popular weld pool monitoring system that has been adopted in References [9,15–23]. The major parts of the system include two point grey cameras, $C1$ and $C2$, and one Lasiris SNF with the wavelength at 635 nm. A linear glass polarizing filter with the wavelength from 400 nm to 700 nm is mounted on camera $C2$ to remove the strong arc light. The structured light laser line pattern is projected by the Lasiris SNF laser generator onto the weld pool surface and reflected onto the diffusive imaging plane $P1$. The calibrated camera $C2$ views the reflected laser lines from the back side of $P1$, which is made up of a piece of glass and a piece of high-quality paper. To facilitate the computation, $P1$, the YZ plane and the image plane of the calibrated camera $C2$ are set to be parallel to each other. During calibration, the laser lines are projected by the Lasiris SNF laser onto a horizontal diffusive plane and camera C1 is used to calculate the length of the straight laser lines in the world coordinate system. Then the horizontal diffusive plane is replaced by a horizontal mirror plane, which reflects the laser lines onto the vertical diffusive plane *P*1. The calibrated camera C2 is used to calculate the length of the imaged straight laser lines in the world coordinate system. Both camera C1 and camera C2 are not attached to the welding platform to avoid the vibration generated during the welding process.

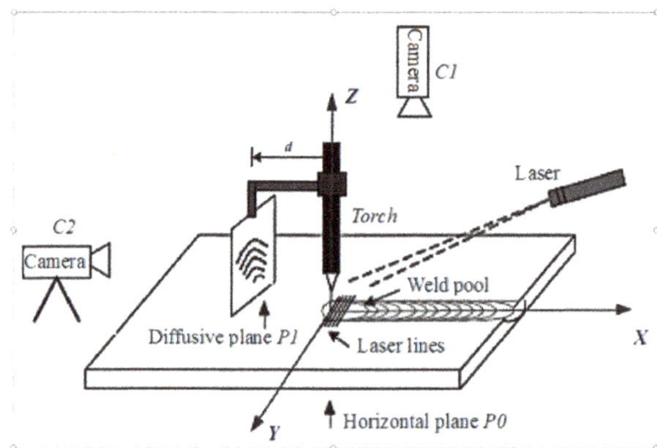

Figure 1. The configuration of the weld pool monitoring system.

3. State of the Art Methods for Laser Line Segmentation

3.1. Top-Hat Transform Based Method

In Reference [19], the uneven background caused by the arc light was removed by the top-hat transform, which is formulated as [24]:

$$T(f) = f - f \cdot e, \tag{1}$$

where · denotes the opening operation, f denotes the image and e denotes the structuring element. e is chosen as a disk with radius 15 in this research work. After the image was enhanced by the top-hat transform, the image was segmented by the threshold selection method. The selected welding parameters are as follows. The wire feed speed is 55 mm/s, the welding speed is 5 mm/s, the peak current is 220 A and the base current is 50 A. As a result, the captured image is very clear for segmentation. However, their segmentation result still misses significant parts for the laser lines.

3.2. FFT Filtering Based Method

In Reference [20], fast Fourier transform [25] was used to remove the uneven arc light caused background and then the image was segmented by threshold selection. The selected welding parameters are as follows. The welding speed is 0 mm/s, the peak current is 160 A and the base current is 80 A. With these welding parameters, the change of the weld pool's geometry is relatively slow compared to that in Reference [19]. As a result, the captured images are relatively easier for automatic image processing. Hence, the authors could segment the captured images by specifying the threshold manually after FFT filtering.

3.3. Difference Operation Based Method

In Reference [21], the difference method was proposed to reduce the unevenly distributed background. The differenced image is obtained by the following operation.

$$f_d(x,y) = f(x + \Delta d, y) - f(x,y),$$ (2)

where Δd denotes the step size of the operation and it is determined by off line analysis of the average width of the laser lines. The step size should be greater than or equal to the width of the laser line in the captured images and it is selected as 10 in this research work. The laser lines are then segmented from the differenced image by the threshold selection method.

3.4. Grey Level Co-Occurrence Matrix Based Method

Grey level co-occurrence matrix (GLCM) [26] computes the frequencies of different combinations of pixel values or grey levels occurring in an image and then forms a matrix to represent these frequencies. Therefore, it is usually used to segment textured images or objects with an unevenly distributed background. The second order GLCM is usually used for segmentation and it is formulated as:

$$P(i,j|d,\theta) = \frac{\#\{k,l \in D | f(k) = i, f(l) = j, \| k - l \| = d, \angle(k - l) = \theta\}}{\#\{m,n \in D | \| m - n \| = d, \angle(m - n) = \theta\}},$$ (3)

where d is the distance of pixel k and pixel l. θ is the angle of vector $(k - l)$ with the horizontal line or vertical line. The combination of d and θ represents the relative position of pixel k and pixel l, which have gray-scale value i and j respectively.

During our implementation, we update all the intensity values in the moving window by subtracting its minimal intensity value and then adding one. The quantization level of the GLCM matrix becomes one and the maximal intensity value of the updated moving window. Following the computation of the GLCM, the contrast measure is used to form the GLCM image. Then, the GLCM image is segmented by a global threshold.

3.5. Combination Method

In Reference [23], a combination approach was used to segment the laser lines. The combined image processing algorithms include a difference operation, a two-dimensional spline fitting enhancement operation, a gradient feature detection filter and the slope difference distribution-based threshold selection. In both [21,23], the laser lines are segmented by the slope difference distribution-based threshold selection method that could be summarized as follows. Firstly, the gray-scales of the original image is rearranged in the interval from 1 to 255 and its normalized histogram distribution $P(x)$ is computed. Secondly, the normalized histogram $P(x)$ is smoothed by the fast Fourier transform (FFT) [25] based low pass filter with the bandwidth $W = 10$. Thirdly, two slopes, the right slope and the left slope, are computed for each point on the smoothed histogram distribution. The slope difference distribution is computed as the differences between the right slopes and their corresponding left slopes. In the slope difference distribution, the position where the valley with the maximum absolute value occurs is selected as the threshold. The slope difference distribution-based threshold selection method is critical in this application. For the comparison with state of the art image segmentation methods, please refer to the related research work [21], where this threshold selection method is compared with the state of the art method using the same type of images. The comparisons with state of the art image segmentation methods showed that the slope difference distribution based threshold selection method is significantly more accurate in segmenting some specific types of images, including the laser line images.

3.6. Performance Evaluation

Figure 2a shows a typical image captured by the structured light monitoring system with the following welding parameters. The wire feed speed is 84.67 mm/s, the welding speed is 3.33 mm/s, the peak current is 230 A and the base current is 70 A. As can be seen, these parameters are significantly higher than those used in References [19,20]. With these parameters, the resultant weld pool surface changes more rapidly and irregularly. Hence, the quality of the captured image in this study is much reduced. The results by top hat method, FFT method, difference method, GLCM method and combination method are shown in Figure 2b–f respectively. Compared to the segmentation result by combination method, the segmentation results by other state of the art methods appear to be very inaccurate. Although the difference method achieves the second best result, there is one big segmented line on the top caused by the edge of the imaging plane. It costs additional effort for the subsequent unsupervised clustering. In addition, there are noise blobs that are hard to be distinguished from the laser line parts.

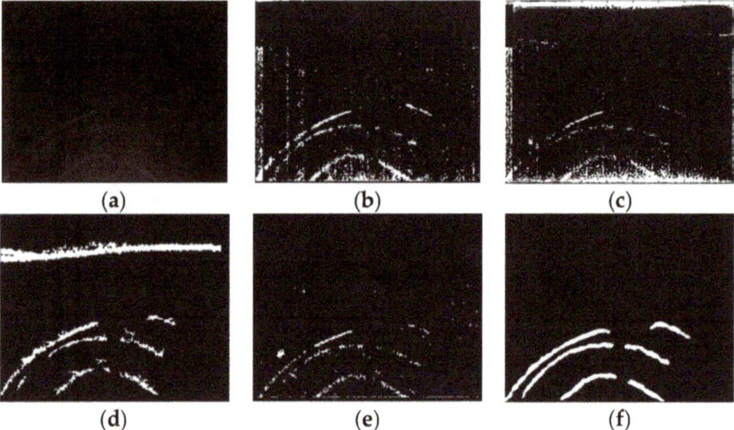

Figure 2. Evaluation of state of the art methods (**a**) one typical captured image; (**b**) segmentation result by the top hat method; (**c**) segmentation result by the FFT method; (**d**) segmentation result by the difference method; (**e**) segmentation result by the GLCM method; (**f**) segmentation result by combination method.

4. Analysis of the Combination Approach

In Reference [23], a combination approach is proposed to segment the reflected laser lines as accurate as possible. The flowchart of this segmentation approach comprises a difference operation, a two-dimensional spline fitting enhancement operation, a gradient feature detection filter and the slope difference distribution-based threshold selection. However, the reasons why the segmentation approach should contain these image processing algorithms were not explained adequately in Reference [23]. Quantitative results to validate the effectiveness of this segmentation approach were also not given. Here, we will theoretically explain why the proposed two-dimensional spline fitting enhancement method and the gradient feature detection filter work well in segmenting the laser lines.

The intensity distribution, I_a of the captured image caused by the arc light could be modeled mathematically by the following equation [21]:

$$I_a(T) = \tau \left| \frac{2hv^3}{c^2} \times \frac{\vec{N} \times cos\beta \times O}{e^{\frac{hv}{\gamma T}} - 1} \right| \times \frac{d}{r^3} = \frac{C_a(T)}{r^3}, \tag{4}$$

where τ is the intensity mapping function of the CCD camera [27]. $\vec{N}$ is the surface normal at position p. β is the angle between the surface normal and the incident light. O denotes the color value. r is the distance between the arc light center and position p. d is the distance from the arc light center to the diffusive imaging plane $P1$. v is the spectral frequency of arc light and h is the planck's constant. c is the speed of the light and γ is the boltzmann's constant. T is the temperature of arc light source, which is determined by the welding current that alternates with the frequency of 10 Hz. The frame rate of the camera C2 is set to 300 frames per second during the experiment. Hence, 30 images are captured with different sampled currents at one period of the current wave. The temperature T is determined by the value of the current and thus it changes with the current from frame to frame in the time domain. In the same frame, the temperature T is a constant. At each image point, both the intensity mapping function τ and the angle α remain the same in all the captured image sequences. Thus, C^a is a constant at each specific image point in one frame and it varies from frame to frame with the value of T. From Equation (4), it becomes much obvious that the intensity distribution produced by the arc light effect on the captured image is inversely proportional to the r^3. Thus, the intensity distribution caused by the arc light can be modeled as:

$$a(x,y) = \frac{C_a(T)}{\left[(x-x_0)^2 + (y-y_0)^2 + d^2\right]^{3/2}},$$ (5)

where (x_0, y_0) denotes the center of arc light distribution and it may lie outside of the image. The arc light is affected by the additional laser line and can be modeled as:

$$f(x,y) = l(x,y) + a(x,y),$$ (6)

where $l(x,y)$ denotes the intensity distribution of the laser line and it is formulated as:

$$l(x,y) = \begin{cases} w\mu(x,y); & (x,y) \in A \\ 0; & (x,y) \notin A \end{cases},$$ (7)

where A denotes the laser line area, w is a constant whose value is higher than the average value of the arc light distribution, I_a. $\mu(x,y)$ is the membership function that represents the fuzziness of the laser line and is formulated as:

$$\mu(x,y) = Exp\left(-\frac{(L(x,y) - \mu_L)^2}{2\sigma_L^2}\right),$$ (8)

where $L(x,y)$ denotes the ideal intensity distribution of the laser line. μ_L is its mean and σ_L is its variance respectively.

Combining Equations (5)–(8), we get the model of the intensity distribution for the image with reflected laser lines.

$$f(x,y) = \begin{cases} w\mu(x,y) + \frac{C_a(T)}{\left[(x-x_0)^2+(y-y_0)^2+d^2\right]^{3/2}}; & (x,y) \in A \\ \frac{C_a(T)}{\left[(x-x_0)^2+(y-y_0)^2+d^2\right]^{3/2}}; & (x,y) \notin A \end{cases}.$$ (9)

From the above model, we see that the fuzziness caused by $\mu(x,y)$ should be reduced effectively to obtain high segmentation accuracy. However, the moving average filter can only reduce the Gaussian noise instead of removing the fuzziness. Thus, a new enhancement method is required.

After the image fuzziness has been added in the derived image model (Equation (9)), we need to theoretically find the reasons that why the fuzziness causes parts of the segmented laser lines missing. To this end, the mathematical explanation of how the gradient detection filter works for segmenting the objects from the uneven background is described at first. According to Equation (9), the gradient value at the position (x, y) that is caused by the arc light background can be formulated as:

$$g^a = \frac{C_a(T)}{\left[\left(\frac{N-1}{2}+x-x_0\right)^2+\left(\frac{N-1}{2}+y-y_0\right)^2+d^2\right]^{3/2}} - \frac{C_a(T)}{\left[\left(x-x_0-\frac{N-1}{2}\right)^2+\left(y-y_0-\frac{N-1}{2}\right)^2+d^2\right]^{\frac{3}{2}}}. \tag{10}$$

The gradient values caused by the laser line can be formulated as:

$$g^l = \begin{cases} w\mu(x,y) + g^a; & \textit{if top part of } K_g \notin A \\ -w\mu(x,y) + g^a & \textit{if bottompart of } K_g \notin A \\ g^a; & \textit{if whole part of } K_g \in A \end{cases}. \tag{11}$$

For the designed filter $K_g(N, \theta)$ in this research work, the following conditions need to be met.

$$w\mu(x,y) \gg g^a. \tag{12}$$

Since g^a is the gradient of a small part of the background with a range of N, its value is much reduced compared to the variation of the total background. Thus, the condition of Equation (12) can be easily satisfied if μ is a constant to make the laser line gradient distinguished from the background variation. However, μ is the membership function and its value is random between 0 and 1. It is impossible to make Equation (12) true for every point. Hence, if the local fuzzy membership $\mu(x, y)$ could be fitted as a global function for the whole image to remove the randomness, Equation (12) could be easily satisfied. A two-dimensional spline function is an ideal global function to the image and thus it is used. The fitting process is implemented by minimizing the following energy function to get an enhanced image, $s(x, y)$ from the original image $f(x, y)$.

$$E = \frac{1}{2} \iint (s(x,y) - f(x,y))^2 dxdy + \frac{1}{2} \iint \left| \frac{d^2 s(x,y)}{dxdy} \right|^2 dxdy. \tag{13}$$

5. The Proposed Approaches

Although the same weld pool monitoring system has been adopted in References [9,15–23], the monitoring algorithms have been proposed differently and divergently. In addition, most proposed monitoring approaches only work for the specifically designed welding scenario. As described in Section 3, either the proposed monitoring approach in Reference [19] or the proposed monitoring approach in Reference [20] could not work for the weld pool scenario in Reference [23]. On the contrary, the proposed monitoring approach in Reference [23] works better for the weld pool scenario in both [19,20] than their proposed monitoring approaches. The reason lies in that the quality of the captured laser line images in References [19,20] is much higher than that of the captured images in Reference [23]. When the quality of the captured image is reduced greatly, the requirement for the monitoring algorithms increases significantly. The proposed monitoring approach is the most robust one for the structured laser line based weld pool monitoring systems shown in Figure 1 up to date. However, it might be redundant for monitoring of simple weld pool scenarios as described in References [19,20]. In Reference [19], GTAW process is used. The wire feed speed is 55 mm/s, the welding speed is 5 mm/s, the peak current is 220A and the base current is 50 A. In Reference [20], GMAW-P process is used. The welding speed is 0 mm/s, the peak current is 160 A and the base current is 80 A. In Reference [23], GMAW-P process is used. The wire feed speed is 84.67 mm/s, the welding speed is 3.33 mm/s, the peak current is 230A and the base current is 70 A. As a result, the produced weld pools in References [19,20] are much more stable than that produced in Reference [23]. Thus, the quality of the captured images in References [19,20] is much better than that of the images captured in Reference [23]. On the other hand, the processing time is also important for on line monitoring. Therefore, the combination approach proposed in Reference [23] is not always optimum when both accuracy and efficiency are considered. The monitoring approach should be designed as fast as possible after the accuracy has been met.

In this paper, we combine the previously proposed image processing algorithms in different groups and form different segmentation approaches. We then evaluate their segmentation accuracy and segmentation efficiency quantitatively. Besides the five image processing algorithms proposed recently, we also add the traditional GLCM as an additional component. We name the segmentation by combining the difference method and threshold selection as *approach 1*, the segmentation by combining GLCM and threshold selection as *approach 2*, the segmentation by combining GLCM, the difference method and threshold selection as *approach 3*, the segmentation by combining the gradient detection filter and threshold selection as *approach 4*, the segmentation by combining the gradient detection filter, the difference method and threshold selection as *approach 5*, the segmentation by combining the gradient detection filter, spline fitting and threshold selection as *approach 6*, the segmentation by combining GLCM, spline fitting, the difference method and threshold selection as *approach 7*, the segmentation by combining the gradient detection filter, spline fitting, the difference method and threshold selection as *approach 8*.

As can be seen, there are five basic image processing methods that consist of (1), the difference method; (2), the GLCM; (3) the spline fitting; (4) the gradient detection filter; and (5), the threshold selection method. The difference method, the GLCM and the spline fitting have been explained by Equations (2), (3) and (13) respectively.

The gradient feature detection filter is formulated as:

$$K_g = R(VH, \theta), \tag{14}$$

where

$$V = [-k; v_1; v_2; \ldots; v_{N-1}; k], \tag{15}$$

$$H = [h_0, h_1, \ldots, h_{N-1}, h_N], \tag{16}$$

$$h_i = w_h(i); i = 0, \ldots, N, \tag{17}$$

$$v_i = w_v(i); i = 1, \ldots, N-1, \tag{18}$$

where N equals the width of the laser line and determines the size of the kernel. k is a constant. w_h and w_v are two weighting functions. As can be seen, the product VH is a N by N matrix. $R(VH, \theta)$ is to rotate the matrix VH by θ degrees in the counterclockwise direction around its center point. θ is orthogonal to the line direction and is chosen as $90°$ in this research.

The threshold selection method is implemented as follows. The histogram distribution of the image is computed, normalized and filtered by a low pass Discrete Fourier Transform (DFT) filter with the bandwidth 10. A line model is fitted with N adjacent points at each side of the sampled point. The slopes of the fitted lines at point i, $a_1(i)$ and $a_2(i)$, are then obtained. The slope difference, $s(i)$, at point i is computed as:

$$s(i) = a_2(i) - a_1(i); i = 16, \ldots, 240, \tag{19}$$

The continuous function of the above discrete function, $s(i)$ is the slope difference distribution, $s(x)$. To find the candidate threshold points, the derivative of $s(x)$ is set to zero.

$$\frac{ds(x)}{dx} = 0, \tag{20}$$

Solving the above equation, the valleys $V_i; i = 1, \ldots, N_v$ of the slope difference distribution are obtained. The position where the valley V_i yields the maximum absolute value is chosen as the optimum threshold in this specific application.

6. Results and Discussion

6.1. Experimental Results

To rank the accuracy and the efficiency the proposed approaches with different algorithm combinations, we show the qualitative results in Figures 3–12 for visual comparison. As can be seen, the performances of the proposed approaches vary depending on the quality of the captured images. For some images (e.g., Figures 11 and 12), most approaches work well. Considering the computation time, *approach 8* is not always the optimum solution in monitoring different welding processes with different welding parameters.

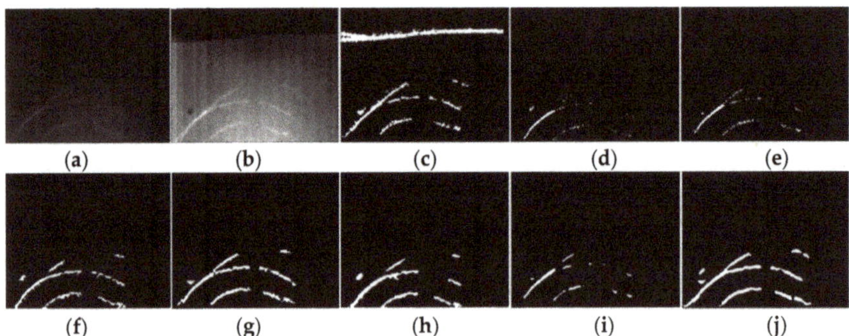

Figure 3. Performance comparison of the eight methods with image 1; (**a**) original image; (**b**) filtered image; (**c**) *approach 1*; (**d**) *approach 2*; (**e**) *approach 3*; (**f**) *approach 4*; (**g**) *approach 5*; (**h**) *approach 6*; (**i**) *approach 7*; (**j**) *approach 8*.

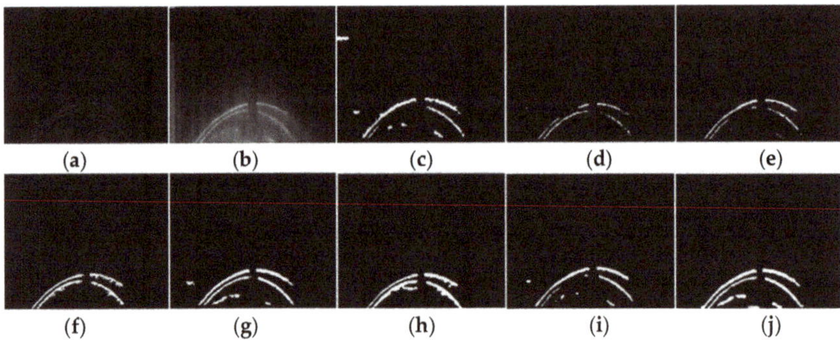

Figure 4. Performance comparison of the eight methods with image 2; (**a**) original image; (**b**) filtered image; (**c**) *approach 1*; (**d**) *approach 2*; (**e**) *approach 3*; (**f**) *approach 4*; (**g**) *approach 5*; (**h**) *approach 6*; (**i**) *approach 7*; (**j**) *approach 8*.

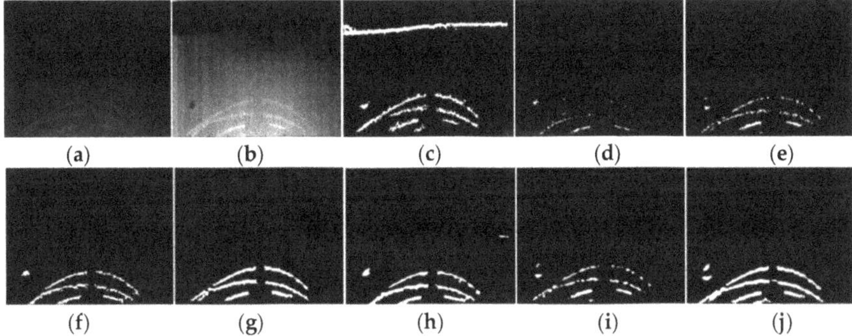

Figure 5. Performance comparison of the eight methods with image 3; (**a**) original image; (**b**) filtered image; (**c**) *approach 1*; (**d**) *approach 2*; (**e**) *approach 3*; (**f**) *approach 4*; (**g**) *approach 5*; (**h**) *approach 6*; (**i**) *approach 7*; (**j**) *approach 8*.

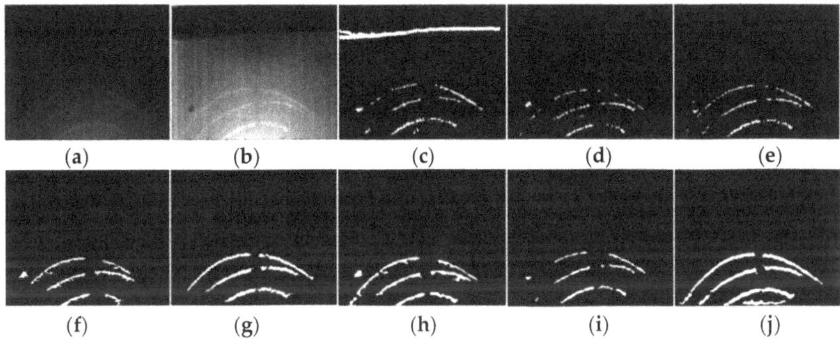

Figure 6. Performance comparison of the eight methods with image 4; (**a**) original image; (**b**) filtered image; (**c**) *approach 1*; (**d**) *approach 2*; (**e**) *approach 3*; (**f**) *approach 4*; (**g**) *approach 5*; (**h**) *approach 6*; (**i**) *approach 7*; (**j**) *approach 8*.

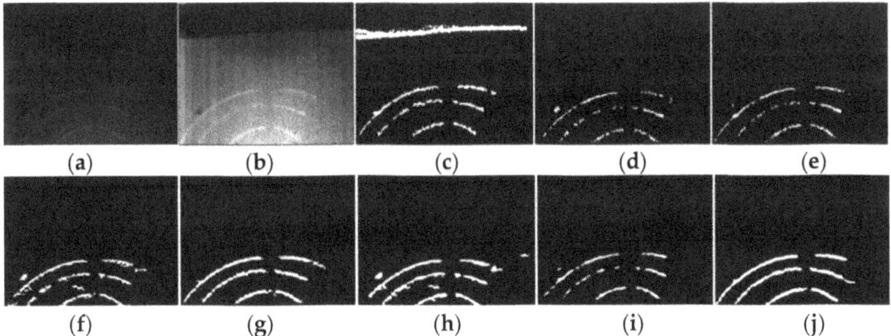

Figure 7. Performance comparison of the eight methods with image 5; (**a**) original image; (**b**) filtered image; (**c**) *approach 1*; (**d**) *approach 2*; (**e**) *approach 3*; (**f**) *approach 4*; (**g**) *approach 5*; (**h**) *approach 6*; (**i**) *approach 7*; (**j**) *approach 8*.

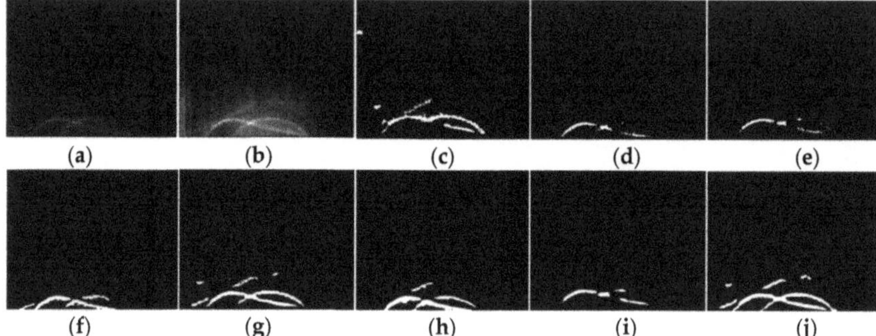

Figure 8. Performance comparison of the eight methods with image 6; (**a**) original image; (**b**) filtered image; (**c**) *approach 1*; (**d**) *approach 2*; (**e**) *approach 3*; (**f**) *approach 4*; (**g**) *approach 5*; (**h**) *approach 6*; (**i**) *approach 7*; (**j**) *approach 8*.

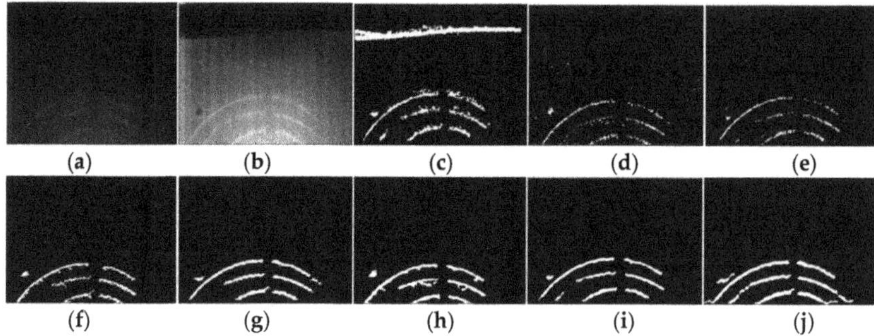

Figure 9. Performance comparison of the eight methods with image 7; (**a**) original image; (**b**) filtered image; (**c**) *approach 1*; (**d**) *approach 2*; (**e**) *approach 3*; (**f**) *approach 4*; (**g**) *approach 5*; (**h**) *approach 6*; (**i**) *approach 7*; (**j**) *approach 8*.

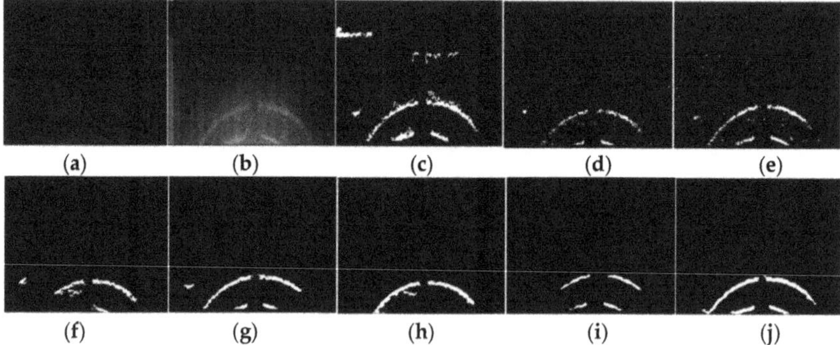

Figure 10. Performance comparison of the eight methods with image 8; (**a**) original image; (**b**) filtered image; (**c**) *approach 1*; (**d**) *approach 2*; (**e**) *approach 3*; (**f**) *approach 4*; (**g**) *approach 5*; (**h**) *approach 6*; (**i**) *approach 7*; (**j**) *approach 8*.

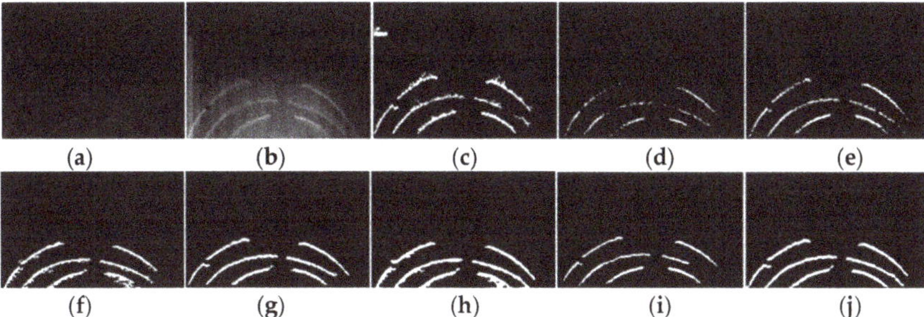

Figure 11. Performance comparison of the eight methods with image 9; (**a**) original image; (**b**) filtered image; (**c**) *approach 1*; (**d**) *approach 2*; (**e**) *approach 3*; (**f**) *approach 4*; (**g**) *approach 5*; (**h**) *approach 6*; (**i**) *approach 7*; (**j**) *approach 8*.

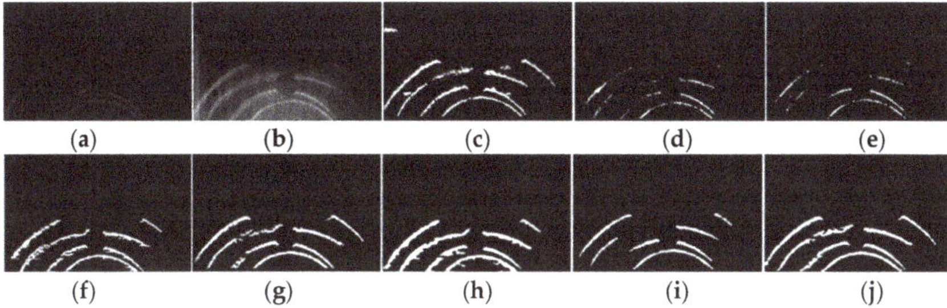

Figure 12. Performance comparison of the eight methods with image 10; (**a**) original image; (**b**) filtered image; (**c**) *approach 1*; (**d**) *approach 2*; (**e**) *approach 3*; (**f**) *approach 4*; (**g**) *approach 5*; (**h**) *approach 6*; (**i**) *approach 7*; (**j**) *approach 8*.

We use 30 images to compare the accuracy of these eight approaches quantitatively and the comparison is shown in Table 1. As can be seen, *approach 8* achieves the best segmentation accuracy. Since the computation time is also critical for on line monitoring, we compare the average computation times of processing the images by these eight approaches programmed with VC++ and Matrox image processing library on the computer with Intel i7-3770 3.4 GHz dualcore CPU. The comparison is shown in Table 2. As can be seen, *approach 1* using the difference operation proposed in Reference [21] is fastest while its segmentation accuracy is not acceptable for some images, e.g., Figure 8. The second fastest is *approach 4* using the gradient detection filter proposed in this paper and its segmentation accuracy is better than *approach 1*. The third fastest is *approach 5*, which combines the gradient dection filter and the difference operation and it achieves adequate segmentation accuracy for subsequent unsupervised processing. On the other hand, *approach 5* has achieved the second-best segmentation accuracy. Hence, *approach 5* is the optimum method for on line processing when the requirement for processing time is strict. In summary, *approach 8* is the best choice to segment the reflected laser lines with the highest accuracy for the developed GMAW weld pool monitoring system in Reference [23] while *approach 1* or *approach 5* is the best choice for the monitoring systems developed in References [19,20]. *approach 1* might be the best choice to segment the reflected laser lines in good quality images. From the quantitative results, we could conclude that the recently proposed image processing algorithms are the most effective steps to form effective segmentation approaches. These image processing algorithms are all proposed based on the analysis of the modeling of the image intensity distribution.

Table 1. Comparison of computation accuracy for eight methods.

Approaches	F-Measure
Approach 1	0.5380
Approach 2	0.2074
Approach 3	0.2276
Approach 4	0.5011
Approach 5	0.8546
Approach 6	0.4743
Approach 7	0.4469
Approach 8	0.9176

Table 2. Comparison of computation time for eight methods.

Approaches	Computation Time
Approach 7	0.05 s
Approach 3	0.045 s
Approach 2	0.042 s
Approach 8	0.031 s
Approach 6	0.028 s
Approach 5	0.0145 s
Approach 4	0.0138 s
Approach 1	0.01 s

6.2. Discussion

The major contributions of the work include:

(1) We combine the recently proposed image processing algorithms and the traditional GLCM to propose different approaches to segment the reflected laser lines. Their performances including accuracy and processing time are evaluated and compared thoroughly in this paper, which is critical in implementing the on-line weld pool monitoring system;

(2) The image processing algorithms proposed previously are explained theoretically in this paper, which serves as a complementation to the previous research [23];

(3) More efficient segmentation approaches for images captured under mild welding parameters with relatively high quality are proposed in this paper.

Image segmentation is fundamental and challenging in many machine vision applications. The most effective methods are usually obtained from the formulated mathematical model and address the characteristics of the captured image sequences well. As a result, the best solution usually needs to combine different image processing algorithms and forms a heuristic approach to achieve the best accuracy and required efficiency. As a typical example of visual intelligent sensing, the research conducted in this work might benefit other researches that need to automatically and robustly extract visual information from the image sequences in different industrial applications.

7. Conclusions

For the image segmentation of the reflected laser lines during on line monitoring of weld pool surface, both accuracy and the efficiency are important. In this paper, we propose eight approaches to segment the reflected laser lines by combining different image processing algorithms. We evaluate their accuracy and efficiency extensively with both qualitative and quantitative results. Experimental results ranked the accuracy and the efficiency of the proposed approaches objectively. The quantitative results showed that the recently proposed image processing methods, including the difference method, the threshold selection method, the gradient detection method and the spline fitting method, are the most effective steps to form the effective segmentation approaches. The quality of the captured image is mainly determined by the welding process. During monitoring weld pool with violent changes, e.g.,

GMAW weld pool, all these recently proposed image processing methods should be combined as an approach to achieve the required accuracy. During monitoring gently changing weld pool, e.g., GTAW weld pool, only the difference method, the gradient detection method and the threshold selection method are required to form approach 5 that could meet the required accuracy while achieving higher efficiency.

Author Contributions: Conceptualization, Z.W. (Zhenzhou Wang); Methodology, Z.W. (Zhenzhou Wang); Software, Z.P. and Z.W. (Zihao Wang); Validation, X.Q., L.L. and J.P.; Formal Analysis, C.Z. and S.M.; Investigation, Z.W. (Zhenzhou Wang); Resources, Z.W. (Zhenzhou Wang); Data Curation, Z.W. (Zhenzhou Wang); Writing-Original Draft Preparation, Z.W. (Zhenzhou Wang); Writing-Review & Editing, C.Z., Z.P., Z.W. (Zihao Wang), X.Q, L.L., J.P. and S.M.; Visualization, Z.W. (Zhenzhou Wang); Supervision, Z.W. (Zhenzhou Wang); Project Administration, Z.W. (Zhenzhou Wang); Funding Acquisition, Z.W. (Zhenzhou Wang).

Conflicts of Interest: The authors declare no conflict of interest.

References

1. Xu, G.; Hu, J.; Tsai, H.L. Three-dimensional modeling of arc plasma and metal transfer in gas metal arc welding. *Int. J. Heat Mass Transf.* **2009**, *52*, 1709–1724. [CrossRef]
2. Wang, X.X.; Fan, D.; Huang, J.K.; Huang, Y. A unified model of coupled arc plasma and weld pool for double electrodes TIG welding. *J. Phys. D Appl. Phys.* **2014**, *47*, 275202. [CrossRef]
3. Chen, X.H.; Dharmawan, A.G.; Foong, S.H.; Soh, G.S. Seam tracking of large pipe structures for an agile robotic welding system mounted on scaffold structures. *Robot. Comput. Integr. Manuf.* **2018**, *50*, 242–255. [CrossRef]
4. Liu, W.; Li, L.; Hong, Y.; Yue, J. Linear Mathematical Model for Seam Tracking with an Arc Sensor in P-GMAW Processes. *Sensors* **2017**, *17*, 591. [CrossRef] [PubMed]
5. Xu, Y.L.; Lv, N.; Zhong, J.Y.; Chen, H.B.; Chen, S.B. Research on the Real-time Tracking Information of Three-dimension Welding Seam in Robotic GTAW Process Based on Composite Sensor Technology. *J. Intell. Robot. Syst.* **2012**, *68*, 89–103. [CrossRef]
6. Zou, Y.B.; Chen, T. Laser vision seam tracking system based on image processing and continuous convolution operator tracker. *Opt. Lasers Eng.* **2018**, *105*, 141–149. [CrossRef]
7. Chen, Z.; Chen, J.; Feng, Z. Monitoring Weld Pool Surface and Penetration Using Reversed Electrode Images. *Weld. J.* **2017**, *96*, 367S–375S.
8. Lv, N.; Zhong, J.Y.; Chen, H.B.; Lin, T.; Chen, S.B. Real time control of welding penetration during robotic GTAW dynamical process by audio sensing of arc length. *Int. J. Adv. Manuf. Technol.* **2014**, *74*, 235–249. [CrossRef]
9. Li, C.K.; Shi, Y.; Du, L.M.; Gu, Y.F.; Zhu, M. Real-time Measurement of Weld Pool Oscillation Frequency in GTAW-P Process. *J. Manuf. Process.* **2017**, *29*, 419–426. [CrossRef]
10. Yang, M.X.; Yang, Z.; Cong, B.Q.; Qi, B. A Study on the Surface Depression of the Molten Pool with Pulsed Welding. *Weld. J.* **2014**, *93*, 312S–319S.
11. Ko, S.H.; Choi, S.K.; Yoo, C.D. Effects of surface depression on pool convection and geometry in stationary GTAW. *Weld. J.* **2001**, *80*, 39.
12. Qi, B.J.; Yang, M.X.; Cong, B.Q.; Liu, F.J. The effect of arc behavior on weld geometry by high-frequency pulse GTAW process with 0Cr18Ni9Ti stainless steel. *Int. J. Adv. Manuf. Technol.* **2013**, *66*, 1545–1553. [CrossRef]
13. Rokhlin, S.; Guu, A. A study of arc force, pool depression and weld penetration during gas tungsten arc welding. *Weld. J.* **1993**, *72*, S381–S390.
14. Zhang, Y.M.; Li, L.; Kovacevic, R. Dynamic estimation of full penetration using geometry of adjacent weld pools. *J. Manuf. Sci. Eng.* **1997**, *119*, 631–643. [CrossRef]
15. Kovacevic, R.; Zhang, Y.M. Real-time image processing for monitoring of free weld pool surface. *J. Manuf. Sci. Eng.* **1997**, *119*, 161–169. [CrossRef]
16. Saeed, G.; Lou, M.J.; Zhang, Y.M. Computation of 3D weld pool surface from the slope field and point tracking of laser beams. *Meas. Sci. Technol.* **2004**, *15*, 389–403. [CrossRef]
17. Zhang, W.J.; Wang, X.W.; Zhang, Y.M. Analytical real-time measurement of three-dimensional weld pool surface. *Meas. Sci. Technol.* **2013**, *24*, 115011. [CrossRef]

18. Zhang, Y.M.; Song, H.S.; Saeed, G. Observation of a dynamic specular weld pool surface. *Meas. Sci. Technol.* **2006**, *17*, 9–12. [CrossRef]

19. Ma, X.J.; Zhang, Y.M. Gas tungsten arc weld pool surface imaging: Modeling and processing. *Weld. J.* **2011**, *90*, 85–94.

20. Shi, Y.; Zhang, G.; Ma, X.J.; Gu, Y.F.; Huang, J.K.; Fan, D. Laser-vision-based measurement and analysis of oscillation frequency in GMAW-P. *Weld. J.* **2015**, *94*, 176–187.

21. Wang, Z.Z. Monitoring of GMAW weld pool from the reflected laser lines for real time control. *IEEE Trans. Ind. Inform.* **2014**, *10*, 2073–2083. [CrossRef]

22. Wang, Z.Z. An imaging and measurement system for robust reconstruction of weld pool during arc welding. *IEEE Trans. Ind. Electron.* **2015**, *62*, 5109–5118. [CrossRef]

23. Wang, Z.Z. Unsupervised recognition and characterization of the reflected laser lines for robotic gas metal arc welding. *IEEE Trans. Ind. Inform.* **2017**, *13*, 1866–1876. [CrossRef]

24. Dougherty, E.R. *An Introduction to Morphological Image Processing*; SPIE International Society for Optical Engine: Bellingham, WA, USA, 1992.

25. Edelman, A.; Mccorquodale, P.; Toledo, S. The future fast Fourier transform. *SIAM J. Sci. Comput.* **1999**, *20*, 1094–1114. [CrossRef]

26. Haralick, R.M.; Shanmugam, K.; Dinstein, I. Textural Features for Image Classification. *IEEE Trans. Syst. Man Cybern.* **1973**, *6*, 610–621. [CrossRef]

27. Grossberg, M.; Nayar, S. Determining the Camera Response from Images: What is Knowable? *IEEE Trans. Pattern Anal. Mach. Intell.* **2003**, *25*, 1455–1467. [CrossRef]

Article

A Novel Discriminating and Relative Global Spatial Image Representation with Applications in CBIR

Bushra Zafar [1,2]*, **Rehan Ashraf** [1], **Nouman Ali** [3,4], **Muhammad Kashif Iqbal** [5], **Muhammad Sajid** [6], **Saadat Hanif Dar** [3] and **Naeem Iqbal Ratyal** [6]

[1] Department of Computer Science, National Textile University, Faisalabad 38000, Pakistan; rehan@ntu.edu.pk
[2] Department of Computer Science, Government College University, Faisalabad 38000, Pakistan;
[3] Department of Software Engineering, Mirpur University of Science & Technology, Mirpur AJK 10250, Pakistan; nali@caa.tuwien.ac.at (N.A.); saadat.dar@gmail.com (S.H.D.)
[4] Computer Aided Automation, Computer Vision Lab, Vienna University of Technology, A-1040 Vienna, Austria
[5] Department of Mathematics, Government College University, Faisalabad 38000, Pakistan; kashifiqbal@gcuf.edu.pk
[6] Department of Electrical Engineering, Mirpur University of Science & Technology, Mirpur AJK 10250, Pakistan; sajid.ee@must.edu.pk (M.S.); naeemratyal@hotmail.com (N.I.R.)
* Correspondence: bkgcuf@gmail.com

Received: 18 October 2018; Accepted: 12 November 2018; Published: 14 November 2018

Abstract: The requirement for effective image search, which motivates the use of Content-Based Image Retrieval (CBIR) and the search of similar multimedia contents on the basis of user query, remains an open research problem for computer vision applications. The application domains for Bag of Visual Words (BoVW) based image representations are object recognition, image classification and content-based image analysis. Interest point detectors are quantized in the feature space and the final histogram or image signature do not retain any detail about co-occurrences of features in the 2D image space. This spatial information is crucial, as it adversely affects the performance of an image classification-based model. The most notable contribution in this context is Spatial Pyramid Matching (SPM), which captures the absolute spatial distribution of visual words. However, SPM is sensitive to image transformations such as rotation, flipping and translation. When images are not well-aligned, SPM may lose its discriminative power. This paper introduces a novel approach to encoding the relative spatial information for histogram-based representation of the BoVW model. This is established by computing the global geometric relationship between pairs of identical visual words with respect to the centroid of an image. The proposed research is evaluated by using five different datasets. Comprehensive experiments demonstrate the robustness of the proposed image representation as compared to the state-of-the-art methods in terms of precision and recall values.

Keywords: image analysis; image retrieval; spatial information; image classification; computer vision

---

1. Introduction

In recent years, with the rapid development of imaging technology, searching or retrieving a relevant image from an image archive has been considered an open research problem for computer vision based applications [1–4]. Higher retrieval accuracy, low memory usage and reduction of semantic gap are examples of common problems related to multimedia analysis and image retrieval [3,5]. The common applications of multimedia and image retrieval are found in the fields of video surveillance, remote sensing, art collection, crime detection, medical image processing and image retrieval in real-time applications [6]. Most of the retrieval systems, both for multimedia and images, rely on the matching of textual data with the desired query [6]. Due to the existing semantic gaps, the performance of these systems suffers [7]. The appearance of a similar view in images belonging to different image categories, results in the closeness of the feature vector values, and degrades the performance of image retrieval [6]. The main focus of the research in Content-Based Image Retrieval (CBIR) is to retrieve images that are in a semantic relationship with a query image [8]. CBIR provides a framework that compares the visual feature vector of a query image to the images places in the dataset [9].

The Bag of Visual Words (BoVW), also known as Bag of Features (BoF) [10], is commonly used for video and image retrieval [11]. The local features or interest point detectors are extracted from a group of training images. To achieve a compact representation, the feature space is quantized to construct a code-book that is also known as visual vocabulary or visual dictionary. The final feature vector, which consists of histograms of visual words, is orderless with respect to the sequence of co-occurrences in the 2D image space. The performance of the BoVW model suffers as the extraction of spatial information is beneficial in image classification and retrieval-based problems [6,12].

Various approaches have been proposed to enhance the performance of image retrieval, such as soft assignments, computation of larger codebooks and visual word fusion [8]. All of these techniques do not contain any information about the visual word's locations in the final histogram-based representation [13]. There are two common techniques that can compute the spatial information from the image. These are based on (1) the construction of histograms from different sub-regions of image, and (2) visual word co-occurrence [13–15]. The first approach is to split the image into different cells for the histogram's computation; it is reported to be robust for content-based image matching applications [16]. Spatial Pyramid Matching (SPM) [16] is considered as a notable contribution for the computation of spatial information for BoVW-based image representation. In SPM, an image is divided into different sizes of rectangular regions for the creation of level-0, level-1 and level-2 histograms of visual words. However, SPM is sensitive to image transformations (i.e., rotation, flipping and translation) and loses its discriminative power, resulting in the misclassification of two similar scene images [17,18].

The second approach to the computation of spatial layout is based on relationships among visual words [19–21]. This paper proposes a novel approach to extracting the image spatial layout based on global relative spatial orientation of visual words. This is achieved by computing the angle between identical visual word pairs with respect to the centroid in the image. Figure 1 provides an illustration to better understand the proposed approach. The image in Figure 1 is rotated at varying angles. It can be seen that the same angle is computed between visual words irrespective of the image orientation.

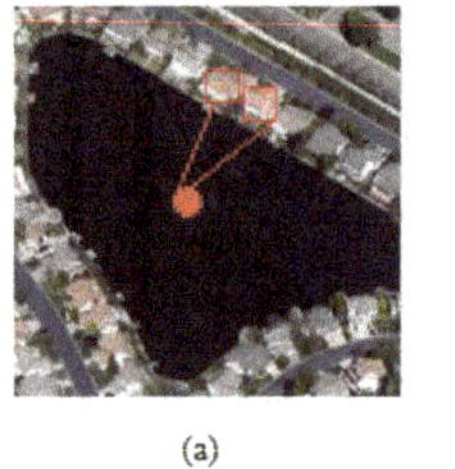
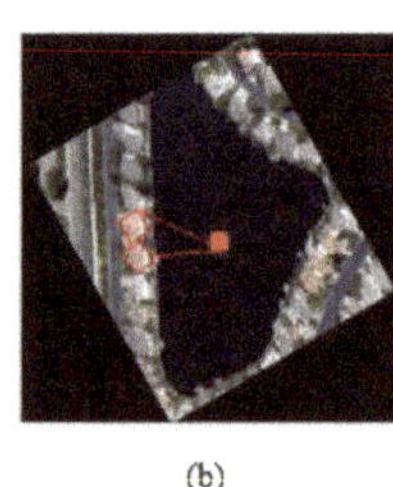

(a) (b) (c)

Figure 1. Angle between identical visual word pairs with respect to the centroid. Here (**a**) represents the original image, (**b**) the image rotated by 120°, and (**c**) the image rotated by 180°.

The main contributions of this research are the following: (1) the addition of the discriminating relative global spatial information to the histogram of BoVW model and (2) reduction of the semantic gap. An efficient image retrieval system must be capable to retrieve images that meet user preferences and their specific requirements. The reduction of the semantic gap specifies that the related categories are given higher similarity scores than unrelated categories. The proposed representation is capable of handling geometric transformations, i.e., rotation, flipping and translation. Extensive experiments on five standard benchmarks demonstrate the robustness of the proposed approach and a remarkable gain in the precision and recall values over the state-of-the-art methods.

The structure of the paper is as follows. Section 2 contains the literature review and related work; Section 3 is about the BoVW model and proposed research; and Section 4 deals with the experimental parameters and image benchmarks, while also presenting a comparison with the existing state-of-the-art techniques. Section 5 provides a discussion, while Section 6 concludes the proposed research with future directions.

2. Related Work

According to the literature [6], SIMPLicity, Blobworld and Query by Image Content (QBIC) are examples of computer vision applications that rely on the extraction of visual features such as color, texture and shape. Image Rover and WebSeek are examples of image search systems that rely on a query-based or keyword-based image search [6]. The main objective of any CBIR system is to search for relevant images that are similar to the query image [22]. Overlapping objects, differences in the spatial layout of the image, changes in illumination and semantic gaps make CBIR challenging for the research community [8]. Wang et al. [23] propose the Spatial Weighing BOF (SWBOF) model to extract the spatial information by using three approaches, i.e., local variance, local entropy and adjacent block distance. This model is based on the concept of the different parts of an image object contributing to image categorization in varying ways. The authors demonstrate significant improvement over the traditional methods. Ali et al. [9] extract the visual information by dividing an image into triangular regions to capture the compositional attributes of an image. The division of the image into triangular cells is reported as an efficient method for histogram-based representation. Zeng et al. [24] propose spatiogram-based image representation that consists of a color histogram that is quantized by using Gaussian Mixture Models (GMMs). The quantized values of GMMs are used as an input for the learning of the Expectation-Maximization (EM). The retrieval is performed on the basis of the closeness of the feature vector values of two spatiograms which are obtained by using the Jensen–Shannon Divergence (JSD) [24]. Yu et al. [25] investigate the impact of the integration of different mid-level features to enhance the performance of image retrieval. They investigate the impact of the integration of SIFT descriptors with LBP and HOG descriptors respectively, in order to address the problem of the semantic gap. Weighed $k-$means clustering is used for quantization, and best performance is reported with SIFT-LBP integration.

To reduce the semantic gap between the low-level features and the high-level image concepts, Ali et al. [8] propose image retrieval based on the visual words integration of Scale Invariant Feature Transform (SIFT) and Speeded−Up Robust Features (SURF). Their approach acquires the strength of both features, i.e., invariance to scale and rotation of SIFT and robustness to illumination of SURF. In another recent work, Ali et al. [26] propose a late fusion of binary and local descriptors i.e., FREAK and SIFT to enhance the performance of image retrieval. Filliat et al. [27] present an incremental and interactive localization and map-learning system based on BoW. Hu et al. [28] propose a real-time assistive localization approach that extracts compact and effective omnidirectional image features which are then used to search a remote image feature-based database of a scene, in order to help indoor navigation.

In another recent work, Li et al. [29] propose a hybrid framework of local (BoW) and global image features for efficient image retrieval. According to Li et al. [29], a multi-fusion based on two lines of image representation can enhance the performance of image retrieval. The authors [29] extract the texture information by using Intensity-Based Local Difference Patterns (ILDP) and by selecting the HSV color space. This scheme is selected to capture the spatial relationship patterns that exist in the images. The global color information is extracted by using the H and S components. The final feature vector is constituted by combining the H, S feature space and ILDP histograms. The experimental result validates that the fusion of color and texture information enhances the performance of image retrieval [29]. According to Liu et al. [30], the ranking and incompatibility of the image feature descriptor is not considered much in the domain of image retrieval. The authors address the problem of incompatibility by using gestalt psychology theory and manifold learning. A combination of gradient direction and color is used to imitate human visual uniformity. The selection of a proposed feature scheme [30] enhances the image retrieval performance. According to Wu et al. [31], ranking and feature representation are two important factors that can enhance the performance of image retrieval and they are considered separately in image retrieval models. The authors propose a texton uniform descriptor and apply an intrinsic manifold structure through visualizing the distribution of image representations on the two-dimensional manifold. This process provides a foundation for subsequent manifold-based ranking and preserves intrinsic neighborhood structure. The authors apply a Modified Manifold Ranking (MMR) to enhance and propagate adjacent similarity between the images [31]. According to Varish et al. [32], a hierarchical approach to CBIR based on a fusion of color and texture can enhance the performance of image retrieval. The color feature vectors are computed on the basis of quantized HSV color space, and texture values are computed to achieve rotation invariance on the basis of Value (V) component of HSV space. The sub-band of various Dual Tree Complex Wavelet Transform (DT-CWT) is applied to compute the principal texture direction.

Zou et al. [33] propose an effective feature selection approach based on Deep Belief Networks (DBN) to boost the performance of image retrieval. The approach works by selecting more reconstructible discriminative features using an iterative algorithm to obtain the optimized reconstruction weights. Xia et al. [34] perform a systematic investigation to evaluate factors that may affect the retrieval performance of the system. They focus the analysis on the visual feature aspect to create powerful deep feature representations. According to Wan et al. [7], a pre-trained deep convolution neural network outperforms the existing feature extraction techniques at the cost of high training computations for large-scale image retrieval. It is important to mention that the approaches based on deep networks may not be an optimal selection as they require large-scale training data with a lot of computations to train a classification-based model [21,35].

3. Proposed Methodology

The basic notations for the BoVW model are discussed in this section. This is then followed by a discussion of the proposed Relative Global Spatial Image Representation (RGSIR) and the details of its implementation.

3.1. BoVW Model

The Bag-of-Words (BoW) methodology was first proposed in textual retrieval systems [11] and was further applied in the form of BoVW representation for image analysis. In BoVW, the final image representation is a histogram of visual words. It is termed a bag, as it counts how many times a word occurs in a document. A histogram does not have any order and does not retain any information regarding the location of visual words in the 2D image space [9,16]. The similarity of two images is determined by histogram intersection. In the case of dissimilar images, the result of the intersection is small.

As a first step in BoVW, the local features are extracted from the image Im, and the image is represented as a set of image descriptors, such as $Im = \{d_1, d_2, d_3, ..., d_I\}$, where d_i denotes the local image features and I represents total image descriptors. The feature extraction can be done by applying some local descriptors such as SIFT descriptors [36]. The key points can be acquired automatically by using interest point detectors or by applying dense sampling [16].

Consequently, there are numerous local descriptors created for each image for a given dataset. The extracted descriptors are vector quantized by applying k-means [11] clustering technique to construct the visual vocabulary, as in

$$v = \{w_1, w_2, w_3, ..., w_K\} \tag{1}$$

where K shows the specified number of clusters or visual words and v denotes the constructed visual vocabulary.

The assignment of each descriptor to the nearest visual word is done by computing the minimum distance as follows:

$$w(d_j) = \underset{w \in v}{argmin}\, Dist(w, d_j) \tag{2}$$

here, $w(d_j)$ represents the visual word mapped to jth descriptor and $Dist(w,d_j)$ depicts the distance between the descriptor d_j and visual word w.

The histogram representation of an image is based on the visual vocabulary. The number of histogram bins equates the number of visual words in the code book or dictionary (i.e., K). Each histogram bin bin_i represents a visual word w_i in v and signifies the number of descriptors mapped to a particular visual word as shown in (3)

$$bin_i = card(D_i)\ \ where\ \ D_i = \{d_j, j \in 1, ..., n \mid w(d_j) = w_i\} \tag{3}$$

D_i is the set of descriptors mapped to a particular visual word w_i in an image, and the cardinality of this set is given by $Card(D_i)$. The final histogram representation for the image is created by repeating the process for each word in the image. The histograms hence created do not retain the spatial context of the interest points.

3.2. The Proposed Relative Global Spatial Image Representation (RGSIR)

In the BoVW model the final image representation is created by mapping identical image patches to the same visual word. In [20], Khan et al. capture the spatial information by modeling the global relationship between identical visual word pairs (PIWs). Their approach exhibits invariance to translation and scaling but is sensitive to rotation [20,37], since the relative relationship between PIWs is computed with respect to the x-axis. Anwar et al. [37] propose an approach to acquire rotation invariance by computing angles between Triplets of Identical Visual Words (TIWs). Although the approach of [37] acquires rotation invariance, it significantly increases computation complexity due to the increase in the number of possible triplet combinations. For instance, if the number of identical visual words is 30, the number of distinct pair combinations is 435 and the number of possible distinct triplet combinations is 4060.

This paper proposes a novel approach to acquiring spatial information for transformation invariance by computing the global geometric relationship between pairs of identical visual words. This is accomplished by extracting the spatial distribution of these pairs with respect to a centroid in an image as shown in Figure 2.

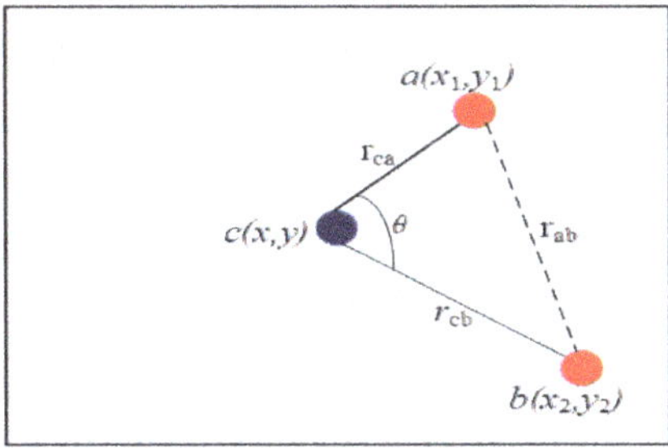

Figure 2. Angle between identical visual word pairs with respect to the centroid.

Hence we define the set of all pairs (PW) of identical visual words related to a visual word w_i as:

$$PW_i = \{(a, b) \mid (d_a, d_b) \in D_i^2, d_a \neq d_b\} \tag{4}$$

where $a(x_1, y_1)$ and $b(x_2, y_2)$ are the spatial locations of the descriptors d_a and d_b, respectively. Since the ith histogram bin signifies the descriptor d_i, its value determines the total occurrences of the word w_i. The cardinality of the set PW_i is $^{b_i}C_2$. The centroid $c = (x, y)$ of an image Im of size $R \times C$ is calculated as

$$x = \frac{1}{\mid Im \mid} \sum_{i=1}^{\mid Im \mid} x_i, \quad y = \frac{1}{\mid Im \mid} \sum_{i=1}^{\mid Im \mid} y_i \tag{5}$$

where $Im = \{(x_i, y_i) \mid 1 \leq x_i \leq R, 1 \leq y_i \leq C\}$ and $\mid Im \mid$ is the number of elements in Im. Let r_{ab} be the Euclidean distance between a and b, then

$$r_{ab} = \sqrt{(x_1 - x_2)^2 + (y_1 - y_2)^2} \tag{6}$$

Similarly, the Euclidean distances of a and b from c are calculated as

$$r_{ca} = \sqrt{(x_1 - x)^2 + (y_1 - y)^2}$$
$$r_{cb} = \sqrt{(x_2 - x)^2 + (y_2 - y)^2}$$

Using the Law of cosines, we have

$$\theta = \arccos \left(\frac{(r_{ca})^2 + (r_{cb})^2 - (r_{ab})^2}{2(r_{ca})(r_{cb})} \right) \tag{7}$$

where $\theta = \angle acb$.

The θ angles obtained are then concatenated to create the histogram representation with bins equally distributed between 0–180°. The optimal number of bins used for histogram representation is determined empirically. The $RGSIR_i$ represents the spatial distribution for a particular visual word w_i. The $RGSIR_i$ obtained from all the visual words in an image are concatenated to create the global image representation. A bin replacement technique is used to transform the BoVW representation to RGSIR. This is achieved by replacing each bin of the BoVW histogram with the associated $RGSIR_i$ related to a particular w_i. To add the spatial information while keeping the frequency information intact, the sum of all bins of $RGSIR_i$ is normalized to the size of the bin bin_i of the BoVW histogram that is being replaced. The image representation for RGSIR is hence formulated as:

$$RGSIR = (\alpha_1 RGSIR_1, \alpha_2 RGSIR_2, ..., \alpha_K RGSIR_K) \tag{8}$$

where α_i, the coefficient of normalization, is given by $\alpha_i = \frac{bin_i}{\lVert RGSIR_i \rVert}$. If the size of the visual vocabulary is K and the number of histogram bins is H, then the dimensions of RGSIR are $K \times H$.

3.3. Implementation Details

The histogram representations for all of the datasets are created by following the same sequence of steps as shown in Figure 3. As a preprocessing step, the images are converted to gray-scale mode by using the available standard resolution, and the dense SIFT features are extracted on six multi-scales, i.e., {2,4,6,8,10,12} for the computation of codebook [38]. The step size of 5 is applied to compute the Dense SIFT features [38]. Dense features are selected, as the dense regular grid has shown to possess better discriminative power [16]. To save computation time for clustering, 40% of the features (per image) are selected by applying a random selection on a training set to compute the codebook.

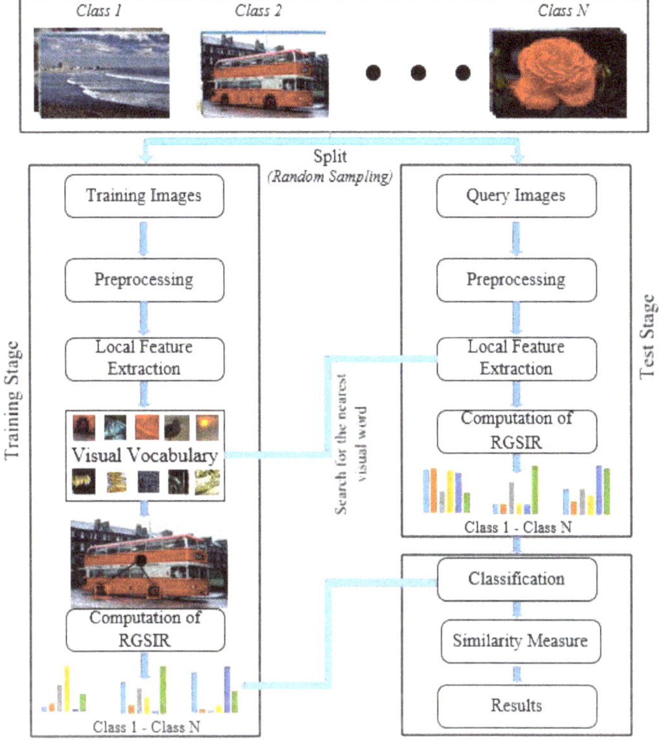

Figure 3. Block diagram of the proposed work.

To quantize the descriptors, *k-means* clustering is applied to generate visual vocabulary. Since the size of the codebook is one of the major factors that affects the performance of image retrieval, the proposed approach is evaluated by using different sizes of codebook to sort out the best retrieval performance. The visual vocabulary is constructed from the training set and the evaluation is done using the test set. The experiments are repeated in 10 trials to remove the ambiguity created by the random initialization of cluster centers by *k-means*. For each trial, the training and test images are stochastically selected and the average retrieval performance is reported in terms of precision and recall values, which are considered as standard image retrieval measures [8,39].

The calculation of RGSIR involves computing subsets of pairs from sets of identical visual words. To accelerate computation, a threshold value is set and a random selection is applied to limit the number of identical words used for creating the pair combinations. We use a nine-bin RGSIR representation for the results presented in Section 4. Figure 4 gives the empirical justification for the number of bins on two different image benchmarks used in our experiments. Support Vector Machine (SVM), a supervised learning technique, is used for classification. The SVM Hellinger Kernel is applied to the normalized RGSIR histograms. The optimal value for the regularization parameter is determined by applying 10-fold cross validation on the training dataset. As we have used a classification-based framework for image retrieval, the class of the image is predicted by using the classifier labels; similarity among the images of the same class is determined on the basis of distance in decision values [8]. The results obtained from the evaluation metrics are normalized and average values are reported in tables in graphs. MATLAB is used to simulate the research by using Corei7, a 7th generation processor with 16 GB RAM.

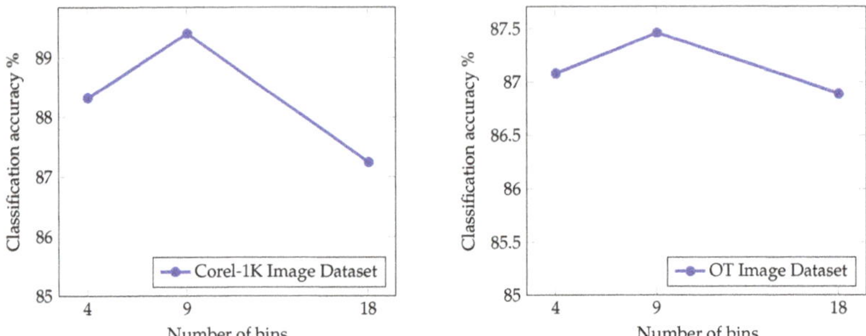

Figure 4. The influence of the number of bins on the performance of RGSIR.

4. Datasets and Performance Evaluation

This section provides a description of the datasets, measures used for evaluation, and the details of the experiments conducted for the validation of the proposed research.

4.1. Dataset Description

To assess the effectiveness of the proposed research for image retrieval, experiments are conducted on the benchmark datasets used extensively in the literature. The first dataset used in our experiments is the Corel-1K [40] image dataset. The Wang's image dataset is comprised of a total of 1000 Corel images from diverse contents such as beach, flowers, horses, mountains, food, etc. The images are grouped into 10 categories with image sizes of 256×384 or 384×256 pixels. The second dataset is the Corel-1.5K image benchmark comprised of 15 classes with 100 images per category [40]. Figure 5 shows sample images from Corel-1K and Corel-1.5K, respectively.

Figure 5. Randomly selected images from each class of Corel-1K and Corel-1.5K image datasets [40].

The third dataset used to validate the efficacy of the proposed RGSIR is the Corel-2K image benchmark. Corel-2K is a subset of Corel image dataset and is comprised of 2000 images classified into 20 semantic categories. Example images from this dataset are shown in Figure 6.

Figure 6. Class representatives from the Corel 2000 image dataset [40].

The forth dataset is the Oliva and Torralba (OT) dataset [41], which includes 2688 images classified into 8 semantic categories. This dataset exhibits high inter and intra-class variability, as the river and forest scenes are all considered as forest. Moreover, there is no specific sky category, since all the images contain the sky object. The average image size is 250×250 pixels and the images are collected from different sources (i.e., commercial databases, digital cameras, websites). This is a challenging dataset as the images are sampled from different perspectives, varying rotation angles, different spatial patterns and different seasons. Figure 7 shows the photo gallery of images for the OT image dataset.

Figure 7. Class representatives from the OT image dataset [41].

The last dataset used is our experiments is the RSSCN image dataset [33], released in 2015, comprised of images collected from Google Earth. It consists of 2800 images categorized into 7 typical scene categories. There are 400 images per class, and each image has a size of 400×400 pixels. It is a challenging dataset, as the images in each class are sampled at 4 different scales, with 100 images per scale under varied imaging angles. Consistent with related work [33], the dataset is stochastically split into two equal image subsets for training and testing, respectively. Example images from this dataset are shown in Figure 8.

Figure 8. Class representatives from the RSSCN image dataset [33].

4.2. Evaluation Measures

Let the database $I_1, ..., I_n, ..., I_N$ be a set of images represented by the spatial attributes. To retrieve an image identical to the query image Q, each image from the database I_n is compared with Q, using the appropriate distance function (Q, I_n). The database images are then sorted based on the distances such that $(d(Q, I_{n_i}) \leq (d(Q, I_{n_i+1})$ holds for each pair images I_{n_i} and $I_{n_{i+1}}$ of distances in the sequence $I_{n_1}, ..., I_{n_i}, ..., I_{n_N}$.

4.2.1. Precision

The performance of the proposed method is measured in terms of precision P and recall R, which are the standard measures used to evaluate CBIR. Precision measures the specificity of the image retrieval, and it gives the number of relevant instances retrieved in response to a query image. The Precision (P) is defined as

$$P = \frac{Number\ of\ relevant\ images\ retrieved}{Total\ number\ of\ images\ retrieved} \tag{9}$$

4.2.2. Recall

The Recall is the fraction of the relevant instances retrieved to the total number of instances of that class in the dataset. It measures the sensitivity of the image and is given by

$$R = \frac{Number\ of\ relevant\ images\ retrieved}{Total\ number\ of\ relevant\ images} \tag{10}$$

4.2.3. Mean Average Precision (MAP)

Based on P and R values, we also report results in terms of precision vs recall curve (P-R curve) and the mean average precision (MAP). The P-R curve represents the tradeoff between precision and recall for a given retrieval approach. It reflects more information about retrieval performance that is determined by the area under the curve. If the retrieval system has better performance, the curve is as far from the origin of coordinates as possible. The area between the curve and the X-Y axes should be larger, which is usually measured and is approximate to MAP [42]. In other words, the most common way to summarize the *P-R* curve in one value is *P-R*. *P-R* is the mean of the average precision (*AP*) scores of all queries and is computed as follows:

$$MAP = \frac{1}{|T|} \sum_{Q \epsilon T} AP(Q) \tag{11}$$

where *T* is the set of test images or queries *Q*. An advantage of MAP is that it contains both precision and recall aspects and is sensitive to the entire ranking [43].

4.3. Performance on Corel-1K Image Dataset

The Corel-1K image benchmark is extensively used to evaluate CBIR research. To ensure fair comparison experiments, the dataset is stochastically partitioned into training and test subsets with a ratio of 0.5:0.5. The image retrieval performance of the proposed image representation is compared with the existing state-of-the-art CBIR approaches. In order to obtain a sustainable performance, the mean average precision of RGSIR is evaluated by using visual vocabulary of different sizes [50, 100, 200, 400, 600, 800]. The best image retrieval performance for Corel-1K is obtained for a vocabulary of size 600, as can be seen in Figure 9.

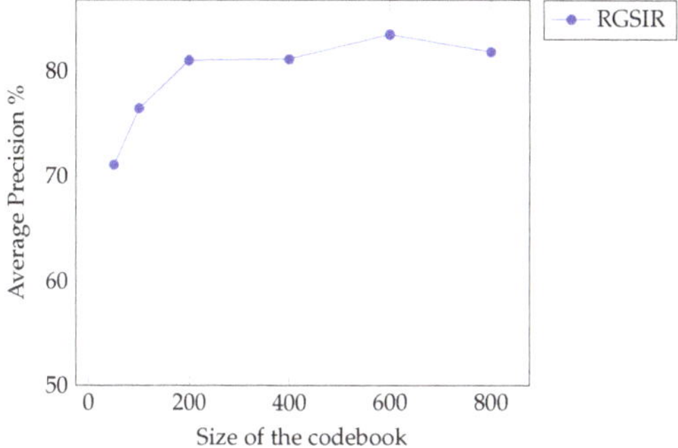

Figure 9. Average Precision as a function of vocabulary size.

The class-wise comparison obtained from the proposed research in terms of precision and recall is presented in Tables 1 and 2. It can be seen that the proposed approach outperforms the state-of-the-art image retrieval approaches. The proposed RGSIR provides 17.7% higher precision compared to Yu et al. [25]. Our proposed representation outperforms SWBOF [23] by {13.7%, 2.74%} in terms of average precision and recall values for the top 20 retrieval. RGSIR yields {8.23%, 1.65%} higher performance compared to [8] in terms of average retrieval precision and recall values.

Table 1. Comparison of precision when using Corel-1K image dataset.

Class Name/ Method	RGSIR	Li et al. [29]	Level-1 RBF-NN [9]	Visual Words Integration SIFT-SURF [8]	SWBOF [23]	SIFT-LBP [25]
African People	72.80	76.55	73.06	60.08	64.00	57.00
Beach	69.40	63.70	69.98	60.39	54.00	58.00
Building	66.20	69.05	76.76	69.66	53.00	43.00
Bus	97.16	87.70	92.24	93.65	94.00	93.00
Dinosaur	100.00	99.40	99.35	99.88	98.00	98.00
Elephant	80.80	91.05	81.38	70.76	78.00	58.00
Flower	94.60	91.70	83.40	88.37	71.00	83.00
Horse	90.80	95.40	82.81	82.77	93.00	68.00
Mountain	76.20	83.40	78.60	61.08	42.00	46.00
Food	86.00	65.80	82.71	65.09	50.00	53.00
Mean	83.40	82.36	82.03	75.17	69.70	65.70

Table 2. Comparison of recall when using Corel-1K image dataset.

Class Name/ Method	RGSIR	Li et al. [29]	Level-1 RBF-NN [9]	Visual Words Integration SIFT-SURF [8]	SWBOF [23]	SIFT-LBP [25]
African People	14.56	15.31	14.61	12.02	12.80	11.4
Beach	13.88	12.74	14.00	12.08	10.80	11.6
Building	13.24	13.81	15.35	13.93	10.30	8.6
Bus	19.43	17.54	18.45	18.73	18.80	18.6
Dinosaur	20.00	19.88	19.87	19.98	19.60	19.6
Elephant	16.16	18.21	16.28	14.15	15.60	11.6
Flower	18.92	18.34	16.68	17.67	14.20	16.6
Horse	18.16	19.08	16.56	16.55	18.60	13.6
Mountain	15.24	16.68	15.72	12.22	8.40	9.2
Food	17.20	13.16	16.54	13.02	10.00	10.6
Mean	16.68	16.48	16.41	15.03	13.94	13.14

The proposed RGSIR results in {1.04%, 0.2%} higher precision and recall values compared to the work of Li et al. [29]. Experimental results validate the robustness of the proposed approach against the state-of-the-art retrieval methods.

The comparative analysis of the proposed research with the existing state-of-the-art verifies the effectiveness of RGSIR for image retrieval. The average precision depends on the total number of relevant images retrieved, and hence is directly proportional to the number of relevant images retrieved in response to a given query image. It is evident from the Figure that the proposed approach attains the highest number of relevant images against a given query image as compared to the state-of-the-art approaches. Similarly, the average recall is directly proportional to the number of relevant images retrieved to the total number of relevant images of that class present in the dataset. The proposed approach outperforms the state-of-the-art methods by attaining the highest precision and recall values.

The *P-R* curve obtained for the Corel-1K image benchmark is shown in Figure 10. The *P-R* curve demonstrates the ability of the retrieval system to retrieve relevant images from the image database in an appropriate similarity sequence. The area under the curve illustrates how effectively different methods perform in the same retrieval scenario. The results indicate that the proposed spatial features enhance the retrieval performance as compared to the state-of-the-art image retrieval approaches.

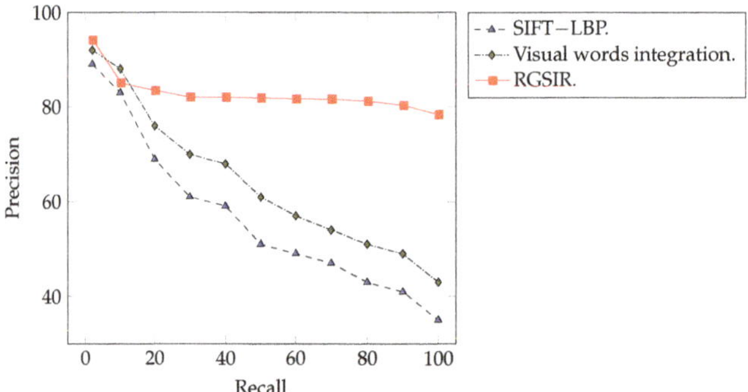

Figure 10. *P-R* curve obtained using Corel-1K image benchmark.

The image retrieval results for the semantic classes of Corel-1K image dataset are shown in Figures 11 and 12 (which reflects the reduction of the semantic gap). The image shown in the first row is the query image and the remaining 20 images are images retrieved by applying a similarity measure that is based on image classification score values. Here a classification label is used to determine the class of the image, while the similarity with-in the same class is calculated on the basis of similarity among classification scores of images of the same class from the test dataset.

Figure 11 shows that, for a given query image, all images of the related semantic category are retrieved. In Figure 12 it can be seen that in a search based on a flower query image, an image from a different semantic category containing flowers is also displayed in the 3rd row in addition to images from the flower image category. The experimental results demonstrate that the proposed approach achieves much higher performance compared to the state-of-the-art complementary approaches [9,23,25].

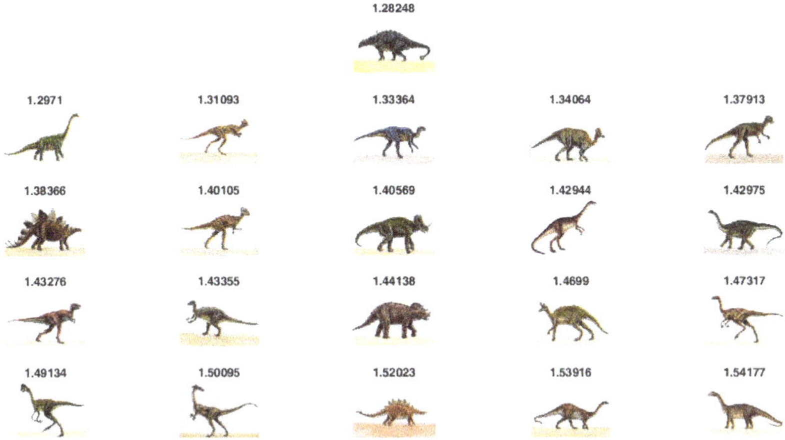

Figure 11. Result of image retrieval for the semantic class "Dinosaurs".

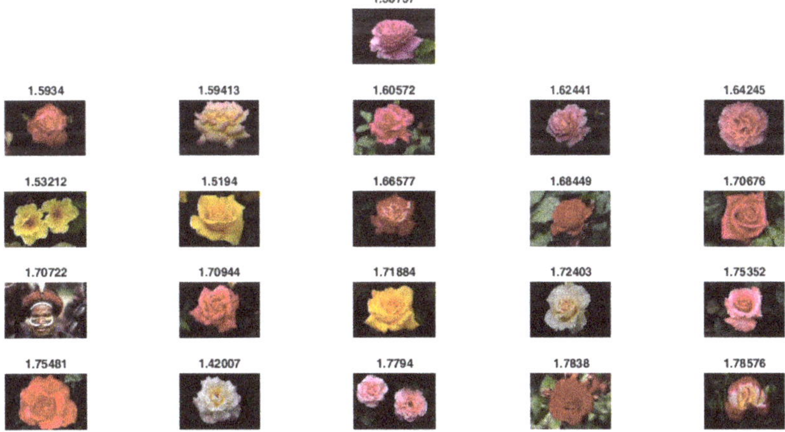

Figure 12. Result of image retrieval for the semantic class "Flowers".

4.4. Performance on Corel-1.5K Image Dataset

To further assess the effectiveness of the proposed method, experiments are conducted on Corel-1.5 image benchmark. The image retrieval performance of Corel-1.5 dataset is analyzed using the visual vocabulary of different sizes. The optimal performance is obtained for a vocabulary size of 400. Table 3 provides a comparison of the mean average precision for the top 20 retrievals with the state-of-the-art image retrieval approaches [8,24,26].

It is evident from the table that the proposed RGSIR provides better retrieval performance compared to the state-of-the-art approaches with higher retrieval precision values than those of the existing research. Experimental results demonstrate that the proposed approach provides {18.9%,

3.78%} better performance compared to the method without soft assignment, i.e., SQ + Spatiogram [24] and {8.75%, 2.77%}, than the probabilistic GMM + mSpatiogram [24] in terms of precision and recall, respectively.

Table 3. Comparison of Average Retrieval Precision and Recall when using Corel-1.5K image dataset.

Performance and Name of Method	RGSIR	Ali et al. [26]	Visual words Integration SIFT-SURF [8]	GMM + mSpatiogram [24]	SQ + Spatiogram [24]
Precision	82.85	72.60	74.95	74.10	63.95
Recall	16.57	14.52	14.99	13.80	12.79

The proposed approach based on relative spatial feature extraction achieves 7.9% higher retrieval precision compared to the image retrieval based on visual words integration of SIFT and SURF [8]. Our proposed approach provides {10.25%, 2.05%} better precision and recall results compared to the late fusion based approach [26]. The experimental results demonstrate that our proposed approach significantly improves the retrieval performance compared to the state-of-the-art image retrieval techniques.

4.5. Performance on Corel-2K image Dataset

The optimal performance for the Corel-2K image dataset is obtained for a vocabulary size of 600. Table 4 provides a comparison of Corel-2K with the state-of-the-art image retrieval approaches. It is evident that the proposed approach yields the highest retrieval accuracy. The proposed approach provides 13.68% highest mean retrieval precision compared to the second best method. Figure 13 illustrates the average precision and recall values for the top 20 image retrievals. The experimental results validate the efficacy of the proposed approach for content-based image retrieval.

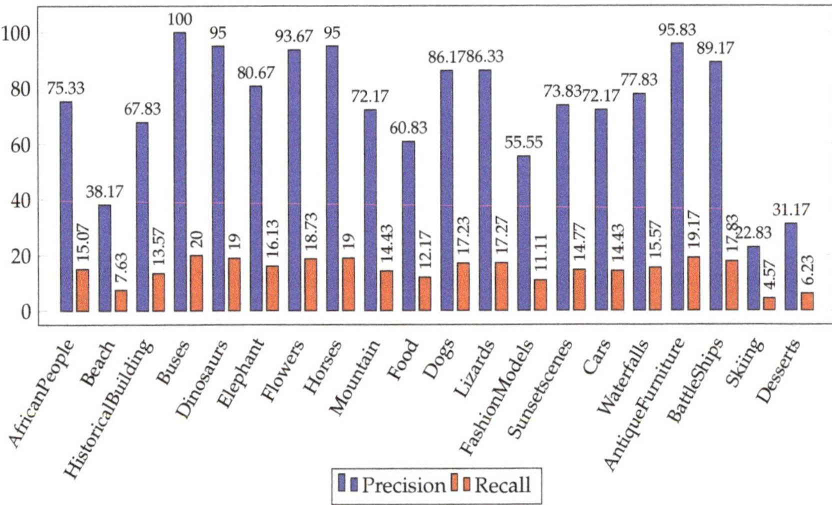

Figure 13. Average Precision and Recall of the proposed RGSIR for the top 20 retrievals using Corel-2K image becnchmark.

Table 4. Comparison of the mean average precision using Corel-2K image benchmark.

Performance/ Method	RGSIR	Visual Words Integration SIFT-SURF [8]	MissSVM [44]	MI-SVM [45]
MAP	79.09	65.41	65.20	54.60

The image retrieval results for the semantic classes of Corel-2K image dataset are shown in Figures 14 and 15 (which reflect the reduction of the semantic gap). The image displayed in the first row is the query image and the remaining images are the results of the top 20 retrievals selected on the basis of the image classification score displayed at the top of each image.

Figure 14. Result of image retrieval for the semantic class "Lizards".

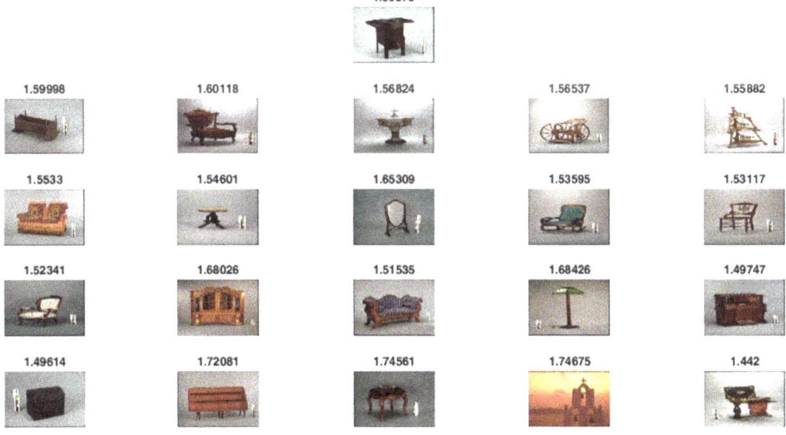

Figure 15. Result of image retrieval for the semantic class "Antique Furniture".

4.6. Image Retrieval Performance While Using Oliva and Torralba (OT-Scene) Dataset

To demonstrate the effectiveness of the proposed research, experiments are performed on the challenging OT image dataset. The best performance for the proposed research is obtained for a vocabulary size of 600. As the proposed approach has been designed on a classification-based framework, Figure 16 provides a class-wise comparison of the classification accuracy of the proposed approach with the recent state-of-the-art classification approaches [46,47]. Shrivastava et al. [46] propose a fusion of color, texture and edge descriptors to enhance the performance of image classification and report an accuracy of 86.4%. Our proposed approach outperforms SPM by 3.85% [48] and yields 1.06% higher accuracy compared to [46]. Zang et al. [47] use the Object Bank (OB) approach to construct powerful image descriptors and boost the performance of OB-based scene image classification. The best mean classification accuracy for the proposed RGSIR is 87.46%, while the accuracy reported by Zang et al. [47] is 86.5%. The proposed approach provides 0.96% higher accuracy compared to their work. It is observed that the performance of the proposed approach is low for the natural coast and the open country category due to high variability in these classes. The proposed approach based on spatial features provides better performance compared to the state-of-the-art retrieval approaches.

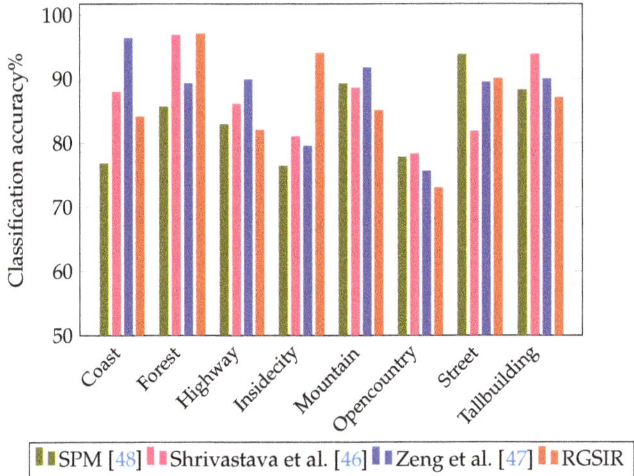

Figure 16. Class-wise comparison between of the proposed research with the state-of-the-art methods for OT scene image dataset.

The comparison of the proposed research with existing research [8] in terms of precision is presented in Table 5. The proposed approach provides 13.17% higher accuracy compared to the second best method in comparison. The experimental results validate the efficacy of the proposed approach for content based image retrieval.

Table 5. Comparison of the mean average precision using OT-Scene image benchmark.

Performance/ Method	RGSIR	Visual Words Integration SIFT-SURF [8]	Log Gabor + OC-LBP Technique [49]	Late Fusion (SIFT + FREAK) [26]	Feature Extraction with Morphological Operators
MAP	82.92	69.75	63.74	63.14	60.70

4.7. Performance on the RSSCN Image Dataset

To evaluate the effectiveness of proposed approach for scene classification, experiments are conducted on the challenging high resolution remote sensing scene image dataset. The training test ratio of 0.5:0.5 is used for the RSSCN image dataset as is followed in the literature [33]. The training set comprises 1400 stochastically selected images and the remaining images are used to assess the retrieval performance. The optimal retrieval performance is obtained for a visual vocabulary size of 200. As we have used a classification based framework for image retrieval, it is important to note here that the classification accuracy for the proposed RGSIR is 81.44% and the accuracy reported by the dataset creator is 77%. Our proposed representation provides 4.44% higher accuracy compared to the deep learning technique, i.e., the DBN adopted by the Zou et al. [33].

Table 6 provides a comparison of the retrieval performance of RSSCN with the state-of-the-art image retrieval approaches. We have computed MAP for the top 100 retrievals using the proposed RGSIR. Xia et al. [34] perform an extensive analysis to develop a powerful feature representation to enhance image retrieval. They consider different CNN representative models, i.e., CaffeNet [50],

VGG-M [51], VGG-VD19 [52] and GoogLeNet [53], in combination with different feature extraction approaches. As our proposed approach is based on mid-level features, we have selected BoW based aggregation methods for comparison. Mid-level features are more resilient to various transformations such as rotation, scale and illumination [34]. The proposed approach provides 16.63% higher accuracy compared to VGG-M (IFK). The proposed RGSIR outperforms the GoogLeNet (BoW), VGG-VD19 (BoW) and CaffeNet (BoW) by 18.56%, 19.2% and 20.88 %, respectively.

Table 6. Comparison of the mean average precision when using RSSCN image benchmark.

Performance/ Method	RGSIR	CaffeNet (BoW) [34]	VGG-VD19 (BoW) [34]	GoogLeNet (BoW) [34]	VGG-M (IFK) [34]
MAP	72.42	51.54	53.22	53.86	55.79

It is important to note here that we have selected the RSSCN image dataset as the images are captured at varying angles and exhibit significant rotation differences. Hence the robustness of the proposed approach to rotation in-variance is also illustrated to some extent. The top 20 retrieval results against the "Forest" and "River & Lake" semantic categories of the RSSCN image dataset are shown in Figures 17 and 18.

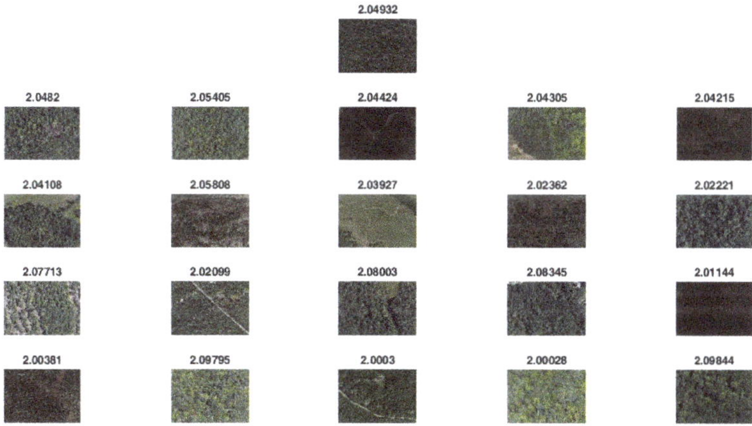

Figure 17. Results of image retrieval for the semantic class "Forest".

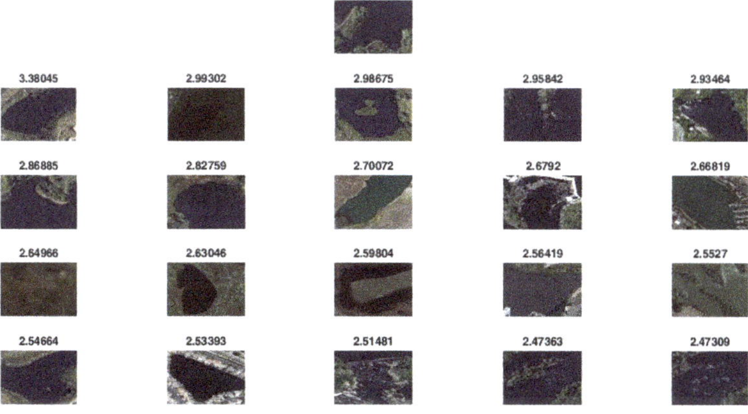

Figure 18. Results of image retrieval for the semantic class "River & Lake".

5. Discussion

In this paper, we have proposed an image retrieval approach based on relative geometric spatial relationships between visual words. Extensive experiments on challenging image benchmarks demonstrate that the proposed approach outperforms the concurrent and the state-of-the-art image retrieval approaches based on feature fusion and spatial feature extraction techniques [8,23,24].

5.1. Factors Affecting the Performance of the System

One of the factors affecting the retrieval performance is the size of the visual vocabulary. We have conducted experiments with visual vocabulary of different sizes to determine the optimal performance of the proposed representation as discussed in the preceding sections. Another factor affecting the performance of the system is the ratio of the training images used to train the classifier. Figure 19 provides a comparison of different training test ratios i.e., 70:30, 60:40, 50:50 for the Corel-1K image dataset. It can be seen that the performance of the system increases at higher training test ratios. However, to be consistent with related approaches [8], 50:50 is used to report the precision and recall retrieval results for the experimental comparisons presented in Section 4.

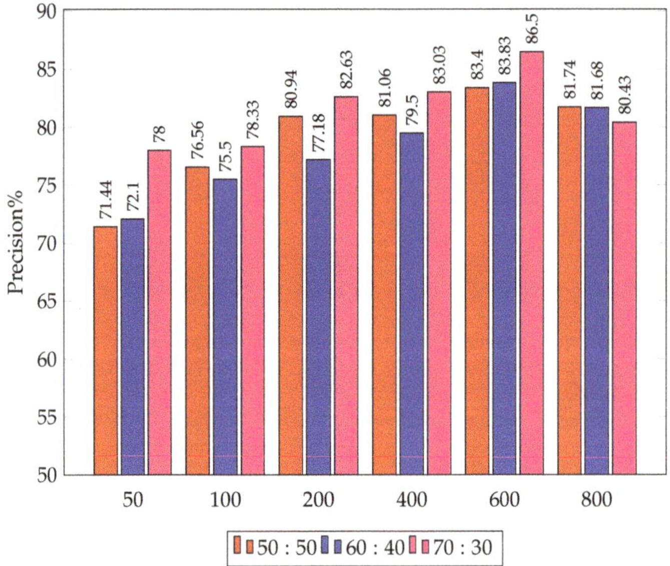

Figure 19. Average precision of the proposed RGSIR on visual vocabulary of different sizes using different training test ratios.

5.2. Invariance to Basic Transformations

Spatial Pyramid Matching (SPM) [16] is the most notable contribution to incorporate spatial context into the BoVW model. SPM captures the absolute spatial distribution of visual words. However, SPM is sensitive to image transformations such as rotation, flipping and translation. For images that are not well-aligned, SPM may lose its discriminative power. An object may rotate by any angle on the image plane (rotation), it may be flipped horizontally or vertically (flipping), or the object may appear anywhere in an image (translation). The proposed approach is capable of addressing various transformations, by encoding the global relative spatial orientation of visual words. This is achieved by computing the angle between identical visual word pairs with respect to the centroid in image. Figure 20 provides an illustration to better understand our approach.

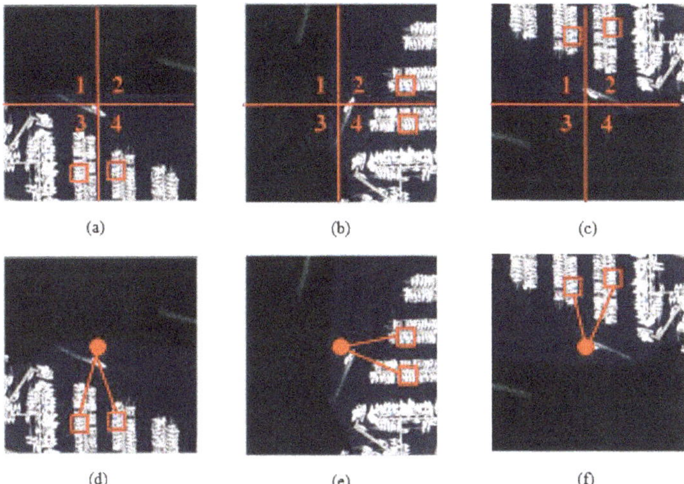

Figure 20. SPM (**a,b,c**) vs.the proposed approach (**d,e,f**). Here (**a,d**) represent the original images, (**b,e**) the images rotated by 90°, and (**c,f**) vertically flipped images.

The upper region of Figure 20a–c represents the idea of histograms constructed with SPM [16], while the lower region demonstrates the proposed approach Figure 20d–f. In the figures, we can see Figure 20a,d the original image, Figure 20b,e the image rotated by 90° and Figure 20c,f the vertically flipped image. The performance of SPM [16] degrades in this case, as the objects occupy different regions in the original and transformed images. In Figure 20a the identical visual words are located in the 3rd and 4th regions, in Figure 20b they are found in the 2nd and 4th regions, while in Figure 20c they are in the 1st and 2nd regions, respectively. Hence the three histogram representations will be different for the same image. In the case of the proposed RGSIR, the same histogram representation will be generated for the original and for the transformed images, as the angle between identical visual words with respect to the centroid remains the same.

Figure 21 presents a graphical comparison of the average precision for the top 20 retrievals with the concurrent state-of-the-art approaches. Chathurani et al. [54] propose a Rotation Invariant Bag of Visual Words (RIBoW) approach to encode the spatial information using circular image decomposition in combination with a simple shifting operation using global image descriptors. They report improved performance to existing BoVW approaches. Although SPM [16] encodes the spatial information, it is sensitive to rotation, translation and scale variance of an image. The circular decomposition approach [54] partitions the image into sub-images, and features which are then extracted from each sub-image are used for feature representation. The proposed RGSIR provides 10.4% higher retrieval precision compared to the second best method.

Experimental results demonstrate the superiority of the proposed approach to the concurrent state-of-the-art approaches. It is important to note here that some approaches incorporate the spatial context prior to the visual vocabulary construction step, while others do so after it [9]. The proposed approach adds this information after the visual vocabulary construction step. In future, we intend to enhance the discriminative power of the proposed approach by extracting rotation-invariant features at the feature extraction step, prior to the construction of the visual vocabulary.

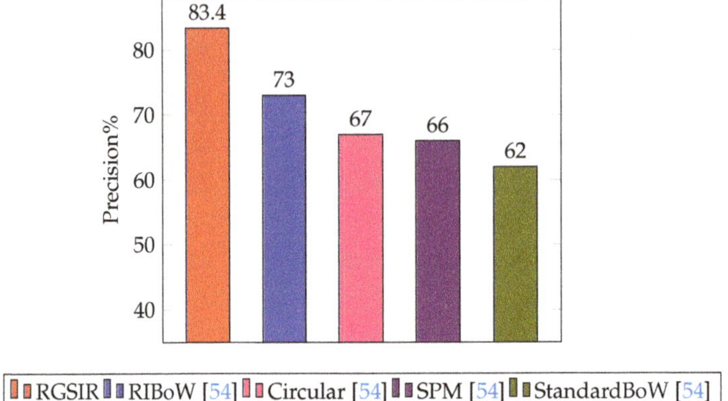

Figure 21. Average precision comparison of the proposed RGSIR with the state-of-the-art approaches for the Corel-1K image benchmark.

6. Conclusions and Future Directions

The final feature vector for the BoVW model contains no information regarding the distribution of visual words in the 2D image space. Due to this reason, the performance of a computer vision application suffers, as spatial information of visual words in the histogram-based feature vector enhances the performance of image retrieval. This paper presents a novel approach to image representation to incorporate the spatial information to the inverted index of the BoVW model. The spatial information is added by calculating the global relative spatial orientation of visual words in a transformation-invariant manner. This is established by computing the geometric relationship between pairs of identical visual words with respect to the centroid of an image. The experimental results and quantitative comparisons demonstrate that our proposed representation significantly improves the retrieval performance in terms of precision and recall values. The proposed approach outperforms other concurrent methods and provides competitive performance as compared with the state-of-the-art approaches.

Furthermore, the proposed approach is not confined to the retrieval task but can be applied to other image analysis tasks, such as object detection. This is because we incorporate the invariant spatial layout information into the BoVW image representation, thereby ensuring seamless application of follow-up techniques.

In future, we would like to enhance the discriminative power of the proposed approach by extracting rotation invariant low-level features at descriptor level. We intend to create a unified representation, tolerant to all kinds of layout variances. As the proposed method has shown excellent results on five image benchmarks, in future we aim to apply a pre-trained deep convolution neural network for the computation of histogram of visual words for learning of classifier to a large scale image dataset. Combining our image representation with a complementary absolute feature extraction method and enriching it with other cues such as color and shape is another possible direction for future research.

Author Contributions: Conceptualization, B.Z., R.A. and N.A.; Data curation, B.Z., R.A. and N.A.; Formal analysis, B.Z., R.A. and N.A. ; Investigation, B.Z., R.A., N.A. and S.H.D.; Methodology, B.Z. and N.A.; Project administration, N.A., M.S., S.H.D. and N.I.R.; Resources, M.K.I., M.S., S.H.D. and N.I.R.; Software, B.Z., N.A. and M.S.; Supervision, R.A. and N.A.; Validation, B.Z., N.A., M.K.I. and N.I.R.; Visualization, B.Z., M.K.I., S.H.D. and N.I.R.; Writing—original draft, B.Z., R.A., N.A., M.K.I., M.S. and N.I.R.; Writing—review and editing, R.A., N.A., M.K.I., M.S., S.H.D. and N.I.R.

Funding: This research received no external funding.

References

1. Irtaza, A.; Adnan, S.M.; Ahmed, K.T.; Jaffar, A.; Khan, A.; Javed, A.; Mahmood, M.T. An Ensemble Based Evolutionary Approach to the Class Imbalance Problem with Applications in CBIR. *Appl. Sci.* **2018**, *8*, 495. [CrossRef]
2. Ye, J.; Kobayashi, T.; Toyama, N.; Tsuda, H.; Murakawa, M. Acoustic Scene Classification Using Efficient Summary Statistics and Multiple Spectro-Temporal Descriptor Fusion. *Appl. Sci.* **2018**, *8*, 1363. [CrossRef]
3. Piras, L.; Giacinto, G. Information fusion in content based image retrieval: A comprehensive overview. *Inf. Fusion* **2017**, *37*, 50–60. [CrossRef]
4. Nazir, A.; Ashraf, R.; Hamdani, T.; Ali, N. Content based image retrieval system by using HSV color histogram, discrete wavelet transform and edge histogram descriptor. In Proceedings of the 2018 International Conference on Computing, Mathematics and Engineering Technologies (iCoMET), Sukkur, Pakistan, 3–4 March 2018; pp. 1–6.
5. Zhu, L.; Shen, J.; Xie, L.; Cheng, Z. Unsupervised visual hashing with semantic assistant for content-based image retrieval. *IEEE Trans. Knowl. Data Eng.* **2017**, *29*, 472–486. [CrossRef]
6. Alzu'bi, A.; Amira, A.; Ramzan, N. Semantic content-based image retrieval: A comprehensive study. *J. Vis. Commun. Image Represent.* **2015**, *32*, 20–54. [CrossRef]
7. Wan, J.; Wang, D.; Hoi, S.C.H.; Wu, P.; Zhu, J.; Zhang, Y.; Li, J. Deep Learning for Content-Based Image Retrieval: A Comprehensive Study. In Proceedings of the ACM International Conference on Multimedia, Orlando, FL, USA, 3–7 November 2014; pp. 157–166. [CrossRef]
8. Ali, N.; Bajwa, K.B.; Sablatnig, R.; Chatzichristofis, S.A.; Iqbal, Z.; Rashid, M.; Habib, H.A. A novel image retrieval based on visual words integration of SIFT and SURF. *PLoS ONE* **2016**, *11*, e0157428. [CrossRef] [PubMed]
9. Ali, N.; Bajwa, K.B.; Sablatnig, R.; Mehmood, Z. Image retrieval by addition of spatial information based on histograms of triangular regions. *Comput. Electr. Eng.* **2016**, *54*, 539–550. [CrossRef]
10. O'Hara, S.; Draper, B.A. Introduction to the bag of features paradigm for image classification and retrieval. *arXiv* **2011**, arXiv:1101.3354.
11. Sivic, J.; Zisserman, A. Video Google: A text retrieval approach to object matching in videos. In Proceedings of the Ninth IEEE International Conference on Computer Vision, Nice, France, 13–16 October 2003; pp. 1470–1477.
12. Liu, P.; Miao, Z.; Guo, H.; Wang, Y.; Ai, N. Adding spatial distribution clue to aggregated vector in image retrieval. *EURASIP J. Image Video Process.* **2018**, *2018*, 9. [CrossRef]
13. Anwar, H.; Zambanini, S.; Kampel, M.; Vondrovec, K. Ancient Coin Classification Using Reverse Motif Recognition: Image-based classification of Roman Republican coins. *IEEE Signal Process. Mag.* **2015**, *32*, 64–74. [CrossRef]
14. Ali, N.; Zafar, B.; Riaz, F.; Dar, S.H.; Ratyal, N.I.; Bajwa, K.B.; Iqbal, M.K.; Sajid, M. A Hybrid Geometric Spatial Image Representation for scene classification. *PLoS ONE* **2018**, *13*, e0203339. [CrossRef] [PubMed]
15. Zafar, B.; Ashraf, R.; Ali, N.; Ahmed, M.; Jabbar, S.; Naseer, K.; Ahmad, A.; Jeon, G. Intelligent Image Classification-Based on Spatial Weighted Histograms of Concentric Circles. *Comput. Sci. Inf. Syst.* **2018**, *15*, 615–633. [CrossRef]
16. Lazebnik, S.; Schmid, C.; Ponce, J. Beyond bags of features: Spatial pyramid matching for recognizing natural scene categories. In Proceedings of the 2006 IEEE Computer Society Conference on Computer Vision and Pattern Recognition, New York, NY, USA, 17–22 June 2006; Voume 2, pp. 2169–2178.
17. Li, X.; Song, Y.; Lu, Y.; Tian, Q. Spatial pooling for transformation invariant image representation. In Proceedings of the 19th ACM International Conference on Multimedia, Scottsdale, AZ, USA, 28 November–1 December 2011; pp. 1509–1512.
18. Karmakar, P.; Teng, S.W.; Lu, G.; Zhang, D. Rotation Invariant Spatial Pyramid Matching for Image Classification. In Proceedings of the 2015 International Conference on Digital Image Computing: Techniques and Applications (DICTA), Adelaide, Australia, 23–25 November 2015; pp. 1–8.
19. Liu, D.; Hua, G.; Viola, P.; Chen, T. Integrated feature selection and higher-order spatial feature extraction for object categorization. In Proceedings of the 2008 IEEE Conference on Computer Vision and Pattern Recognition, Anchorage, AK, USA, 23–28 June 2008; pp. 1–8.

20. Khan, R.; Barat, C.; Muselet, D.; Ducottet, C. Spatial orientations of visual word pairs to improve bag-of-visual-words model. In Proceedings of the British Machine Vision Conference, Surrey, UK, 3–7 September 2012; pp. 89.1–89.11.

21. Zafar, B.; Ashraf, R.; Ali, N.; Ahmed, M.; Jabbar, S.; Chatzichristofis, S.A. Image classification by addition of spatial information based on histograms of orthogonal vectors. *PLoS ONE* **2018**, *13*, e0198175. [CrossRef] [PubMed]

22. Ahmed, K.T.; Irtaza, A.; Iqbal, M.A. Fusion of local and global features for effective image extraction. *Appl. Intell.* **2017**, *47*, 526–543. [CrossRef]

23. Wang, C.; Zhang, B.; Qin, Z.; Xiong, J. Spatial weighting for bag-of-features based image retrieval. In *Integrated Uncertainty in Knowledge Modelling and Decision Making*; Springer: Berlin/Heidelberg, Germany, 2013; pp. 91–100.

24. Zeng, S.; Huang, R.; Wang, H.; Kang, Z. Image retrieval using spatiograms of colors quantized by gaussian mixture models. *Neurocomputing* **2016**, *171*, 673–684. [CrossRef]

25. Yu, J.; Qin, Z.; Wan, T.; Zhang, X. Feature integration analysis of bag-of-features model for image retrieval. *Neurocomputing* **2013**, *120*, 355–364. [CrossRef]

26. Ali, N.; Mazhar, D.A.; Iqbal, Z.; Ashraf, R.; Ahmed, J.; Khan, F.Z. Content-Based Image Retrieval Based on Late Fusion of Binary and Local Descriptors. *arXiv* **2017**, arXiv:1703.08492.

27. Filliat, D. A visual bag of words method for interactive qualitative localization and mapping. In Proceedings of the 2007 IEEE International Conference on Robotics and Automation, Roma, Italy, 10–14 April 2007; pp. 3921–3926.

28. Hu, F.; Zhu, Z.; Mejia, J.; Tang, H.; Zhang, J. Real-time indoor assistive localization with mobile omnidirectional vision and cloud GPU acceleration. *AIMS Electron. Electr. Eng.* **2017**, *1*, 74–99. [CrossRef]

29. Li, L.; Feng, L.; Wu, J.; Sun, M.X.; Liu, S.l. Exploiting global and local features for image retrieval. *J. Cent. South Univ.* **2018**, *25*, 259–276. [CrossRef]

30. Liu, S.; Wu, J.; Feng, L.; Qiao, H.; Liu, Y.; Luo, W.; Wang, W. Perceptual uniform descriptor and ranking on manifold for image retrieval. *Inf. Sci.* **2018**, *424*, 235–249. [CrossRef]

31. Wu, J.; Feng, L.; Liu, S.; Sun, M. Image retrieval framework based on texton uniform descriptor and modified manifold ranking. *J. Vis. Commun. Image Represent.* **2017**, *49*, 78–88. [CrossRef]

32. Varish, N.; Pradhan, J.; Pal, A.K. Image retrieval based on non-uniform bins of color histogram and dual tree complex wavelet transform. *Multimedia Tools Appl.* **2017**, *76*, 15885–15921. [CrossRef]

33. Zou, Q.; Ni, L.; Zhang, T.; Wang, Q. Deep learning based feature selection for remote sensing scene classification. *IEEE Geosci. Remote Sens. Lett.* **2015**, *12*, 2321–2325. [CrossRef]

34. Xia, G.S.; Tong, X.Y.; Hu, F.; Zhong, Y.; Datcu, M.; Zhang, L. Exploiting Deep Features for Remote Sensing Image Retrieval: A Systematic Investigation. *arXiv* **2017**, arXiv:1707.07321.

35. Vassou, S.A.; Anagnostopoulos, N.; Amanatiadis, A.; Christodoulou, K.; Chatzichristofis, S.A. Como: A compact composite moment-based descriptor for image retrieval. In Proceedings of the 15th International Workshop on Content-Based Multimedia Indexing, Florence, Italy, 19–21 June 2017; p. 30.

36. Lowe, D.G. Distinctive image features from scale-invariant keypoints. *Int. J. Comput. Vis.* **2004**, *60*, 91–110. [CrossRef]

37. Anwar, H.; Zambanini, S.; Kampel, M. Encoding spatial arrangements of visual words for rotation-invariant image classification. In Proceedings of the 36th German Conference, GCPR 2014, Münster, Germany, 2–5 September 2014; pp. 443–452.

38. Tuytelaars, T. Dense interest points. In Proceedings of the 2010 IEEE Conference on Computer Vision and Pattern Recognition (CVPR), San Francisco, CA, USA, 13–18 June 2010; pp. 2281–2288.

39. Mehmood, Z.; Anwar, S.M.; Ali, N.; Habib, H.A.; Rashid, M. A novel image retrieval based on a combination of local and global histograms of visual words. *Math. Probl. Eng.* **2016**, *2016*, 8217250. [CrossRef]

40. Li, J.; Wang, J.Z. Real-time computerized annotation of pictures. *IEEE Trans. Pattern Anal. Mach. Intell.* **2008**, *30*, 985–1002. [PubMed]

41. Oliva, A.; Torralba, A. Modeling the shape of the scene: A holistic representation of the spatial envelope. *Int. J. Comput. Vis.* **2001**, *42*, 145–175. [CrossRef]

42. Zhou, J.; Liu, X.; Liu, W.; Gan, J. Image retrieval based on effective feature extraction and diffusion process. *Multimedia Tools Appl.* **2018**, 1–28. [CrossRef]

43. Deselaers, T.; Keysers, D.; Ney, H. Features for image retrieval: An experimental comparison. *Inf. Retr.* **2008**, *11*, 77–107. [CrossRef]

44. Zhou, Z.H.; Xu, J.M. On the relation between multi-instance learning and semi-supervised learning. In Proceedings of the 24th International Conference on Machine Learning, Corvalis, OR, USA, 20–24 June 2007; pp. 1167–1174.

45. Andrews, S.; Tsochantaridis, I.; Hofmann, T. Support vector machines for multiple-instance learning. In *Advances in Neural Information Processing Systems*; MIT Press: Cambridge, MA, USA, 2002; pp. 577–584.

46. Shrivastava, P.; Bhoyar, K.; Zadgaonkar, A. Image Classification Using Fusion of Holistic Visual Descriptions. *Int. J. Image Graph. Signal Process.* **2016**, *8*, 47. [CrossRef]

47. Zang, M.; Wen, D.; Liu, T.; Zou, H.; Liu, C. A pooled Object Bank descriptor for image scene classification. *Expert Syst. Appl.* **2018**, *94*, 250–264. [CrossRef]

48. Yin, H. Scene Classification Using Spatial Pyramid Matching and Hierarchical Dirichlet Processes. MSc Thesis, Rochester Institute of Technology, Rochester, NY, USA, 2010.

49. Walia, E.; Verma, V. Boosting local texture descriptors with Log-Gabor filters response for improved image retrieval. *Int. J. Multimedia Inf. Retr.* **2016**, *5*, 173–184. [CrossRef]

50. Jia, Y.; Shelhamer, E.; Donahue, J.; Karayev, S.; Long, J.; Girshick, R.; Guadarrama, S.; Darrell, T. Caffe: Convolutional architecture for fast feature embedding. In Proceedings of the 22nd ACM International Conference on Multimedia, OR, Florida, USA, 3–7 November 2014; pp. 675–678.

51. Chatfield, K.; Simonyan, K.; Vedaldi, A.; Zisserman, A. Return of the devil in the details: Delving deep into convolutional nets. *arXiv* **2014**, arXiv:1405.3531.

52. Szegedy, C.; Liu, W.; Jia, Y.; Sermanet, P.; Reed, S.; Anguelov, D.; Erhan, D.; Vanhoucke, V.; Rabinovich, A. Going deeper with convolutions. In Proceedings of the IEEE Conference on Computer Vision and Pattern Recognition, Boston, MA, USA, 7–12 June 2015; pp. 1–9.

53. Mousavian, A.; Kosecka, J. Deep convolutional features for image based retrieval and scene categorization. *arXiv* **2015**, arXiv:1509.06033.

54. Chathurani, N.; Geva, S.; Chandran, V.; Cynthujah, V. Content-Based Image (Object) Retrieval with Rotational Invariant Bag-of-Visual Words Representation. In Proceedings of the 2015 IEEE 10th International Conference on Industrial and Information Systems (ICIIS), Peradeniya, Sri Lanka, 18–20 December 2015; pp. 152–157.

Article

Double Low-Rank and Sparse Decomposition for Surface Defect Segmentation of Steel Sheet

Shiyang Zhou, Shiqian Wu *, Huaiguang Liu, Yang Lu and Nianzong Hu

School of Machinery and Automation, Wuhan University of Science and Technology, Wuhan 430081, China;
zhoushiyang@wust.edu.cn (S.Z.); liuhuaiguang@wust.edu.cn (H.L.); luyanglymj@gmail.com (Y.L.);
hnz307517599@gmail.com (N.H.)
* Correspondence: shiqian.wu@wust.edu.cn; Tel.: +86-27-6886-2478

Received: 2 August 2018; Accepted: 9 September 2018; Published: 12 September 2018

Featured Application: **The proposed DLRSD-based segmentation method can be applied for other industrial products, such as glass, fabric, LCD and AMOLED.**

Abstract: Surface defect segmentation supports real-time surface defect detection system of steel sheet by reducing redundant information and highlighting the critical defect regions for high-level image understanding. Existing defect segmentation methods usually lack adaptiveness to different shape, size and scale of the defect object. Based on the observation that the defective area can be regarded as the salient part of image, a saliency detection model using double low-rank and sparse decomposition (DLRSD) is proposed for surface defect segmentation. The proposed method adopts a low-rank assumption which characterizes the defective sub-regions and defect-free background sub-regions respectively. In addition, DLRSD model uses sparse constrains for background sub-regions so as to improve the robustness to noise and uneven illumination simultaneously. Then the Laplacian regularization among spatially adjacent sub-regions is incorporated into the DLRSD model in order to uniformly highlight the defect object. Our proposed DLRSD-based segmentation method consists of three steps: firstly, using DLRSD model to obtain the defect foreground image; then, enhancing the foreground image to establish the good foundation for segmentation; finally, the Otsu's method is used to choose an optimal threshold automatically for segmentation. Experimental results demonstrate that the proposed method outperforms state-of-the-art approaches in terms of both subjective and objective tests. Meanwhile, the proposed method is applicable to industrial detection with limited computational resources.

Keywords: surface defect of steel sheet; image segmentation; saliency detection; low-rank and sparse decomposition

---

1. Introduction

Surface defect detection plays an important role in quality enhancement in industrial product manufacturing. However, traditional defect detection is performed by human eyes, which yields low efficiency and high missing rate. Currently, vision-based automated defect detection has drawn much attention, which has important theoretical and practical value [1–4]. In automatic surface inspection of steel sheet, segmentation of surface defect is a significant step, which generates a binary map to identify defects. In the past two decades, commonly-used segmentation methods can be classified into three categories: statistical-based methods, filter-based methods and model-based methods. Statistical-based methods, such as Otsu's method [5], gray level co-occurrence matrix, local binary pattern, maximum entropy, region growing and morphological watersheds, are used to evaluate the spatial distribution of pixel intensities for segmentation. Filter-based methods, such as discrete Fourier transform [6], discrete Gabor transform [7] and discrete wavelet transform [8,9], apply a bank of filters to the image, in which the energies of the filters response are utilized as features to segment the defects. Model-based approaches obtain certain models with specific feature distributions or other attributes using diverse descriptors, for instance, level set, fuzzy theory, partial differential equations and texture patterns.

Most recently, with the development of saliency detection technology, segmentation methods that use saliency map are gradually rising in the industrial defect inspection field. This method constructs a saliency map that highlights the defect regions standing out from the rest of the image, which provide the good foundation for segmentation. Guan et al. [10] proposed saliency map construction method using Gaussian pyramid decomposition. Then segmentation is conducted with the saliency map. This model exhibits good performance for strip steel defect detection. Li et al. [11] devised a low-rank representation-based saliency detection model for textile fabric defect detection. Zhao et al. [12] also presented a novel saliency detection model, which obviously improve the accuracy of automated defect segmentation.

These methods achieve good results on defect segmentation for a certain and homogeneous texture, but remain a challenging issue for segmentation with miscellaneous textures due to random disturbance. Specially, as the surface defect image of steel sheet has a low signal-to-noise ratio, low contrast between defect object and background, heterogeneous and scattered defect, cluttered and complicated background, these methods still lack of accuracy and suffer from limited adaptability and robustness in industrial practice.

Usually, a defect-free surface in industrial products has consistent texture. The emergence of defects can be regarded as the foreground object superposed in the regular-texture background. As shown in Figure 1, a surface defect image of steel sheet I is decomposed into two parts: relatively homogeneous background image B and a defect foreground image F that is the desired image for the following segmentation.

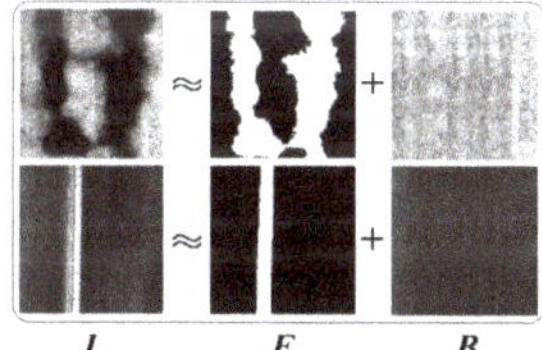

Figure 1. Illustration of surface defect image decomposition.

Inspired by the above analysis, an easy-to-implement method based on double low-rank and sparse decomposition (DLRSD) is proposed in this paper for surface defect segmentation. Considering double low-rank and sparse characteristics of surface defect image, combined with a local consistency constrain among spatially adjacent sub-regions by imposing Laplacian regularization, the feature matrix that form by I can be adaptively decomposed into foreground feature matrix that form by defect foreground image F and background feature matrix that form by background image B in a certain feature space, respectively. Specifically, the foreground image F is served as the source image for segmentation, which can better cope with the intra-class variations and background clutters, leading to a higher performance. Theoretical analysis and experimental results demonstrate the feasibility and effectiveness of the proposed DLRSD-based segmentation method for the surface defect of steel sheet. At the same time, it provides an interesting perspective for the industrial product's surface defect segmentation.

The rest of this paper is organized as follows. In Section 2, we review some existing saliency detection methods, especially the structural matrix decomposition-based methods. In Section 3 we introduce the proposed DLRSD model, including formulation and optimization. Section 4 presents the DLRSD-based defect segmentation method. Also, we give more detail on enhancing the original defect foreground image. Section 5 describes experimental results between our proposed method and some state-of-the-art methods. Finally, conclusions are given in Section 6.

2. Related Work

During the past few years, there are many methods attempting to segment the salient object from the saliency map of an input image [13–19]. The quality and effectiveness of segmentation are decided by the quality of saliency map [20]. Based on the milestone work, some structural matrix decomposition-based methods transform a saliency detection problem into a feature subspace decomposition problem, which can improve detection results in terms of both speed and accuracy. Particularly, many studies conclude that low-rank matrix decomposition-based methods can obtain better saliency detection performance. These methods assume that an image can be represented as a combination of a highly redundant part (e.g., visually consistent background regions) and a sparse part (e.g., salient object foreground regions). Therefore, given the feature matrix of an input image, it can be decomposed into a low-rank matrix corresponding to the non-salient background and a sparse matrix corresponding to the salient foreground objects. Yan et al. [21] employed sparse coding as a feature representation vector of image. Zou et al. [22] designed multi-scale superpixel segmentation to construct the feature matrix and prior matrix. Although Shen et al. [23] adopted learnt linear transformation of the feature space to integrate low-level features and high-level prior knowledge, the learnt transform matrix is correlated for training data set. Unfortunately, the sparsity assumption of the salient objects can't be guaranteed universally, especially when the salient objects with big size occupy most of the image, and then suffer from limited adaptability. Therefore, Peng et al. [24] developed tree-structured sparsity-inducing regularization and Laplacian regularization to disentangle the salient objects and background precisely, and then obtained competitive results. But it may be difficult to suppress some small background regions with distinctive appearances because of the constructed index-tree is not precise enough. Subsequently, Sun et al. [25] presented diversity-induced regularization based on Hilbert–Schmidt independence criterion, which make the background much cleaner in the saliency map and boost the saliency detection performance. But, they don't consider the low-rank characteristic for the foreground regions and background regions simultaneously, and ignore the spatial and pattern relations of image regions, which may lead to very noisy saliency map and influences on the final segmentation performance.

To solve the problems mentioned above, the proposed DLRSD model considers the correlation between defective regions and defect-free regions, which is different from existing methods in essence. Besides, it uses the nuclear norm to depict the low-rank property of defect object rather than consider it as the sparse noises, which can produce more accurate and reliable saliency map that represents the defect foreground image.

3. Double Low-Rank and Sparse Decomposition Model

In this section, we will introduce the proposed DLRSD model and optimization procedure in details.

3.1. Problem Formulation

Let $\{R_1, R_2, \cdots, R_K\}$ be a set of K non-overlapping sub-regions of a surface defect image I, all the feature vectors of sub-regions can construct the feature matrix D. The proposed DLRSD model is to design an effective model to decompose the feature matrix D into a feature matrix S that represents a defect foreground image F and a feature matrix L that represents a background image B:

$$D = S + L \tag{1}$$

In order to separate defect regions and background regions accurately, some constrains are needed for characterizing two feature matrices S and L. According to the surface defect image I that is pre-processed by superpixel segmentation, both defect foreground and background contain multiple homogeneous and highly similar sub-regions, for each defective sub-region, the corresponding locations in saliency map has high probability in larger brightness, indicating that it has higher saliency value. Besides, different defective sub-regions are highly correlated and the corresponding feature vectors lie in a low-dimensional subspace. Therefore, the feature matrix S is expected to be low-rank. Meanwhile, most of background sub-regions tend to have lower saliency value. They are strongly correlated and lie in a low-dimensional feature subspace that is independent of the defect foreground subspace. The strong correlations among the background sub-regions suggest that feature matrix L may have the low-rank property. What is more, in order to reduce the influence of noises and enhance the robustness to uneven illumination, we assume that the background lies in a sparse feature subspace and can be characterized by a sparse matrix.

Based on above analysis, the structured matrix decomposition model can be constructed as follows:

$$\min_{L,S}\left(\mathrm{rank}(L) + \alpha\,\mathrm{rank}(S) + \beta\Theta(S,L) + \gamma||L||_0\right)$$
$$\mathrm{s.t.}\,D = S + L \tag{2}$$

where $\mathrm{rank}(\cdot)$ denotes the rank of matrix; $||\cdot||_0$ denotes l_0 norm of matrix, which equals the number of non-zero element of matrix; $\Theta(S,L)$ denotes the regularization to enlarge the margin and reduce the coherence between the feature subspaces induced by S and L; $D \in \mathbb{R}^{d\times K}$ represents the feature matrix; $\alpha > 0$, $\beta > 0$ and $\gamma > 0$ are regularization parameters.

To separate the defect object from the background easily, spatially adjacent sub-regions with smaller spatial distance and more similar feature vector should be assigned to similar and higher weight values, the local invariance assumption [26] based Laplacian regularization $\Theta(S,L)$ [24] can be defined as follows:

$$\Theta(S,L) = \frac{1}{2}\sum_{i,j=1}^{K}||s_i - s_j||_2^2 w_{ij} = \mathrm{tr}\left(SMS^T\right) \tag{3}$$

where $\mathrm{tr}(\cdot)$ denotes the trace of a matrix; s_i denotes the i-th column of matrix S; the element w_{ij} of affinity matrix $W \in \mathbb{R}^{K\times K}$ denotes the weight that represents the feature similarity between sub-regions R_i and R_j; $M \in \mathbb{R}^{K\times K}$ is a Laplacian matrix.

According to the undirected graph model from a surface defect image, each sub-region is represented by a node, the affinity matrix W is

$$w_{ij} = \begin{cases} \exp\left(\dfrac{-||p_i - p_j||_2^2}{2\sigma_p^2}\right)\exp\left(\dfrac{-||\bar{f}_i - \bar{f}_j||_2^2}{2\sigma_f^2}\right) & R_i \text{ and } R_j \text{ are spatially adjacent} \\ 0 & \text{otherwise} \end{cases} \tag{4}$$

where $p_i \in \mathbb{R}^2$ and $p_j \in \mathbb{R}^2$ denote the central coordinate of R_i and R_j; $\bar{f}_i \in \mathbb{R}^d$ and $\bar{f}_j \in \mathbb{R}^d$ denote the feature vector of R_i and R_j; $\exp\left(\dfrac{-||p_i - p_j||_2^2}{2\sigma_p^2}\right)$ represents spatial connectivity between R_i and R_j, which represents the spatial contiguity; $\exp\left(\dfrac{-||\bar{f}_i - \bar{f}_j||_2^2}{2\sigma_f^2}\right)$ gives the feature similarity between R_i and R_j; σ_p and σ_f are two scalars.

The Laplacian matrix M is

$$M_{ij} = \begin{cases} -w_{ij} & i \neq j \\ \sum_{i\neq j} w_{ij} & \text{otherwise} \end{cases} \tag{5}$$

In particular, the Laplacian regularization $\Theta(S,L)$ can preserve the local consistency and invariance among the spatially adjacent sub-regions with similar saliency values in saliency maps. More specifically, the defect foreground is more uniformly highlighted and the background noise is also better suppressed, and eventually separates the defect from the background as much as possible.

3.2. Optimization

As rank$(\cdot)$ and $||\cdot||_0$ are not convex, Equation (2) is NP-hard problem. A common heuristic criterion is to replace rank$(\cdot)$ and $||\cdot||_0$ are replaced by nuclear norm $||\cdot||_*$ and l_1 norm $||\cdot||_1$ respectively. It has been shown that nuclear norm-based models can obtain the optimal low-rank solution in many kinds of applications [27,28]. Then, Equation (2) can be converted to the following convex surrogate optimization problem:

$$\min_{L,S}\left(||L||_* + \alpha||S||_* + \beta\mathrm{tr}\left(SMS^T\right) + \gamma||L||_1\right)$$
$$\text{s.t.}D = S + L \tag{6}$$

where $||\cdot||_*$ equals the sum of singular values of matrix; $||\cdot||_1$ equals the sum of the absolute values of each element of matrix. For a matrix $A = a_{ij} \in \mathbb{R}^{m\times n}$, $||A||_p = \left(\sum_{i=1}^{m}\sum_{j=1}^{n}|a_{ij}|^p\right)^{\frac{1}{p}}$.

To solve Equation (6) efficiently, the alternating direction method (ADM) algorithm [29] can be adopted. By introducing the auxiliary variables H and J, the augmented Lagrange function is given as follows:

$$\begin{aligned}
O(&L, S, H, J, Y_1, Y_2, Y_3, \mu)\\
= &||L||_* + \alpha||S||_* + \beta\mathrm{tr}\left(HMH^T\right) + \gamma||J||_1\\
&+\mathrm{tr}\left(Y_1^T(D - L - S)\right) + \tfrac{\mu}{2}||D - L - S||_F^2\\
&+\mathrm{tr}\left(Y_2^T(H - S)\right) + \tfrac{\mu}{2}||H - S||_F^2\\
&+\mathrm{tr}\left(Y_3^T(J - L)\right) + \tfrac{\mu}{2}||J - L||_F^2
\end{aligned} \tag{7}$$

where $||\cdot||_F^2$ denotes the Frobenius norm of matrix, which is defined as the sum of squares of each element of matrix; Y_1, Y_2 and Y_3 are Lagrange multipliers; $\mu > 0$ is a penalty parameter.

Therefore, Equation (7) can be converted to the following equivalent optimization problem:

$$\begin{aligned}
O(&L, S, H, J, Y_1, Y_2, Y_3, \mu)\\
= &\tfrac{1}{2}\left\|D - L - S + \tfrac{Y_1}{\mu}\right\|_F^2 + \tfrac{1}{2}\left\|H - S + \tfrac{Y_2}{\mu}\right\|_F^2 + \tfrac{1}{2}\left\|J - L + \tfrac{Y_3}{\mu}\right\|_F^2\\
&+\tfrac{1}{\mu}||L||_* + \tfrac{\alpha}{\mu}||S||_* + \tfrac{\beta}{\mu}\mathrm{tr}\left(HMH^T\right) + \tfrac{\gamma}{\mu}||J||_1
\end{aligned} \tag{8}$$

The above optimization problem can be solved by alternately updating one variable while others fixed. The detailed ADM algorithm for proposed DLRSD model is summarized in Algorithm 1.

(1) Updating H

In order to solve H, the optimal solution can be obtained by Equation (9):

$$\min_{H}\left(\frac{1}{2}\left\|H - S + \frac{Y_2}{\mu}\right\|_F^2 + \frac{\beta}{\mu}\mathrm{tr}\left(HMH^T\right)\right) \tag{9}$$

Differentiating it with respect to H, and let it to be zero, therefore

$$H - S + \frac{Y_2}{\mu} + \frac{2\beta}{\mu}HM = 0 \tag{10}$$

The close-form solution can be obtained as follows:

$$H^* = \left(S - \frac{Y_2}{\mu}\right)\left(I + \frac{2\beta}{\mu}M\right)^{-1} \tag{11}$$

(2) Updating J

In order to solve J, the optimal solution can be obtained by Equation (12):

$$\min_{J} \left(\frac{1}{2}\left\|L - \frac{Y_3}{\mu} - J\right\|_F^2 + \frac{\gamma}{\mu}||J||_1 \right) \tag{12}$$

The solution is

$$J^* = \Psi_{\frac{\gamma}{\mu}}\left(L - \frac{Y_3}{\mu}\right) \tag{13}$$

where $\Psi_{\frac{\gamma}{\mu}}(\cdot)$ denotes soft-thresholding shrinkage operator, which is defined as

$$\Psi_{\frac{\gamma}{\mu}}(T) = \text{sgn}(T)\max\left(|T| - \frac{\gamma}{\mu}, 0\right) = \begin{cases} T_{ij} - \frac{\gamma}{\mu} & T_{ij} > \frac{\gamma}{\mu} \\ 0 & -\frac{\gamma}{\mu} \leq T_{ij} \leq \frac{\gamma}{\mu} \\ T_{ij} + \frac{\gamma}{\mu} & T_{ij} < -\frac{\gamma}{\mu} \end{cases} \tag{14}$$

where T denotes a matrix, T_{ij} denotes the (i, j)-th element of T, $\text{sgn}(T)$ is the matrix whose entries are the signs of those of T.

(3) Updating L

In order to solve L, the optimal solution can be obtained by Equation (15):

$$\min_{L} \left(\frac{1}{2}\left\|D - S + \frac{Y_1}{\mu} - L\right\|_F^2 + \frac{1}{2}\left\|J + \frac{Y_3}{\mu} - L\right\|_F^2 + \frac{1}{\mu}||L||_* \right) \tag{15}$$

It can be rewritten as follows:

$$\min_{L} \left(\frac{1}{2}\left\|\frac{1}{2}\left(D - S + J + \frac{Y_1 + Y_3}{\mu}\right) - L\right\|_F^2 + \frac{1}{4\mu}||L||_* \right) \tag{16}$$

Its solution is

$$L^* = U\Psi_{\frac{1}{4\mu}}(\Sigma)V^T \tag{17}$$

where $(U, \Sigma, V) = \text{svd}\left[\frac{1}{2}\left(D - S + J + \frac{Y_1 + Y_3}{\mu}\right)\right]$, $\text{svd}(\cdot)$ denotes singular value decomposition operator.

(4) Updating S

In order to solve S, the optimal solution can be obtained by Equation (18):

$$\min_{S} \left(\frac{1}{2}\left\|D - L + \frac{Y_1}{\mu} - S\right\|_F^2 + \frac{1}{2}||H + \frac{Y_2}{\mu} - S||_F^2 + \frac{\alpha}{\mu}||S||_* \right) \tag{18}$$

It can be rewritten as follows:

$$\min_{S} \left(\frac{1}{2}\left\|\frac{1}{2}\left(D - L + H + \frac{Y_1 + Y_2}{\mu}\right) - S\right\|_F^2 + \frac{\alpha}{4\mu}||S||_* \right) \tag{19}$$

Its solution is

$$S^* = U\Psi_{\frac{\alpha}{4\mu}}(\Sigma)V^T \tag{20}$$

where $(U, \Sigma, V) = \text{svd}\left[\frac{1}{2}\left(D - L + H + \frac{Y_1 + Y_2}{\mu}\right)\right]$.

(5) Updating Y_1, Y_2 and Y_3

$$\begin{aligned}
Y_1 &= Y_1 + \mu(D - L - S) \\
Y_2 &= Y_2 + \mu(H - S) \\
Y_3 &= Y_3 + \mu(J - L)
\end{aligned} \tag{21}$$

(6) Updating μ

$$\mu = \min(\rho\mu, \mu_{\max}) \tag{22}$$

where $0 < \rho < 1$.

Algorithm 1 Solving DLRSD via ADM.

Input: Data matrix $D \in \mathbb{R}^{d \times K}$, parameters $\alpha > 0, \beta > 0, \gamma > 0$ and $\epsilon > 0$

Output: The optimal solution $L^* \in \mathbb{R}^{d \times K}$ and $S^* \in \mathbb{R}^{d \times K}$

1: Initializing

$L = S = H = J = 0, Y_1 = Y_2 = Y_3 = 0$ $\mu = 10^{-1}, \mu_{max} = 10^{10}, \rho = 1.1, NUM = 100, k = 1$

While $k \leq NUM$ OR $||D - L - S||_F^2 < \epsilon, ||H - S||_F^2 < \epsilon, ||J - L||_F^2 < \epsilon$

    2: Updating H

    $H^* = \left(S - \frac{Y_2}{\mu}\right)\left(I + \frac{2\beta}{\mu}M\right)^{-1}$

    3: Updating J

    $J^* = \Psi_{\frac{\gamma}{\mu}}\left(L - \frac{Y_3}{\mu}\right)$

    4: Updating L

    $L^* = U\Psi_{\frac{1}{4\mu}}(\Sigma)V^T$

    5: Updating S

    $S^* = U\Psi_{\frac{\alpha}{4\mu}}(\Sigma)V^T$

    6: Updating Y_1, Y_2 and Y_3

    $Y_1 = Y_1 + \mu(D - L - S)$

    $Y_2 = Y_2 + \mu(H - S)$

    $Y_3 = Y_3 + \mu(J - L)$

    7: Updating μ

    $\mu = \min(\rho\mu, \mu_{\max})$

    8: Iteration

    $k = k + 1$

End While

---

4. DLRSD-Based Surface Defect Segmentation

In this section, we describe how to apply the proposed DLRSD model to surface defect segmentation. The segmentation method has three stages. In first stage, we use DLRSD model to obtain the defect foreground image F. While in second stage, we utilize regression optimization to enhance F. At last, the segmentation is finished by Otsu's method. The framework of DLRSD-based segmentation method is shown in Figure 2, the detailed procedure is summarized in Algorithm 2.

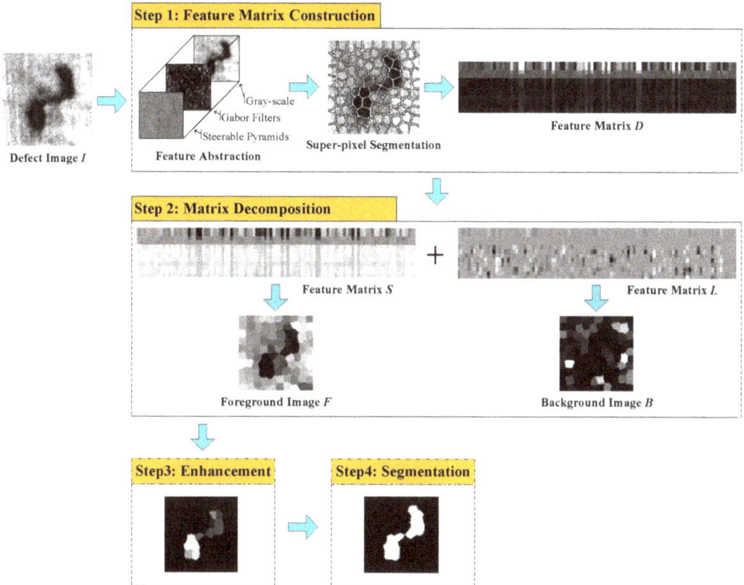

Figure 2. Diagram of the proposed double low-rank and sparse decomposition (DLRSD)-based segmentation method for surface defect image.

4.1. Feature Matrix Construction

According to [23,24], for each pixel $\{I_i\}_{i=1,2,...,N}$ of a surface defect image I, where N denotes the number of pixels, different types of low-level visual features, including gray-scale, Gabor filters and steerable pyramids, are extracted.

(1) Gray-scale

The pixel value of each pixel in defect image I is extracted for gray-scale feature, which is normalized by subtracting its mean value over the entire image.

(2) Gabor filters

Gabor filters responses with eight directions on two different scales are performed on the defect image I, yielding 16 filter responses for each pixel.

(3) Steerable pyramids

Steerable pyramid filters with four directions on two different scales are performed on the defect image I, yielding 8 filter responses for each pixel.

All those 25 features are then stacked vertically to construct a 25-dimension feature vector $\{f_i\}_{i=1,2,...,N} \in \mathbb{R}^d$ for each pixel. Then, in order to improve the efficiency of defect detection and achieve the better structural information about defect image, we conduct superpixel segmentation for image I by adaptive simple linear iterative clustering (ASLIC) algorithm [30]. Each compact, edge-aware and perceptually homogeneous sub-region $\{R_j\}_{j=1,2,...,K}$ can be represented by feature vector $\bar{f}_j \in \mathbb{R}^d$, where $\bar{f}_j$ represents the mean feature vector of all pixels that belong to R_j, where K denotes the number of sub-regions. By arranging $\bar{f}_j$ into a matrix, the feature matrix $D = (\bar{f}_1, \bar{f}_2, \cdots, \bar{f}_K) \in \mathbb{R}^{d \times K}$ of image I is obtained.

4.2. Matrix Decomposition

According to Algorithm 1, the input feature matrix D is decomposed into structured components S and L. According to the obtained $S^* = (s_1, s_2, \cdots, s_K) \in \mathbb{R}^{d \times K}$ and $L^* = (l_1, l_2, \cdots, l_K) \in \mathbb{R}^{d \times K}$, each column of these two matrixes represents the feature vector of corresponding sub-region, respectively. Then, we transfer S^* and L^* from the feature domain to the spatial domain for constructing saliency map. The saliency value of each sub-region in foreground image F and background image B are $\max(s_j)$ and $\max(l_j)$, respectively, where $s_j \in \mathbb{R}^{d \times 1}$ and $l_j \in \mathbb{R}^{d \times 1}$ denotes the j-th column of S^* and L^*, $\max(\cdot)$ denotes the maximum component of the vector, $j = 1, 2, \ldots, K$. After allocating the saliency value to corresponding pixels and normalizing, the defect foreground image F and background image B can be obtained.

4.3. Enhancement

As shown in Figure 2, the original foreground image F can be enhanced in consistency, completeness of defect objects and suppression of background noise. In the paper, the regression optimization method is adopted by combining foreground image F and background image B. The optimization problem can be formulated as follows:

$$\min_{s_i} \left(\sum_{i=1}^{K} w_i^f (s_i - 1)^2 + \sum_{i=1}^{K} w_i^b s_i^2 + \sum_{i,j=1}^{K} w_{ij} (s_i - s_j)^2 \right) \tag{23}$$

where w_i^f denotes saliency value of sub-region in foreground image F, $w_i^f = Val(P_j)$; w_i^b denotes saliency value of sub-region in background image B, $w_i^b = Val(Q_j)$; s_i denotes the optimized saliency value of sub-region in foreground image F.

According to $s = (s_1, s_2, \cdots, s_K)^T \in \mathbb{R}^{K \times 1}$, $W^b = \operatorname{diag}\left[\left(w_1^b, w_2^b, \cdots, w_K^b \right)^T \right] \in \mathbb{R}^{K \times K}$, and $W^f = \operatorname{diag}\left[\left(w_1^f, w_2^f, \cdots, w_K^f \right)^T \right] \in \mathbb{R}^{K \times K}$, the Equation (23) can be reformulated as follows:

$$\min_{s} \left(s^T W^b s + s^T W^f s - 2W^f s + W^f 1 + 2s^T Ms \right) \tag{24}$$

where $1 \in \mathbb{R}^{K \times 1}$ denotes a one vector, $M \in \mathbb{R}^{K \times K}$ denotes the same Laplacian matrix in Equation (5). Differentiating it with respect to s, and let it to be zero, therefore

$$2W^b s + 2W^f s - 2W^f 1 + 4Ms = 0 \tag{25}$$

The solution is

$$s = \left(W^f + W^b + 2M \right)^{-1} W^f 1 \tag{26}$$

Through Equation (26), the sub-regions within the same class (foreground or background) have more similar saliency values while the sub-regions from different classes (foreground and background) have different saliency values. The saliency value of defect sub-region in foreground image is bigger, while the saliency value of background sub-region is smaller, so that the surface defect object can be highlighted further.

4.4. Segmentation

After obtaining the enhanced foreground image F, the high-quality binary image can be obtained through a simple Otsu's method. In binary image of surface defect, white pixel represents surface defect regions, and black pixel represents background regions.

Algorithm 2 DLRSD-based defect segmentation.

Input: Surface defect image I
Output: Binary segmentation image
1: Construct the feature matrix D of I
2: Run Algorithm 1 to get the defect foreground feature matrix S
3: Enhance defect foreground image F
4: Segment enhanced F by Otsu's method

---

5. Experiment

In this section, several experiments are conducted to verify the superiority of our proposed method. We first introduce the experimental setups, which include parameters settings and evaluation metrics. Then, computational complexity, convergence, noise immunity and segmentation results are discussed. At last, the qualitative and quantitative comparisons are presented.

5.1. Experimental Setup

In order to verify and evaluate the effectiveness and robustness of the proposed method, we have adopted the NEU surface defect database established by Kechen Song [12] in our experiments. The size of each surface defect image is 200×200 and the number of image is 300 per class. Two typical surface defect images, such as Patch and Scratch, are selected in the experiments. Our proposed method is compared with eight representative saliency detection methods quantitatively and qualitatively, such as RPCA [28], IS [13], ULR [23], RBD [31], SBD [32], DSR [33], RS [16] and SMF [24], where RPCA, IS, ULR, RBD, SBD, DSR, RS and SMF represent the method of robust principal component analysis, image signature, unified low rank matrix recovery, robust background detection, spaces of background-based distribution, dense and sparse reconstruction, ranking saliency and structured matrix decomposition, respectively. Only a few examples are shown in the paper, the whole segmentation results are uploaded in Baidu Disk (https://pan.baidu.com/s/1QkwFfWsUE9hKL86prlL4nw, Code: iydw).

5.1.1. Parameters Settings

In Equation (6), α represents the redundancy of defect foreground, β represents the uniformity of defect foreground, γ represents the sparsity of background. We conduct some experiments to study the detection performance variation with respect to different α, β and γ, which shows that the detection performance can achieve a high level at $\alpha \in (0.2, 0.4)$, $\beta \in (0.9, 1.3)$ and $\gamma \in (0.05, 0.25)$. In order to achieve the better segmentation results, α, β and γ are set to 0.35, 1.2 and 0.1, respectively. For other methods in our comparison, we use the source codes provided by the authors with default parameters.

5.1.2. Evaluation Metrics

The qualitative evaluation metrics refers to evaluate the detection performance based on human subjective feeling. For example, the boundary of surface defect is clear, and the contrast between defect object and background is obvious.

There are five quantitative evaluation metrics, including precision-recall (P-R) curve, receiver operating characteristic (ROC) curve, average F-Measure (F_ζ), area under ROC (AUC) and mean square error (MAE). They are defined as follows:

$$FPR = \frac{FP}{FP + TN} \tag{27}$$

$$TPR = \frac{TP}{TP + FN} \tag{28}$$

$$F_\zeta = \frac{1}{N} \sum_{i=1}^{N} \frac{(\zeta^2 + 1) \times \text{precision} \times \text{recall}}{\zeta^2 \times \text{precision} + \text{recall}} \tag{29}$$

$$MAE = \frac{\sum_{i=1}^{H} \sum_{j=1}^{W} |S(i,j) - G(i,j)|}{H \times W} \tag{30}$$

where a pixel that belonging to defect is defined as a positive example, and a pixel that belonging to background is defined as a negative example; true positive (TP) indicates that the positive pixel is judged correctly, true negative (TN) indicates that the negative pixel is judged correctly, false positive (FP) indicates that the positive pixel is judged as the negative pixel mistakenly, false negative (FN) indicates that the negative pixel is judged as the positive pixel mistakenly; precision $= TP/(TP + FP)$, recall $= TP/(TP + FN)$; N represents the number of surface defect image samples of the same class, H and W denotes the height and width of surface defect image, respectively; precision is defined as the percentage of defect pixels correctly assigned, while recall is the ratio of correctly detected defect pixels to all true defect pixels. F_ζ represents the weighted harmonic mean of precision and recall. Besides, P-R curve is obtained by binarizing the saliency map using a number of thresholds ranging from 0 to 255; TPR represents true positive rate, FPR represents false positive rate; MAE measures the dissimilarity between the saliency map S and the ground truth G.

5.2. Experimental Results Analysis

5.2.1. Analysis of Computational Complexity

According to Algorithm 1, the main computational load is singular value decomposition operation in updating matrix S and L. As the size of matrix D is $d \times K$, the computational complexity is reduced from (dK^2) to (drK) by the low-rank constraint, where r denotes the rank of matrix. In our experiments, $d = 25$, $K = 100$, so the computational complexity is low.

5.2.2. Analysis of Convergence

According to Algorithm 1 and ADM algorithm, when penalty parameter sequence $\{\mu_k\}$ is increasing monotonically and bounded, the Lagrange multipliers Y_1, Y_2 and Y_3 can converge to the optimal solution linearly; when $\{\mu_k\}$ is increasing monotonically and unbounded, Y_1, Y_2 and Y_3 can converge to the optimal solution super-linearly. As shown in Figure 3, the x-axis denotes the iteration number, and the y-axis is the value of objective function. We can see that the objective function value converges in a very fast manner, usually within 40 iterations, which also proves the fast convergence property of the proposed DLRSD model.

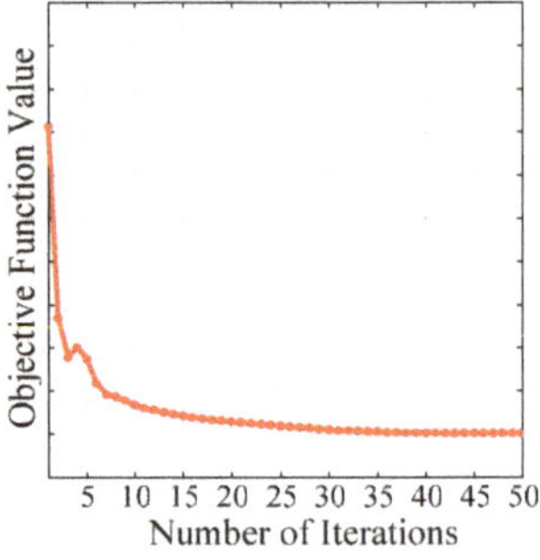

Figure 3. Convergence Curve of DLRSD model.

5.2.3. Analysis of Segmentation Results

From enhanced defect foreground image shown in Figure 4c, it has achieved the goal of "highlight the foreground and suppressing the background". It can accurately extract the entire defect object and assigns nearly uniform saliency values to all sub-regions within the defect objects. Figure 4d shows that the segmentation images are similar to ground truth, the whole defect object can be uniformly highlighted, and boundary of defect object is well-defined. Therefore, we locate the defects accurately.

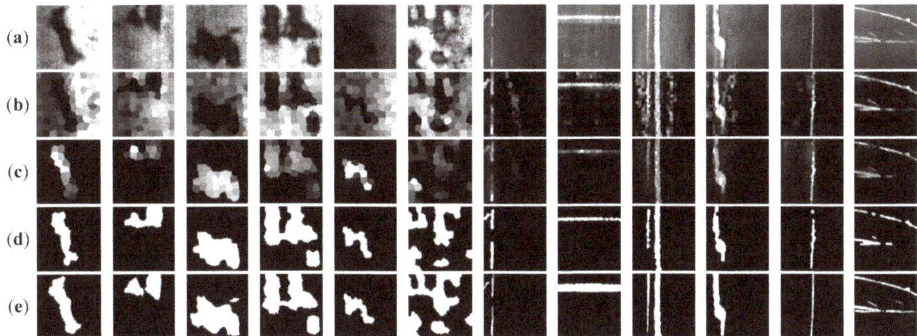

Figure 4. Segmentation results of the proposed DLRSD-based method: (**a**) input image; (**b**) original defect foreground image; (**c**) enhanced defect foreground image; (**d**) segmentation image by Otsu's method; (**e**) manual-labeled ground-truth image.

5.2.4. Analysis of Robustness to Noise

Considering the surface defect image is polluted by Gaussian noise with SNR, including 22 dB, 18 dB, 14 dB and 10 dB, the same experiments are conducted to verify the robustness of the proposed DLRSD model. According to Table 1, when SNR decreases gradually, the AUC and MAE can remain a high level, especially when $SNR = 18$ dB, AUC can remain around 0.8. It's shown that the proposed DLRSD model is robust to noise and can lead to better saliency detection result, which establishes the good foundation for segmentation. The experimental results also indicate that adding sparse constraint for background can reduce the influence from noises, which is a reasonable strategy for surface defect detection.

Table 1. Experimental results with different noise.

Index \ SNR	No Noise	22 dB	18 dB	14 dB	10 dB
AUC	0.8350	0.8216	0.7922	0.7414	0.6918
MAE	0.1584	0.1638	0.1837	0.2114	0.2384

5.3. Comparison with State-of-the-Art Methods

5.3.1. Qualitative Comparison

The qualitative comparison results by the proposed method and other eight methods are shown in Figure 5. It's shown that most saliency detection methods can handle well simple images with relatively homogenous background (e.g., row 4, 5, 7 and 8). They can uniformly highlight the whole defect object and generate high-quality saliency map and segmentation image. However, for some complex defect images containing multiple objects (e.g., row 5, 6, 10 and 11), having a cluttered background (e.g., row 6), and showing there are similarities between the defect objects and background (e.g., row 2 and 9), the whole defect objects could not be uniformly highlighted, and parts of the background being falsely taken as the defect objects. It can be seen that the contrast of saliency maps obtained by RPCA, DSR and RS is low and ambiguous, especially for Patch defects (e.g., row 5 and 6), which is difficult to define a proper threshold to segment the defects. The saliency maps obtained by RBD and SBD miss detecting parts of the defect objects, while some incorrectly include background regions into detection results. Hence, there are some missing defects and fake defects in their final segmentation image. Differently, although IS, ULR and SMF produce the good saliency map, there are many pixels that belonging to the background are misjudged by defect, and some background regions also stand out with the defect regions. By contrast, our proposed method separates the defect objects from the background successfully and locates various defects precisely. It more efficiently highlights the complete defect object with well-defined boundaries and effectively suppresses the backgrounds than the other saliency detection methods. These results illustrate our proposed method not only enhances the contrast between surface defect and background effectively but also improves the robustness to the different illumination conditions, various shapes, scales, directions and locations of surface defect.

Figure 5. Qualitative comparisons: (**a**) ground-truth; (**b**) RPCA; (**c**) IS; (**d**) ULR; (**e**) RBD; (**f**) SBD; (**g**) DSR; (**h**) RS; (**i**) SMF; (**j**) Ours. We can see that our segmentation results, which are produced by simple Otsu's method on the saliency map, are very closer to the ground truth.

5.3.2. Quantitative Comparison

Figure 6 shows the quantitative results of the proposed DLRSD model against eight state-of-the-art methods. It is known that it perform competitively and is both better than the other methods in terms of the P-R curve, ROC curve and F-Measure curve. Especially, the precision can remain above 90% within a large threshold range.

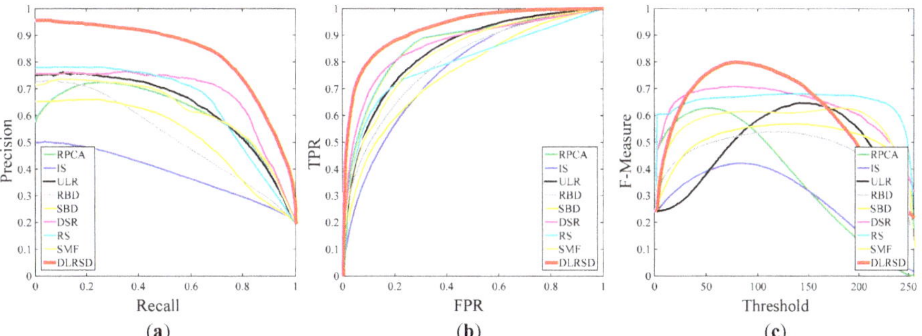

Figure 6. Quantitative comparisons: (**a**) precision-recall (P-R) curve; (**b**) receiver operating characteristic (ROC) curve; (**c**) F-measure curve.

Table 2 summarizes the quantitative results of all the eight methods. We can see that the proposed DLRSD model has achieved the best performance in AUC, F_β and MAE. Compared with SMF, it increased by 8.52% and 4.05% in AUC and F_ζ, respectively, decreased by 5.01% in MAE. All experiments are run in Matlab 2018a on a PC with an Intel Core *i7-4790@2.90GHz* CPU and 8GB RAM, the running time of the proposed DLRSD model is slightly slower than RS but much faster than ULR and SMF.

Table 2. Quantitative comparisons in terms of area under ROC (AUC), F_ζ, mean square error (MAE), and Time.

Method \ Index	AUC	F_ζ	MAE	Time (s)
RPCA [28]	0.7636	0.3633	0.1860	0.1982
IS [13]	0.7140	0.2814	0.2485	**0.0032**
ULR [23]	0.7843	0.4780	0.2976	4.5504
RBD [31]	0.7125	0.4607	0.2090	0.0331
SBD [32]	0.6907	0.5038	0.2390	0.6619
DSR [33]	0.7786	0.6264	0.1626	1.1797
RS [16]	0.7469	0.6454	0.1758	0.1281
SMF [24]	0.7497	0.5655	0.2085	0.4615
Ours	**0.8350**	**0.6060**	**0.1584**	0.1713

Based on the above qualitative and quantitative analyses, it confirms that our proposed method consistently outperforms some state-of-the-art methods and verifies the effectiveness of the proposed structural constraints in separating the low-rank and sparse subspaces.

6. Conclusions

Based on the salient characteristics of the defects in the surface defect image of steel sheet, we formulate the defect segmentation as a problem of saliency detection. We design a double low-rank and sparse decomposition model to obtain high-quality defect foreground image directly, which provides a robust way to segment the surface defect. We experimentally compare our proposed method with some state-of-the-art methods on surface defect images. The experimental results prove that the proposed method performs efficiently and competitively for the surface defect segmentation task and has a strong adaptive ability for the complex and varying surface defects of steel sheet. Our proposed method is an unsupervised framework, which skips the training process and therefore enjoys more flexibility. In the future, we will focus on combining our proposed method with convolutional auto-encoder and expanding the method to other industrial products' defect detection.

Author Contributions: S.Z. designed the DLRSD model and performed the evaluation experiments. S.W. collaborated closely and contributed valuable comments and ideas. H.L. arranged the datasets, as well as reviewed the article. Y.L. and N.H. developed the automatic optical inspection procedure. All authors contributed to writing the article.

Funding: This research was funded by Natural Science Foundation of China, under grant number 61775172 and 51805386, Natural Science Foundation of Hubei Province, under grant number 2017CFC830.

Acknowledgments: The authors would like to thank Kechen Song and Yungang Tan for providing the surface defect images. MATLAB procedure was revised and optimized from [23,24,33].

References

1. Hanbaya, K.; Talub, M.F.; Özgüvenc, Ö.F. Fabric defect detection systems and methods-a systematic literature review. *OPTIK* **2016**, *127*, 11960–11973. [CrossRef]

2. Neogi, N.; Mohanta, D.K.; Dutta, P.K. Review of vision-based steel surface inspection systems. *EURASIP J. Image Video Process.* **2014**, *2014*, 1–19. [CrossRef]

3. Yun, J.P.; Kim, D.; Kim, K.H.; Lee, S.J.; Park, C.H.; Kim, S.W. Vision-based surface defect inspection for thick steel plates. *Opt. Eng.* **2017**, *56*, 1–12. [CrossRef]

4. Madrigal, C.A.; Branch, J.W.; Restrepo, A.; Mery, D. A method for automatic surface inspection using a model-based 3D descriptor. *Sensors* **2017**, *17*, 2262. [CrossRef] [PubMed]

5. Ma, Y.P.; Li, Q.W.; Zhou, Y.Q.; He, F.J.; Xi, S.Y. A surface defects inspection method based on multidirectional gray-level fluctuation. *Int. J. Adv. Robot. Syst.* **2017**, *14*, 1–7. [CrossRef]

6. Aiger, D.; Talbot, H. The phase only transform for unsupervised surface defect detection. In Proceedings of the 2010 IEEE Computer Society Conference on Computer Vision and Pattern Recognition, San Francisco, CA, USA, 13–18 June 2010; pp. 295–302. [CrossRef]

7. Choi, D.C.; Jeon, Y.J.; Kim, S.H.; Moon, S.; Yun, J.P.; Kim, S.W. Detection of pinholes in steel slabs using Gabor filter combination and morphological features. *ISIJ Int.* **2017**, *57*, 1045–1053. [CrossRef]

8. Jeon, Y.J.; Choi, D.C.; Lee, S.J.; Yun, J.P.; Kim, S.W. Defect detection for corner cracks in steel billets using a wavelet reconstruction method. *J. Opt. Soc. Am. A* **2014**, *31*, 227–237. [CrossRef] [PubMed]

9. Liu, K.; Wang, H.Y.; Chen, H.Y.; Qu, E.Q.; Tian, Y.; Sun, H.X. Steel surface defect detection using a new Haar-Weibull-Variance model in unsupervised manner. *IEEE Trans. Instrum. Meas.* **2017**, *66*, 2585–2596. [CrossRef]

10. Guan, S.Q. Strip steel defect detection based on saliency map construction using Gaussian pyramid decomposition. *ISIJ Int.* **2015**, *55*, 1950–1955. [CrossRef]

11. Li, P.; Liang, J.L.; Shen, X.B.; Zhao, M.H.; Sui, L.S. Textile fabric defect detection based on low-rank representation. *Multimed. Tools Appl.* **2017**, 1–26. [CrossRef]

12. Zhao, Y.J.; Yan, Y.H.; Song, K.C. Vision-based automatic detection of steel surface defects in the cold rolling process: Considering the influence of industrial liquids and surface textures. *Int. J. Adv. Manuf. Technol.* **2017**, *90*, 1665–1678. [CrossRef]

13. Hou, X.D.; Harel, J.; Koch, C. Image signature: Highlighting sparse salient regions. *IEEE Trans. Pattern Anal. Mach. Intell.* **2012**, *34*, 194–201. [CrossRef]

14. Perazzi, F.; Krahenbuhl, P.; Pritch, Y.; Hornung, A. Saliency filters: Contrast based filtering for salient region detection. In Proceedings of the 2012 IEEE Conference on Computer Vision and Pattern Recognition, Providence, RI, USA, 16–21 June 2012; pp. 733–740. [CrossRef]

15. Shi, J.P.; Yan, Q.; Xu, L.; Jia, J.Y. Hierarchical image saliency detection on extended CSSD. *IEEE Trans. Pattern Anal. Mach. Intell.* **2016**, *38*, 717–729. [CrossRef] [PubMed]

16. Zhang, L.H.; Yang, C.; Lu, H.C.; Ruan, X.; Yang, M.H. Ranking saliency. *IEEE Trans. Pattern Anal. Mach. Intell.* **2017**, *39*, 1892–1904. [CrossRef] [PubMed]
17. Zhou, Q.Q.; Zhang, L.; Zhao, W.D.; Liu, X.H.; Chen, Y.F.; Wang, Z.C. Salient object detection using coarse-to-fine processing. *J. Opt. Soc. Am. A* **2017**, *34*, 370–383. [CrossRef] [PubMed]
18. Yang, J.M.; Yang, M.H. Top-down visual saliency via joint CRF and dictionary learning. *IEEE Trans. Pattern Anal. Mach. Intell.* **2017**, *39*, 576–588. [CrossRef] [PubMed]
19. Wang, J.D.; Jiang, H.Z.; Yuan, Z.J.; Cheng, M.M.; Hu, X.W.; Zheng, N.N. Salient object detection: A discriminative regional feature integration approach. *Int. J. Comput. Vis.* **2017**, *123*, 251–268. [CrossRef]
20. Peng, Q.M.; Cheung, Y.M.; You, X.G.; Tang, Y.Y. A hybrid of local and global saliencies for detecting image salient region and appearance. *IEEE Trans. Syst. Man Cybern. Soc.* **2017**, *47*, 86–97. [CrossRef]
21. Yan, J.C.; Zhu, M.Y.; Liu, H.X.; Liu, Y.C. Visual saliency detection via sparsity pursuit. *IEEE Signal Process. Lett.* **2010**, *17*, 739–742. [CrossRef]
22. Zou, W.B.; Liu, Z.; Kpalma, K.; Ronsin, J.; Zhao, Y.; Komodakis, N. Unsupervised joint salient region detection and object segmentation. *IEEE Trans. Image Process.* **2015**, *24*, 3858–3873. [CrossRef] [PubMed]
23. Shen, X.H.; Wu, Y. A unified approach to salient object detection via low rank matrix recovery. In Proceedings of the 2012 IEEE Conference on Computer Vision and Pattern Recognition, Providence, RI, USA, 16–21 June 2012; pp. 853–860. [CrossRef]
24. Peng, H.W.; Li, B.; Ling, H.B.; Hu, W.M.; Xiong, W.H.; Maybank, S.J. Salient object detection via structured matrix decomposition. *IEEE Trans. Pattern Anal. Mach. Intell.* **2017**, *39*, 818–832. [CrossRef] [PubMed]
25. Sun, X.L.; He, Z.X.; Xu, C.; Zhang, X.J.; Zou, W.B.; Baciu, G. Diversity induced matrix decomposition model for salient object detection. *Pattern Recogn.* **2017**, *66*, 253–267. [CrossRef]
26. Cai, D.; He, X.F.; Han, J.W.; Huang, T.S. Graph regularized non-negative matrix factorization for data representation. *IEEE Trans. Pattern Anal. Mach. Intell.* **2011**, *33*, 1548–1560. [CrossRef] [PubMed]
27. Bruckstein, A.M.; Donoho, D.L.; Elad, M. From sparse solutions of systems of equations to sparse modeling of signals and images. *SIAM Rev.* **2009**, *51*, 34–81. [CrossRef]
28. Candès, E.J.; Li, X.D.; Ma, Y.; Wright, J. Robust principal component analysis. *J. ACM* **2011**, *58*, 1–37. [CrossRef]
29. Lin, Z.C.; Chen, M.M.; Ma, Y. *The Augmented Lagrange Multiplier Method for Exact Recovery of Corrupted Low-Rank Matrices*; University of Illinois Urbana-Champaign Technical Report; UILU-ENG-09-2215; University of Illinois Urbana-Champaign: Champaign, IL, USA, 2009.
30. Achanta, R.; Shaji, A.; Smith, K.; Lucchi, A.; Fua, P.; Susstrunk, S. SLIC superpixels compared to state-of-the-art superpixel methods. *IEEE Trans. Pattern Anal. Mach. Intell.* **2012**, *34*, 2274–2282. [CrossRef] [PubMed]
31. Zhu, W.J.; Liang, S.; Wei, Y.C.; Sun, J. Saliency optimization from robust background detection. In Proceedings of the 2014 IEEE Conference on Computer Vision and Pattern Recognition, Columbus, OH, USA, 23–28 June 2014; pp. 2814–2821. [CrossRef]
32. Zhao, T.; Li, L.; Ding, X.H.; Huang, Y.; Zeng, D.L. Saliency detection with spaces of background-based distribution. *IEEE Signal Process. Lett.* **2016**, *23*, 683–687. [CrossRef]
33. Lu, H.C.; Li, X.H.; Zhang, L.H.; Ruan, X.; Yang, M.H. Dense and sparse reconstruction error based saliency descriptor. *IEEE Trans. Image Process.* **2016**, *25*, 1592–1603. [CrossRef] [PubMed]

 *applied sciences*

Article

A UAV-Based Visual Inspection Method for Rail Surface Defects

Yunpeng Wu [1,2], Yong Qin [1,3,*], Zhipeng Wang [1,3,*] and Limin Jia [1,3]

[1] State Key Laboratory of Rail Traffic Control and Safety, Beijing Jiaotong University, Beijing 100044, China; 16114225@bjtu.edu.cn (Y.W.); lmjia@bjtu.edu.cn (L.J.)
[2] School of Traffic and Transportation, Beijing Jiaotong University, Beijing 100044, China
[3] Beijing Research Center of Urban Traffic Information Sensing and Service Technologies, Beijing 100044, China
* Correspondence: yqin@bjtu.edu.cn (Y.Q.); zpwang@bjtu.edu.cn (Z.W.)

Received: 29 May 2018; Accepted: 21 June 2018; Published: 24 June 2018

Abstract: Rail surface defects seriously affect the safety of railway systems. At present, human inspection and rail vehicle inspection are the main approaches for the detection of rail surface defects. However, there are many shortcomings to these approaches, such as low efficiency, high cost, and so on. This paper presents a novel visual inspection approach based on unmanned aerial vehicle (UAV) images, and focuses on two key issues of UAV-based rail images: image enhancement and defects segmentation. With regards to the first aspect, a novel image enhancement algorithm named Local Weber-like Contrast (LWLC) is proposed to enhance rail images. The rail surface defects and backgrounds can be highlighted and homogenized under various sunlight intensity by LWLC, due to its illuminance independent, local nonlinear and other advantages. With regards to the second, a new threshold segmentation method named gray stretch maximum entropy (GSME) is presented in this paper. The proposed GSME method emphasizes gray stretch and de-noising on UAV-based rail images, and selects an optimal segmentation threshold for defects detection. Two visual comparison experiments were carried out to demonstrate the efficiency of the proposed methods. Finally, a quantitative comparison experiment shows the LWLC-GSME model achieves a recall of 93.75% for T-I defects and of 94.26% for T-II defects. Therefore, LWLC for image enhancement, in conjunction with GSME for defects segmentation, is efficient and feasible for the detection of rail surface defects based on UAV Images.

Keywords: rail surface defect; UAV image; defect detection; gray stretch maximum entropy; image enhancement; defect segmentation

---

1. Introduction

Rail transportation plays a significant role in the development of economic and industrial growth, and the failures of railway facilities (such as defects on the rail surface) are directly related to catastrophic accidents [1]. With the development of high-speed and high-load rail transit, the probability of rail surface defects is increasing rapidly. In general, rail surface defects which include corrugations and discrete defects due to wheel-rail contact conditions are the most common forms of defects [2]. Corrugations arise from periodic slip of the wheel on the rail as trains run on tracks [3]. The discrete defects are generated on the rail surface in an apparently random manner, i.e., without periodic characteristics, as shown in Figure 1. Those defects might cause serious accidents, or may even result in a catastrophic derailment of vehicles. Thus, this paper mainly discusses the detection of surface discrete defects.

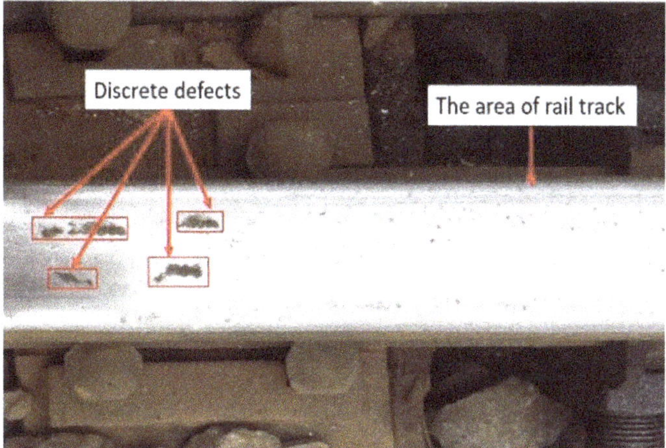

Figure 1. The discrete defects on the rail surface.

Currently, there are regular inspections of tracks in order to maintain safe and efficient operation [4]. Historically, inspection tasks are performed by trained personnel, by walking along the tracks. However, the manual inspection is inappropriate due to its low-efficiency, lack of objectivity, and high false alarm rate. Furthermore, the results are seriously dependent on the capability of the observer to detect possible anomalies and recognize critical conditions [5]. Therefore, automatic and nondestructive inspection methods should be urgently developed.

At present, nondestructive inspection methods have been widely developed in a variety of industry inspection applications, due to their high efficiency and high precision [6]. Several methods have been applied to rail defects inspection, such as acoustic emission inspection [7], electromagnetic inspection [8], ultrasonic surface waves inspection [9], and visual inspection(VI) [10–12]. Particularly, with the development of computer vision techniques, VI (visual inspection) has been widely applied. VI is the most notable method for the surface defect detection because of its high speed and low cost [13]. Some researchers have studied rail surface defects by VI [14–16]. The VI method is an attractive approach for discrete defect detection.

According to the traditional VI approach for surface defects inspection, a high definition (HD) camera is used to capture rail images; it is embedded in a detection system installed under an inspection train. Currently, most related researches are based on this approach. However, it has inevitable drawbacks, such as limited detection range, high cost, and so on. Inspection trains have to run over significant distances to capture rail images for a wide range of detection. The detectable parts and viewing angles are limited, especially for mountainous areas or across rivers.

Unmanned Aerial Vehicles (UAVs) have become a research hotspot in many fields. The UAV-based inspection scheme is efficient and cost-effective, and has become attractive for change inspection in small-scale regions [17]. With the rapid development of UAVs, UAV-based aerial photography has been widely employed for engineering surveying and mapping [18], crop measurements [19], wind turbine blade surface inspection [20], power facilities inspection [21], historical buildings inspection [22], forest fire detection [23], bridge crack detection [24], fault detection in photovoltaic cells [25], and other detection applications such as target tracking [26], tracking and classification of multiple moving objects [27] and object recognition [28], etc. In general, UAV-based aerial photography has been extensively applied in various industries due to its advantages: low cost, ease of control, and flexibility.

As mentioned, an inspection method of rail surface defects based on UAV combined with VI is proposed in this paper. A typical approach to detect surface defects is to automatically extract defects after image enhancement [15]. The most two popular methods for image enhancement are by the histogram equalization (HE) and homomorphic filtering algorithms. However, since the HE is a linear algorithm and only averages the gray level distribution rather than enlarging the gray scale, there are several shortcomings, such as loss of image detail information and noise amplification [29]. Homomorphic filtering algorithm based on the illumination reflectance model is a frequency-domain processing to compress image light regions and enhance contrast. This method makes use of the frequency information of images. However, it often blurs image details, leading to a lack of deliberation of the spatial local characteristics of images [29]. In addition, gray values in the global scope change dramatically because of the uneven illumination and reflectance properties of rail surfaces [14]. Therefore, the two global enhancement methods are not suitable for rail images.

For defect detection, research on models that automatically locate defects after image enhancement has several achievements. For instance, a visual inspection system (VIS) is proposed for discrete defect detection [14]. In the VIS, images are captured by a high-speed digital camera fixed on a train. Subsequently, track extraction based on projection profile (TEBP) algorithm is used to extract the areas of rail track in images, and then the local normalized (LN) method and the defect localization based on projection profile (DLBP) method are applied to detect rail surface defects. The MLC-PEME model is proposed for defect detection [30]. Firstly, the histogram-based track extraction (HBTE) algorithm is used to extract areas of rail track in images which were captured by a camera fixed on an inspection train, then the MLC (Michelson-like contrast) combined with proportion emphasized maximum entropy (PEME) method is applied to detect defects. These methods performed well; however, they are seriously influenced by noise and background points, and consequently, have massive false detection rates [15]. An inverse PM diffusion model is proposed to enhance images [16]. Therefore, an adaptive threshold binarization is able to readily locate surface defects. However, it is seriously influenced by noise points, and yields in high false detection rates [15].

Most inspection models are based on the VI system fixed on inspection trains, and research on the UAV-based inspection of rail surface defects is rarely discussed. In this study, we faced the following serious challenges:

- Rail position variances in UAV images. Unlike inspection trains, the camera angle of HD camera installed on UAV is sensitive to environment aspects (such as wind and turbulence) and operators. Although the UAV can balance itself by using GPS flight mode, rail positions in images captured by UAV aerial photography are extremely variable. Therefore, the variances of rail positions bring difficulties to rail extraction.
- Non-uniform illumination and noise corruption. Due to partial occlusion of infrastructures around the rail (such as catenary etc.), reflectance properties of rail surface and shake of the UAV and other environmental factors, the brightness and contrast of images are uneven and low. According to [17,31], UAV digital images are likely to be corrupted by noises during the acquisition or transmission. In general, the gray levels of surface defects are lower than that of background (non-defect area) [14], but the order of these values is often broken because of non-uniform illumination and noise corruption, as shown in the Figure 1.
- Few characteristics for defects segmentation. A corrugation initiates and develops easily because of the periodic occurrence of contact vibration [32]. However, it is difficult to inspect discrete defects by the VI method due to the lack of periodicity. Surface defects have low grey-level, that distinguishes them from the dynamic background. Therefore, the grey-level is considered to be the most available feature [14]. Therefore, the existing object recognition methods based on sophisticated texture and shape features are unfeasible, due to the limitation of visual features [15].

Also, due to the above challenges, these inspection models based on inspection trains are unable to be used in the case of UAV rail images. To overcome these challenges, this paper presents a novel image enhancement algorithm based on Weber's law and a new threshold segmentation method based on the gray stretch in wavelet domain.

Weber's law was first proposed by German physiologist Weber, and later formulated quantitatively as a mathematical expression referred to as Weber contrast by psychologist Fechner [33]. It reveals the global influence of background stimulus on humans' sensitivity to the intensity increment [34]. Weber contrast is commonly used in cases where small features are present on a large uniform background, i.e., where the average luminance is approximately equal to the background luminance [35]. Due to few defects existing on rail surfaces and the high reflection properties of rail surfaces, the brightness mean of longitudinal line along a track approximates to the background luminance. It is supposed that the Weber's law is suitable for enhancing UAV-based rail images. Therefore, this paper proposes a novel LWLC (Local Weber-like Contrast) algorithm based on Weber's law.

Although the rail surface defects and backgrounds can be highlighted and homogenized by LWLC, respectively, noise points and low contrast of background and defects still exist in UAV-based rail images. This leads to inaccurate segmentation thresholds based on traditional threshold methods, such as Otsu method [36] and maximum entropy (ME) method [37]. An image segmentation method based on gray stretch and threshold algorithm (GSTA) [38] outperforms the Otsu method [38]. This method of GSTA uses the Otsu method to obtain a threshold after image wavelet decomposition, and then grayscale of object and background on this image is extended to a large scale based on this threshold. Subsequently, the Otsu method is also used to get an optimization after the image is reconstructed in wavelet domain. This method has achieved attractive results on image segmentation. In addition, the wavelet transform combined with median/mean filtering is extremely effective for image (or UAV image) de-noising [31,39,40]. Therefore, inspired by these successes, this paper put forward a new threshold method named *gray stretch maximum entropy* (GSME), which utilizes gray stretch in wavelet domain combined with median filtering de-nosing to increase defects detection performance.

The advantages of the proposed methods in this paper are as follows: (1) LWLC algorithm is local, nonlinear, and illuminance independent. This algorithm can adapt to different sunlight illuminance, eliminate the significant changes of gray-scale, and highlight defects of UAV-based rail images. (2) The GSME method emphasizes gray stretch and image de-noising on rail images, and automatically gives more suitable segmentation thresholds for rail surface defects detection. (3) To the surface defects detection based on UAV-based rail images, a LWLC-GSME model that achieves a recall of 93.75% for T-I defects and a recall of 94.26% for T-II defects can provide a feasible solution.

The remaining sections of this paper is organized as follows: details of the LWLC and the GSME algorithms are described in Methodology. The experiment setup and result analysis are presented in Experiment results and performance analysis. A conclusion is provided at the end of the paper.

2. Methodology

The inspection method for rail surface defects based on UAV images is proposed in this paper. The UAV-based rail images are captured by UAV equipped with a high-definition camera. The customized image processing methods are applied to analyze and detect these images. The flow diagram of inspection method for rail surface defects based on UAV image in the study is shown in Figure 2. All images used for this paper are captured by a UAV equipped with HD cameras, with the aircraft was flying at an altitude of 30 m above the rail. There are three subjects discussed in this section: (1) the extraction of area of rail tracks; the pseudocode of the tracks extraction is presented in the appendix, (2) UAV-based rail images enhancement based on LWLC algorithm, and (3) the defects segmentation based on the GSME method.

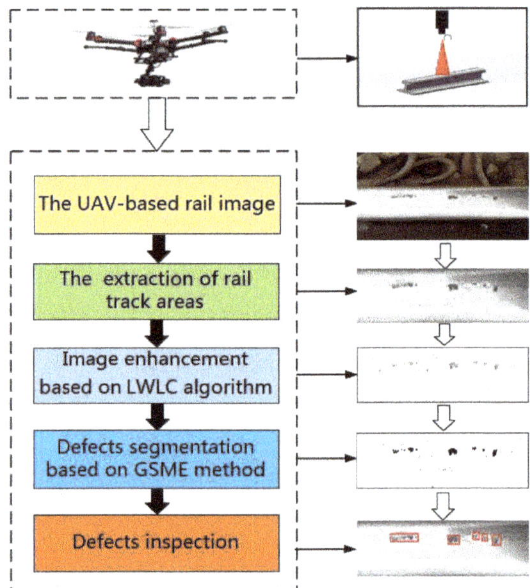

Figure 2. Flow diagram of the inspection method for the rail surface defects based on the UAV images.

2.1. Rail Track Extraction

The rail images captured by UAVs involves redundant areas, as shown in Figure 3. Besides the areas of rail tracks, the rest areas are excluded for the next step. Therefore, Hough Transform and the method based on cumulative gray value of each pixel column are used to extract the area of rail track from rail images. Hough transform is a graphic detection algorithm based on the duality of point and line, and can be applied in the extraction of rail tracks [41].

Considering an image as a $M \times N$ matrix, the N-dimensional matrix Cg that consists of cumulative gray value of each pixel column of the image matrix is determined by:

$$Cg = [\sum_{i=0}^{M-1} D_{i0}, \sum_{i=0}^{M-1} D_{i1}, \ldots\ldots \sum_{i=0}^{M-1} D_{i(N-1)}],$$
$$Cg(n) = \sum_{i=0}^{M-1} D_{in}, \ n \in [0, N-1] \tag{1}$$

where, D_{xy} is the pixel value of the coordinate (x, y). The matrix Cg of the vertical rail image ($M = 550$, $N = 350$) is shown in Figure 3. It should be noted that $Cg(n)$ is mapped to a small range. From this figure, it can be observed that the value $Cg(n)$ of area of the rail track is higher than the rest. This method is based on two factors: (1) Area of rail track has a higher value of $Cg(n)$. (2) The width w_d of the rail track is fixed in the rail image, as shown in Figure 3. The detail procedure about the method for rail track extraction is shown in Appendix A.

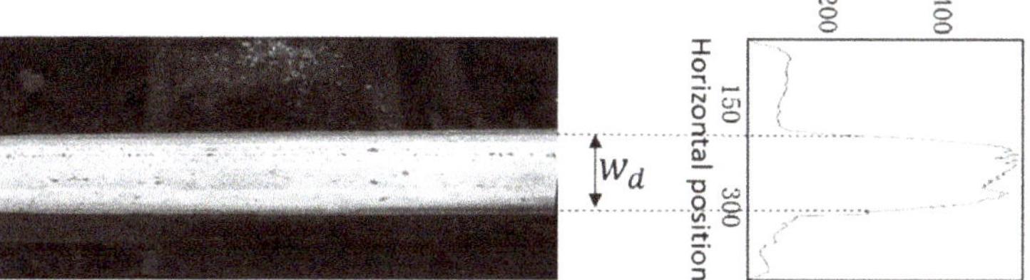

Figure 3. Cumulative gray value of each pixel column for a rail image. Horizontal position denotes location of pixel column.

2.2. The Local Weber-Like Contrast Algorithm for Rail Images Enhancement

The brightness of images is non-uniform because of uneven natural light, the reflectance properties of rail surfaces [42], vibration of UAVs, or any other environmental factors. Therefore, defects and backgrounds are always mixed together. According to our experience, the following characteristics of UAV rail images occur:

- Lower variation range of gray values in local regions. The reflection property and illumination of each longitudinal line in rail images is stable [14]. In the local line window, the variation range of gray values has small variation, and the most obvious features can be used for image enhancement [15].
- Greater variation range of gray values in global scope. In general, the rail images have a large variation range of gray level in global scope due to uneven natural light and the reflectance properties of rail surfaces. The reflected light in smooth parts of rail surfaces is more than the rough parts [42].
- Confused gray values between defects and background. In general, the gray value of surface defects is lower than that of background, but the order is often broken because of illumination non-uniformity and noise corruption, as shown in the Figure 4.
- Consistent features in the same longitudinal direction. Actually, a rail surface shares consistent features in the longitudinal direction as a train runs on a rail, since the friction for the points in the longitudinal direction between the rail surface and train wheels has an almost identical impact on the rail surface. In a rail image, intensity for the pixel points along longitudinal direction is consistent with relatively gray value changes caused by defect points and noise points [15]. Therefore, the surface discrete defects can be derived by the analysis of the information in longitudinal regions.
- Higher gray mean of each longitudinal line for a track. According to our observation, the gray means along longitudinal lines of a UAV rail image are higher under normal conditions. This is because that UAV are supposed to fly in fine weathers and natural light conditions, and the surface reflectivity of rail tracks in operation is high because of its smoothness, as shown in the Figure 4.

As Figure 4 shows, the gray means along longitudinal lines of the image is high and the brightness of defects is quite low [14]. Since there are few defects in the image [30], the mean can be considered as background in the longitudinal direction. This feature completely satisfies a suitable range of Weber Contrast. As one of the most classical luminance contrast statistics, Weber Contrast is popular to cope with small, sharp-edged graphic objects on larger uniform backgrounds [43]:

$$C_w = \frac{L_o - L_b}{L_b} \tag{2}$$

where, L_o is the luminance of the symbol and L_b is the luminance of the immediately adjacent background. When the background is lighter than the object, C_w is negative and ranges from 0 to -1. When the background is darker, C_w is positive and ranges from 0 to potentially very large numbers.

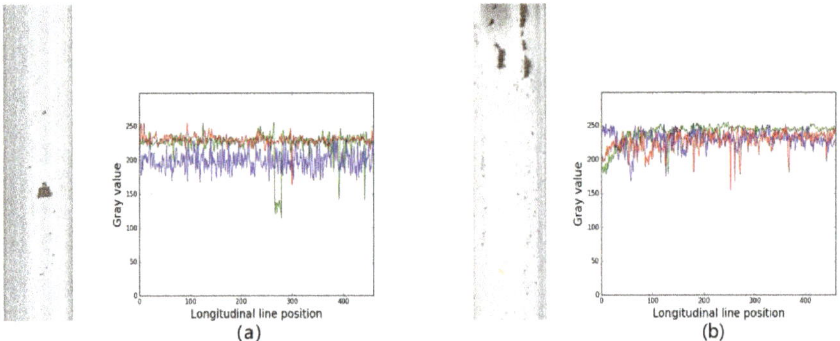

Figure 4. The gray value along longitudinal line of an image. (**a**) The gray value of 10th (red), 80th (blue), 110th longitudinal line (green) of an UAV rail image form slight defects dataset. (**b**) The gray value of 10th, 80th, 110th longitudinal line of an UAV rail image form datasets of serious defects.

Inspired by Weber Contrast and based on these characteristics for UAV rail images, the LWLC algorithm is proposed for adapting to different sunlight illuminance and eliminating the significant changes of gray-scale in this paper. The proposed gray stretch method for defects segmentation will be introduced in the next section. Assuming a pixel (x, y) and its surrounding window T in a rail image I, the intensity $LWLC_{(x,y)}$ of each pixel is given by:

$$LWLC_{(x,y)} = \frac{I(x,y) - E(I(\tilde{x},\tilde{y}))}{E(I(\tilde{x},\tilde{y}))}, \ (\tilde{x},\tilde{y}) \in T \tag{3}$$

where, $I(x, y)$ denotes the gray value of the pixels in the image, E is the mean of $I(\tilde{x}, \tilde{y})$ in T in window T. Figure 5a shows $LWLC$ value with the mean range $[100, 255]$ due to higher brightness of UAV rail images. And Figure 5b,c present the curve with $E = 100$ and the curve with $E = 220$, respectively. In Figure 5b, the low range $[0, 100]$ of I maps to $[-1, 0]$, while the high range $[100, 255]$ of I maps to $(0, 155/100)$. In contrast, the curve with $E = 220$ in Figure 5c shows that the greater low range $[0, 220]$ of I maps to $[-1, 0]$ due to its low slope. Thereby, along with the brightness increasing, the LWLC value is progressively reduced and the stretch of the range of I is weakened. These characteristics are similar to the human vision system, that is likely to discern contrast under the darker illuminance [30,44].

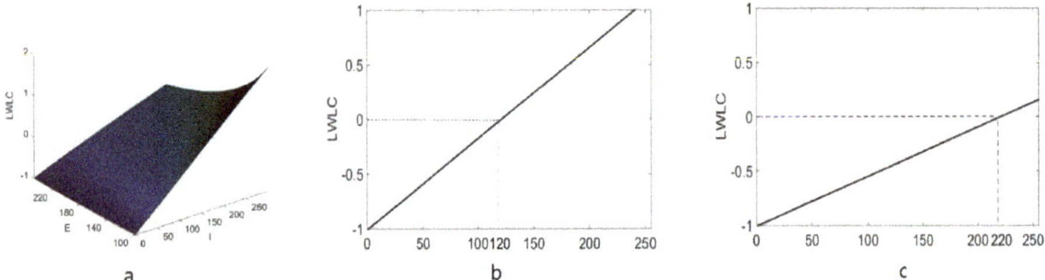

Figure 5. The surface of LWLC (Local Weber-like Contrast) measure. (**a**) The surface of LWLC. (**b**) The curve with $E = 100$. (**c**) The curve with $E = 120$.

The gray value $I(x,y)$ of the pixels in an image can be approximately determined by:

$$I(x,y) = L(x,y) \times R(x,y) \tag{4}$$

where $L(x,y)$ is light source intensity on the camera lens, and $R(x,y)$ is the coefficient of reflection attribute [45]. In a local window T, $L(x,y)$ can be regarded as a constant L due to the fact that the sunlight intensity in this small T reflected by the surface of a rail track is barely change under sunlight illumination. Therefore, Equation (3) can be replaced by:

$$
\begin{aligned}
LWLC_{(x,y)} &= \frac{L \times R(x,y) - L \times \overline{\mu}_R(\widetilde{x},\widetilde{y})}{L \times \overline{\mu}_R(\widetilde{x},\widetilde{y})} \\
&= \frac{R(x,y) - \overline{\mu}_R(\widetilde{x},\widetilde{y})}{\overline{\mu}_R(\widetilde{x},\widetilde{y})} \quad , \; (\widetilde{x},\widetilde{y}) \in T
\end{aligned}
\tag{5}
$$

where $\overline{\mu}_R(\widetilde{x},\widetilde{y})$ is the mean of reflection attribute coefficient $R(x,y)$ in a local window T. From this Equation (5), $LWLC$ is just dependent on $R(x,y)$ and $\overline{\mu}_R(\widetilde{x},\widetilde{y})$ rather than light source intensity $L(x,y)$. As a result, it is supposed that $LWLC$ can keep steady under the change of sunlight illuminance. On the other hand, $R(x,y)$ generally varies less in a local window T. This means that the value of $R(x,y) - \overline{\mu}_R(\widetilde{x},\widetilde{y})$ approximates 0. Thereby, when there are a smooth background window T_1 with a large $\overline{\mu}_R(\widetilde{x},\widetilde{y})$ and a coarse background window T_1 with a small $\overline{\mu}_R(\widetilde{x},\widetilde{y})$ in a rail image, the difference of their $LWLC$ matrix is not obvious. Therefore, a uniform background can be achieved by $LWLC$.

Briefly, rail surface defects and backgrounds can be highlighted and homogenized by $LWLC$. Based on these features, $LWLC$ can enhance the UAV-based rail images of non-uniformity brightness due to the various reflection attribute of rail surfaces under various sunlight intensity.

A transformed image can be obtained by Equation (3), which has contrast enhancement. The choice of window T size is very important, because it affects the quality and efficiency of this algorithm. Based on UAV rail image features in the same longitudinal direction presented in the above characteristics, this study adopted a lined (longitudinal direction) window T (100×1) in this paper. The experiment in [30] also proves the excellent performance of this local line window.

In a local line window, gray value of defects is considered to be lower than the other areas on rail surface, because the light of the window is equal and less light can be reflected by defects. Therefore, if gray value of a pixel is lower than the mean value of all pixels in this window, it may be regarded as a defect point. In contrast, this is regarded as a background point. Based on these factors and Equation (3), the pixels belonging to non-defect (background and irregular) points can be translated into uniform background by setting a dynamic threshold $E(I(\widetilde{x},\widetilde{y}))$ by

$$
LWLC_{(x,y)} = \begin{cases} \frac{I(x,y) - E(I(\widetilde{x},\widetilde{y}))}{E(I(\widetilde{x},\widetilde{y}))}, & if\ I_{(x,y)} < E(I(\widetilde{x},\widetilde{y})) \\ 0, & otherwise. \end{cases}
\tag{6}
$$

In summary, the proposed LWLC algorithm for image enhancement is described as follows:

(i) By convolution with an image matrix I and a designed lined window, calculates LWLC value of each pixel in I by Equation (6), so that a LWLC matrix can be acquired.

(ii) Mapping gray-values of the $LWLC$ matrix to [0, 255].

2.3. Defect Segmentation Mehtod Based on Gray Stretch Maximum Entropy

The GSME algorithm is able to determine an optimal segmentation threshold by stretching gray levels between the objection and background and reduces noise in the image's wavelet domain. The procedure of the algorithm is shown in Figure 6.

Based on one-level 2-D DWT algorithm, the rail image is decomposed into four bands (LL, HL, LH, HH). For the LL band, the ME algorithm is used to obtain a segmentation threshold after reconstructing its coefficient, and then the gray stretch method is used to enhance contrast between background and foreground. For HL, LH and HH bands, the median filtering template of horizontal line, vertical line, and diagonal line is used to eliminate noise of three high frequency wavelet coefficients, respectively.

Subsequently, the ME algorithm is used to select a segmentation threshold after reconstructing the rail image.

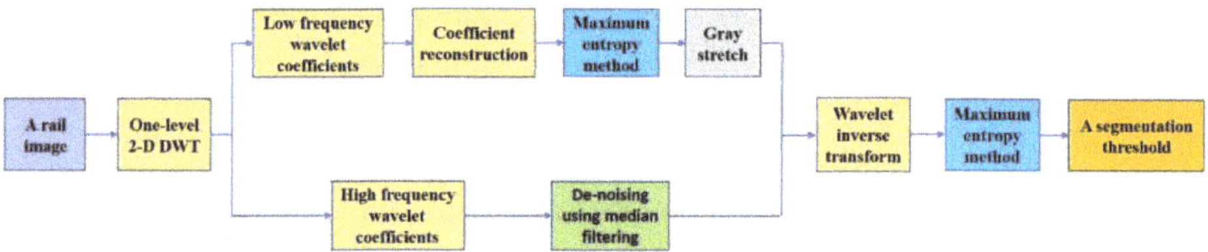

Figure 6. The procedure of the GSME method.

2.3.1. A Brief Introduction for 2-D Discrete Wavelet Transform

The discrete wavelet transform (DWT) can not only express some features of a signal easily and efficiently, but also provides a powerful insight into an image's spatial and frequency characteristics [38,46]. Two-dimensional functions such as images can be expanded from one-dimensional wavelet transform [46]. In two dimensions, a 2-D scaling function and three 2-D wavelets are given by:

$$
\begin{aligned}
\varphi(x,y) &= \varphi(x)\varphi(y)\\
\psi^H(x,y) &= \psi(x)\varphi(y)\\
\psi^V(x,y) &= \varphi(x)\psi(y)\\
\psi^D(x,y) &= \psi(x)\psi(y)
\end{aligned}
\tag{7}
$$

where $\varphi(x,y)$ is a two-dimensional scaling function, ψ^H corresponds to some variations along columns(such as horizontal edges), ψ^V corresponds to variations along rows (for example, vertical edges), and ψ^D responds to variations along diagonals. Each of them can be seen as products of two 1-D functions. If the 2-D scaling and wavelets functions are given, the 1-D DWT can be extended to two dimensions. Firstly, the scaled and translated basis functions are defined as:

$$
\varphi_{j,m,n}(x,y) = 2^{\frac{j}{2}}\varphi(2^j x - m,\ 2^j y - n)
\tag{8}
$$

$$
\psi^i{}_{j,m,n}(x,y) = 2^{\frac{j}{2}}\psi(2^j x - m,\ 2^j y - n),\ \ i = \{H,V,D\}
\tag{9}
$$

where index i is a superscript that assumes the values H, V, and D in Equation (7). The discrete wavelet transform of $M \times N$ image $f(x,y)$ is given by:

$$
W_\varphi(j_0,m,n) = \frac{1}{\sqrt{MN}} \sum_{x=0}^{M-1}\sum_{y=0}^{N-1} f(x,y)\varphi_{j_0,m,n}(x,y)
\tag{10}
$$

$$
W_\psi^i(j,m,n) = \frac{1}{\sqrt{MN}} \sum_{x=0}^{M-1}\sum_{y=0}^{N-1} f(x,y)\psi_{j,m,n}^i(x,y),\ \ \ i = \{H,V,D\}
\tag{11}
$$

where j_0 is an arbitrary starting scale that is set to 0 by default, the $W_\varphi(j_0,m,n)$ coefficients are an approximation of $f(x,y)$ at scale j_0, and the $W_\psi^i(j,m,n)$ coefficients add horizontal, vertical, and diagonal details for scales $j \geq j_0$. The 2-D DWT is achieved by using digital filters and down-sampling, as shown in Figure 7. According to the 2-D DWT scaling and wavelet functions, we can take the 1-D FWT (fast wavelet transform) of the rows of $f(x,y)$, followed by the 1-D FWT of the resulting columns. Therefore, an original 2-D image can be decomposed into four sub-image sets which contain different frequency characteristics by high-pass and low-pass filter: a scaling component W_φ involving low-pass information and three wavelet components, W_ψ^H, W_ψ^D, and W_ψ^V, corresponding respectively to the horizontal, diagonal, and vertical details, as in Figure 5. The 1-level DWT method can also reduce noise and the background disturbances.

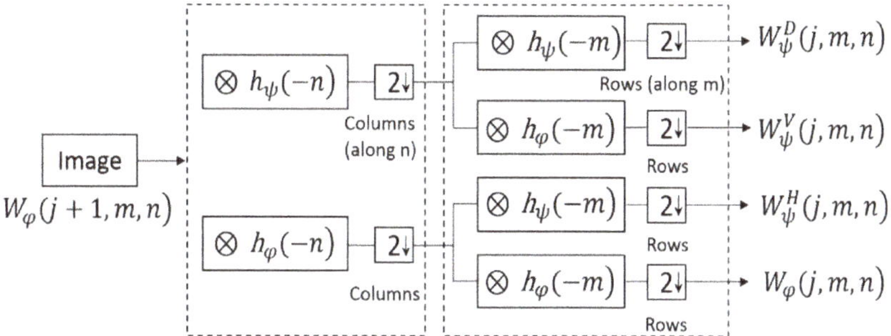

Figure 7. The 2-D DWT diagram for an image by using digital filters and down-sampling. An original 2-D image can be decomposed into four sub-image: a scaling component W_φ and three wavelet components, W_ψ^H, W_ψ^D, and W_ψ^V. In Figure 7, $\otimes$ denotes convolution symbol, h_ψ denotes low pass filter, and h_φ denotes high pass filter.

2.3.2. The Gray Stretch Maximum Entropy Threshold Method

T. Pun et al. proposed the entropy threshold principle [47,48] which uses entropy of image gray histogram to obtain the segmentation threshold. The maximum entropy (ME) algorithm [37] is proposed to optimize a threshold afterwards. ME method can confirm one threshold which maximizes the total content of information provided by cumulative object probability distribution ϕ_o and cumulative background probability distribution ϕ_b. They are given by:

$$P_n = \frac{f_n}{M}, \ n \in [0, 255] \tag{12}$$

$$\phi_o = \sum_{n=0}^{T-1} p_n, \quad \phi_b = 1 - \phi_o \tag{13}$$

where P_n is the probability of gray value n in an image. Given a rail image I that is normalized to 256 gray levels, and the entropy of ϕ_o and ϕ_b is defined as:

$$H_o(T) = -\sum_{n=0}^{T-1} \left(\frac{P_n}{\phi_o(T)} \ln \frac{P_n}{\phi_o(T)} \right), \tag{14}$$

$$H_b(T) = -\sum_{m=T}^{255} \left(\frac{P_n}{\phi_b(T)} \ln \frac{P_n}{\phi_b(T)} \right) \tag{15}$$

where M is the total pixel number of image I, f_n is the frequency of gray value n in I. An optimal threshold T^* can be obtained by:

$$T* = \arg \max \left(H_o(T) + H_b(T) \right), \ T \in [0, 255] \tag{16}$$

The ME method takes into account both the distribution information of image pixel gray and the spatial information of pixels. However, its performance is not perfect for defect segmentation due to the aforementioned characteristics of rail images. Therefore, GSME algorithm is proposed in this paper, as shown in Figure 6.

(i)   Based on one-level 2-D DWT algorithm, the rail image is decomposed into four wavelet coefficients that include approximation (low frequency region), horizontal, vertical, and diagonal details.

(ii) For low frequency region (LL region) of image decomposed by wavelet, the ME algorithm is used to obtain a segmentation threshold after reconstructing its coefficient, and then the gray stretch method is used to enhance contrast between background and foreground, as the following equations:

$$f_\varphi(x,y) = \frac{1}{\sqrt{MN}} \sum_m \sum_n W_\varphi(j_0, m, n)\varphi_{j_0,m,n}(x,y), \tag{17}$$

$$f_\varphi{}^*(x,y) = \begin{cases} f_\varphi(x,y) - af_\varphi(x,y), & if\ f_\varphi(x,y) < T^* \\ f_\varphi(x,y) + af_\varphi(x,y), & otherwise. \end{cases} \tag{18}$$

where $f_\varphi(x,y)$ denotes reconstructing image function, a denotes stretch factor and is set to and value between 0.1 and 0.5 in general.

(iii) For the image, its energy is mainly distributed in the low frequency region. In the high frequency area, the proportion of noise energy is large, so this study focuses on de-noising in this area. In Ref [39], Tang et al. use the filter templates of three different directions for de-noising. For example, the line template of horizontal direction is used for W_ψ^H de-noising, because the wavelet coefficients contain the high-frequency information in the horizontal direction and low-frequency information in the vertical direction of the image signal. Inspired by the median filtering method employed in wavelet domain, this study used the median filtering template of horizontal line, vertical line, and diagonal line to eliminate noise of three high frequency wavelet coefficients, respectively.

(iv) The rail image can be reconstructed based on discrete wavelet inverse transform algorithm. The formula for reconstruction image is given by:

$$f(x,y) = f_\varphi{}^*(x,y) + \frac{1}{\sqrt{MN}} \sum_{i=H,V,D} \sum_{j=j_0}^{\infty} \sum_m \sum_n W_\psi^i(j, m, n)\psi_{j,m,n}^i(x,y) \tag{19}$$

(v)   The ME algorithm is used to select a segmentation threshold after reconstructing the rail image by discrete wavelet inverse transform.

The GSME algorithm performs well for processing rail images in which the contrast between foreground and background is low.

3. Experiment Results

To demonstrate the proposed LWLC-GSME model, experiments were carried out with comparisons with related well-established methods.

3.1. Experiment Setup

3.1.1. A Brief Introduction of the Equipment for UAV Images Acquisition

As shown in Figure 6, the DJI Matrice 600 equipped with Zenmuse Z30 (DJI-Innovations, Shenzhen, China) was used to capture rail images. The Matrice 600 is a six-rotor flying platform designed for professional aerial photography and industrial applications. The aircraft uses six Intelligent Flight Batteries to extend the time of flight. The built-in API Control feature, expandable center frame, and maximum takeoff weight of 15.1 kg make the Matrice 600 ideal for connecting other devices to meet the specific needs of different applications. The Zenmuse Z30 enables non-contact distance detection by a high-performance camera system with a zoom lens. This aerial camera offers

30× optical zoom, 6× digital zoom, and HD 1080P video. The UAV adopts an industrial level Zenmuse platform with a precision of 0.01 degrees, so the problem of image blur caused by jitter is effectively solved.

3.1.2. Experiment Environment

As shown in Figure 8, the UAV equipped with an aerial camera was used to capture image or video information of rails as it flies overhead. Then, customized image processing software was used to analyze the captured UAV images. To avoid interference of obstructions, the flight height was set to 30 m. The images of rail tracks from the UAV were acquired on the freight line near the Baoding railway south station and the freight line near Nansihuan in Beijing. One of the experiment environments is shown in Figure 7. The speed of the UAV and the actual length of the rail track in an UAV image are 2 m/s and 1 m, respectively. Therefore, to cover every part of the rail tracks, the frequency of the camera shutter is set to 2 frames per second. Each image captured by an UAV has corresponding POS information that contains the coordinates of the aircraft at that moment, and therefore, the location of rail defects can be found based on these coordinates.

Figure 8. Experiment environment.

3.1.3. Defects and Evaluation

All images used for this experiment were captured, and several examples of UAV images containing discrete defects are shown in Figure 9. A large dataset was constructed to verify the algorithm, which contains 50 rail images, and each one has a lot of defects (more than 2) on its surface. Rail surface defects are divided into two categories depending on size and maintenance standard of the railway. In general, a defect whose size is larger than 255 mm^2 should be inspected as soon as possible, because it may result in serious accidents [49]. Therefore, in our data set, the defects are divided into two types, according to size Ω of defects, as shown in Table 1.

Figure 9. Examples of UAV images containing discrete defects on rail surfaces.

Table 1. The type of rail surface defects.

Defects Type	T-I Defect	T-II Defect
Area (Ω)	$25\ \text{mm}^2 < \Omega \leq 255\ \text{mm}^2$	$255\ \text{mm}^2 < \Omega$

The dataset includes 208 defects: 126 in T-I and 82 in T-II. These defects are labeled by experts.

Within the Matlab 2014 compile environment, the designed software is achieved by Matlab program languages, and an inspected defect is automatically marked by a rectangle. Then, the inspected defect is accepted as correct if it matches the marked defect in the corresponding image.

In information retrieval and pattern recognition, recall and precision are the basic criteria for evaluation of retrieval quality. The two criteria are used for evaluation of our experiment result in this paper. The precision (P) and recall (R) are respectively given by:

$$P = TP/(TP + FP) \tag{20}$$

$$R = TP/NP \tag{21}$$

where TP is the number of defects that were inspected correctly, FP is the number of wrongly inspected defects, and NP is the number of marked defects for the corresponding defect category (T-I and T-II). Specifically, the recall is more significant than precision, because a defect which is not detected may have severe consequences.

It should be noted that each defect in the dataset is labeled with a minimum enclosing rectangle; thereby, the real region of rail surface defects is approximated by the region of its minimum enclosing rectangle. As to the LWLC-GSME model, all these detected defects are also automatically marked by a minimum enclosing rectangle. This is regard as correct inspection if the minimum enclosing rectangle of a detected defect overlaps the corresponding labeled image more than 85%; otherwise, it is error detection. TP denotes number of defects that were correctly inspected.

3.2. Performance Analysis

Two groups of visual comparison experiments and a qualitative comparison experiment for defects inspection are presented in this section. Every defect in the images is marked by a red rectangle.

3.2.1. Image Enhancement

The effectiveness of image enhancement method was first verified by performing experiments on several randomly selected images. It should be noted that these selected images include defects for types T-I and T-II, and have characteristics of low contrast and varying illumination. In this section, this paper compares the LWLC algorithm with traditional enhancement methods, including histogram equalization (HE), LN [14] and MLC [30] algorithms.

Figure 10 presents comparison results of these methods. The HE method only averages the gray level distribution rather than enlarging the gray scale, and retains a number of irregular (noise) points, as shown in the Figure 10B. The LN method has poor enhancement effect because the image loses a lot of the significant detailed information. Not surprisingly, the MLC algorithm achieves competitive performance and highlights defects on UAV-based rail images. However, to UAV-based rail images having high brightness mean and a great deal irregular points, the MLC algorithm has a poor ability at extending grayscale range between irregular points and defect points. The MLC algorithm can distinguish defects from background, but the distinguishing capability for irregular (noise) points and defect points is not as good as that of the LWLC algorithm. As shown in Figure 10D,E, the enhanced defects by LWLC algorithm are more obviously highlighted than the MLC algorithm.

It is worth noting that the proposed method can effectively remove the influence of uneven illumination. As shown in Figure 10A, for the above two images with several shadows, the global image enhancement method (HE method, as shown in Figure 10B) makes images loss detail information and amplifies irregular points (noise points and shadow points), since the HE is a linear algorithm and only averages the gray level distribution. In addition, although the LN method (as shown in Figure 10C) can remove shadows, it also removes defects, due to the fact that a rail surface contains a small number of defects, and the difference between defects and backgrounds in longitudinal direction is large. However, because of few defects existing on rail surfaces and high reflection properties of rail surfaces, the brightness mean of longitudinal line along a track approximates to the background luminance. Thereby, the proposed LWLC algorithm based on Weber's law can address these issues, and effectively remove the uneven illumination due to the feature of Weber's law presented in Section 1. It can be seen that the proposed LWLC algorithm is superior to other two methods, as shown in the Figure 10E.

For defect segmentation, the LWLC algorithm combined with the GSME method can achieve better performance, and the experiment of defect segmentation will be described in the next section.

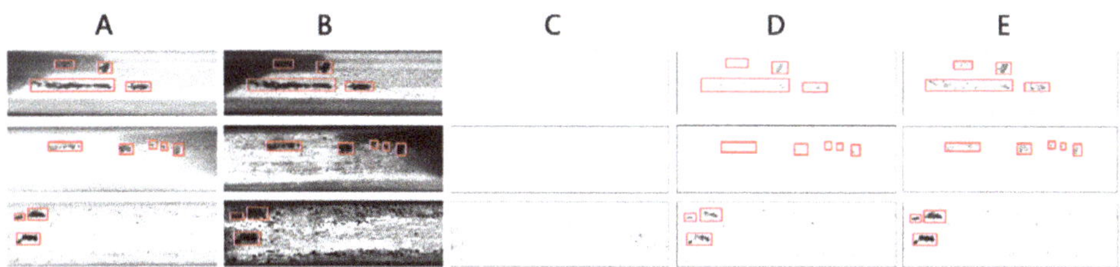

Figure 10. Examples of four enhancement methods for non-uniform illumination rail images. (**A**) Three examples of extracted rail images. (**B**) Three examples of enhancement image by HE method. (**C**) Three examples of enhancement image by LN method. (**D**) Three examples of enhancement image by MLC algorithm. (**E**) Three examples of enhancement image by LWLC algorithm. In Figure 10, the discrete defects on images have been marked by red rectangle to compare enhancement performance of LWLC algorithm with related methods.

3.2.2. Defect Segmentation

On the basis of the LWLC, this paper compares GSME with traditional image segmentation methods including the maximum entropy (ME) algorithm, the proportion emphasized maximum entropy (PEME) algorithm, and the GSTA method. The ME [37] method can confirm one threshold which maximizes the total content of information provided by object distribution and background distribution. After enhancing images by MLC, PEME [30] is used to obtain an optimal segmentation threshold. PEME is an improved ME method which reduces the proportion of the background information and increases an exponent factor in original equation. The GSTA method uses wavelet transform and the Otsu algorithm to enhance image edges, then Otsu to extract the objection of the image. In the comparative experiment, the PEME method adopts MLC for image enhancement, and the other three adopt the same LWLC approach. All methods adopt the same evaluation criterion mentioned above, as shown in Figure 11. According to the Refs. [30,38] and our experiment, the parameter ρ of GSTA method, the parameter β of MLC+PEME and the parameter α of LWLC+GSME are set to 0.2, 2, and 0.3, respectively.

As shown in Figure 11B, LWLC+ME can retain more detailed information of defects than the LWLC+GSTA and MLC+PEME methods, but its performance in terms of noise suppression, non-defect points removal, and defect details preservation is weaker than that of LWLC+GSME. Reasons for the inapplicability of the PEME approach are as follows: on one hand, UAV images have lower contrast, relatively obscure texture features, and more noise due to lighting based on natural light and higher distance between camera and rail; on the other, the variation in natural illumination can't be controlled by a human being, and it has an inevitable effect on defect extraction. If the image is captured as the aircraft flies above the rail at different distances, there is no uniform model to set the exponent factor β of the PEME model. In Figure 11C, it can be seen that defect details including shape and area information can't be effectively retained by the LWLC+GSTA method, and there are more noise and non-defect points than with the other methods. Figure 11D shows that MLC+PEME highlighted defect areas, but that noise and non-defect points in the image can't be restrained. As shown in Figure 11E, it can be seen that defects are remarkably segmented with the least noise and non-defect points based on LWLC+GSME.

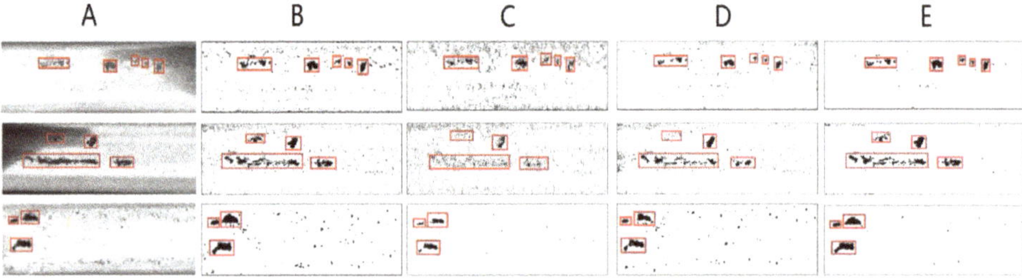

Figure 11. Examples of four defect segmentation methods for rail surface images. (**A**) Three examples of extracted rail images. (**B**) Three examples of defect segmentation by LWLC+ME method. (**C**) Three examples of defect segmentation by LWLC+GSTA. (**D**) Three examples of defect segmentation by MLC+PEME method. (**E**) Three examples of defect segmentation by LWLC+GSME method. In Figure 10, the discrete defects on images have been marked by red rectangle to compare enhancement performance of LWLC+GSME method with related methods.

3.2.3. Qualitative Comparison between LWLC+GSME and Related Methods

Finally, a quantitative analysis for the defect inspection after segmentation is given by Figures 12 and 13. The two figures further explain that the LWLC+GSME method is more suitable for detection of rail defects based on the UAV image.

Although the GSTA method enhances image contrast, it can't repress interference of noise effectively. These three methods acquire poor segmentation effects because they are susceptible to noise and background points. In contrast, LWLC+GSME not only achieves contrast enhancement between defects and background, but also obtains the best segmentation effects. For example, LWLC+GSME achieves a precision of 88.63% for T-I defects and a precision of 90.32% for T-II defects. It should be noted that both of MLC+PEME and LWLC+GSME obtain similar effects for precision, because they see non-defect regions as defect under light disturbance (such as uniform illumination and low contrast).

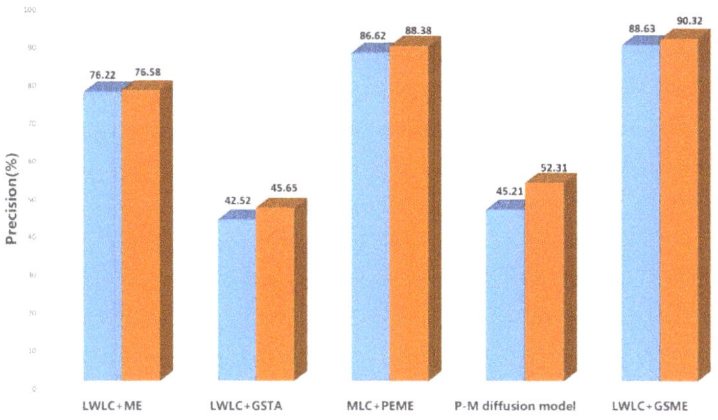

Figure 12. Comparison of defection precision for four detection methods. The blue block and orange block denotes T-I defect and T-II defect, respectively.

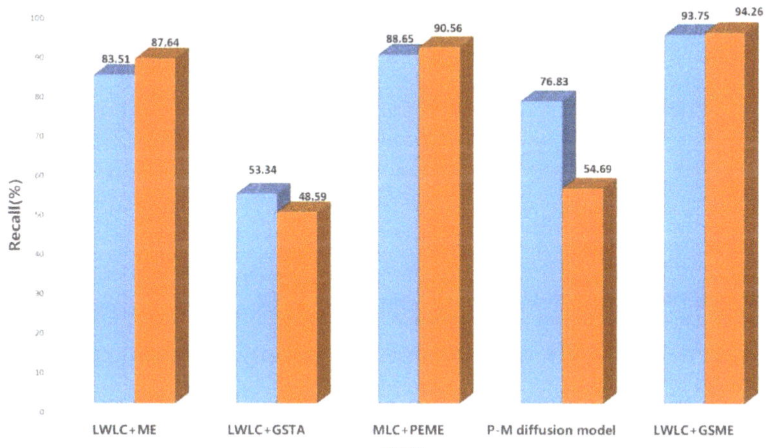

Figure 13. Comparison of defection recall for four detection methods. The blue block and orange block denotes T-I defect and T-II defect, respectively.

In addition, based on the ME principle and characteristics of the defects mentioned above, a suitable segmentation threshold should be relatively small under the condition of complete retention of defects [30]. Three examples of the segmentation threshold with four segmentation methods are shown in Table 2. From this table, a relatively small threshold is obtained based on LWLC+GSME. It further illustrates that the proposed method can select a better segmentation threshold. It should be noted that thresholds based on LWLC+GSTA are the smallest. However, this method achieves poor performance for T-I and T-II defects, as shown in Figures 11 and 12; this is because this method uses Gaussian kernels to enhance object edges of high frequency areas in the wavelet domain, and UAV-based images contain a lot of irregular (noise) points. For this reason, the method also enhances these noises in rail images, thereby yielding poor defects detection performance.

Table 2. Examples of segmentation threshold values with four segmentation methods; LWLC: The proposed Local Weber-like Contrast algorithm; ME: The maximum entropy algorithm; GSTA: The gray stretch and threshold algorithm; MLC: The Michelson-like contrast algrorithm; PEME: The proportion emphasized maximum entropy method; GSME: The gray stretch maximum entropy method.

Defects Inspection Model	LWLC+ME	LWLC+GSTA	MLC+PEME	LWLC+GSME
Original images (A1)	228	140	226	189
Original images (A2)	213	120	225	170
Original images (A3)	188	82	206	161

In Table 2, original images are correspond to the three images in Figure 10A. For example, A1 is the first line image in Figure 10A.

4. Conclusions

To cope with rail surface defects, an inspection approach based on UAV images is proposed in this study. The proposed LWLC algorithm can highlight not only defects and homogenized backgrounds of UAV-based rail images, but also eliminates the adverse effects of non-uniform illumination. Furthermore, we put forward the GSME method for defects segmentation, which reduces irregular points and obtains excellent segmentation effects. The integrated LWLC+GSME method further illustrates great flexibility and effectiveness in detecting discrete defects.

Finally, this study compared LWLC and LWLC+GSME with related methods, and the results of experiments show the significance of the proposed method. The quantitative experimental results show that the proposed method achieves a recall of 93.75% for T-I defects and of 94.26% for T-II defects, and that it is efficient and feasible to detect rail surface defects based on UAV images. It was verified that the proposed model can obtain excellent results.

In future, our research work will focus on the following two aspects: firstly, we will explore new models for rail defect classification based on UAV images and assess the health of rails; secondly, based on the development of high UAV photography, fast detection models in complex environments will be developed to increase detection efficiency.

Author Contributions: Y.W. collected and analyzed the data, made charts and diagrams, conceived and performed the experiments and wrote the paper; Y.Q. conceived the structure and provided guidance; Z.W. provided the UAV equipment and searched the literature; L.J. modified the manuscript.

Funding: This research is supported by the National Key R&D Program of China (No. 2016YFB1200203), National Natural Science Foundation of China (No. 91738303) and State Key Laboratory of Rail Traffic Control and Safety (Contract Nos. RCS2016ZQ003 and RCS2016ZT018)

Acknowledgments: This research is also supported by National Engineering Laboratory for System Safety and Operation Assurance of Urban Rail Transit.

Conflicts of Interest: The authors declare no conflict of interest.

Appendix A

In order to reduce the computation payload, the colored image is transformed into a gray image. The method for rail track extraction is described as follows.

Firstly, the longest line of rail edge is detected by Hough transform, as shown in Figure A1B, and the image is rotated by the angle θ between the line and horizontal direction so that the rails are parallel to the vertical direction, as shown in Figure A1C. And then the followed Algorithm A1 is used to find the most left position of a rail track after the matrix $Cg(n)$ is obtained by Equation (1).

Algorithm A1. The Algorithm A1 for track extraction.

1	**procedure** Algorithm A1 ($Cg(n)$, W_d)
2	**for** $m \leftarrow 1, M - W_d + 1$ **do**
3	**for** $n \leftarrow m, W_d$ **do** /* W_d is the width of the rail track.*/
4	$Cg(n) \leftarrow Cg(n) + Cg(n + 1)$
5	$CumCg(m) \leftarrow Cg(n)$
6	**end for**
7	$maxCumCg \leftarrow -1$
8	$p\_left \leftarrow 0$
9	**for** $m \leftarrow 1, M - W_d + 1$ **do**
10	$p\_CumCg \leftarrow CumCg(m)$
11	**if** $p\_CumCg > maxCumCg$ **then**
12	$maxCumCg \leftarrow p\_CumCg$
13	$p\_left \leftarrow m$
14	**end if**
15	**end for**
16	**return** p_left /* The most left position of a rail track (p_left)*/
17	**end procedure**

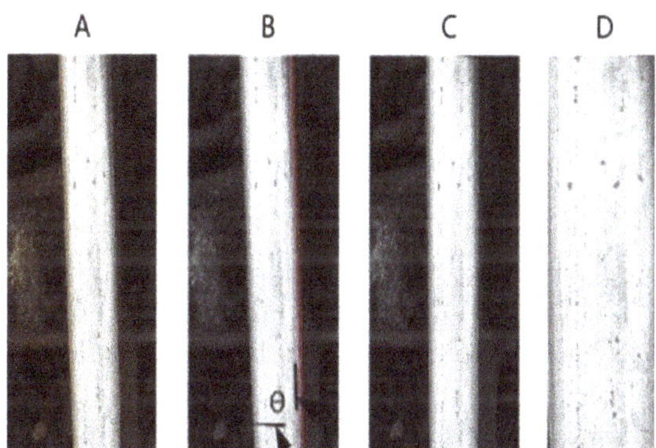

Figure A1. The example of the rail track extraction. (**A**) The original image contains a rail. (**B**) The detection for the longest line and the inclined angle θ of the rail based on Hough transform method. (**C**) The rail image correction by rotating the angle θ. (**D**) The rail track extraction based on integral projection of vertical pixel column for a rail image.

References

1. Arivazhagan, S.; Shebiah, R.N.; Magdalene, J.S.; Sushmitha, G. Railway track derailment inspection system using segmentation based fractal texture analysis. *ICTACT J. Image Video Process.* **2015**, *6*, 1060–1065.
2. Cannon, D.; Edel, K.O.; Grassie, S.; Sawley, K. Rail defects: An overview. *Fatigue Fract. Eng. Mater. Struct.* **2003**, *26*, 865–886. [CrossRef]
3. Grassie, S. Rail corrugation: Characteristics, causes, and treatments. *Proc. Inst. Mech. Eng. Part F J. Rail Rapid Transit* **2009**, *223*, 581–596. [CrossRef]
4. Edwards, J.R.; Hart, J.M.; Sawadisavi, S.; Resendiz, E.; Barkan, C.; Ahuja, N. Advancements in Railroad Track Inspection Using Machine-Vision Technology. In Proceedings of the AREMA Conference on American Railway and Maintenance of Way Association, Chicago, IL, USA, 2–5 August 2009.
5. Marino, F.; Distante, A.; Mazzeo, P.L.; Stella, E. A real-time visual inspection system for railway maintenance: Automatic hexagonal-headed bolts detection. *IEEE Trans. Syst. Man Cybern. Part. C (Appl. Rev.)* **2007**, *37*, 418–428. [CrossRef]

6. Tsai, D.-M.; Wu, S.-C.; Chiu, W.-Y. Defect detection in solar modules using ICA basis images. *IEEE Trans. Ind. Inform.* **2013**, *9*, 122–131. [CrossRef]

7. Zhang, X.; Feng, N.; Wang, Y.; Shen, Y. An analysis of the simulated acoustic emission sources with different propagation distances, types and depths for rail defect detection. *Appl. Acoust.* **2014**, *86*, 80–88. [CrossRef]

8. Liu, Z.; Li, W.; Xue, F.; Xiafang, J.; Bu, B.; Yi, Z. Electromagnetic tomography rail defect inspection. *IEEE Trans. Magn.* **2015**, *51*, 1–7. [CrossRef]

9. Hesse, D.; Cawley, P. The Potential of Ultrasonic Surface Waves for Rail Inspection. In Proceedings of the AIP Conference, Golden, CO, USA, 25–30 July 2004; American Institute of Physics: College Park, MD, USA, 2005; Volume 760, pp. 227–234.

10. Tang, X.-N.; Wang, Y.-N. Visual inspection and classification algorithm of rail surface defect. *Comput. Eng.* **2013**, *9*, 25–30.

11. Resendiz, E.; Hart, J.M.; Ahuja, N. Automated visual inspection of railroad tracks. *IEEE Trans. Intell. Transp. Syst.* **2013**, *14*, 751–760. [CrossRef]

12. Liu, J.; Li, B.; Xiong, Y.; He, B.; Li, L. Integrating the symmetry image and improved sparse representation for railway fastener classification and defect recognition. *Math. Probl. Eng.* **2015**, *2015*, 462528. [CrossRef]

13. Ph Papaelias, M.; Roberts, C.; Davis, C. A review on non-destructive evaluation of rails: State-of-the-art and future development. *Proc. Inst. Mech. Eng. Part F J. Rail Rapid Transit* **2008**, *222*, 367–384. [CrossRef]

14. Li, Q.; Ren, S. A real-time visual inspection system for discrete surface defects of rail heads. *IEEE Trans. Instrum. Meas.* **2012**, *61*, 2189–2199. [CrossRef]

15. Gan, J.; Li, Q.; Wang, J.; Yu, H. A hierarchical extractor-based visual rail surface inspection system. *IEEE Sensors J.* **2017**, *17*, 7935–7944. [CrossRef]

16. He, Z.; Wang, Y.; Yin, F.; Liu, J. Surface defect detection for high-speed rails using an inverse pm diffusion model. *Sens. Rev.* **2016**, *36*, 86–97. [CrossRef]

17. Sieberth, T.; Wackrow, R.; Chandler, J.H. Automatic detection of blurred images in UAV image sets. *ISPRS J. Photogramm. Remote Sens.* **2016**, *122*, 1–16. [CrossRef]

18. Siebert, S.; Teizer, J. Mobile 3d mapping for surveying earthwork projects using an unmanned aerial vehicle (UAV) system. *Autom. Constr.* **2014**, *41*, 1–14. [CrossRef]

19. Pérez-Ortiz, M.; Peña, J.M.; Gutiérrez, P.A.; Torres-Sánchez, J.; Hervás-Martínez, C.; López-Granados, F. Selecting patterns and features for between-and within-crop-row weed mapping using UAV-imagery. *Expert Syst. Appl.* **2016**, *47*, 85–94. [CrossRef]

20. Wang, L.; Zhang, Z. Automatic detection of wind turbine blade surface cracks based on UAV-taken images. *IEEE Trans. Ind. Electron.* **2017**, *64*, 7293–7303. [CrossRef]

21. Liu, C.; Liu, Y.; Wu, H.; Dong, R. A safe flight approach of the UAV in the electrical line inspection. *Int. J. Emerg. Electr. Power Syst.* **2015**, *16*, 503–515. [CrossRef]

22. Kaamin, M.; Idris, N.A.; Bukari, S.M.; Ali, Z.; Samion, N.; Ahmad, M.A. Visual Inspection of Historical Buildings Using Micro UAV. In Proceedings of the MATEC Web of Conferences, Qingdao, China, 25–27 August 2017; EDP Sciences: Paris, France, 2017; Volume 103, p. 07003.

23. Yuan, C.; Liu, Z.; Zhang, Y. UAV-Based Forest Fire Detection and Tracking Using Image Processing Techniques. In Proceedings of the 2015 International Conference on Unmanned Aircraft Systems (ICUAS), Denver, CO, USA, 9–12 June 2015; pp. 639–643.

24. Rau, J.; Hsiao, K.; Jhan, J.; Wang, S.; Fang, W.; Wang, J. Bridge crack detection using multi-rotary UAV and object-base image analysis. *Int. Arch. Photogramm. Remote Sens. Spat. Inf. Sci.* **2017**, *42*, 311. [CrossRef]

25. Arenella, A.; Greco, A.; Saggese, A.; Vento, M. In Real Time Fault Detection in Photovoltaic Cells by Cameras on Drones. In Proceedings of the 14th International Conference on Image Analysis and Recognition, Montreal, QC, Canada, 5–7 July 2017; Springer: New York, NY, USA, 2017; pp. 617–625.

26. Xiao, Q.; Zhang, Q.; Wu, X.; Han, X.; Li, R. Learning Binary Code Features for UAV Target Tracking. In Proceedings of the 3rd IEEE International Conference on Control Science and Systems Engineering (ICCSSE), Beijing, China, 17–19 August 2017; IEEE: Piscataway, NJ, USA, 2017; pp. 65–68.

27. Baykara, H.C.; Bıyık, E.; Gül, G.; Onural, D.; Öztürk, A.S. Real-Time Detection, Tracking and Classification of Multiple Moving Objects in UAV Videos. In Proceedings of the 2017 IEEE 29th International Conference on Tools with Artificial Intelligence (ICTAI), Boston, MA, USA, 6–8 November 2017; IEEE: Piscataway, NJ, USA, 2017; pp. 945–950.

28. Radovic, M.; Adarkwa, O.; Wang, Q. Object recognition in aerial images using convolutional neural networks. *J. Imaging* **2017**, *3*, 21. [CrossRef]

29. Liang, L.; He, W.-P.; Lei, L.; Zhang, W.; Wang, H.-X. Survey on enhancement methods for non-uniform illumination image. *Appl. Res. Comput.* **2010**, *5*, 008.

30. Li, Q.; Ren, S. A visual detection system for rail surface defects. *IEEE Trans. Syst. Man Cybern. Part C (Appl. Rev.)* **2012**, *42*, 1531–1542. [CrossRef]

31. Chen, X.; Xu, J.-P. A denoising algorithm of image for UAV based on wavelet transform and mean-value filtering. *Fire Control Command. Control* **2011**, *8*, 049.

32. Jin, X.; Wen, Z.; Wang, K. Effect of track irregularities on initiation and evolution of rail corrugation. *J. Sound Vib.* **2005**, *285*, 121–148. [CrossRef]

33. Fechner, G. Über ein wichtiges psychophysiches grundgesetz und dessen beziehung zur schäzung der sterngrössen. *Abk. K. Ges. Wissensch. Math.-Phys. K* **1858**, *1*, 4.

34. Shen, J. On the foundations of vision modeling: I. Weber's law and Weberized TV restoration. *Phys. D Nonlinear Phenom.* **2003**, *175*, 241–251. [CrossRef]

35. Available online: https://en.wikipedia.org/wiki/Contrast_(vision) (accessed on 1 May 2018).

36. Otsu, N. A threshold selection method from gray-level histograms. *IEEE Trans. Syst. Man Cybern.* **1979**, *9*, 62–66. [CrossRef]

37. Kapur, J.N.; Sahoo, P.K.; Wong, A.K. A new method for gray-level picture thresholding using the entropy of the histogram. *Comput. Vis. Graph. Image Process.* **1985**, *29*, 273–285. [CrossRef]

38. Liu, L.; Yang, N.; Lan, J.; Li, J. Image segmentation based on gray stretch and threshold algorithm. *Opt.-Int. J. Light Electron. Opt.* **2015**, *126*, 626–629. [CrossRef]

39. Tang, S.-W.; Lin, J. Image denoising with combination of wavelet transform and median filtering. *J. Harbin Inst. Technol.* **2002**, *24*, 1334–1336.

40. Rakheja, P.; Vig, R. Image denoising using combination of median filtering and wavelet transform. *Int. J. Comput. Appl.* **2016**, *141*, 31–35. [CrossRef]

41. Jin, B.; Niu, H.; Hou, T. Study of straight line rail image edge detection based on improved hough. *Video Eng.* **2015**, *39*, 17–19.

42. Li, Q.; Tan, Y.; Huayan, Z.; Ren, S.; Dai, P.; Li, W. A Visual Inspection System for Rail Corrugation Based on Local Frequency Features. In Proceedings of the 2016 IEEE 14th International Conference on Dependable, Autonomic and Secure Computing, 14th International Conference on Pervasive Intelligence and Computing, 2nd International Conference on Big Data Intelligence and Computing and Cyber Science and Technology Congress (DASC/PiCom/DataCom/CyberSciTech), Auckland, New Zealand, 8–12 August 2016; IEEE: Piscataway, NJ, USA, 2016; pp. 18–23.

43. Whittle, P. The psychophysics of contrast brightness. In *Lightness, Brightness, and Transparency*; Gilchrist, A.L., Ed.; Lawrence Erlbaum Associates, Inc.: Hillsdale, NJ, USA, 1994; pp. 35–110.

44. Arend, L.E.; Spehar, B. Lightness, brightness, and brightness contrast: 1. Illuminance variation. *Percept. Psychophys.* **1993**, *54*, 446–456. [CrossRef] [PubMed]

45. Agaian, S.S. Visual Morphology. *Proc. SPIE* **1999**, *3646*, 139–150.

46. Rafael, C.; Gonzalez, R.E.W. *Digital Image Processing*, 3rd ed.; Pearson Education, Inc.: Upper Saddle River, NJ, USA, 2008.

47. Pun, T. A new method for grey-level picture thresholding using the entropy of the histogram. *Signal. Process.* **1980**, *2*, 223–237. [CrossRef]

48. Pun, T. Entropic thresholding, a new approach. *Comput. Graph. Image Process.* **1981**, *16*, 210–239. [CrossRef]

49. Xu, W. Study on Defect Recognition for Rail Surface Based on Machine Vision. Master's Thesis, Beijing Jiaotong University, Beijing, China, 1 March 2015.

Article

Feature-Learning-Based Printed Circuit Board Inspection via Speeded-Up Robust Features and Random Forest

Eun Hye Yuk [1], Seung Hwan Park [2], Cheong-Sool Park [1] and Jun-Geol Baek [1,*]

[1] Department of Industrial Management Engineering, Korea University, Seoul 02841, Korea; eunhyeyuk@korea.ac.kr (E.H.Y.); dumm97@gmail.com (C.-S.P.)
[2] Data Science, SK Hynix Semiconductor, Icheon 17336, Korea; seunghwan1.park@sk.com
* Correspondence: jungeol@korea.ac.kr; Tel.: +82-2-3290-3396

Received: 14 May 2018; Accepted: 31 May 2018; Published: 5 June 2018

Featured Application: The main contribution of this work is to propose an inspection method using image data generated at the actual manufacturing process. This proposed method can help printed circuit board (PCB) manufacturers more effectively detect defects, such as scratches and improper etching, in an automated optical inspection (AOI). Moreover, the proposed method of this work can be also applied to the field of dermatology, where it has to detect skin diseases, as well as in PCB inspection.

Abstract: With the coming of the 4th industrial revolution era, manufacturers produce high-tech products. As the production process is refined, inspection technologies become more important. Specifically, the inspection of a printed circuit board (PCB), which is an indispensable part of electronic products, is an essential step to improve the quality of the process and yield. Image processing techniques are utilized for inspection, but there are limitations because the backgrounds of images are different and the kinds of defects increase. In order to overcome these limitations, methods based on machine learning have been used recently. These methods can inspect without a normal image by learning fault patterns. Therefore, this paper proposes a method can detect various types of defects using machine learning. The proposed method first extracts features through speeded-up robust features (SURF), then learns the fault pattern and calculates probabilities. After that, we generate a weighted kernel density estimation (WKDE) map weighted by the probabilities to consider the density of the features. Because the probability of the WKDE map can detect an area where the defects are concentrated, it improves the performance of the inspection. To verify the proposed method, we apply the method to PCB images and confirm the performance of the method.

Keywords: image inspection; non-referential method; feature extraction; fault pattern learning; weighted kernel density estimation (WKDE)

---

1. Introduction

Because the era of the Internet of Things (IoT) has been accompanied by the rapid development of the semiconductor industry and communication technologies, the use of high-tech products, such as mobile phones and wearable devices, has been spreading widely in our daily lives. The printed circuit board (PCB) is one of the key components of such electronic devices. A PCB is a thin plate made by printing an electrically conductive circuit on an insulator. Figure 1 shows an image of a generic PCB. As shown in Figure 1, a PCB connects to different parts electrically through a point-to-point wiring process.

In the past, handwork wiring led to frequent failures of the wire junctions and a short circuit as the wire insulation began to age. Furthermore, this type of work was conducted manually for inner-connecting components within the board. Such wiring, therefore, required a significant amount of time and effort. With advancements in technology, PCBs were developed for efficient and automated production. Because PCBs allow for the use of mass production, they allow devices to be smaller and lighter. In addition, they allow a high level of reliability at low production costs. Because of these advantages, most electronic devices use PCBs.

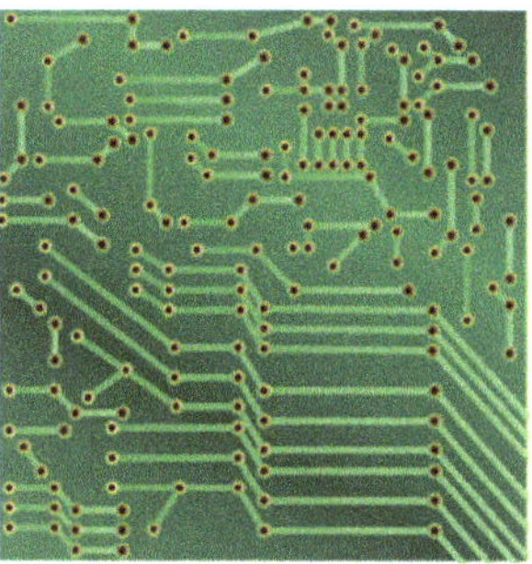

Figure 1. An illustrative image of a printed circuit board (PCB).

According to the recent trend of miniaturized and high-performance electronic devices, the demand for PCBs has increased substantially. The current PCB market has shown an average annual growth rate of 4%. Based on increasing demand, electronics manufacturers require perfect quality and a high level of accuracy from their PCB inspections to assure their competitive edge. To meet these quality requirements, PCB manufacturers conduct an inspection before proceeding to the main process. The PCB manufacturing process consists of many steps: cutting, inner layer etching, an automatic optical inspection (AOI), lay-up, lamination, etching, drilling, solder masking, routing, a bare board test (BBT), quality control, packing, and shipping, in that order. Before proceeding to the main processes, such as lamination and etching, faults in a PCB are detected during the AOI stage. An AOI is an automated visual inspection using an image comparison method. Most PCB manufacturers inspect their PCBs through the AOI process. The types of defects detected during the AOI process include scratches, improper etching, and open circuits. In particular, scratches are fatal defects because they have the potential to change the electrical properties and can result in a malfunction of the completed product.

There are typically three methods applied during the AOI process: a comparison reference (CR), non-reference verification (NV), and a hybrid approach (HA). The CR method compares an inspected image with a reference image. It measures any existing dissimilarities between the reference image and the inspected image. Thus, this method requires a reference image, and inspection is difficult without such an image. The NV approach tests the design rule of the PCB for detecting faults. It essentially verifies the widths of the insulators and conductors. However, this approach makes it difficult for users to design the rules as restrictions on the image features, and it cannot detect faults without such rules [1]. The HA approach combines various types of CR and NV methods. That is, the approach utilizes a reference image or design rules to inspect a PCB image. However, it still does not solve the limitation of being unable to conduct an inspection without reference information. Because the circuits used in a PCB are diverse and complex, the reference information increases significantly, thereby decreasing the inspection's efficiency. Therefore, additional studies are required to detect faults without reference information.

Numerous algorithms have been proposed to improve the accuracy of a PCB inspection. Such algorithms can be categorized into two approaches: referential and non-referential methods.

A referential method is based on a comparison between the inspected image and a reference image. To measure the dissimilarity between these two images, image subtraction and template matching techniques are used. Wu et al. proposed an inspection method based on a subtraction method [2]. The image subtraction method compares both images using an XOR logic operator [3]. The resulting image, which is obtained after this operation, contains only portions of a fault [4]. Therefore, the reference image and the inspected image should be placed in a fixed position to compare both images. In addition, a reference image of the same size is required. Template matching is a technique for identifying the parts of the image that match the reference image. This method extracts the features of both the reference and inspected images. It then calculates the similarities of these features. One of the major disadvantages of template matching is that a large amount of information regarding the reference image must be used. Therefore, this method requires a mass storage device that can store all of the information. Moreover, the inspected images also have to be precisely matched for a comparison with the reference image [5]. Acciani et al. suggested an inspection algorithm that extracts the wavelet and geometric features, and then detects a defect after learning the fault pattern using a neural network and k-nearest neighbors [6]. When extracting the features of a PCB image, this method uses the maximum value of the correlation coefficients between the features of the reference image and the inspected image. This approach of applying machine-learning algorithms is called the learning-based model. The aim of this approach is to automatically detect faults through pattern recognition [7]. In [8], the authors proposed a defect detection method using feature matching for non-repetitive patterned images. The method first extracts features of both the reference image and the inspected image using a modified corner detector. It then detects a fault by finding a correspondence between two feature sets. Thus, the methods [6,8] still require a reference image to detect a fault.

A non-referential method, on the other hand, does not require a reference image. However, this method uses design rules. If the inspected image does not conform to the design specification standards, it is considered defective. Ye and Danielson suggested a verifying algorithm for minimum conductive and insulator trace widths [9]. This algorithm uses morphological techniques, which are methods for processing binary and grayscale images based on shape. Morphological techniques do not require a predefined model of a perfect pattern because we can construct specific shapes in an image by choosing an appropriate neighborhood shape [10]. However, this method has a disadvantage in that we should apply different pre-processing algorithms to check for faults in a PCB. In addition, it automatically increases the inspection time [5]. Tsai et al. proposed a non-referential defect detection approach for bond pads [11]. This approach restores the shape of the bond pads using Fourier image reconstruction. The method then evaluates the similarities and differences in the pad shape. However, the method has difficulty in reconstructing an original image of a PCB in the absence of a reference image because the shape of a circuit is quite varied and complex. Thus, the method is not suitable for detecting faults on a PCB. As mentioned earlier, a non-referential image analysis has certain limitations.

Rosten and Drummon proposed a very fast and high-quality corner detector using machine-learning techniques [12]. This algorithm uses a decision tree to learn the properties of the corner points and determine which point is a corner point instead of directly counting the number of consecutive points of the same type. Pernkopf proposed an approach for the detection of three-dimensional faults on scale-covered steel surfaces [13]. After extracting features, the approach uses a Bayesian network to learn the feature property and classify the faults. Classification research, which is learning the patterns of the extracted features through image processing, is actively carried out to diagnose skin cancer not only in manufacturing but also in dermatology [14]. It is indispensable to extract meaningful features that can explain the properties of shape and color in order to effectively detect a melanoma [15]. Roberta suggested a method to find meaningful features by combining three characteristics of shape, texture, and color [16].

Based on the concept of a learning property, we propose a new non-referential method for fault detection by extracting features based on an image-processing method and learning the fault information using random forests. The proposed method utilizes features obtained through speeded-up robust features (SURF) to describe the fault information. Therefore, the method can detect a fault quickly and accurately without a reference image and without being affected by environmental changes, such as the size, rotation, and location of the PCB. The proposed method first extracts robust features using an image-processing technique and learns the fault pattern using an efficient classification technique utilizing high-dimensional data. We then calculate the probability and draw a weighted kernel density estimation (WKDE) map weighted by the probability and identify whether a fault has occurred.

The remainder of this paper is organized as follows: In Section 2, the background of the proposed method is briefly reviewed. Section 3 describes the procedure of the proposed inspection algorithm. Section 4 presents the experimental results through a comparison of the receiver operating characteristic (ROC) curves. Finally, some concluding remarks are given in Section 5.

2. Background of the Proposed Method

This section describes the two algorithms used in the proposed method. We first introduce SURF, which is used for extracting the features of the PCB. These features include important information in the image, such as the size, angles, coordinates, and color. By digitizing the image, they enable calculations related to the pattern recognition to be applied during image processing. We then describe the random forests used to learn the fault pattern and calculate the probability.

2.1. Speeded-Up Robust Features (SURF)

SURF first digitizes a PCB image into a vector to enable computational calculations through image processing for learning the fault pattern. Among the various types of image processing techniques, including scale invariant feature transform (SIFT) and orientation by intensity centroid (ORB), we use SURF, which is the most robust algorithm with regard to environmental factors such as size, shape, and color. The performance of SIFT does not differ significantly compared to that of SURF. However, SIFT is not able to extract the features in a small defective area. ORB is specialized in extracting the features of a rounded or curved edge, rather than a straight line, in a circuit. Because a PCB is mainly composed in a straight formation, ORB is inappropriate for extracting PCB features. On the other hand, SURF is robust and useful for feature detection. It also improves the computational speed compared to SIFT [17]. In addition, SURF provides robust features and distinctive descriptors to the size and rotation transform. Thus, it results in a strong performance regardless of the fault size and shape of the circuit. For this reason, the proposed method applies SURF for the feature extraction. The following subsections describe the SURF algorithm, which has two phases. First, a point of interest is detected in an image using a Hessian matrix. In the second phase, SURF generates descriptors with 64 dimensions, which describe the characteristics of the features.

2.1.1. Interest Point Detection Based on a Hessian Detector

SURF uses a Hessian matrix for detecting interest points, which represent the characteristics of the image and include useful information for identification, including corner points. An integral image is used in an approximated Hessian matrix to improve the calculation speed of box-type convolution filters. The integral image $J(x,y)$ at coordinate $x = (x,y)$ $x = (x,y)$ is the sum of all pixels within a rectangular area formed by the origin and coordinate x in input image I, the equation of which is as follows:

$$J(x,y) = \sum_{i=0}^{i \leq x} \sum_{j=0}^{j \leq y} I(i,j). \tag{1}$$

Given point $x = (x,y)$ in input image I, a Hessian matrix $H(x,\sigma)$ at a scale of σ is defined through Equation (2).

$$H(x,\sigma) = \begin{bmatrix} L_{xx}(x,\sigma) & L_{xy}(x,\sigma) \\ L_{xy}(x,\sigma) & L_{yy}(x,\sigma) \end{bmatrix}, \tag{2}$$

where $L_{xx}(x,\sigma)$, $L_{xy}(x,\sigma)$, and $L_{yy}(x,\sigma)$ represent the convolution of the Gaussian second-order derivative space with the image at coordinate $x = (x,y)$.

Bay et al. proposed the use of 9×9 box filters to approximate the second-order Gaussian partial derivatives and rapidly compute the image convolutions using integral images [17]. An approximation of the second-order derivatives is denoted by $D_{xx}(x,\sigma)$, $D_{xy}(x,\sigma)$, and $D_{yy}(x,\sigma$. Bay et al. suggested using a σ of 1.2, which is the lowest scale [17]. Therefore, the determinant of an approximated Hessian matrix is calculated using Equation (3).

$$\det\big(H_{approx}(x,\sigma)\big) = D_{xx}(x,\sigma)D_{yy}(x,\sigma) - \big(0.9D_{xy}(x,\sigma)\big)^2. \tag{3}$$

To obtain the robust features to scale, SURF forms a pyramid scale space with various scales on the original image. The size of the box filter expands along with the pyramid scale space while the size of the original image remains fixed. Consequently, we can have an effect of enlarging or reducing the size of the image without scaling the original. Moreover, we can apply box filters of any size at the same speed directly to the original image, which can improve the computational speed.

After the determinant of the approximated Hessian matrix is calculated at each scale, non-maximum suppression in a $3 \times 3 \times 3$ neighborhood is applied to find the maxima, which describe the edge of the features best. The maxima are then interpolated in terms of both the scale and image space [18]. Finally, we can detect the stable location of the feature through one of the maxima.

2.1.2. Descriptor Generation

After the features are detected as mentioned above, SURF generates a descriptor, which describes the characteristics of the features, such as the shape and color, using the sum of the Haar wavelet responses [19]. The Haar wavelets enable SURF to increase the robustness and decrease the computational time. To generate the descriptor, the first step is constructing a square region around the points of interest and assigning a reproducible orientation based on the orientation-selection method introduced in [17]. The region is then split equally into 4×4 sub-regions to retain some of the spatial information as shown in Figure 2. In each sub-region, we compute the Haar wavelet responses at regularly 5×5 spaced sample points. Bay et al. suggested that the level of performance is best at the each of the above sizes [17]. The wavelet responses are calculated in the x- and y-axis directions (d_x and d_y) and summed over each sub-region. The wavelet responses are then weighted with a Gaussian centered on the point of interest to increase the robustness toward geometric deformations and localization errors. Moreover, to obtain information regarding the polarity of the intensity changes, the absolute values of the responses $|d_x|$ and $|d_y|$ are also summed. Thus, each sub-region has a four-dimensional descriptor vector as shown in Equation (4).

$$v = \big|\sum d_x, \sum d_y, \sum |d_x|, \sum |d_y|\big|. \tag{4}$$

As a result, the overall descriptor has a 64-dimension vector because 4 vectors are created for each 4×4 sub-region. Figure 2 shows a descriptor vector obtained by summing the wavelet response in a sub-region.

Finally, the descriptor vectors for each sub-region are normalized to reduce the impact on the environment, such as an external light or illumination. Eventually, we can obtain the robust features and descriptors of the PCB image using SURF. The descriptor for a single feature is generated, as shown in Figure 3, where m is the number of extracted features.

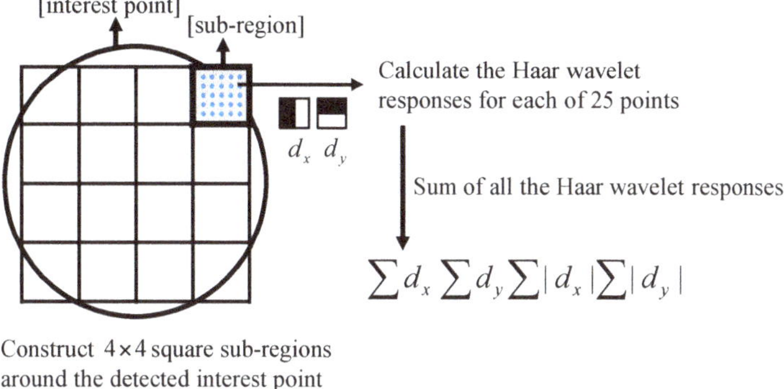

Figure 2. Diagram for generating the descriptors.

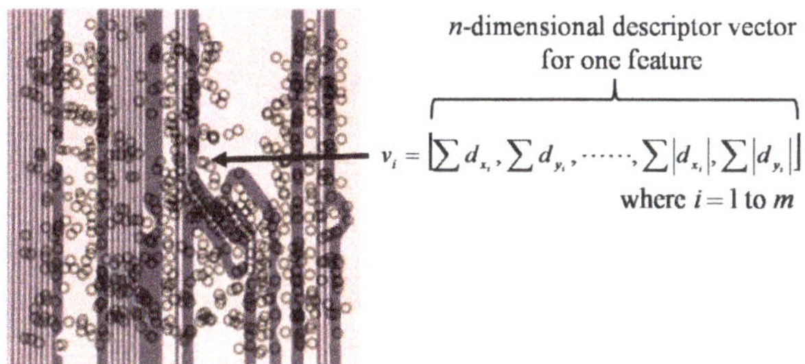

Figure 3. Features and descriptors obtained in a PCB image using speeded-up robust features (SURF).

2.2. Random Forests

Random forests is a classification method proposed by Breiman and is a type of ensemble learning method that constructs a multitude of decision trees and combines the predictions from them [20]. The random forests method is relatively accurate and fast for high-dimensional data. Furthermore, it prevents overfitting problems by voting on multiple trees and shows good predictive performance for noisy data. Because the descriptor vector obtained from SURF is composed of high-dimensional data, the algorithm is efficient for feature classification. Therefore, we use random forests to learn the fault pattern and calculate the probability of a fault feature being classified.

This algorithm begins by drawing many bootstrap samples from the training set. Classification trees are then built for each bootstrap sample using a decision tree learning method. After the forests are formed, the new data are put into each classification tree for classification. Each tree votes on the result of the tree's decision regarding the class of data. The forests then predict the class by taking the majority vote from the classification trees [21]. Because the random forests method generates many classification trees using bootstrapping rather than pruning, we can obtain low variation trees. It can therefore avoid an overfitting and thereby improve the performance. However, the performance can decrease when the data are extremely imbalanced. The class of PCB features is mostly normal, and the number of fault features is low. Thus, the performance inevitably decreases. To complement this disadvantage, we suggest utilizing the probability of being classified as a fault rather than predicting the class using the random forests and by considering the density of the features. By considering the density, the probability of being a defective feature increases, whereas the probability of being a normal feature decreases. Therefore, the proposed method detects faults more effectively than a method using only random forests. We next describe the process for calculating the probability of the test data being classified as a fault when using random forests. Then, in Section 4, we describe how to use the calculated probability to detect a fault.

The probability $p(y = k)$ of being classified as k is $p(y = k) = \pi_k$, where $0 \leq \pi_k \leq 1$. Here, n is the number of classes. The density function $f_k(x)$ for the features in class k is $f_k(x) = f(x|y = k)$. Thus, the density function $f(x)$ for all features is calculated through Equation (5) [22].

$$f(x) = \sum_{k=1}^{n} \pi_k f_k(x).$$

(5)

Finally, we can estimate the probability that the class of new data x_{new} will be predicted as k using a Bayesian probability. The equation of this probability is denoted through Equation (6).

$$p(y = k|x = x_{new}) = \frac{\pi_k f_k(x_{new})}{f(x_{new})}$$

(6)

Thus, the probability p_i that a new feature x_i will be classified as a fault or as normal is calculated using Equation (7).

$$p_i = \begin{cases} \frac{\pi_F f(x_i)}{f(x_i)}, & p(y = Fault|x = x_i) \\ 1 - \frac{\pi_F f(x_i)}{f(x_i)}, & p(y = Normal|x = x_i) \end{cases}$$

(7)

3. Proposed Method

Based on the algorithms described in Section 2, we propose a scratch fault detection method, which is depicted in Figure 4. First, SURF extracts the features of the PCB image. The extracted features are composed of a multivariate vector representing the properties found in the PCB image, such as an edge, blob, or curve. Because SURF also extracts the features in a scratch-defective edge, it is difficult to distinguish between a normal edge and a fault edge using only these features.

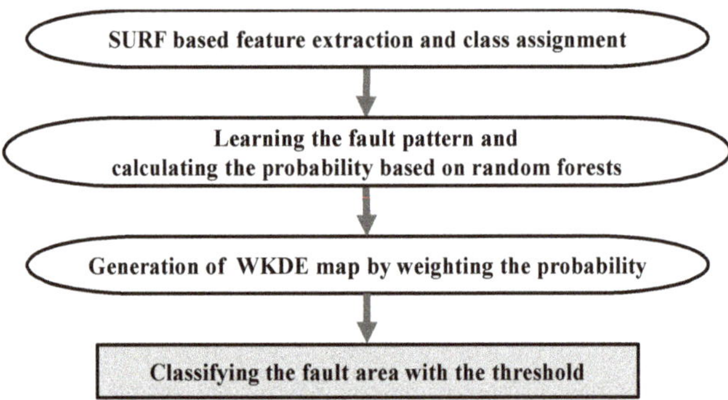

Figure 4. Flowchart of the proposed PCB inspection method. WKDE, weighted kernel density estimation.

Hence, we assign a normal or fault class to the features using the coordinates of the features within a defective or normal area. We then learn the normal and fault properties using the multivariate vector and feature class and calculate the probability of each feature being classified as a fault class. We then generate a new weighted kernel density estimation (WKDE) map. Kernel density estimation (KDE) is a technique for estimating the density based on the coordinates. That is, KDE is a technique that considers the density of the features. To maximize the effectiveness of the features, the probability of affecting the spatial density is given as a weight to the KDE. Thus, we consider not only the properties but also the densities of the features. After generating a WKDE map, we can predict the fault area based on the WKDE value. The proposed method is described in more detail in the following subsection.

3.1. SURF-Based Feature Extraction and Class Assignment

As the first step of the proposed method, SURF digitizes a PCB image as a vector. SURF then detects the features and extracts the descriptors in the form of a vector as mentioned in the previous section. Figure 5a shows an example of an original PCB image. The circuits in the area inside the dotted line have a defective edge, whereas the other circuits have a normal edge. The extracted features are shown in the circles in Figure 5b. As Figure 5b illustrates, the features are obtained identically without discriminating between a defective edge and a normal edge. In other words, SURF cannot distinguish between a defective and a normal edge. Thus, we should learn the fault pattern of the features by setting the normal and fault areas and assigning classes to each feature.

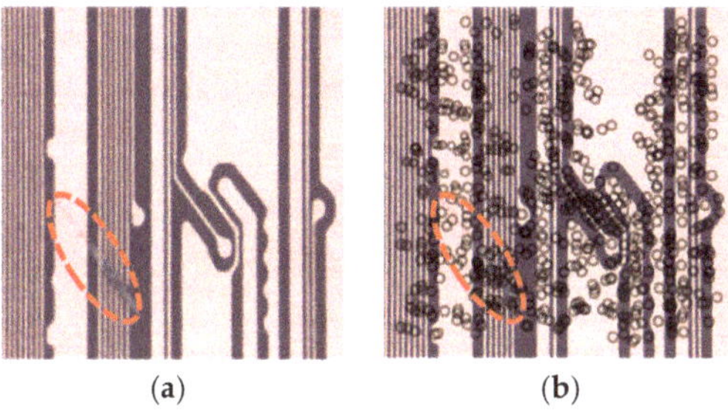

(a)   (b)

Figure 5. An example of SURF-based feature extraction (**a**) Original image; (**b**) A resultant image after extracting the features of the PCB.

As shown in Figure 6a, we define a lozenge area with four points so that the fault area can be visually represented in the clearest manner. We can then consider the fault occurring in the lozenge area. We assign the features within the lozenge area in Figure 6b to the fault class. The remaining features from areas other than the lozenge area are designated as being of a normal class through Equation (8).

$$c(x_i) = \begin{cases} \textit{Fault class, if } x_i \in X_L \\ \textit{Normal class, otherwise} \end{cases}, \tag{8}$$

where x_i indicates each feature composed using a multivariate vector and X_L is a set of features within the lozenge area.

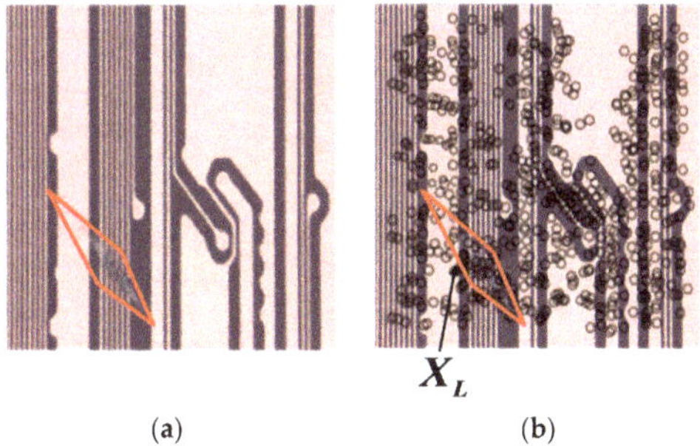

(a)   (b)

Figure 6. An area selection for a class assignment: (**a**) A fault occurring in the lozenge area; (**b**) The features within the lozenge area are the fault class and the remaining features are the normal class.

3.2. Learning the Fault Pattern and Calculating the Probability Based on Random Forests

Given the class-assigned features, the following step is to handle fault patterns using a machine-learning algorithm. This method is based on the random forests method, which is a well-known ensemble learning model for classification. We first learn the fault pattern of the features. We use a 136-dimension input vector, X_i, composed using extended descriptors and the information of each feature, such as its location and class, for a more precise description. The extended descriptors are computed by summing d_x and $|d_x|$ separately for $d_y < 0$ and $d_y \geq 0$. Similarly, $\sum d_y$ and $\sum |d_y|$ are split for $d_x < 0$ and $d_x \geq 0$. Thus, the number of vector dimensions of the descriptors doubles to 128. After building the learning model by training the properties of the features, we can find the probability p_i that each feature of an inspected image will be classified as a fault, which is calculated using Equation (7). Because the classes of the features are extremely imbalanced, most features are classified as normal. That is, the probability that the features will be estimated as a fault is considerably less than the probability that they will be predicted as normal. Consequently, because it is difficult to detect a fault area using only the results of the random forests, we should consider the density in addition to the map of the probability of being predicted as a fault.

3.3. Generation of WKDE Map by Weighting the Probability

This section describes how to apply the probability calculated in the previous section to generate a weighted kernel density estimation (WKDE) map, a technique that reflects the kernel density estimation (KDE). KDE is a probability density estimation method using a kernel function. In addition, it is a nonparametric method that is even applicable to high-dimensional data [23]. The representative kernel functions include Gaussian, uniform, and Epachenikov functions [24]. This paper uses a Gaussian kernel function to estimate the density.

After calculating the probability, we generate a WKDE map by giving the weight, w_i, to the kernel function in order to complement any disadvantages as mentioned earlier. Because the PCB features have x- and y-axis coordinates, this method uses the weighted multivariate kernel density estimation.

Let $\{x_1, y_1\}, \{x_2, y_2\}, \cdots, \{x_m, y_m\}$ be the coordinates of the PCB features. The WKDE value is denoted through Equation (9) [25].

$$\hat{f}(x,y) = \frac{1}{mh} \sum_{i=1}^{m} w_i K\left\{ \frac{(x,y) - (x_i, y_i)}{h} \right\}, \tag{9}$$

where m is the total number of PCB features, h is a scaled kernel that controls the smoothness of the estimate, and w_i is the probability of being classified as a fault. In addition, K is a Gaussian kernel function.

KDE and WKDE maps were drawn using the features extracted from the PCB image in Figure 5 as shown in Figure 7. The x and y axes in the KDE and WKDE maps represent the coordinates of the features. The z axis of the KDE map indicates the KDE value, which is calculated based on the coordinates of each feature, whereas the z axis of the WKDE map is the WKDE value obtained by weighting the probability to the KDE. Because the KDE considers only the density, it has a high KDE value in areas where many features are extracted regardless of the defective area as shown in Figure 7a.

Consequently, it is difficult to distinguish between a normal area and a fault area on the KDE map. On the other hand, the WKDE value, which reflects the probability, increases when the probability of each feature being classified as a fault is higher than the probability of the other features. Otherwise, the WKDE value decreases because there is a relative difference between the probabilities. The area marked with a circle in Figure 7b has a relatively high WKDE value owing to the weight. That is, because the marked area is densely concentrated with fault features, the WKDE value of this area is higher than the KDE value. In addition, the KDE values in the remaining areas are randomly distributed, but decrease significantly after being weighted. Thus, we can predict the marked area as a fault area.

Figure 8 shows a three-dimensional (2-D) WKDE map overlapping the original image (Figure 5a). In the 3-D WKDE map, it is difficult to intuitively recognize the coordinates of a feature. Thus, we convert the 3-D image into a two-dimensional (2-D) image to prove that the marked area in Figure 7b is consistent with the real fault area. The area within the dotted line in Figure 8 indicates where the actual faults occur, whereas the shaded area is an area predicted as a fault because the WKDE value exceeds the threshold. As Figure 8 shows, we can confirm that the shaded area matches the dotted area. Therefore, we monitor the WKDE value and predict the features as being a fault if the WKDE value exceeds the threshold. In this paper, we set a WKDE value of higher than 1% as the threshold. However, the threshold changes according to the process yield or to past experience.

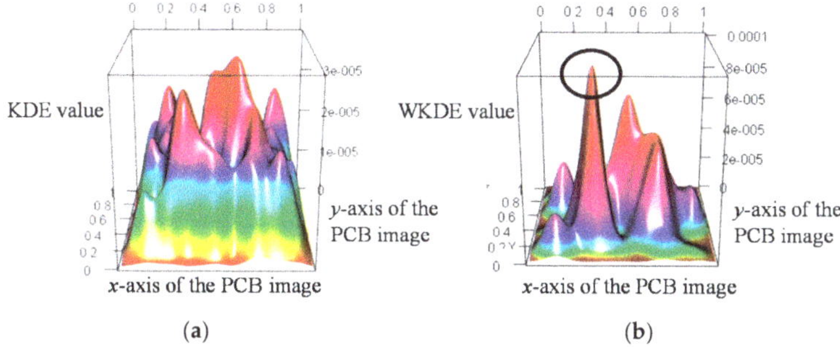

Figure 7. The kernel density estimation (KDE) and WKDE map in three-dimensions (3-D) for a PCB image: (**a**) KDE map; (**b**) Proposed WKDE map.

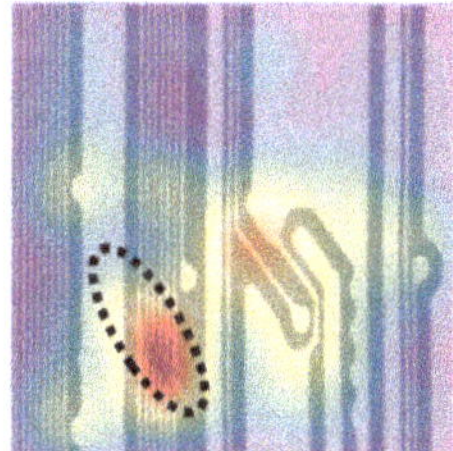

Figure 8. Two-dimensional (2-D) WKDE map on an original image.

4. Experimental Results

The present experiment used PCB image data to verify the proposed method. PCB images were collected from a battery manufacturer in Korea. We obtained 10 PCB images by selecting a fault image with a particular scratch property. As shown in Figure 9, the scratch fault, which is indicated by the dotted line, has a fine and thin form, indicating that the circuit may be broken. Because a scratch defect may prevent a current flow within the circuit, it has a significant impact on the PCB quality. Therefore, this paper analyses PCB images with a scratch defect.

We first extracted the features in the 10 PCB images using SURF. We then assigned a class to each feature by designating the defective area and organized the 136-dimension input vector including information such as the coordinates and class. The data were then divided into the test set and training set for learning purposes. This experiment applied a 10-fold cross validation using the 10 images. For example, nine images are used for training, and the remaining image is used for testing. Next, we obtained the WKDE value for each of the 10 PCB images as mentioned in Section 3.

To diagnose a fault area, statistical process control (SPC) can be applied to monitor the WKDE value. SPC is a method of quality control that uses statistical methods based on the distribution of data [26]. However, there is no assumption regarding the distribution of the features. For this reason, this research cannot utilize SPC. Hence, a binary classification was applied based on a single continuous WKDE value to examine whether a feature is a fault. The features were classified by comparing the WKDE value with a threshold x^*, which is called the cut-off value. A feature was classified as a fault if $\hat{f}(x, y) \geq x^*$ and was classified as normal otherwise.

The appropriate threshold can be chosen based on various criteria [27]. Thus, we conducted the experiments using a variety of thresholds, and adopted an appropriate threshold based on a receiver operating characteristic (ROC) curve, which is a tool for demonstrating the performance of a classifier. An ROC curve is a threshold-independent technique for evaluating the performance of a model and represents the relationship between the model's sensitivity and specificity [27]. In this study, the ROC curve was drawn based on the true- and false-negative rates. The ROC curve for a good model achieves a high true-negative rate whereas the false-negative rate is relatively small. Therefore, the performance of a model can be evaluated by comparing the area under the ROC curve (AUROC). Robust models have an AUROC of close to 1.0, whereas poorer models have an AUROC near 0.5, and worthless models have a value of less than 0.5 [28].

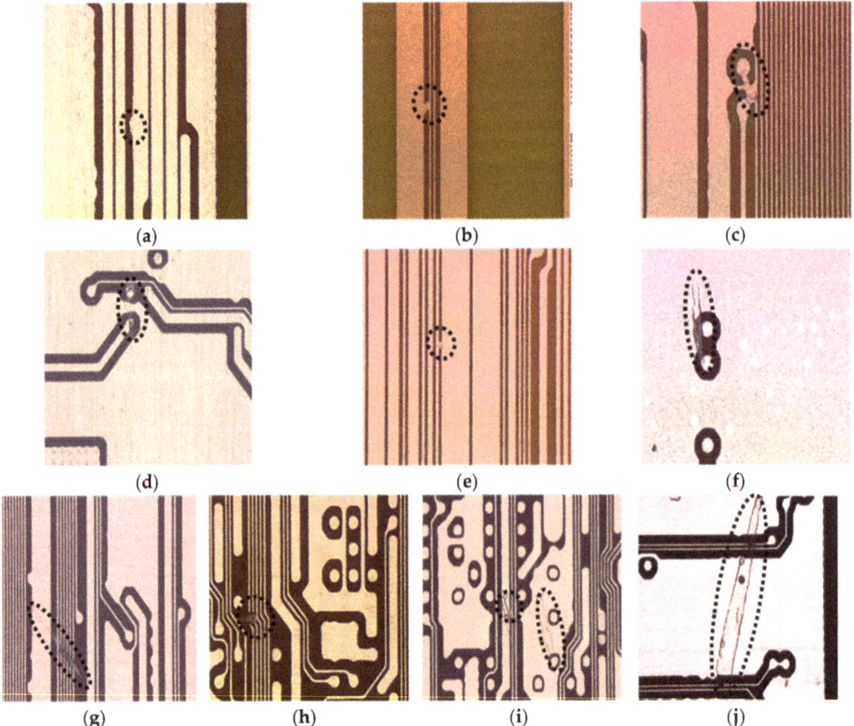

Figure 9. Example images of the 10 PCB images used: (a)–(j) The area marked with a circle or an ellipsoid on the dotted line is where the scratch-fault occurred.

To demonstrate that both the probability and density of the features have to be considered for detecting a fault, this paper compares the performances of three different methods: (1) monitoring only the probability (Method 1), (2) using a KDE map (Method 2), and (3) using the proposed WKDE map (Method 3). Each method applies binary classification to detect a fault. Method 1 detects a fault area by monitoring only the probability obtained by the random forests. An area is determined as a fault area when the probability of the feature being a fault exceeds a certain threshold. That is, this method does not consider the density of the features. Method 2 identifies a defect by considering only the density of the features without taking into account the probability. Thus, it monitors the KDE value. Finally, Method 3, the proposed method, detects a fault by monitoring the WKDE value. This method considers both the properties of the features and their density. Finally, we draw an ROC curve to compare the performances of the three methods. Figure 10 shows the ROC curves for the three methods. A method in which the AUROC is close to 1.0 detects a PCB fault more precisely. As can be seen, the AUROC of Method 1 is larger than that of Method 2 for 6 of the 10 images. However, the AUROC of Method 1 is less than that of Method 3 for all 10 images. Thus, the performance of Method 1 is not proper for detecting faults. Although Method 2 outperforms Method 1 for certain images, it has a poorer level of performance than Method 3. On the other hand, Method 3 outperforms the other methods, and its AUROC is close to 1.0 for all images. Therefore, Method 3 is appropriate for detecting a scratch fault on a PCB.

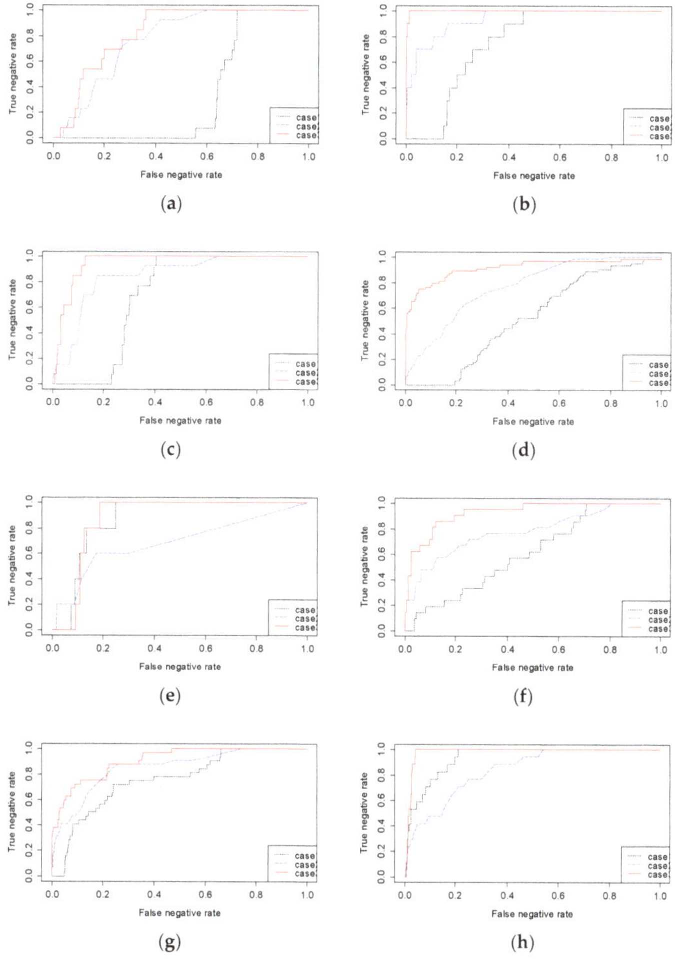

Figure 10. *Cont.*

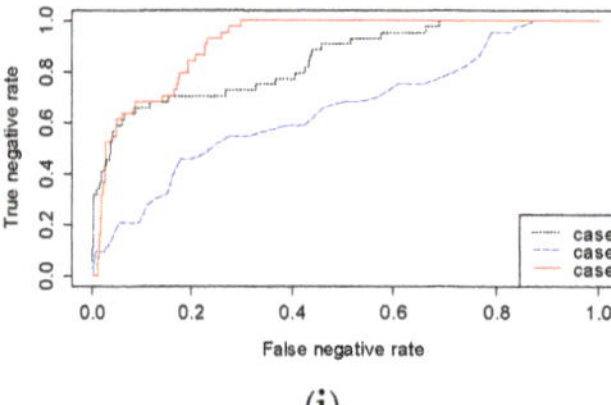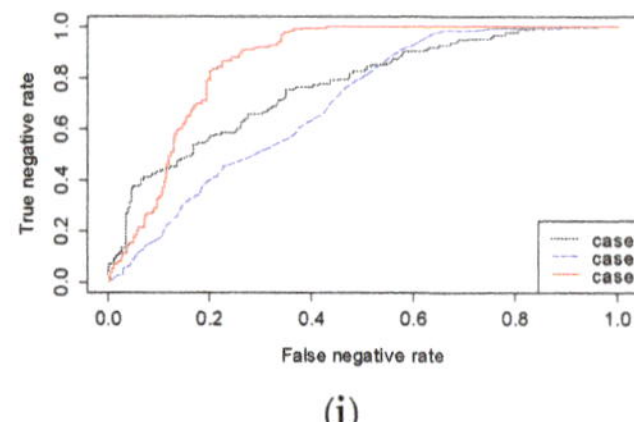

(i)   (j)

Figure 10. Performance comparison of three different methods for 10 scratch-fault images: (**a**)–(**j**) For the 10 PCB images, the mean AUROC of the Method 1 is 0.70, the Method 2 is 0.78, and the Method 3 is 0.91. The Method 3 considering both probability and density has better detection performance in all 10 images than the other method considering only one property.

5. Conclusions

In this paper, we proposed a new non-referential method by learning the fault pattern and generating a WKDE map. This method can learn various fault patterns regardless of the type of defect. Thus, the proposed method allows for a flexible PCB inspection without limitations regarding the specific type of fault. Furthermore, it can be extended to deal with unknown faults for PCB inspection. The performance of the proposed method is demonstrated by comparing the ROC curve for the three methods presented above. In addition, it was found that considering both the probability and density of the features is effective for detecting a scratch fault. Thus far, our method has dealt only with the 10 scratch faults due to security issues, and it needs to be applied to other PCB images with more data and various fault patterns. In addition, using a clustering technique, further work needs to be conducted to redefine the fault type when it is not specified, and an appropriate detection method for each type of defect should be studied. Although this paper used SURF to detect robust features, other image processing techniques can be applied. In other words, the classification accuracy can be further enhanced using another algorithm that includes more detailed edge information. The classification performance of the proposed algorithm is proved by the ROC curve. In the future, it is necessary to study the threshold selection method and its performance evaluation method for practical application in the future.

Author Contributions: E.Y. designed and implemented the algorithm to solve the defined problem. S.H.P. and C.-S.P. performed the experiments and data analyses. J.-G.B. validated the proposed algorithm and guided the whole research. All authors read and approved the final manuscript.

Acknowledgments: This work was supported by the National Research Foundation of Korea (NRF) grant funded by the Korea government (MSIP) (NRF-2016R1A2B4013678). This work was also supported by BK21 Plus (Big Data in Manufacturing and Logistics Systems, Korea University).

Conflicts of Interest: The authors declare no conflict of interest.

References

1. Malge, P.S.; Nadaf, R.S. A survey: Automated visual PCB inspection algorithm. *Int. J. Eng. Res. Technol.* **2014**, *3*, 1. Available online: https://www.ijert.org/browse/volume-3-2014/january-2014-edition?start=20 (accessed on 2 June 2018).
2. Wu, W.Y.; Wang, M.J.J.; Liu, C.M. Automated inspected of printed circuit boards through machine vision. *Comput. Ind.* **1996**, *28*, 103–111. [CrossRef]
3. Sundaraj, K. PCB inspection for missing or misaligned components using background subtraction. *WSEAS Trans. Inf. Sci. Appl.* **2009**, *6*, 778–787. Available online: https://dl.acm.org/citation.cfm?id=1558809 (accessed on 2 June 2018).
4. Chauhan, A.P.S.; Bhardwaj, S.C. Detection of bare PCB defects by image subtraction method using machine vision. In Proceedings of the World Congress on Engineering, London, UK, 6–8 July 2011; Volume 2, pp. 103–111.

5. Moganti, M.; Ercal, F.; Dagli, C.H.; Tsunekawa, S. Automatic PCB inspection algorithms: A survey. *Comput. Vis. Image Understand.* **1996**, *63*, 287–313. [CrossRef]
6. Acciani, G.; Brunetti, G.; Fornarelli, G. Application of neural networks in optical inspection and classification of solder joints in surface mount technology. *Int. IEEE Trans. Ind. Inform.* **2006**, *2*, 200–209. [CrossRef]
7. Huang, S.H.; Pan, Y.C. Automated visual inspection in the semiconductor industry: A survey. *Comput. Ind.* **2015**, *66*, 1–10. [CrossRef]
8. Kim, H.W.; Yoo, S.I. Defect detection using feature point matching for non-repetitive patterned images. *Pattern Anal. Appl.* **2014**, *17*, 415–429. [CrossRef]
9. Ye, Q.-Z.; Danielsson, P.E. Inspection of printed circuit boards by connectivity preserving shrinking. *IEEE Trans. Pattern Anal. Mach. Intell.* **1988**, *10*, 737–742. [CrossRef]
10. Elbehiery, H.; Hefnawy, A.; Elewa, M. Surface defects detection for ceramic tiles using image processing and morphological techniques. *WEC* **2007**, *1*, 1488–1492. Available online: https://scholar.waset.org/1307-6892/15176 (accessed on 2 June 2018).
11. Tsai, D.M.; Su, Y.J. Non-referential, self-compared shape defect inspection for bond pads with deformed shapes. *Int. J. Prod. Res.* **2009**, *47*, 1225–1244. [CrossRef]
12. Rosten, E.; Drummond, T. Machine learning for high-speed corner detection. In Proceedings of the 9th European Conference on Computer Vision-ECCV, Graz, Austria, 7–13 May 2006; pp. 430–443.
13. Pernkopf, F. Detection of surface defects on raw steel blocks using Bayesian network classifiers. *Pattern Anal. Appl.* **2004**, *7*, 333–342. [CrossRef]
14. Oliveira, R.B.; Papa, J.P.; Pereira, A.S.; Tavares, J.M.R. Computational methods for pigmented skin lesion classification in images: Review and future trends. *Neural Comput. Appl.* **2018**, *29*, 613–636. [CrossRef]
15. Ma, Z.; Tavares, J.M.R. Effective features to classify skin lesions in dermoscopic images. *Expert Syst. Appl.* **2017**, *84*, 92–101. [CrossRef]
16. Oliveira, R.B.; Pereira, A.S.; Tavares, J.M.R. Computational diagnosis of skin lesions from dermoscopic images using combined features. *Neural Comput. Appl.* **2018**. [CrossRef]
17. Bay, H.; Ess, A.; Tuytelaars, T.; Van Gool, L. Speeded-up robust features (SURF). *Comput. Vis. Image Understand.* **2008**, *110*, 346–359. [CrossRef]
18. Su, J.; Xu, Q.; Zhu, J. A scene matching algorithm based on SURF feature. In Proceedings of the International Conference on Image Analysis and Signal Processing (IASP), Huangzhou, China, 9–11 April 2010; pp. 434–437.
19. Strang, G.; Nguyen, T. *Wavelets and Filter Banks*, 2nd ed.; Wellesely-Cambridge Press: Wellesley, MA, USA, 1996.
20. Breiman, L. Random forests. *Mach. Learn.* **2001**, *45*, 5–32. [CrossRef]
21. Ohn, S.Y.; Chi, S.D.; Han, M.Y. Feature Selection for Classification of Mass Spectrometric Proteomic Data Using Random Forest. *J. Korea Soc. Simul.* **2013**, *22*, 139–147. [CrossRef]
22. Li, C. Probability Estimation in Random Forests. Master's Thesis, Utah State University, Logan, UT, USA, 2013.
23. Thurstain-Goodwin, M.; Unwin, D. Defining and delineating the central areas of towns for statistical monitoring using continuous surface representations. *Trans. GIS* **2000**, *4*, 305–317. [CrossRef]
24. Kurata, E.; Mori, H. Short-term load forecasting using informative vector machine. *Electr. Eng. Jpn.* **2009**, *166*, 23–31. [CrossRef]
25. Anderson, T.K. Kernel density estimation and K-means clustering to profile road accident hotspots. *Accid. Anal. Prev.* **2009**, *41*, 359–364. [CrossRef] [PubMed]
26. Fadel, H.K.; Holloway, L.E. Using SPC and template monitoring method for fault detection and prediction in discrete event manufacturing systems. In Proceedings of the IEEE International Symposium on Intelligent Control/Intelligent Systems and Semiotics, Cambridge, MA, USA, 17 September 1999.
27. Kang, S.; Cho, S.; An, D.; Rim, J. Using wafer map features to better predict die-level failures in final test. *IEEE Trans. Semicond. Manuf.* **2015**, *28*, 431–437. [CrossRef]
28. Freeman, E.A.; Moisen, G.G. A comparison of the performance of threshold criteria for binary classification in terms of predicted prevalence and kappa. *Ecol. Model.* **2008**, *217*, 48–58. [CrossRef]

Review

Research Progress of Visual Inspection Technology of Steel Products—A Review

Xiaohong Sun [1,2], Jinan Gu [1,*], Shixi Tang [1] and Jing Li [1]

[1] School of Mechanical Engineering, Jiangsu University, Zhenjiang 212013, China; jxxsxh_1002@163.com (X.S.); tangsx@yctu.edu.cn (S.T.); aygxylj@163.com (J.L.)

[2] School of Mechanical Engineering, Anyang Institute of Technology, Anyang 455000, China

* Correspondence: gujinan@tsinghua.org.cn

Received: 15 September 2018; Accepted: 2 November 2018; Published: 8 November 2018

Abstract: The automation and intellectualization of the manufacturing processes in the iron and steel industry needs the strong support of inspection technologies, which play an important role in the field of quality control. At present, visual inspection technology based on image processing has an absolute advantage because of its intuitive nature, convenience, and efficiency. A major breakthrough in this field can be achieved if sufficient research regarding visual inspection technologies is undertaken. Therefore, the purpose of this article is to study the latest developments in steel inspection relating to the detected object, system hardware, and system software, existing problems of current inspection technologies, and future research directions. The paper mainly focuses on the research status and trends of inspection technology. The network framework based on deep learning provides space for the development of end-to-end mode inspection technology, which would greatly promote the implementation of intelligent manufacturing. .

Keywords: defect inspection; image processing; feature extraction; classification methods

---

1. Introduction

China's iron and steel industry has made tremendous contributions to the development of its national economy. In recent years, the rapid rise in the output of steel products has been accompanied by a large number of defects, which could bring significant economic losses to enterprises, and ultimately affect their brand image. Therefore, it is necessary to study detection methods, particularly since artificial detection methods no longer meet the enterprise requirements regarding time, cost, and precision.

Visual detection technology based on image processing has been widely used in various fields, such as medicine [1], the iron and steel industry [2,3], art [4], the textile industry [5], and the automobile industry [6] for its unique advantages of intuition, accuracy, and convenience. Early detection methods for steel defects are classified as contact detection and non-contact detection [7]. The former receives information through direct contact with the sample surface by the sensing element of a contact-detection device. The latter is based on the technology of photoelectricity, and electromagnetism to obtain the parameter information of the sample surface without contacting it.

Contact-detection methods include magnetic particle testing (MPT) and liquid penetration testing (LPT). Although intuitive images can be quickly obtained via these methods, to do so is not practicable. Non-contact methods include ultrasonic scanning, electromagnetic testing, etc., in which ultrasound or electromagnetic signals are converted to optical signals. Results are not intuitive, but need to be judged by professionals.

Visual detection based on image sensors stands out from the range of non-contact detection technologies, because it is an effective combination of the high speed achieved with contact detection methods and the independence of non-contact detection methods. The key feature is that it can be implemented by using only a universal computer and a dedicated image processor. Indeed, the number of publications related to defect detection has grown rapidly over the past decade, which is a trend that may be due to the rapid advance of computing capacity, the enhancement of sensor performance, and the great improvement of image processing technology.

The basic components of a typical visual system are an image acquisition unit, an image processing unit, and a control execution unit (Figure 1). The image acquisition unit is the component of the system hardware, and its main task is to obtain high-quality images, since low-quality images lead to algorithm burden. Excellent visual software can quickly and accurately detect the target features in the image and minimize dependence on the system hardware. The sorting mechanism can adopt an electromechanical system or hydraulic system, but the dynamic characteristics (i.e., rapidity and stability) of the system are important. This paper focus on image acquisition and image processing units.

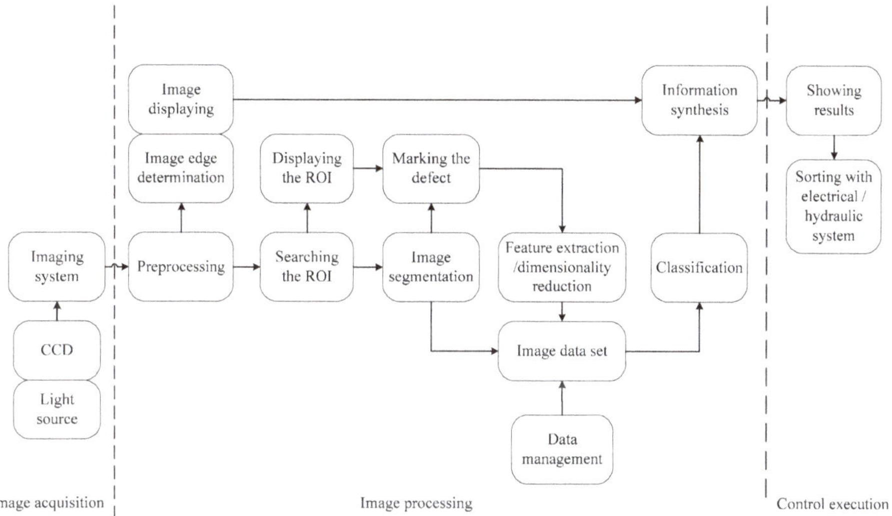

Figure 1. Visual inspection system. Note: CCD: Charge coupled devices; ROI: Region of interest.

With the maturity of the basic theory of image analysis, the development of the detection field has advanced by leaps and bounds, and several reviews of defect detection have been undertaken. Commercially available inspection equipment and visual inspection systems, as well as practical applications of visual inspection, are summarized in Chin [8], Newman, [9]. Amongst recent literature reviews, a large number of methods and techniques for the free surface detection of parts are studied in Li [10]. Development trends of visual inspection are presented in article Shirvaikar [11], which mainly focuses on the introduction of visual detection systems relating to hardware and software, but detailed algorithm comparisons are not provided. Reviews of visual detection in the manufacture of textiles Hanbay [12], food, and agriculture Jfs [13] have also contributed to the development of detection technology. However, because the reflective properties of steel products (55–65%) [14] differ between foods and fabrics, these testing methods are for reference only. Notably, a comprehensive review of defect detection in steel surfaces has been conducted Neogi [15], and is a valuable article for researchers in the field. However, the chronological distribution of the references in Neogi [15] suggests that it is somewhat dated, with 13.82% of references from before 2000, 65.5% from 2001–2010, and 20.68% from after 2010. In contrast, the chronological distribution of references in the present paper—10.41% from prior to 2000, 37.5% from 2000–2010, and 52.08% from after 2010—indicates that it is more up-to-date with the latest technological developments. Thus, it is intended that the present study provide a supplement to Neogi [15].

In Section 2, the types of steel products and common defects are presented, so as to understand the complexity and diversity of visual detection. In Section 3, the hardware composition of inspection systems is explained. Detection and classification methods are reviewed according to different theories in Sections 4 and 5. In Section 6, a literature analysis is conducted, which includes not only an analysis of detection technology, but also an analysis of the scale of the detection market. Conclusions and further prospects are provided in Section 7.

2. Types of Defects in Steel Products

The wide variety of steel products can be roughly divided into two categories: flat products and long products (Figure 2).

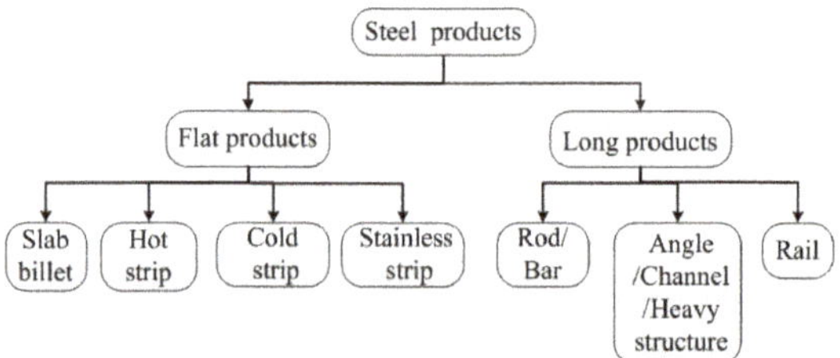

Figure 2. Types of steel products (reproduced from Neogi [15]).

Steel products have been identified as having 55 types of defects Neogi [15]; Figure 3 shows some examples.

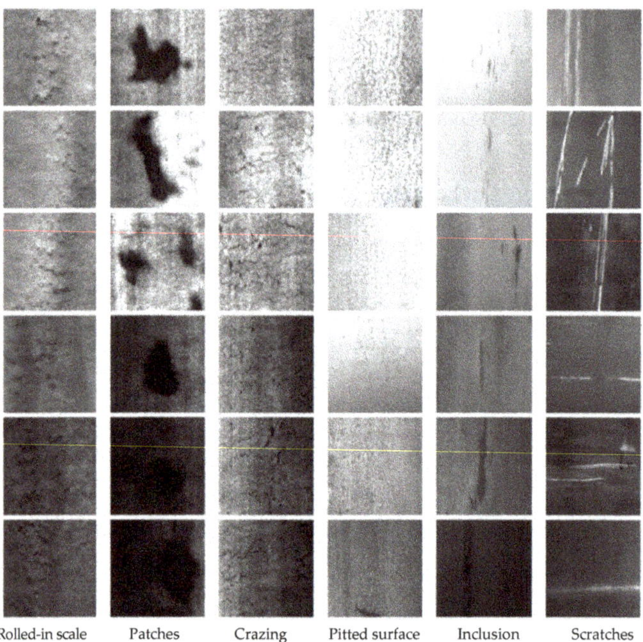

Figure 3. Examples of steel defects (reproduced from Song [16]).

The wide range of defects can be classified according to the general type of steel product as follows in Table 1.

Table 1. Common defect types of different steel products.

Steel Type	Type of Defect	References
Billet	Cracks, scratches	[17–21]
Hot strip	Cracks, longitudinal scratches, transverse scratches, scales, delamination, roll marks, pit defects, seams, inclusions	[16,22–30]
Cold strip	Bruises, slags, inclusions, seams, oxide scales, cracks, holes, feather roll marks, Latex marks	[31–43]
Stainless steel	Slag inclusions, scratches in pickling, scratches	[44]
Rod/bar	Cracks, scratches	[45–49]

Establishing a defect detection system is not easy. In order to create a reliable and repeatable test system, produ ct manufacturers often need to work with test engineers to conduct qualitative and quantitative analyses of potential defects. Defects can be roughly divided into three situations:

(1) In most cases, defects can be easily detected by using standard imaging tools. For example, pinholes Liu [31] and certain impurities are usually round, and pixels that appear bright on an image fall on a dark background, or vice versa, so are easily discernible.

(2) A slightly more complex situation is that the defect definition, including its size and shape, are not very clear, but it can still be distinguished from the underlying background. This situation mainly includes wear, or slender, low-contrast linear defects Yun [17]. Examples are scratches Dupont [18] and cracks Choi [50] on products. These types of defects may require more advanced imaging detection tools.

(3) The most complicated case is a defect in which its definition, size, and shape are not clear, and there is no recognition mode, so that it is difficult to distinguish the defect from the underlying background. This kind of situation mainly includes printing defects and some random medium "impurities" which pose significant challenges to detection technology.

3. The Hardware Composition of the Inspection System

3.1. Camera

An industrial camera is at the core of the system hardware. The frame rate (the rate at which the camera collects and transmits images) and resolution are two important parameters of the camera. The frame rate must be greater than the detection speed; 10 fps is usually sufficient to meet industrial requirements. The required resolution depends on the size of the features relative to the overall image. For example, suppose the surface scratch of an object is detected, the size of the object to be photographed is $a \times b$ mm, and the detection accuracy is 0.01 mm. Then, the minimum resolution formula of the camera can be determined as:

$$(a/0.01) \times (b/0.01) = Resolution \tag{1}$$

Industrial cameras are divided into two types according to the differences of the image sensors: charge coupled devices (CCD) and complementary metal oxide semiconductor devices (CMOS) Koller [51]. The differences are as follows.

(1) Different imaging processes: A CCD utilizes a small number of output nodes to output data uniformly, thereby ensuring good signal consistency. By contrast, each pixel in a CMOS chip has its own signal amplifier, with charge conversion done separately, so output signal consistency

is poorer and more greatly affected by signal noise than with a CCD. However, a significant advantage of CMOS is low power consumption.

(2) Different integration: The CCD manufacturing process is complex, and the output of a CCD consists only of an analog electrical signal, which requires a decoder, analog converter, and image signal processor. As a result, a CCD has low integration. A CMOS, on the other hand, can collect signals with an analog-to-digital converter on a chip with high integration and low cost. With the advancement of CMOS imaging technology, CMOS will have greater applications in the future.

(3) Different image output speed: A CCD adopts photosensitive outputs sequentially, relatively slowly. With a CMOS, each charge element has its own switch controller, and the readout speed is very fast. Most high-speed cameras with a frame rate greater than 500 fps use CMOS.

(4) Different noise levels: CCD technology is mature, and the imaging quality is superior to that of CMOS. CMOS has a higher degree of integration, a closer spacing distance, and more interference.

As part of so-called Industry 4.0, factories around the world are developing automation and intelligence, in which smart sensors play an important role. The smart camera Lee [52] has the functions of processor, memory, communication interface, operating system, etc., which can process a large amount of data in advance and assist subsequent automatic detection and judgment. Nguyen et al. [53] noted an ultra-high-speed silicon image sensor. The test chip of this image sensor realizes a temporal resolution of 10 ns. For a silicon image sensor, the limit is 11.1 fps. Considering the theoretical derivation, this high-speed image sensor can reach a frame rate near the theoretical limit.

3.2. Light Selection

Lighting devices will vary because of different operating environments. For hot rolled steel, the strip itself is a luminous heating element. In order to reduce the interference of internal light sources, the intensity of the light source should be much higher than that of the steel strip. Thus, the light source can only have high-power, long-distance characteristics, so that it can provide high-intensity light at a long distance. For the cold rolling environment, although the relative distance between the light source and the steel strip is short, a more continuous light is needed to attenuate unstable infrared light, in order to ensure the highest sensitivity of the lens to the visible light spectrum. Overall, high strength, life span, design freedom, heat radiation, and response speed should be considered during lighting arrangement.

At present, some classical light sources are optical fiber, LED (light-emitting diode) lights, and stroboscopic xenon lamps. Among the latter, the strobe xenon lamp is mainly used in the area array CCD detection system as shown in Figure 4a, since it can effectively deal with adverse environmental conditions, such as fog Luo [22].

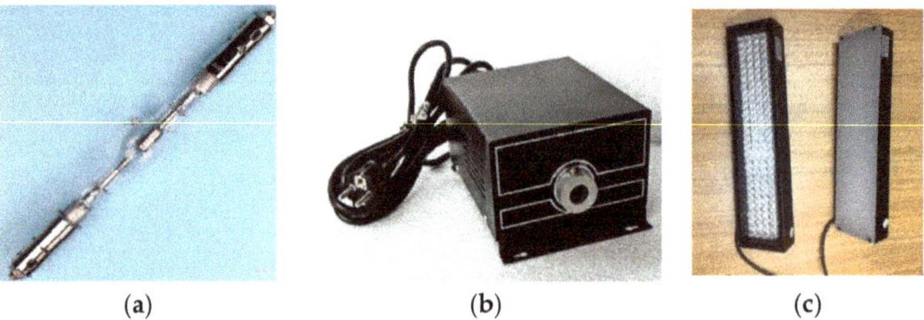

(a)   (b)   (c)

Figure 4. (a) Stroboscopic xenon lamps; (b) Halogen lamps; (c) Light-emitting diode (LED) lamps.

An example of an optical fiber light source is the halogen lamp (Figure 4b), which is used in conjunction with a color filter adapter. The output of light through the cylinder prism can avoid overflow and increase light intensity by 10% Wu [54], which can result in an ideal performance if high-power halogen lamps are used in industrial sites. However, halogen lamps are not suitable for hot-rolled steel in poor environments, and are mainly used in the testing environments of cold-rolled steel and finished products because of their high price and susceptibility to damage.

LED lamps (Figure 4c) are a spontaneous radiation source. Spontaneous radiation is a process in which an excited atom spontaneously transitions from a high-energy state to a low-energy state, emitting a photon at the same time. LED lamps have the advantages of non-related light, no optical resonator, long lifespan, and easy maintenance. However, the weaknesses of the lamps are their narrow spectral range and that the wavelength is affected by the materials. The lifespan of LED lamps will also shorten as the ambient temperature increases, so they are not suitable for high-temperature applications. In addition, LEDs cannot be directly connected in parallel. Therefore, a method of a single-channel serial multi-channel parallel connection method is used to form an array LED, which then forms a light source through the prism. Owing to its low cost and long lifespan, LED light sources are usually equipped with cooling devices for hot-rolled inspection.

3.3. Lighting Method Selection

In addition to the influence of the light source and CCD sensor on detection effects, the lighting mode has a greater impact on the detection effect of steel products. Usually, lighting can be divided into light and dark lighting.

The bright field lighting method is shown in Figure 5a. In this mode, the light source and the CCD are on the same side of the strip. The light emitted by the light source enters the camera after being reflected by the detection target. The reflection angle β is equal to the incident angle α, and the line between the CCD sensor and the image of the light source must be on the same line as the reflected light.

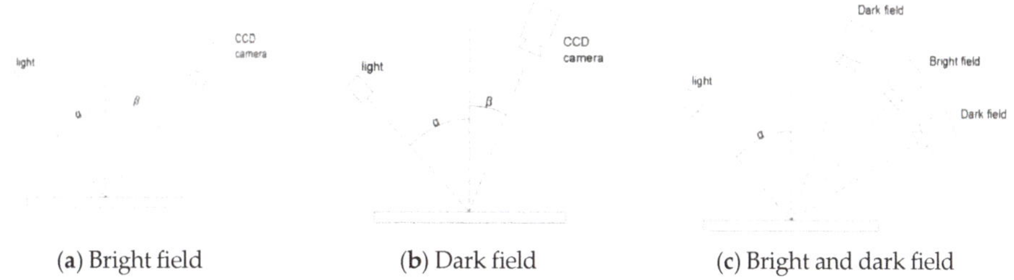

| (a) Bright field | (b) Dark field | (c) Bright and dark field |

Figure 5. Lighting methods.

The reflected light is evenly distributed on each area of the CCD sensor when there is no defect on the surface. However, reflected light at the defect position will change when a defect exists, and the illuminance entering the CCD sensor will be weakened. Therefore, the reflection of light in the defect area is altered for three-dimensional defects, and the illuminance of the defect position into the CCD sensor is less than the background light entering the CCD sensor; that is to say, the defect image is darker than the background image. As for two-dimensional defects, the reflection of the defect does not change, but two-dimensional defects are usually of different shades. The light in the darker region absorbs more light; therefore, the gray value of the defect image is higher than the background image when the defect color is lighter than the surface color of the strip. Conversely, the grayscale value of the defect image is lower than the background image when the color of defects is darker than the color of non-detects.

Therefore, the use of bright field lighting can not only detect two-dimensional (2D) defects, it can also detect three-dimensional (3D) defects. However, it is worth noting that the results will be significantly affected if a large fluctuation of the steel strip leads to exceeding the range of the reflection angle. Overall, the bright field method is more appropriate for detecting a type of defect that reflects and absorbs light, especially dark targets with a bright background, such as scales, oxide skins, pits, water marks, etc.

In the dark field lighting method (Figure 5b), the light source and the CCD sensor are also on the same side of the strip. In this case, the reflection angle β is not equal to the incident angle α, and the line between the CCD sensor and the image of the light source is not on the same line as the reflected light; therefore, it is difficult for light to enter the CCD sensor. Only when three-dimensional defects exist on the strip surface will the defect change the reflective nature of the light into a diffuse reflection. Then, the camera is able to collect some diffuse light, and the light from the defect position will be stronger than from areas without a defect. In addition, the light source itself has a collection effect of high-intensity light. Even if the incident angle is changed, it has little effect on the illumination of reflected light on CCD sensors. As a result, the CCD can still effectively detect a surface defect when the surface of the strip steel generates vibration. On the whole, the dark field lighting method is more suitable for the type of defects that can emit diffuse reflected light on the surface of a bright steel plate, and, in particular, bright targets with dark backgrounds, such as skins, pits, and indentations. There is a certain degree of tolerance to the vibration of the detection point.

The bright and dark double field lighting mode (Figure 5c) addresses the problems of detection of two-dimensional defects in dark field lighting and vibration at the detection point. However, this method has a high requirement for a high-intensity concentrated light effect of the light source; that is to say, the part of the dark field detection that is included will not be ideal if the concentrated light effect is not strong.

Overall, the appropriate light source and lighting mode allow us to capture the features of the object more accurately and improve the contrast between the object and the background. In this way, high-quality images can be obtained, and good detection results can be achieved.

4. Detection Methods

The performance of the software algorithm directly determines the result of detection. The detection task is arduous and challenging due to similarities between classes of defects and diversity within the classes. Due to this, domestic and foreign scholars have conducted significant research regarding steel inspection. Publications of the past 30 years can be classified according to basic theories, as shown in Figure 6.

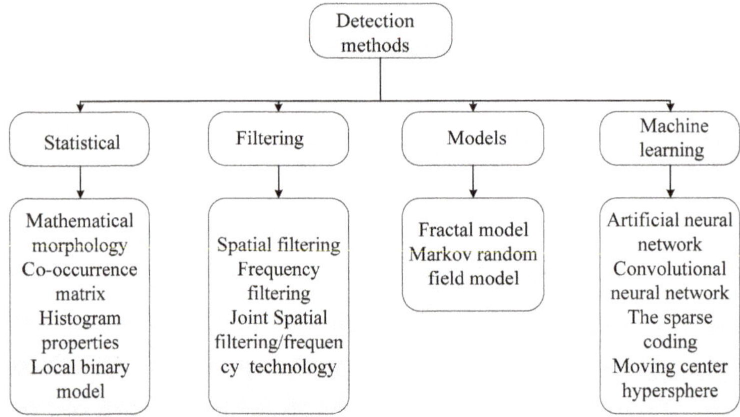

Figure 6. Classification map of steel surface defect detection methods.

4.1. Statistics

The statistical method is to establish a mathematical model using probability theory and mathematical statistics, which can be used to infer, predict, quantitatively analyze, and summarize the spatial distribution data of pixels. As a result, it can provide a basis and reference for subsequent decision-making. In the statistical method, the spatial distribution of gray values is defined with various forms of characterization, such as mathematical morphology, the co-occurrence matrix, histogram properties, and local binary models. These methods are widely used in the field of visual inspection. Table 2 compares defect detection methods based on statistics.

Table 2. Strengths and weaknesses of statistical detection methods for steel defects.

Method	Strengths	Weaknesses
Mathematical morphology	Computational simplicity.	Morphological operations are only implemented on non-periodic steel defects.
	Geometric representation of texture images.	
	Highly suitable for random or natural textures.	
Co-occurrence matrix	Extracting spatial relationship of pixels with different statistical computations.	Difficult to judge the optimal displacement vector.
	High accuracy rate.	
Histogram properties	Translation and rotation invariance.	Low detection rate (50–70%) for irregular textures.
	Calculation is simple.	Sensitive to noise.
Local binary pattern	Calculation is simple.	Too dependent on the gray value of the center point pixel.
	Recognition ability is strong.	
	Rotation invariance and gray invariance.	

4.1.1. Mathematical Morphology

Mathematical morphology is a subject of image analysis based on lattice theory and topology. The basic operations include: corrosion and expansion, open and closed operations, skeleton extraction, limit corrosion, hit-and-miss transformation, morphological gradient, top-hat transformation, particle analysis, and watershed transformation. The cost matrix theory based on mathematical morphology, combined with the K-nearest neighbor (KNN) classifier, has been shown to detect eight defects of flat steel products in Dopont [18].

In Yun [19], Liu [23], Zheng [55], a genetic algorithm combined with mathematical morphology was adopted to realize the defect detection of steel products. In particular, Liu [23] studied an enhancement operator based on mathematical morphology (EOBMM), combined with a binarization method based on genetic algorithm (BMBGA), which can effectively overcome the effects of non-uniform illumination and enhance the detailed information of the image. A method using mathematical morphology in combination with filtering methods was described in Wu [24]. There are other, similar combinations; for example, mathematical morphology can be combined with the curvelet transform or the Gabor transform. Mathematical morphology in conjunction with the curvelet transform has been used for the detection of metallic surfaces Cord [56]. A morphological operation combined with an optimized Gabor filter method was derived to address the problem of detection performance decreasing due to billet shape, multiple defects, and scales Yun [57]. There are also studies on morphological methods Nguyen [53], Wu [54]. .

4.1.2. Co-Occurrence Matrix

The spatial gray level co-occurrence matrix was first proposed by Haralick [58], and it is a popular texture analysis method belonging to second-order statistics, which is defined by the joint probability density of two positional pixels. Texture features derived from the co-occurrence matrix (energy, entropy, contrast, uniformity, deficit moment, and correlation) have been used in various surface defects detection methods. The detection and classification of defects can be realized by extracting the spatial features of a gray-level co-occurrence matrix (GLCM) in combination with a classifier Yu [32].

4.1.3. Histogram Properties

Image histograms are widely used in various fields of image processing, because they have low computational cost and many other advantages, such as image translation, rotation, and scale invariance, specifically in the fields of threshold segmentation of grayscale images, image retrieval, and image classification based on color. There are many histogram statistics, four of which (mean, standard deviation, variance, and median) are the most frequently used as texture features. Liu et al. [59] performed a multivariate discriminant function based on a statistical histogram to model, and used three statistical characteristics—the deviation (Dg), the mean (mg), and the variance (Vg)—to represent the shape of a point. In Luo [22], they also adopted a method that selected a suitable threshold based on the histogram to extract features. In Martins [33], a study of principal component analysis combined with histogram statistics were presented.

4.1.4. Local Binary Pattern

The local binary pattern (LBP) is an operator that describes the local texture features of the image with rotation invariance and grayscale invariance. It is worth mentioning that the application of the LBP in classification recognition generally uses the statistical histogram of the LBP feature spectrum as the feature vector rather than the LBP feature spectrum itself. In order to improve the recognition rate, a new feature descriptor known as the adjacent evaluation completed local binary pattern (AECLBP) was proposed by Song et al. [16] for hot-rolled steel strip detection. In a recent study, an LBP operator with symbol and size, combined with a histogram and shape and distance statistic features, were developed by Chu et al. [60]. On one hand, this method can solve multi-class classification problems; on the other, it also has an anti-noise ability and high classification efficiency. On the whole, the LBP method performs better than the co-occurrence matrix and filtering methods in accurately detecting the surface texture defects of steel.

4.2. Methods Based on Filtering

Most of the methods discussed in this section have the common feature that they apply a filter bank to an image to calculate the energy of the filter response. These methods can be divided into the spatial domain, frequency domain, and joint spatial/frequency analysis methods. Table 3 compares filtering-based detection methods.

Table 3. Strengths and weaknesses of filtering-based detection methods for steel defects.

Name	Strengths	Weaknesses
Spatial domain	A more centralized text-based approach (in which the segmentation of the text file is separate from the image).	Difficult to determine the optimal filter parameters.
		High computation cost.
Frequency analysis	Spatial frequency spectrum is invariant to shift, rotation, and scaling.	Lack ability of spatial orientation.
	Suitable for the detection of global and local defects.	Not suitable for random texture detection.
	FFT (Fast Fourier transform.) calculation time is short (600 pixels with 2.2 ms).	
Gabor Transform	Suitable for high dimensional feature space.	Difficult to determine the optimal filter parameters.
	An adaptive filter selection method is implemented to reduce the computational complexity.	No rotation invariance.
	Suitable for defect detection in airspace and frequency domain.	
Wavelet transform	Suitable for multi-scale image analysis.	Easily to be affected by feature correlations between the scales.
	High detection rate (83–97%).	
	Efficient image compression with less information loss.	
Multiscale Geometric Analysis	Suitable for the optimal and sparse representation of high-dimension data.	Redundancy problem (i.e., repeated data in a data set) cannot be solved.
	Good at image processing of strong noise background.	
	Compression with less information.	

4.2.1. Spatial Domain

Spatial filtering is an enhancement method based on neighborhood processing that directly conducts operations in the two-dimensional space where the image is located. The most common operation of spatial filtering is template arithmetic, and the basic idea is to use the value of a pixel as a function of its own gray value and the gray value of its neighboring pixels. In spatial filtering, the gradient filters are mainly used to detect edges, lines, and isolated points. Sobel, Robert, Canny, Laplacian, and Deriche filters are popular tools for measuring edge density. Dupont et al. provided a method that used the Prewitt filter to extract edge information and realize defects of sheet products [18]. Guo et al. [61] used the Sobel gradient edge detection operator combined with the Fisher discriminant to detect defects on steel surfaces. Spatial filtering methods are also discussed in the literature [25, 34,35,45,46,62–64] for the defect detection of various steel products. For the optical properties of highly reflective surfaces of cold-rolled strip, Zhao et al. [36] adopted a kind of homomorphic filtering algorithm based on a partial differential equation (PDE). Recently, the application of filter banks has been expressed in Bulnes [46], Liu [65], Li [47], and particularly in Li [47], which used mean filtering combined with a local annular contrast (LAC) detection method, which led to better performance.

4.2.2. Frequency Analysis

To address the limitation of spatial filtering methods (i.e., the kernel cannot be found in the defect image), the frequency domain analysis method was derived. This method firstly converts the image into frequency domain signals using the Fourier transform, and secondly performs a filtering analysis of the signals, and finally converts the signals back to the spatial domain to be stored by inverse Fourier transform. Three articles [25,37,38] were published successively, outlining the methods of frequency domain analysis for the defect detection of cold-rolled and hot-rolled steel. Among them, the method proposed in Wu [25] is the most effective, with a detection rate of up to 92.68%. They proposed a method of fast Fourier transform (FFT) combined with a local border search algorithm (LBSA) for the detection of hot-rolled steel strips.

4.2.3. Joint Spatial/Frequency Analysis Methods

Gabor Transform

Since the Fourier transform lacks spatial localization ability, it has a poor performance in practical applications. In order to solve this problem, the windowed Fourier transform was developed in 1946; this is known as the Gabor transform if the window function is a Gaussian function. Yun et al. [19] applied a Gabor filter optimized by genetic algorithm (GA) for the detection of corner cracking and thin cracking defects. Jeon et al. [66] adopted the Gabor filter to perform edge-pair detection in order to reduce the influence of lighting conditions, with satisfactory effects. D. Choi et al. [67] used a Gabor filter with morphological defect detection algorithms to detect pinholes on the surface of the steel plate. Choi [50] employed Gabor filtering and dual-threshold segmentation detection methods for the crack detection of steel plates. Among these methods, Choi [50] achieved the best results, with a detection rate of up to 94.43%.

Wavelet Transform

The Gabor transform is not adaptive, because the sliding window function is fixed once selected. On the contrary, the wavelet transform has a time-frequency window that can be adjusted; that is, the width of the window changes with the frequency. Thus, it overcomes the limitation of the Gabor transform. Wavelet transform was first put forward in 1974. The method was first used for defect detection Kaya [44] in 1995, because of the poor performance in diagonal detection using traditional edge-detection methods. Soon after, there were a lot of extensions of the method based on wavelet transform, such as the snake projection wavelet algorithm Li [68], undecimated wavelet transform algorithm [17], wavelet transform to obtain the approximate sub-image method Zhang [48], three-layer Haar wavelet feature set method Ghorai [26], discrete wavelet transform combined with adaptive local binarization method Yun [49], wavelet filtering in combination with center-surrounding difference method Xu [69] and, recently, anisotropic diffusion filter based on wavelet transform method [31]. The characteristics of wavelet transform are presented incisively and vividly in a large number of publications.

Multiscale Geometric Analysis

The excellent characteristics of wavelet transform in one-dimensional data analysis cannot be simply extended to two-dimensional or multi-dimensional data, because it cannot make full use of the unique geometric features of the data itself in the case of higher dimensions. Thus, it is not the optimal or the sparsest method for function representation. The multi-scale geometric analysis (MGA) method arose in response to the proper time and conditions, and typical representations of multi-scale geometric analysis appeared. Ridgelet transform Candès [70], wedgelet Claypoole [71], beamlet Donoho [72], curvelet Candès [73], bandelet Pennec [74], and contourlet Do [75] were successively proposed.

Zhang [27] studied a new image fusion method using bandelet transform based on MGA. In this method, a low-pass subband coefficient of a source image by bandelet transform is inputted into a pulse-coupled neural network (PCNN), and the fused image can then be obtained through inverse bandelet transform using the coefficient and geometric flow parameters. Ai et al. [20] applied the curvelet transform to decompose the image combined with Fourier transform to extract features. Xu et al. [76] explored a method of MGA based on the non-symmetry and anti-packing model (NAM). This method can adaptively be applied to three types of steel products—continuous casting slabs, hot-rolled steel, and cold-rolled steel—that cannot be assessed by the traditional method.

It can be seen from the theoretical development of signal analysis methods that Fourier analysis is especially suitable for analyzing stable signals over a long period of time. The Gabor transform has its own application, but its effect depends on the window function. Wavelet analysis is especially suitable for analyzing mutated and singular signals. Multi-scale geometric analysis is suitable for the "sparse" function representation of high-dimension data.

4.3. Method Based on Model

4.3.1. Fractal Model (FM)

The fractal model (FM) was first derived by Mandelbrot [77] in 1983. Fractal dimension and porosity are the most important metrics in a fractal model. The former is a measure of complexity and irregularity, while the latter represents structural change or unevenness. In 2008, Blackledge et al. [78] used a membership function to analyze the partial structure and fractal features of images for extracting new information. Yazdchi et al. [79] researched a multifractal-based segmentation method to locate defects, and then extracted 10 features, such as multi-dimensional fractal dimension, variance, mean value, and maximum value in the principal component vector to achieve defect detection. The method achieved accuracies of 97.9%.

4.3.2. Markov Random Field Model

One of the main uses of the Markov random field (MRF) model in image processing is image segmentation, which is the technology and process of dividing the image into several specific and unique regions and extracting the target of interest; hence, it is a key step between image processing and image analysis.

In MRF, two random fields are often used to describe the image. One is the labeling field, which is often called the implicit random field. The prior distribution is used to describe the local correlation of the label field. The other is the grayscale field or feature field. The distribution function is often used to describe the distribution of observation data or feature vectors under the condition of the labeling field. The process of obtaining feature vectors is the process of detection.

Based on Bayesian theory, MRF turns the image segmentation problem into a process of obtaining the maximum probability density. The formula is as follows:

$$P(W|S) = \frac{P(S|W)P(W)}{P(S)} \tag{2}$$

where $P(W)$, $P(S)$, and $P(S|W)$ are the prior probability, fixed value based on the observed value, and conditional probability distribution based on the observation S (also called the likelihood function), respectively. Then, the problem is converted into finding the maximum value of $P(S|W)P(W)$. The Markov random field was used as a texture analysis method, which was combined with a KNN classifier to achieve six kinds of steel surface defects detection, with classification rates of 79.36–91.36% Ünsalan [80]. Table 4 shows the comparison of model-based detection methods.

Table 4. Strengths and weaknesses of model-based detection methods for steel defects.

Name	Strengths	Weaknesses
Fractal model (FM)	Remain invariant to large geometric transformations and lighting variations.	Low characteristic dimensions lead to weak judgment.
Markov random field model	Can be used with statistical and spectral methods for segmentation applications.	Not invariant to rotation and scaling.
		Cannot detect small defects.
	Captures the local texture orientation information.	Not suitable for global texture analysis.
		Strong spatial constraint.

4.4. Method Based on Machine Learning

4.4.1. Artificial Neural Networks

Since 1980, artificial neural networks (ANN) have been a hotspot in the field of artificial intelligence. ANN abstracts the human brain neural network from the perspective of information processing to establish a simple model, and can then be used to form different networks according to different connection modes to achieve various functions (Figure 7). As early as 2000, Caleb [28] proposed an adaptive learning classification for surface defects of hot-rolled steel, and the average percentage classification accuracy was 84% for training data and 64% for test data. Later, an improved BP (Back propagation) algorithm based on error function was mentioned in Peng [39] to conduct the surface quality inspection of cold-rolled strip. Zhao [81] came up with an improved BP algorithm based on singular value decomposition and a generalized inverse matrix for five common defects (cracks, oxide, skin, holes, and scratches) of steel plate, overcoming the slow training of the traditional BP algorithm, with results showing that it could meet real-time requirements.

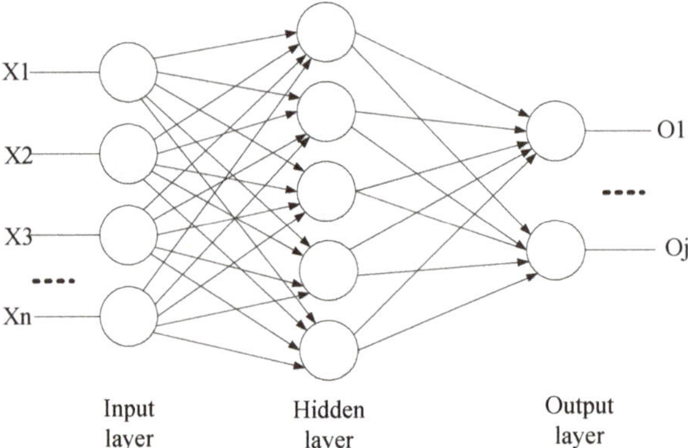

Figure 7. Artificial neural networks (ANN) model.

4.4.2. Convolutional Neural Network

Convolutional neural networks (CNNs) belong to a branch of ANN. As a network structure with fewer layers, ANNs have limited representation ability for complex functions, and generalization ability for complex classification problems is restricted to some extent. However, CNNs can realize complex function approximation by learning a deep nonlinear network structure. The deep neural network (DNN) has more layers (8–152 layers) than ANN, and needs more training data (4000–10,000 images).

An end-to-end detection network model was outlined in Yi [40]. Since the feature detection layer of CNNs is learned by training data, explicit feature extraction is avoided when using CNNs, while implicit learning is carried out from the training data. Moreover, since the weights of neurons on the same feature mapping surface are the same, the network can learn in parallel. It is worth mentioning that the detection accuracy is 99.29%. The CNN's structure for surface defect recognition model is presented in Table 5. In a recent study, Park et al. [82] employed CNNs to detect several types of defects on textured and non-textured surfaces, which was difficult to achieve by traditional machine learning methods. The recognition rate was 98%, and the time consumed for single image recognition was 0.01135 s. Masci et al. [21] proposed a max-pooling convolutional neural network method for the classification of steel defects. Compared with the commonly used support vector machine (SVM) classifier for feature descriptor training, this method can not only obtain better detection effects, it can also be directly used to detect the original image and segmentation defects, avoiding further time consumption and difficulty in optimizing adaptive preprocessing. However, changes in image size in a particular classification task have not yet been addressed by standard CNNs;

nevertheless Masci et al. [41] put forward the multi-scale pyramidal pooling network, which had three characteristics: (1) a pyramidal pooling layer that made the net independent of input image size; (2) multi-scale feature extraction; and, (3) an encoding layer emulating standard dictionary-based encoding strategies. Hence, the problem of image scale in traditional CNNs was solved.

Table 5. Convolutional neural networks (CNNs) structure for surface defect recognition model (reproduced from Yi [40]).

Layer Type	Filter Size	Volume Size
Input	N/A	(3,144,144)
Convolution	(5,5)	(32,140,140)
Max pooling	(2,2)	(32,70,70)
Convolution	(5,5)	(32,66,66)
Max pooling	(2,2)	(32,33,33)
Convolution	(4.4)	(64,30,30)
Max pooling	(2,2)	(64,15,15)
Convolution	(4,4)	(64,12,12)
Max pooling	(2,2)	(64,6,6)
Convolution	(3,3)	(128,4,4)
Max pooling	(2,2)	(128,2,2)
Fully connected	N/A	(256)
Fully connected	N/A	(512)
Softmax	N/A	(7)

4.4.3. Moving Center Hypersphere

The moving center hypersphere (MCH) is a way to compress a reference sample. The basic idea of the MCH is to use a hypersphere to represent a cluster of points to approximate each sample with a number of hyperspheres. The center of the hypersphere is then moved, and its radius is expanded so that it should contain as many sample points as possible, and ultimately contain all of the sample points in the space. Two recent articles fully illustrate the novelty of this method. In 2017, a method with quantile hypersphere based on machine learning (QH-ML) was employed in Chu [83] for six kinds of defects on a steel surface. Soon after, in 2018, a defect classification model was established in Gong [84], which was a multi-hypersphere support vector machine (MHSVM) with additional information. It is not hard to see that this method has good generalization ability.

4.4.4. Sparse Coding

A sparse coding algorithm is an unsupervised learning method. The purpose of the sparse coding algorithm is to find a set of overcomplete base vectors φ, so that we can represent the input vector x as a linear combination of these base vectors:

$$x = \sum_{i=1}^{k} a_i \phi_i \tag{3}$$

where a is the weight. In reference Liu [85], the method of sparse coding was adopted to achieve defect detection. Table 6 shows the comparison of defect detection methods based on machine learning.

Table 6. Strengths and weaknesses of machine learning-based detection methods for steel defects.

Name	Strengths	Weaknesses
Artificial neural networks	Real-time performance suitable for industrial application.	Large-scale feature vectors lead to high calculation cost.
	Can learn complex nonlinear input-output relationships.	Need lots of data (100–4000 images).

Table 6. *Cont.*

Name	Strengths	Weaknesses
Convolutional neural network	End-to-end mode (raw image input, classification results output).	Large data sets are required (4000–10,000 images).
	High detection rate (95–100%).	
Moving center hypersphere	Not sensitive to noise.	Optimal choice of parameters is difficult.
	High classification accuracy and efficiency (93–96%).	
Sparse coding	Can be used not only in the input phase, but also in the output phase.	The calculation time is too long (more than 45.6 s [85]) to allow real-time detection.

5. Classifier

The overall framework of the classifier is shown in Figure 8. There are two types of classifiers that are commonly used: supervised and unsupervised. The supervised classifier is a method of pattern recognition that is based on the samples provided by known training areas to find the characteristic parameters as decision rules, and then to establish the discriminant function to classify unknown sample images. The unsupervised classifier is an image classification method without a priori category standard, which is based on the characteristic differences of different image categories in the feature space. Based on the cluster theory, the decision rule of classification is established according to the statistical characteristics of the samples, and the classification is then presented.

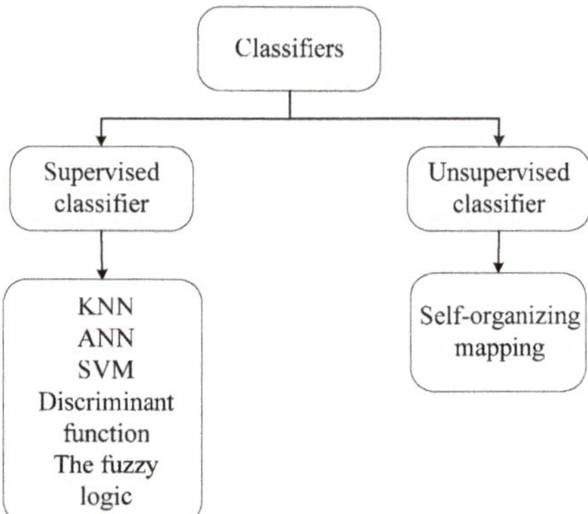

Figure 8. Kinds of classifiers. Note. KNN: K-nearest neighbor, ANN: artificial neural network, SVM: support vector machine.

5.1. Supervised Classifier

5.1.1. K-Nearest Neighbor (KNN)

The KNN method is uses the following steps. First, it extracts the characteristics of new data, and compares them with each data feature in the test set. Then, it extracts the nearest K data point's feature labels from the test set. Finally, the most frequently occurring category of the nearest K data points is counted as the category of the new data.

The KNN algorithm is the simplest and most effective classification algorithm; it is simple and easy to implement. When the training data set is large, a large amount of storage space is required, and the distance between the samples to be measured and all of the samples in the training data set needs to be calculated, so it is very time-consuming, and time complexity is O(n) (which is a level of time complexity).

Ünsalan et al. [80] developed a texture analysis method combined with the K-nearest neighbor classifier to achieve satisfactory recognition accuracy; however, the method could not meet real-time requirements.

5.1.2. Artificial Neural Network

Artificial neural networks are free of the restrictions of early discrete transfer functions, and use continuous functions, such as sigmoid or hyperbolic tangent functions, to imitate the response of the neuron to excitation. The training process adopts the back propagation algorithm. The ANN resolves matters that could not be simulated or solved with logic problems before. Further, more layers allow the network to implement complex situations in practice. Moreover, this method can automatically construct nonlinear features, so it can be used to solve the problem of nonlinear partitions. Examples of practical applications include Martins [33], Wu [37], Kang [42], Tang [62], Li [68], Yazdchi [79], Yazdchi [86], among which Yazdchi [86] employed a three-layer feed forward neural network, with training by the error back-propagation method. Classification accuracy reached 97.89%. The publication of literatures using this classifier was mostly concentrated in 2000–2010, and its status in the mainstream has been gradually replaced since 2010.

5.1.3. Support Vector Machine

Since neural network training requires a large number of samples and there are multiple local optimums, the expression ability of shallow neural networks for feature learning is limited. However, there are many parameters in deep neural networks, which may lead to an overfitting problem. Support vector machines (SVMs) can overcome this problem. SVMs have the following advantages over neural networks (ANNs): (1) their cost function is convex, and there is a global optimal value; (2) they are able to cope with small sample sets; (3) they have good generalization performance and robustness; (4) the introduction of a kernel function solves the nonlinear problem; and, (5) they can also avoid the dimension disaster. In Neogi [87], Yu [32], Song [16], Chu [60], Wu [25], Liu [65], Jia [88], Zhao [36], Ghorai [26], Choi [43], Agarwal [29], the excellent performance of support vector machines is demonstrated. We found that there has been a large amount of literature based on support vector machine classifiers since 2010.

5.1.4. Discriminant Function (DF)

Pattern classification using discriminant functions not only depends on the geometric properties of the discriminant function (i.e., linear and nonlinear functions), it also depends on the coefficients of the discriminant function. As long as the samples that are being studied are separable, the coefficients of the discriminant function can be determined using a given set of samples [56,59].

5.1.5. Fuzzy Logic (FL)

Fuzzy logic based on the concept of a membership function makes use of fuzzy sets and fuzzy reasoning rules, and can represent transitional boundaries or qualitative knowledge experience. Therefore, fuzzy logic is good at expressing qualitative knowledge and experience with unclear boundaries. For example, an information extraction technique based on fuzzy logic and membership function theory to design decision rules is discussed in Blackledge [78].

5.1.6. Learning Vector Quantizer (LVQ)

The learning vector quantizer (LVQ) is a kind of supervised learning algorithm for pattern classification that was put forward in 1988, which is an extension of the unsupervised self-organizing map (SOM) algorithm. The basic idea of LVQ is to use a small number of weight vectors representing the topology of the data. Compared with the unsupervised self-organizing neural network algorithm, the LVQ algorithm has a wider application in the field of pattern recognition because of the introduction of supervised signals during the process of updating weight vectors. Olsson et al. [89] developed a statistical feature extraction technology combined with an LVQ classifier to complete defect inspection. Subsequently, Wu et al. [30] employed an FFT-based extraction feature combined with the LVQ classifier for the detection of surface defects in hot-rolled strips.

5.2. Unsupervised Classifier

The self-organizing map (SOM) is an important type of neural network based on unsupervised learning methods that was first put forward in 1981. Since then, with the rapid development of neural networks in the mid to late 1980s, self-organizing map theory and its applications have also made considerable progress. The self-organizing map network conducts classification by finding the optimal set of reference vectors. Compared with the traditional pattern clustering method, the clustering center can be mapped to a surface or a plane while keeping the topology unchanged. Hence, the problem of discriminating unknown cluster centers can be solved by using self-organizing maps. For example, in Kang [42], the authors researched an adaptive classification technique based on a combination of supervised learning neural network with error back-propagation (NN-BP) and unsupervised learning (SOM).

Table 7 shows the performance comparison of classification methods for steel defects recognition.

Table 7. Strengths and weaknesses of classifiers for steel defect recognition.

Name of Classifier		Strengths	Weaknesses
Supervised classifier	KNN	The algorithm is simple, clear, and easy to achieve.	Time consuming (time complexity is O(n)).
			Not suitable for the unbalanced distribution of samples.
	ANN	Suitable for nonlinear separable problems.	Too many layers (i.e., more than eight layers) make it prone to overfitting.
		High classification accuracy.	
	SVM	Suitable for small sample sets.	Difficult to find the kernel function.
		Good generalization performance and robustness.	
		Suitable for nonlinear problems.	Solving quadratic programming of functions requires a lot of storage space.
		Suitable for high-dimensional situations.	
	DF	Suitable for multi-class classification.	Difficult to determine the coefficients of functions.
		Accurate and efficient.	
	FL	Suitable for expressing qualitative knowledge and experience in cases of unclear boundaries.	Poor classification accuracy (60%–85%).
	LVQ	Simple structure.	Input of heterogeneous samples will prevent convergence of weight vectors.
		Adaptive ability.	Information of each dimension of the input sample is not fully utilized.
Unsupervised classifier	SOM	Suitable for identifying unknown cluster centers.	"Dead nodes" can appear when the number of neuron nodes is more than the number of categories.
		Fault tolerant function.	
		Self-associative function.	

Note: KNN: K-nearest neighbor, ANN: Artificial neural network, SVM: Support vector machine, DF: Discriminant function, FL: Fuzzy logic, LVQ: Learning vector quantizer, SOM: Self-organizing map.

6. Analysis

The following is an analysis of visual detection from the perspective of scientific literature to the perspective of market size.

6.1. Literature Analysis

From the review of detection methods, we can see that a large number of publications over the past 30 years have been related to statistics and filtering methods, as shown in Figure 9.

From the point of development trends, both statistical methods and filtering methods have shown a significant downward trend since 2010, while the discussion of learning-based methods has steadily improved. This has much to do with the upsurge of deep learning in recent years. Model-based detection methods have always been out of the mainstream.

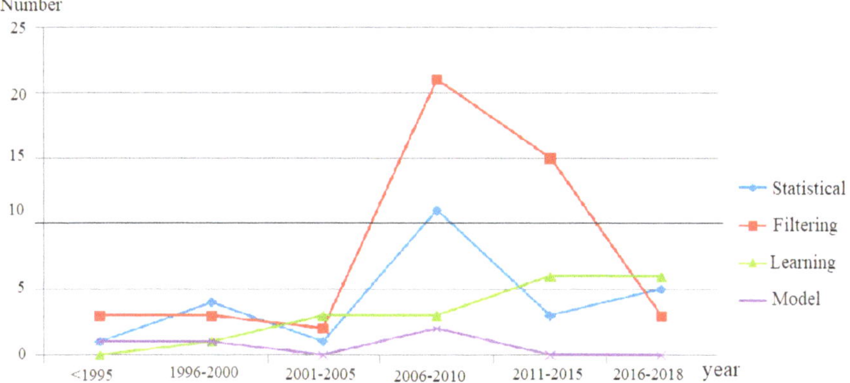

Figure 9. Distribution map detection method.

A detailed analysis of filtering methods is provided in Figure 10, because these methods have attracted much attention. From the 36 papers collected (relating to space filtering and frequency domains), it can be seen that although the best method of defect detection cannot be determined, it is clear that the joint spatial/frequency analysis methods (i.e., Gabor transform, wavelet transform, and MGA) have increased since 2010, which shows that these methods have been increasingly recognized by a majority of researchers.

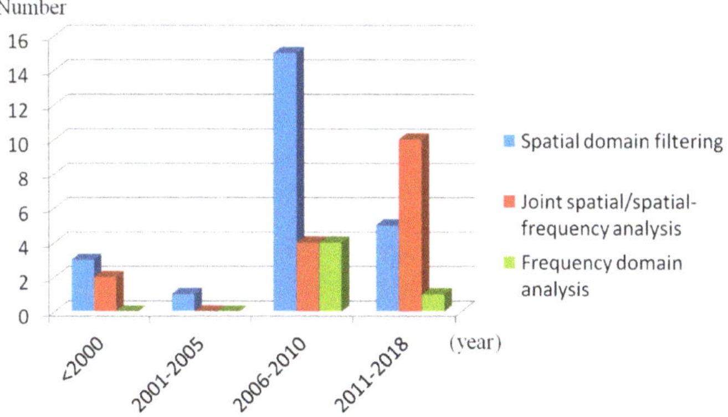

Figure 10. Distribution of filtering detection methods.

In terms of classification methods, supervised classification methods have always dominated compared to unsupervised classification methods. As the knowledge set of defect models is imperfect, the supervised classification method is preferred if prior knowledge is available, since this method can achieve superior results.

Support vector machines and neural networks based on back propagation (NN-BP) are the mainstream supervised classification methods.

It can be seen from the Figure 11 that the NN-BP method was a classifier that was commonly discussed in the literature prior to 2010, and that the frequency of discussion of SVMs has increased sharply since 2010.

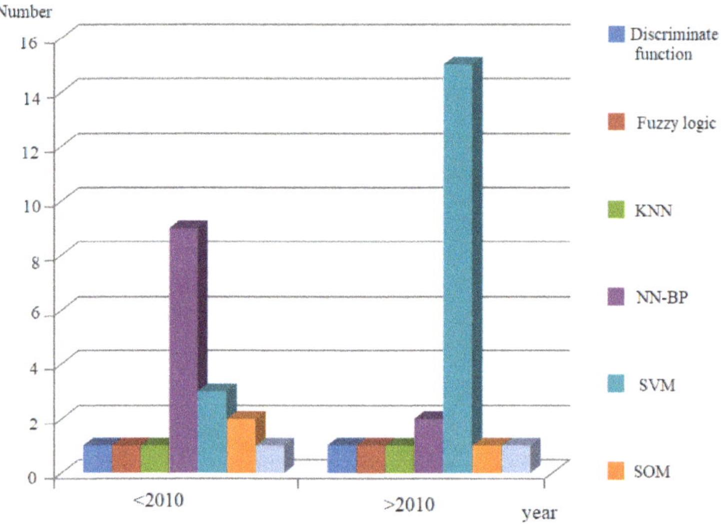

Figure 11. Distribution of classification methods.

6.2. Market Size Analysis of Visual Inspection

In 2017, the size of the global machine vision market was about USD $7.2 billion, growing 6.8% year-on-year. The market size is expected to be USD $7.7 billion in 2018, and could break through USD $9 billion in 2021, with an expected average annual compound growth rate of around 7.5% for 2018–2021. Germany and the United States are the world's two largest national machine vision markets, accounting for more than 30% of the worldwide market in 2017. China's machine vision industry has emerged since 2010, and is now in a period of rapid development. China's market size in 2017 was CNY ¥2.9 billion (about USD $42.64 million), accounting for 6.41% of the global market, and up 18.3% year-on-year. With the deepening of automation and the intellectualization of various industries, it is estimated that the average annual growth rate of China's machine vision market will be around 20% in 2018–2021, which was higher than the global average growth rate, as shown in Figure 12 [90].

The world's major machine vision manufacturers include Keenshi, Konrad, Darsa, Panasonic, and Omron. In 2016, their combined market share was about 38.0%. Typical Chinese enterprises are Daheng, New Epoch Technology, and Shenzhen JT Automation Equipment, which are less competitive compared with international well-known players, and each made up less than 1.5% of the global market in 2016.

At present, Chinese machine vision products are mainly used in semiconductor, electronic manufacturing, automobile, and other fields. The demand for machine vision in these fields accounted for nearly 60% of total demand in 2017.

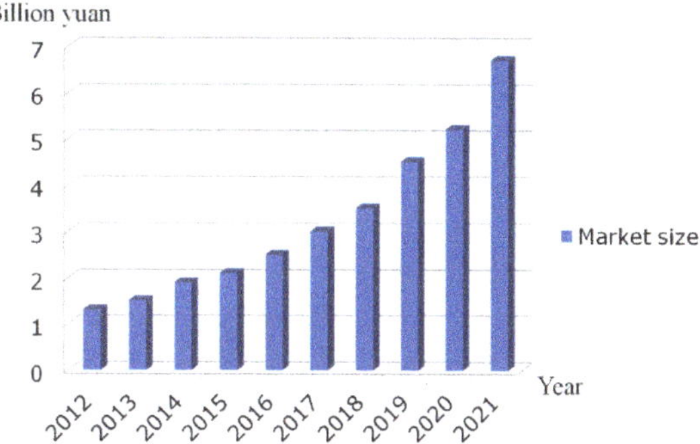

Figure 12. China machine vision market size.

7. Conclusions

In this paper, studies of software and hardware for visual detection from 90 papers are reviewed. The discussion of hardware includes coverage of cameras, light sources, and lighting modes, and a basis of selection is provided. In the software discussion, detection methods are divided into the categories of statistics, filtering, models, and machine learning according to basic theories of image processing. Classification methods are divided into supervised and unsupervised learning. The main ideas, advantages, and disadvantages of these methods are discussed, which can help users choose the most appropriate methods for different application environments.

Recommendations relating to the key technologies of visual detection, cameras, light sources, and image-processing algorithms can be summarized as follows:

1. The linear array camera is an inevitable choice for the selection of industrial cameras, because area-array cameras cannot achieve the resolution and frame rate required in conditions of high detection accuracy and fast motion. The frame rate of the camera must be greater than the speed of the object. Therefore, large frame rate, small pixel size line array cameras have good development prospects.

2. LED light sources have good color performance, a wide spectrum range (i.e., they can cover the whole range of visible light), high luminous intensity, and a long period of stability. As their manufacturing processes and technology matures, and prices fall, LED lamps will be used more widely.

3. It is difficult to select one kind of detection algorithm to meet the range of needs of accurate detection for multiple types of unbalanced defects; therefore, the fusion of multiple technologies is an expected trend.

4. The conventional detection process starts with feature extraction, followed by classification and a result output. The feature extraction process adopts artificial design features, and is tedious and complicated. However, the end-to-end approach combines feature extraction and the classification process into one body through deep learning neural networks, and features are extracted automatically through the learning of training sets (Figure 13), as seen in Yi [40], Park [82], Masci [21]. This method is simple and achieves high detection accuracy. Moreover, it can be readily generalized. However, its biggest disadvantage is that it needs a large number of training images, with specific needs of training sets (e.g., the training set must cover sufficient defect types); otherwise, detection results are not ideal. The excellent performance of convolutional networks based on deep learning in the field of image processing makes it inevitable that it will be developed further in the future. The convolutional neural network algorithm with small and zero samples will be the focus of future research in the field of visual detection.

5. For industrial applications, it is important that real-time performance meets production requirements. However, detection accuracy depends on the complexity of the deep network, while the complexity of the network can restrict the production process. Therefore, it is a direction

of future efforts to find a balance between algorithm complexity, detection accuracy, and time taken for detection.

6. A well-recognized standard data set and a good communication protocol for experimental data is required for the detection of defects on steel surfaces. Only in this way can fair, comparative analysis be realized.

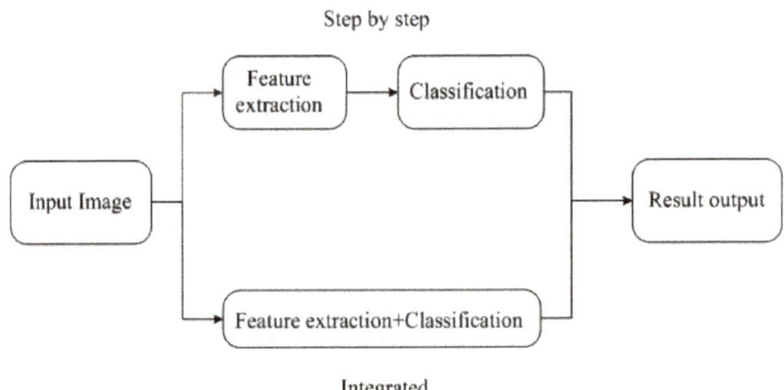

Figure 13. Two models of image processing.

In addition, our future work will pay attention to research progress on the detection of surface defects of steel products based on image processing in order to continuously enrich and update the relevant literature review.

Author Contributions: X.S. conducted the literature review with the discussion, correction, and guidance provided by J.G., S.T., and J.L. at every stage of the process: from the general structure to the specific details. All authors have read and approved the final manuscript.

Funding: This research was funded by the National Natural Science Foundation of China (No. 51875266), Jiangsu Province Graduate Research and Innovation Program (No. KYCX18-2227). Research Innovation Program for College Graduates of Jiangsu under Grand (No. KYLX16-0880). Industry-Academia Prospect Research Foundation of Jiangsu under Grand (No. BY2016066-06).

Conflicts of Interest: The authors declare no conflict of interest.

References

1. Pinho, E.; Costa, C. Unsupervised learning for concept detection in medical images: A comparative analysis. *Appl. Sci.* **2018**, *8*, 1213. [CrossRef]
2. Kong, X.; Li, J. Image Registration-Based Bolt Loosening Detection of Steel Joints. *Sensors* **2018**, *18*, 1000. [CrossRef] [PubMed]
3. Kong, X.; Li, J. Vision-Based Fatigue Crack Detection of Steel Structures Using Video Feature Tracking. *Comput.-Aided Civ. Inf.* **2018**, *9*, SI783–SI799. [CrossRef]
4. Calvo-Zaragoza, J.; Castellanos, F.J.; Vigliensoni, G.; Fujinaga, I. Deep neural networks for document processing of music score images. *Appl. Sci.* **2018**, *8*, 654.
5. Cho, C.S.; Chung, B.M.; Park, M.J. Development of real-time vision-based fabric inspection system. *IEEE Trans. Ind. Electron.* **2005**, *52*, 1073–1079. [CrossRef]
6. Gao, X.; Wu, Y.; Yang, K. Vehicle bottom anomaly detection algorithm based on sift. *Optik* **2015**, *126*, 3562–3566.
7. Jang, T.S.; Lee, S.S.; Kwon, I.B.; Lee, W.J.; Lee, J.J. Noncontact detection of ultrasonic waves using fiber optic sagnac interferometer. *IEEE Trans. Ultrason. Ferroelectr. Frequency Control* **2002**, *49*, 767–775. [CrossRef]
8. Chin, R.T.; Harlow, C.A. Automated visual inspection: A survey. *IEEE Trans. Pattern Anal. Mach. Intell.* **1982**, *4*, 557–573. [CrossRef] [PubMed]
9. Newman, T.S.; Jain, A.K. A survey of automated visual inspection. *Vis. Image Underst.* **1995**, *61*, 231–262. [CrossRef]
10. Li, Y.; Gu, P. Free-form surface inspection techniques state of the art review. *Comput. Aided Des.* **2004**, *36*, 1395–1417.

11. Shirvaikar, M. Trends in automated visual inspection. *J. Real-Time Image Process.* **2006**, *1*, 41–43. [CrossRef]
12. Hanbay, K.; Talu, M.F.; Özgüven, Ö.F. Fabric defect detection systems and methods—A systematic literature review. *Optik* **2016**, *127*, 11960–11973. [CrossRef]
13. Jfs, G.; Leta, F.R. Applications of computer vision techniques in the agriculture and food industry: A review. *Eur. Food Res. Technol.* **2012**, *235*, 989–1000.
14. Eui Jae, R.I. Dependences of the reflectivity and adhesion of thin metal films on the various process parameters for polyester substrates during physical vapor depositions. *Met. Mater. Int.* **2002**, *8*, 591–599.
15. Neogi, N.; Mohanta, D.K.; Dutta, P.K. Review of vision-based steel surface inspection systems. *EURASIP J. Image Video Process.* **2014**, *1*, 50. [CrossRef]
16. Song, K.; Yan, Y. A noise robust method based on completed local binary patterns for hot-rolled steel strip surface defects. *Appl. Surf. Sci.* **2013**, *285*, 858–864. [CrossRef]
17. Yun, J.P.; Choi, S.H.; Jeon, Y.; Choi, D.; Kim, S.W. Detection of line defects in steel billets using undecimated wavelet transform. In Proceedings of the 2008 IEEE International Conference on Control, Automation and Systems, Seoul, Korea, 14–17 October 2008; pp. 1725–1728.
18. Dupont, F.; Odet, C.; Cartont, M. Optimization of the recognition of defects in flat steel products with the cost matrices theory. *NDT E Int.* **1997**, *30*, 3–10. [CrossRef]
19. Yun, J.P.; Choi, S.H.; Seo, B.; Chang, H.P.; Sang, W.K. Defects detection of billet surface using optimized gabor filters. *IFAC Proc. Vol.* **2008**, *41*, 77–82. [CrossRef]
20. Ai, Y.H.; Xu, K. Surface Detection of Continuous Casting Slabs Based on Curvelet Transform and Kernel Locality Preserving Projections. *J. Iron. Steel Res. Int.* **2013**, *20*, 80–86. [CrossRef]
21. Masci, J.; Meier, U.; Ciresan, D.; Schmidhuber, J. Steel defect classification with Max-Pooling Convolutional Neural Networks. International Joint Conference on Neural Networks, Brisbane, QLD, Australia, 10–15 June 2012; IEEE: Piscataway, NJ, USA, 2012; Volume 20, pp. 1–6.
22. Luo, Q.; He, Y. A cost-effective and automatic surface defect inspection system for hot-rolled flat steel. *Robot. Cim.-Int. Manuf.* **2016**, *38*, 16–30. [CrossRef]
23. Liu, M.; Liu, Y.; Hu, H.; Nie, L. Genetic algorithm and mathematical morphology based binarization method for strip steel defect image with non-uniform illumination. *J. Vis. Commun. Image Represent.* **2016**, *37*, 70–77. [CrossRef]
24. Wu, X.; Xu, K.; Xu, J. Application of Undecimated Wavelet Transform to Surface Defect Detection of Hot Rolled Steel Plates. In Proceedings of the 2008 IEEE Conference on Image and Signal, Sanya, China, 27–30 May 2008; pp. 528–532.
25. Wu, G.; Kwak, H.; Jang, S.; Xu, K.; Xu, J. Design of online surface inspection system of hot rolled strips. In Proceedings of the 2008 IEEE International Conference on Automation and Logistics, Qingdao, China, 1–3 September 2008; pp. 2291–2295.
26. Ghorai, S.; Mukherjee, A.; Gangadaran, M. Automatic defect detection on hot-rolled flat steel products. *IEEE Trans. Instrum. Meas.* **2013**, *62*, 612–621. [CrossRef]
27. Zhang, X. Image Fusion Method for Strip Steel Surface Detect Based on Bandelet-PCNN. *Adv. Mater. Res.* **2012**, *546–547*, 806–810. [CrossRef]
28. Caleb, P.; Steuer, M. Classification of surface defects on hot rolled steel using adaptive learning methods. In Proceedings of the International Conference on Knowledge-Based Intelligent Engineering Systems and Allied Technologies, Brighton, UK, 30 August–1 September 2000; IEEE: Piscataway, NJ, USA, 2000; pp. 103–108.
29. Agarwal, K.; Shivpuri, R.; Zhu, Y.; Chang, T.S.; Huang, H. Process knowledge based multi-class support vector classification (PK-MSVM) approach for surface defects in hot rolling. *Expert. Syst. Appl.* **2011**, *38*, 7251–7262. [CrossRef]
30. Wu, G.; Zhang, H.; Sun, X. A Bran-new Feature Extraction Method and its application to Surface Defect Recognition of Hot Rolled Strips. In Proceedings of the IEEE International Conference on Automation and Logistics, Jinan, China, 18–21 August 2007; pp. 2069–2074.
31. Liu, W.; Yan, Y. Automated surface defect detection for cold-rolled steel strip based on wavelet anisotropic diffusion method. *Int. J. Ind. Syst. Eng.* **2014**, *17*, 224–239. [CrossRef]
32. Yu, Y.W.; Yin, G.F.; Du, L.Q. Image classification for steel strip surface defects based on support vector machines. *Adv. Mater. Res.* **2011**, *217–218*, 336–340. [CrossRef]

33. Martins, L.A.O.; PaáDua, F.L.C.; Almeida, P.E.M. Automatic detection of surface defects on rolled steel using Computer Vision and Artificial Neural Networks. In Proceedings of the IECON 2010—36th Annual Conference on IEEE Industrial Electronics Society, Glendale, AZ, USA, 7–10 Novrmber 2010; pp. 1081–1086.
34. Toshihiro, S.; Hideki, T.; Yasuo, T. Automatic surface inspection system for tin mill black plate (TMBP). *JFE Tech. Rep.* **2007**, *9*, 60–63.
35. Yun, S.W.; Kong, N.W.; Lee, G.; Park, P.G. Development of defect detection algorithm in cold rolling. *Int. J. Control Autom.* **2008**, *1*, 1729–1733.
36. Zhao, J.; Zhao, J.; Yang, Y.; Li, G. The Cold Rolling Strip Surface Defect On-Line Inspection System Based on Machine Vision. In Proceedings of the Second Pacific-Asia Conference on IEEE Circuits, Communications and System (PACCS), Beijing, China, 1–2 August 2010; pp. 402–405.
37. Wu, G.F. *Online Surface Inspection Technology of Cold Rolled Strips, Multimedia*; InTech: London, UK, 2010.
38. Wu, G.; Xu, K.; Xu, J. Application of a new feature extraction and optimization method to surface defect recognition of cold rolled strips. *Int. J. Min. Met. Mater.* **2007**, *14*, 437–442. [CrossRef]
39. Peng, K.; Zhang, X. Classification Technology for Automatic Surface Defects Detection of Steel Strip Based on Improved BP Algorithm. In Proceedings of the International Conference on Natural Computation IEEE, Tianjin, China, 14–16 August 2009; Volume 1, pp. 110–114.
40. Yi, L.; Li, G.; Jiang, M. An End-to-End Steel Strip Surface Defects Recognition System Based on Convolutional Neural Networks. *Steel Res. Int.* **2016**, *88*, 176–187. [CrossRef]
41. Masci, J.; Meier, U.; Fricout, G.; Schmidhuber, J. Multi-scale pyramidal pooling network for generic steel defect classification. In Proceedings of the International Joint Conference on Neural Networks, Dallas, TX, USA, 4–9 August 2013; Piscataway, NJ, USA, 2014; pp. 1–8.
42. Kang, G.W.; Liu, H.B. Surface defects inspection of cold rolled strips based on neural network. In Proceedings of the International Conference on Machine Learning and Cybernetics IEEE, Guangzhou, China, 18–21 August 2005; Volume 8, pp. 5034–5037.
43. Choi, K.; Koo, K.; Jin, S.L. Development of Defect Classification Algorithm for POSCO Rolling Strip Surface Inspection System. In Proceedings of the International Joint Conference IEEE, SICE-ICASE, Busan, Korea, 18–21 October 2006; Volume 2007, pp. 2499–2502.
44. Kaya, K.; Bilgutay, N.M.; Murthy, R. Flaw detection in stainless steel samples using wavelet decomposition. In Proceedings of the 1994 IEEE Conference on Ultrasonics Symposium, Cannes, France, 31 October–3 November 1994; pp. 1271–1274.
45. Choi, S.H.; Yun, J.P.; Seo, B.; Sang, W.K. Real-time defects detection algorithm for high-speed steel bar in coil, world academy of science. *Eng. Technol.* **2007**, *25*, 66–70.
46. Bulnes, F.G.; Usamentiaga, R.; García, D.F.; Molleda, J. Vision-based sensor for early detection of periodical defects in web materials. *Sensors* **2012**, *12*, 10788–10809. [CrossRef] [PubMed]
47. Li, W.B.; Lu, C.H.; Zhang, J.C. A local annular contrast based real-time inspection algorithm for steel bar surface defects. *Appl. Surf. Sci.* **2012**, *258*, 6080–6086. [CrossRef]
48. Zhang, J.; Kang, D.; Won, S. Detection of scratch defects for wire rod in steelmaking process. In Proceedings of the 2010 IEEE International Conference on Control, Automation and Systems, Gyeonggi-do, Korea, 27–30 October 2010; pp. 319–323.
49. Yun, J.P.; Choi, D.C.; Jeon, Y.J. Defect inspection system for steel wire rods produced by hot rolling process. *Int. J. Adv. Manuf. Technol.* **2014**, *70*, 1625–1634. [CrossRef]
50. Choi, D.C.; Jeon, Y.J.; Lee, S.J.; Yun, J.P.; Kim, S.W. Algorithm for detecting seam cracks in steel plates using a gabor filter combination method. *Appl. Opt.* **2014**, *53*, 4865–4872. [CrossRef] [PubMed]
51. Koller, N.; O'Leary, P.; Lee, P. Comparison of cmos and ccd cameras for laser profiling. In Proceedings of the SPIE the International Society for Optical Engineering, San Jose, CA, USA, 3 May 2004; pp. 108–115.
52. Lee, S.H.; Yang, C.S. A real time object recognition and counting system for smart industrial camera sensor. *IEEE Sens. J.* **2017**, *17*, 2516–2523. [CrossRef]
53. Nguyen, A.Q.; Vts, D.; Shimonomura, K.; Takehara, K.; Etoh, T.G. Toward the ultimate-high-speed image sensor: From 10 ns to 50 ps. *Sensors* **2018**, *18*, 2407. [CrossRef] [PubMed]
54. Wu, L.; Zhu, J.; Xie, H. A modified virtual point model of the 3d dic technique using a single camera and a bi-prism. *Meas. Sci. Technol.* **2014**, *25*, 115008. [CrossRef]
55. Zheng, H.; Kong, L.X.; Nahavandi, S. Automatic inspection of metallic surface defects using genetic algorithms. *J. Mater. Process. Technol.* **2002**, *125*, 427–433. [CrossRef]

56. Cord, A.; Bach, F.; Jeulin, D. Texture classification by statistical learning from morphological image processing: Application to metallic surfaces. *J. Microsc.* **2010**, *239*, 159–166. [CrossRef] [PubMed]

57. Yun, J.P.; Park, C.; Bae, H.; Hwang, H.; Choi, S. Vertical Scratch Detection Algorithm for High-speed Scale-covered Steel BIC (Bar in Coil). In Proceedings of the 2010 IEEE International Conference on Control Automation and Systems, Gyeonggi-do, Korea, 27–30 October 2010; Volume 6245, pp. 342–345.

58. Haralick, R.M. Statistical and structural approaches to texture. *Proc. IEEE* **2005**, *67*, 786–804. [CrossRef]

59. Liu, W.; Yan, Y.; Li, J.; Zhang, Y.; Sun, H. Automated On-Line Fast Detection for Surface Defect of Steel Strip Based on Multivariate Discriminant Function. In Proceedings of the 2008 IEEE Second International Symposium on Intelligent Information Technology Application, Computer Society, Shanghai, China, 20–22 December 2008; Volume 2, pp. 493–497.

60. Chu, M.; Gong, R.; Gao, S.; Zhao, J. Steel surface defects recognition based on multi-type statistical features and enhanced twin support vector machine. *Chemometr. Intell. Lab.* **2017**, *171*, 140–150. [CrossRef]

61. Guo, J.H.; Meng, X.D.; Xiong, M.D. Study on Defection Segmentation for Steel Surface Image Based on Image Edge Detection and Fisher Discriminant. *J. Phys. Conf. Ser.* **2006**, *48*, 364–368. [CrossRef]

62. Tang, B.; Kong, J.Y.; Wang, X.D.; Chen, L. Surface Inspection System of Steel Strip Based on Machine Vision. In Proceedings of the First International Workshop on Database Technology and Applications, Wuhan, China, 25–26 April 2009; pp. 359–362.

63. Yang, S.S.; He, Y.H.; Wang, Z.L.; Zhao, W.S. A method of steel strip image segmentation based on local gray information. In Proceedings of the 2008 IEEE International Conference on Industrial Technology, Chengdu, China, 21–24 April 2008; pp. 1–4.

64. Guan, S. Strip Steel Defect Detection Based on Saliency Map Construction Using Gaussian Pyramid Decomposition. *ISIJ Int.* **2015**, *55*, 1950–1955. [CrossRef]

65. Liu, Y.C.; Hsu, Y.L.; Sun, Y.N.; Tsai, S.J.; Ho, C.Y.; Chen, C.M. A computer vision system for automatic steel surface inspection. In Proceedings of the fifth IEEE conference on Industrial Electronics and Applications (ICIEA), Taichung, Taiwan, 15–17 June 2010; pp. 1667–1670.

66. Jeon, Y.J.; Choi, S.H.; Yun, J.P. Detection of scratch defects on slab surface. In Proceedings of the 1994 IEEE International Conference on Control, Automation and Systems, Gyeonggi-do, Korea, 26–29 October 2011; pp. 1274–1278.

67. Choi, D.; Yun, J.P.; Sang, W.K.; Jeon, Y. Pinhole detection in steel slab images using gabor filter and morphological features. *Appl. Opt.* **2011**, *50*, 5122–5129. [CrossRef] [PubMed]

68. Li, J.; Shi, J.; Chang, T.S. On-line seam detection in rolling processes using snake projection and discrete wavelet transform. *J. Manuf. Sci.-Trans. ASME* **2006**, *129*, 926–933. [CrossRef]

69. Xu, S.H.; Guan, S.Q.; Chen, L.L. Steel Strip Defect Detection based on Human Visual Attention Mechanism Model. *Appl. Mech. Mater.* **2014**, *530–531*, 456–462. [CrossRef]

70. Candès, E.J.; Donoho, D.L. Ridgelets: A key to higher-dimensional intermittency? *Philos. Trans. Math. Phys. Eng. Sci.* **1999**, *357*, 2495–2509.

71. Claypoole, R.L., Jr.; Baraniuk, R.G. Multiresolution wedgelet transform for image processing. In Proceedings of the SPIE—The International Society for Optical Engineering, San Diego, CA, USA, 4 December 2000; Volume 4119, pp. 253–262.

72. Donoho, D.L.; Huo, X. Beamlets and Multiscale Image Analysis. *Multiscale Multiresolut. Methods* **2001**, *20*, 149–196.

73. Candès, E.J.; Donoho, D.L. Recovering edges in ill-posed inverse problems: Optimality of curvelet frames. *Ann. Stat.* **2002**, *30*, 784–842. [CrossRef]

74. Pennec, E.L.; Mallat, S. Bandelet representations for image compression. In Proceedings of the 2001 IEEE International Conference on Image Processing, Thessaloniki, Greece, 7–10 October 2001; p. 12.

75. Do, M.N.; Vetterli, M. Contourlets: A Directional Mulitresolution Image Representation. In Proceedings of the 1994 IEEE International Conference on Image Processing, Rochester, NY, USA, 22–25 September 2002; pp. 357–360.

76. Xu, K.; Xu, Y.; Zhou, P.; Wang, L. Application of RNAMlet to surface defect identification of steels. *Opt. Laser Eng.* **2018**, *105*, 110–117. [CrossRef]

77. Mandelbrot, B.B. *The Fractal Geometry of Nature*; W. H. Freeman: New York, NY, USA, 1983.

78. Blackledge, J.; Dubovitskiy, D. A Surface Inspection Machine Vision System that Includes Fractal Texture Analysis. *Dublin Inst. Technol.* **2008**, *3*, 76–89.

79. Yazdchi, M.; Yazdi, M.; Mahyari, A.G. Steel Surface Defect Detection Using Texture Segmentation Based on Multifractal Dimension. In Proceedings of the International Conference on Digital Image Processing, Bangkok, Thailand, 7–9 March 2009; pp. 346–350.

80. Ünsalan, C.; Erçil, A. *Automated Inspection of Steel Structures, Recent Advances in Mechatronics*; Springer Ltd.: Singapore, 1999.

81. Zhao, X.Y.; Lai, K.S.; Dai, D.M. An improved bp algorithm and its application in classification of surface defects of steel plate. *J. Iron Steel Res. (Int.)* **2007**, *14*, 52–55. [CrossRef]

82. Park, J.K.; Kwon, B.K.; Park, J.H.; Kang, D.J. Machine learning-based imaging system for surface defect inspection. *Int. J. Precis. Eng. Manuf. Green Technol.* **2016**, *3*, 303–310. [CrossRef]

83. Chu, M.; Zhao, J.; Liu, X.; Gong, R. Multi-class classification for steel surface defects based on machine learning with quantile hyper-spheres. *Chemometr. Intell. Lab.* **2017**, *168*, 15–27. [CrossRef]

84. Gong, R.; Wu, C.; Chu, M. Steel surface defect classification using multiple hyper-spheres support vector machine with additional information. *Chemometr. Intell. Lab.* **2018**, *172*, 109–117. [CrossRef]

85. Liu, Z.; Hu, J.; Hu, L.; Zhang, X.L.; Kong, J.Y. Research on on-line surface defect detection for steel strip based on sparse coding. *Adv. Mater. Res.* **2012**, *548*, 749–752. [CrossRef]

86. Yazdchi, M.R.; Mahyari, A.G.; Nazeri, A. Detection and Classification of Surface Defects of Cold Rolling Mill Steel Using Morphology and Neural Network. In Proceedings of the 2008 IEEE International Conference on Computational Intelligence for Modelling Control & Automation, Vienna, Austria, 24 July 2009; pp. 1071–1076.

87. Neogi, N.; Mohanta, D.K.; Dutta, P.K. Defect detection of steel surfaces with global adaptive percentile thresholding of gradient image. *J. Inst. Eng.* **2017**, *98*, 557–565. [CrossRef]

88. Jia, H.; Yi, L.M.; Shi, J.; Chang, T.S. An Intelligent Real-time Vision System for Surface Defect Detection. In Proceedings of the International Conference on Pattern Recognition IEEE, Cambridge, UK, 26 August 2004; pp. 239–242.

89. Olsson, J.; Gruber, S. Web process inspection using neural classification of scattering light. *IEEE Trans. Ind. Electron.* **1993**, *40*, 228–234. [CrossRef]

90. Global and China Machine Vision Industry Report. Available online: http://www.pday.com.cn/Htmls/Report/201704/24516162.html (accessed on 24 April 2017).

Article

Fine-Grain Segmentation of the Intervertebral Discs from MR Spine Images Using Deep Convolutional Neural Networks: BSU-Net

Sewon Kim [1], Won C. Bae [2,3], Koichi Masuda [4], Christine B. Chung [2,3] and Dosik Hwang [1,*]

[1]   School of Electrical and Electronic Engineering, Yonsei University, Seoul 03722, Korea; sewon.kim@yonsei.ac.kr
[2]   Department of Radiology, VA San Diego Healthcare System, San Diego, CA 92161-0114, USA; wbae@ucsd.edu (W.C.B.); cbchung@ucsd.edu (C.B.C.)
[3]   Department of Radiology, University of California-San Diego, La Jolla, CA 92093-0997, USA
[4]   Department of Orthopedic Surgery, University of California-San Diego, La Jolla, CA 92037, USA; koichimasuda@ucsd.edu
*   Correspondence: dosik.hwang@yonsei.ac.kr; Tel.: +82-2-2123-5771

Received: 13 August 2018; Accepted: 12 September 2018; Published: 14 September 2018

Featured Application: The application of this research aims to provide clinicians with a robust deep learning model for fine-grain segmentation of tissues in medical images, and therefore to provide accurate quantitative information of intervertebral discs in magnetic resonance spine images, which can be useful for diagnosis, surgical planning, and treatment monitoring.

Abstract: We propose a new deep learning network capable of successfully segmenting intervertebral discs and their complex boundaries from magnetic resonance (MR) spine images. The existing U-network (U-net) is known to perform well in various segmentation tasks in medical images; however, its performance with respect to details of segmentation such as boundaries is limited by the structural limitations of a max-pooling layer that plays a key role in feature extraction process in the U-net. We designed a modified convolutional and pooling layer scheme and applied a cascaded learning method to overcome these structural limitations of the max-pooling layer of a conventional U-net. The proposed network achieved 3% higher Dice similarity coefficient (DSC) than conventional U-net for intervertebral disc segmentation (89.44% vs. 86.44%, respectively; $p < 0.001$). For intervertebral disc boundary segmentation, the proposed network achieved 10.46% higher DSC than conventional U-net (54.62% vs. 44.16%, respectively; $p < 0.001$).

Keywords: intervertebral disc; segmentation; convolutional neural network; fine grain segmentation; U-net; deep learning; magnetic resonance image; lumbar spine

---

1. Introduction

Low back pain is a common disease in modern society. It can be caused by disorders of lumbar components such as an intervertebral disc, paraspinal muscle, and vertebral body. Therefore, it is important to examine the specific components of the lumbar spine for accurate diagnosis and treatment. Assessment of the intervertebral disc is particularly important since its shape is liable to physiological (age-related) and pathological changes [1,2]. Magnetic resonance (MR) imaging is a very effective non-invasive imaging modality for obtaining such information. However, segmentation of intervertebral discs in MR spine images is typically challenging for the following reasons: (1) object shapes are deformed and rotated; (2) the contrast between an object and its surroundings can be very low, which renders the boundary unclear; (3) the intensity within an object is not uniform.

Segmentation of intervertebral discs in MR spine images has been extensively studied. Ayed et al. [3] studied the application of graph-cut method for intervertebral disc segmentation and Michopoulou et al. [4] sought to detect and segment intervertebral discs using atlas-based and fuzzy clustering methods. Law et al. [5] proposed a detection and segmentation method for intervertebral discs using anisotropic oriented flux, while Rabia et al. [6] proposed a 3D intervertebral disc segmentation algorithm using a simplex active surface model using weak shape prior. However, performance of these conventional methods, which depend on mathematical algorithms with hand-crafted features, is limited by the challenges mentioned above.

Recent years have witnessed remarkable advances in the field of machine learning, especially with the use of deep-learning techniques. Convolutional neural networks (CNNs) effectively extract image features and perform effective classification based on these features. Several intelligent techniques, such as computer aided diagnoses that employ CNNs, have been reported in the field of medical imaging [7]. Ji et al. [8] attempted segmentation of intervertebral discs in MR spine images using a classification network by splitting the entire image into small patches.

The most common and effective CNN in medical image segmentation is the U-network (U-net) proposed by Ronneberger et al. [9]. As shown in Figure 1, a U-net is composed of an encoding part and a decoding part. The encoding part of conventional U-net is composed of convolutional layers and pooling layers and the decoding part is composed of convolutional layers and up-convolutional layers. Conventional U-net performs efficient feature extraction and segmentation using a large receptive field obtained through this structure [8]. However, since conventional U-net is based on feature extraction network for image classification, information pertaining to fine details of the image may disappear during the pooling process in the encoding part. For example, max-pooling layers, which is commonly used in U-nets, retains a pixel with the largest value among the neighboring four pixels and removes the information of the other pixels. Therefore, the pooling layer helps to efficiently detect the dominant information representing image characteristics, albeit with a loss of detailed information. The missing detail is not restored during up-convolutional layers. A skip connection can be added to this network to overcome this problem; however, it cannot completely recover the finer details. As a result, low-frequency information of the image is generally emphasized [10,11]. Figure 2 displays a comparison between the results of the conventional U-net segmentation and manually segmented labels. Dice similarity coefficient (DSC) [12] of segmentation for a whole area of intervertebral discs is 87.49%, while the DSC at the boundaries of the discs is as low as 40.87%. This suggests that it is difficult to achieve fine grain segmentation with conventional U-net and it may lead to unsatisfactory results for complex objects, such as intervertebral discs.

Dilated convolution is a way to overcome this limitation. Dilated convolution uses filters of various sizes with various rates. It allows users to control the resolution in the feature extraction process and to enlarge the field of view (FOV) without increasing parameter and cost [13,14].

In this paper, we propose a new network which can effectively perform fine grain segmentation for intervertebral discs. In our proposed network, pooling layers are modified to compensate for the aforementioned drawbacks. Convolutional layers and network structure are also improved to maximize the efficiency of the overall segmentation network. A preliminary study of this method was partially presented at the annual meeting of International Society for Magnetic Resonance in Medicine (ISMRM) in 2018 [15].

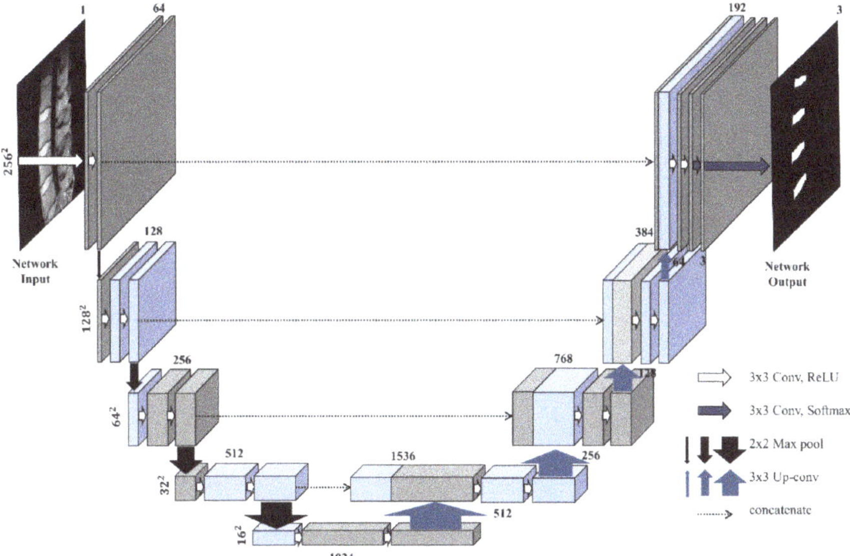

Figure 1. Structure of conventional U-network (U-net).

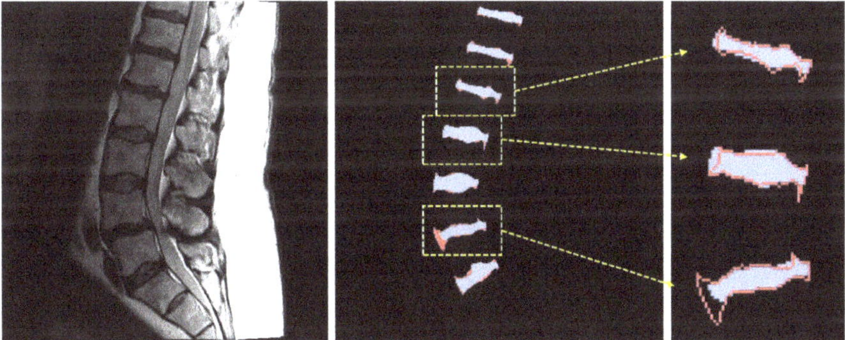

Figure 2. Intervertebral disc segmentation results from the conventional U-net. Blue areas are the results from the conventional U-net and red areas are manually segmented labels. Red lines are the boundaries of the labels.

2. Materials and Methods

2.1. Network Design: Boundary Specific U-Network (BSU-Net)

The purpose of this paper is to design a new network architecture based on U-nets, which can overcome the problems encountered in the detailed segmentation tasks. Hence, we propose a boundary specific U-network (BSU-net). The proposed network has a complex form of pooling layers and convolutional layers which are referred to as BSU-pooling layers and residual blocks respectively, and has a cascaded structure that uses preliminary outcomes of conventional U-net for efficient network learning. A schematic illustration of BSU-net is shown in Figure 3.

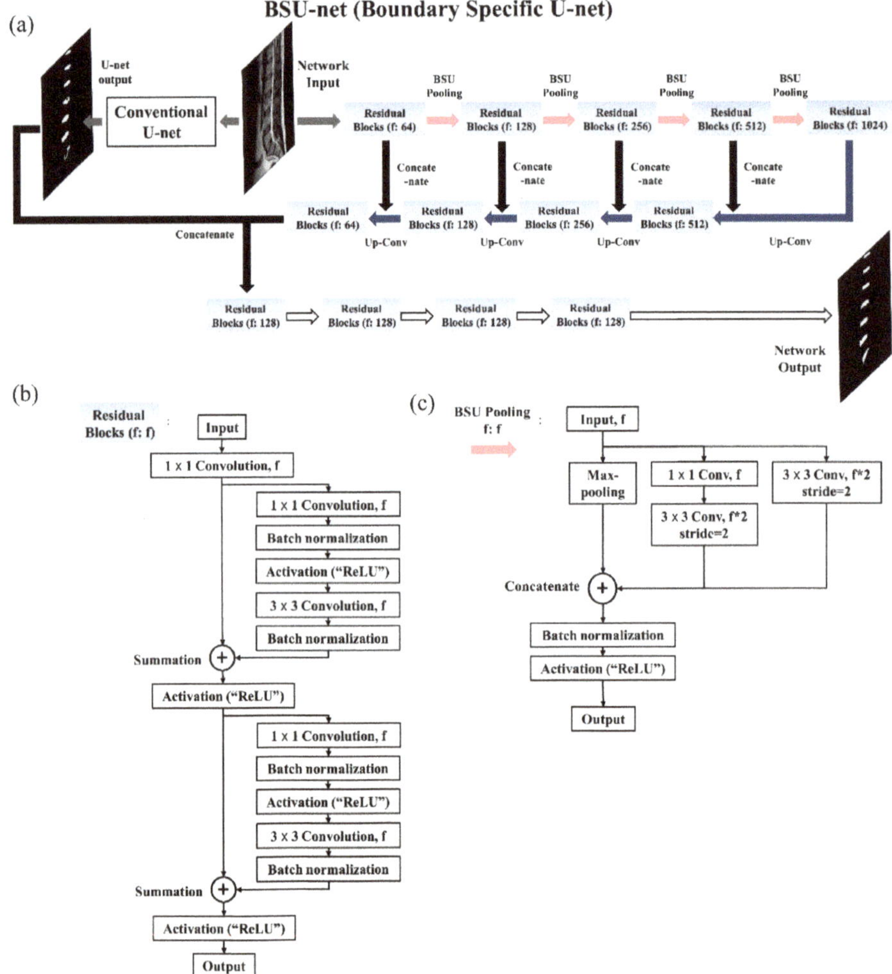

Figure 3. Whole structure of the proposed network. (**a**) Structure of the boundary specific U-network (BSU-net). (**b**) Structure of residual block. (**c**) Structure of BSU-pooling layer.

2.1.1. BSU-Pooling Layer

BSU-net has three components. The first is the advanced pooling process. Conventional max-pooling layer used in conventional U-net discards rest of the pixels in a calculation field except for one pixel with maximum value. This process contributes to the efficiency of feature extraction; however, the loss of the information contained in the discarded pixels during the pooling process results in an inaccurate estimation of boundaries of target object in detailed segmentation tasks. Therefore, there is a need for an advanced pooling layer scheme that can minimize the loss of information while increasing the efficiency of feature extraction. The proposed BSU-pooling layer shown in Figure 3c uses both a max-pooling layer that increases the efficiency of feature extraction and convolutional layers that compute the neighboring information without discarding it. In this case, the stride of the convolutional layers is set to 2, so that down-sampling effect as in the max-pooling layer is possible. Furthermore, the inputs of the layer are preserved through multiple paths: a path passing through

3×3 convolutional layer and a path passing through 1×1 convolutional layer and another subsequent 3×3 convolutional layer (Figure 3c).

2.1.2. Residual Block

The second component of BSU-net is the application of residual learning. Residual learning is applied to improve the efficiency of the convolutional layer. Conventional U-net is a very deep neural network with a large number of convolutional layers. Conventional U-net used in this study has a total of 38 convolutional layers and 62,803,650 learning parameters. Use of such a large number of consecutive convolutional layers can lead to the problem of gradient vanishing, which can degrade learning efficiency. The concept of residual learning was introduced to solve this problem [16]. Suppose we have a simple network $\mathcal{H}$ which is a part of a certain deep neural network. When $\mathcal{H}$ consists of two convolutional layers $\mathcal{F}_n$ and $\mathcal{F}_{n+1}$ and activation functions σ as shown in Figure 4a, output for the network with an input vector x is defined as $\mathcal{H}(x) = \sigma_{n+1}(\mathcal{F}_{n+1}(\sigma_n(\mathcal{F}_n(x)))), x \in \mathbb{R}^{w \times h \times c}$ where w, h, and c, respectively, denote the width, height, and the number of channels. During back propagation, gradient vanishing can occur if the weights of $\mathcal{F}_n$ or $\mathcal{F}_{n+1}$ are close to zero [16]. But if we change the network output $\mathcal{H}(x)$ to $\mathcal{H}(x) - x$, gradient vanishing is avoided. The changed network $\mathcal{S}$ is defined as $\mathcal{S}(x) = \mathcal{H}(x) - x$ and is also expressed as $\mathcal{H}(x) = \mathcal{S}(x) + x$. $\mathcal{H}$ is converted to $\mathcal{S}$ with "shortcut connection" between input and output as shown in Figure 4b. In this case, gradient vanishing rarely occurs because 1 is added to $\frac{\partial \mathcal{S}(x)}{\partial x}$. This change improves learning efficiency and allows the network to respond appropriately to small changes in input [16]. Residual block embeds this residual learning in BSU-net as displayed in Figure 3b. The first 1×1 convolutional layer immediately after the input is arranged to match filter size.

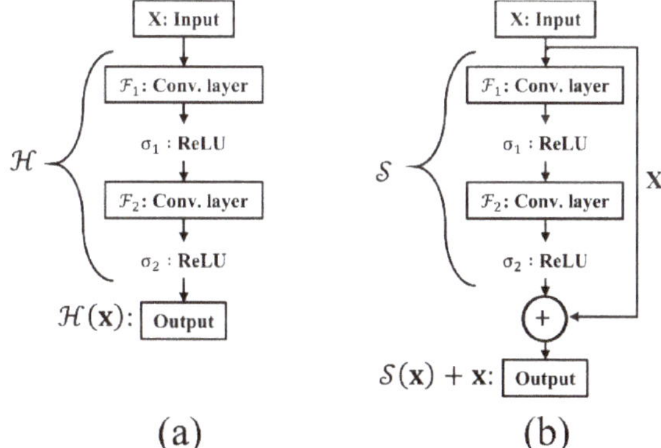

Figure 4. Introduction of residual learning. (**a**) Conventional neural network layers. (**b**) A learning network of residual function $\mathcal{S}$.

2.1.3. Cascaded Network

Several studies have revealed that cascaded learning of networks improves learning efficiency and network performance [17–19]. It is an efficient way to improve performance of an entire network to provide outcomes from other networks or to combine outcomes from multiple networks like ensemble networks [20–22]. As shown in Figure 3a, conventional U-net outcomes are used to guide the learning of the entire BSU-net. This augments overall segmentation and fine grain segmentation and results in improved overall performance of the network.

2.2. Experimental Materials

The dataset used in the experiments comprised of 3D MR spine images of 20 patients sourced from Spineweb dataset 10 [23,24]. Among this dataset, the images used in actual experiments are 1 to 3 mid-sagittal images per patient, totaling 25. The pixel size of images is 1.5×1.5 mm. Label data were made manually by a spine MR researcher and reviewed by a radiologist with an experience of more than 10 years. The experiments were implemented using 5-fold cross validation and each experiment had 5 test images and 20 training images. For fair validation of the network, all images from a single patient were used exclusively for either training or test.

The segmentation accuracy was evaluated using a DSC [12], and to assess the accuracy of measurement of fine details the evaluation was divided into the following three parts: (1) whole area; (2) boundary area; (3) boundary area with 2 pixels' thickness. The first part evaluates segmentation accuracy of the entire area of intervertebral discs. The second and third parts evaluate the accuracy of the boundaries of the intervertebral discs whose boundary thickness was defined as 1 pixel and 2 pixels, respectively. A modified Hausdorff distance (MHD) was also used to evaluate the segmentation accuracy [25]. Smaller MHD indicates the better segmentation performance. Paired *t*-test [26] was used to compare the results for three types of measurements; p-values below 0.05 were considered statistically significant.

Conventional U-net and dilated U-net were compared with BSU-net. Dilated U-net is a network in which dilated convolution is applied to conventional U-net. In the structure of dilated U-net used in this study, max-pooling layers used in conventional U-net are replaced with convolutional layers with stride 2, and dilated convolution blocks are placed before each convolutional layer with stride 2. Dilated convolution blocks are composed of three concatenated dilated convolutional layers whose rate is 1, 2, and 3 respectively, and a convolutional layer placed after them. Activation function (rectified linear unit (ReLU)) and batch normalization were used after each convolutional or dilated convolutional layer.

The proposed network and all the neural networks used in our experiments were trained and tested using Google tensorflow library based on python 2.7 (Google, Mountain View, CA, USA) [27]. The computing hardware used in the experiments were as follows: GPU, NVIDIA GeForce GTX 1080 (NVIDIA Corp., Santa Clara, CA, USA); CPU, 3.60 GHz Octa core (Xeon, Intel, Santa Clara, CA, USA); Memory, 32 GB. Hyper parameters applied to the experiments were as follows: Learning rate was 10^{-3}, total training epoch was 200, and optimizer was Adam. All images used as input for the networks were resized to 256×256 size matrix and normalized to values between 0 and 1.

3. Results

As shown in Table 1, both dilated U-net and BSU-net showed better results than conventional U-net in all DSC measurements. Furthermore, BSU-net showed better results than dilated U-net. As observed from these common trends, application of cascaded learning, BSU-pooling, and residual learning improved segmentation performance. In DSC measurement 1 (whole area segmentation), dilated U-net showed 2.02% higher DSC than conventional U-net and BSU-net showed a 3.00% higher DSC than conventional U-net. In DSC measurement 2 (boundary segmentation, thickness = 1 pixel), dilated U-net showed 8.29% higher DSC than conventional U-net and BSU-net showed 10.45% higher DSC than conventional U-net. In DSC measurement 3 (boundary segmentation, thickness = 2 pixels), dilated U-net showed 5.66% higher DSC than conventional U-net and BSU-net showed 7.34% higher DSC than conventional U-net. MHD results for three different networks showed similar trends (Table 2). Dilated U-net showed 0.03 mm lower MHD than conventional U-net and BSU-net showed 0.08 mm lower MHD than conventional U-net. Figure 5 compares the distributions of results according to the three DSC measurements and MHD measurement. In three DSC measurements, dilated U-net and BSU-net showed significant improvement in performance over conventional U-net. In DSC measurement 1, dilated U-net showed significantly increased DSC compared to conventional U-net ($p < 0.01$) and BSU-net showed significantly higher DSC compared to conventional U-net ($p < 0.001$).

In DSC measurements 2 and 3, both dilated U-net and BSU-net showed significantly higher DSC than conventional U-net ($p < 0.001$) On the other hand, in MHD measurement, dilated U-net showed no statistical difference compared to conventional U-net ($p > 0.05$), while BSU-net showed statistically significant MHD compared to conventional U-net ($p < 0.05$). Figure 6 shows the comparisons between three networks. It is noticeable that under-segmented area in the boundaries of intervertebral discs decreased in order of Figure 6b–d and correctly segmented area increased in order of Figure 6b–d. This indicates that BSU-net segmented more accurately than the other two networks.

Table 1. Dice similarity coefficient (DSC) measurements for the three different models. Accuracy for boundary area is very limited.

		Mean (%)	SD (%)
	U-net	86.44	2.24
Whole area segmentation	Dilated U-net	88.46	2.63
	BSU-net	89.44	2.14
Boundary segmentation (thickness = 1 pixel)	U-net	44.16	4.18
	Dilated U-net	52.45	4.08
	BSU-net	54.62	4.59
Boundary segmentation (thickness = 2 pixels)	U-net	67.51	3.59
	Dilated U-net	73.17	3.70
	BSU-net	74.85	3.20

Table 2. Modified Hausdorff distance (MHD) measurements for the three different models.

	Mean (mm)	SD (mm)
U-net	0.89	0.14
Dilated U-net	0.86	0.14
BSU-net	0.81	0.10

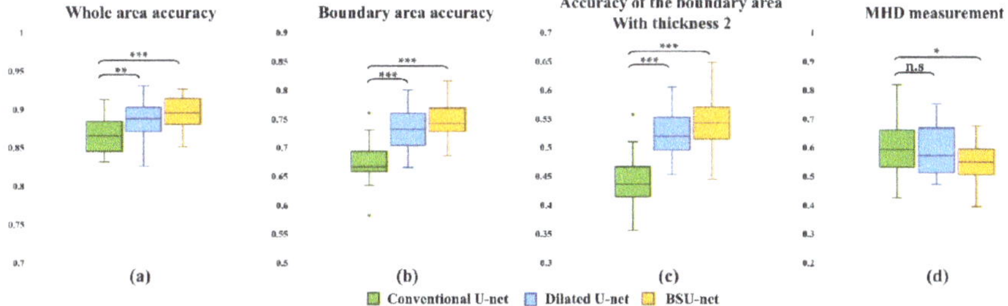

Figure 5. Segmentation results of networks. (**a**) Dice coefficients for whole area of intervertebral discs. (**b**) Dice coefficients of the boundaries of intervertebral discs whose thickness is defined as 1 pixel. (**c**) Dice coefficients of the boundaries of intervertebral discs whose thickness is defined as 2 pixels. (**d**) MHDs of intervertebral discs. A paired *t*-test was performed to calculate *p*-values. * denotes $p < 0.05$, ** denotes $p < 0.01$, *** denotes $p < 0.001$, and n.s. denotes not significant ($p > 0.05$).

BSU-net has three components: BSU-pooling layer, residual block, and cascaded network. Table 3 shows the results of five different networks including U-net, BSU-net and three different networks applying several BSU-net components (BSU-pooling layer, BSU-pooling layer and residual block, and cascaded learning network). When pooling layers of U-net were replaced with BSU-pooling layers, the results of three DSC measurements and MHD measurement were improved compared to conventional U-net. The applications of residual blocks and BSU-pooling layers (i.e., BSU-layers) to U-net improved the results of all DSC measurements compared to conventional U-net while there was little increasement of MHD result. Cascaded U-net has a similar structure to BSU-net, but conventional convolutional layers and pooling layers are used instead of BSU-layers. Cascaded U-net showed higher DSC and smaller MHD compared to conventional U-net. The application of each component improved the segmentation performance in most cases.

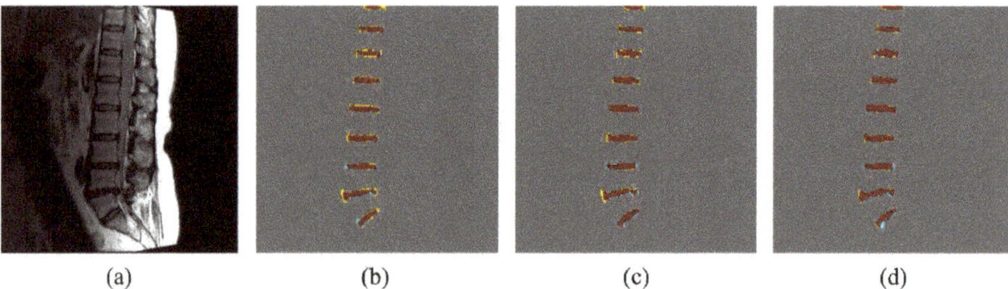

(a)   (b)   (c)   (d)

Figure 6. Segmentation result from networks. Brown area, yellow area, and blue area denote correctly segmented area, under-segmented area, and over segmented area, respectively. (**a**) Input image. (**b**) U-net result. (**c**) Dilated U-net result. (**d**) BSU-net result.

Table 3. DSC and MHD measurements for five different networks including conventional U-net, BSU-net and three different networks applying several components of BSU-net.

	DSC (%)			MHD (mm)
	Measurement 1	**Measurement 2**	**Measurement 3**	
Conventional U-net	86.44 ± 2.24	44.16 ± 4.18	67.51 ± 3.59	0.89 ± 0.14
U-net + BSU-pooling layer	87.30 ± 3.16	50.68 ± 5.50	71.68 ± 4.76	0.88 ± 0.14
U-net + BSU-layer	87.19 ± 2.67	51.88 ± 5.67	71.68 ± 5.48	0.90 ± 0.18
Cascaded U-net	87.70 ± 4.00	50.25 ± 8.68	71.33 ± 7.63	0.86 ± 0.17
BSU-net	89.44 ± 2.14	54.62 ± 4.59	74.85 ± 3.20	0.81 ± 0.10

Figures 7–9 show the results of the five different networks in Table 3. Figure 7b–d shows segmentation results of conventional U-net, U-net applying BSU-layers, and BSU-net, respectively. U-net applying BSU-layers segmented more delicately than conventional U-net, but there are some incorrectly segmented areas. On the other hand, the results of BSU-net have detailed boundaries and no incorrectly segmented area. Figure 8b–d shows segmentation results of conventional U-net, cascaded U-net, and BSU-net, respectively. The white pixels represent estimated boundary pixels that are perfectly matched with true boundary labels. It is easily noticeable that cascaded U-net found a higher number of true boundary pixels than conventional U-net, and BSU-net detected the most among the three different networks. The enlarged views at the bottom of Figure 8 clearly show the results from each and demonstrate the improved performance of BSU-net. Figure 9b–d also shows segmentation results of conventional U-net, cascaded U-net, and BSU-net, respectively. In this case, cascaded U-net did not properly segment intervertebral disc, and its results are worse than those of conventional U-net. In some cases of cascaded U-net, it segmented intervertebral discs smaller than their actual size. On the other hand, BSU-net showed successful performance in these cases. Standard deviations in Table 3 shows the stability of BSU-net. Standard deviations of BSU-net are the lowest in most accuracy measurements while those of cascaded U-net are the highest in most accuracy measurements.

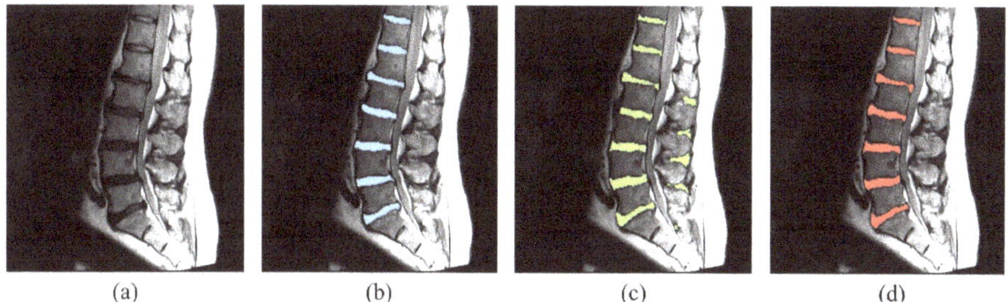

Figure 7. Segmentation results of the networks overlaid on the input image. (**a**) The input magnetic resonance (MR) image. (**b**) The input MR image with U-net segmentation result. (**c**) The input MR image with the result from the modified U-net which is U-net whose convolutional and pooling layers are replaced with BSU-layers. (**d**) The input MR image with BSU-net result.

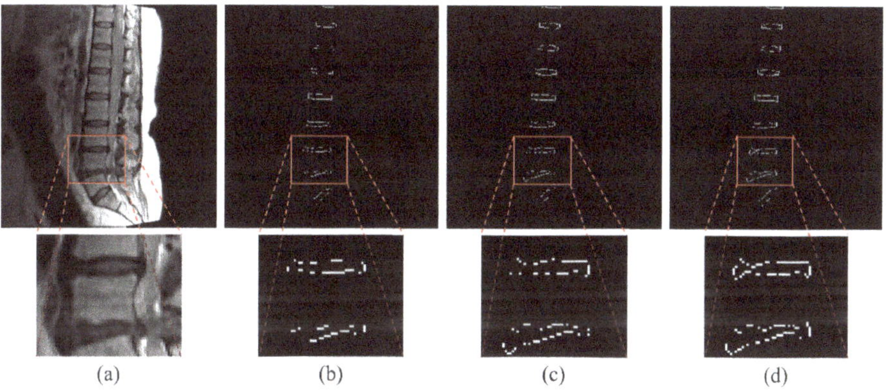

Figure 8. Segmentation results. (**a**) Input MR spine image. (**b**) Boundary segmentation result from U-net. (**c**) Boundary segmentation result from cascaded U-net. (**d**) Boundary segmentation result from BSU-net. White pixels correspond to boundary pixels that were perfectly matched with true boundary labels. BSU-net preserved more boundaries than other models.

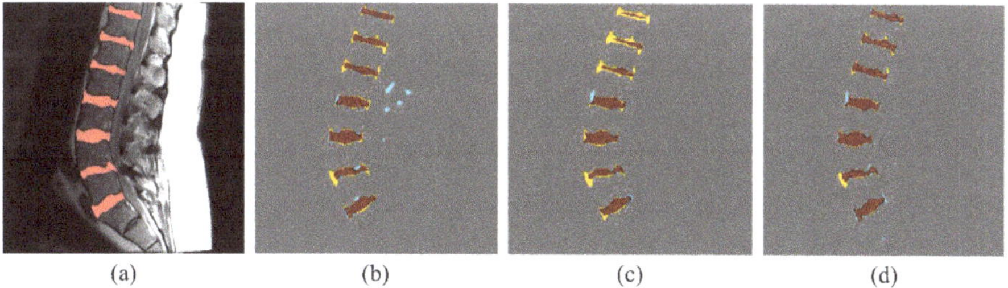

Figure 9. Segmentation results from all networks illustrating the outlier case of cascaded U-net. Brown area, yellow area, and blue area denote correctly segmented area, under-segmented area, and over segmented area, respectively. (**a**) Input image with label. (**b**) U-net result. (**c**) Cascaded U-net result. (**d**) BSU-net result.

4. Discussion

Conventional U-net is a commonly used deep learning network that displays good performance in various kinds of studies. It is used for segmentation of organs and cancers in various types of medical images [28–30], and it is also used for object segmentation of optical images [31].

However, conventional U-net has limited ability for detailed boundary segmentation [10] due to the structural limitations of a max-pooling layer that plays a key role in feature extraction process. It is not suitable for segmentation of objects with complex boundaries, such as intervertebral discs. The purpose of our proposed network, BSU-net, is to improve the pooling layer of conventional U-net. In this paper, BSU-net showed a better performance than conventional U-net for intervertebral disc segmentation in MR spine images. This indicates that BSU-net can perform more precise and fine-grain segmentation than conventional U-net. BSU-net will be of value in MR studies where quantitative MR values of disc need to be determined.

As shown in Tables 1 and 2 and Figure 5, dilated U-net performed better than conventional U-net and BSU-net showed better performance than dilated U-net. In most accuracy measurements, dilated U-net showed statistically significant performance improvement, but the improvement in MHD measurement was quite small. MHD indicates the accuracy of boundaries because it is based on the distances between obtained boundaries and reference boundaries. This indicates that the results of dilated U-net have many incorrectly segmented areas. Figure 10 shows the results of dilated U-net and BSU-net. There are some incorrectly segmented areas in the results of dilated U-net while the results of BSU-net have no incorrectly segmented areas. This is because the feature extraction process of dilated U-net did not remove unnecessary information compared to BSU-net. The number of trainable parameters used in BSU-net is 53,740,674 which is approximately 22% lower than dilated U-net (69,048,584) and approximately 14% lower than conventional U-net (62,803,650). This indicates that BSU-net performed successful fine-grain segmentation efficiently.

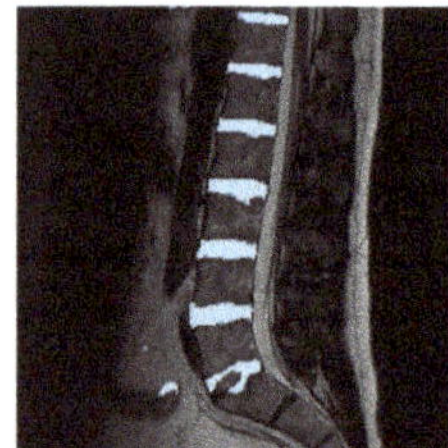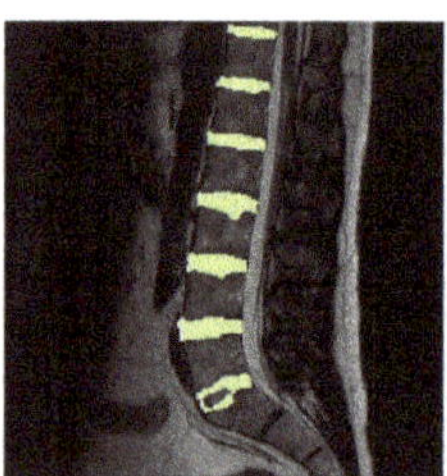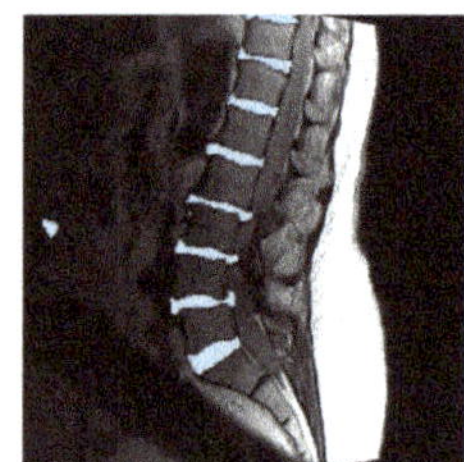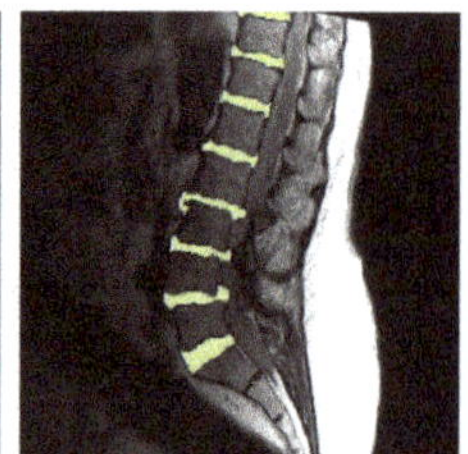

Figure 10. Comparison between dilated U-net and BSU-net. Blue area denotes segmentation results of dilated U-net and green area denotes segmentation results of BSU-net.

The components of the BSU-net are the BSU-pooling layer and residual block, and cascaded network. As shown in Table 3, the application of each component contributed to performance enhancement. The performance improvement of applying residual blocks is much smaller than those of applying other components. However, the number of trainable parameters were approximately 12% decreased. Therefore, the application of residual blocks brought efficiency to the entire learning.

When BSU-layers were applied to U-net, the result of DSC measurement 1 was only 0.74% higher than conventional U-net. The application of BSU-layers brought improved performance in terms of fine-grain segmentation, given the fact that the result of DSC measurement 2 was 7.72% higher than conventional U-net and the result of accuracy measurement 3 was 4.18% higher than conventional U-net. However, the MHD result of U-net applying BSU-layers is worse than conventional U-net. These results indicate that the results of U-net applying BSU-layers had many incorrectly segmented areas. Figure 7 shows many incorrectly segmented areas in the results of U-net applying BSU-layers and they decreased the accuracy of whole segmented areas. These incorrectly segmented areas occurred because BSU-layers preserved the detailed information which was discarded in the feature extraction process in conventional U-net. The retention of this information affected the performance of the network. Therefore, in order to fully utilize the advantages of BSU-layers, there is a need for a guiding mechanism that can discard unnecessary parts and narrow the target area into proper regions. Cascaded learning method can use the outcomes of conventional U-net to effectively guide BSU-layers to focus on the proper regions. This is the reason why BSU-net, which combines cascaded learning method and BSU-layers at the same time, can achieve a high performance. Figure 7d shows the successful segmentation results of BSU-net without incorrectly segmented area. Appropriate guidance for BSU-layers improved the efficiency of the entire network.

In general, cascaded learning uses the outcomes of former networks as inputs at the beginning of following networks [17–19]. However, cascaded learning applied to BSU-net puts the outcomes of conventional U-net at the back-end rather than the beginning of the following network. This is because detailed information of conventional U-net outcomes disappeared during the pooling process in the encoding part of the network. A network showed 1.67%, 4.01%, and 2.92% lower accuracy for three DSC measurements respectively when the outcomes of conventional U-net were put into the initial part of the following network.

As shown in Table 3, standard deviations of cascaded U-net are highest in most accuracy measurements. Figure 9 also shows the unstable performance of cascaded U-net. For eight out of the 25 cases, cascaded U-net showed over 1% lower accuracy than conventional U-net in all eight cases; two of these showed more than 7% lower accuracy. Contrastingly, BSU-net showed lower accuracy than conventional U-net in just one case where the difference is smaller than 1%. This is because important information pertaining to the boundary areas was discarded during the feature extraction process in cascaded U-net. The loss of important information in the max-pooling process is a noticeable problem. On the other hand, BSU-net distinguished most intervertebral disc areas correctly, while unsegmented areas and over-segmented areas did not deviate much from the actual boundaries. These results also indicate that the application of BSU-layers to cascaded U-net provides stability and generality to the network. Furthermore, the use of BSU-layers enables efficient training of the network. Cascaded U-net used in our experiments has 63,912,898 trainable parameters in a total of 42 convolutional layers (3×3 convolutional layers: 41 and 1×1 convolutional layer: 1), while BSU-net has 53,740,674 trainable parameters, approximately 16% less than that in cascaded U-net, in a total of 79 convolutional layers (3×3 convolutional layers: 35 and 1×1 convolutional layer: 44).

5. Conclusions

Intervertebral disc segmentation in MR images is challenging owing to their complex shapes and non-uniform intensity. This study introduces a robust deep-learning segmentation network, boundary specific U-net (BSU-net), which can successfully segment intervertebral discs with complex boundaries.

Conventional U-net is a deep learning segmentation algorithm for image segmentation which is commonly used in various fields. However, conventional U-net is not suitable for intervertebral disc segmentation because its performance with respect to the details of segmentation (such as the boundaries) is still limited owing to the structural limitations of the max-pooling layer that plays a key role in the feature extraction process in conventional U-net. The proposed BSU-net can overcome the limitations of conventional U-net and achieve fine-grain segmentation. BSU-net uses modified convolutional and pooling layers and applies cascaded learning method to overcome the structural limitations of conventional U-net. BSU-net performed intervertebral discs segmentation in MR spine images with higher accuracy than conventional U-net, especially in the boundary areas.

Obtaining specific information about intervertebral discs is of great help for the diagnosis and treatment of lumbar diseases. In many translational studies with real patients, quantitative MRI such as T_2 mapping is used to show treatment efficiency or track subtle changes over time. BSU-net, though not clinically applicable at this time, will be of great value in translational MR studies where quantitative MR values of the disc need to be determined using regions of interest. Our finding of 89% Dice similarity coefficient of BSU-net against human annotator compares favorably with inter-observer agreement of about 80% [32].

Author Contributions: W.C.B., K.M., and C.B.C. proposed the idea and contributed to data acquisition and performed manual segmentation. S.K. contributed to performing data analysis, algorithm construction,

Appl. Sci. **2018**, *8*, 1656

and writing the article. D.H. technically supported the algorithm and evaluation and also professionally reviewed and edited the paper.

Funding: This research was supported in parts by the National Research Foundation of Korea (NRF) grant funded by the Korean government (MSIP) (2016R1A2B4015016) in support of Dosik Hwang, and National Institute of Arthritis and Musculoskeletal and Skin Diseases of the National Institutes of Health in support of Won C. Bae (Grant Number R01 AR066622). The contents of this paper are the sole responsibility of the authors and do not necessarily represent the official views of the sponsoring institutions.

Acknowledgments: The authors thank Yohan Jun, and Hyungseob Shin for their professional preliminary reviews.

Conflicts of Interest: The authors declare no conflict of interest.

References

1. Luoma, K.; Riihimäki, H.; Luukkonen, R.; Raininko, R.; Viikari-Juntura, E.; Lamminen, A. Low back pain in relation to lumbar disc degeneration. *Spine* **2000**, *25*, 487–492. [CrossRef] [PubMed]
2. Modic, M.T.; Steinberg, P.M.; Ross, J.S.; Masaryk, T.J.; Carter, J.R. Degenerative disk disease: Assessment of changes in vertebral body marrow with MR imaging. *Radiology* **1988**, *166*, 193–199. [CrossRef] [PubMed]
3. Ayed, I.B.; Punithakumar, K.; Garvin, G.; Romano, W.; Li, S. Graph cuts with invariant object-interaction priors: Application to intervertebral disc segmentation. In Proceedings of the Biennial International Conference on Information Processing in Medical Imaging, Kloster Irsee, Germany, 3–8 July 2011; Springer: Berlin/Heidelberg, Germany, 2011; pp. 221–232.
4. Michopoulou, S.K.; Costaridou, L.; Panagiotopoulos, E.; Speller, R.; Panayiotakis, G.; Todd-Pokropek, A. Atlas-based segmentation of degenerated lumbar intervertebral discs from MR images of the spine. *IEEE Trans. Biomed. Eng.* **2009**, *56*, 2225–2231. [CrossRef] [PubMed]
5. Law, M.W.; Tay, K.; Leung, A.; Garvin, G.J.; Li, S. Intervertebral disc segmentation in MR images using anisotropic oriented flux. *Med. Image Anal* **2013**, *17*, 43–61. [CrossRef] [PubMed]
6. Haq, R.; Besachio, D.A.; Borgie, R.C.; Audette, M.A. Using shape-aware models for lumbar spine intervertebral disc segmentation. In Proceedings of the 22nd International Conference on Pattern Recognition (ICPR), Stockholm, Sweden, 24–28 August 2014; IEEE: Piscataway, NJ, USA, 2014; pp. 3191–3196.
7. Mansour, R.F. Deep-learning-based automatic computer-aided diagnosis system for diabetic retinopathy. *Biomed. Eng. Lett.* **2018**, *8*, 41–57. [CrossRef]
8. Ji, X.; Zheng, G.; Belavy, D.; Ni, D. Automated intervertebral disc segmentation using deep convolutional neural networks. In Proceedings of the International Workshop on Computational Methods and Clinical Applications for Spine Imaging, Athens, Greece, 17 October 2016; Springer: Cham, Switzerland, 2016; pp. 38–48.
9. Ronneberger, O.; Fischer, P.; Brox, T. U-net: Convolutional networks for biomedical image segmentation. In Proceedings of the International Conference on Medical Image Computing and Computer-Assisted Intervention, Munich, Germany, 5–9 October 2015; Springer: Cham, Switzerland, 2015; pp. 234–241.
10. Ye, J.C.; Han, Y.; Cha, E. Deep convolutional framelets: A general deep learning framework for inverse problems. *SIAM J. Imaging Sci.* **2018**, *11*, 991–1048. [CrossRef]
11. LeCun, Y.; Bengio, Y.; Hinton, G. Deep learning. *Nature* **2015**, *521*, 436. [CrossRef] [PubMed]
12. Dice, L.R. Measures of the amount of ecologic association between species. *Ecology* **1945**, *26*, 297–302. [CrossRef]
13. Chen, L.C.; Papandreou, G.; Kokkinos, I.; Murphy, K.; Yuille, A.L. Deeplab: Semantic image segmentation with deep convolutional nets, atrous convolution, and fully connected CRFs. *IEEE Trans. Pattern Anal. Mach. Intell.* **2018**, *40*, 834–848. [CrossRef] [PubMed]
14. Yu, F.; Koltun, V. Multi-scale context aggregation by dilated convolutions. *arXiv*, 2015; arXiv:1511.07122.
15. Kim, S.; Bae, W.C.; Hwang, D. Automatic delicate segmentation of the intervertebral discs from MR spine images using deep convolutional neural networks: ICU-net. In Proceedings of the 26th Annual Meeting of ISMRM, Paris, France, 16–21 June 2018; p. 5401.
16. He, K.; Zhang, X.; Ren, S.; Sun, J. Deep residual learning for image recognition. In Proceedings of the IEEE Conference on Computer Vision and Pattern Recognition, Las Vegas, NV, USA, 26 June–1 July 2016; pp. 770–778.

17. Qin, H.; Yan, J.; Li, X.; Hu, X. Joint training of cascaded CNN for face detection. In Proceedings of the IEEE Conference on Computer Vision and Pattern Recognition, Las Vegas, NV, USA, 26 June–1 July 2016; pp. 3456–3465.

18. Eo, T.; Jun, Y.; Kim, T.; Jang, J.; Lee, H.J.; Hwang, D. KIKI-net: Cross-domain convolutional neural networks for reconstructing undersampled magnetic resonance images. *Magn. Reson. Med.* **2018**. [CrossRef] [PubMed]

19. Christ, P.F.; Elshaer, M.E.A.; Ettlinger, F.; Tatavarty, S.; Bickel, M.; Bilic, P.; Rempfler, M.; Armbruster, M.; Hofmann, F.; D'Anastasi, M.; et al. Automatic liver and lesion segmentation in CT using cascaded fully convolutional neural networks and 3D conditional random fields. In Proceedings of the International Conference on Medical Image Computing and Computer-Assisted Intervention, Athens, Greece, 17–21 October 2016; Springer: Cham, Switzerland, 2016; pp. 415–423.

20. Liu, M.; Zhang, D.; Shen, D. Alzheimer's Disease Neuroimaging Initiative. Ensemble sparse classification of Alzheimer's disease. *NeuroImage* **2012**, *60*, 1106–1116. [CrossRef] [PubMed]

21. Liao, R.; Tao, X.; Li, R.; Ma, Z.; Jia, J. Video super-resolution via deep draft-ensemble learning. In Proceedings of the IEEE International Conference on Computer Vision, Santiago, Chile, 11–18 December 2015; pp. 531–539.

22. Deng, L.; Platt, J.C. Ensemble deep learning for speech recognition. In Proceedings of the Fifteenth Annual Conference of the International Speech Communication Association, Singapore, 14–18 September 2014; pp. 1915–1919.

23. Cai, Y.; Osman, S.; Sharma, M.; Landis, M.; Li, S. Multi-modality vertebra recognition in arbitrary views using 3d deformable hierarchical model. *IEEE Trans. Med. Imaging* **2015**, *34*, 1676–1693. [CrossRef] [PubMed]

24. Spineweb. Available online: http://spineweb.digitalimaginggroup.ca/ (accessed on 13 September 2018).

25. Dubuisson, M.P.; Jain, A.K. A modified Hausdorff distance for object matching. In Proceedings of the 12th International Conference on Pattern Recognition, Jerusalem, Israel, 9–13 October 1994; IEEE: Piscataway, NJ, USA, 1994; pp. 566–568.

26. McDonald, J.H. *Handbook of Biological Statistics*, 2nd ed.; Sparky House: Baltimore, MD, USA, 2009; Volume 2, pp. 173–181.

27. TensorFlow. Available online: http://www.tensorflow.org/ (accessed on 13 September 2018).

28. Yu, L.; Yang, X.; Chen, H.; Qin, J.; Heng, P.A. Volumetric ConvNets with Mixed Residual Connections for Automated Prostate Segmentation from 3D MR Images. In Proceedings of the Thirty-First AAAI Conference on Artificial Intelligence, San Francisco, CA, USA, 4–9 February 2017; pp. 66–72.

29. Christ, P.F.; Ettlinger, F.; Grün, F.; Elshaera, M.E.A.; Lipkova, J.; Schlecht, S.; Ahmaddy, F.; Tatavarty, S.; Bickel, M.; Bilic, P.; et al. Automatic liver and tumor segmentation of CT and MRI volumes using cascaded fully convolutional neural networks. *arXiv*, 2017; arXiv:1702.05970.

30. Yuan, Y.; Chao, M.; Lo, Y.C. Automatic skin lesion segmentation using deep fully convolutional networks with jaccard distance. *IEEE Trans. Med. Imaging* **2017**, *36*, 1876–1886. [CrossRef] [PubMed]

31. Oliveira, G.L.; Burgard, W.; Brox, T. Efficient deep models for monocular road segmentation. In *Intelligent Robots and Systems (IROS), Proceedings of the IEEE/RSJ International Conference on Intelligent Robots and Systems (IROS), Daejeon, Korea, 9–14 October 2016*; IEEE: Piscataway, NJ, USA, 2016; pp. 4885–4891.

32. Claudia, C.; Farida, C.; Guy, G.; Marie-Claude, M.; Carl-Eric, A. Quantitative evaluation of an automatic segmentation method for 3D reconstruction of intervertebral scoliotic disks from MR images. *BMC Med. Imaging* **2012**, *12*, 26. [CrossRef] [PubMed]

Article

Semi-Automatic Segmentation of Vertebral Bodies in MR Images of Human Lumbar Spines

Sewon Kim [1], Won C. Bae [2,3], Koichi Masuda [4], Christine B. Chung [2,3] and Dosik Hwang [1,*]

[1] School of Electrical and Electronic Engineering, Yonsei University, Seoul 06974, Korea;
    sewon.kim@yonsei.ac.kr
[2] Department of Radiology, VA San Diego Healthcare System, San Diego, CA 92161-0114, USA;
    wbae@ucsd.edu (W.C.B.); cbchung@ucsd.edu (C.B.C.)
[3] Department of Radiology, University of California-San Diego, La Jolla, CA 92093-0997, USA
[4] Department of Orthopedic Surgery, University of California-San Diego, La Jolla, CA 92037, USA;
    koichimasuda@ucsd.edu
* Correspondence: dosik.hwang@yonsei.ac.kr; Tel.: +82-2-2123-5771

Received: 13 August 2018; Accepted: 5 September 2018; Published: 7 September 2018

Abstract: We propose a semi-automatic algorithm for the segmentation of vertebral bodies in magnetic resonance (MR) images of the human lumbar spine. Quantitative analysis of spine MR images often necessitate segmentation of the image into specific regions representing anatomic structures of interest. Existing algorithms for vertebral body segmentation require heavy inputs from the user, which is a disadvantage. For example, the user needs to define individual regions of interest (ROIs) for each vertebral body, and specify parameters for the segmentation algorithm. To overcome these drawbacks, we developed a semi-automatic algorithm that considerably reduces the need for user inputs. First, we simplified the ROI placement procedure by reducing the requirement to only one ROI, which includes a vertebral body; subsequently, a correlation algorithm is used to identify the remaining vertebral bodies and to automatically detect the ROIs. Second, the detected ROIs are adjusted to facilitate the subsequent segmentation process. Third, the segmentation is performed via graph-based and line-based segmentation algorithms. We tested our algorithm on sagittal MR images of the lumbar spine and achieved a 90% dice similarity coefficient, when compared with manual segmentation. Our new semi-automatic method significantly reduces the user's role while achieving good segmentation accuracy.

Keywords: semi-automatic segmentation; MR spine image; vertebral body; graph-based segmentation; correlation

1. Introduction

Low back pain is a common disease in modern society [1,2]. As a multifactorial disease, numerous lumbar components including intervertebral discs, paraspinal muscle, or alterations of the vertebral body may contribute to low back pain. Magnetic resonance (MR) imaging is a noninvasive imaging modality that is widely used for both morphological and quantitative evaluation of the human lumbar spine. Evaluation of the vertebral bodies in MR images (Figure 1a) plays a key role in the diagnosis and establishing treatment strategies. Vertebral body segmentation on MR images provides clinically useful information including quantitative biomarkers, volume, and shape. Previous methods for vertebral body segmentation are inherently challenging, owing to the similar signal intensity of the vertebral body and anatomically contiguous tissues and the inconsistent boundaries [3,4].

The segmentation of grayscale medical images has been extensively researched [5]. Some of the classical methods are described here. Histogram-based segmentation is one of the most commonly used methods [6,7]; it uses only pixel intensity to segment the image by applying a threshold. However, this method faces many limitations, because it does not include shape or position information. Another conventional method is the region-growing method that begins by placing a seed [8–11]. This seed point grows on the basis of the similarity of its neighboring pixels and extends out to fill a region of interest (ROI). This method is also based on pixel intensity, which is disadvantageous in the case of noisy images. Another approach employs an edge detection algorithm. For example, the Canny edge-detection algorithm obtains vertical and horizontal gradients using various kernels after limitation of noise via medial or Gaussian filters [12]. Subsequently, non-maximum suppression is used to make the edge thinner and more precise, whereas thresholding and edge tracking further improves the edge. Clustering is another commonly used method. The K-means clustering algorithm determines the number of clusters, k, with the initial center position of the clusters and repeats the process until the center position of the clusters and their corresponding data converge [13–15]. The fuzzy C-means algorithm is another clustering method that sets the number of clusters and the objective function with initial values and repeats the calculation until the objective function is minimized [16–19].

The actively investigated segmentation method in recent medical imaging is the graph-cut algorithm. This algorithm transforms an image into a graph G = (V, E), where V (node) is the actual spatial element to be segmented and E (edge) is the similarity of each pixel [20–24]. Depending on how V and E are specified, various graphs can be created. The generated graph is transformed into a similarity matrix to perform division. Various algorithms, such as the min-cut max-flow algorithm, are used for the division [25,26]. Use of the graph-cut algorithm for medical image segmentation has shown good results. Egger and Kapur reported a 91% Dice similarity coefficient (DSC) [27], whereas Schwarzenberg and Freisleben reported an 81% DSC using the graph-cut method for vertebral body segmentation [3]. However, in both these studies, each vertebral body was individually segmented, and the method required repeated user inputs to assign a seed point at the center of each vertebral body. Furthermore, because the performance of the algorithm depends on the number of nodes and the connection method, the method requires the user to calibrate multiple parameters for each segmentation.

The most recent approaches are the deep-learning-based methods, which exhibit superior performance over existing mathematical algorithms. Ronneberger reported successful segmentation of medical images using a U-net image segmentation network [28]. Korez also showed the possibility of a deep learning segmentation for vertebral bodies in MR lumbar spine images [29]. However, these deep neural networks require a large amount of data and use a large amount of memory, which is a key limitation. For example, it might be difficult for a single graphics processing unit (GPU), sold in the market, to learn and use the U-net with a 512×512 size image matrix. Furthermore, a repeated heavy training processes may be required for the segmentation of different types of images (e.g., use of different contrasts, sequences, or different MR scanners). On the other hand, mathematical algorithms do not require a large amount of training datasets and can be applied to large sized images. Unlike common optical images, the acquisition of a large amount of data is a key limitation in medical imaging. Therefore, the segmentation methods that employ mathematical algorithms for medical image processing are still needed.

Our proposed method aims to reduce user inputs to a minimum while still achieving high accuracy in a short execution time during the segmentation of vertebral bodies. We first reduce the number of user inputs for vertebral body selection (i.e., a rectangular ROI containing a single vertebral body) to just one for the whole image. The remaining vertebral bodies are automatically detected using a correlation algorithm without any further user inputs. Subsequently, the boundary of the vertebral body inside each ROI is automatically segmented using the graph-cut and line-based methods together with the incorporation of the Hough transform and edge-detection algorithm. No additional user input is required. A preliminary study of this paper was partially presented at the annual meeting of International Conference on Electronics, Information, and Communication in 2017 [30].

This paper comprises four sections. Section 2 describes the details of our proposed algorithm. Section 3 presents the experimental results. Finally, Section 4 discusses the study as a whole and concludes the article.

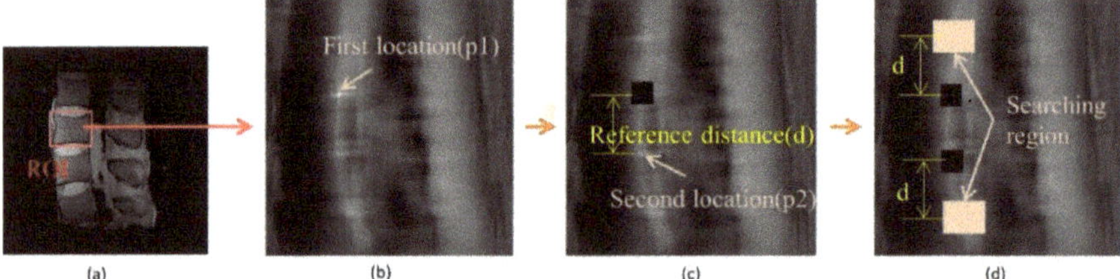

Figure 1. Protocol to find vertebral body region of interest (ROIs) using minimum user input. (**a**) T2-weighted spin echo MR image of a cadaveric lumbar spine and a user-defined ROI placement. (**b**) Correlation map and the location of the user-defined ROI (the first location, p1 by the maximum value). (**c**) Determination of the second location, p2, of the next closest vertebral body and the reference distance, d. (**d**) Detection of the other remaining ROI locations utilizing the reference distance, p1, and p2. Black pixels in (**c,d**) are disregarded pixels for the next ROI search.

2. Materials and Methods

2.1. Materials

This cadaveric study was exempted from institutional review board approval. Nineteen lumbar spines (L1 through L5) from cadaveric donors (age range 46–60 years) were obtained from a local tissue bank. MR imaging was performed on a 3-Tesla system (General Electric Signa HDx, San Diego, CA, USA). The cadaveric spines were placed in the supine position to ensure vertical alignment of the vertebral bodies in the MR images. Our segmentation algorithm was developed on the basis of this position, which is typically used in many MR imaging protocols. A T2-weighted fast spin echo sequence was used with the following parameters: Mid-sagittal plane; repetition time (TR) = 2000 ms; echo time (TE) = 7.6 ms; field of view (FOV) = 180 − 220 mm; pixel spacing = 0.39 mm; slice thickness = 3 mm; flip angle = 90°; and bandwidth = ±62.5 kHz. Image processing and test experiments were conducted using Matlab R2012b (The Mathworks Inc., Natick, MA, USA).

2.2. ROI Detection

The first part of our algorithm requires minimal user inputs regarding the location of one of the vertebral bodies and the size of the rectangular ROI window (width: lx and height: ly); the automated algorithm then searches other locations with the same window for the remaining vertebral bodies. This ROI detection step is necessary to increase the efficiency of the subsequent segmentation process by setting uniform ROIs, each of which contains only one vertebral body. When the user defines a rectangular area containing a single vertebral body along with some of the surrounding tissues, as shown in Figure 1a, the correlation algorithm [31] obtains a correlation map (Figure 1b) between the user-defined ROI and other candidate ROIs throughout the image.

When examining the correlation map (Figure 1b), the highest correlation value is easily found at the center position (p1) of the user specified ROI. The next vertebral body location is searched by finding the second highest correlation value. However, the pixels in the immediate vicinity of p1 tend to have high correlation values that are close to the correlation value of p1. Therefore, correlation values in the immediate neighborhood of p1 (equivalent to 1/3 of the ROI size) are disregarded. Then, the search continues for the next highest correlation value and the corresponding location of the next vertebral body, p2. After finding p2, the neighborhood of p2 in the correlation map is also disregarded to search for the next vertebral body locations. In this case, we use a reference distance, d, defined as the distance between p1 and p2 (Figure 1c) for subsequent searches. The searching area can be reduced to the small area whose width and height are lx and ly, respectively, because the reference distance provides the approximate locations of the upper or lower vertebral bodies. Through this process, vertebral body positions are found in the entire image and each ROI for the individual vertebral body is set for subsequent fine segmentation, as described in the next section.

The information on the intervertebral disc can also be used to further refine the vertebral body location. The searched locations are corrected again using these intervertebral discs that are sandwiched between the vertebral bodies. Assuming that an intervertebral disc ROI exists in the lx/2 portion just above the vertebral body ROI, locations of other intervertebral discs can also be obtained in the same manner (Figure 1). The horizontal locations of the vertebral bodies and the intervertebral discs are generally similar. Therefore, if the horizontal location of a searched vertebral body ROI is significantly different from the horizontal location of the surrounding intervertebral disc, the horizontal location of a searched vertebral body ROI needs to be adjusted. When the difference between the horizontal locations of the searched vertebral body and the intervertebral disc is larger than 10% of ly, the horizontal location of the searched vertebral body ROI is replaced with that of the searched intervertebral disc ROI.

2.3. Roi Fine Tuning—Hough Transform and Canny Edge Filtering

The vertebral bodies are slightly different in size and these may be rotated with respect to each other due to kyphosis or lordosis [32]. Therefore, after identifying ROIs for the vertebral bodies, the ROIs need to be fine-tuned to make the orientation and size of the vertebral bodies similar to each other for ease of segmentation. Canny edge filtering and the Hough transform [12,33] are used for this purpose.

2.3.1. ROI Fine Tuning—Orientation Adjustment

First, Canny edge filtering is applied to the vertebral body region **R** to extract the edge components **E** of the region. Second, a Hough transform is applied to **E** to estimate the rotation angle of the vertebral body. **H** is the Hough transformed image of **E**, as shown in Figure 2c. The x-axis of **H** is a rotation angle, and the y-axis of **H** is the distance from origin. Assuming that the shape of the vertebral body is similar to a rectangle, the y-axis value of **H** has the smallest value at $0°$. However, when the vertebral body rotates, the x-coordinates of the smallest **H** deviates from $0°$; the degree of deviation indicates the rotation angle of the vertebral body. Therefore, the ROI can be adjusted by rotating it in the direction opposite to that of the estimated rotation angle.

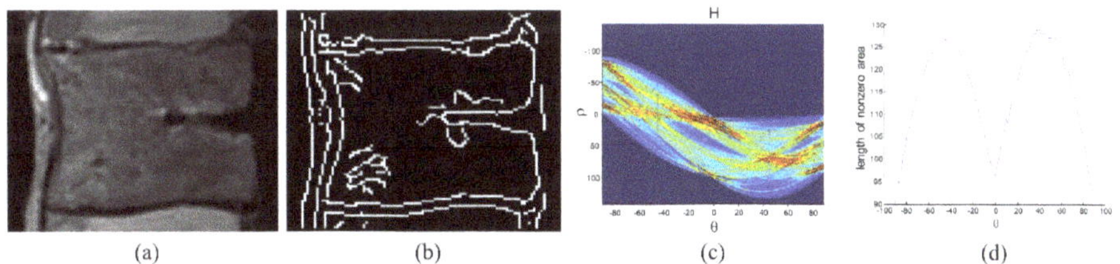

(a)   (b)   (c)   (d)

Figure 2. Analysis for the region of interest (ROI) refinement of the vertebral body. (**a**) Vertebral body ROI. (**b**) Edge of the vertebral body ROI. (**c**) Hough transformed image of the vertebral body ROI. (**d**) Length of nonzero area at each angle.

2.3.2. ROI fine Tuning—Boundary Adjustment

Define $\mathbf{Px}(x)$ as the result of projecting **E** on the x-axis, and $\mathbf{Py}(y)$ as the result of projecting **E** on the y-axis. and represent the size of the vertebral body ROI. As the vertebral body is rectangular, the minimum values of **Px** and **Py** are $\mathbf{Px}\left(\frac{lx}{2}\right)$ and $\mathbf{Py}\left(\frac{ly}{2}\right)$. On the other hand, the projection of the left and right sides of the vertebral body edge will be at both ends of **Px**, having considerably larger values than $\mathbf{Px}\left(\frac{lx}{2}\right)$. Similarly, the point at which the upper and lower sides of the vertebral body edge are projected will be at both ends of $\mathbf{Py}$, having larger values than $\mathbf{Py}\left(\frac{ly}{2}\right)$. Therefore, if $max\left(\mathbf{Px}\left(2 < i < \frac{lx}{2}\right)\right) < \alpha\mathbf{Px}\left(\frac{lx}{2}\right)$, it implies that the ROI does not include the left side of the vertebral body, and hence, the ROI should extend to the left. Likewise, if $max\left(\mathbf{Px}\left(\frac{lx}{2} < i < lx - 2\right)\right) < \alpha\mathbf{Px}\left(\frac{lx}{2}\right)$, the ROI should extend to the right. This rule is also applied to the relationships between $\mathbf{Py}$ and both the upper and lower sides of the ROI, such that the ROI can contain the vertebral body within a specified interval. This process allows us to adjust vertebral body ROIs so that they contain all the necessary boundaries of the vertebral body, thereby increasing the effectiveness of the subsequent segmentation.

2.4. Segmentation

During the ROI fine tuning process, we can determine the approximate position of the vertebral body boundary.

$$bL = \underset{i}{argmax}\left(\mathbf{Px}\left(1 < i < \frac{lx}{2}\right)\right) \tag{1}$$

$$bR = \underset{i}{argmax}\left(\mathbf{Px}\left(\frac{lx}{2} < i \leq lx\right)\right) \tag{2}$$

$$bU = \underset{i}{argmax}\left(\mathbf{Py}\left(1 < j < \frac{ly}{2}\right)\right) \tag{3}$$

$$bD = \underset{i}{argmax}\left(\mathbf{Py}\left(\frac{ly}{2} < j \leq ly\right)\right) \tag{4}$$

where $(bL, bU), (bR, bU), (bL, bD)$, and (bR, bD) are approximate vertices of the vertebral body boundary. Then, we draw lines connecting each dot at points spaced inward and outward from the approximate vertices, as shown in Figure 3a. Using these points, the vertebral body ROI is divided into eight areas, as illustrated in Figure 3b, and different segmentation methods are applied to these areas. Figure 4 shows the flowchart of the segmentation process. Areas $\mathbf{U}$ and $\mathbf{D}$ detect boundaries using the graph-cut based method. $\mathbf{U}$ and $\mathbf{D}$ are converted to graph $\mathbf{G} = (\mathbf{V}, \mathbf{E})$. As $\mathbf{G}$ should be divided into the top and bottom, the nodes $n \in \mathbf{V}$ and edges $e \in \mathbf{E}$ are set along the x-axis. Intervals between each pixel along the x-axis are set to the nodes $n \in \mathbf{V}$ and pixel intensities between the nodes are set to the edges $e \in \mathbf{E}$ [25]. As a result, boundaries that can properly distinguish the top and bottom in the x-direction of $\mathbf{U}$ and $\mathbf{D}$ are detected.

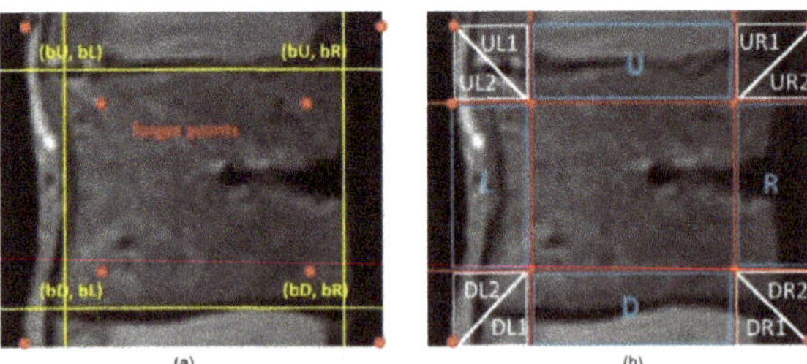

Figure 3. Partitioning area. (**a**) Pointing base vertices. (**b**) Partitioned area.

Because $\mathbf{L}$ and $\mathbf{R}$ contain the background, applying the graph-cut to find the minimum route is difficult. In this case, segmentation is performed by capturing boundary points for each line of the ROI. First, k-means clustering is performed to divide the total ROI into 10 clusters. Let $\mathbf{YL}$ be the clustered $\mathbf{L}$, whose size is (lx, ly) and $\mathbf{YL_i}$ be the i-th line of $\mathbf{YL}$. At this time, the base value BL_i is ly or the peak point nearest to ly, and the reference value $RL_i = ceil(\mathbf{YL_i}(BL_i)/2)$. $\mathbf{YL_i}$ can be divided into three cases. Case 1 has one or more deep valleys and peaks between BL_i and 1, and the peak value far from

BL_i is larger than RL_i. Case 2 is similar to the case 1, but the peak values are smaller than RL_i. Case 3 is the case without a valley. For each case, the boundary point EL_i is as follows:

$$\text{Case 1}: \ EL_i \ = \ \underset{j}{argmin}(\mathbf{YL_i}(PL_i < j < ly)) \tag{5}$$

$$\text{Case 2,3}: \ EL_i \ = \ ML_i \tag{6}$$

PL_i denotes the peak point closest to BL_i which has a valley smaller than RL_i towards BL_i and having a larger value than the valley by 2 cluster stages or more. ML_i is the point closest to BL_i among the points having the value of $ceil(\mathbf{YL_i}(BL_i)/3)$. The case of region $\mathbf{R}$ is similar to the case of $\mathbf{L}$, but performed in the opposite direction. $\mathbf{YR}$ is the clustered $\mathbf{R}$, whose size is $(rx,\ ry)$ and let $\mathbf{YR_i}$ be the i-th line of $\mathbf{YR}$. BR_i is 1 or the peak point nearest to 1, and the reference value $RR_i \ = \ ceil(\mathbf{YR_i}(BR_i)/2)$. Case 1 has one or more deep valleys and peaks between BR_i and ry, and the peak value far from BR_i is larger than RR_i. Case 2 is similar to case 1, but the peak values are smaller than RR_i. Case 3 is the case without a valley. For each case, the boundary point ER_i is as follows:

$$\text{Case 1}: \ ER_i = \ \underset{j}{argmin}(\mathbf{YR_i}(1 < j < PR_i)) \tag{7}$$

$$\text{Case 2,3}: \ ER_i \ = \ MR_i \tag{8}$$

PRi denotes the peak point closest to BR_i, which has a valley smaller than RR_i towards BR_i and having larger value than the valley by 2 cluster stages or more. MR_i is the point closest to BR_i among the points having the value of $ceil(\mathbf{YR_i}(BR_i)/3)$.

Before the segment the **UL1**, **UL2**, **UR1**, **UR2**, **DL1**, **DL2**, **DR1**, and **DR2** have to change their triangular shape to parallelograms by copying points. Subsequently, the segmentation method of U and D is applied to **UL1**, **UR1**, **DL1**, and **DR1** areas, and the segmentation method of **L** and **R** is applied to the **UL2**, **UR2**, **DL2**, and **DR2** areas.

Finally, when all edge points are found, the cubic function is fitted for each side of the boundary to remove the edge points above the error range. The points that deviate by a regular distance from the cubic function curve are replaced with points on the function curves, and finally the segmentation process is ended by connecting all the points.

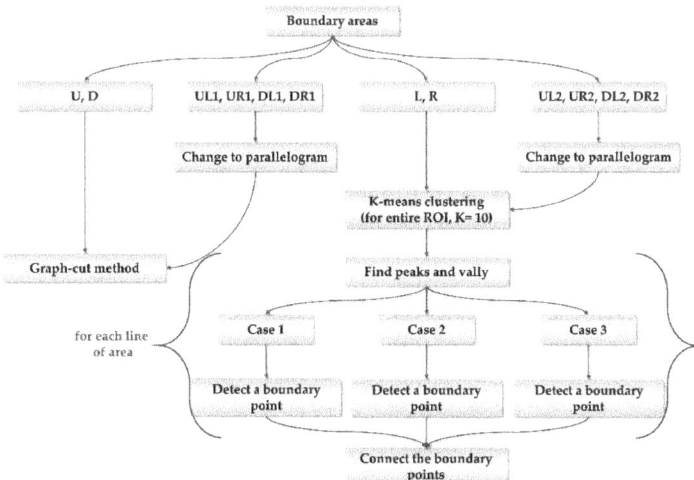

Figure 4. Flow chart of segmentation process.

3. Results and Discussions

3.1. ROI Detection

We tested the ROI detection algorithm using the whole image and user-defined ROI as the minimum user input. The first test was performed with only vertebral body correlation maps, and the second test was performed with both correlation maps of the vertebral body and intervertebral disc. For all 85 vertebral bodies in 19 MR spine images, the number of detection failure cases in the vertebral body test was three, whereas the vertebral body and intervertebral disc test was zero. Detection error of the vertebral body and the intervertebral disc test was 10.41 pixels, which was 19.36% lower than that of the vertebral body test (12.91 pixels). Therefore, it is clear that the use of both ROIs yields better results and that the intervertebral disc exhibits more precise information of location as compared with the vertebral body. Figure 5 also shows that the use of the intervertebral discs adjusts the ROI detection more closely. Detection of ROIs using the vertebral body ROI offered by the user poses a difficulty only in finding other vertebral bodies of the MR spine images; however, more precise detection can be performed if the intervertebral disc ROI is additionally used for ROI detection. In addition, when the user ROI was set to the central vertebral body, there was no detection failure case. On the other hand, when the ROI was set to the uppermost vertebral body, the detection failure case occurred 0.93 times per image, and when the ROI was set to the undermost vertebral body, the detection failure case occurred 0.77 times per image.

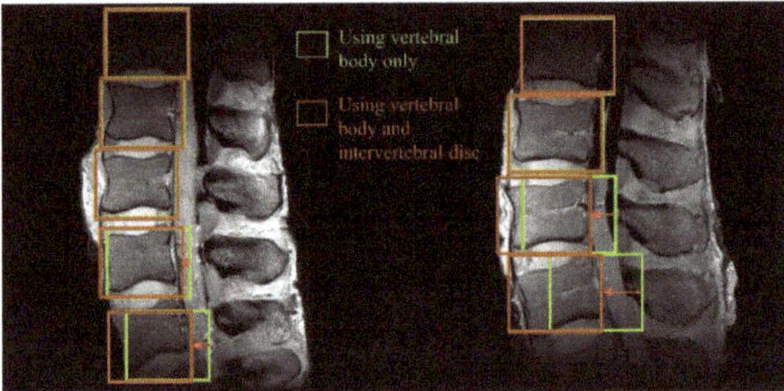

Figure 5. Results of the region of interest (ROI) detection. Green boxes are the results of using only vertebral body ROI, and brown boxes are the results of using both vertebral body ROI and intervertebral disc ROI.

In most cases, the vertebral bodies and intervertebral discs are at a similar position in the horizontal direction. In other words, knowing the position of the intervertebral discs helps to understand the horizontal position of the vertebral bodies. As the intervertebral discs have more compact and unique shapes than the vertebral body, their correlation with other tissues is quite small. Therefore, more accurate locations of vertebral bodies can be found using correlation maps of the intervertebral discs.

Aligning the ROIs with a fixed size cannot completely cover all the vertebral bodies because the vertebral bodies have various sizes and shapes. The ROI fine tuning process adjusts the ROIs to be fitted for each of these various vertebral bodies and aligns them to the center of the ROIs. When the ROI fine tuning process was not applied, there were several cases in which a part of the vertebral body was not included within the ROI, on average, of 1.31 cases per image.

3.2. Segmentation Results

Segmentation experiments were performed on 19 MR spine images including 85 vertebral bodies. All user ROIs were manually specified vertebral body regions. The DSC was used to evaluate the accuracy of segmentation, and the references were manually obtained by an expert [34]. DSC is defined as

$$DSC = \frac{2|\mathbf{F}_{manual} \cap \mathbf{F}_{result}|}{|\mathbf{F}_{manual}| + |\mathbf{F}_{result}|} \tag{9}$$

where $\mathbf{F}_{manual}$ represents the manual references and $\mathbf{F}_{result}$ represents the segmentation results of the proposed method. Table 1 shows the DSC results for 19 MR spine images.

Table 1. Dice similarity coefficient (DSC) results of the experiments.

No.	Volume of Vertebral Body (mm^3)		Number of Voxels		DSC (%)
	The Proposed	Manual	The Proposed	Manual	
1	2347.58	2134.11	5723.00	5379.40	90.96
2	2069.50	2059.26	5298.25	5265.50	94.11
3	2904.11	2731.39	7435.00	6882.25	90.50
4	1363.59	1468.81	3491.00	4037.00	88.11
5	2028.70	2034.23	5193.80	5263.20	93.77
6	2422.79	2360.53	6202.75	6003.50	91.01
7	1858.32	1843.76	4757.60	4864.00	86.94
8	2247.85	2136.49	5231.20	4847.00	87.79
9	1762.97	1520.00	4102.80	3548.40	86.58
10	1658.64	1686.57	3860.00	3933.60	85.30
11	2412.74	2172.44	6177.00	5408.00	88.82
12	2376.49	2240.58	6084.20	5695.20	90.20
13	2148.50	2127.33	5500.50	5432.75	92.80
14	3222.66	2872.19	8689.50	7508.25	86.43
15	2667.37	2445.22	7192.25	6443.50	88.81
16	2264.05	2365.08	6104.75	6445.25	93.33
17	2212.97	2296.04	5967.00	6247.00	93.51
18	2494.64	2662.20	6726.50	7231.25	91.34
19	2385.87	2425.71	6433.20	6694.60	87.92
$\mu \pm \sigma$	2255.23 ± 429.94	2188.52 ± 377.83	5798.44 ± 1265.49	5641.56 ± 1110.66	89.91 ± 2.77

The average DSC was 89.91% (Table 1). This is a valuable result as it is comparable to the 90.97% achieved by Egger et al. who used the square-cut algorithm that requires more user inputs than the proposed method [27]. Further, it is higher than the 81.33% reported by Schwarzenberg et al. who used the cube-cut algorithm [3]. The standard deviation score 2.77% is similar to that reported by Egger et al. (2.2%) [27] and much smaller than that reported by Schwarzenberg et al. (5.07%) [3]. For the number of voxels and the vertebral body volume, the square-cut algorithm showed a difference of 3.83% on both, compared with manual segmentation results [27], and the cube-cut algorithm showed differences of 15.95% and 5.97%, respectively, compared with manual segmentation results [3]. Our method, on the other hand, showed differences of 3.05% and 2.78% respectively, compared with manual segmentation results. Our results are lower than the cube-cut algorithm and square-cut algorithm. While a direct comparison of the results obtained with different algorithms is difficult, owing to differences with respect to image contrast type, resolution, image quality, and the number of test datasets used in each study, it generally suggests a comparable performance range of our algorithm. The square-cut and cube-cut algorithms, which are based on the graph-cut method, have no ROI detection processes; therefore, these algorithms require the locations of all vertebral bodies. Furthermore, because the results depend on the cost of constructing the graph, these algorithms require users to adjust the parameters such as the number of nodes and edges according to the circumstances. On the other hand, the proposed method requires the user to draw just one vertebral body ROI. Table 2 shows the comparison of the user inputs between Square-cut and Cube-cut algorithms and our method. Square-cut and Cube-cut algorithms requires four parameters to adjust and seed points for all vertebral bodies, but our method only needs a single vertebral body ROI. This suggests that our method can maintain or improve the segmentation performance despite reduction in the number of user inputs, and the automated portion of the whole segmentation process increases, as compared with the existing methods.

Table 2. Comparisons of the user inputs between existing algorithms and the proposed method.

	Square-Cut & Cube-Cut	The Proposed Method
Parameter tuning	The number of rays The number of nodes Maximum length of the rays Delta value	None
Necessary inputs	Seed points for each vertebral body	A single vertebral body ROI

Figure 6 shows a representative segmentation result, wherein Figure 6b shows the manually segmented areas, and Figure 6c shows the results of the proposed method. Figure 6d shows the difference between the manual references and the result of the proposed method. The results of the proposed method are generally similar with the manual references. It can be seen that four vertebral bodies are correctly identified and segmented and that most regions are recognized as vertebral bodies. Nonetheless, there are some limitations. The results of the proposed method have non-smooth boundaries, as compared with the manual references. When examining each vertebral body, the top and bottom boundaries are smooth and match well with the actual reference, whereas the left and right boundaries are not entirely smooth and have protruding parts. Based on these experiments and evaluations, it seems that the results of the line-based algorithm are not as good as those obtained with the graph-based algorithm.

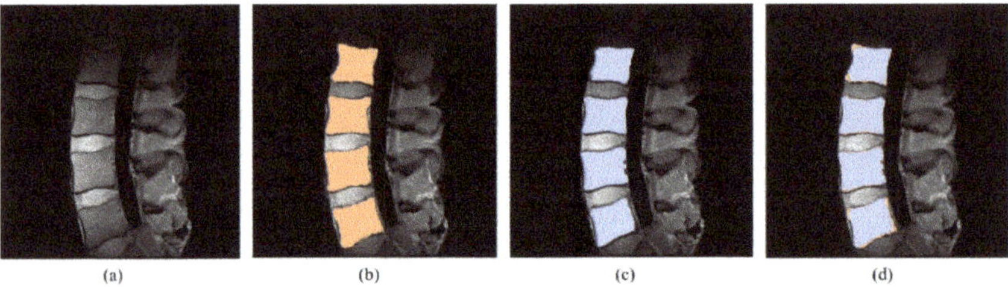

Figure 6. Result of the segmentation algorithm. (**a**) Original image. (**b**) Manual segmentation result. (**c**) The result of the proposed segmentation algorithm. (**d**) The comparison between (**b**) and (**c**).

Figure 7 shows eight different examples of the results of the proposed method. It can be observed that the overall performance looks good in terms of ROI detection and the subsequent segmentation has an average DSC of 89.91 (Table 1). These results show that the proposed method has successfully reduced the user's role while maintaining a performance comparable to existing methods. There are slight differences between the manual references and the results of the proposed method with respect to the upper and lower parts of the images. The proposed method performs segmentation depending on the relative contrast of pixel intensities in the ROI. Therefore, the proposed method may not be able to achieve accurate segmentation if the contrast of the images is not good as shown in the upper and lower parts of the images in Figure 7. This problem may arise with most segmentation algorithms depending on the image contrast.

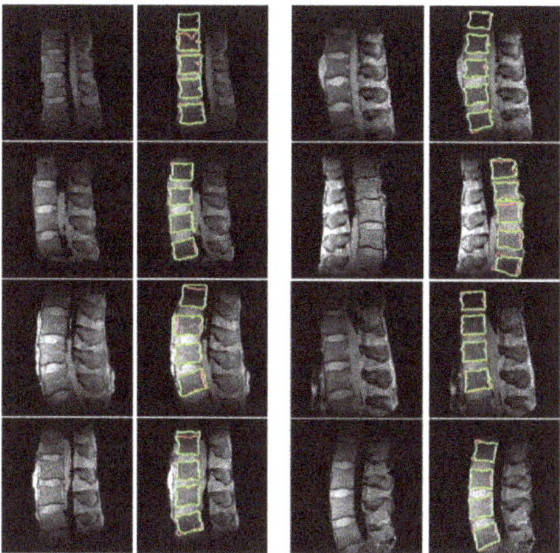

Figure 7. Segmentation results for other magnetic resonance (MR) spine images. Manual segmentation results (red line). The proposed method results (green line).

For additional experiments, the data augmentation technique using a rotation process was applied to the original sample images. 19 MR images were rotated by $10°$, $20°$, and $30°$ with respect to the vertical axis. Therefore, a total of 51 rotated images were obtained. Table 3 shows the segmentation results for them. The average DSC scores were 89–90 in all the cases, and the standard deviations of the DSC scores were about three, which was similar to the results of the original sample MR images. Both results in Tables 1 and 3 demonstrate the robustness of our method.

Table 3. Dice similarity coefficient (DSC) results of the additional experiments.

	Images Rotated by $10°$	Images Rotated by $20°$	Images Rotated by $30°$
μ	89.15	89.70	89.55
σ	3.45	2.96	3.18

4. Conclusions

Obtaining information about the vertebral bodies in the lumbar spine is very important for the diagnosis and treatment of low back pain [35]. In this study, we proposed a semi-automatic algorithm for vertebral body segmentation that requires the user to specify only a single ROI. Our method overcomes the disadvantages of the existing algorithms that require multiple user inputs, such as clicking on each vertebral body, assigning ROIs for each vertebral body, or adjusting multiple parameters depending on circumstances. In our method, only one user-defined-ROI is required to identify the ROIs for all vertebral bodies in the image. Additionally, the detected ROIs were finely adjusted using the edge-detection algorithm and Hough transformation to consider the orientation and size of the vertebral bodies. The subsequent automatic segmentation was performed by combining both the line-based and graph-cut-based methods depending on the parts of the bodies. The experimental results demonstrate comparable or an even higher performance than the existing methods, even though the automated portion of the entire segmentation process is increased. Furthermore, our method does not require a heavy training dataset necessary for deep-learning -based methods and can be applied regardless of the image size.

In future studies, we will apply this technique to in vivo datasets and evaluate its performance for various types of contrast images obtained from living patients. The in vivo datasets may have different noise levels, field of views, contrasts, and resolutions. Motion-related artifacts and abnormality of the vertebral bodies and discs with spinal injuries or diseases should be further investigated.

The application of dorsal vertebrae, which was not included in this study, can be a good future topic to expand and evaluate our technique. A three-dimensional (3D) extension of this technique is also one of our future studies. 3D extension of other approaches has yielded better results than those obtained via two-dimensional (2D) methods [28,36]. Our 3D extension is also expected to improve the overall segmentation performance.

Author Contributions: W.C.B.; K.M. and C.B.C. proposed the idea and contributed to data acquisition and performed manual segmentation; S.K. contributed to performing data analysis, algorithm construction, and writing the article; D.H. technically supported the algorithm and evaluation and also professionally reviewed and edited the paper.

Funding: This research was supported in parts by the National Research Foundation of Korea (NRF) grant funded by the Korean government (MSIP) (2016R1A2B4015016) in support of Dosik Hwang, and National Institute of Arthritis and Musculoskeletal and Skin Diseases of the National Institutes of Health in support of Won C. Bae (Grant Number R01 AR066622). The contents of this paper are the sole responsibility of the authors and do not necessarily represent the official views of the sponsoring institutions.

Conflicts of Interest: The authors declare no conflict of interest.

References

1. Murray, C.J.; Abraham, J.; Ali, M.K.; Alvarado, M.; Atkinson, C.; Baddour, L.M.; Bartels, D.H.; Benjamin, E.J.; Bhall, K.; Birbeck, G.; et al. US Burden of Disease Collaborators the State of US Health, 1990–2010: Burden of diseases, injuries, and risk factors. *JAMA* **2013**, *310*, 591–608. [CrossRef] [PubMed]

2. Gatchel, R.J. The continuing and growing epidemic of chronic low back pain. *Healthcare* **2015**, *3*, 838–845. [CrossRef] [PubMed]

3. Schwarzenberg, R.; Freisleben, B.; Nimsky, C.; Egger, J. Cube-cut: Vertebral body segmentation in MRI-data through cubic-shaped divergences. *PLoS ONE* **2014**, *9*, e93389. [CrossRef] [PubMed]

4. Larhmam, M.A.; Mahmoudi, S.A.; Benjelloun, M.; Mahmoudi, S.; Manneback, P. A portable multi-CPU/multi-GPU based vertebra localization in sagittal MR images. In Proceedings of the 11th International Conference Image Analysis and Recognition, Vilamoura, Portugal, 22–24 October 2014; Volume 8815, pp. 209–218.

5. Pham, D.L.; Xu, C.; Prince, J.L. Current methods in medical image segmentation. *Annu. Rev. Biomed. Eng.* **2000**, *2*, 315–337. [CrossRef] [PubMed]

6. Otsu, N. A Threshold Selection Method from Gray-level Histograms. *IEEE Trans. Syst. Man Cybern.* **1979**, *9*, 23–27. [CrossRef]

7. Schneiderman, H.; Kanade, K. A histogram-based method for detection of faces and cars. In Proceedings of the 2000 International Conference on Image Processing, Vancouver, BC, Canada, 10–13 September 2000; Volume 3, pp. 504–507.

8. Zhu, S.C.; Lee, T.S.; Yuille, A.L. Region competition: Unifying snakes, region growing, energy/bayes/MDL for multi-band image segmentation. In Proceedings of the IEEE International Conference on Computer Vision, Cambridge, MA, USA, 20–23 June 1995; pp. 416–423.

9. Pohle, R.; Toennies, K.D. Segmentation of medical images using adaptive region growing. In Proceedings of the Medical Imaging 2001: Image Processing, San Diego, CA, USA, 3 July 2001; Volume 4322, pp. 1337–1346.

10. Chang, Y.L.; Li, X. Adaptive image region-growing. *IEEE Trans. Image Process.* **1994**, *3*, 868–872. [CrossRef] [PubMed]

11. Adams, R.; Bischof, L. Seeded region growing. *IEEE Trans. Pattern Anal. Mach. Intell.* **1994**, *16*, 641–647. [CrossRef]

12. Canny, J. A computational approach to edge detection. *IEEE Trans. Pattern Anal. Mach. Intell.* **1986**, *6*, 679–698. [CrossRef]

13. Hartigan, J.A.; Wong, M.A. Algorithm AS 136: A k-means clustering algorithm. *J. R. Stat. Soc. Ser. C (Appl. Stat.)* **1979**, *28*, 100–108. [CrossRef]

14. Jain, A.K.; Dubes, R.C. *Algorithms for Clustering Data*; Prentice-Hall: Upper Saddle River, NJ, USA, 1988; ISBN 0-13-022278-X.

15. Alsabti, K.; Ranka, S.; Singh, V. An efficient k-means clustering algorithm. *Electr. Eng. Comput. Sci.* **1997**, *43*.

16. Bezed, J.C.; Ehrlich, R.; Full, W. FCM: The fuzzy c-means clustering algorithm. *Comput. Geosci.* **1984**, *10*, 191–203.
17. Chuang, K.S.; Tzeng, H.L.; Chen, S.; Wu, J.; Chen, T.J. Fuzzy c-means clustering with spatial information for image segmentation. *Comput. Med. Imaging Graph.* **2006**, *30*, 9–15. [CrossRef] [PubMed]
18. Pham, D.L.; Prince, J.L. Adaptive fuzzy segmentation of magnetic resonance images. *IEEE Trans. Med. Imaging* **1999**, *18*, 737–752. [CrossRef] [PubMed]
19. Clark, M.C.; Hall, L.O.; Goldgof, D.B.; Clarke, L.P.; Velthuizen, R.P.; Silbiger, M.S. MRI segmentation using fuzzy clustering techniques. *IEEE Eng. Med. Boil. Mag.* **1994**, *13*, 730–742. [CrossRef]
20. Camilus, K.S.; Govindan, V.K. A review on graph-based segmentation. *Int. J. Image Graph. Signal Process.* **2012**, *4*, 1–13. [CrossRef]
21. Wu, Z.; Leahy, R. An optimal graph theoretic approach to data clustering: Theory and its application to image segmentation. *IEEE Trans. Pattern Anal. Mach. Intell.* **1993**, *15*, 1101–1113. [CrossRef]
22. Felzenszwalb, P.F.; Huttenlocher, D.P. Efficient graph-based image segmentation. *Int. J. Comput. Vis.* **2004**, *59*, 167–181. [CrossRef]
23. Shi, J.; Malik, J. Normalized cuts and image segmentation. *IEEE Trans. Pattern Anal. Mach. Intell.* **2000**, *22*, 888–905.
24. Egger, J.; O'Donnel, T.; Hopfgartner, C.; Freisleben, B. Graph-based tracking method for aortic thrombus segmentation. In Proceedings of the 4th European Conference of the International Federation for Medical and Biological Engineering (IFMBE), Antwerp, Belgium, 23–27 November 2008; Springer: Berlin/Heidelberg, Germany, 2009; pp. 584–587.
25. Boykov, Y.; Kolmogorov, V. An experimental comparison of min-cut/max-flow algorithms for energy minimization in vision. *IEEE Trans. Pattern Anal. Mach. Intell.* **2004**, *26*, 1124–1137. [CrossRef] [PubMed]
26. Boykov, Y.; Veksler, O.; Zabih, R. Fast approximate energy minimization via graph cuts. *IEEE Trans. Pattern Anal. Mach. Intell.* **2001**, *23*, 1222–1239. [CrossRef]
27. Egger, J.; Kapur, T.; Dukatz, T.; Kolodziej, M.; Zukić, D.; Freisleben, B.; Nimsky, C. Square-cut: A segmentation algorithm on the basis of a rectangle shape. *PLoS ONE* **2012**, *7*, e31064. [CrossRef] [PubMed]
28. Ronneberger, O.; Fischer, P.; Brox, T. U-Net: Convolutional Networks for Biomedical Image Segmentation. In Proceedings of the 18th International Conference on Medical Image Computing and Computer-Assisted Intervention, Munich, Germany, 5–9 October 2015; Volume 9351, pp. 234–241.
29. Korez, R.; Likar, B.; Pernuš, F.; Vrtovec, T. Model-Based Segmentation of Vertebral Bodies from MR Images with 3D CNNs. In Proceedings of the 19th International Conference on Medical Image Computing and Computer-Assisted Intervention, Athens, Greece, 17–21 October 2016; Volume 9901, pp. 433–441.
30. Kim, S.; Bae, W.C.; Hwang, D. Semi-automatic segmentation algorithm for vertebral body in MR spine image. In Proceedings of the ICEIC 2017 International Conference on Electronics, Information, and Communication, Phuket, Thailand, 11–14 January 2017; pp. 810–812.
31. Weisstein, E.W. Cross-Correlation Theorem, 2014. MathWorld—A Wolfram Web Resource. Available online: http://mathworld.wolfram.com/Cross-CorrelationTheor-em.htm (accessed on 13 August 2018).
32. Tüzün, C.; Yorulmaz, I.; Cindaş, A.; Vatan, S. Low back pain and posture. *Clin. Rheumatol.* **1999**, *18*, 308–312. [CrossRef] [PubMed]
33. Duda, R.O.; Hart, P.E. Use of the Hough transformation to detect lines and curves in pictures. *Commun. ACM* **1972**, *15*, 11–15. [CrossRef]
34. Dice, L.R. Measures of the amount of ecologic association between species. *Ecology* **1945**, *26*, 297–302. [CrossRef]
35. Zukic, D.; Vlasák, A.; Dukatz, T.; Egger, J.; Horínek, D.; Nimsky, C.; Kolb, A. Segmentation of Vertebral Bodies in MR Images. *Vis. Model. Vis.* **2012**, *12*, 135–142.
36. Štern, D.; Likar, B.; Pernuš, F.; Vrtovec, T. Parametric modelling and segmentation of vertebral bodies in 3D CT and MR spine images. *Phys. Med. Biol.* **2011**, *56*, 7505. [CrossRef] [PubMed]

Article

Data Balancing Based on Pre-Training Strategy for Liver Segmentation from CT Scans

Yong Zhang [1,2], Yi Wang [1,*], Yizhu Wang [2], Bin Fang [1], Wei Yu [1], Hongyu Long [1] and Hancheng Lei [1]

[1] College of Computer Science, Chongqing University, No.174 Shazhengjie, Shapingba, Chongqing 400044, China; zhangyong7630@163.com (Y.Z.); bf@cqu.edu.cn (B.F.); yu_wei0811@163.com (W.Y.); leihancheng97@163.com (H.L.); leihancheng97@163.com (H.L.)
[2] Ziwei king star Digital Technology Co., Ltd., Nine Floors of G4 A Block, Phase 2 Innovation Industrial Park, Hefei High-tech Zone, Hefei 230000, China; wyizhu@ziweidixing.com
*   Correspondence: YiWang@cqu.edu.cn; Tel.: +86-13062351083

Received: 2 March 2019; Accepted: 29 April 2019; Published: 2 May 2019

Abstract: Data imbalance is often encountered in deep learning process and is harmful to model training. The imbalance of hard and easy samples in training datasets often occurs in the segmentation tasks from Contrast Tomography (CT) scans. However, due to the strong similarity between adjacent slices in volumes and different segmentation tasks (the same slice may be classified as a hard sample in liver segmentation task, but an easy sample in the kidney or spleen segmentation task), it is hard to solve this imbalance of training dataset using traditional methods. In this work, we use a pre-training strategy to distinguish hard and easy samples, and then increase the proportion of hard slices in training dataset, which could mitigate imbalance of hard samples and easy samples in training dataset, and enhance the contribution of hard samples in training process. Our experiments on liver, kidney and spleen segmentation show that increasing the ratio of hard samples in the training dataset could enhance the prediction ability of model by improving its ability to deal with hard samples. The main contribution of this work is the application of pre-training strategy, which enables us to select training samples online according to different tasks and to ease data imbalance in the training dataset.

Keywords: data imbalance; Contrast Tomography (CT); pre-training strategy; segmentation

---

1. Introduction

Accurate segmentation of the liver can greatly help the subsequent segmentation of liver tumors, as well as assisting doctors in making accurate disease condition assessment and treatment planning of patients [1]. Traditionally, liver delineation relies on the slice-by-slice manual segmentation of Contrast Tomography (CT) or Magnetic Resonance Imaging (MRI) by radiologists, which is time-consuming and prone to influence by internal and external variations. With the rapid increase of CT and MRI data, traditional manual segmentation method has become increasingly unable to meet the clinical needs. Therefore, automatic segmentation tools are required for practical clinical applications.

Automatic segmentation methods such as region growing, intensity thresholding, and deformable model-based methods have achieved automatic or semi-automatic segmentation to a certain extent, with good segmentation results. However, these models rely on hand-crafted features and have limited feature extraction ability. Recently, methods of deep learning, especially full convolutional networks (FCNs), have achieved great success on a broad array of recognition problems [2–4]. Many researchers advance this stream using deep learning methods in segmentation tasks such as liver [1,5–7], kidney [8], vessel [9–11] and pancreas [12–14]. All the models mentioned above are based on a large amount of data. However, there often are two kinds of data imbalance problems in the training process for the segmentation of CT scans: (i) data imbalance in images: the imbalance between background voxels and target voxels, as shown in Figure 1; (ii) data imbalance between images: the imbalance of hard or easy predicted examples in training datasets (the easily segmented slices are called easy samples or easy slices, while the difficult samples are defined as hard samples or hard slices) in training dataset. As shown in Figure 2A,B, the features of some slices are obvious and easy to segment. However, in some others, as shown in Figure 2C,D), the features of liver are not obvious, which may be due to poor quality of CT image or the liver self-defect (e.g., liver morphological variation, liver lesions, etc.), and it is difficult to accurately segment liver from these slices. Moreover, it is easy to qualitatively divide hard samples and easy samples according to the segmentation results, but it is difficult or almost impossible to describe the characteristics of hard samples and easy samples, and accurately distinguish them in training dataset before training process.

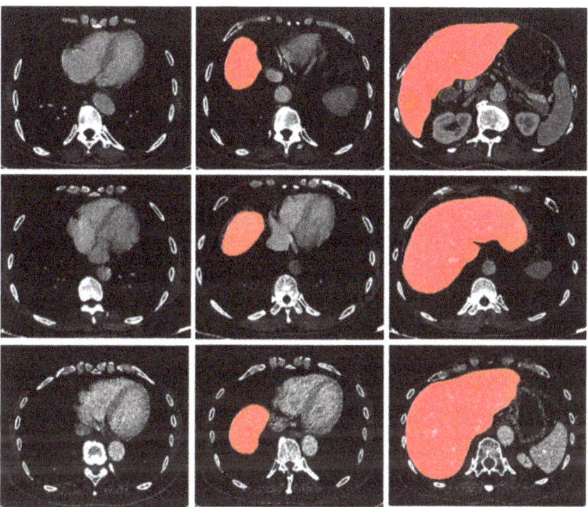

Figure 1. Examples of the imbalance between background voxels and target voxels in images. Each row shows a CT scan from individual patients. The read regions denote the liver.

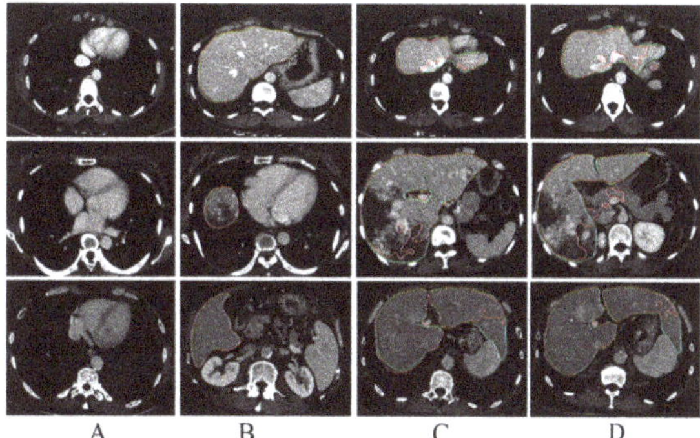

Figure 2. Examples of easy and hard predicted slices in CT scans. The predicted results are based on the FCN model with 10×10^5 iterations. (**A**,**B**) display the easy samples; (**C**,**D**) display the hard slices. Blue and red lines denote ground truth and prediction results. Each row shows results acquired from an individual case.

Using Dice coefficient [15] as the loss function in training process can solve the first kind of data imbalance by reducing or even ignoring the contribution of background voxels. However, due to the similarity between adjacent slices in medical images, and different training tasks (for example the same slice may be a hard example in liver segmentation task but an easy example in kidney or spleen segmentation task), it is difficult to classify medical images in training dataset automatically using traditional methods before the training process. When there are many easy samples, the contribution of hard slices will be overwhelmed in the training process, which could cause a significant reduction in the prediction ability of the model for difficult samples, and may even lead to overfitting. Therefore, it is necessary to classify the training samples and increase the proportion of hard samples in training datasets.

Recently, focal loss, which could automatically adjust the contribution of easy-negative samples in training process and rapidly focus on hard examples in every batch training process, has achieved great success in one-stage detector objection [16]. However, focal loss failed to change the imbalance between hard samples and easy samples in training dataset, the contribution of hard slices may still be overwhelmed in the training process. In order to solve or alleviate this imbalance problem, we introduce an online hard example enhancement method to increase the proportion of the hard samples in the training dataset. Frist, we use partial slices in the whole training dataset to train a pre-training model according to the needs of segmentation task, and then the pre-training model is used to distinguish hard samples and easy samples in the rest slices of the whole training dataset, i.e., adding the identified hard samples to the training datasets used in the pre-training processes. Second, the hard slices identified by pre-training model are selected and enhanced by flipping, and then these slices are added to the dataset used in pre-training process to enhance the ratio of hard slices in training dataset, and improve the contribution of hard slices in training process. Therefore, the basic purpose of pre-training strategy is to get a sample classifier, which could distinguish hard/easy slices according to actual task need.

To demonstrate the effectiveness of the proposed method, we adopt a classical 2D FCN model based on VGG-16 [17] and 2D U-Net [3], as shown in Figures A1 and A2 respectively, for the task of the liver segmentation, kidney segmentation and spleen segmentation from Computed Tomography (CT) scans.

2. Materials and Methods

2.1. Dataset and Processing

We test our method on datasets acquired from different scanners of different medical institutions. The collected dataset composes of 260 CT scans, with a largely varying in slice spacing from 0.45 mm to 5 mm. And 220 CT scans were randomly selected for training, the rest 40 cases for testing. For images pre-processing, the image intensity values were truncated to the range of $[-150, 250]$ hounsfield unit (HU) to remove the irrelevant details [9].

2.2. Selection of Training Samples

Inspired by pre-training strategy, in this work, a pre-training model is used as a sample classifier to classify hard samples and easy samples in training dataset. Frist, the whole training dataset was divided into two parts (A and B) based on their simple statistics information (e.g., the number of slices in volume, the proportion of positive and negative samples in volume). In this way, the ratio of positive and negative slices in two subsets (A and B) can be guaranteed the same as that of the whole training dataset. Part A is used for the later sample classification and screening, while part B is used for model pre-training. Second, slices in part B are enhanced by flipping and mirroring, and then these enhanced slices are used in model pre-training process. And we get a pre-training model when model is trained to a set iteration (such as 5×10^5 iterations in this work). Third, the pre-training model is used to predict slices in part A, and all slices in part A are simply divided into two categories, hard samples, and easy samples, by their Dice score. Next, the hard slices in part A are enhanced by flipping, and then added to the training dataset (part B) used in pre-training process. Finally, we continue the training process until reaching to the set 10×10^5 iterations, and then get the final segmentation model. Just 5×10^5 iterations are needed in the final training process if 5×10^5 iterations were done in the pre-training process and the pre-training model structure is consistent with the final model, while 10×10^5 iterations are needed in the final training stage if the pre-training model structure is inconsistent with the final model. In this study, we use the same model structure in pre-training process and final training stage.

2.3. Evaluation Metrics

Dice coefficient, which measures the amount of an agreement between two image regions, was used to evaluate the segmentation performance on the test dataset.

2.4. Implementation Details

Classical 2D FCN model structure and 2D U-Net are used for segmentation tasks from CT scans using the TensorFlow package [9]. We use stochastic gradient descent (SGD) with a mini-batch size 16. Inspired by [1], the "poly" learning rate policy where the current learning rate equals to the initial learning rate multiplying (1−(iterations)/(total_iterations))^power. We set the initial learning rate to 0.001 and the power to 0.9 and the models are trained for up to 10×10^5 iterations. We use the Dice coefficient as the loss function in the training process. For data augmentation, we adopt a random mirror, flip for all datasets. We use the aforementioned training strategy in the pre-training process and the final training stage.

3. Results

As for the strong similarity between adjacent slices in CT scans, we assume that the contribution of some slices could be replaced by others in the training process. To test this idea, we select partial cases at a certain ratio from the whole training dataset, based on their simple statistical information (e.g., the number of slices in volume, proportion of positive samples and negative samples in volume). And then the selected cases were enhanced by flipping and mirroring.

As shown in Table 1, reducing the number of training samples within a certain range has less influence on the segmentation ability of FCN. However, the prediction ability of FCN decreases significantly when the selection of training samples is further reduced. The max value, which refers to the best segmentation results of the model, has little change in different selection ratio (the ratio of training scans in part B to total number of scans in training dataset) experiments. Meanwhile, the min value, referring to the worst segmentation results of the model, decreases significantly when training samples decrease substantially. These phenomena are also observed in kidney segmentation and spleen segmentation from CT scans using FCN model, as shown in Tables A1 and A2. Moreover, the same results were also discovered in liver segmentation, kidney segmentation and spleen segmentation tasks using U-Net model, as shown in Tables A3–A5. These results suggest that there is redundancy in the training dataset, and that too little training data is harmful in the model training process.

Table 1. Liver segmentation results on test dataset based on different selection ratio using FCN model.

Model	Dice Score		
	Mean	**Min**	**Max**
selection ratio = 1, baseline	0.9705 ± 0.011	0.923	0.9856
selection ratio = 0.8	0.9706 ± 0.011	0.921	0.9853
selection ratio = 0.5	0.9702 ± 0.013	0.9124	0.9843
selection ratio = 0.3	0.9512 ± 0.035	0.8635	0.9844
selection ratio = 0.2	0.9475 ± 0.039	0.812	0.9817
proposed model	0.9789 ± 0.012	0.947	0.9854

As for the performance of FCN model begins to decline significantly when the selection ratio is less than 0.5, so we set selection ratio as 0.5 in the proposed model, and divide the training dataset into two parts (A and B) in liver segmentation. Slices in part B are used for model training, and we get the pre-training model after 5×10^5 iterations. Using the pre-training model to predict slices in part A, we then simply classify these slices in part A into two categories, i.e., hard and easy simples, based on their Dice score. In the liver segmentation task using FCN model, we set the threshold to 0.923, the min Dice scores of baseline. Six thousand, two hundred and sixty-eight slices are classified as hard samples; however, 35,984 slices are classified as easy samples, almost 6-fold the number of hard samples. Hard samples in part A were enhanced by flipping and added to the dataset (part B) used in pre-training process. Then, we continue the training process until model reaching 10×10^5 iterations.

As shown in Table 1, the proposed model performs slightly better than the baseline in liver segmentation with a smaller training dataset. Moreover, adding hard examples has almost no effect on the max value of Dice score, but it can significantly increase the min value compared with the baseline. This indicates that increasing the ratio of hard samples in the training dataset has little influence on easily segmented cases, but could greatly improve the segmentation ability of model on hard samples.

The segmentation results display in 3D form in Figure 3A,B show that the proposed method could enhance liver segmentation results, especially in some details. Liver segmentation results of hard examples have been greatly improved compared with the baseline, as shown in Figure 3C,D, which may be attributed to the increase of the number of hard samples in training dataset. The above results suggest that enhancing the proportion of hard samples in the training dataset could improve the prediction performance of FCN model in the liver segmentation task, as well as model's ability to deal with hard samples.

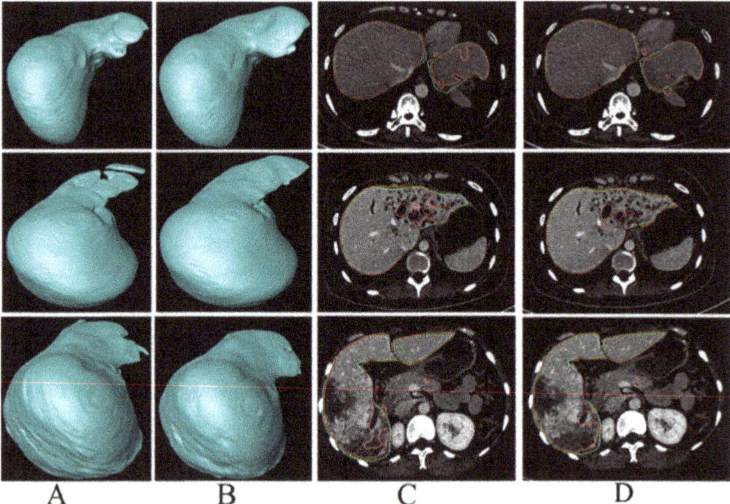

Figure 3. Results of Liver segmentation using FCN model. (**A**,**B**) display the 3D liver segmentation result of the baseline and proposed a model, respectively; (**C**,**D**) display the hard samples liver segmentation results of the baseline and proposed a model, respectively; Blue and red lines in **C** and **D** denote ground truth and prediction results. Each row shows results acquired from an individual case.

4. Discussion

It is often thought that the more data, the better the performance in deep learning. However, in this work, we observed that a proper reduction of training samples in training process had little effect on the segmentation performance of model. This may be due to the strong similarity between two adjacent slices in CT images, which makes it difficult to ensure each image in the training dataset is independent from others; in other words, the contribution of some samples can be replaced by others in the training process. However, it is hard to screen out which one may be redundant. The relatively shallow network structure, which has relatively weak deep feature extraction capability, may be another reason for the phenomenon observed in this work. Meanwhile, the significantly reduced performance of the model in the case of a large reduction in training dataset also supports the point that the more data, the better the performance in deep learning.

Additionally, the same slices may play different roles in different segmentation tasks. For example, the positive-hard samples in the liver segmentation task may be negative-easy ones in kidney or spleen segmentation. Therefore, it is difficult to classify samples with the traditional unsupervised method. Inspired by the pre-training strategy, we use a pre-training method as a sample classifier to classify hard samples and easy samples in training dataset. We obtained better performance from the model after adding the enhanced hard examples.

Author Contributions: Conceptualization, Y.W. (Yi Wang) and B.F.; Data curation, Y.Z., Y.W. (Yizhu Wang), W.Y. and H.L. (Hancheng Lei); Funding acquisition, Y.W. (Yizhu Wang); Writing–original draft, Y.Z. and Y.W. (Yi Wang); Writing–review & editing, H.L. (Hongyu Long).

Funding: This work was supported by the National Natural Science Foundation of China (61876026, 61672120) and The Social Livelihood Science and Technology Innovation Special Project of CSTC (no. CSTC2015shmszx120002).

Conflicts of Interest: The authors declare no conflict of interest.

Appendix A

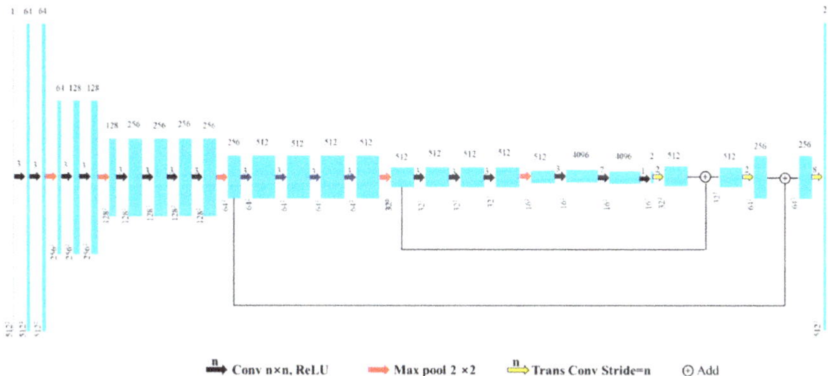

Figure A1. FCN architecture. Each blue box corresponds to a multi-channel feature map. The number of channels is denoted on top of the box. The x-y-size is provided at the lower left edge of the box.

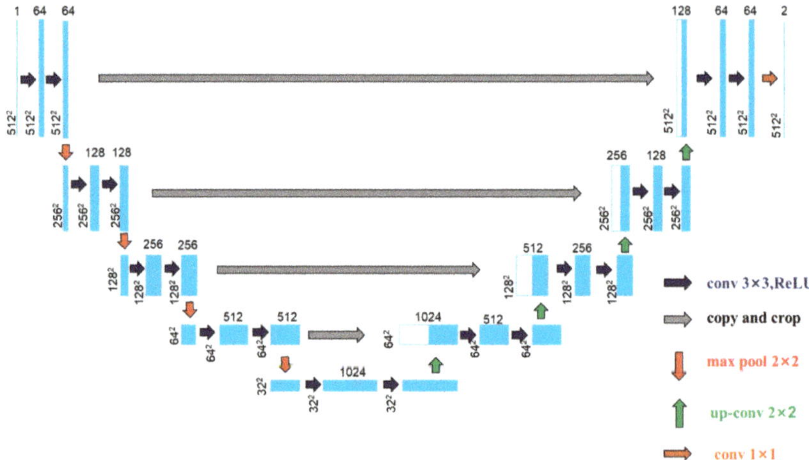

Figure A2. U-net architecture. Each blue box corresponds to a multi-channel feature map. The number of channels is denoted on top of the box. The x-y-size is provided at the lower left edge of the box. White boxes represent copied feature maps. The arrows denote the different operations.

Table A1. Kidney segmentation results on test dataset based on different selection ratio using FCN model.

Model	Dice Score		
	Mean	**Min**	**Max**
selection ratio = 1, baseline	0.9167 ± 0.13	0.8262	0.9698
selection ratio = 0.8	0.9106 ± 0.089	0.8213	0.9703
selection ratio = 0.5	0.9078 ± 0.114	0.822	0.9693
selection ratio = 0.3	0.8835 ± 0.035	0.7935	0.9687
selection ratio = 0.2	0.8511 ± 0.039	0.7724	0.9684
proposed model	0.9258 ± 0.067	0.8547	0.9693

Table A2. Spleen segmentation results on test dataset based on different selection ratio using FCN model.

Model	Dice Score		
	Mean	**Min**	**Max**
selection ratio = 1, baseline	0.9773 ± 0.016	0.9364	0.9973
selection ratio = 0.8	0.9762 ± 0.012	0.9379	0.9957
selection ratio = 0.5	0.9767 ± 0.014	0.9359	0.9969
selection ratio = 0.3	0.9714 ± 0.035	0.9358	0.997
selection ratio = 0.2	0.8981 ± 0.139	0.7724	0.9965
proposed model	0.9801 ± 0.007	0.9563	0.9975

Table A3. Liver segmentation results on test dataset based on different selection ratio using U-Net model.

Model	Dice Score		
	Mean	**Min**	**Max**
selection ratio = 1, baseline	0.9532 ± 0.035	0.9017	0.9875
selection ratio = 0.8	0.9501 ± 0.031	0.908	0.9863
selection ratio = 0.5	0.9486 ± 0.023	0.898	0.9867
selection ratio = 0.3	0.9107 ± 0.057	0.834	0.9821
selection ratio = 0.2	0.8932 ± 0.063	0.796	0.9806
proposed model	0.9604 ± 0.022	0.912	0.987

Table A4. Kidney segmentation results on test dataset based on different selection ratio using U-Net model.

Model	Dice Score		
	Mean	**Min**	**Max**
selection ratio = 1, baseline	0.9206 ± 0.073	0.8345	0.9732
selection ratio = 0.8	0.9188 ± 0.092	0.8351	0.9725
selection ratio = 0.5	0.9158 ± 0.113	0.8298	0.973
selection ratio = 0.3	0.8735 ± 0.127	0.7653	0.9563
selection ratio = 0.2	0.8621 ± 0.153	0.7549	0.9517
proposed model	0.9287 ± 0.028	0.8591	0.9708

Table A5. Spleen segmentation results on test dataset based on different selection ratio using U-Net model.

Model	Dice Score		
	Mean	**Min**	**Max**
selection ratio = 1, baseline	0.9795 ± 0.009	0.9473	0.9971
selection ratio = 0.8	0.9803 ± 0.007	0.9482	0.9962
selection ratio = 0.5	0.9780 ± 0.015	0.9367	0.9969
selection ratio = 0.3	0.9704 ± 0.023	0.9289	0.9972
selection ratio = 0.2	0.9057 ± 0.089	0.8124	0.9963
proposed model	0.9857 ± 0.005	0.9579	0.9969

References

1. Li, X.; Chen, H.; Qi, X.; Dou, Q.; Fu, C.W.; Heng, P.A. H-Dense UNet: hybrid densely connected UNet for liver and tumor segmentation from CT volumes. *IEEE Trans. Med. Imaging* **2018**, *37*, 2663–2674. [CrossRef] [PubMed]
2. Shaoqing, R.; Kaiming, H.; Girshick, R.; Xiangyu, Z.; Jian, S. Object detection networks on convolutional feature maps. *IEEE Trans. Pattern Anal. Mach. Intell.* **2017**, *39*, 1476–1481.
3. Ronneberger, O.; Fischer, P.; Brox, T. U-net: Convolutional networks for biomedical image segmentation. In Proceedings of the International Conference on Medical Image Computing and Computer-Assisted Intervention, Munich, Germany, 5–9 October 2015; pp. 234–241.
4. Hwang, H.; Rehman, H.Z.U.; Lee, S. 3D U-Net for skull stripping in brain MRI. *Appl. Sci.* **2019**, *9*, 569. [CrossRef]
5. Vorontsov, E.; Tang, A.; Pal, C.; Kadoury, S. Liver lesion segmentation informed by joint liver segmentation. In Proceedings of the 2018 IEEE 15th International Symposium on Biomedical Imaging (ISBI), Washington, DC, USA, 4–7 April 2018; pp. 1332–1335.
6. Lu, F.; Wu, F.; Hu, P.J.; Peng, Z.Y.; Kong, D.X. Automatic 3D liver location and segmentation via convolutional neural network and graph cut. *Int J. CARS* **2017**, *12*, 171–182. [CrossRef] [PubMed]
7. Kaluva, K.C.; Khened, M.; Kori, A.; Krishnamurthi, G. 2D-Densely Connected Convolution Neural Networks for automatic Liver and Tumor Segmentation. *arXiv* **2018**, arXiv:1802.02182.
8. Zhao, F.; Gao, P.; Hu, H.; He, X.; Hou, Y.; He, X. Efficient kidney segmentation in micro-CT based on multi-atlas registration and random forests. *IEEE Access* **2018**, *6*, 43712–43723. [CrossRef]
9. Moccia, S.; Momi, E.; Hadji, S.E.; Mattos, L.S. Efficient kidney segmentation in micro-CT based on multi-atlas registration and random forests. *Comput. Meth. Prog Bio.* **2018**, *158*, 71–91. [CrossRef] [PubMed]
10. Jin, Q.G.; Meng, Z.P.; Tuan, D.P.; Chen, Q.; Wei, L.Y.; Su, R. DUNet: A deformable network for retinal vessel segmentation. *arXiv* **2018**, arXiv:1811.01206.
11. Tetteh, G.; Efremov, V.; Forkert, N.D.; Schneider, M.; Kirschke, J.; Weber, B.; Zimmer, C.; Piraud, M.; Menze, B.H. DeepVesselNet: vessel segmentation, centerline prediction, and bifurcation detection in 3-D angiographic volumes. *arXiv* **2018**, arXiv:1803.09340.
12. Cai, J.L.; Lu, L.; Xing, F.Y.; Yang, L. Pancreas segmentation in CT and MRI images via domain specific network designing and recurrent neural contextual learning. *arXiv* **2018**, arXiv:1803.11303.

13. Roth, H.R.; Lu, L.; Lay, N.; Harrison, A.P.; Farag, A.; Sohn, A.; Summers, R.M. Spatial aggregation of holistically-nested convolutional neural networks for automated pancreas localization and segmentation. *Med. Image Anal.* **2018**, *45*, 94–107. [CrossRef] [PubMed]

14. Yu, Q.H.; Xie, L.X.; Wang, Y.; Zhou, Y.Y.; Fishman, E.K.; Yuille, A.L. Recurrent Saliency Transformation Network: Incorporating Multi-Stage Visual Cues for Small Organ Segmentation. In Proceedings of the 2018 IEEE Conference on Computer Vision and Pattern Recognition (CVPR), Salt Lake City, UT, USA, 18–22 June 2018; pp. 8280–8289.

15. Taha, A.A.; Hanbury, A. Metrics for evaluating 3D medical image segmentation: Analysis, selection, and tool. *BMC Med. Imaging* **2015**, *15*, 29. [CrossRef] [PubMed]

16. Lin, T.Y.; Goyal, P.; Girshick, R.; He, K.; Dollar, P. Focal loss for dense object detection. In Proceedings of the IEEE International Conference on Computer Vision (ICCV), Venice, Italy, 22–29 October 2017; pp. 2980–2988.

17. Long, J.; Shelhamer, E.; Darrell, T. Fully Convolutional Networks for Semantic Segmentation. In Proceedings of the IEEE Conference on Computer Vision and Pattern Recognition (CVPR), Boston, MA, USA, 7–12 June 2015; pp. 3431–3440.

Article

A Joint Training Model for Face Sketch Synthesis

Weiguo Wan [1] and Hyo Jong Lee [1,2,*]

[1] Division of Computer Science and Engineering, Chonbuk National University, Jeonju 54896, Korea; wanwgplus@gmail.com
[2] Center for Advanced Image and Information Technology, Chonbuk National University, Jeonju 54896, Korea
* Correspondence: hlee@chonbuk.ac.kr

Received: 18 March 2019; Accepted: 24 April 2019; Published: 26 April 2019

Featured Application: The proposed face sketch synthesis method can be applied for various applications, such as law enforcement and digital entertainment.

Abstract: The exemplar-based method is most frequently used in face sketch synthesis because of its efficiency in representing the nonlinear mapping between face photos and sketches. However, the sketches synthesized by existing exemplar-based methods suffer from block artifacts and blur effects. In addition, most exemplar-based methods ignore the training sketches in the weight representation process. To improve synthesis performance, a novel joint training model is proposed in this paper, taking sketches into consideration. First, we construct the joint training photo and sketch by concatenating the original photo and its sketch with a high-pass filtered image of their corresponding sketch. Then, an offline random sampling strategy is adopted for each test photo patch to select the joint training photo and sketch patches in the neighboring region. Finally, a novel locality constraint is designed to calculate the reconstruction weight, allowing the synthesized sketches to have more detailed information. Extensive experimental results on public datasets show the superiority of the proposed joint training model, both from subjective perceptual and the FaceNet-based face recognition objective evaluation, compared to existing state-of-the-art sketch synthesis methods.

Keywords: face sketch synthesis; face sketch recognition; joint training model

1. Introduction

Face sketch synthesis is a key branch of face style transformation, which generates face sketches for given input photos with the help of face photo-sketch pairs as the training dataset [1]. It has achieved wide applications in both law enforcement and digital entertainment. For example, sketches drawn according to the description of victims or witnesses can help identify a suspect by matching the sketch against a mugshot dataset from a police department. Face sketch synthesis reduces the texture discrepancy between photos and sketches for the face recognition procedure [2] and thus increases the recognition accuracy [3]. In digital entertainment, people are increasingly preferring to use face sketches as their portrait in social media; the sketch synthesis technique can also simplify animation production [4].

During the past two decades, various sketch synthesis methods have been proposed. The exemplar-based method is an important category of existing synthesis approaches. It synthesizes sketches for test photos by utilizing photo-sketch pairs as training data. The exemplar-based method mainly consists of neighbor selection and reconstruction weight representation [5]. In the neighbor selection process, K nearest training photo patches are selected for a test photo patch. In the reconstruction weight representation, a weight vector between the test photo patch and the selected photo patches is calculated. The target sketch patch can be obtained using weighted averaging of the K training sketch patches corresponding to the selected photo patches with the calculated weight vector. The final sketch is obtained by averaging all the generated sketch patches.

Exemplar-based face sketch synthesis originates from the Eigen-transformation research by Tang et al. [6]. In their work, all training photo-sketch pairs were used to generalize the target sketch. Principal component analysis was adopted to learn the weight coefficients by projecting the input test photo onto the training photos.

It is difficult to represent nonlinear relationships between face photos and sketches by only learning one holistic reconstruction model. Thus, Liu et al. [7] proposed a locally linear embedding (LLE)-based sketch synthesis method to estimate the nonlinear mapping with piecewise linear mappings. The LLE method works at the image patch level, in which K nearest training photo patches are searched in terms of Euclidean distance for each test photo patch. However, the LLE method suffers from a serious noise problem. To resolve this problem, Song et al. [8] formulated face sketch synthesis into a spatial sketch denoising problem and calculated the reconstruction weight using the conjugate gradient solver.

To describe the dependency relationship between neighboring sketches, Wang et al. [9] introduced a multi-scale Markov random fields (MRF) model to represent the neighboring constraint between adjacent sketch patches using a compatibility function. However, this method only chose the best single sketch patch from the training data for the test photo patch, meaning it could not synthesize new sketch patches. Additionally, the optimization process in the MRF model is an NP-hard problem. To overcome these limitations, Zhou et al. [10] extended the MRF model by introducing the linear combination of nearest neighbors to structure a Markov weight fields (MWF) model, which is capable of synthesizing new sketch patches that do not exist in the training dataset. In reference [11], a sparse representation-based face sketch synthesis method was proposed by Gao et al. Peng et al. [12] proposed a multiple representation-based face sketch synthesis method, which is able to obtain high-quality sketch images. However, this method is time-consuming because of the online neighbor selection.

Recently, Wang et al. [13] proposed a state-of-the-art face sketch synthesis method, based on random sampling and locality constraint (RSLCR). They randomly sampled the training photo and sketch patches in place of a neighbor search, and then employed the locality constraint to model the distinct correlations between the test photo patch and sampled photo patches while calculating the reconstruction weight coefficients. However, the target sketch patch reconstruction obtained by weighted-averaging the hundreds of sampled sketches can be regarded as a low-pass filter process, which results in blurred synthesized sketches. In addition, the Bayesian inference was utilized in reference [14] to incorporate the neighboring constraint in both the neighbor selection and reconstruction weight representation, which can obtain impressive performance.

Apart from the exemplar-based method, deep learning techniques were also applied to face sketch synthesis, generating new trends. Zhang et al. [15] first proposed a fully convolutional network (FCN) to learn end-to-end mapping from photos to sketches. It consisted of six convolutional layers with rectified linear units as activation functions. Zhang et al. [16] utilized the branched FCN to structure a decomposition representation learning framework for sketch synthesis. Additionally, the generative adversarial network (GAN) [17] was developed for image style transformation (e.g., photo-to-sketch generation or vice versa). The deep learning-based sketch synthesis methods can preserve textural structure well; however, serious noise effects occur in the synthesized results. This is mainly because of the limited available of training data, which is insufficient to train large networks robustly [18].

A common problem with these exemplar-based approaches is that they ignore the role of training sketches when calculating reconstruction weights. This is because the basic assumption of these exemplar-based methods is that a photo patch and its corresponding sketch patch have a similar geometric manifold structure. If two photo patches are similar, then their sketch patch counterparts are also similar. However, owing to potential misalignment, the reconstruction weight obtained from the test photo patch and selected training photo patches may not be suitable for sketch patches reconstruction [19].

We propose a new exemplar-based method, the joint training model, to solve the problem. In our method, the training photo and the sketch patches are concatenated. Instead of directly using the sketch patches, the high-pass filtered component of the training sketches are adopted to reduce the effect of the modality difference between a photo and a sketch. Then, we employ the offline random sampling method to select the joint training photo and the sketch patches for the test photo patch. Moreover, a modified locality constraint is designed to calculate the reconstruction weight. With the obtained reconstruction weight, the target sketch patch can be synthesized. Experimental results indicate that the proposed joint model significantly eliminates noise and improves the synthesized sketch quality. It also preserves the detail information of the test photo, which other methods cannot do. Therefore, synthesized face sketch images using our method achieve higher accuracy in face sketch recognition.

The contributions of this paper are summarized as below.

(1) To consider the training sketches during the reconstruction weight representation process, a joint training model is proposed to integrate the training photo and sketch information.
(2) We design a modified locality constraint that modulates the reconstruction weight through the distance between the high-pass filtered images of test patches and the sampled training sketch patches.
(3) The proposed method yields high quality sketches with more detail information and less noise over the wide range of datasets, promoting the accuracy of the sketch-based suspect identification.

The organization of the rest of the paper is as follows. Section 2 introduces the relevant works, including some example-based sketch synthesis methods, which are the basic works of the proposed method. The proposed model is described in detail in Section 3. Section 4 provides a comparison of experiments and their results. Conclusions are then given in Section 5.

2. Technical Backgrounds

In this paper, excepted as noted, a bold uppercase letter and a bold lowercase letter represent a matrix and a column vector, respectively; regular uppercase and lowercase letters denote scalars. Given a test photo $\mathbf{T}$, it is divided into patches $\mathbf{t}^{(i,j)}$ with r pixels overlapping between neighboring patches. (i, j) denotes the location of the patch at the i-th row and the j-th column, $i \in \{1, \cdots, m\}$, $j \in \{1, \cdots, n\}$. Notice that each patch is represented as a q-dimensional column vector, where $q = p^2$, and p is the size of the patch. Similarly, the target sketch is denoted as $\mathbf{S}$. $\mathbf{s}^{(i,j)}$ denotes the target sketch patch corresponding to the testing patch $\mathbf{t}^{(i,j)}$. The training dataset, which consists of M photo-sketch pairs, are similarly divided into patches. Let $\mathbf{X}^{(i,j)} = \{\mathbf{x}_k^{(i,j)}\}_{k=1}^K$ and $\mathbf{Y}^{(i,j)} = \{\mathbf{y}_k^{(i,j)}\}_{k=1}^K$ denote the set of K selected training photo patches and the corresponding sketch patches of the test photo patch $\mathbf{t}^{(i,j)}$, respectively. The weight coefficients $\mathbf{w}^{(i,j)} = (w_1^{(i,j)}, \cdots, w_K^{(i,j)})^T$ are calculated to linearly combine the candidate sketch patches.

2.1. The LLE Method

In the LLE method [6], K nearest patches are first obtained for each test photo patch $\mathbf{t}^{(i,j)}$. Then, the linear reconstruction coefficients $\mathbf{w}^{(i,j)}$ can be calculated by resolving the following minimization problem:

$$\min_{\mathbf{w}^{(i,j)}} \|\mathbf{t}^{(i,j)} - \mathbf{X}^{(i,j)}\mathbf{w}^{(i,j)}\|_2^2, \ s.t. \ \mathbf{1}^T\mathbf{w}^{(i,j)} = 1, \tag{1}$$

Then the target sketch patch $\mathbf{s}^{(i,j)}$ can then be synthesized.

$$\mathbf{s}^{(i,j)} = \sum_{k=1}^{K} w_k^{(i,j)} y_k^{(i,j)} = \mathbf{Y}^{(i,j)} \cdot \mathbf{w}^{(i,j)}, \tag{2}$$

After all target sketch patches are generated, the final sketch can be achieved by averaging the overlapped pixel intensities.

2.2. The RSLCR Method

Instead of searching the nearest neighbor photo patches, the RSLCR method [13] proposed to randomly sample K photo-sketch patch pairs from training data in a predicted neighbor region for the test patch $\mathbf{t}^{(i,j)}$. Additionally, to consider the correlation between different sampled patches, a locality constraint [20] was introduced to impose a weight to the distances of the test photo patch and random sampled photo patches. The reconstruction weight representation model of the RSLCR method can be written as follows.

$$\min_{\mathbf{w}^{(i,j)}} \|\mathbf{t}^{(i,j)} - \mathbf{X}^{(i,j)}\mathbf{w}^{(i,j)}\|_2^2 + \lambda \|\mathbf{d}^{(i,j)} \odot \mathbf{w}^{(i,j)}\|, \; s.t.\, \mathbf{1}^T \mathbf{w}^{(i,j)} = 1, \tag{3}$$

where $\odot$ denotes element-wise multiplication, λ balances the reconstruction error and the locality constraint, and $\mathbf{d}^{(i,j)}$ is the Euclidean distance vector between the test photo patch $\mathbf{t}^{(i,j)}$ and sampled training photo patches $\mathbf{X}^{(i,j)}$.

3. Joint Training Model for Face Sketch Synthesis

In most exemplar-based face sketch synthesis methods, only the test photo and training photos are considered for selecting the candidate photo patches and calculating the reconstruction weight. This strategy cannot achieve an optimal result when the training photo and sketch are misaligned. In this paper, we put forward a novel exemplar-based face sketch synthesis approach, which takes the training sketch into account by joining it with its corresponding training photo for reconstruction weight representation.

Figure 1 shows the illustration of the proposed face sketch synthesis method. First, the high-pass filtered components of the sketch images in the training data are extracted by using the Laplacian of Gaussian (LoG) filter. Then, the extracted high-pass filtered components are attached to their corresponding photo and sketch images to form the joint training data. After that, the candidate joint patch pairs are sampled with the offline random sampling strategy. In the test phase, the high-pass filtered component of the input photo is extracted with the same filter and attached to the input photo to obtain the joint test photo. For each joint test patch, a reconstruction weight can be calculated by approximating the joint test patch and joint training photos. Finally, the target patch can be constructed by linearly combining the sampled joint training sketches with the reconstruction weight and the final sketch image can be synthesized by averaging the overlapping sketch patches. The strategy to structure the joint training model and the way to calculate the reconstruction weight coefficients of the proposed method will be explained next in detail.

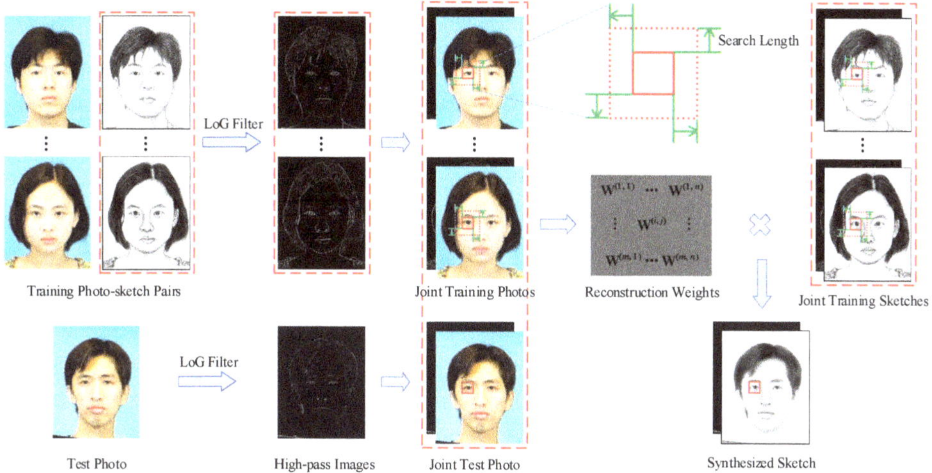

Figure 1. Illustration of the proposed joint model for face sketch synthesis.

3.1. Joint Training Model

For each training sketch, the corresponding high-pass filtered component is first obtained with a LoG filter. Let $\mathbf{Z}^{(i,j)} = \{\mathbf{z}_k^{(i,j)}\}_{k=1}^{K}$ denote the set of K randomly selected high-pass filtered image patches of the training sketches corresponding to a test patch $\mathbf{t}^{(i,j)}$. Then, the sampled K joint training photo and sketch patches for $\mathbf{t}^{(i,j)}$ can be denoted as Equations (4) and (5), respectively.

$$\mathbf{U}^{(i,j)} = \begin{pmatrix} \mathbf{X}^{(i,j)} \\ \mathbf{Z}^{(i,j)} \end{pmatrix} = \left\{ \begin{pmatrix} \mathbf{x}_k^{(i,j)} \\ \mathbf{z}_k^{(i,j)} \end{pmatrix} \right\}_{k=1}^{K}, \tag{4}$$

$$\mathbf{V}^{(i,j)} = \begin{pmatrix} \mathbf{Y}^{(i,j)} \\ \mathbf{Z}^{(i,j)} \end{pmatrix} = \left\{ \begin{pmatrix} \mathbf{y}_k^{(i,j)} \\ \mathbf{z}_k^{(i,j)} \end{pmatrix} \right\}_{k=1}^{K}, \tag{5}$$

For the test photo $\mathbf{T}$, the high-pass image $\mathbf{H}$ is obtained with a LoG filter. Let $\mathbf{h}^{(i,j)}$ denotes the high-pass image patch corresponding to the test photo patch $\mathbf{t}^{(i,j)}$. We concatenate these two patches as the joint test patch $\mathbf{t}_1^{(i,j)}$:

$$\mathbf{t}_1^{(i,j)} = \begin{pmatrix} \mathbf{t}^{(i,j)} \\ \mathbf{h}^{(i,j)} \end{pmatrix}. \tag{6}$$

After modeling the joint training photos and sketches, the reconstruction weight can be calculated by approximating the joint test patch and joint training photos, which will be discussed next.

3.2. Face Sketch Synthesis

Assuming there are M pairs of joint training photos and sketches that are geometrically aligned, they are divided into patches of fixed size ($p \times p \times 2$). Each joint patch is reshaped to a $2q$-dimensional column vector. For each joint test patch location, we extend the search region around the patch with c pixels. Thus, there are $(2c + 1)^2$ patches in the search region for one patch location, and there are $(2c + 1)^2 M$ joint training photo/sketch patch-pairs. Among these patch-pairs, K joint training photo patches $\mathbf{U}^{(i,j)} \in \mathbb{R}^{2p^2 \times K}$ and joint training sketch patches $\mathbf{V}^{(i,j)} \in \mathbb{R}^{2p^2 \times K}$ are randomly and simultaneously selected.

For each joint test patch $\mathbf{t}_1^{(i,j)}$, the reconstruction weight is calculated as follow:

$$\min_{\mathbf{w}^{(i,j)}}\|\mathbf{t}_1^{(i,j)} - \mathbf{U}^{(i,j)}\mathbf{w}^{(i,j)}\|_2^2 + \lambda\|\mathbf{d}_1^{(i,j)} \odot \mathbf{w}^{(i,j)}\| + \mu\|\mathbf{d}_2^{(i,j)} \odot \mathbf{w}^{(i,j)}\|, \quad s.t.\, 1^T\mathbf{w}^{(i,j)} = 1, \tag{7}$$

where $\mathbf{w}^{(i,j)} \in \mathbb{R}^{K\times 1}$ is the weight coefficient for the joint test patch $\mathbf{t}_1^{(i,j)}$. $\mathbf{d}_1^{(i,j)} \in \mathbb{R}^{K\times 1}$ is the Euclidean distance vector between the joint test patch $\mathbf{t}_1^{(i,j)}$ and sampled joint training photo patches $\mathbf{U}^{(i,j)}$, and $\mathbf{d}_2^{(i,j)} \in \mathbb{R}^{K\times 1}$ is the Euclidean distance vector between the LoG test patch image $\mathbf{h}^{(i,j)}$ and K sampled LoG training sketch patches $\mathbf{Z}^{(i,j)}$.

Equation (7) has the closed-form solution:

$$\mathbf{aw\prime}^{(i,j)} = (\mathbf{C}^{i,j} + \lambda\mathrm{diag}(\mathbf{d}_1^{(i,j)}) + \mu\mathrm{diag}(\mathbf{d}_2^{(i,j)}))\backslash 1, \tag{8}$$

$$\mathbf{w}^{(i,j)} = \mathbf{w\prime}^{(i,j)}/1^T\mathbf{w\prime}^{(i,j)}, \tag{9}$$

where $\mathbf{1}$ is a column vector in which all elements are 1. $\mathbf{C}^{i,j} = (\mathbf{U}^{(i,j)} - 1\mathbf{t}_1^{(i,j)^T})(\mathbf{U}^{(i,j)} - 1\mathbf{t}_1^{(i,j)^T})^T$ denotes the data covariance matrix, and $\mathrm{diag}(\cdot)$ extends the vector into a diagonal matrix.

The target joint sketch patch $\mathbf{s}^{(i,j)}$ can be synthesized by linearly combining the sampled joint training sketches with the reconstruction weight coefficients $\mathbf{w}^{(i,j)}$.

$$\mathbf{s}^{(i,j)} = \mathbf{V}^{(i,j)}\mathbf{w}^{(i,j)}. \tag{10}$$

The obtained joint sketch patch is a $2q$-dimensional vector. Thus, we extract the first half and reshape it to a $p \times p$ patch, which is the target sketch patch. After obtaining all target sketch patches, the final target sketch can be achieved with an averaged overlapping area.

4. Evaluation Experiments

4.1. Datasets

We evaluated the performance of the proposed method on two publicly available datasets: The Chinese University of Hong Kong (CUHK) face sketch (CUFS) dataset [9] and the CUHK face sketch FERET (CUFSF) dataset [21]. The CUFS dataset includes three sub-datasets: The CUHK student dataset (188 subjects) [22], the AR dataset (123 subjects) [23], and the XM2VTS dataset (295 subjects) [24]. The CUFSF dataset includes 1194 subjects from the FERET dataset [25]. Artist drew sketches corresponding to each face photo in these datasets. In our experiments, all face images were normalized into the size of 200×250 by centering the coordinates on two eyes and the mouth. Some face photo-sketch pairs from these two datasets are shown in Figure 2.

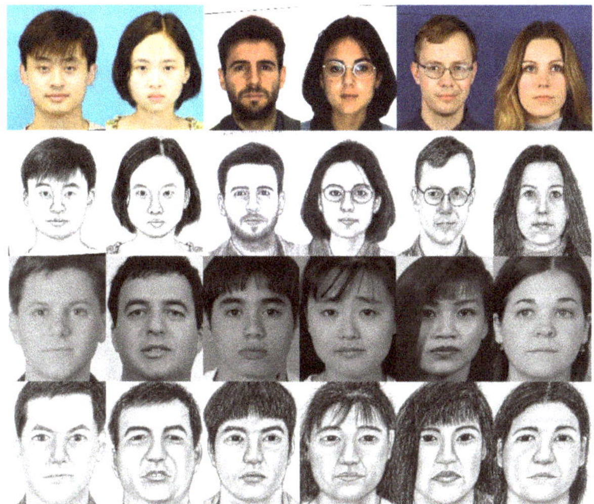

Figure 2. Example of face sketch-photo pairs in the CUFS dataset (first two rows) and the CUFSF dataset (last two rows). The first and the third row are face photos and the second and the last rows are corresponding face sketches drawn by the artist.

4.2. Experimental Setting

In this section, the data distribution and parameter settings are introduced. For fair comparison, we employed the same cross validation technique used in most exemplar-based sketch synthesis works. For the CUFS dataset, 88 face photo-sketch pairs were taken as the training dataset and the remaining 100 pairs were taken for test in the CUHK student dataset; 80 pairs were chosen for training and the remaining 43 pairs for test in AR dataset; 100 pairs for training and the remaining 195 pairs for test in XM2VTS dataset. For the CUFSF dataset, 250 face photo-sketch pairs were randomly chosen for training and the rest of the 944 pairs for test.

Parameters were set as follow. Patch size was $p = 20$, overlap size was $o = 14$, search length was $c = 5$, the number for random sampling was $K = 800$, and the regularization parameter λ and μ were both set to 0.5. The window size and standard deviation of LoG filter were 5 and 0.5, respectively.

To evaluate performance, four traditional exemplar-based methods (LLE [7], MRF [9], MWF [10], and RSLCR [13]) and two deep learning-based methods (FCN [15] and GAN [17]) were compared. These six state-of-the-art face sketch synthesis results are released by Wang et al. [13].

4.3. Synthesizesd Sketch Results Comparison

Figure 3 shows some synthesized face sketches from different methods on the CUFS dataset. The first two rows are from the CUHK student dataset, the middle two rows are from the AR dataset, and the last two rows are from the XM2VTS dataset. Moreover, the corresponding local blocks of the synthesized sketches are displayed in Figure 4. From Figure 3, it can be seen that the sketches synthesized by the LLE and MRF methods suffer serious block effects. The MWF method obtains better performance than the LLE and MRF methods in the CUHK student and AR datasets. However, the results are unsatisfying in the XM2VTS dataset, because it contains more face variations, such as aging, race, and hair styles. The RSLCR method generates fine textures and structures, because more candidate patches are incorporated via random sampling and the locality constraint. However, this method results in blurred outputs. The FCN and GAN methods overcome the blurring effect by using pixel-to-pixel mapping from a photo to a sketch. However, they tend to have undesirable artifacts because of instabilities in training, while generating high-resolution images. Our proposed method achieves much better performance than these six benchmarked methods in all the three datasets. More detailed information is preserved in the synthesized sketches, such as the double-fold eyelids and hair grain. This illustrates that the proposed method is capable of generating identity-preserved sketches.

Figure 3. Synthesized sketches on the CUFS dataset by locally linear embedding (LLE) [7], Markov random fields (MRF) [9], Markov weight fields (MWF) [10], random sampling and locality constraint (RSLCR) [13], fully convolutional network (FCN) [15], generative adversarial network (GAN) [17], and our proposed method, respectively. Face photos in the first two rows are from the CUHK student dataset; second two rows are from the AR dataset; and the last two rows are from the XM2VTS dataset, respectively.

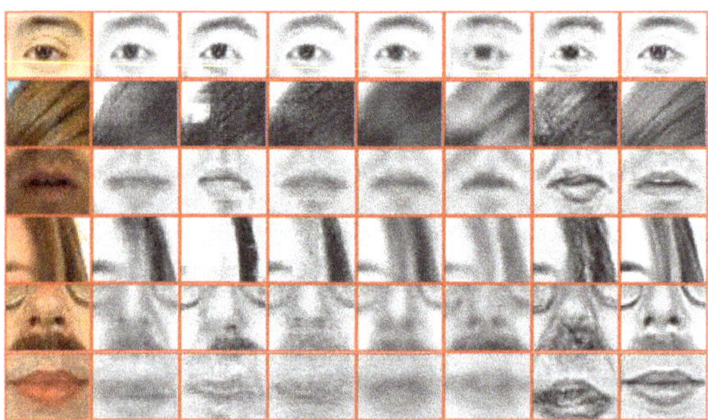

Figure 4. Local cut-out effect on Figure 3; same marshalling sequence as Figure 3.

We also investigated the robustness of the proposed method against shape exaggeration and illumination variations on the CUFSF dataset. Figure 5 shows some synthesized face sketches from different methods used on the CUFSF dataset. The block effect still exists in the LLE and MRF results. The MWF and RSLCR methods obtain similar performance on the CUFSF dataset. The FCN method suffers from serious noise and artifacts. The GAN results show much improvement, but distortion occurs in the synthesized sketches by this method. By comparison, the proposed method achieves the most vivid and clear sketches, reflecting the robustness of our proposed method.

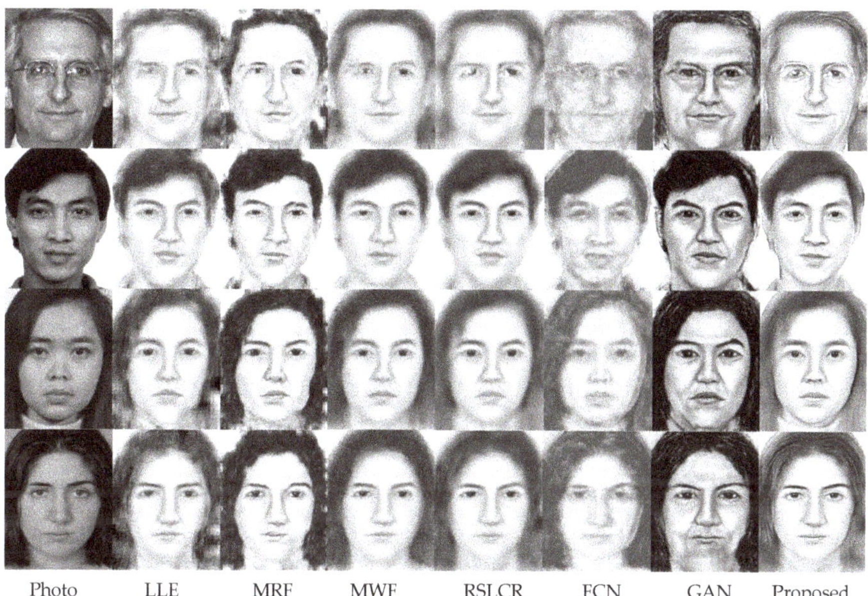

Figure 5. Synthesized sketches on the CUFSF dataset by LLE [7], MRF [9], MWF [10], RSLCR [13], FCN [15], GAN [17], and the proposed method, respectively.

Overall, the synthesized face sketch images on the CUFS and CUFSF datasets by our method have the following superiorities compared to the benchmarked methods: (1) Less block artifacts and noise, because we calculated the target sketch patch by weighted averaging hundreds randomly sampled training sketch patches but not few nearest patches; (2) rich facial detail, due to the high-pass filtered components were adopted to build the joint training model, which is able to generate high-quality face sketch patches; (3) complete facial structures, as a result of the training sketch images were took into consideration when computing the reconstruction weights, which weakens the influence by the misalignment in training images.

Table 1 shows the average time consumption of different methods on different datasets. Here, only exemplar-based methods are compared, because it takes a very long time to train the neural network for deep learning-based methods, though the test time is quite fast once the model is trained. From Table 1, it can be seen that the LLE, MRF, and MWF methods have no scalability of training data. With the amplification of training data, the running time increased radically, such as the CUFSF dataset. The RSLCR and our proposed method were less susceptible to the size of training data, owing to the random sampling strategy. Although the proposed method is not the fastest, it still has comparable time consumption as the other methods.

Table 1. Average running time (s) to generate one sketch by different methods.

Methods	LLE	MRF	MWF	RSLCR	Proposed
CUHK	536.34	8.60	16.10	18.79	20.80
AR	496.47	8.40	15.33	19.10	19.75
XM2VTS	642.50	10.40	18.80	18.14	20.78
CUFSF	1591.95	24.25	45.20	17.66	20.17

4.4. Face Sketch Recognition

Face sketch recognition is commonly used to quantitatively evaluate the face sketch synthesis methods and to collectively compare the synthesized sketch images [8,15,26]. A higher face sketch recognition rate means that the corresponding sketch synthesis method is more effective and the synthesized sketch images are better. In this study, FaceNet [27] was employed to conduct the face sketch recognition experiments. To demonstrate the recognition performance of the synthesized sketches using our proposed method, we used the sketches synthesized using different methods as probe images to match the gallery images, consisting of the corresponding artist-drawn sketches. The 338 synthesized sketches in the CUFS dataset were taken as the probe set, and the corresponding ground-truth sketches drawn by the artist were taken as the gallery set. For the CUFS dataset, 944 synthesized sketches were taken as the probe set and the corresponding sketches drawn by the artist were taken as the gallery set.

Figure 6 shows the face sketch recognition accuracies of FaceNet on the CUFS and the CUFSF dataset, respectively. The proposed method achieved the best accuracy on both datasets, 97.04% in CUFS and 87.18% in CUFS, at rank-100, respectively. Table 2 shows the rank-1, rank-5, and rank-10 recognition rates, where rank-n measures the accuracy of the top-n best matches. The FCN and GAN methods got higher recognition accuracy on CUFS dataset, but similar accuracy on CUFSF dataset compared with the traditional exemplar-based methods. It indicates that the performance of deep learning-based methods degrades when the dataset has challenging variations. The synthesized sketches of our method obtained the highest rate in rank-1, rank-5, and rank-10. The recognition results indicated that the better generated texture features and more detailed information preserved by the proposed method contributes to the face sketch recognition. This further demonstrates the superiority of our proposed method.

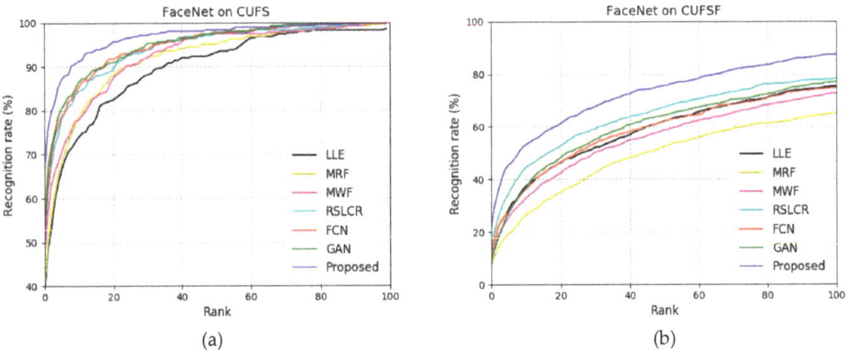

Figure 6. Face sketch recognition accuracies of FaceNet on the CUFS (**a**) and the CUFSF (**b**) datasets.

Table 2. Recognition accuracies (%) of FaceNet on the CUFS and CUFSF datasets.

Methods	CUFS			CUFSF		
	rank-1	rank-5	rank-10	rank-1	rank-5	rank-10
LLE	38.7	63.6	72.8	11.3	26.3	36.2
MRF	39.6	65.1	76.3	7.2	18.0	25.3
MWF	47.3	68.0	77.5	10.6	23.6	31.5
RSLCR	51.7	74.2	83.4	14.3	32.4	43.1
FCN	51.4	76.3	84.0	11.4	26.4	35.2
GAN	52.6	78.7	84.9	9.2	24.9	35.6
Proposed	65.4	85.2	90.5	21.2	43.6	52.1

5. Conclusions

Without considering the training sketches in the reconstruction weight representation process, the exemplar-based face sketch synthesis method had difficulty generating ideal results. This paper proposed a joint training model by concatenating the original training photos and sketches with high-pass filtered image patches of the training sketches. Additionally, we constructed a new locality constraint in the reconstruction weight process. With these improvements, more detailed information was preserved in the synthesized sketches. Experimental results demonstrated that the proposed method not only reduced the noise, but it also increased the definition of the synthesized sketches. Thus, the proposed joint training model is a practical and effective technique for face sketch synthesis. As analyzed and discussed previously, the deep learning-based face sketch synthesis methods are immature. In the future, a deep learning-based synthesis approach will be explored by employing more training data and designing optimized networks.

Author Contributions: Conception and design of the proposed method: H.J.L and W.W.; performance of the experiments: W.W.; writing of the paper: W.W.; paper review and editing: H.J.L.

Funding: This research was supported by "Research Base Construction Fund Support Program" funded by Chonbuk National University in 2018. This research was also supported by Basic Science Research Program through the National Research Foundation of Korea (NRF) funded by the Ministry of Education (GR 2016R1D1A3B03931911). This study was also financially supported by the grants of China Scholarship Council (CSC No.2017 08260057).

Conflicts of Interest: The authors declare no conflict of interest

References

1. Wang, N.; Zhang, S.; Gao, X.; Li, J.; Song, B.; Li, Z. Unified framework for face sketch synthesis. *Signal Process.* **2017**, *130*, 1–11. [CrossRef]
2. Li, J.; Yu, X.; Peng, C.; Wang, N. Adaptive representation-based face sketch-photo synthesis. *Neurocomputing* **2017**, *269*, 152–159. [CrossRef]
3. Wan, W.; Lee, H.J. FaceNet Based Face Sketch Recognition. In Proceedings of the 2017 International Conference on Computational Science and Computational Intelligence, Las Vegas, NV, USA, 14–16 December 2017; pp. 432–436.
4. Zhang, Y.; Wang, N.; Zhang, S.; Li, J.; Gao, X. Fast face sketch synthesis via KD-tree search. In Proceedings of the European Conference on Computer Vision, Amsterdam, The Netherland, 8–16 October 2016; Springer: Cham, Switzerland, 2016; pp. 64–77.
5. Wang, N.; Zhu, M.; Li, J.; Song, B.; Li, Z. Data-driven vs. model-driven: Fast face sketch synthesis. *Neurocomputing* **2017**, *257*, 214–221. [CrossRef]
6. Tang, X.; Wang, X. Face sketch recognition. *IEEE Trans. Circuits Syst. Video Technol.* **2004**, *14*, 1–7. [CrossRef]
7. Liu, Q.; Tang, X.; Jin, H.; Lu, H.; Ma, S. A nonlinear approach for face sketch synthesis and recognition. In Proceedings of the IEEE Conference on Computer Vision and Pattern Recognition, San Diego, CA, USA, 20–25 June 2005; pp. 1005–1010.

8. Song, Y.; Bao, L.; Yang, Q.; Yang, M. Real-time exemplar-based face sketch synthesis. In Proceedings of the European Conference on Computer Vision, Zurich, Switzerland, 5–12 September 2014; Springer: Cham, Switzerland, 2014; pp. 800–813.

9. Wang, X.; Tang, X. Face photo-sketch synthesis and recognition. *IEEE Trans. Pattern Anal. Mach. Intell.* **2009**, *31*, 1955–1967. [CrossRef] [PubMed]

10. Zhou, H.; Kuang, Z.; Wong, K. Markov weight fields for face sketch synthesis. In Proceedings of the IEEE Conference on Computer Vision and Pattern Recognition, Rhode Island, USA, 18–20 June 2012; pp. 1091–1097.

11. Gao, X.; Wang, N.; Tao, D.; Li, X. Face sketch–photo synthesis and retrieval using sparse representation. *IEEE Trans. Circuits Syst. Video Technol.* **2012**, *22*, 1213–1226. [CrossRef]

12. Peng, C.; Gao, X.; Wang, N.; Tao, D.; Li, X.; Li, J. Multiple representations-based face sketch–photo synthesis. *IEEE Trans. Neural Netw. Learn. Syst.* **2016**, *27*, 2201–2215. [CrossRef] [PubMed]

13. Wang, N.; Gao, X.; Li, J. Random sampling for fast face sketch synthesis. *Pattern Recognit.* **2018**, *76*, 215–227. [CrossRef]

14. Wang, N.; Gao, X.; Sun, L.; Li, J. Bayesian face sketch synthesis. *IEEE Trans. Image Process.* **2017**, *26*, 1264–1274. [CrossRef] [PubMed]

15. Zhang, L.; Lin, L.; Wu, X.; Ding, S.; Zhang, L. End-to-end photo-sketch generation via fully convolutional representation learning. In Proceedings of the 5th ACM on International Conference on Multimedia Retrieval, Shanghai, China, 23–26 June 2015; pp. 627–634.

16. Zhang, D.; Lin, L.; Chen, T.; Wu, X.; Tan, W.; Izquierdo, E. Content-adaptive sketch portrait generation by decompositional representation learning. *IEEE Trans. Image Process.* **2017**, *26*, 328–339. [CrossRef] [PubMed]

17. Goodfellow, I.J.; Pouget-Abadie, J.; Mirza, M.; Xu, B.; Warde-Farley, D.; Ozair, S.; Courville, A.; Bengio, Y. Generative adversarial nets. In Proceedings of the International Conference on Neural Information Processing Systems, Montreal, QC, Canada, 8–13 December 2014; MIT: Cambridge, MA, USA; pp. 2672–2680.

18. Jiang, J.; Yu, Y.; Wang, Z.; Liu, X.; Ma, J. Graph-Regularized Locality-Constrained Joint Dictionary and Residual Learning for Face Sketch Synthesis. *IEEE Trans. Image Process.* **2019**, *28*, 628–641. [CrossRef] [PubMed]

19. Wang, N.; Gao, X.; Sun, L.; Li, J. Anchored neighborhood index for face sketch synthesis. *IEEE Trans. Circuits Syst. Video Technol.* **2018**, *28*, 2154–2163. [CrossRef]

20. Wang, J.; Yang, J.; Yu, K.; Lv, F.; Huang, T.; Gong, Y. Locality-constrained linear coding for image classification. In Proceedings of the IEEE Conference on Computer Vision and Pattern Recognition, San Francisco, CA, USA, 13–18 June 2010; pp. 3360–3367.

21. Zhang, W.; Wang, X.; Tang, X. Coupled information-theoretic encoding for face photo-sketch recognition. In Proceedings of the IEEE Conference on Computer Vision and Pattern Recognition, Colorado Springs, CO, USA, 21–25 June 2011; pp. 513–520.

22. Tang, X.; Wang, X. Face photo recognition using sketch. In Proceedings of the IEEE International Conference on Image Processing, New York, NY, USA, 22–25 September 2002; pp. 257–260.

23. Martinez, A.; Benavente, R. *The AR Face Database*; Technical Report; CVC: Barcelona, Spain, 1998.

24. Messer, K.; Matas, J.; Kittler, J.; Luettin, J.; Maitre, G. XM2VTSDB: The extended M2VTS database. In Proceedings of the International Conference on Audio- and Video-Based Biometric Person Authentication, Washington, DC, USA, 22–24 March 1999; pp. 965–966.

25. Phillips, P.; Moon, H.; Rauss, P.; Rizvi, S. The FERET evaluation methodology for face recognition algorithms. *IEEE Trans. Pattern Anal. Mach. Intell.* **2000**, *22*, 1090–1104. [CrossRef]

26. Chen, C.; Tan, X.; Wong, K.K. Face sketch synthesis with style transfer using pyramid column feature. In Proceedings of the IEEE Winter Conference on Applications of Computer Vision, Lake Tahoe, NV, USA, 12–15 March 2018; pp. 485–493.

27. Schroff, F.; Kalenichenko, D.; Philbin, J. Facenet: A unified embedding for face recognition and clustering. In Proceedings of the IEEE Conference on Computer Vision and Pattern Recognition, Boston, MA, USA, 8–12 June 2015; pp. 815–823.

Article

Evaluating the Overall Accuracy of Additional Learning and Automatic Classification System for CT Images

Hiroyuki Sugimori

Faculty of Health Sciences, Hokkaido University, Sapporo 060-0812, Japan; sugimori@hs.hokudai.ac.jp; Tel.: +81-11-706-3410

Received: 28 December 2018; Accepted: 14 February 2019; Published: 17 February 2019

Featured Application: This article describes the evaluation of the automatic classification system for computed tomography (CT) images using a deep learning technique. Additional learning for automatic training will help to create various classification models in the medical fields. The results in this study will be useful for creating new classification models.

Abstract: A large number of images that are usually registered images in a training dataset are required for creating classification models because training of images using a convolutional neural network is done using supervised learning. It takes a significant amount of time and effort to create a registered dataset because recently computed tomography (CT) and magnetic resonance imaging devices produce hundreds of images per examination. This study aims to evaluate the overall accuracy of the additional learning and automatic classification systems for CT images. The study involved 700 patients, who were subjected to contrast or non-contrast CT examination of brain, neck, chest, abdomen, or pelvis. The images were divided into 500 images per class. The 10-class dataset was prepared with 10 datasets including with 5000–50,000 images. The overall accuracy was calculated using a confusion matrix for evaluating the created models. The highest overall reference accuracy was 0.9033 when the model was trained with a dataset containing 50,000 images. The additional learning for manual training was effective when datasets with a large number of images were used. The additional learning for automatic training requires models with an inherent higher accuracy for the classification.

Keywords: deep learning; medical image classification; additional learning; CT image; automatic training; GoogLeNet

---

1. Introduction

Deep learning techniques [1–3], including deep convolutional neural networks (CNNs), are being employed widely in the field of image processing to conduct image classification [4–6], object detection [7,8], and image segmentation [9–12] tasks. Recently, many studies [4–17] have investigated the applications of deep learning techniques in medical imaging, which now serve as an expansion to this field.

Image diagnosis using computed tomography (CT) and magnetic resonance imaging (MRI) is currently becoming indispensable in the medical field. Although a large number of CT and MRI images are being generated from daily medical examinations, these images are referred to as a follow-up for only a few specific patients. There are many existing models [4–7,13] for the classification of medical images; however, these models are not usually updated since they are created only when needed. Thus, it is not possible to improve such models because they lack procedures and feasibility to retrain the additional medical images. Additionally, creating models requires a large number of images that usually are registered images in a training dataset because training images for CNN are processed using supervised learning algorithms. Herein, we focus on additional learning and automatic learning for CT images because a current CT scanner has the ability to generate a large number of images per examination. Although there is an existing report [13] on the classification of CT images including contrast enhancement data, there are no reports on the classification of medical images based on the evaluation of the additional learning and automatic image learning system. This study aims to evaluate the overall accuracy of the additional learning and the automatic classification systems for CT images.

2. Materials and Methods

2.1. Subjects and CT Images

The study included 700 patients (male: 371, female: 329; mean age $\pm$ standard deviation (SD): 59.2 $\pm$ 19.5 years), who were subjected to either a contrast or non-contrast CT examination of the brain, neck, chest, abdomen, or pelvis in January, 2016. This study was approved by the ethics committee of the Hokkaido University Hospital. The CT images were obtained on a 320-detector-row CT scanner (Aquilion ONE; Canon Medical Systems, Otawara, Japan), an 80-detector-row CT scanner (Aquilion PRIME; Canon Medical Systems, Otawara, Japan), and a 64-detector-row Light Speed VCT (GE Medical Systems, Milwaukee, WI, USA).

2.2. Datasets

The dataset of CT images for creating models for classification was divided in 10 classes for brain, neck, chest, abdomen, and pelvis with contrast-enhanced (CE) and non-contrast-enhanced examination, defined as plain (P). The number of images in each class was 500, 1000, 1500, 2000, 2500, 3000, 3500, 4000, 4500, and 5000 images from the earliest date that they were acquired from the 700 patients; the names of the corresponding datasets were defined as 5 K, 10 K, 15 K, 20 K, 25 K, 30 K, 35 K, 40 K, 45 K, and 50 K, respectively, where the letter K represents one thousand; e.g., the 5 K dataset includes a total of 5000 images, 500 images each of the 10 classes in that dataset. For the validation dataset, another three different datasets (A, B, and C) of 1000 images for each class were prepared (a total of 30,000 images), which were exclusive from the above datasets. The names and details of each dataset are listed in Table 1.

Table 1. Names of datasets and the number of images in each label.

Class Name	Dataset										Validation Dataset		
	5 K	10 K	15 K	20 K	25 K	30 K	35 K	40 K	45 K	50 K	A	B	C
Brain (P)	500	1000	1500	2000	2500	3000	3500	4000	4500	5000	1000	1000	1000
Brain (CE)	500	1000	1500	2000	2500	3000	3500	4000	4500	5000	1000	1000	1000
Neck (P)	500	1000	1500	2000	2500	3000	3500	4000	4500	5000	1000	1000	1000
Neck (CE)	500	1000	1500	2000	2500	3000	3500	4000	4500	5000	1000	1000	1000
Chest (P)	500	1000	1500	2000	2500	3000	3500	4000	4500	5000	1000	1000	1000
Chest (CE)	500	1000	1500	2000	2500	3000	3500	4000	4500	5000	1000	1000	1000
Abdomen (P)	500	1000	1500	2000	2500	3000	3500	4000	4500	5000	1000	1000	1000
Abdomen (CE)	500	1000	1500	2000	2500	3000	3500	4000	4500	5000	1000	1000	1000
Pelvis (P)	500	1000	1500	2000	2500	3000	3500	4000	4500	5000	1000	1000	1000
Pelvis (CE)	500	1000	1500	2000	2500	3000	3500	4000	4500	5000	1000	1000	1000
Total number of images	5000	10,000	15,000	20,000	25,000	30,000	35,000	40,000	45,000	50,000	10,000	10,000	10,000

P: plain, CE: contrast enhanced.

The image range of each class was defined as follows. Brain: slice from the anterior tip of the parietal bone to the foramen magnum; neck: slice from the foramen magnum to the pulmonary apex; chest: slice from the pulmonary apex to the diaphragm; abdomen: slice from the diaphragm to the top of an iliac crest; pelvis: slice from the top of an iliac crest to the distal end of the ischium (Figure 1). The range of each class was the same as that of a previous report [13] for the classification of CT images.

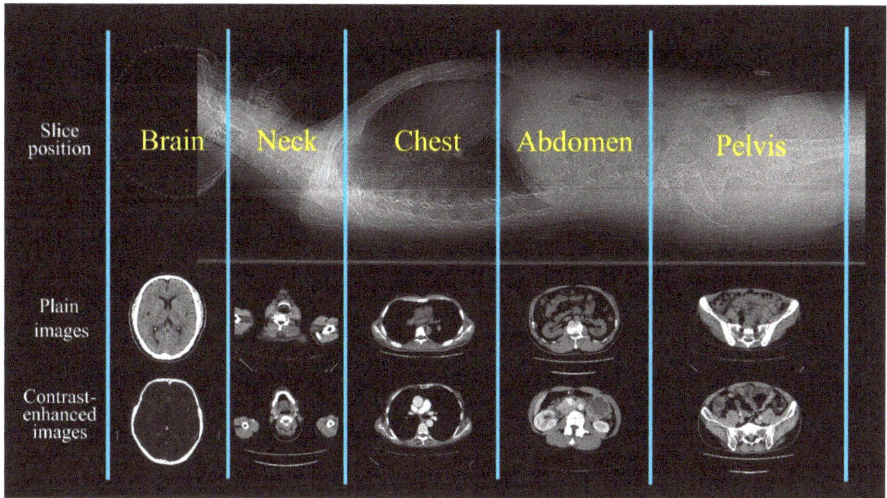

Figure 1. Slice position and sample CT images of the 10 classes.

CE examination involved the intravascular injection of contrast media before examination. The timing of the scan from injection was not considered. Exclusion criteria of CT images for the datasets were images with excessive magnification, images with the reconstruction kernel of bone or lung, images with nothing above the anterior tip of the parietal bone, and images with only arms or legs.

2.3. Preprocessing of Images for Creating the Models

The CT images were retrieved from the picture archiving and communication system. To convert the images for use by the training database, the CT images were converted from digital imaging and communications in medicine (DICOM) format to joint photographic experts group (JPEG) format using a dedicated DICOM software (XTREK view, J-MAC SYSTEM Inc., Sapporo, Japan). The window width and level of DICOM image were used to preset values in the DICOM tag. The DICOM images were converted to JPEG images with a size of 512 × 512 pixels. JPEG files were sorted into folders according to the class that each image belonged to.

2.4. Manual Training of the Images for Creating the Models

The outline of the training performed for creating the models is shown in Figure 2. The authoring software for deep learning was performed via in-house MATLAB (The Mathworks Inc., Natick, MA, USA) software, and a deep learning optimized machine with two GTX1080 Ti GPUs with 11.34 TFlops of single precision, 484 GB/s of memory bandwidth, and 11 GB of memory per board were used. Herein, GoogLeNet [3] with 22 layers was used as the CNN architecture (Figure 3). The hyper-parameters of the training models are as follows: Maximum training epochs were 10 and an initial learning rate was 0.0001. The learning rate was fixed throughout the training. The overall accuracy was calculated using the confusion matrix in the software. The results were evaluated using the validation datasets. Each dataset for training was sorted by a radiological technologist with 17 years of experience. Datasets were divided into 500 images per class for every 5000 images and 1000 images per class for every 10,000 images to create a model for each dataset. Additional learning processes, which were repeated up to the 50 K dataset, were performed to evaluate the accuracy after additional effects.

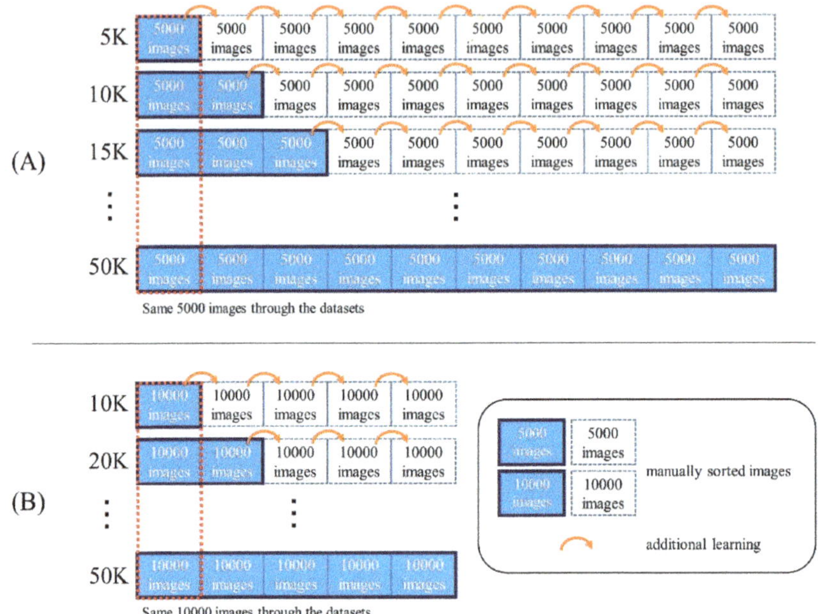

Figure 2. Outline of the training performed for creating the models. Additional learning for every (**A**) 5000 images (**B**) 10,000 images.

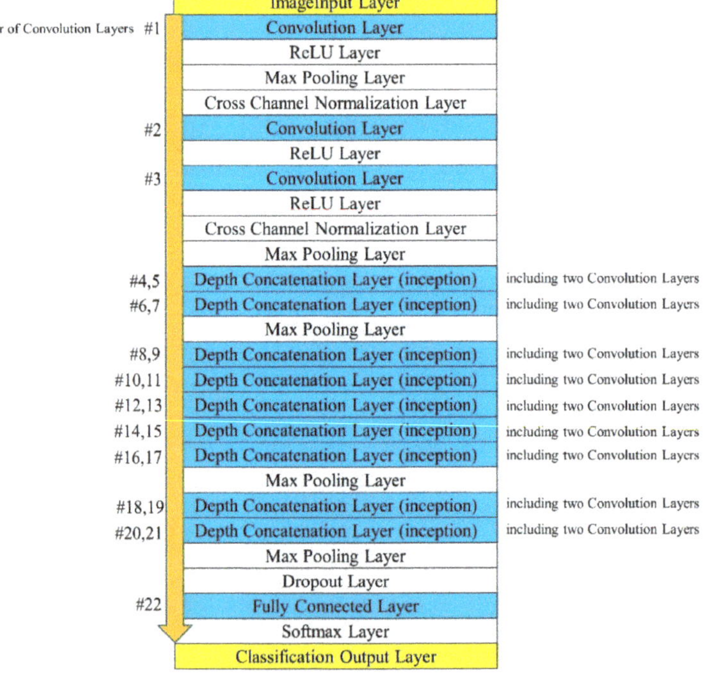

Figure 3. The CNN architecture, which has 22 convolutional layers.

2.5. Automatic Training for Creating Models

The outline of the training for creating the models is shown in Figure 3. The authoring software, machine, CNN architecture, and hyper-parameters of training models were the same as those discussed in Section 2.4. The automatic training system was developed with MATLAB software because supervised learning usually requires images that were classified by humans. Differing from manual training, the following functions were added to the software. (i) The created models with each dataset were used to automatically classify new images into the classes to which they should belong. (ii) The classified JPEG files were sorted into each folder according to their image classes. (iii) The classified images were used for the training to create new models. (iv) The automatic classification and creation of a model was repeated up to the 50 K dataset (Figure 4). The new images provided were divided into 500 images per class for every 5000 images and 1000 images per class for every 10,000 images.

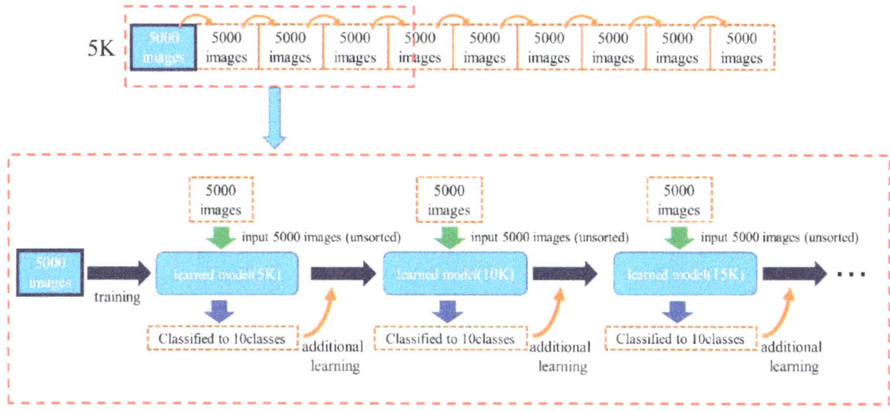

Figure 4. Details of the additional learning process for automatic training (Example case using 5 K datasets).

2.6. Evaluation of the Created Models

The confusion matrix obtained using each dataset, shown in Figure 5, is an indicator of the performance of the created models. The training performed with 10 image classes is shown as a 10×10 table and all performances were based on numbers obtained by applying the classifier to the validation dataset. The overall accuracy was obtained as a ratio of the number of correctly classified images in all validation images to the total number of images. The overall accuracies in each dataset were calculated as reference accuracy. Accuracies of the manual and automatic training were calculated for each dataset. Furthermore, the overall accuracies were evaluated three times with each validation dataset and presented at mean regardless of the dataset.

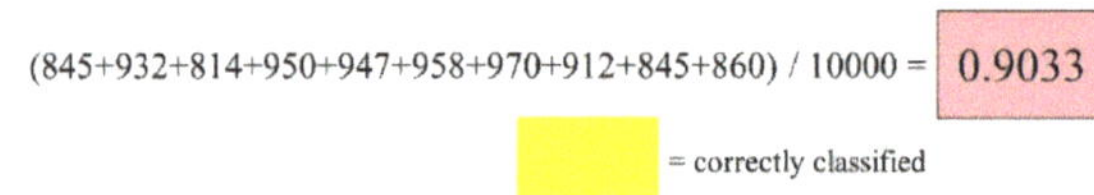

Figure 5 confusion matrix (the full matrix is rendered as a table below):

	Brain (P)	Brain (CE)	Neck (P)	Neck (CE)	Chest (P)	Chest (CE)	Abdomen (P)	Abdomen (CE)	Pelvis (P)	Pelvis (CE)
Brain (P)	845	0	1	0	0	0	0	0	0	0
Brain (CE)	80	932	9	16	0	0	0	0	0	0
Neck (P)	40	8	814	34	11	0	0	0	2	0
Neck (CE)	26	60	165	950	2	12	0	0	0	1
Chest (P)	0	0	5	0	947	27	11	0	0	0
Chest (CE)	0	0	0	0	26	958	0	17	0	0
Abdomen (P)	8	0	0	0	12	0	970	54	22	0
Abdomen (CE)	0	0	1	0	0	3	5	912	1	17
Pelvis (P)	1	0	5	0	0	0	9	0	845	122
Pelvis (CE)	0	0	0	0	2	0	5	17	130	860

(845+932+814+950+947+958+970+912+845+860) / 10000 = **0.9033**

= correctly classified

Figure 5. Confusion matrix for evaluating the overall accuracy, which was calculated using the validation dataset A with 50 K dataset.

3. Results and Discussions

3.1. Reference Accuracy

Table 2 shows the overall accuracy for each dataset. With an increase in the size of image datasets, the overall accuracy became higher. The highest overall accuracy for the datasets used was 0.9033 and the model was trained using the 50 K dataset.

Table 2. Overall accuracy for each dataset.

Dataset Type	Group	Dataset									
		5 K	10 K	15 K	20 K	25 K	30 K	35 K	40 K	45 K	50 K
Validation dataset	A	0.6028	0.6532	0.7293	0.7914	0.8334	0.8369	0.8615	0.8947	0.8986	0.9033
	B	0.4833	0.5352	0.5713	0.6166	0.6789	0.7056	0.7693	0.7877	0.7884	0.8422
	C	0.4927	0.5101	0.5899	0.613	0.7121	0.7397	0.8208	0.8472	0.8633	0.8974
	mean	0.5263	0.5662	0.6302	0.6737	0.7415	0.7607	0.8172	0.8432	0.8501	0.881

3.2. Manual Training

Figure 6 shows the relation between datasets and the overall accuracy of the created model for manual training. For the additional learning of every 5000 images, the overall accuracy when additional learning started from 5 K to 20 K increased continuously up to 25 K. However, after exceeding the 30 K dataset, the overall accuracy fluctuated. For the additional learning of every 10,000 images, the overall accuracy increased continuously up to 40 K. However, the overall accuracy of the dataset of 40 K slightly declined compared to that of 30 K.

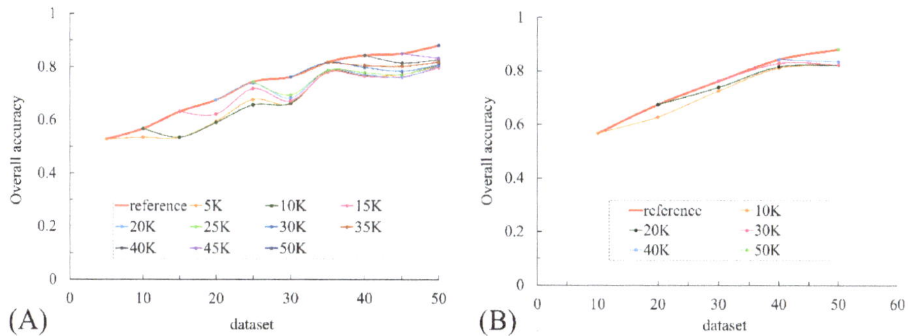

Figure 6. Relation between datasets and overall accuracy for manual training. (**A**) Additional learning for every 5000 images, (**B**) additional learning for every 10,000 images.

3.3. Automatic Training

Figure 7 shows the relation between datasets and the overall accuracy of the created model for automatic training. For the additional learning of every 5000 images, there was little increase in the overall accuracy when the additional learning started from 5 K to 20 K. There was a gradual decrease in the overall accuracy when the additional learning started from 25 K to 35 K and over 40 K dataset. There were no subsequent data when the additional learning started from 5 K to 20 K because some created models could not classify new images up to 10 classes because they had incomplete classification models. For the additional learning for every 10,000 images, there was little increase in the overall accuracy. However, when the additional learning started from 40 K, the overall accuracy was maintained at a high value.

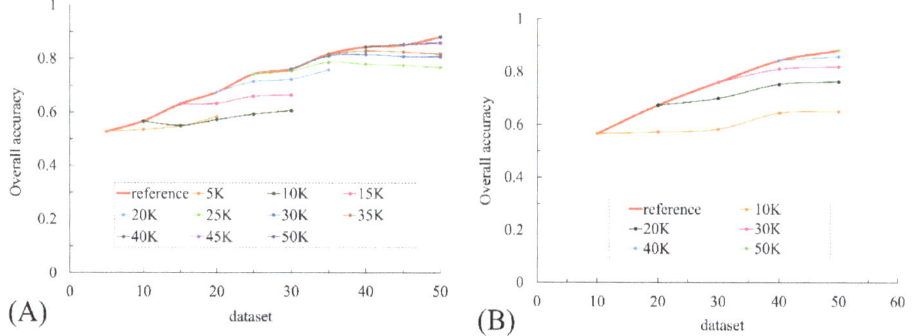

Figure 7. Relation between datasets and the overall accuracy for automatic training. (**A**) Additional learning for every 5000 images, (**B**) additional learning for every 10,000 images.

This study evaluated the overall accuracy of the additional learning and automatic classification system for CT images. From the viewpoint of additional learning, there was a significant improvement of the overall accuracy for the manual training. However, the additional dataset to be added should be prepared with a large number of images because the training for every 5000 images might be affected by specific feature amount. One of the reasons for the fluctuating accuracy, as shown in Figure 6A, might be insufficient feature information in the dataset. For the additional learning of every 5000 images, the number of images for the additional training was small perhaps because, as shown by a previous report [13], the number of CT images affected the accuracy of training the dataset. If the additional dataset included specific patients' data (for instance, the patient who suffered serious traffic accident), the feature amount through the training may be changed dramatically. Therefore, additional images with a variety of features should be prepared by using a high enough number of images for additional learning. On the contrary, automatic training showed no improvement in the overall accuracy, one reason being that the inherent accuracy is not affected by the created models. As the reference overall accuracy, the datasets between 5 K and 20 K were under 0.8 of the overall accuracy. Inaccurate classification affected the models created for automatic training. As a result, there was no further improvement in the overall accuracy. However, when additional learning started from the 40 K and larger datasets, the reference accuracy around 0.9 maintained the overall accuracy at this value. This means that automatic training with a model of higher inherent accuracy might be effective in performing accurate classifications.

The limitations of this study are as follows. First, the hyper-parameters of the training models used are fixed parameters. Although a previous study [13] showed that the hyper-parameters and CNN architecture affected the overall accuracy, the CNN architecture of GoogLeNet is suitable for performing classification in many fields owing to its high accuracy; thus, we used fixed parameters. Second, the process of training accuracy and loss were not showed in this study because the ability to generalize was most important for the intended application [18] in the training process; thus, we only focused on the overall accuracy. However, the overfitting would hardly cause problems during training in this study because GoogLeNet adopted the inception module [19] and global average pooling [20] for preventing overfitting. Third, the additional image data was fixed at 500 images per class. Actual human CT images are often taken from a specific region such as from the lung or liver. The number of images in each class was unstable and imbalanced, as observed during the daily routine examinations. Therefore, the standard of the additional images was required to be set to the number of images and not patients because the additional learning needs to be evaluated in the same situation. In the future, we plan to investigate the effects of an imbalanced number of images when creating an additional model. As for the images, the CT images were converted from DICOM to JPEG images in this study. The CT images have Hounsfield Units (HUs, CT-specific numbers); by definition, water is zero HU and air is −1000 HU. A previous study [21] showed the strong correlation between HUs and grayscales though the JPEG images have no information of absolute values. We supposed the classification of the slice position might not be affected in this study.

4. Conclusions

Herein, we evaluated the overall accuracy of the additional learning and the automatic classification system for CT images. It was found that additional learning for manual training was effective when a large number of images were used. The additional learning for automatic training requires models with the inherent higher accuracy for the classification.

Author Contributions: H.S. proposed the idea, contributed to data acquisition, performed manual classification, data analysis, algorithm construction, wrote the article, and edited the paper.

Funding: This study was supported in part by Grants-in-Aid for Regional R&D Proposal-Based Program from Northern Advancement Center for Science & Technology of Hokkaido Japan.

Acknowledgments: The author thanks the laboratory students Kazuya Sasaki for his help.

Conflicts of Interest: The author declares no conflict of interest.

References

1. LeCun, Y.; Bottou, L.; Bengio, Y.; Haffner, P. Gradient-Based Learning Applied to Document Recognition. *Proc. IEEE* **1998**, *86*, 2278–2324. [CrossRef]
2. Krizhevsky, A.; Sutskever, I.; Hinton, G.E. Imagenet classification with deep convolutional neural networks. In Proceedings of the Advances in Neural Information Processing Systems, Lake Tahoe, NV, USA, 3–6 December 2012; pp. 1097–1105.

3. Szegedy, C.; Liu, W.; Jia, Y.; Sermanet, P.; Reed, S.; Anguelov, D.; Erhan, D.; Vanhoucke, V.; Rabinovich, A. Going deeper with convolutions. In Proceedings of the IEEE Conference on Computer Vision and Pattern Recognition, Boston, MA, USA, 7–12 June 2015.

4. Lakhani, P.; Sundaram, B. Deep Learning at Chest Radiography: Automated Classification of Pulmonary Tuberculosis by Using Convolutional Neural Networks. *Radiology* **2017**, *284*, 574–582. [CrossRef] [PubMed]

5. Qayyum, A.; Anwar, S.M.; Awais, M.; Majid, M. Medical image retrieval using deep convolutional neural network. *Neurocomputing* **2017**, *266*, 8–20. [CrossRef]

6. Gao, X.W.; Hui, R.; Tian, Z. Classification of CT brain images based on deep learning networks. *Comput. Methods Programs Biomed.* **2017**, *138*, 49–56. [CrossRef] [PubMed]

7. Masood, A.; Sheng, B.; Li, P.; Hou, X.; Wei, X.; Qin, J.; Feng, D. Computer-Assisted Decision Support System in Pulmonary Cancer detection and stage classification on CT images. *J. Biomed. Inform.* **2018**, *79*, 117–128. [CrossRef] [PubMed]

8. Zhao, X.; Liu, L.; Qi, S.; Teng, Y.; Li, J.; Qian, W. Agile convolutional neural network for pulmonary nodule classification using CT images. *Int. J. Comput. Assist. Radiol. Surg.* **2018**, *13*, 585–595. [CrossRef] [PubMed]

9. Wachinger, C.; Reuter, M.; Klein, T. DeepNAT: Deep convolutional neural network for segmenting neuroanatomy. *Neuroimage* **2018**, *170*, 434–445. [CrossRef] [PubMed]

10. Akkus, Z.; Galimzianova, A.; Hoogi, A.; Rubin, D.L.; Erickson, B.J. Deep Learning for Brain MRI Segmentation: State of the Art and Future Directions. *J. Digit. Imaging* **2017**, *30*, 449–459. [CrossRef] [PubMed]

11. Ren, X.; Xiang, L.; Nie, D.; Shao, Y.; Zhang, H.; Shen, D.; Wang, Q. Interleaved 3D-CNNs for joint segmentation of small-volume structures in head and neck CT images. *Med. Phys.* **2018**, *45*, 2063–2075. [CrossRef] [PubMed]

12. Avendi, M.R.; Kheradvar, A.; Jafarkhani, H. A combined deep-learning and deformable-model approach to fully automatic segmentation of the left ventricle in cardiac MRI. *Med. Image Anal.* **2016**, *30*, 108–119. [CrossRef] [PubMed]

13. Sugimori, H. Classification of Computed Tomography Images in Different Slice Positions Using Deep Learning. *J. Healthc. Eng.* **2018**, *2018*, 9. [CrossRef] [PubMed]

14. Kim, K.H.; Choi, S.H.; Park, S.-H. Improving Arterial Spin Labeling by Using Deep Learning. *Radiology* **2017**, *287*, 658–666. [CrossRef] [PubMed]

15. Liu, F.; Jang, H.; Kijowski, R.; Bradshaw, T.; McMillan, A.B. Deep Learning MR Imaging–based Attenuation Correction for PET/MR Imaging. *Radiology* **2017**, *286*, 676–684. [CrossRef] [PubMed]

16. Yasaka, K.; Akai, H.; Kunimatsu, A.; Abe, O.; Kiryu, S. Liver Fibrosis: Deep Convolutional Neural Network for Staging by Using Gadoxetic Acid–enhanced Hepatobiliary Phase MR Images. *Radiology* **2017**, *287*, 146–155. [CrossRef] [PubMed]

17. Chen, M.C.; Ball, R.L.; Yang, L.; Moradzadeh, N.; Chapman, B.E.; Larson, D.B.; Langlotz, C.P.; Amrhein, T.J.; Lungren, M.P. Deep Learning to Classify Radiology Free-Text Reports. *Radiology* **2017**, *286*, 845–852. [CrossRef] [PubMed]

18. Zheng, Q.; Yang, M.; Yang, J.; Zhang, Q.; Zhang, X. Improvement of Generalization Ability of Deep CNN via Implicit Regularization in Two-Stage Training Process. *IEEE Access* **2018**, *6*, 15844–15869. [CrossRef]

19. Szegedy, C.; Vanhoucke, V.; Ioffe, S.; Shlens, J.; Wojna, Z. Rethinking the Inception Architecture for Computer Vision. In Proceedings of the IEEE Conference on Computer Vision and Pattern Recognition, Boston, MA, USA, 7–12 June 2015.

20. Lin, M.; Chen, Q.; Yan, S. Network in Network. *arXiv*, 2013; arXiv:1312.4400.

21. Kamaruddin, N.; Rajion, Z.A.; Yusof, A.; Aziz, M.E. Relationship between Hounsfield unit in CT scan and gray scale in CBCT. *AIP Conf. Proc.* **2016**, *1791*, 020005. [CrossRef]

Article

Computer-Aided Design and Manufacturing Technology for Identification of Optimal Nuss Procedure and Fabrication of Patient-Specific Nuss Bar for Minimally Invasive Surgery of Pectus Excavatum

Yoon-Jin Kim [1], Jin-Young Heo [1], Ki-Hyun Hong [1], Hoseok I [2,3], Beop-Yong Lim [4] and Chi-Seung Lee [3,4,5,*]

[1] S-ONE Bio CORP., Busan 49241, Korea; yjkim@s-one.co.kr (Y.-J.K.); jy.hue@s-one.co.kr (J.-Y.H.); khh@s-one.co.kr (K.-H.H.)

[2] Department of Thoracic and Cardiovascular Surgery, Pusan National University School of Medicine, Busan 49241, Korea; ihoseok@pusan.ac.kr

[3] Biomedical Research Institute, Pusan National University Hospital, Busan 49241, Korea

[4] Department of Biomedical Engineering, School of Medicine, Pusan National University, Busan 49241, Korea; lbrcj1220@pusan.ac.kr

[5] School of Medicine, Pusan National University, Busan 49241, Korea

* Correspondence: victorich@pusan.ac.kr; Tel.: +82-51-240-6867

Received: 26 October 2018; Accepted: 17 December 2018; Published: 22 December 2018

Abstract: The Nuss procedure is one of the most widely used operation techniques for pectus excavatum (PE) patients. It attains the normal shape of the chest wall by lifting the patient's chest wall with the Nuss bar. However, the Nuss bar is for the most part bent by a hand bender according to the patient's chest wall, and this procedure causes various problems such as the failure of the operation and a decreased satisfaction of the surgeon and patient about the operation. To solve this problem, we proposed a method for deriving the optimal operation result by designing patient-specific Nuss bars through computer-aided design (CAD) and computer-aided manufacturing (CAM), and by performing auto bending based on the design. In other words, a three-dimensional chest wall model was generated using the computed tomography (CT) image of a pectus excavatum patient, and an operation scenario was selected considering the Nuss bar insertion point and the post-operative chest wall shape. Then, a design drawing of the Nuss bar that could produce the optimal operation result was derived from the operation scenario. Furthermore, after a computerized numerical control (CNC) bending machine for the Nuss bar bending was constructed, the Nuss bar prototype was manufactured based on the derived design drawing of the Nuss bar. The Nuss bar designed and manufactured with the proposed method has been found to improve the Haller index (HI) of the pectus excavatum patient by approximately 37% (3.14 before to 1.98 after operation). Moreover, the machining error in the manufacturing was within ±5% compared to the design drawing. The method proposed and verified in this study is expected to reduce the failure rate of the Nuss procedure and significantly improve the satisfaction of the surgeon and patient about the operation.

Keywords: pectus excavatum; nuss procedure; patient-specific nuss bar; minimally invasive surgery; computerized numerical control bending machine; computer-aided design; computer-aided manufacturing

1. Introduction

Pectus excavatum is one of the most well-known chest wall deformities, in which the entire chest including the costal cartilage and sternum is depressed due to the overgrowth of the costal cartilage. The exact cause of PE has not been accurately identified, and it affects about one in every 300 children worldwide. For these patients, the PE operation is strongly recommended since the major organs in the chest such as the heart and lungs can be subjected to pressure, and problems such as the degradation of cardiopulmonary function, growth, and physical activities are likely to be caused by PE [1,2]. Furthermore, the PE operation is also required to solve potential problems related to cosmetic and mental aspects, such as the avoidance of interpersonal relationships.

The PE operations include the Ravitch procedure, sternal turnover operation, Silastic molding method, and Nuss procedure. The Ravitch procedure and the sternal turnover operation, which are called open surgery, require a resection of all deformed costal cartilage. They can correct the chest wall effectively, but it has disadvantages such as a large operation range, a long operating time, and a less cosmetic effect. The Silastic molding method is a minimally invasive surgery that is used for cosmetic effect, but this method cannot solve the problems of physical function. [3,4].

On the other hand, the Nuss procedure, one of the minimally invasive surgeries, was introduced in 1998 by Donald Nuss, a thoracic surgeon. In this method, a metal bar called a Nuss bar is inserted through the ribs and placed below the sternum to lift the depressed chest. It is widely popular worldwide owing to a small surgical wound, a low risk of infection, and a better cosmetic advantage [5,6]. Hence, the Nuss procedure was selected in the present study.

There are several types of PE, for example, symmetrical, asymmetrical, eccentric, and unbalanced types. Accordingly, the proper type of the Nuss bar should be fabricated prior to operation in order to correct the PE accurately and effectively [7]. However, most Nuss bars are manufactured in a straight shape and provided as such to surgeons. For this reason, the surgeons must manually bend the Nuss bar based on their intuition. However, it is extremely difficult to make a patient-specific shape of the Nuss bar during operation. In addition, the Nuss bar can be damaged due to the repeated bending process, and the adjacent tissues can be damaged if the Nuss bar has a sharp angle.

In order to overcome the above obstacle, Lin et al. [8] adopted the three-dimensional (3D) printing method to fabricate a patient-specific Nuss bar. In their research, the polylactic acid-based 3D printed Nuss bar was developed and implemented into 10 PE patients. They called this novel operation the 3DPMAN (3D Printed Model-Assisted Nuss) procedure, and the initial results of the 3DPMAN procedure was feasible, easy, convenient, and satisfactory.

There are many advantages in fabricating the Nuss bar using polymer 3D printing such as the low fabrication cost, the short fabrication time, and others. However, the polymer Nuss bar has severe shortcomings. The Nuss bars are usually made of titanium alloys or stainless steel since they can maintain high stiffness and strength until they are removed from the body. Accordingly, the chest wall, including the Nuss bar, may be free from sudden failure or fracture due to unexpected excessive external forces. On the other hand, the polymer Nuss bars, such as polylactic acids, have low strength and stiffness compared to titanium alloys or stainless steel. Therefore, the mechanical evaluation of material/structural safety should be sufficiently carried out prior to clinical applications.

If the Nuss bar is made using metal 3D printing, the procedure is costly and time-consuming. In other words, the metal powder (titanium alloy powder) is quite expensive, and the 3D printing takes a great amount of time. In addition, the metal 3D printed Nuss bar must be properly surface treated, and thus requires much time and cost. If the surface is poorly treated, the metal powder can be absorbed into the human body and is extremely dangerous.

Due to the various problems described above, we adopted a method for bending a titanium alloy Nuss bar to the expected normal chest wall shape after the patient's operation. For this, we proposed a method for deriving the optimal operation result of PE patients based on (1) technology to design the patient-specific Nuss bar using a three-dimensional (3D) chest wall model (CAD method) and (2) technology to fabricate the patient-specific Nuss bar using an auto bending machine (CAM method).

2. Materials and Methods

2.1. Fabrication Procedure for the Patient-Specific Nuss Bar via CAD and CAM

The fabrication procedure for the patient-specific Nuss bar can be divided into a design process for the patient-specific Nuss bar (CAD) and a machining process for fabricating the designed Nuss bar (CAM). The overall flow for the fabrication procedure is shown in Figure 1. The detailed procedures are described sequentially in the following chapters.

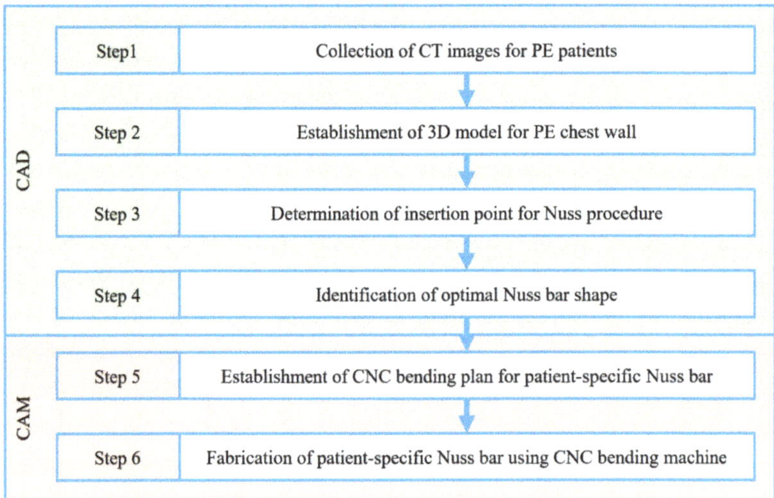

Figure 1. Fabrication procedure for the patient-specific Nuss bar.

2.2. Modeling of the Three-Dimensional (3D) Chest Wall for a Pectus Excavatum Patient

2.2.1. Collection of CT Images for PE Patients

The first step for deriving the 3D chest wall of a PE patient is to collect the medical imaging data of the PE patients. We collected computed tomography (CT) images of 15 PE patients from the Pusan National University Hospital, and classified them based on the chest wall shape into symmetrical and asymmetrical types. Ten patients with a symmetrical depression based on the center of the sternum were included in the symmetric group, and five patients with a depression deviated from the center of the sternum or an asymmetrical depression were included in the asymmetric group.

In general, CT, magnetic resonance imaging (MRI), ultrasound, and other methods can be adopted to fabricate the 3D surface and finite element (FE) model for several organs of the human body. For example, Bonacina et al. [9] have developed a novel algorithm that automatically extracts the facial surface from ultrasound images. Using this method, they fabricated a 3D foetal face model without any human intervention or training procedure.

However, in this study, the CT image was adopted from among various medical images. The reason is that in the case of MRI, the image can be distorted due to the patient's breathing, and in the case of ultrasound, it is not easy to generate a 3D surface and FE model using general-purpose medical image processing software such as Materialise MIMICS due to the noise.

2.2.2. Establishment of 3D Model for PE Chest Wall

For our study, 3D chest wall models were fabricated based on the obtained patient CT images. The models were then used to design patient-specific Nuss bars. The image processing software MIMICS version 17 from Materialise (Leuven, Belgium) was used to produce the 3D chest wall model. From the original CT images, 12 ribs, sternums, and costal cartilages were extracted in the following four steps (see Figure 2).

- Step 1: The CT file of a patient is loaded into the program (MIMICS). The loaded CT image is displayed in black and white depending on the density of each tissue. Thus, the density differences between tissues are used to select the desired body organ on the screen. The density range is specified using the "Thresholding" feature. The 3D surface can be obtained by selecting the rib and sternum tissues easily using the default value (bone region) of each tissue provided by the "Region growing" function of the program.

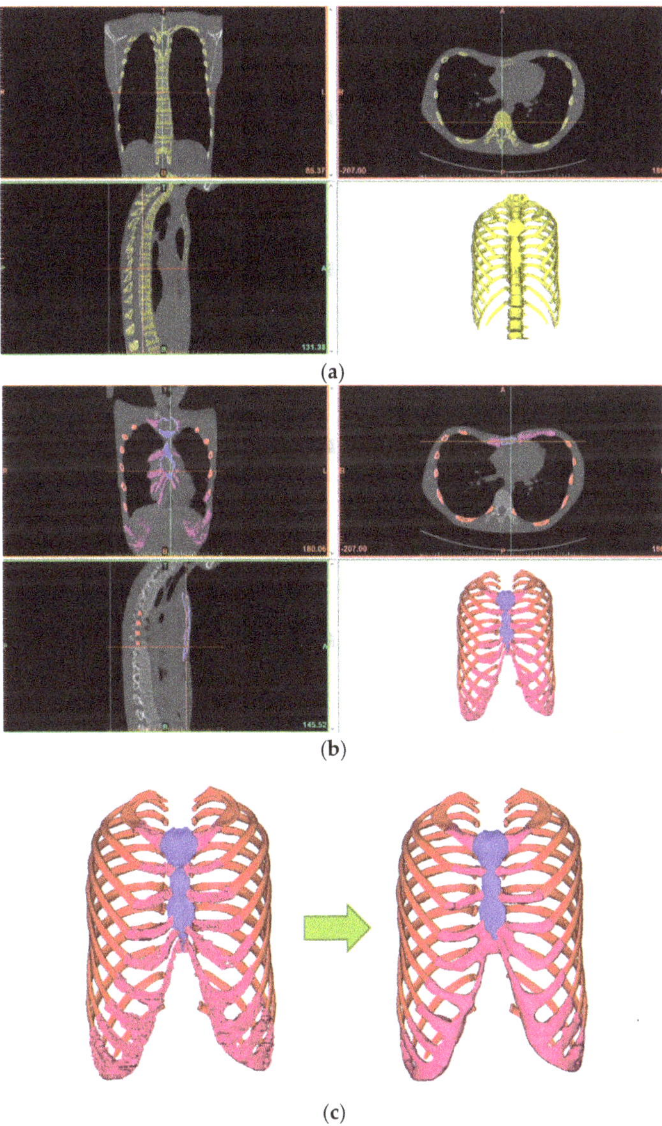

Figure 2. Fabrication procedure for the 3D chest wall model for a PE patient: (**a**) Step 1; (**b**) Steps 2 and 3; (**c**) Step 4.

- Step 2: A 3D model is created from the obtained 3D surface using the "3D calculation" feature. The spine or any other unnecessary tissues are removed using "Edit mask" or similar features because the Nuss bar is not applied to them.
- Step 3: A separate selection task must be performed to obtain the 3D surface of costal cartilage because the costal cartilage region cannot be taken with the Bone default value due to its low density compared to the rib and sternum. A 3D model is created from the 3D surface of the costal cartilage obtained through this separate process.
- Step 4: A smooth-shaped chest wall model is produced finally by modifying the 3D chest wall models of the rib, sternum, and costal cartilage using the "Wrap" and "Smoothing" features.

2.3. Virtual Surgery Scenario for Nuss Procedure

2.3.1. Definition of Haller Index

Two conditions must be considered when performing the Nuss procedure. The first condition is that the Haller index (HI) must be an HI value of a normal chest wall. The second condition is that the damage of the Nuss bar and its adjacent tissues must be minimized. In order to satisfy these two conditions, the insertion point and shape of the Nuss bar must be determined before performing the Nuss procedure. In this study, the optimal operation position and shape were found by quantitatively deriving the Haller index values before and after operation for each Nuss bar insertion point based on the actual CT images of the patients.

Figure 3 shows the pectus indices. The Haller index is a simple mathematical method for measuring and representing PE with a known pectus index. It is calculated by the ratio of the maximum transverse diameter (the maximum length inside the thoracic cage, A) measured on the axial CT section of the chest with the largest deformation and the minimum anterior-posterior (AP) distance (minimum distance between the spine and the sternum, B) [10–12].

Equation (1) is the Haller index equation before the Nuss operation:

$$\text{Pre-operation: HI} = A/B. \tag{1}$$

The Haller index expected after the PE operation is called the ideal chest index (ICI), which is determined by dividing the corrected maximum transverse diameter (**A'**) by the corrected minimum AP distance (C) [13].

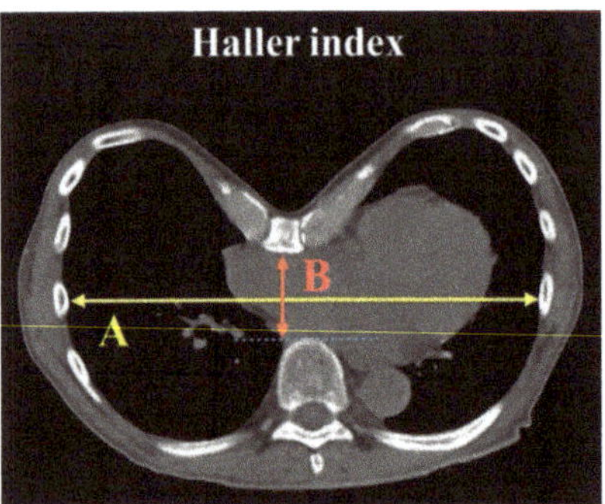

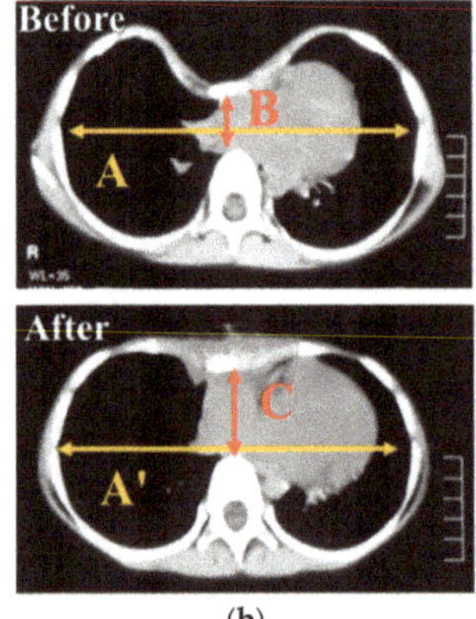

Figure 3. Illustration of pectus indices calculated from the computed tomography (CT) axial image with greatest sternum depression: (**a**) description of the Haller index; (**b**) measurement of the Haller index before and after the Nuss operation.

Equation (2) is the ideal chest index (ICI) equation, which is used instead of the Haller index after the Nuss operation:

$$\text{Post-operation: ICI} = \mathbf{A'}/\mathbf{C}. \tag{2}$$

The severity degree of PE is classified by the HI value. Dr. Mark Thurston classified the degree of PE by HI value as is shown in Table 1 [14].

Table 1. Classification of the Haller index.

Degree of Pectus Excavatum	Range of Haller Index
Normal chest	<2.0
Mild excavatum	2.0–3.2
Moderate excavatum	3.2–3.5
Severe excavatum	>3.5

According to the results of many studies that investigated the correlations between HI and PE, the PE operation is required if the HI is equal to or greater than 3.2 [11,15].

2.3.2. Virtual Surgery Scenario for Insertion Point and Shape Design of Nuss Bar

The virtual surgery scenario for selecting the optimal insertion point and shape was set as follows [16]. Three points were selected based on the ribs around the sternum with the largest depression using the CT image and 3D model of Patient 1. As is shown in Figure 4, insertion point A is the sternum between the second and third ribs, insertion point B is between the third and fourth ribs, and insertion point C is between the fourth and fifth ribs.

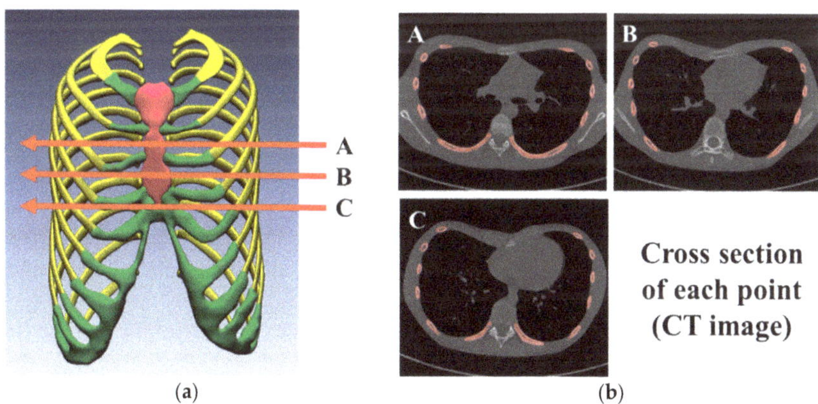

Figure 4. Virtual surgery scenario for insertion point and shape design of Nuss bar: (**a**) Nuss bar is inserted at A, B, and C points from 3D chest wall model; (**b**) CT axial images with insertion points A, B, and C.

Figure 5 shows the chest wall shape expected after the Nuss procedure for each insertion point. The blue line on the CT image is the optimal Nuss bar shape, which was designed by referring to the PE shape type and the shape of a similar chest wall.

To fabricate a patient-specific Nuss bar based on the patient's CT image, a computerized numerical control (CNC) bending machine that can machine every curved surface within the range of the Nuss bar curves must be first constructed. In this study, a CNC bending machine was constructed with the machining purpose.

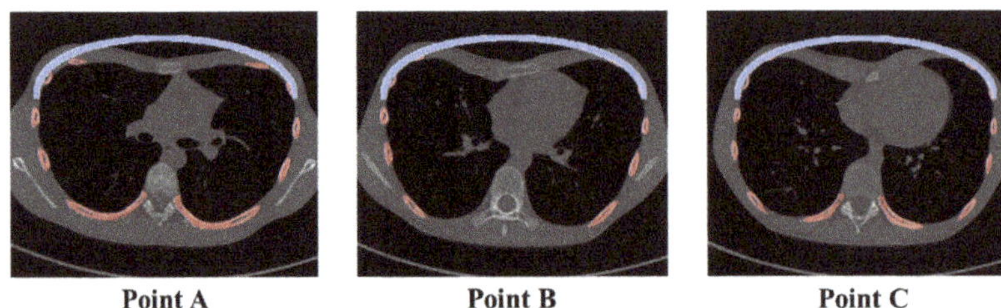

Point A   **Point B**   **Point C**

Figure 5. Illustrations of corrected sternum after Nuss bar insertion. The same shape of the Nuss bar was located on the CT axial image with insertion points.

2.4. Establishment of Computerized Numerical Control (CNC) Bending Machine for Patient-Specific Nuss Bar

2.4.1. Components and Modification of CNC Bending Machine

Figure 6 shows the components of the equipped CNC bending machine which consists of a main body, control cabinet, and air pump. Of these, the main body where the material shaping occurs is largely divided into feeding and rotation units and a bending unit.

Figure 6. Establishment of CNC bending machine. In the figure, components of the CNC bending machine are (1) main body, (2) control cabinet, and (3) air pump.

Figure 7 shows the feeding and rotation units in the main body. The feeding and rotation unit assembly (Figure 7a) consists of a feeder module which pushes the material to the bending head, a set of clamps which holds the material, and a rotation module that rotates the material. The existing set of clamps is designed for round materials such as wire and rod, and cannot hold the Nuss bar material with the bar shape. Therefore, a customized jig for the Nuss bar was designed, fabricated, and mounted onto the clamps (Figure 7b).

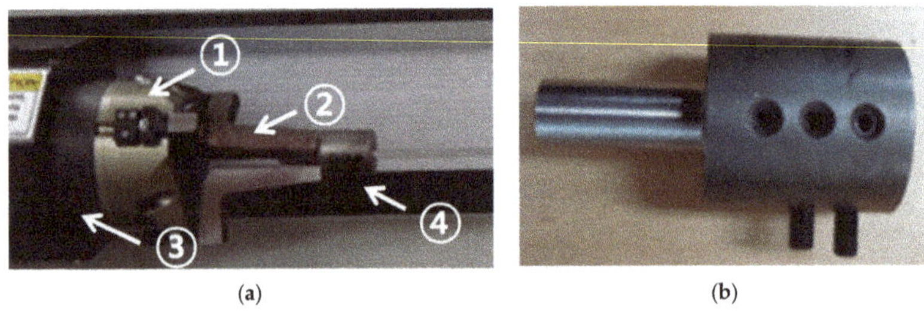

(a)   **(b)**

Figure 7. Images of feeding and rotation parts of the CNC bending machine: (**a**) image of the feeding and rotation assembly, composed of the (1) rotation module, (2) set of clamps, (3) feeding module, and (4) Nuss bar jig; (**b**) image of the customized jig tool for Nuss bar processing.

The design drawing of the jig for the Nuss bar is shown in Figure 8, which was designed considering the specifications of the machined material (width of metal bar: 13 mm, thickness of metal bar: 3 mm).

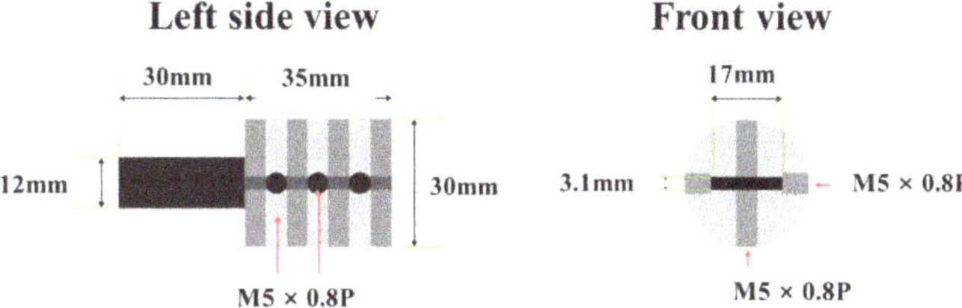

Figure 8. Left and front view of Nuss bar jig design. Designed jig is holding metal bar during the production process of patient-specific Nuss bar by CNC bending machine.

Figure 9 shows the bending unit for bending of the material. The bending unit is composed of multiple tools and is designed to be able to machine 180° rotations at the maximum with 90° in two directions (clockwise and counterclockwise) considering the characteristics of the material (for metal bar, rotation by the rotation unit is impossible). The Nuss bar machining limits of existing tools were overcome by manufacturing a tool that could perform bidirectional bending.

Figure 9. Bending unit for Nuss bar manufacturing by CNC bending machine. (**a**) This picture shows the maximum range of movement of the bending tool. (**b**) This picture shows the Nuss bar (straight) being inserted into the machine; the Nuss bar makes a 90-degree angle with the bending unit.

The control cabinet, which is another component of the CNC bending machine, is used to input the machining data and manage the device-operating options, and the CAM software, which is the machining program, is embedded in it. In addition, a separate air pump must be installed for material feeding.

2.4.2. Installation and Specification of CNC Bending Machine

The metal body frame of the CNC bending machine must be installed on a concrete floor and maintain horizontal balance. The control cabinet must be fixed by wheel brake pedals and all cables connected to the main body must be protected. The specifications of the machine including power consumption, electrical requirement, and air requirement are shown in Table 2.

Table 2. Specification of the CNC bending machine.

Average Power Consumption	Electrical Requirement	Installed Power	Air Requirement
1.9 KW/h	230 Volts/single-phase/50–60 Hz	5 kVA	Dry air 100 psi (min. 80 psi)

2.4.3. Operation Parameters of CNC Bending Machine

The parameters for machine operation such as home position and initial position can be set using the parameters function of the CAM program. Table 3 shows the parameter list and values of the CNC bending machine set for the fabrication of the patient-specific Nuss bar.

Table 3. Operating parameter values to operate the CNC bending machine for the Nuss bar.

Index	Description	Value
0	Default Units	0.0000
1	Stop Machine if done	0.0000
6	Return Bender Speed	100.0000
7	Delay at the end of program	0.0000
12	Cut at the end of program	3.0000
16	Negative Z-axis limit	−200.0000
17	Positive Z-axis limit	200.0000
18	Initial Feeder position *	756.750
19	Initial Bender position	0.0000
20	Output ON Delay	80.0000
21	Output OFF Delay	80.0000
23	Feeder Clamp move to grip	0.0000
24	Arm Collision detection Rotation Default	5.0000

* Initial feeder position is defended on length of material (Nuss bar length)

The accurate bending of materials requires the setting of the tool geometry. After selecting the material and tool in the material selection choice, the Bender Geometry window (click on tool definitions icon) is activated and the appropriate tool geometry is set for the Nuss bar machining.

In this study, the mandrel type tool was selected as the default tool. Table 4 shows the main setting values of the tool geometry for accurate machining of a linear Nuss bar.

Table 4. Tool geometry setting values for increasing accuracy of bending result.

Index	Description	Value
0	Inner bending roller diameter (mm)	19.050
1	Outer bending roller diameter (mm)	19.050
3	Upper Roller Center to X (mm)	27.305
10	Tool type (#)	5 (=Mandrel cluster)
11	Tool Cluster Diameter	75.001

Notes: Setting values for all items except for index numbers 0, 1, 3, 10, 11 on the Tool Geometry window are "0".

3. Results

3.1. Validation of CAD-Based Patient-Specific Nuss Bar Design Technology

To fabricate the patient-specific Nuss bar, we reviewed the CT images of 15 patients and chose one symmetric case called Patient No. 1 from the symmetric group. Then, we proposed three surgery scenarios (Nuss bar insertion points) to derive the patient-specific Nuss bar shape and select the insertion point for Patient No. 1.

The HI value before and after inserting the Nuss bar was calculated for each scenario. The pre- and post-operative maximum transverse diameters (A, A') and the minimum AP distances (B, C), and HI values are summarized in Table 5.

According to a previous study, the post-operative maximum transverse diameter is 95% of the pre-operative value [17]. The post-operative maximum transverse diameter (A') was derived by applying the results of the corresponding study.

Table 5. Haller index values before and after the Nuss procedure in Patient No. 1.

Point	A (mm): Pre-Op.	A′ (mm): Post-Op.	B (mm): Pre-Op.	C (mm): Post-Op.	HI: Pre-Op.	ICI: Post-Op.
A	217.86	206.97	73.01	101.11	2.98	2.05
B	215.22	204.46	68.61	103.49	3.14	1.98
C	217.57	206.70	73.30	104.15	2.97	1.98

Figure 10 shows the pre- and post-operative minimum AP distances according to the Nuss bar insertion point.

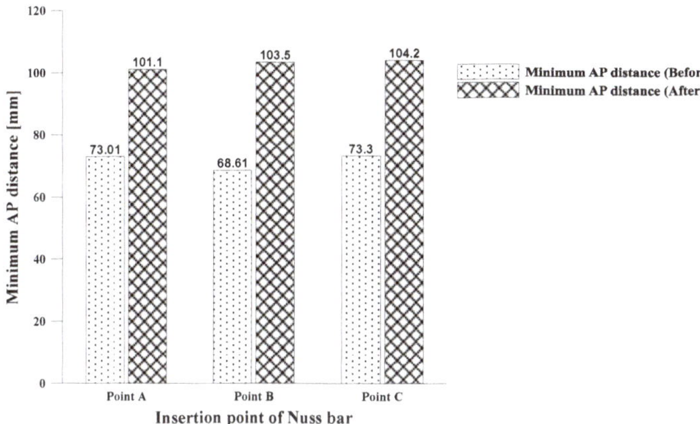

Figure 10. Minimum AP distance before and after the Nuss procedure. Values of minimum AP distance with Nuss bar insertion points are shown by histogram.

The B (Pre-operative) values for each insertion point (points A, B, and C) were measured at 73.01 mm, 68.61 mm, and 73.30 mm, respectively. The smallest value was obtained at position B. The C values for each insertion point were 101.11 mm, 103.49 mm, and 104.15 mm, respectively. The largest value was obtained at position C. The minimum AP distance increased by 38.49%, 50.84%, and 42.09% at the insertion points, respectively. Thus, position B showed the largest increase.

Figure 11 shows the changes in HI values before and after the Nuss procedure.

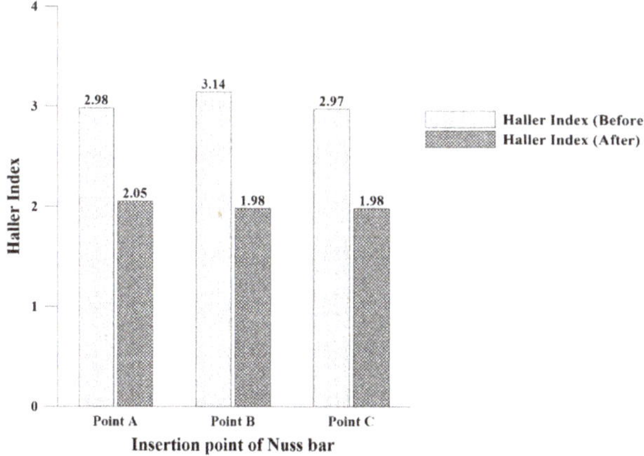

Figure 11. Haller index before and after the Nuss procedure. The values of the Haller index with insertion points of the Nuss bar are shown in the histogram.

The post-operative HI value decreased by 31.31% from 2.98 to 2.01 at insertion point A, by 36.94% from 3.14 to 1.98 at insertion point B, and by 33.33% from 2.97 to 1.98 at insertion point C. At every insertion point, the HI value improved by more than 30%. Furthermore, the improved HI values were near the target HI value of a normal chest wall (HI < 2.5).

The data analysis results revealed that the minimum AP distance and HI improved the most at point B that had the largest HI value before operation. Therefore, the optimal insertion point was confirmed to be B which had a corrected HI value lower than 2.0 and showed the greatest change.

Using the Nuss bar design technology which is described in Section 2.3, we derived the Nuss bar design for asymmetric PE patients (Figure 12). Patients Nos. 2 and 3 represent the eccentric and unbalanced PE patients, respectively.

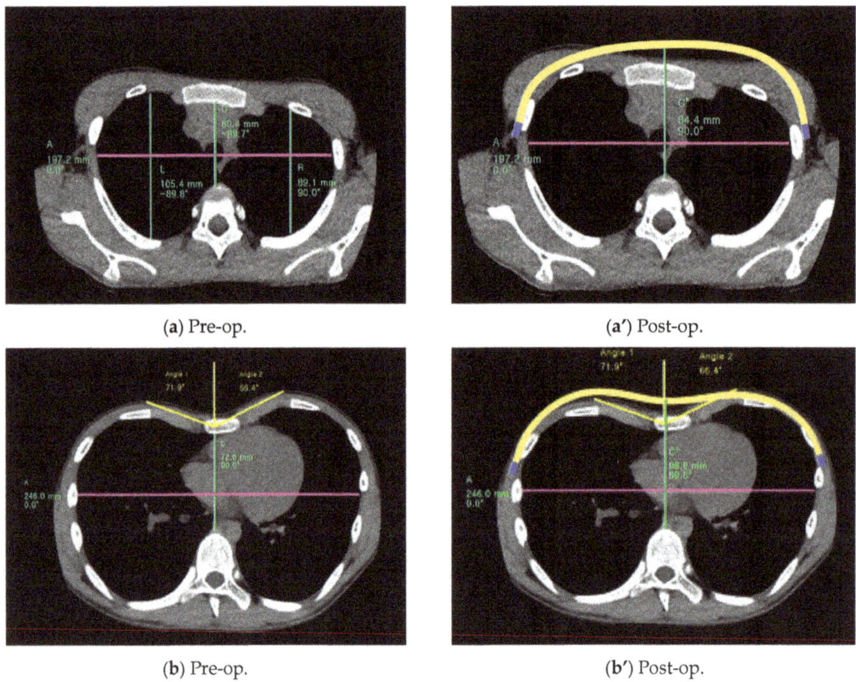

<table>
<tr><td align="center">(a) Pre-op.</td><td align="center">(a′) Post-op.</td></tr>
<tr><td align="center">(b) Pre-op.</td><td align="center">(b′) Post-op.</td></tr>
</table>

Figure 12. CT images of asymmetric patients. (**a**) and (**b**) are CT images taken before the Nuss operation; (**a′**) and (**b′**) are images taken after the Nuss operation. Yellow lines in (**a′**) and (**b′**) show the patient-specific Nuss bar.

The insertion position and shape of the patient-specific Nuss bar are derived using corrected HI values. Additionally, the angle and height on the left and right side of chest have similar values after the Nuss procedure, which is for the cosmetic aspect. Table 6 shows the corrected HI values after Nuss bar insertion.

Table 6. Haller index before and after the Nuss procedure in Patients Nos. 2 and 3.

Type	A (mm): Pre-Op.	A′ (mm): Post-Op.	C (mm): Pre-Op.	C′ (mm): Post-Op.	HI: Pre-Op.	ICI: Post-Op.
Eccentric (Patient No. 2)	197.2	187.34	60.4	84.4	3.26	2.22
Unbalanced (Patient No. 3)	246.0	233.7	72.6	98.6	3.39	2.37

After Nuss bar insertion, the HI values of the eccentric case decreased by 31.9% from 3.26 to 2.22 in the insertion position A. In the unbalanced case, the HI values after Nuss bar insertion decreased by

30% from 3.39 to 2.37 in the insertion position C. The corrected HI value of both patients (eccentric and unbalanced) are below 2.5 (normal range of HI).

3.2. Validation of CAM-Based Patient-Specific Nuss Bar Fabrication Technology

3.2.1. Bending Test for Patient-Specific Nuss Bar Fabrication

Before fabricating the patient-specific Nuss bar, a bending test was performed to verify the bending range and machining accuracy of the equipped CNC bending machine.

- Drawing of Test Design: Central Arc and Transition Value of Nuss Bar

A product drawing for machining test considering the maximum/minimum values of the arc and transition parts of the Nuss bar components was created. Figure 13 shows the components of the Nuss bar [18].

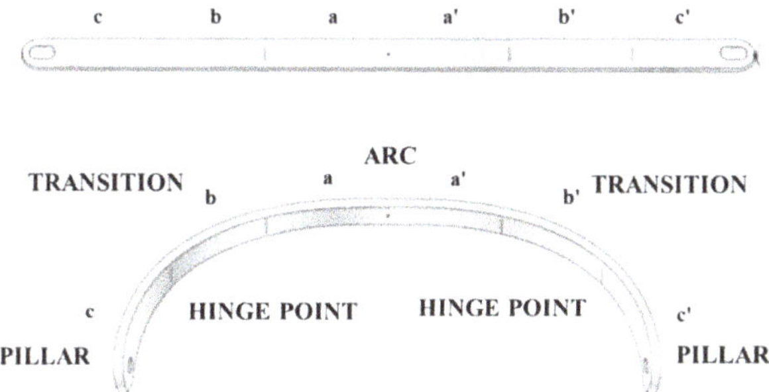

Figure 13. Three components of the Nuss bar. Arc and transition were used to draw the test sample for the bending test.

(1) Central arc design: This part lifts the depressed sternum and is divided into five steps by selecting 200–400 mm for the maximum/minimum lengths of the transverse diameter (50 mm intervals). The maximum/minimum AP distances are determined by using the HI [19]. Five drawings were created by using the circumference within 160° from the center of the ellipse shape. (2) Transition part design: The machining occurs predominantly in this part and the applied value varies by the degree of the chest wall deformation [18]. The maximum/minimum radius range of curvature of the transition part was set to 20–100 mm, and was divided into five steps (20 mm intervals). Five drawings were created with the representative angles of 30°, 60°, and 90° for each curvature.

- Computer-Aided Manufacturing (CAM) Data and Fabrication of Test Designs

The test products were fabricated using the biocompatible metal Titanium-6Al-4V ELI (Ti-Gr5) and SUS 316 LVM (SUS) with a thickness of 3 mm and a width of 13 m. The bending method varied according to the product shape. The feeding method was applied when the curvature of an ellipse shape was low and machining over a wide range was required such as for the central arch. The multi-bending method was applied if different bending methods had to be performed for each narrow point such as for the transition type.

The desired machining result could not be obtained from the initial CNC bending test where the design specifications were applied. In particular, Ti-Gr5 generated greater machining errors due to a strong springback phenomenon compared to SUS.

The springback, which is a property that makes the material return to its original shape, is affected by the material properties (yield strength, modulus of elasticity) and thickness, machining angle, and bending radius [20–22].

Therefore, the CAM data were established considering the springback phenomenon of materials to perform accurate bending. First, the initial springback factor, *Ks*, of each material was calculated as is shown in Figure 14. To calculate the initial *Ks* value, the data for each factor in Table 7 were input into the Bending Springback Calculator (Figure 14) of CUSTOMPART.NET.

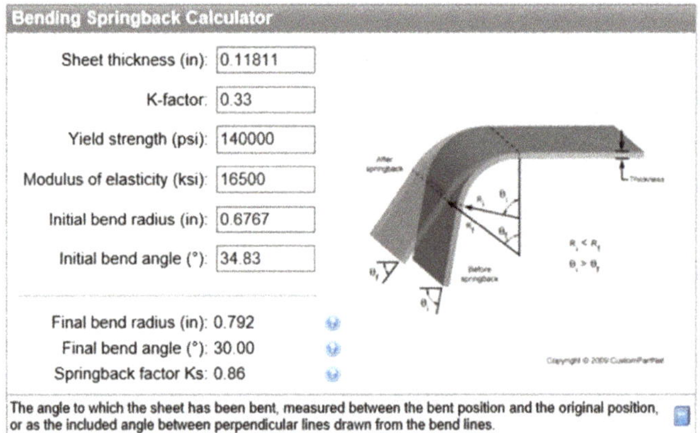

Figure 14. Bending Springback Calculator from CUSTOMPART.NET [23]; springback factor (*Ks*), final bend radius (*FR*), and final bend angle are changed by input values which are sheet thickness (*Mt*), K-factor, yield strength, elasticity, initial bend angle, and initial bend radius (*IR*).

Table 7. Data of springback factors.

Material	Sheet Thickness (mm)	K-Factor	Yield Strength (psi *)	Elastic Modulus (psi) *	Initial Bend Radius (mm)	Initial Bend Angle (°)
Ti-Gr5	3	0.33	140,000	16,500	Up to transition values	
SUS	3	0.33	116,000	28,000		

* The material properties of Titanium-6Al-4V ELI (Ti-Gr5) and SUS (SUS 316 LVM) were referenced from material property data (MetWeb.com).

The CAM data (machining angle and radius value) for the final angle and radius values were calculated reversely by applying the initial *Ks* value to the springback factor (*Ks*) equation, Equation (3) [24], as follows:

$$Ks = \frac{\text{Initial angle (°)}}{\text{Final angle (°)}} = \left(\frac{2 \times IR}{Mt} + 1 \right) \Big/ \left(\frac{2 \times FR}{Mt} + 1 \right), \tag{3}$$

where *IR* and *FR* are the initial and final bend radii, respectively, and *Mt* is the sheet thickness.

The CAM data according to the material, angle, and bending radius were determined through multiple bending trials and errors. Different CAM values were required depending on the material even if a Nuss bar of the same shape was fabricated. The Ti-Gr5 product required additional bending between 2% to 20% as compared to the SUS product.

The prototypes of the central arc and transition values were fabricated using the appropriate machining method and the derived CAM values.

- Results of Bending Test

The machining accuracies of 60 prototypes in total were evaluated through measurement of dimensions. For the dimension measurement test, the maximum width and height were specified as the major measurement indices. Figure 15 shows the dimension measuring points for machining accuracy.

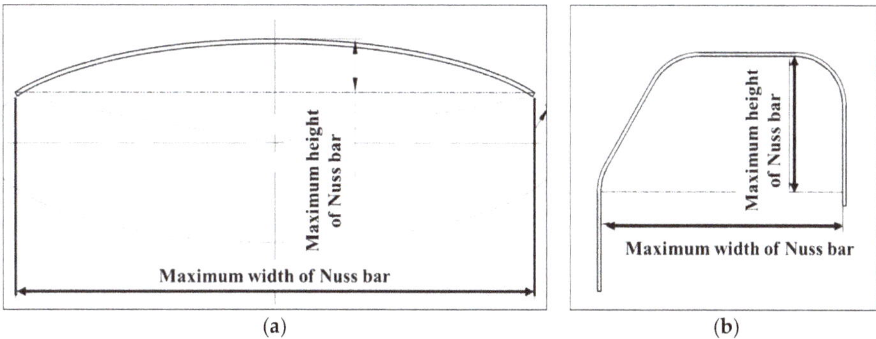

Figure 15. Designs of test sample: (**a**) central arc design with measuring point of maximum width and height; (**b**) transition Nuss bar design with measuring points.

The dimension accuracy results of the five design prototypes considering the central arc were plotted as a graph. Figure 16 shows a graph for the measured width values of central arc designs 1–5 which were fabricated with SUS and Ti-Gr5. The SUS and Ti-Gr5 products have the dimension (width) errors of ±0.41% and ±0.53%, respectively.

Figure 17 shows the height measurements of the central arc design products. SUS and Ti-Gr5 products have the dimension (height) errors of ±2.61% and ±1.48%, respectively. The major dimension measuring results of Ti-Gr5 and SUS products considering the central arc confirmed that the dimension accuracies of all the prototypes were within ±5%.

The dimension accuracy values of the five designs considering the transition values were plotted as a graph. Figure 18 shows the width of the transition Nuss bars 1–5 fabricated with SUS and Ti-Gr5. The width error is ±1.19% for the SUS product and ±0.57% for the Ti-Gr5 product.

Figure 19 shows the measured height, and the height errors of the SUS and Ti-Gr5 products are ±1.73% and ±0.9%, respectively. This result confirms that the dimension accuracies of Ti-Gr5 and SUS products considering the transition value are within ±5%.

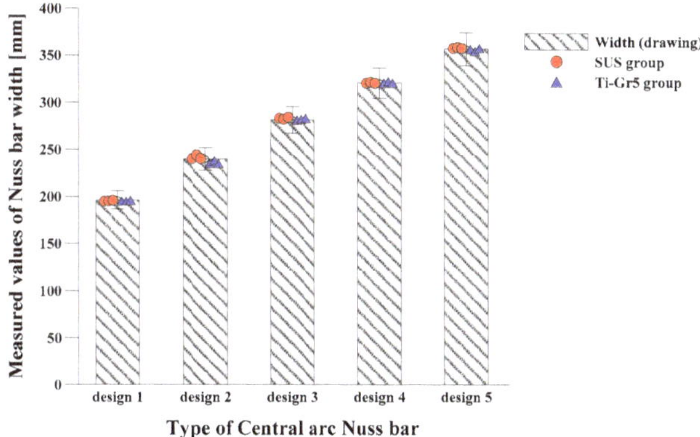

Figure 16. This histogram shows the dimensions of the Nuss bar. The maximum width of SUS and Ti-Gr5 central arc Nuss bars was measured and compared with original width.

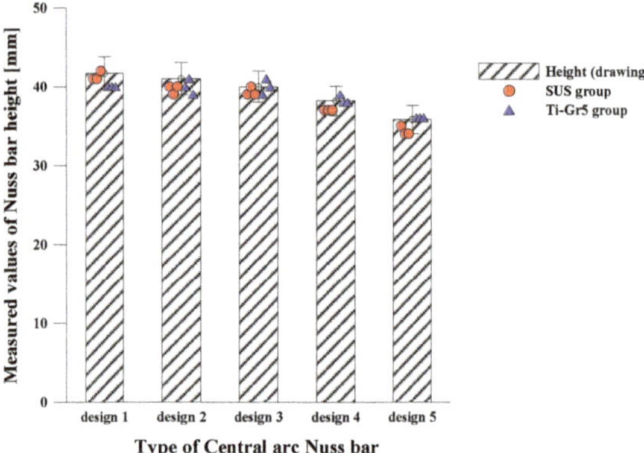

Figure 17. Results for the dimension accuracy of the central arc Nuss bar. This histogram shows the maximum height of the central arc Nuss bars (SUS and Ti-Gr5). Heights were measured and compared with original height.

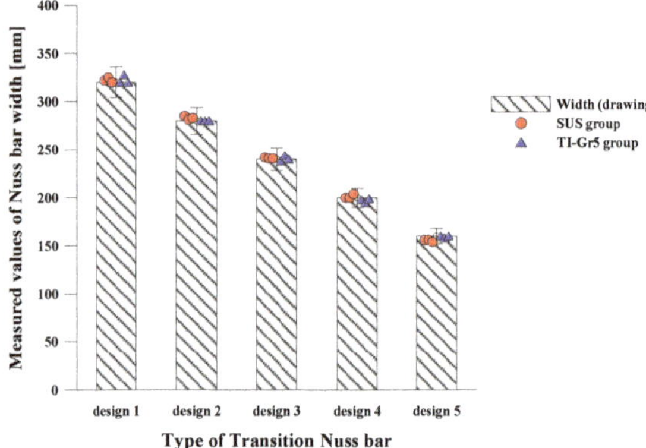

Figure 18. Results for the dimension accuracy of the SUS and Ti-Gr5 transition Nuss bar. This histogram shows the maximum width of the SUS and Ti-Gr5 transition Nuss bars. Widths were measured and compared with original width.

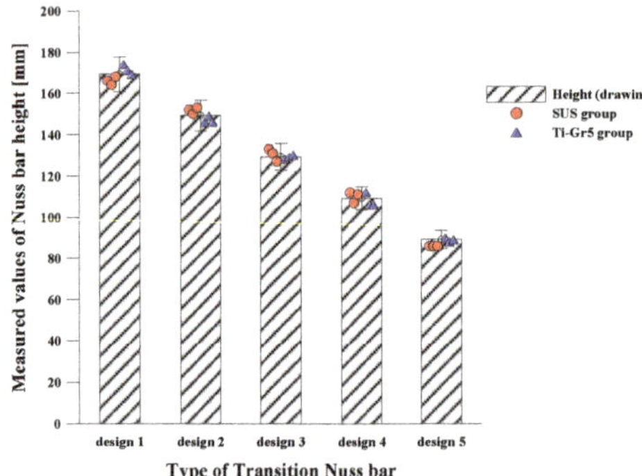

Figure 19. Results for the dimension accuracy of the SUS and Ti-Gr5 transition Nuss bar. This histogram shows the maximum height of the SUS and Ti-Gr5 transition Nuss bars. Heights were measured and compared with original height.

All values within the central arc and transition values of the Nuss bar can be machined using the auto CNC bending machine, and the bending error range of the bending machine was found to be within ±5%. In particular, it is expected that more accurate machining will be possible for products fabricated with the Ti-Gr5 material.

3.2.2. Manufacture of Patient-Specific Nuss Bar

The patient-specific Nuss bars were fabricated with SUS and Ti-Gr5. The design of the patient-specific Nuss bar was derived using CAD-based design technology, and the product design drawings were created for symmetrical, eccentric, and unbalanced Nuss bar shapes in accordance with the morphological classification of PE [2,7,18,25]. Each type of Nuss bar was designed by collecting CT data and using virtual surgery scenarios, and they represent a different patient group (symmetric, eccentric, and unbalanced group).

Figure 20a–c shows the product design drawings of the three types.

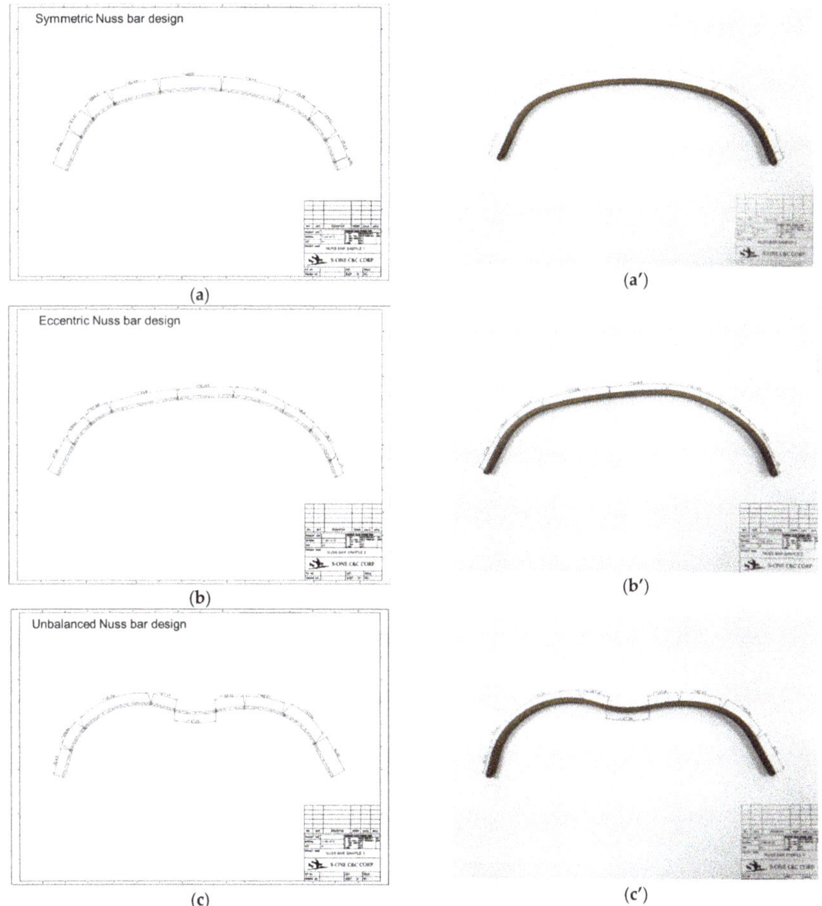

Figure 20. Three types of patient-specific Nuss bar drawings and products with Ti-Gr5: (**a**) symmetric Nuss bar design and (**a′**) symmetric Nuss bar product for Patient No. 1; (**b**) eccentric Nuss bar design, which was fitted for an asymmetric chest wall patient (Patient No. 2), and (**b′**) eccentric Nuss bar product; (**c**) unbalanced Nuss bar design, which was fitted for an asymmetric chest wall patient (Patient No. 3), And (**c′**) unbalanced Nuss bar product.

A total of 18 patient-specific Nuss bar prototypes were created with three prototypes for each design using the CAM data obtained using the same method. Figure 20a′–c′ shows the fabrication outputs of the patient-specific Nuss bars. Major dimensions were measured to verify the accuracy of the fabricated patient-specific Nuss bar prototypes, and the measurement data and distribution are shown in Table 8 and Figure 21.

Table 8. Data table for patient-specific Nuss bar dimension accuracy.

		Symmetric		Eccentric		Unbalanced	
		SUS	Ti-Gr5	SUS	Ti-Gr5	SUS	Ti-Gr5
Drawing (mm)	Width (A)	298.1		304.2		297.8	
	Height (B)	86.3		84.4		78.6	
Specimen 1 (mm)	Width (A)	297.5	298	303	305	299	297
	Height (B)	85	86	86	85	82	80.5
Specimen 2 (mm)	Width (A)	298.5	298	302	304	296	298
	Height (B)	86	85	86	86	82	79.5
Specimen 3 (mm)	Width (A)	300	300	303	302	297	299
	Height (B)	86	85	87	85	82.5	81.5

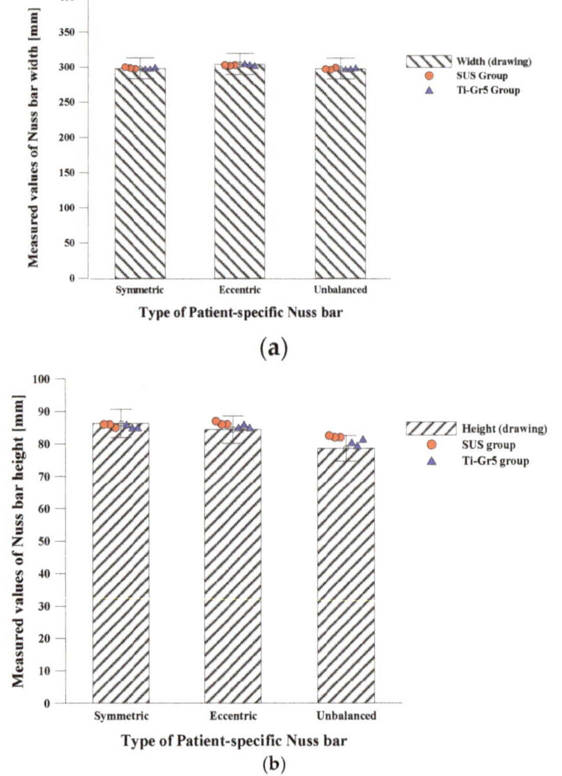

Figure 21. Dimension accuracy histogram of patient-specific Nuss bar: (**a**) maximum width of patient-specific Nuss bars was measured after Nuss bar processing and compared with original value; (**b**) maximum height of patient-specific Nuss bars was measured and compared with original data.

The average dimension accuracies of the titanium and SUS prototypes were ±0.85 % and ±1.47%, respectively. These results show that the patient-specific Nuss bar fabricated with Ti-Gr5 has a higher machining accuracy.

4. Discussion

Since the Nuss procedure was announced by Dr. Nuss in 1997, many Nuss procedures have been carried out and various related studies have been conducted. However, the Nuss bars have been formed by naked eye and experience in the operation room. This study verified the effectiveness of the CAD-based patient-specific Nuss bar design technology for the optimal Nuss procedure.

The CAD-based patient-specific Nuss bar design technology was verified by measuring the Haller index before and after the Nuss procedure. In particular, an increase of the minimum AP distance after the procedure shows that the depressed sternum that pressured the organs inside the chest has been successfully corrected by the patient-specific Nuss bar designed with CAD technology and sufficient internal space has been attained. This method is expected to improve the accuracy and reliability of surgery because the optimal insertion point and correction result of the Nuss bar can be predicted before the actual Nuss procedure.

Furthermore, this study demonstrates the possibility and utility of fabricating the patient-specific Nuss bar through a CNC bending machine using the CAM-based manufacturing technology, and it shows that patient-specific Nuss bars with a dimension error range within ±5% can be fabricated. Using the CAM data for correcting the springback phenomenon for each material and shape will improve productivity because the same products with a certain performance can be manufactured accurately within a short time. However, new CAM data must be constructed if the product shape is changed, or if the material thickness, bending degree, and components are changed even if the product has the same shape. This characteristic will act as a disadvantage when manufacturing diverse products.

5. Conclusions

The CAD/CAM-based patient-specific Nuss bar design and fabrication technology verified in this study will provide a good solution to solve the problems and inconveniences of the current Nuss procedure.

However, the commercialization of the patient-specific Nuss bars fabricated with CAD/CAM-based design and manufacturing technology is still problematic because the CAD-based patient-specific Nuss bar design using the Haller index has limitations with respect to reliability verification and the CAM-based manufacturing method has limitations related to the precise machining of various shapes without establishing the CAM data.

Nevertheless, the proposed method has a positive value in that it can dramatically solve the problem of Nuss bar formation during surgery. In the future, we should obtain clinical data to build the reliability of patient-specific Nuss bars, carry out research based on computer-aided engineering (CAE) to predict the prognosis of the Nuss procedure (e.g., the fixing point of the Nuss bar and the load distribution according to shape)(, and build the CAM database to expand the manufacturing scope of the Nuss bars. Then, we could not only enter the market through the patient-specific Nuss bars, but also secure the possibility of manufacturing various patient-specific orthopedic implants for knees, joints, and spines.

Author Contributions: B.-Y.L. made the 3D model. H.I. developed the general idea of the study, C.-S.L. reviewed and edited paper and supervised the project, J.-Y.H. conceived and designed the study, K.-H.H. performed experiments and collected data, Y.-J.K. wrote the paper and analyzed the data. All authors have read and revised the manuscript.

Funding: This research was supported by a grant from the University Research Park Project of Busan National University funded by the Busan Institute of S & T Evaluation and Planning.

Conflicts of Interest: The authors declare no conflict of interest. The funders had no role in the study design, data collection and analyses, writing of the manuscript, or in the decision to publish the results.

References

1. Kilda, A.; Basevicius, A.; Barauskas, V.; Lukosevicius, S.; Ragaisis, D. Radiological assessment of children with pectus excavatum. *Indian J. Pediatr.* **2007**, *74*, 143–147. [CrossRef] [PubMed]
2. Park, H. Minimally Invasive Surgery for Pectus Excavatum&58; Park Technique. *J. Clin. Anal. Med.* **2011**, *2*, 84–90.
3. Lee, J.H.; Kim, S.J.; Kang, J.H.; Chung, W.S.; Kim, H.; Chon, S.H. Silastic molding method for pectus excavatum correction using a polyvinyl alcohol (Ivalon) sponge. *Korean J. Thorac. Cardiovasc. Surg.* **2012**, *45*, 418–420. [CrossRef] [PubMed]
4. Uemura, S.; Nakagawa, Y.; Yoshida, A.; Choda, Y. Experience in 100 cases with the Nuss procedure using a technique for stabilization of the pectus bar. *Pediatr. Surg. Int.* **2003**, *19*, 186–189. [PubMed]
5. Yoon, Y.S.; Kim, H.K.; Choi, Y.S.; Kim, K.; Shim, Y.M.; Kim, J. A modified Nuss procedure for late adolescent and adult pectus excavatum. *World J. Surg.* **2010**, *34*, 1475–1480. [CrossRef] [PubMed]
6. Nagasao, T.; Miyamoto, J.; Tamaki, T.; Ichihara, K.; Jiang, H.; Taguchi, T.; Yozu, R.; Nakajima, T. Stress distribution on the thorax after the Nuss procedure for pectus excavatum results in different patterns between adult and child patients. *J. Thorac. Cardiovasc. Surg.* **2007**, *134*, 1502–1507. [CrossRef] [PubMed]
7. Park, H.J.; Song, C.M.; Her, K.; Jeon, C.W.; Chang, W.; Park, H.-G.; Lee, S.Y.; Lee, C.S.; Youm, W.; Lee, K.R. Minimally Invasive Repair of Pectus Excavatum Based on the Nuss Principle:An Evolution of Techniques and Early Results on 322 Patients. *Korean J. Thorac. Cardiovasc. Surg.* **2003**, *36*, 164–174.
8. Lin, K.-H.; Huang, Y.-J.; Hsu, H.-H.; Lee, S.-C.; Huang, H.-K.; Chen, Y.-Y.; Chang, H.; Chen, J.-E.; Huang, T.-W. The Role of Three-Dimensional Printing in the Nuss Procedure: Three-Dimensional Printed Model-Assisted Nuss Procedure. *Ann. Thorac. Surg.* **2018**, *105*, 413–417. [CrossRef] [PubMed]
9. Bonacina, L.; Froio, A.; Conti, D.; Marcolin, F.; Vezzetti, E. Automatic 3D foetal face model extraction from ultrasonography through histogram processing. *J. Med. Ultrasound* **2016**, *24*, 142–149. [CrossRef]
10. Archer, J.E.; Gardner, A.; Berryman, F.; Pynsent, P. The measurement of the normal thorax using the Haller index methodology at multiple vertebral levels. *J. Anat.* **2016**, *229*, 577–581. [CrossRef] [PubMed]
11. Robbins, L.P. Pectus excavatum. *Radiol. Case Rep.* **2011**, *6*, 460. [CrossRef] [PubMed]
12. Khanna, G.; Jaju, A.; Don, S.; Keys, T.; Hildebolt, C.F. Comparison of Haller index values calculated with chest radiographs versus CT for pectus excavatum evaluation. *Pediatr. Radiol.* **2010**, *40*, 1763–1767. [CrossRef] [PubMed]
13. Poston, P.M.; Patel, S.S.; Rajput, M.; Rossi, N.O.; Ghanamah, M.S.; Davis, J.E.; Turek, J.W. The correction index: setting the standard for recommending operative repair of pectus excavatum. *Ann. Thorac. Surg.* **2014**, *97*, 1176–1180. [CrossRef] [PubMed]
14. Radiopaedia: Haller Index. Available online: https://radiopaedia.org/articles/haller-index (accessed on 5 March 2018).
15. Daunt, S.W.; Cohen, J.H.; Miller, S.F. Age-related normal ranges for the Haller index in children. *Pediatr. Radiol.* **2004**, *34*, 326–330. [CrossRef] [PubMed]
16. Ewert, F.; Syed, J.; Wagner, S.; Besendoerfer, M.; Carbon, R.T.; Schulz-Drost, S. Does an external chest wall measurement correlate with a CT-based measurement in patients with chest wall deformities? *J. Pediatr. Surg.* **2017**, *52*, 1583–1590. [CrossRef] [PubMed]
17. Rha, E.Y.; Kim, J.H.; Yoo, G.; Ahn, S.; Lee, J.; Jeong, J.Y. Changes in thoracic cavity dimensions of pectus excavatum patients following Nuss procedure. *J. Thorac. Dis.* **2018**, *10*, 4255–4261. [CrossRef] [PubMed]
18. Wall, C.; Group, I.; Chest, O.N.; Diseases, W. Effect of radiotherapy after mastectomy and axillary surgery on 10-year recurrence and 20-year breast cancer mortality: meta-analysis of individual patient data for 8135 women in 22 randomised trials. *Lancet* **2014**, *383*, 2127–2135.
19. Rebeis, E.B.; de Campos, J.R.M.; Fernandez, Â.; Moreira, L.F.P.; Jatene, F.B. Anthropometric index for pectus excavatum. *Clinics* **2007**, *62*, 599–606. [CrossRef] [PubMed]

20. Damián-Noriega, Z.; Pérez-Moreno, R.; Villanueva-Pruneda, S.A.; Domínguez-Hernández, V.M.; Puerta-Huerta, J.P.A.; Huerta-Muñoz, C. A new equation to determine the springback in the bending process of metallic sheet. In Proceedings of the ICCES: International Conference on Computational & Experimental Engineering and Sciences, Crete, Greece, 25–30 September 2008; Volume 8, pp. 25–30.
21. Narita, K.; Niinomi, M.; Nakai, M.; Akahori, T.; Tsutsumi, H.; Oribe, K. Bending fatigue and spring back properties of implant rods made of β-type titanium alloy for spinal fixture. In *Advanced Materials Research*; Trans Tech Publications: Stafa-Zurich, Switzerland, 2010; Volume 89, pp. 400–404.
22. Adamus, J.; Lacki, P.; Motyka, M.; Nitkiewicz, Z. Analysis of titanium sheet bending process. *Inż. Mater.* **2010**, *31*, 716–719.
23. CUSTOMPART.NET: Spring Back Calculator. Available online: https://www.custompartnet.com/calculator/bending-springback (accessed on 12 April 2018).
24. SM: Spring Back. Available online: http://sheetmetal.me/tooling-terminology/spring-back/ (accessed on 2 April 2018).
25. Brown, A.L.; Cook, O. Cardio-respiratory studies in pre and post operative funnel chest (*Pectus excavatum*). *Dis. Chest* **1951**, *20*, 378–391. [CrossRef] [PubMed]

 *applied sciences*

Article

An Efficient Automatic Midsagittal Plane Extraction in Brain MRI

Hafiz Zia Ur Rehman [1] and Sungon Lee [2],*

[1] Department of Mechatronics Engineering, Hanyang University, Ansan 15588, Korea; hzia05@gmail.com
[2] School of Electrical Engineering, Hanyang University, Ansan 15588, Korea
* Correspondence: sungon@hanyang.ac.kr

Received: 16 October 2018; Accepted: 6 November 2018; Published: 9 November 2018

Abstract: In this paper, a fully automatic and computationally efficient midsagittal plane (MSP) extraction technique in brain magnetic resonance images (MRIs) has been proposed. Automatic detection of MSP in neuroimages can significantly aid in registration of medical images, asymmetric analysis, and alignment or tilt correction (recenter and reorientation) in brain MRIs. The parameters of MSP are estimated in two steps. In the first step, symmetric features and principal component analysis (PCA)-based technique is used to vertically align the bilateral symmetric axis of the brain. In the second step, PCA is used to achieve a set of parallel lines (principal axes) from the selected two-dimensional (2-D) elliptical slices of brain MRIs, followed by a plane fitting using orthogonal regression. The developed algorithm has been tested on 157 real T_1-weighted brain MRI datasets including 14 cases from the patients with brain tumors. The presented algorithm is compared with a state-of-the-art approach based on bilateral symmetry maximization. Experimental results revealed that the proposed algorithm is fast (<1.04 s per MRI volume) and exhibits superior performance in terms of accuracy and precision (a mean z-distance of 0.336 voxels and a mean angle difference of 0.06).

Keywords: medical image registration; image alignment in medical images; misalignment correction in MRI; midsagittal plane extraction; symmetry detection; PCA

---

1. Introduction

Segmentation of brain in magnetic resonance images (MRIs) is one of the difficult and crucial steps of clinical diagnostic tools in medical images. The brain is the most complex organ in the human body that can be split into two approximately symmetrical hemispheres using a plane. This plane is known as the midsagittal plane (MSP) [1]. In brain symmetric/asymmetric analysis, automatic MSP extraction that is independent for symmetrical and asymmetrical brain regions is an essential brain segmentation task [2]. Enormous research reflects that the symmetrical structure of the brain deteriorates due to psychological and physical ailments in the brain [3]. Clinical experts use the symmetry of the brain to identify qualitatively asymmetric patterns that signify an ample range of pathologies, such as brain tumors [4,5], brain infections [6], metabolic disorders [7], brain injury [8], and perinatal brain lesions [9]. Similarly, the computer-aided diagnostic and image analysis systems can use the symmetry and asymmetry information as a prior knowledge to embellish the system efficiency in the analysis of altered brain anatomy [10].

Moreover, the detection of MSP is required in registration [11] of medical images as the first step for spatial normalization [12] and anatomical standardization [13] of the brain images. However, legitimate evaluation of symmetric and asymmetric patterns in brain images is possible only when the symmetry axis or the symmetry plane (MSP) is accurately aligned and appropriately oriented within the coordinate system of the MRI scanner [14]. This permits the system to adjust the possible misalignment of brain MRIs. A general phenomenon in brain MRI scanning is that many neuroimaging scanners produce tilted and distorted brain images. The tilt of the head is not always detectable, due to many reasons such as the health conditions, immobility of patients, imprecision of the data calibration systems, and the inexperience of the technicians. Consequently, the slices of the brain MRIs are no more alike within the same orientation, at either the axial or coronal level [15]. Disoriented and misaligned brain MRIs can betray visual inspection and prevalently yield erroneous clinical perception [16]. In summary, assessment of brain MRIs for any anomaly based on cross-referencing of brain hemispheres (left and right), either by a human expert or computer-based software could be affected by false geometrical representation. Consequently, it is essential to correct the tilt and realign the brain MRIs data before further analysis.

Manual misalignment correction is extremely time-consuming and laborious to perform on a huge scale. It also demands an urbane knowledge of brain anatomy. Therefore, it is neither sufficient nor efficient. Alignment or tilt correction of brain MRIs is tantamount to realigning the MSP with the center of the image matrix or image coordinate system [14]. If the MSP is computed precisely, the orientation problem of the MRI volume can be resolved. Thus, the tilt of the head volume can be assessed and adjusted. An ideal MSP can be defined as a virtual geometric plane passing through the interhemispheric fissure (IF) [17], about which the three-dimensional (3-D) anatomical structure of the brain (such as the ventricles, anterior/posterior commissures, corpus callosum, thalamus) exhibits maximum bilateral symmetry [18].

Previously, several approaches that considered the problem of computing the MSP in brain MRIs and other brain image modalities (Computed Tomography (CT), Positron Emission Tomography (PET), Single Photon Emission Computed Tomography (SPECT)) have been published. These approaches can be divided into two distinct groups, varying in their exclusive interpretation of prescribed MSP: (1) shape-based algorithms that identify the location of cerebral IF using features of the head images to estimate MSP; and (2) content-based algorithms that considered MSP as the plane which maximizes the bilateral symmetry of the brain. A comprehensive survey of all the existing MSP extraction methods can be found in a recent review [19].

Shape-based algorithms first segment the longitudinal fissure of the brain MRIs and employ it as a landmark for symmetry analysis and MSP extraction. For instance, Brummer [20] utilized Hough transform for straight line identification on each coronal slice and computed the MSP using interpolation. Guillemaud et al. [21] exploited linear snakes to find the control points on IF lines and estimate MSP plane through these lines using orthogonal regression. Volkau and Nowinski [17,22] and Kuijf et al. [23] proposed simple and accurate methods based on Kullback and Leibler's (KL) measure. These approaches are computationally efficient and independent of internal asymmetries. However, they became unstable in the presence of strong mass effect near IF or invisibility of IF, which is common in some imaging protocols (CT, PET or SPECT).

Content-based algorithms, also known as the similarity-based methods, maximize some similarity measure between the two halves (hemispheres) of the 3-D head volume. Ardekani et al. [24] proposed an iterative local search-based algorithm that uses the cross-correlation between the voxels of either side of the estimated MSP. This method failed on images having asymmetries due to pathological effects. Liu et al. [18] computed the MSP by extracting the two-dimensional (2-D) symmetry axes on each slice using cross-correlation from an edge image, followed by plane fitting. Another technique based on the similarity between two sides of the head volume using block matching was given by Prima et al. [25]. These methods are computationally intensive due to their iterative nature and optimization scheme. Ruppert et al. [26,27] improved the efficiency of similarity and symmetric-based methods, and developed an algorithm using 3-D Sobel edge operator, downsampling, and a multiscale scheme. Although the algorithm used the sagittal orientation for MSP extraction, it can be applied to other orientations (axial, and coronal) as well. The authors tested the algorithm on limited imaging protocols and it is also sensitive to noise. The MSP extraction technique based on 3-D scale invariant feature transform (SIFT) was formulated by Wu et al. [28]. The authors determined the MSP by parallel

3-D SIFT matching and voting, followed by least median of square (LMS) regression. The paper also compared the results of the algorithm with three other MSP extraction methods [16,27,29]. The authors reported that the algorithm is sensitive to noise, blur, and asymmetry, greater than a certain threshold. Moreover, the parameter setting of the algorithm is somehow complex. A computationally simple and robust MSP extraction algorithm was presented by [30] using curve fitting. The method depends on skull stripping in brain images and the authors reported that the algorithm may fail to identify the MSP correctly if the image slices have a rotation angle of greater than 15° or unsuccessful skull stripping.

Recently, Ferrari et al. [31] devised a new MSP extraction algorithm using a sheetness measure obtained from 3-D phase congruency (PC) responses. The authors reported results on synthetic and real brain MRIs. A comparison study of three MSP extraction algorithms (symmetry-based [27], phase congruency [31], and Hessian-based [32]) is presented in [33]. In spite of the enormous variety of algorithms published on MSP extraction, there is no unanimity among the researchers about the best algorithm, due to the ambiguous longitudinal fissure lines, low-contrast brain images, mass effect, and absence of intensity standardization. Moreover, MSP extraction becomes more difficult and challenging when the brain MRIs having a pathological disorder [18,25].

In this article, we have combined the advantages of both aforementioned techniques (to some extent) and developed a new principal component analysis (PCA) and symmetric feature-based approach to automatically compute and reorient the MSP in T_1-weighted MRIs. In fact, the pathological disorder and variations, such as stroke, brain tumor, bleedings, and brain injury, only alter the local intensities and symmetries of brain MRIs. They do not affect the overall shape topological properties of the 3-D head. Furthermore, when the head volume demonstrates a low signal-to-noise ratio (SNR) and significant artifacts, the segmentation of external surfaces is easier as compared to that of internal structures.

Therefore, by considering all these observations and assuming that the head is an ellipsoid-like 3-D solid object, a PCA-based algorithm is designed for MSP extraction. PCA is a fundamental and prevailing statistical technique also known as Hotelling transform substantially used in digital image processing for data dimension reduction [34], feature pattern recognition [35], quality control [36], data decorrelation [37], data compression [38], and segmentation [39]. It is also acknowledged as a low-level digital image processing tool for tasks such as the orientation assessment and alignment of particular shape objects [40,41]. In this paper, PCA has been used for determining the rotation angle (yaw angle) of the bilateral symmetric axis of the brain. The parameters of MSP (yaw angle, roll angle, and offset) are estimated in two steps. In the first step, a coarse value of yaw angle has been estimated using PCA. The angle value is further refined using a cross-correlation method. After thresholding and elliptical area extraction, PCA is used to achieve a set of parallel lines (principal axes) from the selected 2-D slices of brain MRIs. In the second step, the roll angle and the plane offset (a perpendicular distance of MSP from the origin) have been computed by fitting a plane to these parallel lines using orthogonal regression [42]. Initial slices in brain MRIs, where no or very small brain is present (in size), show ambiguous symmetry features as compared to the slices near the center of the brain. Therefore, selected slices have been used for MSP extraction and automatically discarded the ambiguous slices based on semi-axes (major and minor of the ellipse). Similar to the work by Liu [18] who used a weighted mean due to biasing in mean by the initial slices as compared to the superior slices, the removal of ambiguous slices of brain MRIs makes this technique to perform robustly and efficiently. Finally, an affine transformation has been applied to rotate the 3-D head volume to realign (recenter) within the required coordinate system (scanner coordinate system). The proposed technique is insensitive to pathological asymmetries, acquisition noises, and bias fields.

The rest of the paper is categorized as follows: Section 2 describes the methodology of MSP extraction algorithm. Implementation of the algorithm, the description of datasets used for evaluation, and results are reported in Section 3. Section 4 discusses some limitations of the developed algorithm and concludes the proposed technique.

2. Materials and Methods

2.1. Geometry of MSP

Generally, MRI of the brain consists of 3-D volumetric data in three orientations: axial, coronal, and sagittal orientations. In the proposed algorithm, only the axial orientation was considered as an input to the algorithm. The head coordinate system is defined as the ideal head coordinate system (X_o, Y_o, Z_o) and the imaging coordinate system (X, Y, Z), as described by Liu et al. [18]. The origin of the ideal coordinate is the center of the brain with positive X_o pointing to the right. Anterior and superior directions represent positive Y_o and Z_o axes from the center of the brain, respectively. Mathematically, $X_o = 0$ is defined to be the MSP with respect to the ideal coordinate system. Practically, the imaging coordinate system (blue) varies from the ideal coordinate system (black) due to translations and three rotations (pitch ω, roll φ, and yaw θ with respect to X_o, Y_o, and Z_o axes, respectively) of the patient head as portrayed in Figure 1. Therefore, the main objective of MSP extraction is to determine the transformation between the two planes, i.e., $X_o = 0$ and $X = 0$.

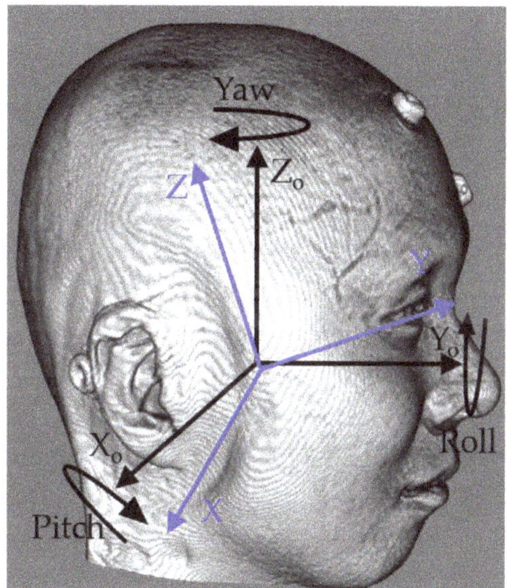

Figure 1. An ideal coordinate system $X_oY_oZ_o$ (black) versus an imaging coordinate system XYZ (blue).

MSP in an image coordinate system can be defined as:

$$aX + bY + cZ + d = 0 \tag{1}$$

where parameters (a, b, c) are not all zero and can be scaled by any non-zero scalar, vector (a, b, c) is the normal vector of the MSP, and $d / \sqrt{a^2 + b^2 + c^2}$ is the perpendicular distance of the plane from the origin.

The intersection of MSP with each axial slice (slice cut perpendicularly to the Z_o) is always a vertical line in the ideal coordinate system. The rth axial slice is represented by a plane equation as:

$$Z = Z_r \tag{2}$$

The intersection of Equations (1) and (2) is the vertical line (ideally a line of bilateral symmetry of brain) on the rth slice and can be written as:

$$aX + bY + (cZ_r + d) = 0 \tag{3}$$

This represents a normal equation of the 2-D line in the XY plane. By comparing Equation (3) with the standard normal equation of the line, the orientation θ_r of the line (vertical line) can be computed as:

$$\theta_r = \tan^{-1}\left(\frac{b}{a}\right) \tag{4}$$

Equation (4) exhibits that the orientation of all the 2-D symmetric lines should be the same, irrespective of the position of slice i.e., Z_r. This angle corresponds to the yaw angle of the patient's head. This angle is estimated using the PCA and cross-correlation technique described in the succeeding paragraph.

Similarly, the perpendicular distance (translation offset) p_r of the line (Equation (3)) from the point $(0, 0, Z_r)$ can be calculated as:

$$p_r = cZ_r + d \tag{5}$$

Equation (5) demonstrates that the offset p_r of the symmetric line on the rth slice ($Z = Z_r$) is linearly related to slice position as a function of plane parameters c and d. This represents an overdetermined set of linear equations in p_r and Z_r. It can be solved by fitting a plane to a set of parallel lines having the orientation θ_r. We use an orthogonal regression [42] using PCA to fit the plane to these lines in 3-D Euclidian space.

To completely determine the MSP parameter (a, b, c, d), we need to assess the transformation between the two planes $X_0 = 0$ (MSP in the ideal coordinate system) and $X = 0$ (MSP in the image coordinate system). The derivation of this transformation can be found in [18]. The final expression for MSP $X_0 = 0$ can be written in terms of the imaging coordinate system as:

$$X \cos \varphi \cos \theta + Y \cos \varphi \sin \theta - Z \sin \varphi - (n \cdot \Delta) = 0 \tag{6}$$

where $n = [\cos \varphi \cos \theta, \cos \varphi \sin \theta, \sin \theta]^T$ is the unit normal vector of the plane, $(\cdot)$ indicates the standard dot product of the vectors, and $\Delta = [\Delta X_0, \Delta Y_0, \Delta Z_0]^T$ is the translation vector.

Dividing by $\cos \varphi$ and $abs(\varphi) \neq 90°$, and comparing the result with Equation (1), we obtain:

$$a = \cos \theta, b = \sin \theta, c = -\tan \varphi, d = -\frac{n \cdot \Delta}{\cos \varphi} \tag{7}$$

Therefore, φ is the roll angle and $\theta = \theta_r$ is the yaw angle of each axial slice of the head's imaging coordinate system.

2.2. Estimation of Yaw Angle (θ_r)

The yaw angle is estimated in two stages. A coarse value (θ_1) is estimated in the first stage using PCA after the region of interest (ROI) extraction. Then, the image is vertically aligned by θ_1. The value of the yaw angle is further refined in the second stage and a cross-correlation technique is exploited to measure θ_2. The sum of θ_1 and θ_2 completes the procedure of calculating yaw angle (θ_r).

2.2.1. Region of Interest Extraction

Due to the fact that the human brain is nearly elliptical, PCA is used, as it aligns the data in the direction of maximum variance (data spread). A reference 2-D slice I_0 is selected from the 3-D volume of the brain MRIs. The selection of the slice is important. As in fact, the higher slice (before and after the half of the total slices present in the volume), the brain in the 2-D image becomes more elliptical in shape. It can give a good estimate of the angle of brain symmetric axis as compared to other slices [43].

Therefore, a 1/55th slice of the total slices present in the volume of brain MRIs is considered as a reference slice. Next, the image is binarized [44] and noise is removed by a mathematical morphological filtering operation called area opening [45] (pp. 112–114). In this filtering procedure, small connected pixels of which the area (in a number of pixels) is less than a specified threshold (η) are removed. The value of "η" can be varied depending upon the type of brain MRIs and the extent of noise. After several experiments, a value of 100 pixels has been considered a good estimate for η (area threshold) in all the brain MRIs including tumor datasets. Mathematical, area opening is represented as:

$$\gamma_\eta = \bigcup_i \left\{ \gamma_{B_i} \big|_i \text{ is connected and } card(_i) = \eta \right\}, i = 1, 2 \ldots, k. \tag{8}$$

where γ_η is the area opening, η is the value of threshold area of the connected pixels, and *card*(B) is the number of elements (cardinal number) of B.

Area opening is equivalently defined as the union of all the morphological opening (erosion followed by dilation) with the connected structuring element of which the size is equal to η. Then, a rectangular area is achieved by searching for the first and last nonzero pixels along the rows (top and bottom) and columns (left and right) of the noise-free binary image, as shown in Figure 2.

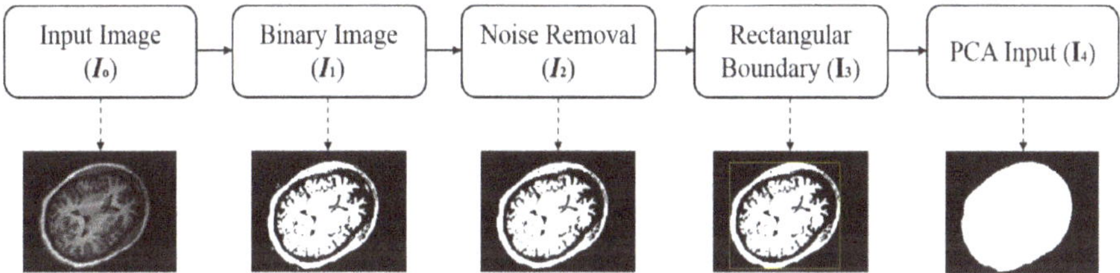

Figure 2. Steps for noise removal and region of interest (ROI) extraction.

Some of the ROI pixels inside the rectangular boundary (image I_3) have value 0. To make all the ROI pixels in the image I_3 to one, the complement of noise-free binary image I_2 is multiplied (logical AND) with the rectangular boundary, such that all the pixels inside the boundary are equal to 1 (Figure 3a). Largest connected component (LCC) is chosen (Figure 3b) from the resulted image and added (logical OR) to I_3 (Figure 3c). Lastly, morphological flood-fill operation [45] (p. 208) is used to fill the holes (0-valued pixels) and achieved the binary image I_4 with a single ROI. This image is exploited as an input for PCA (Figure 3d).

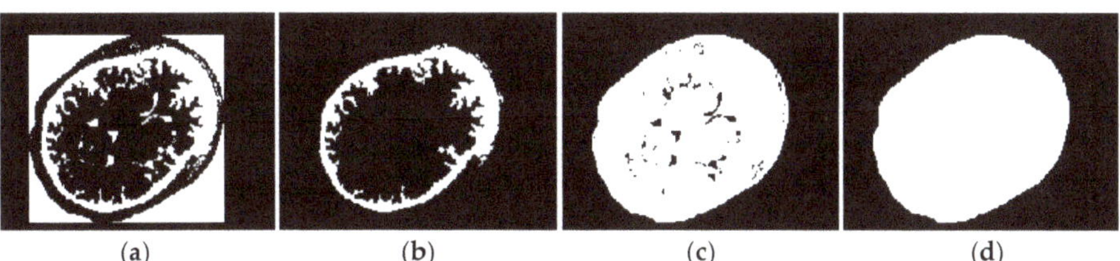

Figure 3. Intermediate steps between images I_3 and I_4, (**a**) logical AND of I_2 with the rectangular boundary, (**b**) largest connected component, (**c**) logical OR of (**b**) with I_3, (**d**) holes filling using morphological flood-fill operation.

2.2.2. Principal Component Analysis

After noise removal and ROI extraction steps, the method uses 2-D coordinates (column and row) of the non-zero pixels of the object's region as data points and an array X is formed as:

$$X = \begin{bmatrix} c_1 & r_1 \\ c_2 & r_2 \\ \vdots & \vdots \\ c_n & r_n \end{bmatrix} \tag{9}$$

where r, c, and n represent row, column, and the total number of data points (1-valued pixels) in the ROI, respectively.

The size of X is $n \times 2$, and rows of X are the location coordinate values of each non-zero pixel. Mean of X is a row vector m_x (mean of the elements in each column of X), and can be computed as:

$$m_x = \frac{1}{n} \sum_{i=1}^{n} X_i \tag{10}$$

Similarly, a covariance matrix P_x can be calculated as:

$$P_x = \frac{1}{n-1} \sum_{i=1}^{n} (X_i - m_x)(X_i - m_x)^T \tag{11}$$

The mean subtraction is vital for performing PCA to ensure that the first principal component represents the direction of maximum variance. The covariance matrix P_x is a real and symmetric matrix with a size of 2×2. Thus, finding a pair of orthonormal eigenvectors is always possible [46].

Suppose the elements of covariance matrix as:

$$P_x = \begin{bmatrix} f & g \\ g & h \end{bmatrix} \tag{12}$$

where $f = \mathrm{var}(x_{ii})$, $h = \mathrm{var}(x_{jj})$, and $g = \mathrm{cov}(x_{ij})$.

The principal components can be determined by solving an eigenvalue problem as:

$$(P_x - \lambda I)e = 0 \tag{13}$$

where λ, I, and e are the eigenvalue, identity matrix, and eigenvector, respectively.

If Equation (13) is to have a solution other than vector zero, then $(P_x - \lambda I)$ must be a nonsingular matrix. Therefore, it leads to a characteristics equation as:

$$\det(P_x - \lambda I)e = 0 \tag{14}$$

After expansion of Equation (14), it becomes a second-degree equation as:

$$\lambda^2 - \lambda(f + h) + \left(fh - g^2\right) = 0 \tag{15}$$

The Equation (15) can be solved using a quadratic formula as:

$$\lambda_1, \lambda_2 = \frac{tr(P_x) \pm \sqrt{tr(P_x)^2 - 4|P_x|}}{2} \tag{16}$$

where $tr(P_x) = f + h$ and $|P_x| = fh - g^2$.

The corresponding eigenvectors of the eigenvalues (λ_1, λ_2) can be calculated as:

$$e_j = \frac{1}{\sqrt{g^2 + (\lambda_j - f)}} \begin{bmatrix} g \\ \lambda_j - f \end{bmatrix} \tag{17}$$

where $e_j(j = 1, 2)$, and λ_j denote the eigenvectors and eigenvalues of P_x, respectively.

The eigenvalues represent the length of the semi-axes of the ellipse and the eigenvectors specify the directions of these axes, as illustrated in Figure 4.

The rotation angle can be computed using the eigenvector corresponding to the largest eigenvalue as:

$$\theta_1 = \tan^{-1}\left(\frac{v_y}{v_x}\right), \quad \theta_1 \in \left(-90°, 90°\right) \tag{18}$$

where v_x, v_y are the horizontal and vertical components of the eigenvector associated with the largest eigenvalue, respectively.

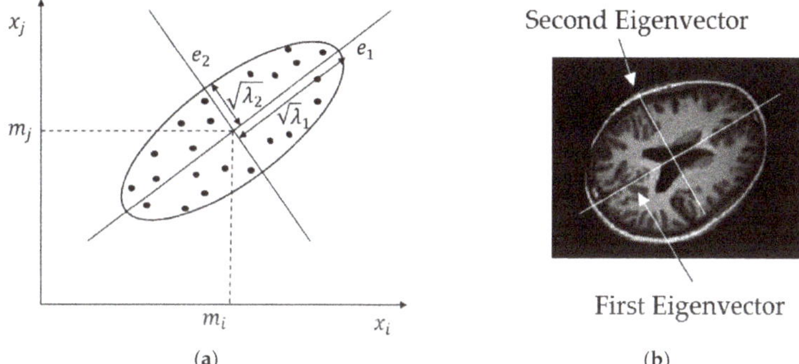

Figure 4. Orthonormal eigenvector by principal component analysis (PCA). (a) Eigenvector e_1 through the center of the elliptical area represents the maximum variance and e_2 is perpendicular to e_1, and (b) extracted eigenvectors from brain brain magnetic resonance images (MRI) are imposed on the image.

This procedure provides a rough estimate of brain bilateral symmetric axis orientation θ_1. Next, the slice is realigned with the vertical axis of the image using the angle θ_1. Generally, it was observed that PCA aligned the brain bilateral symmetric axis with the vertical axis of the image with an error of less than $1°$, as displayed in Figure 5.

The first row of Figure 5 indicates that bilateral symmetry axis of the brain is successfully aligned with the vertical axis (white vertical line) of the image by PCA, but sometimes as the second row in Figure 5 depicts, the bilateral symmetric axis of the brain is not completely aligned with the vertical axis of the image. Therefore, to ensure the accurate alignment of the bilateral symmetric axis of the brain with the vertical axis of the image, another fine alignment step is applied using a cross-correlation technique to find the angle θ_2 with the vertical axis of the aligned image F (output of the previous step).

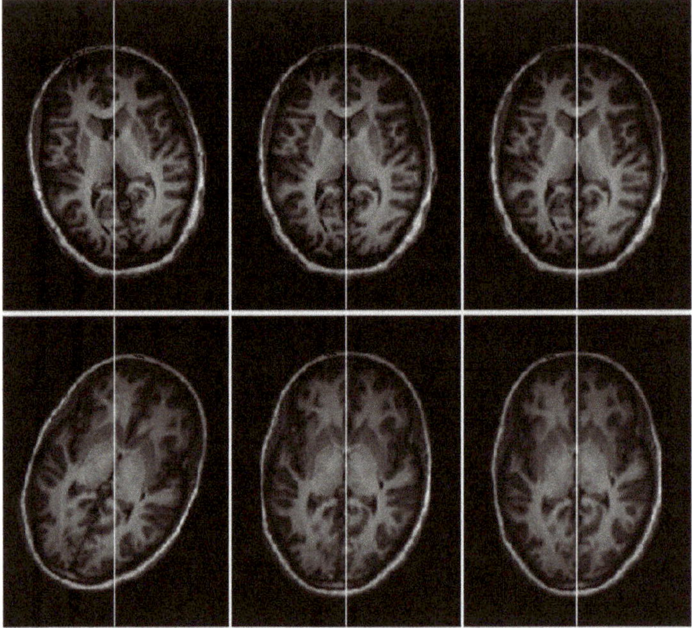

Figure 5. Estimation of symmetry axis orientation θ_r. First column: input images; second column: alignment of the brain bilateral symmetry axis with the vertical axis of the image using θ_1; last column: brain bilateral symmetry axis alignment using $\theta_r = \theta_1 + \theta_2$. Note that $\theta_2 = 0$ in the first row.

2.2.3. Cross-Correlation

In the cross-correlation technique, the aligned image F acquired in the last step is reflected about the current vertical center line to produce a new reflected image F'. The reflected image F' is rotated by $2\theta_2$ ($\theta_2 \in [-10°, 10°]$) with a 0.5° interval about the center of the image, cross-correlated with the image F, and the maximum correlation score is noted. The value of θ_2 at which the cross-correlation score is maximum will be the required angle of bilateral symmetry with the vertical axis of the image. The cross-correlation is accomplished in the frequency space for greater efficiency. The detail of this method can be found in [18].

If θ_2 is zero, it means that the PCA accurately aligns the brain bilateral symmetric axis with the vertical axis of the image, otherwise θ_r (yaw angle) will be:

$$\theta_r = \theta_1 + \theta_2 \tag{19}$$

where θ_1 is the angle obtained from PCA and θ_2 is the angle yielded from the cross-correlation technique.

After applying the cross-correlation technique, the brain bilateral symmetric axis is completely aligned with the vertical axis of the image, as displayed in Figure 5 (last column). Now, the angle θ_r can be used to estimate the first two parameters of the normal of required MSP plane using Equation (7), i.e., $a = \cos\theta$, and $b = \sin\theta$.

2.3. Fitting of Plane in Three Dimensions

According to Equation (4), the angle θ_r will be same for all the 2-D axial symmetric lines on each axial slice, irrespective of the position of slice Z_r. Therefore, each image in the 3-D volume of the brain MRIs is rotated by an angle of $-\theta_r$, so that bilateral symmetry axis of the brain becomes vertically oriented in each image. Normally, the translation offset p_r of each symmetric axis can be computed using a cross-correlation of the aligned image with its vertical reflection about the center of the image [18,25]. This technique has two complications. The first one is the existence of outlier in the translation offset due to pathology effects, image artifacts, and symmetry axis ambiguity in the superior slices of brain. The second one is that this technique is also computationally intensive as it takes approximately 10 s for each slice of size (rows, column) 512×512 [18].

We have introduced a fast approach to estimating the translation offset independent of these constraints. The technique is based on the straightforward and effective observation that the head in the slices of brain MRIs is shaped like an ellipse (ellipsoid in three dimensions). Moreover, a trapezoid area on the surface of the head, center upon $+x$ and $-x$ (from ear to ear) directions of a height (30–$45°$) and a width (45–$70°$), has the most significant geometrical features as described by [16]. Fitting of plane consists of the following three steps:

1. Elliptical Area Extraction

The main objective to extract the elliptical area is to make the offset estimation independent of pathological effects, image artifacts, symmetry axis ambiguity in the higher slices, and computationally efficient. The aligned image is binarized [44] and noise is removed by the same procedure as described in the previous Section 2.2.1. Then, a rectangular area is achieved by searching for the first and last nonzero pixels along the rows (top and bottom) and columns (left and right) of the noise-free binary image (Figure 6d). First and last nonzero columns and rows are denoted by c_1, c_2, r_1, and r_2, respectively. Accordingly, the vertices of the rectangle are labeled as: $A(c_1, r_1)$, $B(c_2, r_1)$, $C(c_2, r_2)$, and $D(c_1, r_2)$, respectively.

Now, the parameters of an ellipse can be determined as:

$$Center\ of\ ellipse:\ O = \left(\frac{c_1 + c_2}{2}, \frac{r_1 + r_2}{2} \right) \tag{20}$$

$$Semi-major\ axis:\ a = \left(\frac{r_2 - r_1}{2} \right) \tag{21}$$

$$Semi-minor\ axis:\ b = \left(\frac{c_2 - c_1}{2} \right) \tag{22}$$

Now, ellipse in parametric form can be written as:

$$x = \frac{c_1 + c_2}{2} + b \cos \theta \tag{23}$$

$$y = \frac{r_1 + r_2}{2} + a \sin \theta \tag{24}$$

Now, within $I(i, j)$, which consists of all the pixels inside the elliptical boundary, the pixels are set to "1" based on Equation (25), as shown in Figure 6e,f, respectively.

$$I(i,j) = \begin{cases} 1 & \text{if } (i,j) \text{ lies inside elliptical area} \\ 0 & \text{otherwise} \end{cases} \tag{25}$$

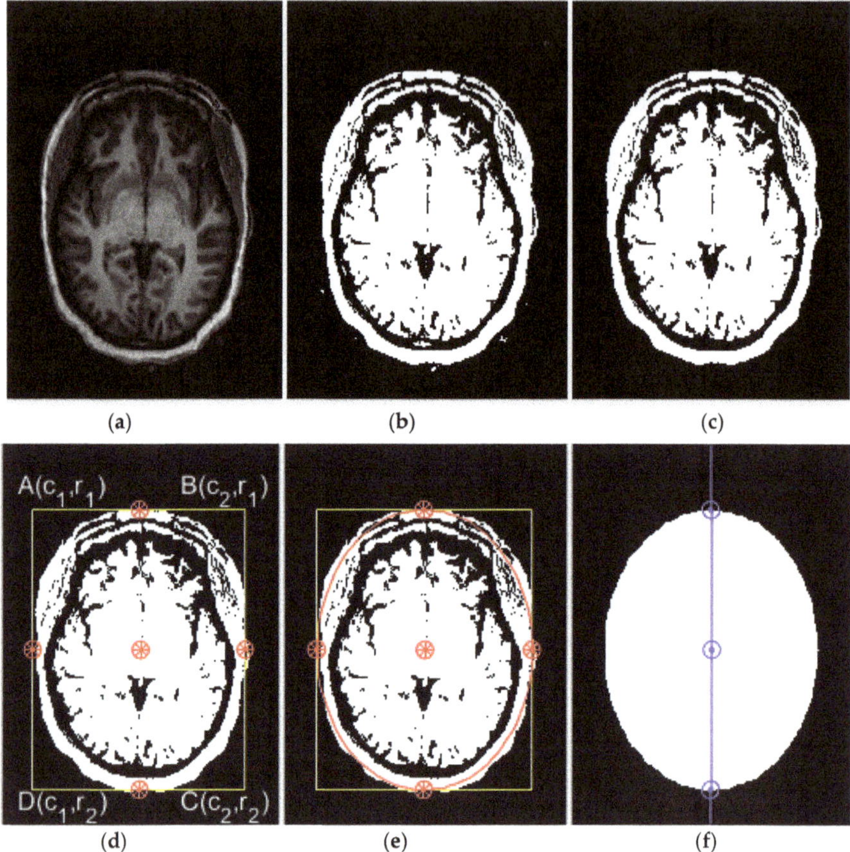

Figure 6. Elliptical area extraction. (**a**) Input image, (**b**) binary image, (**c**) noise removal, (**d**) rectangular boundary extraction, (**e**) elliptical area extraction, and (**f**) symmetric axis (midline) extraction using PCA.

2. Set of Mid-Parallel Lines Extraction

In the axial orientation, inferior and superior slices of the brain volume have ambiguous symmetry axes as compared to the slices near the center of the brain. Secondly, slices higher in the brain are almost ovals [16]. Therefore, ambiguous symmetry slices are automatically eliminated based on the ratio of the semi-axes of the ellipse, and only those slices of which this ratio is greater than 1.2 are extracted. This significantly improves the accuracy and efficiency of the algorithm. Selected sample slices with their respective elliptical area are illustrated in Figure 7.

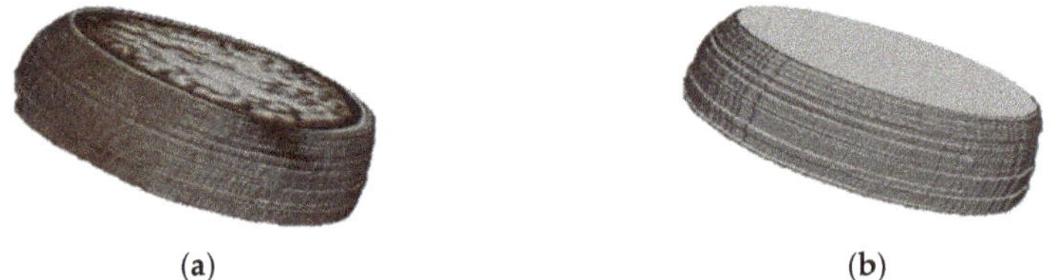

Figure 7. Selected slices from a 3-D volume of brain MRIs. (**a**) Selected slices based on semi-axes, and (**b**) extracted elliptical area from slices.

The midline (principal axis) on each elliptical area slice is extracted using PCA as described in the preceding Section 2.2.2. An array $A_{r,s}$ is formed from the location coordinates of three points (top, center, and bottom) on each midline (Figure 6f) as:

$$A_{r,s} = \left[\begin{array}{ccc} x_{r,s} & y_{r,s} & z_r.1_s \end{array} \right]_{rs \times 3} \qquad r = 1, 2, \ldots, N. \tag{26}$$

where r denotes the rth brain slice, s is the number of points considered on each midline, 1_s is a column vector (s-dimensional) with all its elements equal to one, and N is equal to the total number of slices used (automatically selected).

The first column of $A_{r,s}$ consists of p_r (offset) values of the symmetric axis and the last column contains the indices of the respective slice. According to Equation (5), this makes an overdetermined set of linear equations in p_r and Z_r as a function of plane parameters c and d, as displayed in Figure 8. It can be solved by fitting a plane in 3-D Euclidean space to a set of parallel lines (midlines) having the orientation θ_r.

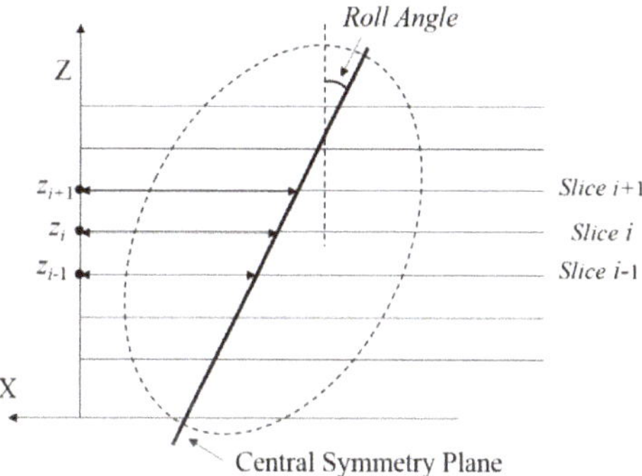

Figure 8. The relationship between the brain slice position (Z_r) and the symmetric axis offset (p_r). The illustration is adapted from [47].

3. Fitting of Plane Using Orthogonal Regression

Orthogonal regression is employed using PCA to fit a plane to these midlines. PCA minimizes the orthogonal distances from the data point to the fitting plane (fitting model). In the linear case, it is also known as total least squares [42]. It is appropriate when all the variables are measured with errors. In contrary to the usual regression, where the assumption is that the predictor variables are measured precisely, and only the response variables have the component of error. Singular value decomposition (*svd*) is used to find the principal components of the PCA. The first step is to center the location data matrix $A_{r,s}$. This can be achieved by subtracting each data point of the matrix from its column mean. The resultant matrix is labeled as B. Then, the *svd* of B can be expressed as:

$$B = USV^T \tag{27}$$

where U is a left-singular vector, S is a diagonal matrix of singular values, and V is a right-singular matrix.

The columns of V will be the required principal components. First two columns (first two principal components) of V define vectors that form a basis for the plane, and the third column (third principal component) is orthogonal to the first two principal components. The coefficients of the third principal component define the normal vector (n) of the MSP having an orientation of θ_r (yaw angle). The third coefficient of the normal vector n can be exploited to estimate the roll angle φ using Equation (7) as:

$$\varphi = -\tan^{-1}(c) \tag{28}$$

where c is the third coefficient of the normal vector.

Similarly, the parameter d of the MSP can be calculated by taking the scalar product of the normal vector n and the mean of $A_{r,s}$.

2.4. Transformation for Tilt Correction

After MSP normal vector computation, the translation matrix and the rotation matrix can be easily determined to correct the tilt (recenter and reorientation) of 3D-volume of brain MRIs. Let R_{req} = yaw(θ)roll(φ)pitch(ω) be the required rotation matrix, which can be written as:

$$R_{req} = \begin{bmatrix} c\varphi c\theta & c\theta s\omega s\varphi - c\omega s\theta & c\omega c\theta s\varphi + s\omega s\theta \\ s\theta c\varphi & c\omega c\theta + s\theta s\omega s\theta & c\omega s\theta s\varphi - s\omega c\theta \\ -s\varphi & s\omega c\varphi & c\omega c\varphi \end{bmatrix} \tag{29}$$

where $c\theta \equiv \cos\theta, s\theta \equiv \sin\theta$, and so on.

The pitch (ω) angle will be zero for MSP, and yaw (θ) and roll (φ) angles are calculated using Equations (19) and (28), respectively. The translation vector between the two coordinate systems, i.e., the center of the volume and centroid of the image grid, can be written as:

$$\Delta = [\Delta X_o, \Delta Y_o, \Delta Z_o]^T \tag{30}$$

Trilinear interpolation is used to reslice the head volume after realignment and tilt correction.

3. Results and Discussion

The presented algorithm has been implemented in MATLAB 2018a on a PC with Intel(R) Core (TM) i5-6600 CPU @ 3.30 GHz, 8 GB RAM. Total time for all algorithmic steps is 1.04 s on average. No attempt has been made on optimization of the code. The developed algorithm has been tested on 157 real T_1-weighted brain MRI datasets including 14 cases from the patients with the brain tumors. All the brain MRI datasets are publicly available. The details of the sample dataset images and parameters are given in Table 1. The first dataset is from Neurofeedback Skull-stripped (NFBS) repository [48] containing 125 volumes of brain MRI having several clinical and subclinical psychiatric syndromes. There is no ground truth (GT) for MSP available for this database. The second database is from the Internet Brain Segmentation Repository (IBSR) [49], containing 18 volumes of T_1-weighted brain MRI with manually segmented hemispheres of the brain. The third dataset is from Montreal Neurological Institute's Brain Images of Tumors for Evaluation (MNI BITE) database [50], which consists of 14 patients, 5 women and 9 men with a mean age of 52 years. The dataset includes 4 patients with low-grade gliomas (brain tumors) and 10 patients with high-grade gliomas (brain tumors). The mean tumor volume calculated manually by the experts was 30 cm^3.

Table 1. Datasets used for evaluation.

Datasets	Detail of Images
NFBS [48]	There are 125 T_1-weighted MRI scans, 77 females and 48 males in the 21–45 age range (average: 31) with a variety of clinical and subclinical psychiatric symptoms. The size of the individual scan is $256 \times 256 \times 192$ and each voxel size is $1 \times 1 \times 1$ mm^3. The first two dimensions in each scan size indicate the individual image size (rows, columns) and the third dimension represents the number of images in the scan.
IBSR [49]	Eighteen volumes of T_1-weighted brain MRI from all age groups from juvenile to adult are available online with ground truth. The size of the individual scan is $256 \times 256 \times 128$ and each voxel size is $1.5 \times 1.5 \times 1.5$ mm^3. Most of the scans in this database have low-contrast images.
MNI BITE [50]	Real T_1-weighted brain MRI of 14 patients with brain tumors (gliomas). We have used scans from Group 2 (pre-operative MRIs) and Group 3 (post-resection MRIs). The size of each scan in Group 2 is $394 \times 466 \times 378$. Group 3 contains scans of different sizes and dimensions.

3.1. Evaluation on Real Datasets

The MSP is extracted from each dataset using the proposed algorithm and some slices perpendicular to the estimated MSP are snipped and displayed in Figure 9. The green line in each image is the intersecting line between the estimated MSP and the corresponding orthogonal slice. The first row (Figure 9a) represents the images from the NFBS database with extracted MSP by the proposed algorithm. The images of the same dataset (NFBS) are synthetically degraded by adding zero-mean Gaussian noise of several levels. Proposed algorithm breaking points can be found by incrementally adding the noise until the algorithm fails to detect the accurate MSP plane. The proposed algorithm successfully estimated the MSP at levels of noise up to SNR = -10.09 decibel (dB). The second row (Figure 9b) indicates representative resulting slices and the estimated MSP from noisy images. Images from the IBSR database are portrayed in Figure 9c. Manual delineation (GT) of brain hemispheres is available only for the IBSR database. The manual delineation boundary is superimposed on the input image with the white pixels by using a morphological gradient and binary skeletonization [51], as shown in Figure 9c. The proposed algorithm successfully extracts the MSP in all the volumes of the IBSR database and no obvious error is detected. The last two rows in Figure 9 contain the MSP-extracted results from the MNI BITE database. Group 2 (Figure 9d) comprises pre-operative MRIs and Group 3 (Figure 9e) includes MRIs acquired at different intervals of time, i.e., before and after surgery. Obvious asymmetries can be seen in the two groups of brain MRIs. The proposed algorithm extracted the symmetry axes from these volumes robustly and accurately.

3.2. Evaluation and Comparison on Synthetic Datasets

The accuracy of the proposed algorithm for extracting MSP is also evaluated by creating a set of 50 symmetrical scans of the real brain MRI from the NFBS database [48]. Each head scan is manually adjusted and perfectly aligned followed by reflecting one half of the head volume about the known MSP to form the other half. The two mirror halves are stitched together to create a symmetrical head scan with known GT MSP. The reason for creating such volumes is that perfectly symmetric head volumes with GT can be manipulated and transformed arbitrarily. In this way, we can avoid typical subjective factors of human visual inspection. Moreover, in reality, no human head scan exhibits perfect digital symmetry [52]. Therefore, GT in real brain MRIs cannot be used directly for MSP algorithm evaluation [18].

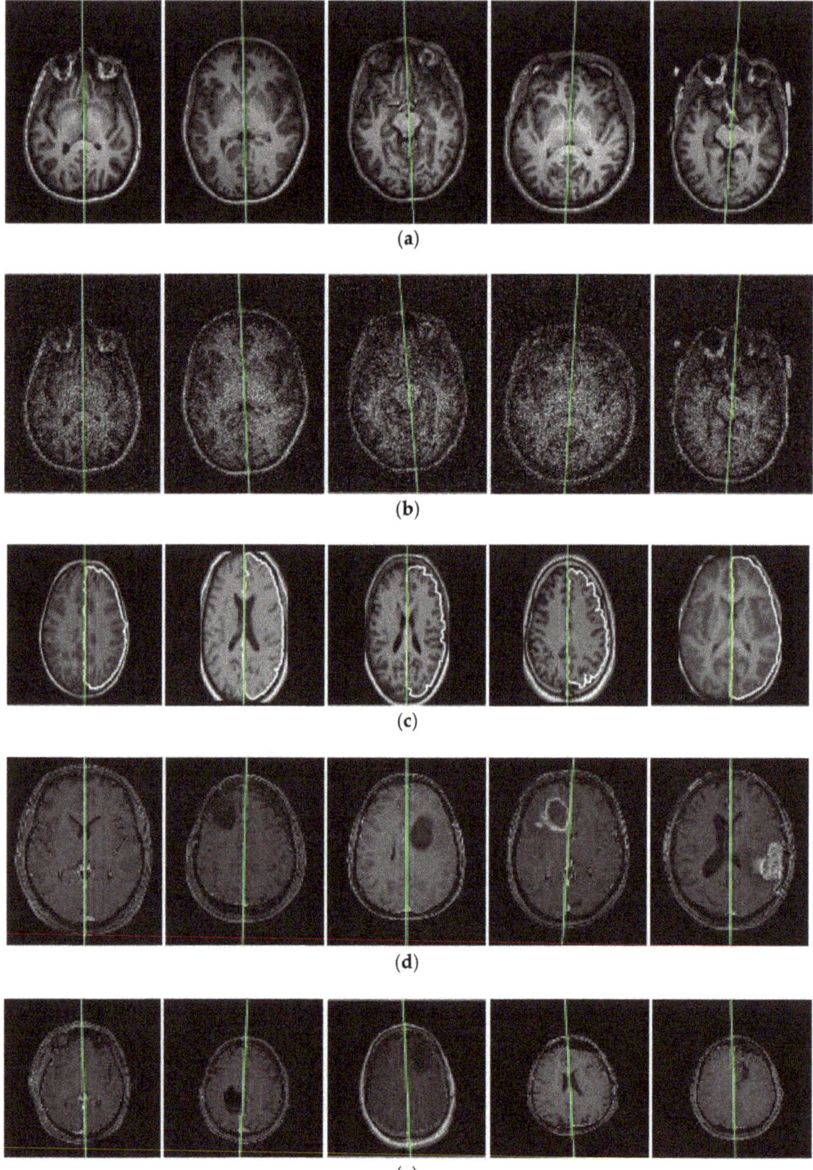

Figure 9. Visual comparison of the proposed algorithm in extracting the symmetric axis (MSP) from real brain MRIs, (**a**) the NFBS database [38], (**b**) images of the same subjects with Gaussian noise, (**c**) the IBSR database [39], (**d**) the MNI BITE database Group 2, and (**e**) the MNI BITE database Group 3 [40].

The presented algorithm results have been compared with a state-of-the-art MSP extraction method proposed by Ruppert et al. [27]. The algorithm is based on maximization of bilateral symmetry using 3-D Sobel edge operator, thresholding, downsampling, and a multiscale scheme. To improve the quantitative analysis of MSP identification, the authors also introduced a new MSP estimation error metric called average z-distance. The detail of this metric is discussed in the succeeding paragraph.

Metrics: Two metrics are measured to assess how accurately the algorithms detected the MSP. One is the angle difference (in degree) and is defined as the angle between normal vectors of GT MSP and estimated MSP. Mathematically, it can be computed using the inner product as:

$$\alpha = \cos^{-1}\left(\frac{<u,v>}{\|u\|\|v\|}\right) \times \frac{180}{\pi} \tag{31}$$

where α is the angular difference, u is the normal vector of GT MSP, and v is the normal vector of calculated MSP.

When the two planes are parallel, this error metric is not sufficient and may be misleading due to translation between the estimated MSP and GT MSP. This problem is circumvented by Ruppert et al. [27] who proposed a new, simple, and fast metric known as average z-distance or z-score (in voxels), to measure MSP estimation error as a function of the distance between the two planes. This distance can be measured [27,33] by computing z coordinate from the plane equations, for the GT plane and the estimated plane, using each "x" and "y". It can be written as:

$$z\ distance = \frac{\sum\limits_{(x,y)}\left(|z_{coord.}(GT) - z_{coord.}(Est.)|\right)}{dim(x) \times dim(y)} \tag{32}$$

where dim is the image dimension along x and y axes, and GT, and $Est.$ stands for ground truth and estimated MSPs, respectively.

Both the algorithms (the proposed algorithm and Ruppert et al. algorithm) are evaluated for all the slices in the perfectly symmetrical head volume by determining their average z-distance with respect to each corresponding GT MSP. The mean z-scores obtained for both algorithms on 50 perfectly symmetric datasets are shown in Table 2.

Table 2. Quantitative results comparison for perfectly symmetric datasets.

	Ruppert et al. Algorithm			Proposed Algorithm		
	z Score (Voxels)	Angle Difference (°)	Time (s)	z-Score (Voxels)	Angle Difference (°)	Time (s)
Mean	1.246	0.10	35.02	0.336	0.06	1.04
Std.	2.041	0.22	1.12	0.324	0.21	0.02
Median	0.50	0.00	34.86	0.250	0.00	1.01

The plot of average z-distance for the individual scan is illustrated in Figure 10. Ruppert et al. approach was unable to truly estimate the MSP in some scans and showed substantially large values for average z-distances, as indicated by the black square in Figure 10. These scans were not considered when measuring the mean z-score and the mean of angle differences, since they would synthetically intensify the z-score and standard deviation results of Ruppert et al. algorithm.

The values of z-score are almost similar as they reported in their paper, except for some scans having a rotation angle (either yaw or roll) of greater than 5°, as displayed in Figure 11. The situations at which their algorithm relied, i.e., smoothing, Sobel operator followed by thresholding of 5% brightest voxels, and symmetric score, are not satisfied in the presence of noise, image artifacts, and intensity inhomogeneity. This causes the offsets of plane underestimated. From Table 2, it is evident that Ruppert et al. algorithm accurately estimated the orientation of the MSP but failed to correctly calculate the offset of MSP. On the other hand, the proposed algorithm results are more consistent and accurate in orientations as well as in offsets.

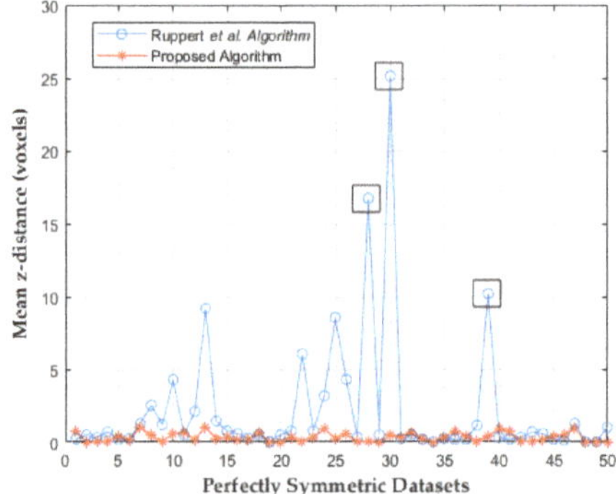

Figure 10. Average z-distances of perfectly symmetric datasets (z-distances indicated by the square are not included for mean z-score calculation).

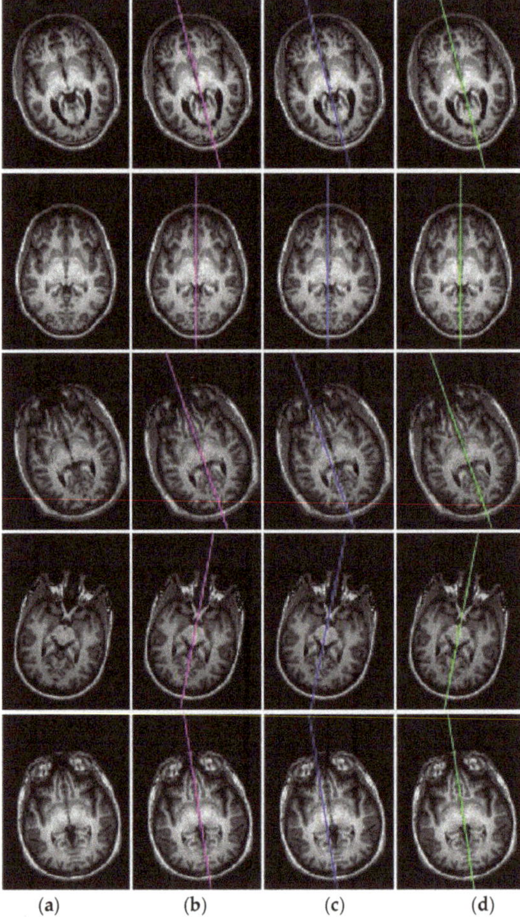

Figure 11. Visual comparison of the proposed algorithm with Ruppert et al. algorithm for extracting the symmetric axis (MSP) from perfectly symmetric datasets: (**a**) input slice, (**b**) ground-truth slice, (**c**) Ruppert et al. [27] algorithm results, and (**d**) proposed algorithm results.

To precisely inspect the accuracy of the proposed algorithm, many illustrative slices orthogonal to the estimated MSP are displayed in Figure 11. The lines in different colors (magenta, blue and green) are the intersecting lines between the extracted MSP and the respective orthogonal slice. The first column represents the input images and the second column shows the input images with the GT MSP intersection line (magenta color line). Similarly, third and fourth columns contain the images of Ruppert et al. and proposed algorithm results, respectively. Visual comparison in Figure 11 also reveals that the proposed algorithm outperformed Ruppert et al. algorithm in terms of accuracy, both in orientation and offsets. Ruppert et al. algorithm could not always achieve a rigorous estimate of MSP, particularly when the brain MRIs underwent a considerable transformation (rotation, translation, noise).

3.3. Evaluation and Comparison on Real Datasets

The proposed method has also been compared with Ruppert et al. method on real brain MRIs. Both techniques have been tested on 125 real head scans of the NFBS database [48] and 18 real head volumes of the IBSR database [49]. No GT for MSP is available for NFBS. Therefore, only visually comparison of the MSP extraction results has been reported for both the algorithms. The first row of Figure 12 shows some of the input image slices orthogonal to MSP.

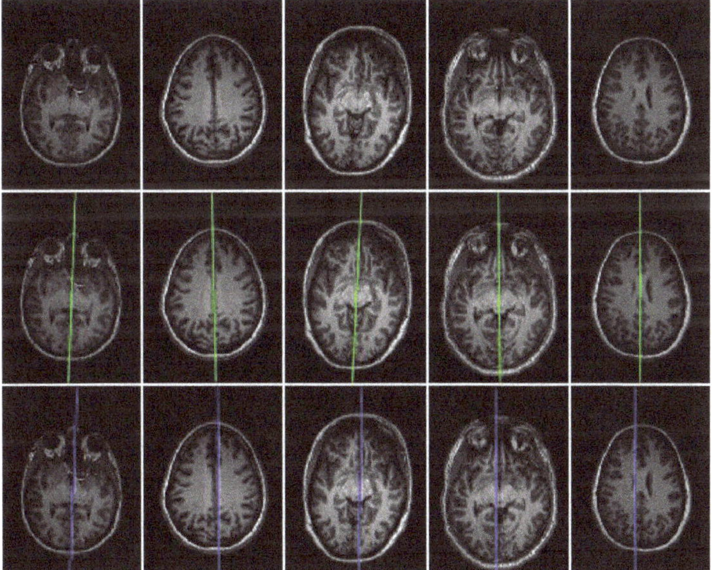

Figure 12. Visual comparison of the proposed algorithm with Ruppert et al. algorithm for extracting the symmetric axis (MSP) from real head volumes (the NFBS database [48]). First row, second row, and third row display the input images, the proposed algorithm results, and Ruppert et al. algorithm results, respectively.

Extracted MSP (green lines) using the proposed algorithm is displayed in the second row of Figure 12. Blue lines in the third row of Figure 12 displays the MSP extraction results of Ruppert et al. algorithm. The same pattern of the results in detecting MSP is shown by Ruppert et al. algorithm in real brain volumes. It estimates the orientation of the MSP more accurately but fails to correctly calculate the offset of MSP. On the other hand, the proposed algorithm detects the MSP more precisely and consistently.

The IBSR database contains the manual delineation (GT) of brain hemispheres. For illustration, the input image from a brain volume is shown in Figure 13a with its manual delineation (mask) of the left (dark) and the right (bright) brain regions (Figure 13b). For comparison purpose, the boundary of the right region is superimposed on the input image with the white pixels using a morphological gradient and the binary skeletonization [51], as shown in Figure 13c. Therefore, this image is used as a GT for evaluation of MSP extraction results. For the precise and accurate result, the intersection line (between the estimated MSP and the corresponding orthogonal slice) should be overlapped or coincided with the white pixels boundary in the GT image.

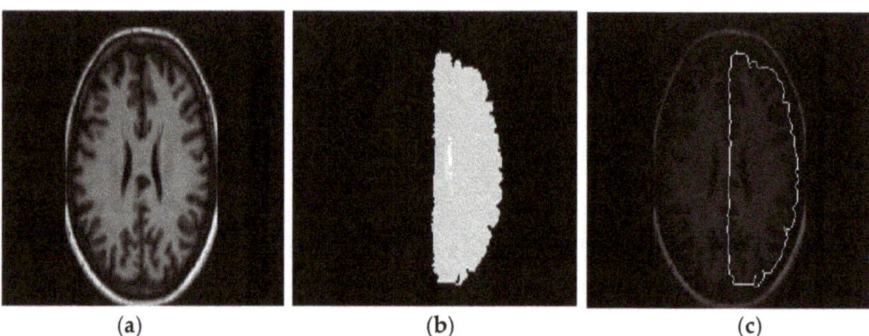

(a) (b) (c)

Figure 13. Superimposition of the ground truth (GT) image on the input image of the IBSR dataset [49]: (a) input image; (b) GT image; and (c) boundary of the right brain region is imposed on the input image with the white pixels.

The first row in Figure 14 consists of all the input images with GT delineation of the right brain. Since the MSP divides the brain into two roughly symmetrical regions, the detected MSP (green line) by the proposed algorithm is almost completely overlapped with the boundary of the GT delineation of the right brain, as shown in the second row of Figure 14. In contrary, the MSP (blue line) estimated by Ruppert et al. algorithm is deteriorated significantly from the real boundary of the GT delineation of the right brain. Note that we did not compare results of brain tumors datasets with Ruppert et al. algorithm because they did not report results on such datasets in their paper.

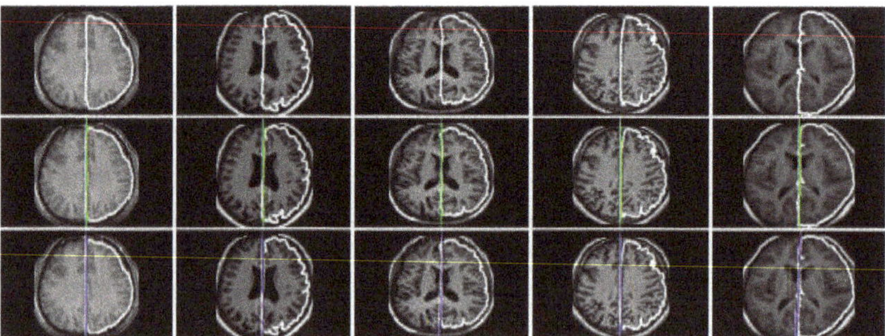

Figure 14. Visual comparison of the proposed algorithm with Ruppert et al. algorithm for extracting the symmetric axis (MSP) from real head volumes of the IBSR database [49]. First row, second row, and third row display the input images, the proposed algorithm results, and Ruppert et al. algorithm results, respectively.

In short, all the promising results given by the proposed algorithm indicates that the developed technique has the highest accuracy and consistency in extracting the MSP. Finally, the results of automatic MSP detection and tilt correction (recenter and reorientation) in brain MRIs are displayed in Figure 15, where the first row represents the tilted slices of the three distinct brain volumes. After computing the parameters of the MSP and affine transformation, we reoriented and recentered the brain volumes. The corrected volume images are displayed in the second row of Figure 15.

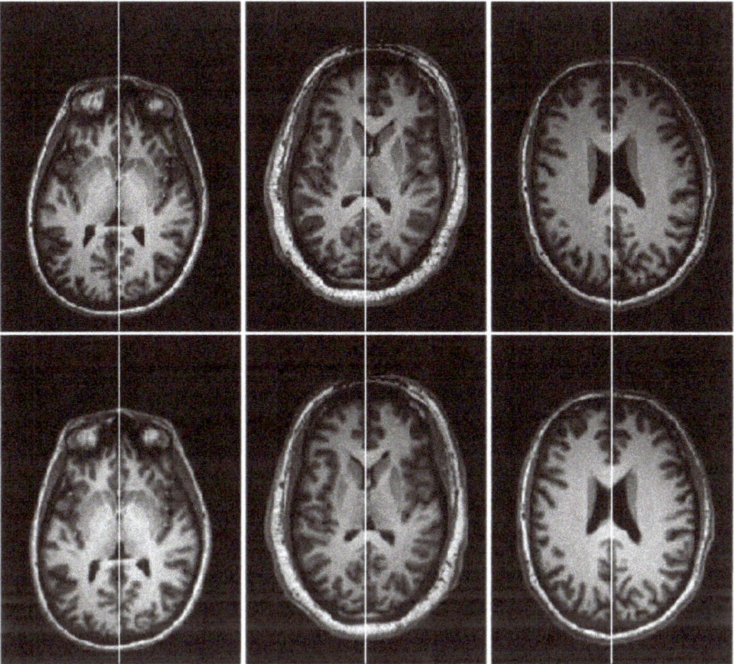

Figure 15. The results of symmetric detection and tilt correction (realignment of the brain head volume). The input head volumes images are in the first row and reoriented (recentered) head volumes images are in the second row.

4. Conclusions

In this paper, we have presented a fully automatic and computationally efficient MSP extraction and tilt correction technique in brain MRIs. The proposed method is based on PCA and symmetric features followed by plane fitting using orthogonal regression. Experimental results on 157 real heterogeneous brain MRIs including 14 datasets with brain tumors and comparison with a state-of-the-art method have confirmed that the proposed technique provides consistent performance with the highest accuracy. Moreover, it is 30 times faster than the competitor algorithm and takes only 1.04 s (on average) for all algorithmic steps per MRI volume. It is also robust on pathological brain MRIs having various intensity inhomogeneities, noises and image artifacts.

Some limitation of the proposed algorithm should be taken into consideration. The algorithm can only take T_1-weighted MRI of the brain in axial orientation as an input. Although one can convert from one orientation (sagittal or coronal) to other using existing algorithms, the selection of brain slice in the first step can affect the yaw angle due to the assumption that it should be the same in each axial slice (see Equation (4)). The reason is that in the true anatomical structure of the brain, MSP is not exactly the plane but a curved surface even for a normal brain. Even though the planer estimation is adequate for many applications such as registration and symmetric/asymmetric analysis of brain images, the results obtained from the PCA in estimating the yaw angle can also affect the performance of the algorithm in the presence of high level of noise. When the SNR is less than -10.09 dB, estimates of the yaw angle cannot be trusted. This problem can be circumvented by increasing the angle range in the cross-correlation step. Our future work will include evaluation of the algorithm on brain images of various sources and modalities such as T_2-weighted, Proton Density (PD) weighted, Fluid Attenuated Inversion Recovery (FLAIR), PET, and SPECT.

Author Contributions: H.Z.U.R. proposed the idea, implemented it and wrote the manuscript. S.L. supervised the study and manuscript-writing process.

Funding: This work was funded by the Korean Government (MSIP) under Grant No. 2015R1C1A1A01056013 and Grant No. 2012M3A6A3055694.

Acknowledgments: The author (H.Z.U.R.) is extremely thankful to the Higher Education Commission (HEC) of Pakistan for HRDI-UESTPs scholarship.

Conflicts of Interest: The authors declare no conflicts of interest.

References

1. Davarpanah, S.H.; Liew, A.W.-C. Brain mid-sagittal surface extraction based on fractal analysis. *Neural Comput. Appl.* **2018**, *30*, 153–162. [CrossRef]
2. Crow, T. Schizophrenia as an anomaly of cerebral asymmetry. In *Imaging of the Brain in Psychiatry and Related Fields*; Springer: Berlin, Germany, 1993; pp. 3–17.
3. Oertel-Knochel, V.; Linden, D.E. Cerebral asymmetry in schizophrenia. *Neuroscientist* **2011**, *17*, 456–467. [CrossRef] [PubMed]
4. Yu, C.-P.; Ruppert, G.C.; Nguyen, D.T.; Falcao, A.X.; Liu, Y. Statistical Asymmetry-based Brain Tumor Segmentation from 3D MR Images. *Biosignals* **2012**, *15*, 527–533.
5. Roy, S.; Bandyopadhyay, S.K. Detection and Quantification of Brain Tumor from MRI of Brain and it's Symmetric Analysis. *Int. J. Inf. Commun. Technol. Res.* **2012**, *2*, 477–483.
6. Hermes, G.; Ajioka, J.W.; Kelly, K.A.; Mui, E.; Roberts, F.; Kasza, K.; Mayr, T.; Kirisits, M.J.; Wollmann, R.; Ferguson, D.J.P.; et al. Neurological and behavioral abnormalities, ventricular dilatation, altered cellular functions, inflammation, and neuronal injury in brains of mice due to common, persistent, parasitic infection. *J. Neuroinflamm.* **2008**, *5*, 48. [CrossRef] [PubMed]
7. Schulte, T.; Muller-Oehring, E.M.; Rohlfing, T.; Pfefferbaum, A.; Sullivan, E.V. White Matter Fiber Degradation Attenuates Hemispheric Asymmetry When Integrating Visuomotor Information. *J. Neurosci.* **2010**, *30*, 12168–12178. [CrossRef] [PubMed]
8. Kumar, A.; Schmidt, E.A.; Hiler, M.; Smielewski, P.; Pickard, J.D.; Czosnyka, M. Asymmetry of critical closing pressure following head injury. *J. Neurol. Neurosurg. Psychiatry* **2005**, *76*, 1570–1573. [CrossRef] [PubMed]
9. Roussigné, M.; Blader, P.; Wilson, S.W. Breaking symmetry: The zebrafish as a model for understanding left-right asymmetry in the developing brain. *Dev. Neurobiol.* **2012**, *72*, 269–281. [CrossRef] [PubMed]
10. Doi, K. Computer-aided diagnosis in medical imaging: Historical review, current status and future potential. *Comput. Med. Imaging Graph.* **2007**, *31*, 198–211. [CrossRef] [PubMed]
11. Alves, R.S.; Tavares, J.M.R. Computer image registration techniques applied to nuclear medicine images. In *Computational and Experimental Biomedical Sciences: Methods and Applications*; Springer: Berlin, Germany, 2015; pp. 173–191.
12. Lancaster, J.L.; Glass, T.G.; Lankipalli, B.R.; Downs, H.; Mayberg, H.; Fox, P.T. A modality-independent approach to spatial normalization of tomographic images of the human brain. *Hum. Brain Mapp.* **1995**, *3*, 209–223. [CrossRef]
13. Minoshima, S.; Koeppe, R.A.; Frey, K.A.; Kuhl, D.E. Anatomic standardization: Linear scaling and nonlinear warping of functional brain images. *J. Nucl. Med.* **1994**, *35*, 1528–1537. [PubMed]
14. Liu, S.X. Symmetry and asymmetry analysis and its implications to computer-aided diagnosis: A review of the literature. *J. Biomed. Inform.* **2009**, *42*, 1056–1064. [CrossRef] [PubMed]
15. Prima, S.; Ourselin, S.; Ayache, N. Computation of the Mid-Sagittal Plane in 3D Medical Images of the Brain. In Proceedings of the 6th European Conference on Computer Vision-Part II, Dublin, Ireland, 26 June–1 July 2000; pp. 685–701.
16. Liu, S.X.; Kender, J.; Imielinska, C.; Laine, A. Employing symmetry features for automatic misalignment correction in neuroimages. *J. Neuroimaging* **2011**, *21*, e15–e33. [CrossRef] [PubMed]

17. Volkau, I.; Prakash, K.B.; Ananthasubramaniam, A.; Aziz, A.; Nowinski, W.L. Extraction of the midsagittal plane from morphological neuroimages using the Kullback–Leibler's measure. *Med. Image Anal.* **2006**, *10*, 863–874. [CrossRef] [PubMed]

18. Liu, Y.; Collins, R.T.; Rothfus, W.E. Robust midsagittal plane extraction from normal and pathological 3-D neuroradiology images. *IEEE Trans. Med. Imaging* **2001**, *20*, 175–192. [CrossRef] [PubMed]

19. Kalavathi, P.; Senthamilselvi, M.; Prasath, V.B.S. Review of Computational Methods on Brain Symmetric and Asymmetric Analysis from Neuroimaging Techniques. *Technologies* **2017**, *5*, 16. [CrossRef]

20. Brummer, M.E. Hough transform detection of the longitudinal fissure in tomographic head images. *IEEE Trans. Med. Imaging* **1991**, *10*, 74–81. [CrossRef] [PubMed]

21. Guillemaud, R.; Marais, P.; Zisserman, A.; Mc Donald, T.; Crow, B. A 3-Dimensional midsagittal plane for brain asymmetry measurement. *Schizophr. Res.* **1995**, *18*, 183–184. [CrossRef]

22. Nowinski, W.L.; Prakash, B.; Volkau, I.; Ananthasubramaniam, A.; Beauchamp, N.J., Jr. Rapid and automatic calculation of the midsagittal plane in magnetic resonance diffusion and perfusion images. *Acad. Radiol.* **2006**, *13*, 652–663. [CrossRef] [PubMed]

23. Kuijf, H.J.; van Veluw, S.J.; Geerlings, M.I.; Viergever, M.A.; Biessels, G.J.; Vincken, K.L. Automatic extraction of the midsagittal surface from brain MR images using the Kullback–Leibler measure. *Neuroinformatics* **2014**, *12*, 395–403. [CrossRef] [PubMed]

24. Ardekani, B.A.; Kershaw, J.; Braun, M.; Kanno, I. Automatic detection of the mid-sagittal plane in 3-D brain images. *IEEE Trans. Med. Imaging* **1997**, *16*, 947–952. [CrossRef] [PubMed]

25. Prima, S.; Ourselin, S.; Ayache, N. Computation of the mid-sagittal plane in 3-D brain images. *IEEE Trans. Med. Imaging* **2002**, *21*, 122–138. [CrossRef] [PubMed]

26. Bergo, F.P.; Ruppert, G.C.; Pinto, L.F.; Falcao, A.X. Fast and Robust Mid-Sagittal Plane Location in 3D MR Images of the Brain. In Proceedings of the BIOSIGNALS, Madeira, Portugal, 28–31 January 2008; pp. 92–99.

27. Ruppert, G.C.; Teverovskiy, L.; Yu, C.-P.; Falcao, A.X.; Liu, Y. A new symmetry-based method for mid-sagittal plane extraction in neuroimages. In Proceedings of the 2011 IEEE International Symposium on Biomedical Imaging: From Nano to Macro, Chicago, IL, USA, 30 March–2 April 2011; pp. 285–288.

28. Wu, H.; Wang, D.; Shi, L.; Wen, Z.; Ming, Z. Midsagittal plane extraction from brain images based on 3D SIFT. *Phys. Med. Biol.* **2014**, *59*, 1367–1387. [CrossRef] [PubMed]

29. Zhang, Y.; Hu, Q. A PCA-based approach to the representation and recognition of MR brain midsagittal plane images. In Proceedings of the 30th Annual International Conference of the IEEEEngineering in Medicine and Biology Society, Vancouver, BC, Canada, 20–24 August 2008; pp. 3916–3919.

30. Kalavathi, P.; Prasath, V.B.S. Automatic segmentation of cerebral hemispheres in MR human head scans. *Int. J. Imaging Syst. Technol.* **2016**, *26*, 15–23. [CrossRef]

31. Ferrari, R.J.; Pinto, C.H.V.; Moreira, C.A.F. Detection of the midsagittal plane in MR images using a sheetness measure from eigenanalysis of local 3D phase congruency responses. In Proceedings of the 2016 IEEE International Conference on Image Processing (ICIP), Phoenix, AZ, USA, 25–28 September 2016; pp. 2335–2339.

32. Descoteaux, M.; Audette, M.; Chinzei, K.; Siddiqi, K. Bone enhancement filtering: Application to sinus bone segmentation and simulation of pituitary surgery. *Comput. Aided Surg.* **2006**, *11*, 247–255. [CrossRef] [PubMed]

33. de Lima Freire, P.G.; da Silva, B.C.G.; Pinto, C.H.V.; Moreira, C.A.F.; Ferrari, R.J. Midsaggital Plane Detection in Magnetic Resonance Images Using Phase Congruency, Hessian Matrix and Symmetry Information: A Comparative Study. In Proceedings of the International Conference on Computational Science and Its Applications, Melbourne, Australia, 2–5 July 2018; pp. 245–260.

34. Toro, C.; Gonzalo-Martín, C.; García-Pedrero, A.; Menasalvas Ruiz, E. Supervoxels-Based Histon as a New Alzheimer's Disease Imaging Biomarker. *Sensors* **2018**, *18*, 1752. [CrossRef] [PubMed]

35. Caggiano, A. Tool Wear Prediction in Ti-6Al-4V Machining through Multiple Sensor Monitoring and PCA Features Pattern Recognition. *Sensors* **2018**, *18*, 823. [CrossRef] [PubMed]

36. Cristalli, C.; Grabowski, D. Multivariate Analysis of Transient State Infrared Images in Production Line Quality Control Systems. *Appl. Sci.* **2018**, *8*, 250. [CrossRef]

37. Zhang, J.; Feng, X.; Liu, X.; He, Y. Identification of Hybrid Okra Seeds Based on Near-Infrared Hyperspectral Imaging Technology. *Appl. Sci.* **2018**, *8*, 1793. [CrossRef]

38. Wang, J.; Zhao, X.; Xie, X.; Kuang, J. A Multi-Frame PCA-Based Stereo Audio Coding Method. *Appl. Sci.* **2018**, *8*, 967. [CrossRef]

39. Maalek, R.; Lichti, D.D.; Ruwanpura, J.Y. Robust Segmentation of Planar and Linear Features of Terrestrial Laser Scanner Point Clouds Acquired from Construction Sites. *Sensors* **2018**, *18*, 819. [CrossRef] [PubMed]

40. Solomon, C.; Breckon, T. *Fundamentals of Digital Image Processing: A Practical Approach with Examples in Matlab*; John Wiley & Sons: Hoboken, NJ, USA, 2011; pp. 247–262.

41. Mudrová, M.; Procházka, A. Principal component analysis in image processing. In Proceedings of the MATLAB Technical Computing Conference, Prague, Czech Republic, 4–8 July 2005.

42. Petras, I.; Bednarova, D. Total Least Squares Approach to Modeling: A Matlab Toolbox. *Acta Montan. Slovaca* **2010**, *15*, 158–170.

43. Minovic, P.; Ishikawa, S.; Kato, K. Symmetry Identification of a 3-D Object Represented by Octree. *IEEE Trans. Pattern Anal. Mach. Intell.* **1993**, *15*, 507–513. [CrossRef]

44. Otsu, N. A threshold selection method from gray-level histograms. *IEEE Trans. Syst. Man Cybern.* **1979**, *9*, 62–66. [CrossRef]

45. Soille, P. *Morphological Image Analysis: Principles and Applications*, 2nd ed.; Springer Science & Business Media: Berlin/Heidelberg, Germany, 2013.

46. Noble, B.; Daniel, J.W. *Applied Linear Algebra*; Prentice-Hall: Upper Saddle River, NJ, USA, 1988; Volume 3.

47. Liu, Y.; Collins, R.T.; Rothfus, W.E. *Automatic Extraction of the Central Symmetry (Mid-Sagittal) Plane from Neuroradiology Images*; Carnegie Mellon University, The Robotics Institute: Pittsburgh, PA, USA, 1996.

48. Puccio, B.; Pooley, J.P.; Pellman, J.S.; Taverna, E.C.; Craddock, R.C. The preprocessed connectomes project repository of manually corrected skull-stripped T1-weighted anatomical MRI data. *Gigascience* **2016**, *5*, 45. [CrossRef] [PubMed]

49. Internet Brain Segmentation Repository (IBSR). Massachusetts General Hospital. Available online: http://www.nitrc.org/projects/ibsr/ (accessed on 26 September 2018).

50. Mercier, L.; Del Maestro, R.F.; Petrecca, K.; Araujo, D.; Haegelen, C.; Collins, D.L. Online database of clinical MR and ultrasound images of brain tumors. *Med. Phys.* **2012**, *39*, 3253–3261. [CrossRef] [PubMed]

51. Gonzalez, R.C.; Woods, R.E. *Digital Image Processing (Global Edition)*, 4th ed.; Pearson: New York, NY, USA, 2018; pp. 747–750.

52. Zhao, L.; Ruotsalainen, U.; Hirvonen, J.; Hietala, J.; Tohka, J. Automatic cerebral and cerebellar hemisphere segmentation in 3D MRI: Adaptive disconnection algorithm. *Med. Image Anal.* **2010**, *14*, 360–372. [CrossRef] [PubMed]

Article

Registration of Dental Tomographic Volume Data and Scan Surface Data Using Dynamic Segmentation

Keonhwa Jung, Sukwoo Jung, Inseon Hwang, Taeksoo Kim and Minho Chang *

Department of Mechanical Engineering, Korea University, Seoul 02841, Korea; rhslrk14@korea.ac.kr (K.J.);
james24@korea.ac.kr (S.J.); his0177@korea.ac.kr (I.H.); pqowep@korea.ac.kr (T.K.)
* Correspondence: mhchang@korea.ac.kr; Tel.: +82-2-3290-3379

Received: 24 August 2018; Accepted: 25 September 2018; Published: 29 September 2018

Abstract: Over recent years, computer-aided design (CAD) has become widely used in the dental industry. In dental CAD applications using both volumetric computed tomography (CT) images and 3D optical scanned surface data, the two data sets need to be registered. Previous works have registered volume data and surface data by segmentation. Volume data can be converted to surface data by segmentation and the registration is achieved by the iterative closest point (ICP) method. However, the segmentation needs human input and the results of registration can be poor depending on the segmented surface. Moreover, if the volume data contains metal artifacts, the segmentation process becomes more complex since post-processing is required to remove the metal artifacts, and initially positioning the registration becomes more challenging. To overcome these limitations, we propose a modified iterative closest point (MICP) process, an automatic segmentation method for volume data and surface data. The proposed method uses a bundle of edge points detected along an intensity profile defined by points and normal of surface data. Using this dynamic segmentation, volume data becomes surface data which can be applied to the ICP method. Experimentally, MICP demonstrates fine results compared to the conventional registration method. In addition, the registration can be completed within 10 s if down sampling is applied.

Keywords: local registration; iterative closest points; multimodal medical image registration

---

1. Introduction

In the dental computer-aided design and computer-aided manufacturing (CAD/CAM) industry, volumetric computed tomography (CT) images and scan surfaces are most commonly used. However, the two data types are very different, because their measurement techniques fundamentally differ. The volume data contains intensity information of the internal organs of the human body, while the surface data contains only the visible surfaces, that is, the teeth and the gingiva. Because of their different features, volume data and surface data are used for different dental applications. However, there are many applications which require both volume data and surface data and for the accurate registration of the volume and surface data is necessary. To achieve this we propose a novel registration method of volume data and surface data.

1.1. Backgrounds

1.1.1. Volumetric Computed Tomography (CT) Data

Volumetric computed tomography (CT) data features a voxel structure, with each voxel having an intensity value. The standard data format for this volume data is the Digital Imaging and Communications in Medicine (DICOM) format, which contains more than 90 valuable information fields such as intensity values, patient details, modality, and manufacturer, acquisition data, and so on [1–3]. The volume data is obtained by X-ray computed tomography (CT) scanning. In practice, the volume data is divided into three parallel planes, the sagittal, axial, and coronal planes, and is used in the analysis of many operations. For dental applications, cone beam computed tomography (CBCT) is used [4–6]. While a 'fan-shaped' X-ray beam is used in medical CT, a 'cone' X-ray beam is used in cone beam CT. Because medical CT features higher X-ray exposure than CBCT, the resolution of the volume data provided by medical CT is higher than that provided by CBCT [7]. However, CBCT is used in many fields of dentistry due to the low X-ray exposure associated with it. Also, CBCT data is much easier to use with 3D interpolation since, due to its X-ray geometry, it forms isotropic voxels, whereas medical CT forms anisotropic voxels.

1.1.2. Dental Surface Data

3D scanners are well established and widely used in industry and dental 3D scanners, which are optimized to scan plaster models, are also widely used in dentistry. The standard data format for the surface data is standard triangle language (STL) [8] and polygon file format (PLY). This surface data contains vertices, faces, normal vectors, and so on. To obtain the surface data, 3D optical scanners using structured light are generally used because they are fast and precise [9,10]. The 3D optical scanner is composed of two cameras for epipolar geometry and one projector for pattern projection. The surface data features much better resolution and accuracy than the volume data.

1.2. Related Works

Volume data and surface data have different features and there are various dental CAD/CAM applications which use both volume and surface data. Therefore, the registration of volume data and surface data is necessary.

Before approaching the registration problem, the intrinsic errors of each data should be considered numerically. 3D dental scanners (Identica Blue, MEDIT Corp., Seongbuk-gu, Seoul, Korea) are accurate to 0.007 mm. However, considering the whole process of making the impression and the plaster model for measurement, the total intrinsic error of the surface data is around 0.06 mm in practice [11,12]. On the other hand, the accuracy of CBCT (MercuRay, HITACHI, Chiyoda, Japan) is approximately 0.20 mm [13]. Dental prostheses cannot be designed using CBCT volume data because of this relatively low accuracy. Thus, the intrinsic error of the scan-derived data is generally negligible and only that of the volume data is a cause for concern.

Usually the registration problem concerns the same types of data and that is the basic premise in 2D images or 3D data registration. However, the registration problem in this paper concerns different types of data, volume data and surface data. Data must be converted to identical data types before the registration process, and most previous works convert the volume data to surface data. This type of conversion process that extracts dental surface data from volume data is called segmentation. Surface data registration can be performed on the resulting segmented dental surface data. Generally, surface registration is done by the iterative closest point (ICP) method, which needs good initial conditions [14–17] and is widely used in the dental CAD/CAM industry. The flow chart for the ICP method is shown in Figure 1.

Although the established registration framework (ICP) is currently used clinically for dental applications, some drawbacks still exist; the requirement of human inputs and the metal artifact problem.

Human input is needed to set the initial positioning. Although there is much research on global registration, which obtains the initial conditions automatically for ICP, applying this algorithm to dental model registration is challenging because dental surface data suffers from ambiguity due to teeth shape characteristics [18]. Segmentation also requires human input. The most commonly used segmentation methods are thresholding, region growing, and active contour methods such as a level set. Thresholding is the most straightforward and basic segmentation method and teeth volume data is segmented by giving lower and upper intensity values [19]. The region growing method starts with a set of seed points and regions are grown based on the similarity of intensity [20,21]. Level set segmentation is performed using 2D axial direction sliced images and stacked for a 3D segmentation result [22–24]. To improve the result of segmentation, combining the above segmentation methods has also been studied [25]. However, every segmentation method mentioned above needs a human input; lower and upper threshold values must be defined for thresholding segmentation, seed points must be defined for region growing segmentation, and initial contours must be defined for level set segmentation. In the established framework for the registration of volume and surface data, this human input stage for segmentation represents the most time-consuming step. Although some automatic tooth segmentation methods have been studied, each study has constraints and cannot be used in a wide variety of applications [26,27]. Also, once the segmentation has been done, post processing, such as surface smoothing and island removing, is required. The result of segmentation may even differ from person to person.

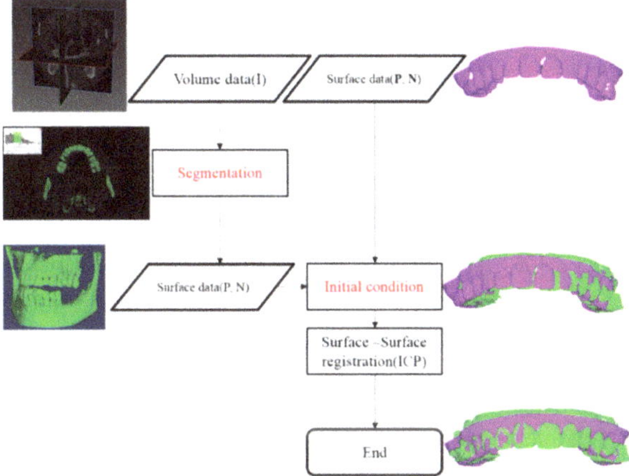

Figure 1. Flow chart: the iterative closest point (ICP)-based method, the established registration method.

Another drawback of the established registration procedure is the metal artifact problem [28]. If a patient has a prosthetic appliance made of metal, the volume data is seriously affected by white saturation. The quality of the resulting segmented surface is also affected. Initial conditioning for ICP also becomes more difficult because non-artifact points on the segmented surface must be selected manually. Many studies have considered the metal artifact problem [29,30] but they have resulted in a reduction rather than an elimination of the metal artifact effect so the problem remains unsolved.

1.3. Motivation and Contribution of the Thesis

ICP is the most useful fine registration algorithm and produces accurate results. However, volume data is composed of voxel structures with intensity values and contains no points or normal data. To find corresponding points between volume and surface data to apply to ICP, point data should be segmented from the volume data. In the established registration procedure, the segmented surface is used as a target surface for the ICP algorithm. Hence, the previous works must make considerable effort to ensure good quality segmented surface data, and this requires human input. We are motivated to try and overcome this fundamental limitation of the established registration procedure. Registration does not require fully segmented surface information, but only the corresponding points for the ICP method. Obtaining these points has proven the most challenging step in previous works on segmentation. The proposed method, the modified iterative closest point (MICP) obtains the corresponding points by dynamic segmentation defined by an intensity profile analysis. The remainder of this paper is organized as follows. In Section 2 detailed explanations of the proposed method are given. In Section 3 the results of the proposed method are shown and a comparison of its effectiveness with that of the established method is provided. Finally, the concluding remarks of the paper are given in Section 4.

2. Proposed Method: MICP

2.1. Overview

The overall process of the algorithm can be written in pseudo-code as follows:

Pseudo-code: The overall process of the proposed method

Data: V, P
Result: T_f
while (e < ending criteria)
$\{V'\} \leftarrow$ interpolation(V)
$\{E\} \leftarrow$ step_edge(V)
$\{M\} \leftarrow$ match(E,P',N)
$\{T_{init}\} \leftarrow$ minimizing point-to-plane distance metric
$\{P'\} \leftarrow \{T_{init}P\}$
end
$\left\{T_f\right\} \leftarrow ICP(T_{init}\cdot P_s,\ P_t)$

To align the volume data (V) and the scan data (P), the two were initially manually placed proximally. For a single point and normal vector from the surface data, an intensity profile can be defined in the volume data. The intensity profile has several new points aligned with the normal vector to the surface data. These points are defined with uniform intervals and new intensity values are given to these points by 3D interpolation (V′). Because the volume data and surface data are initially positioned well, a single intensity step edge is apparent in the intensity profile (Figure 2).

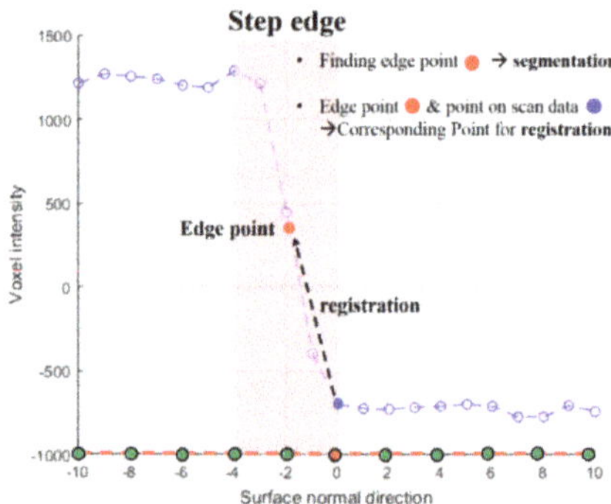

Figure 2. Intensity profile and step edge point.

The step edge represents the boundary of the teeth in the volume data and provides valuable information for both segmentation and registration. For most existing segmentation algorithms that extract surface data from volume data, points of the segmented surface must be positioned in the step edge(E). In other words, an edge point on an intensity profile should relate directly to a 3D segmented point. From the registration aspect, if the volume data and scan data are aligned properly, this step edge point must converge to the origin (x = 0).

The intensity profile is calculated for every point so the step edge can be determined. If the step edge exists in the intensity profile, the interpolated edge point in the 3D vertex can be obtained and the edge point becomes a segmented point. These points are the corresponding points used for the ICP algorithm (M). The rigid transformation matrix can be calculated by minimizing the distance between the matching points. If the average distance between the two data sets becomes less than the ending criteria, the dynamic segmentation process terminates.

Unlike the previous methods that use only volume data for segmentation, we use both volume data and surface data. The proposed registration procedure does not need any human input except for initial positioning and works automatically. In addition, the proposed method uses the edge points of the teeth. Therefore, it is robust to the artifact problem. The overall flow chart of the proposed registration procedure is shown in Figure 3.

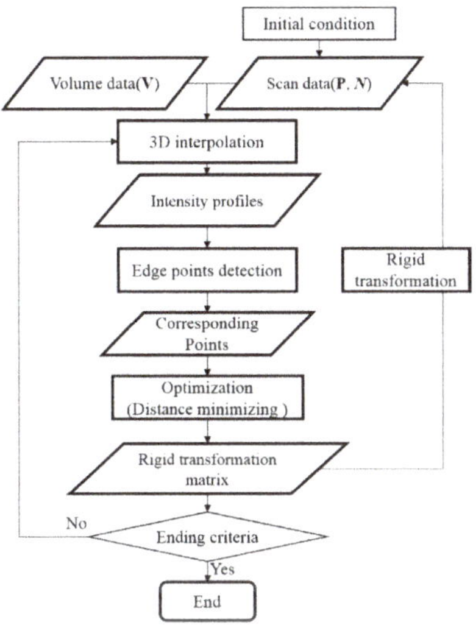

Figure 3. Overall procedure of the proposed method.

2.2. Automatic Registration Process

2.2.1. Defining an Edge Point on Intensity Profile

Based on a single point of scan data, an intensity profile can be generated along the normal direction of the point by 3D interpolating with a uniform interval. For this intensity profile generating process, two parameters are needed, the maximum distance and the interval. In this study, we used 10 voxels as the maximum distance and 1 voxel as the interval, and a total of 21 interpolation values are calculated for 1 intensity profile. With this intensity profile, the presence of step edges can be determined. The first derivative of the line profile can be used to determine the presence of a step edge. The determined step edge could correspond to an intensity increasing shape or an intensity decreasing shape.

Because the normal direction of the surface data and the gradient direction of the volume data are opposite, the sign of the first gradient of the step edge must be negative. In conclusion, if an intensity profile has a high negative value for the first derivatives, that intensity profile has a step edge and the point of the surface data becomes a corresponding point for ICP. Then, the minimum value of the first derivatives is defined. Many studies have evaluated the HU values of materials in CBCT volume data [31–34]. There are two step edges which define the teeth boundaries, whether bone—air or bone—skin. To detect and use both step edges, the step edge defined by the relatively low first derivatives value, the bone—skin value, becomes the reference. Based on the previous studies, −800 was defined as a reasonable slope value of the first derivatives for defining the step edge [19]. All experiments in Section 3 used this slope value to define step edges. This edge point is physically the same as a zero-crossing edge point [35]. The detected edge point on the intensity profile is a 3D point because the intensity profile is defined by 3D interpolation from volume data.

Volume data is highly complex data containing not only teeth, but also bones and tissues. Therefore, more than two step edges could be found. These intensity profiles are not used for the registration process. The proposed registration works based on only the defined points of surface data which are near the step edges. Because of this strict standard, the proposed registration method is robust to cases with metal artifacts.

Once the edge point is found, the sub-voxel level edge point is defined by local 3D interpolation back and forth along the intensity profile. This is the final step for a single intensity profile. The sub-voxel level edge point becomes a corresponding point to the point defining the intensity profile.

2.2.2. Dynamic Segmentation

For a single point on the surface data, a corresponding point in the volume data can be found by intensity profile analysis. If this intensity profile analysis is applied to all points on the surface data, a set of corresponding points for ICP can be obtained. The surface data and volume data are initially positioned. All detected edge points can be visualized as surface data. This surface data is a segmented surface representing the tooth volume data and can be a corresponding point for ICP. In the proposed method, if correspondence is obtained, the conventional ICP step for finding corresponding points becomes unnecessary. Now, points with edge points are set as moving surface data and the segmented points are set as the target surface data. Then, the sum of the distance between the corresponding points is minimized using the singular value decomposition (SVD) method and a rigid transformation matrix can be obtained [36]. The moving surface data is transformed by the obtained transformation matrix in a process which is a single iteration process under the proposed procedure. Even if the data is down-sampled, ICP still uses several thousands of points for which it needs to find correspondences. Repeating SVD iteratively increases the computational costs of ICP. In contrast, the proposed algorithm uses only hundreds of points from the interpolated data and already knows the correspondences. Therefore, the computational costs incurred by using SVD are substantially lower than those of ICP.

This whole process is performed iteratively just the same as for the conventional iterative closest point algorithm. The procedure of the established registration method for volume data and surface data uses only one segmented surface data. However, the proposed registration method includes the segmentation process within the iteration which is why this segmentation method is termed 'dynamic' segmentation. The segmented surface used as the target surface differs every iteration. For the proposed dynamic segmentation method using intensity profile analysis, the volume data can be used directly as input data. Above all, the dynamic segmentation works automatically without needing any human input. During the iteration process, the edge point on the intensity profile gets closer to the point on the surface.

2.3. Factors to Consider in Proposed Method

2.3.1. Normal Correction

Fundamentally, the 3D vertex points are the raw data obtained from the 3D scanning and are not positioned regularly due to the geometry of the model. From this point cloud, a mesh model is generated from various meshing algorithms and a surface normal can be calculated. From the near surface normal directions of a point, a vertex normal can be calculated. However, the mesh model generated from raw scan data is not good mesh data because of its point irregularity. There are many long-edge triangle faces on the raw mesh data and the calculated normal data is noisy. The dynamic segmentation that is proposed for MICP is sensitive to the normal direction of the scan data. Using raw scan data works fine but more accurate registration results can be achieved by correcting the normal data. There are two ways to correct normal data, normal smoothing and remeshing [37,38]. Normal smoothing can remove the high-frequency noise in the normal data and generates more reliable intensity profiles. However, the input normal for smoothing is basically inaccurate because of the irregular mesh data from the raw point cloud data. To overcome the irregularity, we remeshed the mesh data. After remeshing from the original mesh data with equal edge lengths, points on the surface are realigned and regular mesh data is generated (Figure 4). More accurate face normal data can be calculated from the regular mesh data and more accurate vertex normal data can be obtained sequentially.

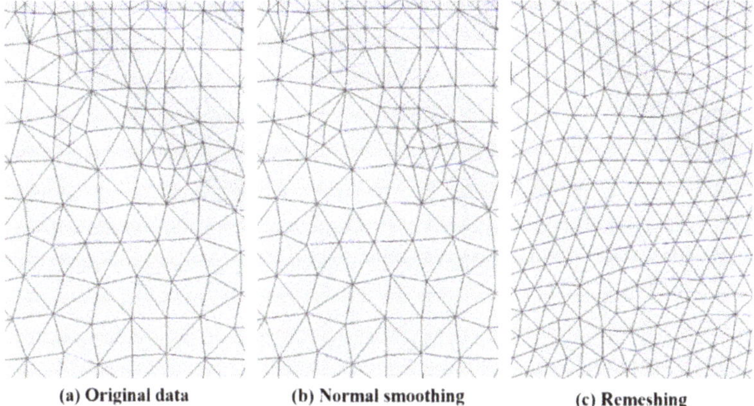

Figure 4. Results of normal correction methods for scan data. (**a**) raw scan data, (**b**) normal smoothing, (**c**) remeshing.

2.3.2. Length Value to Generate Intensity Profile

To generate the intensity profile through point normal direction, a limitation length should be set before the process. To achieve accurate registration, a proper length value is needed. If the intensity profile is set to a short length value, no edge points can be detected and registration cannot be performed because there is no correspondence between the scan data and the CBCT volume data. Alternatively, if the intensity profile is set to a long length value, unintended edge points can be detected and wrong correspondences lead to inaccurate registration results. Edge points of gums, tissue regions or metal artifact regions can be ignored automatically by the proper length value. With the proper distance value, the edge points of teeth regions are segmented and can be used to achieve good correspondences.

2.3.3. Down Sampling

One of the differences between volume data and surface data is their resolution. Surface data have a much higher resolution than CBCT volume data. The resolution of volume data is not as high as that of surface data even if 3D interpolated intensity values are given to all the points. A voxel in the volume data may even correspond to more than dozens of points in the surface data. Thus, using all points of the surface data is ineffective. To improve the efficiency of MICP registration, input surface data was down sampled to match the volume data. The conventional registration process, always contains a segmentation process that takes at least 20 min, so the expected time reduction is low. However, the expected elapsed time of the proposed registration method is dramatically lowered because segmentation is contained in the iteration. Generally, there are 4 down sampling algorithms that are widely used in 3D data handling; uniform sampling, random sampling, normal sampling, and covariance sampling [39]. The volume data has uniform resolution along the x, y, and z directions. Given this feature of the volume data, it is reasonable to use uniform sampling to adjust the resolution of the input data.

3. Results & Discussion

In this chapter, the experimental results of MICP are presented. Also, the proposed method is compared to the conventional registration method in terms of the average distance of points (D value). The proposed algorithm was implemented in MATLAB R2017a (The MathWorks Inc., Natick, MA, USA) on a personal computer with an i7-4770K processor with 8GB memory and a Windows 7 operating system (Microsoft Cop., Redmond, Washington, DC, USA). To visualize the results, MITK 2016.11.0 [40] and Meshlab 2016 [41] software were used.

3.1. Result

3.1.1. Data Sets

To be able to register the CBCT volume data and the dental scan surface data, naturally both data types should be obtained from the same patient. To obtain the experimental result, four sets of volume data and surface data were used for the registration (Figure 5). The volume data were obtained from CBCT (CB MercuRay, HITACHI, Chiyoda, Japan) and the surface data was obtained from a 3D optical dental scanner (Identica blue, MEDIT Corp., Seongbuk-gu, Seoul, Korea). The dimension of all volume data is $512 \times 512 \times 512$. The pixel spacing of the volume data of set1 and set2 is 0.2920 and of set3 and set4 is 0.2.

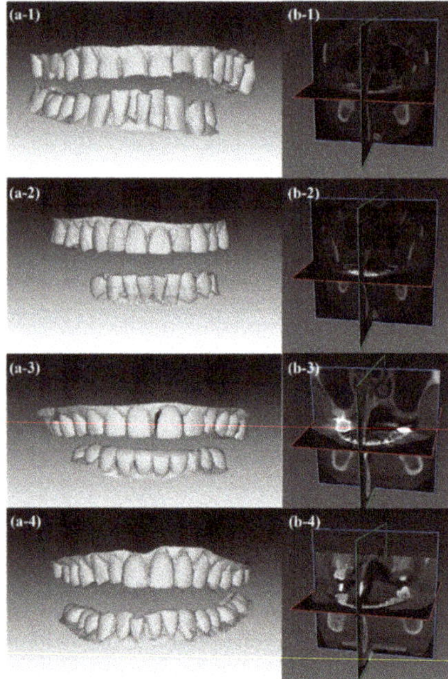

Figure 5. Input data ((**a**): surface data, (**b**): volume data). (**1**) set no.1 (**2**) set no.2 (**3**) set no.3 (**4**) set no.4.

3.1.2. Registration by the Conventional Method

Although there is no ground truth for the registration result of volume data and scan data, conventional registration using segmentation and the iterative closest point has been used in the dental field for a long time. Therefore, to compare registration performance, the ground truth used was the result of conventional registration. For the comparison, the conventional registration process was performed on all input data sets. The region of interest (ROI) was set to the teeth region in the CBCT volume data. In the ROI, thresholding and region growing was performed for segmentation. After the volume data has been segmented, the segmented region is extracted as surface mesh data. Then, human input is used to select a 3-point pair for the initial condition. ICP is performed as the last registration step. The result of ICP is shown in Figure 6 and Table 1. It is hard to measure the exact time cost of this conventional registration because it varies with the skills of the user and the computing power of the hardware. The conventional registration process takes at least 20 min for a no-artifact case and even more for an artifact case. The most time-consuming steps are segmentation and exportation of the surface data.

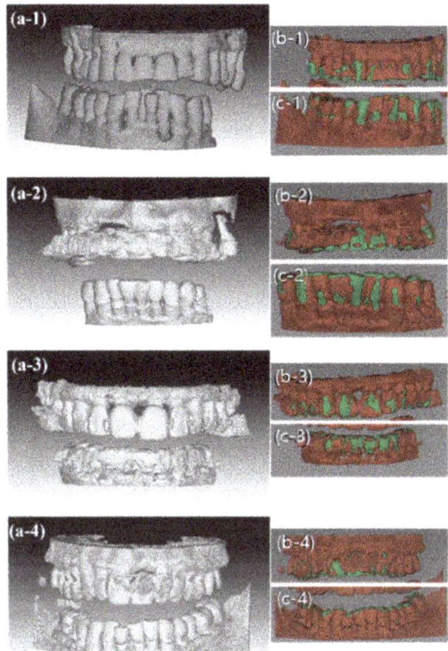

Figure 6. Conventional registration result. ((**a**): segmented surface data, (**b**): maxillary, (**c**): mandible) (**1**) set no.1 (**2**) set no.2 (**3**) set no.3 (**4**) set no.

Table 1. The result of conventional registration.

Input Data		Metal Artifact	E_{ICP} (mm)	SD_{ICP} (mm)
Set no.1	Maxillary	No	0.3612	0.1308
	Mandible	No	0.3357	0.1140
Set no.2	Maxillary	Yes	0.4170	0.1574
	Mandible	No	0.3812	0.1380
Set no.3	Maxillary	Yes	0.3178	0.1242
	Mandible	Yes	0.3464	0.1420
Set no.4	Maxillary	Yes	0.5025	0.1914
	Mandible	no	0.4834	0.1823

3.1.3. Modified Iterative Closest Point (MICP) Registration Results: Normal Correction

Modified iterative closest point (MICP) was performed on the four data sets to register the volume data and surface data. The volume data was set as the fixed data and the surface data was set as the moving data. The result of the MICP method is a 3D rigid transformation from dental scan data to CBCT volume data. To compare the results to those of the conventional registration method, the average point distance of the two surface data after registration was calculated (D).

The results of the MICP using row scan data converged well and show good results compared to those of the conventional registration process. After normal correction, even more accurate registration results are obtained (Figure 7). Table 2 shows the registration numbers in detail both before and after normal correction.

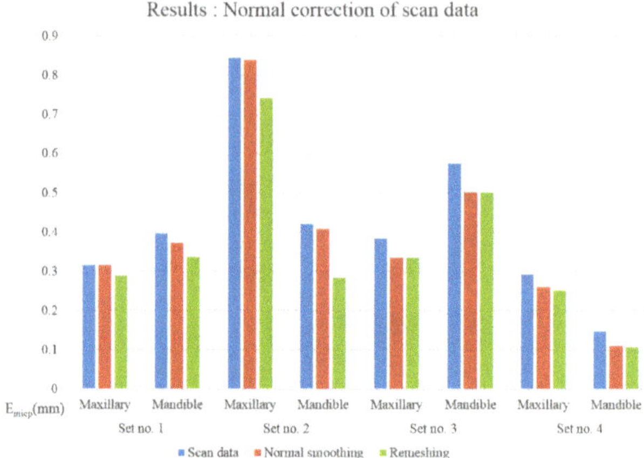

Figure 7. Registration result of modified iterative closest point (MICP) after normal correction.

Table 2. Registration result of modified iterative closest point (MICP) after normal correction.

Input Data		Normal Correction	N_{MICP}	E_{MICP} (mm)	SD_{MICP} (mm)	D_{MICP} (mm)
Set no.1	Maxillary	No	212,125	0.3145	0.7173	0.6178
		Normal sampling		0.3146	0.7198	0.6172
		Remeshing	180,186	0.2882	0.6703	0.6219
	Mandible	No	160,112	0.3953	1.0815	1.1811
		Normal sampling		0.3723	1.0094	1.1774
		Remeshing	131,545	0.3357	0.9217	1.1679
Set no.2	Maxillary	No	199,456	0.8431	1.8762	1.2145
		Normal sampling		0.8374	1.8500	1.2461
		Remeshing	185,546	0.7412	1.6113	1.2730
	Mandible	No	113,333	0.4203	1.2217	1.1116
		Normal sampling		0.4071	1.1760	1.1070
		Remeshing	95,438	0.2830	0.8040	1.0955
Set no.3	Maxillary	No	78,442	0.3828	1.4925	0.4458
		Normal sampling		0.3339	1.2376	0.4350
		Remeshing	51,398	0.3337	1.2289	0.4466
	Mandible	No	34,081	0.5730	1.9817	0.5957
		Normal sampling		0.5000	1.6143	0.5366
		Remeshing	78,657	0.4998	1.6196	0.5387
Set no.4	Maxillary	No	69,563	0.2920	1.1811	0.6204
		Normal sampling		0.2602	1.0053	0.5959
		Remeshing	56,335	0.2515	0.9704	0.5891
	Mandible	No	56,817	0.1457	0.5928	0.9164
		Normal sampling		0.1084	0.3837	0.9146
		Remeshing	44,472	0.1059	0.3801	0.9298

3.1.4. MICP Registration Results: Different Length Values

To identify the optimal length for generating the intensity profiles of the four data sets, MICP is performed as the length values are varied from 1 to 10 (Table 3). The length values 1 and 2 are too short to find edge points on the intensity profiles. Also, the D value increases with the length value. The registration result changes depending on the length used, and 3–6 seem to be good lengths for generating the intensity profile due to the lower D values obtained. Note that, based on this length test, all MICP registration results in this paper use 4 as the length when generating the intensity profile.

Table 3. Modified iterative closest point (MICP) results with different length values for the intensity profiles.

Input Data		Metal Artifact	D Value Respect to Length of Intensity Profile									
			1	2	3	4	5	6	7	8	9	10
Set no.1	Maxillary	No	X	0.7009	0.5681	0.5512	0.5409	**0.5402**	0.5410	0.5442	0.5446	0.5453
	Mandible	No	X	X	**1.1733**	1.1743	1.1735	1.1734	1.1741	1.1776	1.1780	1.1738
Set no.2	Maxillary	Yes	X	1.0571	1.1947	1.2525	**1.0014**	1.0817	1.0813	1.3574	1.3294	1.2866
	Mandible	No	X	1.2707	1.1302	1.1186	1.1215	**1.1185**	1.1186	1.1185	1.1186	1.1184
Set no.3	Maxillary	Yes	X	0.7850	0.4375	0.4101	0.4241	0.4533	0.4557	0.4725	0.4781	0.4817
	Mandible	Yes	X	X	**0.5127**	0.5345	0.5482	0.5703	0.5593	0.5508	0.5530	0.5501
Set no.4	Maxillary	Yes	X	X	0.5614	0.5604	0.5696	0.5844	0.5826	0.5838	0.5717	0.5824
	Mandible	No	X	X	0.9393	0.9370	0.9371	0.9371	0.9371	0.9372	0.0973	0.9371

3.1.5. MICP Registration Results: Down Sampling

Down sampling can be applied to the proposed MICP registration method and the expected time saving is much higher than for the conventional registration procedure. Figure 8 shows the down sampled point cloud using uniform sampling with grid sizes of 1, 5, and 10. The result of MICP using the down sampled surface data is shown in Table 4.

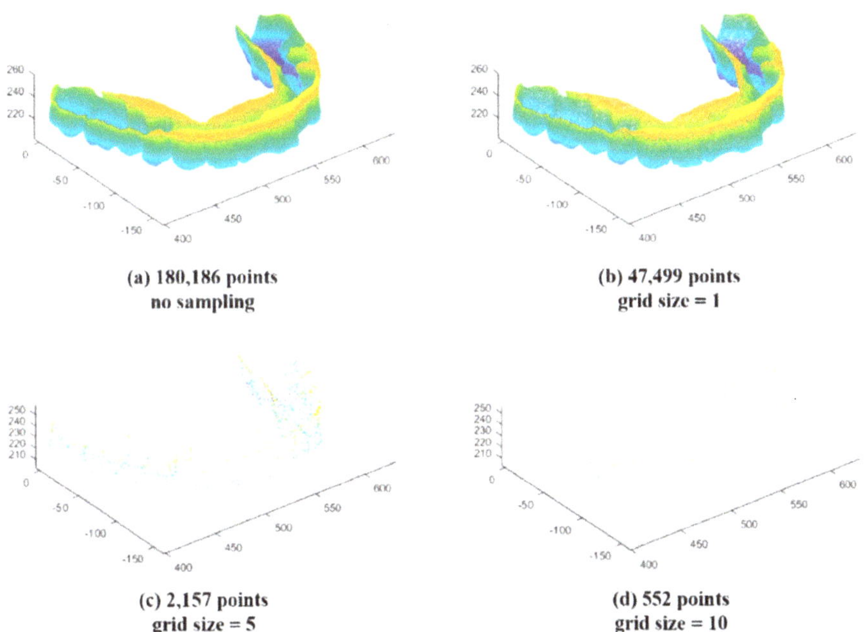

(a) 180,186 points
no sampling

(b) 47,499 points
grid size = 1

(c) 2,157 points
grid size = 5

(d) 552 points
grid size = 10

Figure 8. Uniform sampling results. (**a**) No sampling, (**b**) uniform sampling (grid size = 1 voxel) (**c**) uniform sampling (grid size = 5 voxels), (**d**) uniform sampling (grid size = 10 voxels).

Table 4. MICP registration results using uniform down sampled data.

Input Data		Sampling	N_{Point}	E_{MICP} (mm)	SD_{MICP} (mm)	D_{MICP} (mm)	Time (s)
Set no.1	Maxillary	No	180,186	0.2882	0.6703	0.6219	1062.8
		1	47,499	0.2860	0.6622	0.6159	274.8
		5	2157	0.2528	0.5933	0.5881	12.4
		10	552	0.2434	0.5635	0.5299	3.5
	Mandible	No	131,545	0.3210	0.8730	1.1677	793.4
		1	34,781	0.3131	0.8470	1.1760	210.4
		5	1593	0.2609	0.6881	1.1882	9.9
		10	395	0.2401	0.5460	1.2847	2.9
Set no.2	Maxillary	No	173,449	0.7386	1.6003	1.2689	1142.4
		1	46,158	0.7279	1.5659	1.2766	307.0
		5	2155	0.6971	1.4932	1.2698	14.1
		10	539	0.5489	0.9869	1.3935	4.0
	Mandible	No	84,735	0.2716	0.7618	1.0742	550.4
		1	22,285	0.2636	0.7348	1.0698	146.4
		5	1021	0.1592	0.3407	1.0895	6.9
		10	227	0.1789	0.4136	1.1974	1.7
Set no.3	Maxillary	No	163,714	0.3386	1.2525	0.4448	1149.0
		1	77,997	0.3358	1.2416	0.4464	537.8
		5	3982	0.3337	1.2247	0.4827	27.7
		10	1013	0.3210	1.1381	0.4661	7.4
	Mandible	No	78,657	0.4991	1.6165	0.5390	581.9
		1	40,812	0.5004	1.6167	0.5388	303.0
		5	2097	0.4993	1.6027	0.5468	15.9
		10	543	0.5131	1.6708	0.5920	4.3
Set no.4	Maxillary	No	163,714	0.2597	1.0095	0.5843	1092.8
		1	82,483	0.2589	1.0015	0.5862	580.0
		5	4232	0.2576	0.9843	0.5753	29.4
		10	1081	0.2462	0.9398	0.5856	7.6
	Mandible	No	125,617	0.1093	0.3891	0.9128	871.6
		1	65,067	0.1088	0.3857	0.9130	444.3
		5	3384	0.1067	0.3621	0.9166	23.4
		10	892	0.1248	0.4066	0.9329	6.4

3.1.6. MICP Registration Results

Detailed results of proposed MICP registration are shown in Figures 9 and 10. Surface data with volume rendered volume data is shown. Also, both the axial and sagittal views are shown. The red surface data is the initial condition, the blue surface data is the result of conventional registration, and the green surface data is the result of the proposed MICP registration.

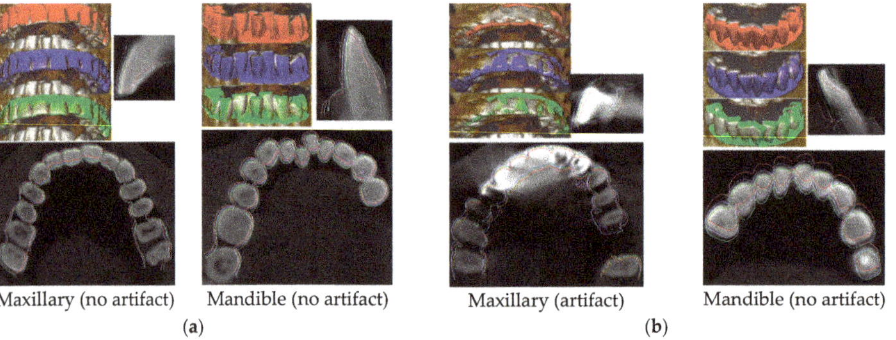

Maxillary (no artifact) Mandible (no artifact) Maxillary (artifact) Mandible (no artifact)

(a) (b)

Figure 9. *Cont.*

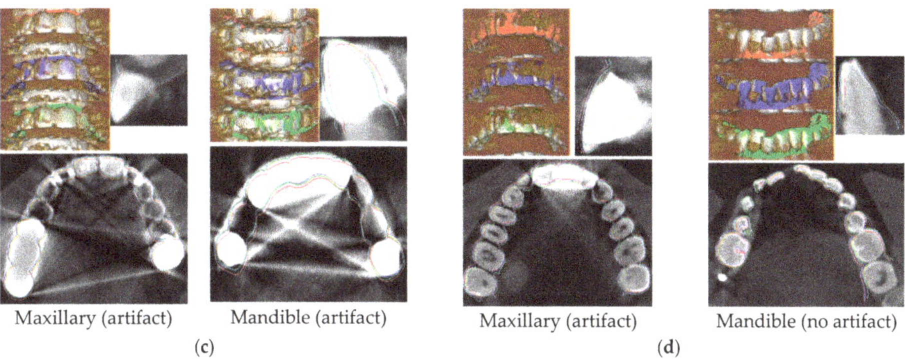

Figure 9. Modified iterative closest point (MICP) registration result. (**a**) Set no.1; (**b**) Set no.2; (**c**) Set no.3; (**d**) Set no. 4.

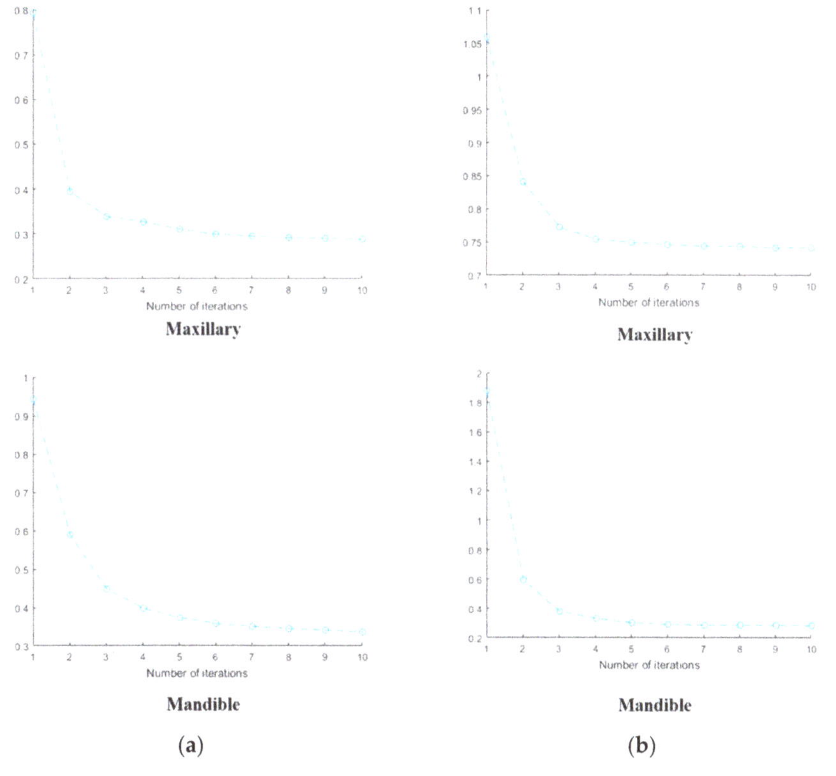

Figure 10. *Cont.*

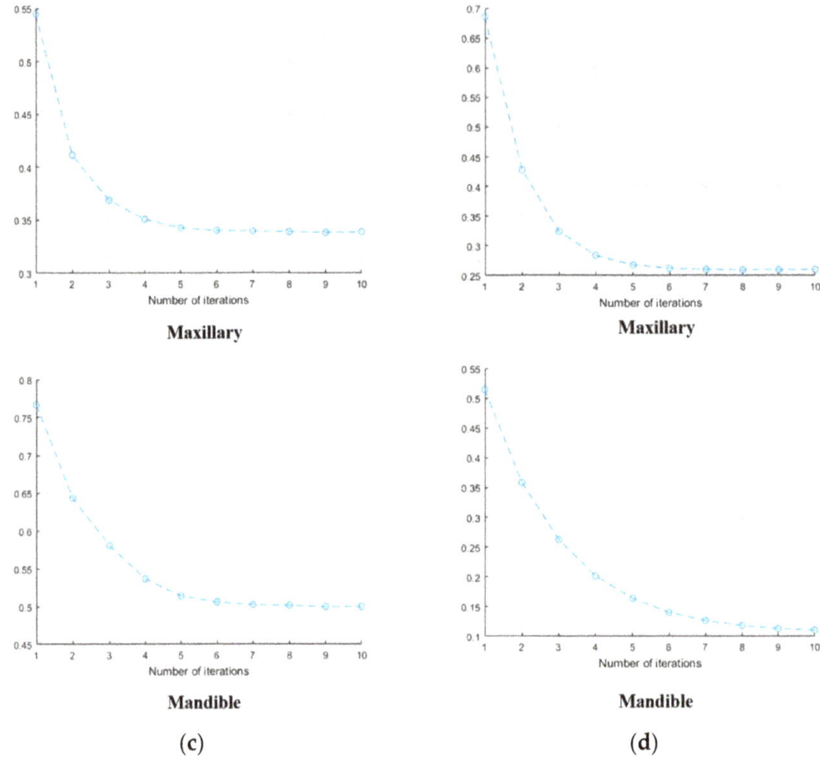

Maxillary

Maxillary

(c)

Mandible

(d)

Mandible

Figure 10. Average point distance of the modified iterative closest point (MICP) result. (**a**) set no.1; (**b**) set no.2; (**c**) set no.3; (**d**) set no.4.

3.2. Discussion

3.2.1. Evaluation of MICP Registration Results

Some patients who take both CBCT volume data and dental scanning surface data may have uneven teeth geometry as in our data set. Even with this uneven and poor teeth condition, the proposed algorithm showed fine results, even for the case with artifacts without needing any extra processing. The convergence error of MICP is similar to that of ICP.

As the proposed algorithm selects reliable points, it is robust to artifact cases. In the proposed procedure, registration and segmentation work complementarily; the better registration result causes the dynamic segmented points to increase, and the increased dynamic segmented points cause a better registration result. For artifact cases, the ratio of the increasing number of segmented points is lower.

The D value is computed in order to compare the result with the conventional registration result. As mentioned earlier, there is no ground truth for the registration of volume data and surface data. In other words, D = 0 does not exactly correspond to a perfect result. However, it is taken as the ground truth on the basis that the conventional registration is currently used in all dental applications by experts in this field. Maximum tolerance in the registration of volume data and surface data for dental applications ranges from 1.0 mm to 2.0 mm. In most cases, the distance values D between the conventional registration result and the result of the proposed method were less than 2.0 mm. This means that the proposed MICP registration result has fine registration accuracy and can be used for conventional applications without any problem. Also, from the early iterations of MICP, the D values decreased in all test cases. This proves that the surface is moving in the right direction.

While the established registration framework takes more than 20 min to register the volume data and the surface data, the proposed MICP registration takes less than 10 min for all cases because the segmentation process became automatic.

3.2.2. Evaluation of MICP Registration Results by Down Sampling

MICP using full surface data is two times faster than conventional registration and down sampling makes MICP even faster. The MICP registration result using the most down sampled surface data took less than 20 s and varying the D values caused almost no significant difference to this time. If the faster registration of volume data and surface data is needed for some applications, MICP with down sampling is a realistic solution, especially considering that it guarantees fine registration results.

3.2.3. Limitations

In the overall MICP process, segmentation is done automatically but setting the initial condition still needs a human input. Like general ICP, MICP could suffer from poor initial conditions. To determine the maximum effective limits of the initial condition, translations and rotations through x, y, z directions are applied to each data set and used as the initial position of the MICP input data. The D value is used to judge whether the registration works or not with the initial conditions.

To set an initial condition for MICP in real applications, landmarks selected by human input are necessary. In medical image registration, landmark-based registration is widely used instead of total manual registration by picking arbitrary point pairs. Consistency of the landmarks on medical images is about 1.64 mm and this is the initial condition for the registration [42]. This 1.64 mm can be considered directly as the initial condition error. From these initial condition tests, acceptable registration results are obtained with 2 mm differences of translation and rotation.

4. Conclusions

In this paper, modified iterative closest point (MICP), an automatic segmentation method for CBCT volume data and dental scan data is proposed. The proposed registration algorithm is based on a classic local registration algorithm, the iterative closest point (ICP). To find corresponding points for registration of CBCT volume data and dental scan data, previous methods had to extract full surface data from the volume data by segmentation. In the proposed method, the step for finding corresponding points was modified to a dynamic segmentation and the volume data could be directly used as input data. The whole registration process, except for the initial condition setting, is automatic and the registration result of the proposed method differs from conventional registration result by less than 2 mm, which is an acceptable tolerance in the dental CAD/CAM industry. With normal correction, more accurate registration results can be achieved and proper distance values for generating the intensity profile are provided. The registration speed is at least two times faster than the conventional method. With down sampling, MICP works much faster and registration is completed within only 10 s.

Author Contributions: K.J. proposed the method of the research, designed the experiments, and wrote the manuscript; S.J. wrote and revised the manuscript; I.H. and T.K. performed the experiments; M.C. provided the expertise in 3D measurement. All authors approved the final version of the manuscript.

Funding: This research was supported by the Technology Innovation Program (10065150, Development for Low-Cost and Small LIDAR System Technology Based on 3D Laser scanning for 360 Real-time Monitoring), funded by the Ministry of Trade, Industry & Energy (MOTIE, Korea) and the Korea Evaluation Institute of Industrial Technology (KEIT, Korea).

Conflicts of Interest: The authors declare no conflict of interest.

References

1. Mildenberger, P.; Eichelberg, M.; Martin, E. Introduction to the DICOM standard. *Eur. Radiol.* **2002**, *12*, 920–927. [CrossRef] [PubMed]

2. Gueld, M.O.; Kohnen, M.; Keysers, D.; Schubert, H.; Wein, B.B.; Bredno, J.; Lehmann, T.M. Quality of DICOM header information for image categorization. *Med. Imaging* **2002**, *4685*, 280–288.

3. Mustra, M.; Delac, K.; Grgic, M. Overview of the DICOM standard. In Proceedings of the 2008 50th International Symposium ELMAR, Zadar, Croatia, 10–12 September 2008; Volume 1, pp. 39–44.

4. Mozzo, P.; Procacci, C.; Tacconi, A.; Tinazzi Martini, P.; Bergamo Andreis, I.A. A new volumetric CT machine for dental imaging based on the cone-beam technique: Preliminary results. *Eur. Radiol.* **1998**, *8*, 1558–1564. [CrossRef] [PubMed]

5. Scarfe, W.C.; Farman, A.G.; Sukovic, P. Clinical applications of cone-beam computed tomography in dental practice. *J. Can. Dent. Assoc.* **2006**, *72*, 75–80. [CrossRef] [PubMed]

6. Scarfe, W.C.; Farman, A.G. What is Cone-Beam CT and How Does it Work? *Dent. Clin. N. Am.* **2008**, *52*, 707–730. [CrossRef] [PubMed]

7. Suomalainen, A.; Vehmas, T.; Kortesniemi, M.; Robinson, S.; Peltola, J. Accuracy of linear measurements using dental cone beam and conventional multislice computed tomography. *Dentomaxillofac. Radiol.* **2008**, *37*, 10–17. [CrossRef] [PubMed]

8. Hiller, J.D.; Lipson, H. STL 2.0: A Proposal for a Universal Multi-Material Additive Manufacturing File Format. In Proceedings of the 20th Solid Freeform Fabrication Symposium (SFF), Austin, TX, USA, 3–5 August 2009; pp. 266–278. [CrossRef]

9. Morano, R.A.; Ozturk, C.; Conn, R.; Dubin, S.; Zietz, S.; Nissanov, J. Structured light using pseudorandom codes. *IEEE Trans. Pattern Anal. Mach. Intell.* **1998**, *20*, 322–327. [CrossRef]

10. Salvi, J.; Fernandez, S.; Pribanic, T.; Llado, X. A state of the art in structured light patterns for surface profilometry. *Pattern Recognit.* **2010**, *43*, 2666–2680. [CrossRef]

11. Reza Rokn, A.; Hashemi, K.; Akbari, S.; Javad Kharazifard, M.; Barikani, H.; Panjnoosh, M. Accuracy of Linear Measurements Using Cone Beam Computed Tomography in Comparison with Clinical Measurements. *J. Dent.* **2016**, *13*, 333.

12. Patcas, R.; Müller, L.; Ullrich, O.; Peltomäki, T. Accuracy of cone-beam computed tomography at different resolutions assessed on the bony covering of the mandibular anterior teeth. *Am. J. Orthod. Dentofac. Orthop.* **2012**, *141*, 41–50. [CrossRef] [PubMed]

13. Van Assche, N.; Quirynen, M. Tolerance within a surgical guide. *Clin. Oral Implant Res.* **2010**, *21*, 455–458. [CrossRef] [PubMed]

14. Besl, P.; McKay, N. A Method for Registration of 3-D Shapes. *IEEE Trans. Pattern Anal. Mach. Intell.* **1992**. [CrossRef]

15. Jung, S.; Song, S.; Chang, M.; Park, S. Range image registration based on 2D synthetic images. *CAD Comput. Aided Des.* **2018**, *94*, 16–27. [CrossRef]

16. Gelfand, N.; Mitra, N.J.; Guibas, L.J.; Pottmann, H. Robust global registration. *Symp. Geom. Process.* **2005**, *2*, 5. [CrossRef]

17. Aiger, D.; Mitra, N.J.; Cohen-Or, D. 4-Points Congruent Sets for Robust Pairwise Surface Registration. *ACM Trans. Graph.* **2008**, *27*, 1. [CrossRef]

18. SanthaKumar, R.; Vidhya, S. Three-Dimensional Reconstruction of Cone Beam Computed Tomography Using Splines Interpolation Technique for Dental Application. *J. Med Devices* **2016**, *10*, 030927. [CrossRef]

19. Rumboldt, Z.; Huda, W.; All, J.W. Review of portable CT with assessment of a dedicated head CT scanner. *Am. J. Neuroradiol.* **2009**, *30*, 1630–1636. [CrossRef] [PubMed]

20. Revol, C.; Jourlin, M. A new minimum variance region growing algorithm for image segmentation. *Pattern Recognit. Lett.* **1997**, *18*, 249–258. [CrossRef]

21. Yau, H.T.; Lin, Y.K.; Tsou, L.S.; Lee, C.Y. An adaptive region growing method to segment inferior alveolar nerve canal from 3d medical images for dental implant surgery. *Comput. Aided Des. Appl.* **2008**, *5*, 743–752. [CrossRef]

22. Hosntalab, M.; Aghaeizadeh Zoroofi, R.; Abbaspour Tehrani-Fard, A.; Shirani, G. Segmentation of teeth in CT volumetric dataset by panoramic projection and variational level set. *Int. J. Comput. Assist. Radiol. Surg.* **2008**, *3*, 257–265. [CrossRef]

23. Ji, D.X.; Ong, S.H.; Foong, K.W.C. A level-set based approach for anterior teeth segmentation in cone beam computed tomography images. *Comput. Boil. Med.* **2014**, *50*, 116–128. [CrossRef] [PubMed]

24. Gao, H.; Chae, O. Individual tooth segmentation from CT images using level set method with shape and intensity prior. *Pattern Recognit.* **2010**, *43*, 2406–2417. [CrossRef]

25. Păvăloiu, I.B.; Vasilăţeanu, A.; Goga, N.; Marin, I.; Ilie, C.; Ungar, A.; Pătraşcu, I. 3D dental reconstruction from CBCT data. In Proceedings of the 2014 International Symposium on Fundamentals of Electrical Engineering, Bucharest, Romania, 28–29 November 2014. [CrossRef]

26. Mortaheb, P.; Rezaeian, M.; Soltanian-Zadeh, H. Automatic dental CT image segmentation using mean shift algorithm. In Proceedings of the 2013 8th Iranian Conference on Machine Vision and Image Processing (MVIP), Zanjan, Iran, 10–12 September 2013; pp. 121–126. [CrossRef]

27. Thariat, J.; Ramus, L.; Maingon, P.; Odin, G.; Gregoire, V.; Darcourt, V.; Malandain, G. Dentalmaps: Automatic dental delineation for radiotherapy planning in head-and-neck cancer. *Int. J. Radiat. Oncol. Boil. Phys.* **2012**, *82*, 1858–1865. [CrossRef] [PubMed]

28. Schulze, R.; Heil, U.; Groß, D.; Bruellmann, D.D.; Dranischnikow, E.; Schwanecke, U.; Schoemer, E. Artefacts in CBCT: A review. *Dentomaxillofac. Radiol.* **2011**, *40*, 265–273. [CrossRef] [PubMed]

29. Wang, G.; Snyder, D.L.; O'Sullivan, J.; Vannier, M.W. Iterative Deblurrinf for CT metal artifact reduction. *IEEE Trans. Med. Imaging* **1996**, *15*, 657–664. [CrossRef] [PubMed]

30. Watzke, O.; Kalender, W.A. A pragmatic approach to metal artifact reduction in CT: Merging of metal artifact reduced images. *Eur. Radiol.* **2004**, *14*, 849–856. [CrossRef] [PubMed]

31. Cann, C.E. Quantitative CT for determination of bone mineral density: A review. *Radiology* **1988**, *166*, 509–522. [CrossRef] [PubMed]

32. Norton, M.R.; Gamble, C. Bone classification: An objective scale of bone density using the computerized tomography scan. *Clin. Oral Implant Res.* **2001**, *12*, 79–84. [CrossRef]

33. Turkyilmaz, I.; Ozan, O.; Yilmaz, B.; Ersoy, A.E. Determination of bone quality of 372 implant recipient sites using hounsfield unit from computerized tomography: A clinical study. *Clin. Implant Dent. Relat. Res.* **2008**, *10*, 238–244. [CrossRef] [PubMed]

34. Rafic, M.; Ravindran, P. Evaluation of on-board imager cone beam CT hounsfield units for treatment planning using rigid image registration. *J. Cancer Res. Ther.* **2015**, *11*, 690. [CrossRef] [PubMed]

35. Grimson, W.E.L.; Hildreth, E.C. Comments on "Digital Step Edges from Zero Crossings of Second Directional Derivatives". *IEEE Trans. Pattern Anal. Mach. Intell.* **1985**, *PAMI-7*, 121–127. [CrossRef]

36. Sorkine, O.; Rabinovich, M. Least-Squares Rigid Motion Using SVD. Technical Notes. 2009, pp. 1–6. Available online: http://www.igl.ethz.ch/projects/ARAP/svd_rot.pdf (accessed on 24 February 2009).

37. Botsch, M.; Kobbelt, L. An intuitive framework for real-time freeform modeling. *ACM Trans. Graph.* **2004**, *23*, 630. [CrossRef]

38. Kobbelt, L.P.; Bareuther, T.; Seidel, H.P. Multiresolution shape deformations for meshes with dynamic vertex connectivity. *Proc. Eurographics 2000* **2000**, *19*, 249–260. [CrossRef]

39. Gelfand, N.; Ikemoto, L.; Rusinkiewicz, S.; Levoy, M. Geometrically stable sampling for the ICP algorithm. In Proceedings of the Fourth International Conference on 3-D Digital Imaging and Modeling, Banff, AB, Canada, 6–10 October 2003; pp. 260–267. [CrossRef]

40. Wolf, I.; Vetter, M.; Wegner, I.; Böttger, T.; Nolden, M.; Schöbinger, M.; Meinzer, H.P. The medical imaging interaction toolkit. *Med. Image Anal.* **2005**, *9*, 594–604. [CrossRef] [PubMed]

41. Cignoni, P.; Corsini, M.; Ranzuglia, G. MeshLab: An Open-Source Mesh Processing Tool. In Proceedings of the Eurographics Italian Chapter Conference, Salerno, Italy, 2–4 July 2008; pp. 129–136. [CrossRef]

42. Schlicher, W.; Nielsen, I.; Huang, J.C.; Maki, K.; Hatcher, D.C.; Miller, A.J. Consistency and precision of landmark identification in three-dimensional cone beam computed tomography scans. *Eur. J. Orthod.* **2012**, *34*, 263–275. [CrossRef] [PubMed]

Article

3-D Point Cloud Registration Algorithm Based on Greedy Projection Triangulation

Jian Liu [1,*], Di Bai [1] and Li Chen [2]

[1]   Institution of Information and Control Engineering, Shenyang Jianzhu University, Shenyang 110168, China;
     baidi922@163.com

[2]   Architectural Design and Research Institute, Shenyang Jianzhu University, Shenyang 110168, China;
     xxliujian@sjzu.edu.cn

*   Correspondence: jeanliu10@163.com; Tel.: +86-024-2469-0042

Received: 21 August 2018; Accepted: 20 September 2018; Published: 30 September 2018

Abstract: To address the registration problem in current machine vision, a new three-dimensional (3-D) point cloud registration algorithm that combines fast point feature histograms (FPFH) and greedy projection triangulation is proposed. First, the feature information is comprehensively described using FPFH feature description and the local correlation of the feature information is established using greedy projection triangulation. Thereafter, the sample consensus initial alignment method is applied for initial transformation to implement initial registration. By adjusting the initial attitude between the two cloud points, the improved initial registration values can be obtained. Finally, the iterative closest point method is used to obtain a precise conversion relationship; thus, accurate registration is completed. Specific registration experiments on simple target objects and complex target objects have been performed. The registration speed increased by 1.1% and the registration accuracy increased by 27.3% to 50% in the experiment on target object. The experimental results show that the accuracy and speed of registration have been improved and the efficient registration of the target object has successfully been performed using the greedy projection triangulation, which significantly improves the efficiency of matching feature points in machine vision.

Keywords: machine vision; point cloud registration; greedy projection triangulation; local correlation

---

1. Introduction

With the rapid development of optical measurement technology and three-dimensional (3-D) imaging [1–3], point cloud data has received substantial attention as a special information format that contains complete 3-D spatial data. The application of the 3-D image information is widespread in the fields of 3-D reconstruction for medical applications [4], 3-D object recognition, reverse engineering of mechanical components [5], virtual reality, and many others such as image processing and machine vision [6,7].

There have been many efforts to achieve point cloud registration. The classic algorithm for this purpose is the iterative closest point [8], proposed by Besl and Mckay. This algorithm can be efficiently applied to registration problems for simple situations. However, if there is significant variance in the initial position of the two cloud points, it is easy to fall into a local optimum and thus increase the possibility of inaccurate registration. In order to provide improved initial parameters, it is necessary to perform the initial registration before accurate registration using algorithms such as the sampling consistency initial registration algorithm [9]. Due to the large capacity and complexity of point cloud data models, describing feature points is one of the most important and decisive steps in the processing for initial registration. Various methods have been developed to obtain feature information, such as local binary patterns (LBP) [10], local reference frame (LRF) [11], signatures of histogram of orientations

(SHOT) [12], and point feature histograms (PFH) [13]. These feature operators can only provide a single description for feature information with high feature dimensions and high computational complexity.

Other efforts have been made in terms of feature matching. Scale-invariant feature transform (SIFT) [14–16] utilizes difference of gaussian (DOG) images to calculate key points. It describes local features of images and obtains the corresponding 3-D feature points through mapping relationships. It has certain stability in terms of the change of view and affine transformation; however, the matching speed for this algorithm is the main limitation. The speeded-up robust features (SURF) algorithm can be used to extract the feature points of the image [17–20] and implement image matching according to the correlation. However, this algorithm relies too much on the gradient direction of the pixels in the local area, which yields unsatisfactory feature matching results. The intrinsic shape signature (ISS) algorithm has been proposed for feature extraction to complete the initial registration process [21,22]; however, wide range in searching feature point pairs and low computational efficiency are the limitations for this algorithm. The method for interpolating point cloud models using basis functions has been proposed for establishing local correlation to reduce computational complexity [23]. There are some limitations in traditional methods, such as the inability to comprehensively describe feature information and slow matching of feature point pairs. These issues limit the accuracy and speed of 3-D point cloud registration and significantly impacts its application in practical fields. Based on the traditional sampling consistency initial registration, and iterative closest point accurate registration, a new point cloud registration algorithm is proposed herein. The proposed algorithm combines fast point feature histograms (FPFH) feature description with greedy projection triangulation. The FPFH feature descriptor describes feature information accurately and comprehensively, and greedy projection triangulation reflects the topological connection between data points and its neighbors, establishes local optimal correlation, narrows the search scope, and eliminate unnecessary matching times. The combination solves the problems of the slow speed and the low accuracy in traditional point cloud registration, which leads to improvements in the optical 3-D measurement technology. The effectiveness of the proposed algorithm is experimentally verified by performing point cloud registration on a target object.

The contents of the paper consist of four sections. In Section 2, the specifications of the point cloud registration algorithm are discussed. In Section 3, experiments and analysis performed using the point cloud library (PCL) are presented. Finally, the conclusions are presented in Section 4.

2. Point Cloud Registration Algorithm

Regarding the complexity of the target, integral information can only be obtained by scanning multiple stations from different directions. The data scanned by each direction is based on its own coordinate system, and then unify them to the same coordinate system. Control points and target points are set in the scan area such that there are multiple control points or control targets with the same name on the map of the adjacent area. Thus, the adjacent scan data has the same coordinate system through the forced attachment of control points. The specific algorithm is as follows.

First, the FPFH of the point cloud is calculated and the local correlation is established to speed-up the search for the closest eigenvalue using the greedy projection triangulation network. Because of the unknown relative position between the two point cloud models, sample consensus initial alignment is used to obtain an approximate rotation translation matrix to realize the initial transformation. In addition, the iterative closest point is further refined to obtain a more accurate matrix with the initial value. The point cloud registration chart is shown in Figure 1.

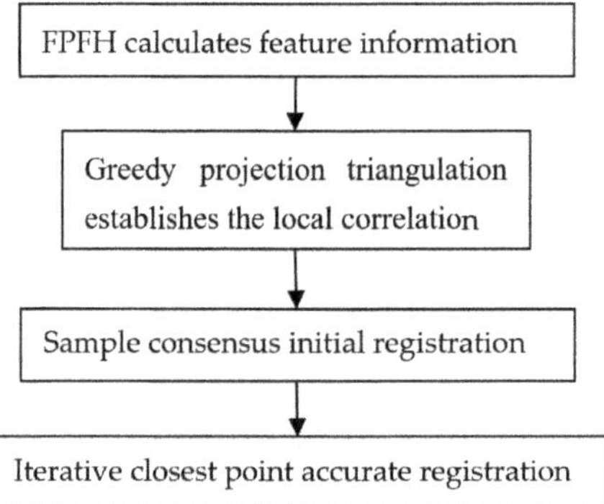

Figure 1. Point cloud registration chart. FPFH (fast point feature histograms).

2.1. Feature Information Description

FPFH is a simplification algorithm for point feature histograms (PFH), which is a histogram of point features reflecting the local geometric features around a given sample point. All neighboring points in the neighborhood K of the sample point P are examined and a local UVW coordinate system is defined as follows:

$$\begin{cases} u = \mathbf{n}_s \\ v = u \times \frac{(P_t - P_s)}{\|P_t - P_s\|} \\ w = u \times v \end{cases} . \tag{1}$$

The relationship between pairs of points in the neighborhood K is represented by the parameters (α, β, θ) and can be obtained as follows:

$$\begin{cases} \alpha = v \times \mathbf{n}_s \\ \beta = u \times \frac{(P_t - P_s)}{\|P_t - P_s\|} \\ \theta = \arctan(w \cdot \mathbf{n}_s, u \cdot \mathbf{n}_t) \end{cases} , \tag{2}$$

where P_s and P_t ($s \neq t$) denote the point pairs and $\mathbf{n}_s$ and $\mathbf{n}_t$ denote their corresponding normals in the sample point neighborhood K.

The eigenvalues of all point pairs are then calculated and the PFH of each sample point P_c is then statistically integrated. Next, the neighborhood K of each point is determined to form a simplified point feature histogram (SPFH), which is then integrated into the final FPFH. Hence, each sample point is uniquely represented by the FPFH feature descriptor. The eigenvalues of FPFH can be calculated using the following equation:

$$FPFH(P_c) = SPFH(P_c) + \frac{1}{k} \sum_{i=1}^{k} \frac{1}{w_k} \cdot SPFH(P_i), \tag{3}$$

where w_k denotes the distance between the sample point P_c and the neighboring point P_k in the known metric space.

2.2. Greedy Projection Triangulation

Greedy projection triangulation bridges computer vision and computer graphics. It converts the scattered point cloud into an optimized spatial triangle mesh, thereby reflecting the topological connection relationship between data points and their neighboring points, and maintaining the global information of the point cloud data [24]. The established triangulation network reflects the topological structure of the target object that is represented by the scattered data-set. The triangulation process is shown in Figure 2. The specific steps are given as follows:

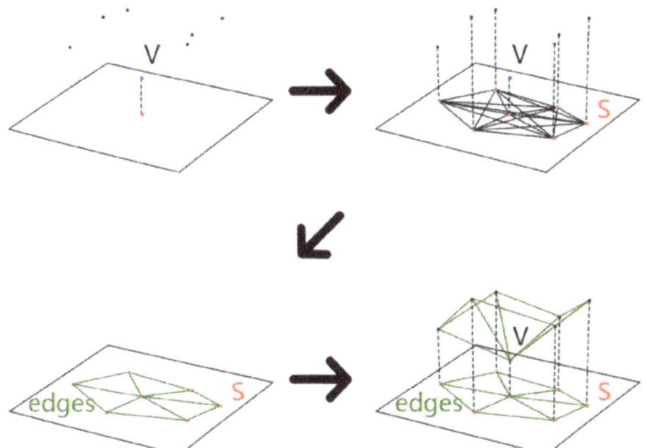

Figure 2. Greedy projection triangulation schematic.

Step 1: A point V and its normal vector exist on the surface of the three-dimensional object. The tangent plane perpendicular to the normal vector must first be determined.

Step 2: The point V and its vicinity are projected to the tangent plane passing through V, denoted as the point set $\{S\}$, and the point set $\{S\}$, which forms all N/2 edges between the two points, is linearly arranged in order of distance from small to large.

Step 3: The local projection method is used to add the shortest edge at each stage and remove the shortest edge from the memory. If the edge does not intersect any of the current triangulation edges, then it is added to the triangulation, otherwise, it is removed. When the memory is empty, the triangulation process ends.

Step 4: Triangulation is used to obtain the connection relationship of the points and return it to the three-dimensional space, which forms the space triangulation of the point V and its nearby points.

Greedy projection triangulation can establish a reasonable data structure for a large number of scattered point clouds in the 3-D space. When positioning a point, the path is unique, and the tetrahedron can be located accurately and quickly, thereby narrowing the search range and eliminating unnecessary matching. This fundamentally improves the overall efficiency of matching feature points.

2.3. Sample Consensus Initial Registration

The sample consensus initial alignment is used for initial registration. Assuming that there exists a source cloud $O_s = \{P_i\}$ and a target cloud $O_t = \{Q_j\}$, then the specific steps are as follows:

Step 1: Based on the FPFH feature descriptor of each sample point, greedy projection triangulation is performed on the target point cloud to establish local correlation of the scattered point cloud data.

Step 2: A number of sampling points are selected in the source point cloud O_s. In order to ensure that the sampling points are representative, the distance between two sampling points must be greater than the preset minimum distance threshold d.

Step 3: Search for the feature points in the target point cloud O_t, whose feature value are close to the sample points in the source point cloud O_s. Given that the greedy projection triangulation establishes a reasonable data structure for the target point cloud and then performs feature matching, it directly locates the tetrahedron with a large correlation and searches for the corresponding point pairs within the local scope.

Step 4: The transformation matrix between the corresponding points is obtained. The performance of registration is evaluated according to the total distance error function by solving the corresponding point transformation, which is expressed as follows:

$$H(l_i) = \begin{cases} \frac{1}{2}l_i^2 & \|l_i\| < m_i \\ \frac{1}{2}m_i(2\|l_i\| - m_i) & \|l_i\| > m_i \end{cases}, \tag{4}$$

in which, m_i is the specified value and l_i is the distance difference after the corresponding point transformation. When the registration process is completed, the one with the smallest error in all the transformations is considered as the optimal transformation matrix for initial registration.

2.4. Iterative Closest Point Accurate Registration

The initial transformation matrix is the key to improved matching for accurate registration. An optimized rotational translation matrix $[\mathbf{R}_0, \mathbf{T}_0]$ was obtained by initial registration, which is used as an initial value for accurate registration to obtain a more accurate transformation relationship by the iterative closest point algorithm.

Based on the optimal rotation translation matrix obtained from the initial registration, the source point cloud O_s is transformed into O_s', and it is used together with O_t as the initial set for accurate registration. For each point in the source point cloud, the nearest corresponding point in the target point cloud is determined to form the initial corresponding point pair and the corresponding point pair with the direction vector threshold is deleted. The rotation matrix $\mathbf{R}$ and translation vector $\mathbf{T}$ are then determined. Given that $\mathbf{R}$ and $\mathbf{T}$ have six degrees of freedom while the number of points is huge, a series of new $\mathbf{R}$ and $\mathbf{T}$ are obtained by continuous optimization. The nearest neighbor point changes with the position of the relevant point after the conversion; therefore, it returns to the process of continuous iteration to find the nearest neighbor point. The objective function is constructed as follows:

$$f(R, T) = \frac{1}{N_P} \sum_{i=1}^{N_P} |O_t^i - R \cdot O_s^i - T|^2, \tag{5}$$

when the change of the objective function is smaller than a certain value, it is believed that the iterative termination condition has been satisfied. More precisely, accurate registration has been completed.

3. Experiment and Analysis

During the experiment, Kinect was used as a 3-D vision sensor to realize point cloud data acquisition. The original point cloud data that was collected was processed on the Geomagic Studio 12 (Geomagic Corporation, North Carolina, the United States) platform and the experiment was completed in Microsoft Visual C++ (Microsoft Corporation, Washington, the United States). The traditional algorithm collects two point cloud data under different orientations of the same object, performs initial registration and fine registration without applying greedy projection triangulation. The greedy projection triangulation is added to address the limitations in terms of registration speed and accuracy, and the superiority of the proposed algorithm is analyzed by comparing with the traditional algorithm.

3.1. Point Cloud Registration Experiment for Simple Target Object

In this experiment, a cup is used as an example for registration. Figure 3a shows the original point cloud data and Figure 3b shows cup point cloud data after removing the background. Figure 3c shows the registration result obtained by the traditional algorithm while Figure 3d shows the registration result obtained using the registration algorithm proposed in this paper. The red regions in the point cloud represents the source point cloud data while the green and the blue regions represent the target point cloud and the rotated cloud point data, respectively.

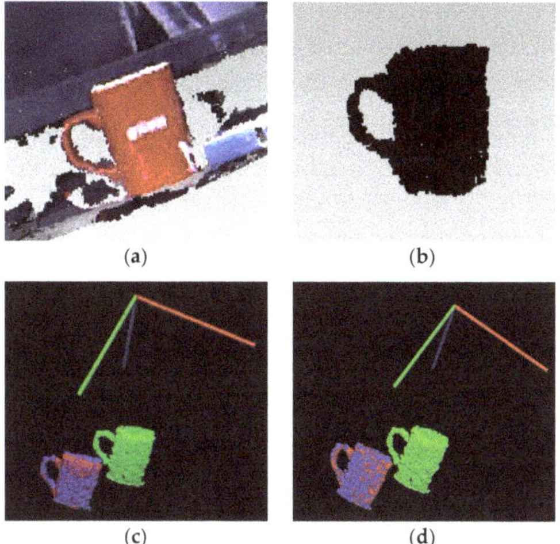

Figure 3. (**a**) Original point cloud; (**b**) Processed point cloud; (**c**) Traditional algorithm; and (**d**) Proposed algorithm.

Table 1 lists the experimental parameters. Table 2 lists the results of the target point cloud conversion obtained using the traditional point cloud registration algorithm and the results obtained using the proposed algorithm, which reflect the relative transformation relationship of the target object. Table 3 compares the registration times of different algorithms.

Table 1. The experimental parameters.

The Number of Point Clouds		Iterative Closest Point Accurate Registration Parameters			
Source point cloud	Target point cloud	Threshold (m)	The maximum number of iterations	Transform matrix difference (m)	Mean square error (m)
7009	5566	0.01	500	1×10^{-10}	0.1

Table 2. Point cloud conversion results of different algorithms.

Algorithms	Transormation matrix of Initial Registration (m)	Transormation matrix of Accurate Registration (m)
Traditional algorithm	$\begin{bmatrix} 0.999 & 0.008 & -0.031 & -0.049 \\ -0.003 & 0.985 & 0.170 & -0.040 \\ 0.032 & -0.170 & 0.985 & 0.027 \\ 0 & 0 & 0 & 1 \end{bmatrix}$	$\begin{bmatrix} 0.999 & 0.007 & -0.017 & -0.057 \\ -0.004 & 0.983 & 0.185 & -0.049 \\ 0.018 & -0.185 & 0.983 & 0.026 \\ 0 & 0 & 0 & 1 \end{bmatrix}$
Proposed algorithm	$\begin{bmatrix} 0.999 & 0.008 & -0.031 & -0.049 \\ -0.030 & 0.985 & 0.170 & -0.040 \\ 0.032 & -0.170 & 0.985 & 0.028 \\ 0 & 0 & 0 & 1 \end{bmatrix}$	$\begin{bmatrix} 0.999 & 0.012 & -0.005 & -0.062 \\ -0.011 & 0.980 & 0.201 & -0.059 \\ 0.007 & -0.200 & 0.980 & 0.025 \\ 0 & 0 & 0 & 1 \end{bmatrix}$

Table 3. Registration time of different algorithms.

Algorithm	Total Registration Time (s)	Initial Registration Time (s)	Accurate Registration Time (s)
Traditional algorithm	0.347	0.340	0.007
Proposed algorithm	0.257	0.252	0.005

When analyzing the above experiments, the attitude of the source cloud is considered as a reference and the attitude of the target object is decomposed into three directions, namely X, Y, and Z. The rotation angle in three directions and the matching error distance between the source cloud and the transformed point cloud are considered as the evaluation indices. The rotation angle and the registration error distance in this experiment are shown in Table 4.

Table 4. Experimental results of different algorithms.

Algorithms	X-Direction Rotation Angle (rad)	Y-Direction Rotation Angle (rad)	Z-Direction Rotation Angle (rad)	Average Error Distance (cm)
Traditional algorithm	0.186	0.018	−0.625	0.158
Proposed algorithm	0.202	0.007	−0.617	0.149

From Table 3, it can be observed that for the same point cloud sample with the same experimental parameters, the initial registration time using the traditional algorithm is 0.340 s. Because of the combination of FPFH feature description and greedy projection triangulation, the initial registration time obtained using the proposed algorithm is 0.252 s. Table 4 shows a comparison of the two algorithms. The average error distance obtained is 1.58 mm and 1.49 mm using the traditional algorithm and the proposed algorithm, respectively. As shown in Table 2, the point cloud is transformed by the different transformation matrix, and the average error distance obtained is smaller by the proposed algorithm.

3.2. Point Cloud Registration Experiment for Complex Target Object

In this experiment, the point cloud models of the same person in different orientations are collected and then registered using different algorithms. Figure 4a shows the results of 3-D reconstruction. Figure 4b,c show the registration results of the traditional algorithm and the proposed algorithm, respectively. As can be seen from the figure, the blue point cloud and the red point cloud are more highly integrated in Figure 4c than that in Figure 4b, which can be known the proposed algorithm is more accurate than the traditional algorithm. Table 5 shows a comparison of the registration time of different algorithms. Table 6 shows the obtained rotation angle and the registration error distance in this experiment. The eight groups affine transformations are performed on the input point cloud data, which verify the reliability of the algorithm. Table 7 shows a comparison of the average registration error distance and the total registration time of the eight groups of experiments. The ratio of average registration error reduction is between 27.3% and 50%, and the ratio of total registration time reduction is about 1.1%. It can be seen that the average registration error distance of the proposed algorithm is smaller and the total registration time is shorter than the traditional one, which verifies the reliability of the proposed algorithm.

Table 5. Registration time of different algorithms.

Orientation	Algorithms	Total Registration Time (s)	Initial Registration Time (s)	Accurate Registration Time (s)
1	Traditional algorithm	11.680	11.428	0.252
	Proposed algorithm	11.553	11.336	0.217
2	Traditional algorithm	8.287	8.196	0.091
	Proposed algorithm	8.029	7.955	0.074

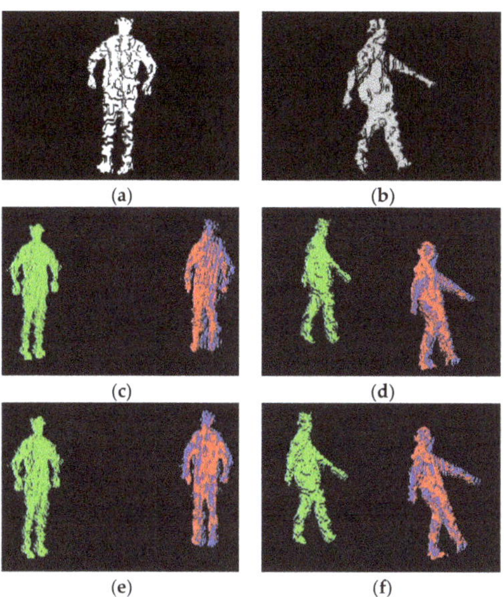

Figure 4. (a,b) 3-D reconstruction; (c,d) Traditional algorithm; (e,f) Proposed algorithm.

Table 6. Experimental results of different algorithms.

Orientation	Algorithms	X-Direction Rotation Angle (rad)	Y-Direction Rotation Angle (rad)	Z-Direction Rotation Angle (rad)	Average Error Distance (cm)
1	Traditional algorithm	0.283	0.702	−0.172	0.015
	Proposed algorithm	0.139	0.561	−0.002	0.011
2	Traditional algorithm	0.053	0.469	−0.587	0.009
	Proposed algorithm	0.003	0.495	−0.625	0.005

Table 7. Comparison of average error distance and total registration time of multiple sets experiment.

Group	Registration Error of Traditional Algorithm (cm)	Registration Error of Proposed Algorithm (cm)	Percentage of Average Registration Error Reduction (%)	Total Registration Time of Traditional Algorithm (s)	Total Registration Time of Proposed Algorithm (s)	Percentage of Total Registration Time Reduction (%)
1	0.015	0.011	36.4	11.680	11.553	1.1
2	0.014	0.011	27.3	11.669	11.549	1.0
3	0.014	0.010	40.0	11.684	11.559	1.1
4	0.015	0.011	36.4	11.681	11.556	1.1
5	0.013	0.010	30.0	11.685	11.558	1.1
6	0.015	0.010	50.0	11.673	11.551	1.1
7	0.015	0.011	36.4	11.678	11.551	1.1
8	0.014	0.011	27.3	11.683	11.558	1.1

Compared with the results obtained from the traditional algorithm, it is concluded that the proposed algorithm has higher registration accuracy and faster registration speed. Its advantages can be attributed to the following factors:

(a) The FPFH feature descriptor describes feature information accurately and comprehensively and avoids the errors in matching feature point pairs.

(b)   Greedy projection triangulation reflects the topological connection between data points and its neighbors, establishes local optimal correlation, narrows the search scope, and reduce unnecessary matching times.

(c)   The combination of the FPFH feature description and the greedy projection triangulation can match similar point pairs accurately and quickly, which is the key to efficient registration.

4. Conclusions

Based on the traditional sample consensus initial alignment and iterative closest point algorithms, a new point cloud registration algorithm based on the combination of the FPFH feature description and the greedy projection triangulation was proposed herein. The 3-D point cloud data is used to improve the information regarding the two-dimensional image, and the data information is completely preserved. The FPFH comprehensively describes the local geometric feature information around the sample point. This simplifies the complexity of feature extraction and improves the accuracy of feature description. Greedy projection triangulation solves the problem that the feature points have a wide search range during the registration process. Thus, the number of matching processes is reduced.

In the registration experiment for target object, the registration speed increased by 1.1% and the registration accuracy improved by 27.3% to 50%. The results show that the optimized spatial triangular mesh established by greedy projection triangulation narrows the search range of feature points, which improved the registration speed and accuracy. The initial registration determines an approximate rotational translation relationship between the two point cloud models. Using it as the initial value, accurate registration is performed to obtain a more precise relative change relationship. The greedy projection triangulation optimizes the traditional registration algorithm, thereby making the registration process faster and more accurate.

Author Contributions: For the research articles, the three authors distributed the responsibilities as J.L. carried out the theoretical algorithm research; D.B. carried out the experimental research and performed the analysis; L.C. conducted the experimental data acquisition and analysis; J.L. and D.B. wrote the paper.

Funding: This research was funded by [the scientific research projects in National Natural Science Foundation of China] grant number [11704263], [Liaoning Province Natural Science Foundation] grant number [201602616], [Liaoning Province Department of Education Scientific Research Project] grant number [2015443].

Acknowledgments: The authors would like to thank Ziyi Meng and Xudong Wang for insightful discussions.

Conflicts of Interest: The authors declare no conflict of interest.

References

1.   Huang, Y.; Da, F.P.; Tao, H.J. An automatic registration algorithm for point cloud based on feature extraction. *Chin. J. Lasers* **2015**, *42*, 250–256. [CrossRef]

2.   Chen, K.; Zhang, D.; Zhang, Y. Point cloud data processing method of cavity 3D laser scanner. *Acta Opt. Sin.* **2013**, *33*, 125–130. [CrossRef]

3.   Wei, S.B.; Wang, S.Q.; Zhou, C.H. An Iterative closest point algorithm based on biunique correspondence of point clouds for 3D reconstruction. *Acta Opt. Sin.* **2015**, *35*, 252–258. [CrossRef]

4.   Logozzo, S.; Kilpelä, A.; Mäkynen, A.; Zanetti, E.M.; Franceschini, G. Recent advances in dental optics—Part II: Experimental tests for a new intraoral scanner. *Opt. Lasers Eng.* **2014**, *54*, 187–196. [CrossRef]

5.   Calì, M.; Oliveri, S.M.; Ambu, R.; Fichera, G. An Integrated Approach to Characterize the Dynamic Behaviour of a Mechanical Chain Tensioner by Functional Tolerancing. *Int. J. Mech. Eng. Educ.* **2018**, *64*, 245–257.

6.   Won-Ho, S.; Jin-Woo, P.; Jung-Ryul, K. Traffic Safety Evaluation Based on Vision and Signal Timing Data. In Proceedings of Engineering and Technology Innovation. *Proc. Eng. Technol. Innov.* **2017**, *7*, 37–40.

7.   Hsu, H.C.; Chu, L.M.; Wang, Z.K.; Tsao, S.C. Position Control and Novel Application of SCARA Robot with Vision System. *Adv. Technol. Innov.* **2017**, *2*, 40–45.

8.   Besl, P.J.; McKay, N.D. A method for registration of 3-D shapes. *IEEE Trans. Pattern Anal. Mach. Intell.* **1992**, *15*, 239–256. [CrossRef]

9. Qiu, L.; Zhou, Z.; Guo, J.; Lv, J. An Automatic Registration Algorithm for 3D Maxillofacial Model. *3D Res.* **2016**, *7*, 20. [CrossRef]

10. Xu, J.F.; Liu, Z.G.; Han, Z.W.; Liu, Y. 3D Reconstruction of Contact Network Components Based on SIFT and LBP Point Cloud Registration. *J. China Railw. Soc.* **2017**, *39*, 76–81. [CrossRef]

11. Chen, X. *Research on Feature Extraction and Recognition of Imaging Lidar Target*; National University of Defense Technology: Changsha, China, 2015.

12. Jia, Y.J.; Xiong, F.G.; Han, X.; Kuang, L.Q. SHOT-based multi-scale key point detection technology. *J. Laser Opt. Prog.* **2018**. Available online: http://kns.cnki.net/kcms/detail/31.1690.TN.20180227.1658.018.html (accessed on 20 September 2004).

13. Huang, J.J. *Research on Real-Time 3D Reconstruction of Mobile Scenes Based on PFH and Information Fusion Algorithm*; Donghua University: Shanghai, China, 2014.

14. Wu, S.G.; He, S.; Yang, X. The Application of SIFT Method towards Image Registration. *Adv. Mater. Res.* **2014**, *1044*, 1392–1396. [CrossRef]

15. Liu, J.F. Feature Matching of Fuzzy Multimedia Image Based on Improved SIFT Matching. *Recent Adv. Electr. Electron. Eng.* **2016**, *9*, 34–38. [CrossRef]

16. Shen, Y.; Pan, C.K.; Liu, H.; Gao, B. Point Cloud Registration Method Based on Improved SIFT-ICP Algorithm. *Trans. Chin. Soc. Agric. Mach.* **2017**, *48*, 183–189.

17. Manish, I.P.; Vishvjit, K.T.; Shishir, K.S. Image Registration of Satellite Images with Varying Illumination Level Using HOG Descriptor Based SURF. *Proc. Comput. Sci.* **2016**, *93*, 382–388. [CrossRef]

18. Li, J.F.; Wang, G.; Li, Q. Improved SURF Detection Combined with Dual FLANN Matching and Clustering Analysis. *Appl. Mech. Mater.* **2014**, *556*, 2792–2796. [CrossRef]

19. Huang, L.; Chen, C.; Shen, H.; He, B. Adaptive registration algorithm of color images based on SURF. *Measurement* **2015**, *66*, 118–124. [CrossRef]

20. Yan, W.D.; She, H.W.; Yuan, Z.B. Robust Registration of Remote Sensing Image Based on SURF and KCCA. *J. Indian Soc. Remote Sens.* **2014**, *42*, 291–299. [CrossRef]

21. Renzhong, L.; Man, Y.; Yu, T.; Yangyang, L.; Huanhuan, Z. Point Cloud Registration Algorithm Based on the ISS Feature Points Combined with Improved ICP Algorithm. *J. Laser Opt. Prog.* **2017**, *54*, 312–319. [CrossRef]

22. Liu, W.Q.; Chen, S.L.; Wu, Y.D.; Cai, G.R. *Fast CPD Building Point Cloud Registration Algorithm Based on ISS Feature Points*; Jimei University: Xiamen, China, 2016; Volume 21, pp. 219–227.

23. Wang, L.H. *Technical Research on 3D Point Cloud Data Processing*; Beijing Jiaotong University: Beijing, China, 2011.

24. Xie, F.J. *Research and Implementation of Key Techniques of 3D Point Cloud Surface Reconstruction on Vehicle Dimensions Measuring System*; Hefei University of Technology: Hefei, China, 2017.

MDPI
St. Alban-Anlage 66
4052 Basel
Switzerland
Tel. +41 61 683 77 34
Fax +41 61 302 89 18
www.mdpi.com

Applied Sciences Editorial Office
E-mail: applsci@mdpi.com
www.mdpi.com/journal/applsci